Thermal barrier coatings

Related titles:

Surface engineering of light alloys: Aluminium, magnesium and titanium alloys
(ISBN 978-1-84569-537-8)
This authoritative book provides a comprehensive review of the various surface engineering techniques employed to improve the properties of light alloys, focusing on titanium, magnesium and aluminium alloys. It examines surface related degradation of light alloys and covers surface engineering technologies in detail. The book includes chapters on corrosion behaviour of Mg alloys, anodising treatments of Mg alloys, micro-arc oxidation of light alloys, physical vapour deposition of light alloys, PIII/PSII of light alloys, laser surface modification of Ti alloys, plasma nitriding of Ti and Al alloys, duplex surface treatments of light alloys and biomedical devices using Ti alloys.

Electroless copper and nickel–phosphorus plating: Processing, characterisation and modelling
(ISBN 978-1-84569-808-9)
Compared with electroplating, electroless plating allows uniform deposits over different surfaces. Electroless copper and nickel–phosphorus deposits provide protective and functional coatings in industries as diverse as electronics, automotive, aerospace and chemical engineering. Written by leading experts in the field, this important book reviews the deposition process and the key properties of electroless copper and nickel–phosphorus deposits as well as their practical applications.

Nanostructured metals and alloys: Processing, microstructure, mechanical properties and applications
(ISBN 978-1-84569-670-2)
Nanostructured metals and alloys have enhanced tensile strength, fatigue strength and ductility and are suitable for use in applications where strength or strength-to-weight ratios are important. Part I of this important book reviews processing techniques for bulk nanostructured metals and alloys. Parts II and III discuss microstructure and mechanical properties, whilst Part IV outlines applications of this new class of material.

Details of these and other Woodhead Publishing materials books can be obtained by:

- visiting our web site at www.woodheadpublishing.com
- contacting Customer Services (e-mail: sales@woodheadpublishing.com; fax: +44 (0) 1223 832819; tel.: +44 (0) 1223 499140 ext. 130; address: Woodhead Publishing Limited, 80 High Street, Sawston, Cambridge CB22 3HJ, UK)

If you would like to receive information on forthcoming titles, please send your address details to: Francis Dodds (address, tel. and fax as above; e-mail: francis.dodds@woodheadpublishing.com). Please confirm which subject areas you are interested in.

Thermal barrier coatings

Edited by
Huibin Xu and Hongbo Guo

WOODHEAD PUBLISHING

Oxford Cambridge Philadelphia New Delhi

Published by Woodhead Publishing Limited,
80 High Street, Sawston, Cambridge CB22 3HJ, UK
www.woodheadpublishing.com

Woodhead Publishing, 1518 Walnut Street, Suite 1100, Philadelphia,
PA 19102-3406, USA

Woodhead Publishing India Private Limited, G-2, Vardaan House,
7/28 Ansari Road, Daryaganj, New Delhi – 110002, India
www.woodheadpublishingindia.com

First published 2011, Woodhead Publishing Limited

British Library Cataloguing in Publication Data
A catalogue record for this book is available from the British Library.

ISBN 978-0-08-101501-8 (print)
ISBN 978-0-85709-082-9 (online)

The publisher's policy is to use permanent paper from mills that operate a sustainable forestry policy, and which has been manufactured from pulp which is processed using acid-free and elemental chlorine-free practices. Furthermore, the publisher ensures that the text paper and cover board used have met acceptable environmental accreditation standards.

Typeset by Toppan Best-set Premedia Limited, Hong Kong

Contents

Contributor contact details

(* = main contact)

Editors

H. Xu and H. Guo
School of Materials Science and Engineering
Beihang University
37 Xueyuan Road
Beijing 100191
China

E-mail: xuhb@buaa.edu.cn
guo.hongbo@buaa.edu.cn

Chapter 1

D. Zhang
School of Energy, Power and Mechanical Engineering
North China Electric Power University
Beijing 102206
China

E-mail: zhangdb@ncepu.edu.cn

Chapter 2

W. Ma* and H. Dong
School of Materials Science and Engineering
Inner Mongolia University of Technology
49 Aimin Road
Xincheng district
Hohhot 010051
China

E-mail: wma@imut.edu.cn

Chapter 3

X. Peng
State Key Laboratory for Corrosion and Protection
Institute of Metal Research
Chinese Academy of Sciences
72 Wenhua Road
Shenhe district
Shenyang 110016
China

E-mail: xpeng@imr.ac.cn

Chapter 4

C. G. Zhou* and Q. H. Yu
School of Materials Science and Engineering
Beihang University
37 Xueyuan Road
Beijing 100191
China

E-mail: cgzhou@buaa.edu.cn

Chapter 5

J. Xu* and K. A. Khor
School of Mechanical and Aerospace Engineering
Nanyang Technological University
50 Nanyang Avenue
Singapore 639798

E-mail: jlxu@ntu.edu.sg

Chapter 6

S. Gong* and Q. Wu
School of Materials Science and Engineering
Beihang University
37 Xueyuan Road
Beijing 100191
China

E-mail: gongsk@buaa.edu.cn

Chapter 7

L. Chen
Department of Material Science and Engineering
China University of Petroleum
18 Fuxue Road
Changping
Beijing 102249
China

E-mail: liqiang_chen@sohu.com

Chapter 8

H. Guo*, R. Yao and L. Zhou
School of Materials Science and Engineering
Beihang University
37 Xueyuan Road
Beijing 100191
China

E-mail: guo.hongbo@buaa.edu.cn

Chapter 9

P. Ke*
Surface Engineering
Ningbo Institute of Materials Technology and Engineering
Chinese Academy of Sciences
519 Zhuangshi Road
Zhenhai district
Ningbo 315201
China

E-mail: plkeimr@hotmail.com
kepl@nimte.ac.cn

Y. Wu and C. Sun*
State Key Laboratory for Corrosion and Protection
Institute of Metal Research
Chinese Academy of Sciences
72 Wenhua Road
Shenhe district
Shenyang 110016
China

E-mail: yingnawu@gmail.com
csun@imr.ac.cn

Chapter 10

C. G. Zhou and Y. X. Song
School of Materials Science and Engineering
Beihang University
37 Xueyuan Road
Beijing 100191
China

E-mail: cgzhou@buaa.edu.cn

Chapter 11

C. Zhang*
Baosteel Group Corporation
Pu Dian Road 370
Pudong New District
Shanghai 200122
China

E-mail: zcx_buaa@163.com

C. Chen
AVIC Commercial Aircraft Engine Co., Ltd
Building 3
Dongchuan Road 555
Minhang District
Shanghai 200241
China

E-mail: chenchen@mse.buaa.edu.cn

Chapter 12

G. Chen
Shaanxi Key Laboratory of Friction Welding Technologies
Northwestern Polytechnical University
127 Youyi Xilu
Xi'an
Shaanxi Province 710072
China

E-mail: gfchen@gmail.com

Chapter 13

R. T. Wu
International Center for Young Scientists (ICYS)
National Institute for Materials Science (NIMS)
1-2-1 Sengen
Tsukuba
Ibaraki 305-0047
Japan

E-mail: wu.rudder@nims.go.jp

Chapter 14

H. Y. Qi* and X. G. Yang
School of Jet Propulsion
Beihang University
37 Xueyuan Road
Beijing 100191
China

E-mail: qhy@buaa.edu.cn

Chapter 15

H. Xu* and J. Wu
School of Materials Science and Engineering
Beihang University
37 Xueyuan Road
Beijing 100191
China

E-mail: xuhb@buaa.edu.cn

Preface

Thermal barrier coatings (TBCs) are finding increasing application in the most demanding high-temperature environment of aircraft and industrial engines. They have the capability to improve the durability of engines by reducing the surface temperature of the underlying components. The use of TBCs, along with internal cooling of the underlying superalloy components, has thus enabled advanced gas turbine engines to operate at temperatures even above the melting temperature of the superalloy, thus achieving a remarkable increase in the efficiency and performance of engines. This technology has been regarded as one of the most important and effective developments in efforts to improve the propulsion efficiency of advanced aero engines.

The TBC components must withstand the harsh environment coupled with high temperature, large temperature gradient, complex stress condition and corrosive atmosphere. No single coating component is able to satisfy these multifunctional requirements. As a result, a complex TBC structure has been developed. Research since the 1970s has been focused on a preferred coating system primarily comprising three separate layers on the substrate, so as to achieve long-term effectiveness in oxidative and corrosive environment at high temperatures, i.e. a ceramic top coat layer with low thermal conductivity to provide excellent thermal insulation for the hot components, a bond coat layer above the substrate to hold the ceramic coat on the substrate, and a thin thermally grown oxide (TGO) layer to provide adhesion between the ceramic top coat and the bond coat. The TBC component is expected to last over thousands of take-offs and landings in aero engines. However, the complexity and diversity of TBC structures and the severity of operating conditions give rise to premature TBC failure. This premature failure has slowed down the use of TBCs and compromised the ability to take full advantage of the energy efficiency and service lifetime potentially offered by TBCs. In view of this, fundamental investigations into TBC materials and processing principles and assessments of TBC performance, lifetime prediction and failure mechanisms have been systematically conducted. The multidisciplinary subjects of materials science, physics, chemistry, mechanics, computation science and thermodynamics are involved in this research field.

At present, the TBC industry is key in the field of aero engines worldwide. It is necessary to make the theory and the applications of TBCs known to both researchers and graduate students, and this is the purpose of this book. The aim is not only to provide a clear and thorough presentation of both fundamental principles and applications of TBC systems in the aero engine industry, but also to prepare the way for future TBC systems by reviewing previous achievements. There are 15 independent chapters in this book, in which progress in TBC materials, advanced processing technologies, physical and chemical phenomena such as oxidation, diffusion, fatigue, thermal conduction and sintering, failure mechanisms and non-destructive evaluation of TBCs and their life prediction have been described.

A general description of the background, processing technology of electron beam physical vapor deposition (EB–PVD), oxidation and thermal cycling performance of TBCs manufactured by EB–PVD is given in Chapter 1. The selection principles and the progress of ceramic top coat materials in TBC systems are introduced in Chapter 2. Several proposals on the development trends in ceramic top coats are raised in this chapter. In Chapter 3, the emphasis is placed on a comparison of the processing and oxidation resistance of various metallic coatings. As a newly developed TBC system, the nanostructured TBCs are introduced in Chapter 4, including the fundamental principle of formation of nanostructure, microstructure features and associated properties, and several potential applications of nanostructured TBCs are suggested.

Two main processing technologies are generally used in the manufacture of TBCs, namely plasma spraying (PS) and electron beam physical vapor deposition (EB–PVD). The principles, microstructure characteristics and applications of PS and EB–PVD are addressed in Chapters 5 and 6, respectively, giving guidance in the selection of processing technology for the production of TBCs to meet different industrial purposes. In Chapter 7, plasma-sprayed TBCs are presented with the emphasis on thermal physical properties, mechanical properties, durability and failure mechanisms. The properties of plasma-sprayed TBCs can be further improved by modifications to the TBC structure. This fact was particularly evident in Chapter 8, in which a modified TBC structure with high segmentation crack density is proposed. In Chapter 9, the structure, characteristics and properties of TBC systems prepared by a so-called detonation plasma spray method are presented.

One of the primary tasks for TBCs is to effectively protect the underlying superalloy substrate from hot corrosion and high-temperature oxidation. The high-temperature oxidation and hot-corrosion behaviors of TBCs are described in Chapter 10. The emphasis is focused on failure mechanisms of TBCs associated with cracking of TGO grown on the metallic bond coat and comparison of several kinds of metallic bond coating materials which are practical use in engineering design. In Chapter 11, the behaviors of TBCs under thermal-mechanical coupled loads are presented, and a failure mechanism is also put forward. Recently, non-destructive evaluation (NDE) methods have been used for the premature failure detection of TBC systems. Chapter 12 reviews the fundamentals and application of different NDE methods with particular emphasis on the latest results in the promising area of emission spectroscopy NDE used for in situ TBCs failure inspection. The inter-diffusion between the bond coat and the underlying substrate under high temperatures has significant influence on the mechanical properties of superalloy, oxidation resistance and durability of TBCs. This is clearly stated in Chapter 13.

Chapter 14 deals with life prediction of TBCs. The ability to foresee, theoretically, the lifetime of TBCs is fundamental in improving the safety of TBC components in industrial applications. In this chapter, three visco-plastic constitutive models are developed for TBC systems. Failure mechanisms of TBCs are discussed and a fatigue life model is set up for life prediction of TBC vanes.

In the final chapter, new TBC materials and processing technologies are reviewed and some possible directions for the development of TBCs in the future are suggested.

The author acknowledges all the contributors to the successful completion of this book. The author would also like to thank Dr Dongbo Zhang in North China Electric Power University (NEPU) for his contribution to Chapter 1, Dr Wen Ma in Inner Mongolia University of Technology (IMUT) for his contribution to Chapter 2, Dr Xiao Peng in Institute of Metal Research (IMR) in Chinese Academy of Science (CAS) for his contribution to Chapter 3, Professor Chungen Zhou and Dr Qinghe Yu in BUAA for their contribution to Chapter 4, Dr Jinlin Xu and Professor K.A. Khor in Nanyang Technological University (NTU) for their contribution to Chapter 5, Professor Shengkai Gong and Dr Qiong Wu in Beihang University (BUAA) for their contribution to Chapter 6, Dr Liqiang Chen in China University of Petroleum (CUP) for his contribution to Chapter 7, Professor Hongbo Guo, Miss Rui Yao and L. Zhou in BUAA for their contribution to Chapter 8, Drs Yingna Wu, Peiling Ke and Chao Sun in IMR of CAS for their contribution to Chapter 9, Professor Chungen Zhou and Dr Yinxue Song for their contribution to Chapter 10, Dr Chunxia Zhang in Baosteel Group Corporation (BGC) and Dr Chen Chen in BUAA for their contribution to Chapter 11, Dr Guofeng Chen in the Northwestern Polytechnical University, China, for his contribution to Chapter 12, Dr Rudder Wu in National Institute for Materials Science (NIMS) Japan for his contribution to Chapter 13, Professors Hongyu Qi and Xiaoguang Yang in BUAA for their contribution to Chapter 14, and Mrs J. Wu for her contribution to Chapter 15. The authors will be grateful for all readers' comments.

Huibin Xu
Beihang University, China

Part I
Materials and structure

1
Thermal barrier coatings prepared by electron beam physical vapor deposition (EB–PVD)

D. ZHANG, North China Electric Power University, China

Abstract: Thermal barrier coatings (TBCs), which offer excellent properties in terms of high-temperature operating, thermal barrier and flow- and wear-resistance, have been developed for application in hot-section components of gas turbines. Research in this area has become a new direction in the surface engineering field. Electron beam physical vapor deposition (EB–PVD) has received much research attention in recent years since it offers many advantages for TBCs.

This chapter introduces the structure and preparation processes of TBCs. Two-layer structured TBCs were prepared by EB–PVD on superalloy IC10/IC6. The behavior of TBCs under thermal cycling was investigated. The roughness of the bond coat was found to have a significant effect on the lifetime of TBCs, the lifetime clearly decreasing during thermal cyclic testing as the roughness average (Ra) of the bond coat increased. The thickness of the pre-oxide layer also affected the lifetime of TBCs, the lifetime decreasing during thermal cyclic testing as the pre-oxide layer thickness increased. Graded thermal barrier coatings were prepared by EB–PVD on superalloy and the thermal cycling behavior was investigated.

Key words: thermal barrier coating, EB–PVD, thermal cycled, two-layer structured TBCs, GTBCs.

1.1 Introduction

There are many methods of preparing thermal barrier coatings (TBCs), such as electron beam physical vapor deposition (EB–PVD), plasma spray, high-velocity oxy-fuel (HVOF), sol-gel, laser chemical vapor deposition, and so on.[1–8] At the time of writing EB–PVD and plasma spray have become popular preparation methods. Plasma spray has a simpler equipment structure than EB–PVD, and thermal spray has a higher spray rate than EB–PVD.

For plasma spray, it is not necessary to heat the substrate, and the deposition rate is much faster than that of EB–PVD. However, coatings prepared by EB–PVD are different from those prepared by plasma spray. EB–PVD TBCs have better surface roughness than those produced by plasma spray, which means that EB–PVD TBCs have good aerodynamics. EB–PVD TBCs have a columnar crystal microstructure and those produced by plasma spray a laminated structure. Due to good resistance to spallation from the substrate, TBCs with the columnar structure obtained by EB–PVD have a longer lifespan than those with the laminated structure obtained by plasma spray during thermal cycling. Therefore, EB–PVD TBCs are likely to become popular in the future.

1.1.1 Electron beam physical vapor deposition (EB–PVD) technology and its equipment

Since the 1980s, EB–PVD has been investigated for preparation of TBCs. A significant development in the 1990s was the invention by Paton Electric Institute of Welding in Ukraine of a new EB–PVD technology which decreased the preparation cost of EB–PVD TBCs. This technology has been widely applied in the production of TBCs.

The electron beam has a very important role as the thermal source of EB–PVD technology. EB–PVD is capable of depositing any type of material. The deposition mechanism consists of an electron beam formed at 2000 °C in an electron gun (either a Pierce gun or an electromagnetism deflexion gun), followed by acceleration of thermal electrons under high voltage. When the high-speed electrons hit materials in the ingot, energy is generated sufficient to melt and convert material into vapor and then deposit it onto the substrate. Compared to other preparation methods, this method has a higher deposition rate. Moreover, deposition parameters can be easily controlled and the surface is very smooth with a Ra surface value of about 1–2 μm. There is a strong adhesion between the substrate and coatings after deposition by EB–PVD. Process parameters are easily modified and the deposition rate can be accurately controlled.

The EB–PVD equipment is more complex compared to other methods. Figure 1.1 gives a diagram of EB–PVD equipment (UE205). There are four electron guns, two sets of sample fixtures (one vertical and one horizontal), three ingots, and two chambers (one of which is a pre-vacuum chamber) in this EB–PVD system. Three electron guns are used to heat the materials in the ingots, respectively, and one electron gun to heat the substrate (workpiece) to achieve the right deposition condition which is able to ensure good adhesion between substrate and coatings. During depositing, the sample fixture keeps rotating to get a stable substrate temperature and uniform coating thickness. Three ingots can deposit from one to three types of material and prepare coatings of different structures. There is a shaft below the material in the ingot. During depositing, the shaft pushes the material up to the ingot, thus increasing the deposition rate. EB–PVD equipment of different types with different structures can prepare TBCs of different shapes and structures. For example, some EB–PVD equipment (UE-206, UE204U) has six electron guns and more power.

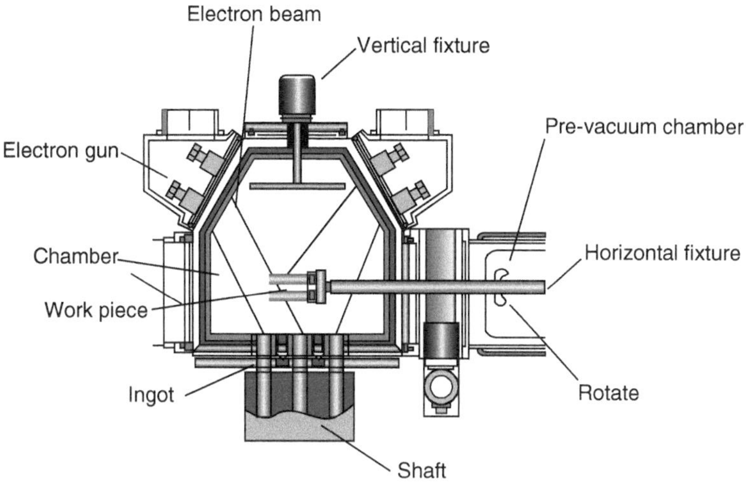

1.1 EB–PVD equipment (UE205).

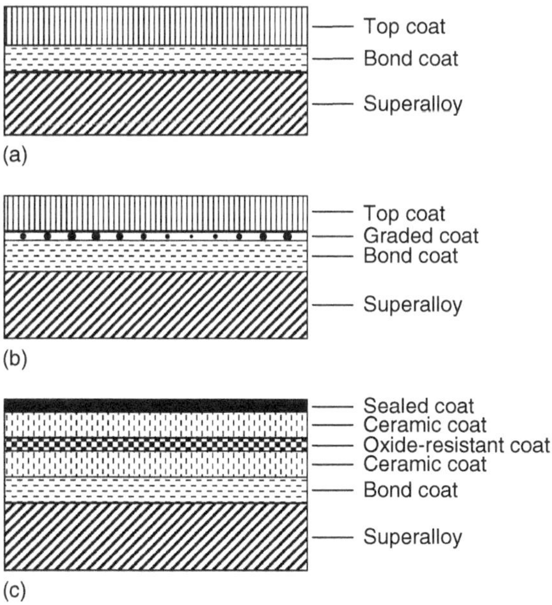

1.2 TBC structures: (a) two-layered; (b) graded; (c) multi-layered.

1.1.2 Thermal barrier coatings (TBCs): structure

There are three types of EB–PVD TBCs: two-layered, graded, and multi-layered. As shown in Fig. 1.2, two-layered structured TBCs are traditional TBCs comprising a bond coat with MCrAlY(M = Ni, Co) and a ceramic top coat with 6–8 wt% Y_2O_3 stabilized ZrO_2. Graded TBCs (GTBCs) can relax the stress between the top coat and the bond coat which is brought by different coefficients of thermal expansion. In the GTBCs system, there is a transition layer with Al_2O_3 between the bond coat and the top coat. Multi-layered TBCs are usually used for specialist applications. In the future, it is likely that nanostructure TBCs will have an important role because they have the capacity to enhance the effectiveness of the thermal barrier.

1.2 Preparation process and parameters

TBCs should be prepared under optimum preparation conditions and parameters. Many preparation parameters can affect the quality of TBCs. Amonst these, three – vacuum pressure, substrate temperature, and deposition rate – can have a significant effect.

1.2.1 Vacuum pressure

O_2 pressure dictates the vacuum pressure in the chamber. If O_2 exists in the chamber, it will decrease the background vacuum. This is because O_2 can result in filament oxidation of the electron gun which will affect its lifetime. If the background vacuum pressure is low during MCrAlY depositing, oxygen will enter into the bond coat, inducing inter-oxidation which will decrease the effectiveness of the bond coat. The background vacuum pressure should be in the range 10^{-2}–10^{-3} Pa during deposition.

1.2.2 Substrate temperature

When the electron beam strikes the materials, the resultant transfer of high energy will generate sufficient heat to create vapor which rises and is deposited on the substrate. There are two possible states during solidification of liquid phase on the substrate. One is that material vapor deposits, adheres, and then diffuses into the substrate, and the fine coating forms after deposition. The other is that material vapor deposits onto but debonds with the substrate. In that case, the coating cannot combine with the substrate. Substrate temperature is an important factor in obtaining the fine coating. To increase the adhesion between coatings and the substrate, the substrate should be heated to a high temperature during depositing. Substrate temperature can also affect the crystal structure. Palatnik *et al.* investigated the relationship of substrate temperature and coating crystal structure. If the substrate temperature (T_s) is lower than 2/3 of the melting point (T_m) of the material, the depositing material solidifies immediately. When it is higher than 2/3 of the melting point, the depositing material vapor changes into the liquid state, then crystallizes, and finally changes into the solid state.[9] Movchan and Demchishin investigated the relationship between substrate temperature and coating crystal structure, then created a model of substrate temperature and coating crystal structure[10,11] as shown in Fig. 1.3.

When the relationship of substrate temperature and the melting point of the depositing material is $T_s/T_m < 0.3$ (Zone I), the coating has a structure of columnar crystals with a round top surface. Because of the 'self-shadow effect' and uniform diffusion, there are a lot of holes between the grain boundaries. If the substrate temperature is low, the coating has a large cooling rate during depositing. The result is that the size of the columnar crystals in the coating system is small. As shown in Zone I in Fig. 1.3, there is a transitional region between zone I and Zone II, which is composed of columnar crystals with fibrillation.

When the relationship of substrate temperature and the melting point of the depositing material is $0.3 < T_s/T_m < 0.5$ (Zone II), the coating has a structure of compacted columnar crystals. Due to diffusion and concreting on the surface, compacted columnar crystals are formed as shown in Zone II in Fig. 1.3. In the range $0.3 < T_s/T_m < 0.5$, the columnar crystals will grow with the increasing T_s.

When the relationship between substrate temperature and the melting point of the depositing material is $0.5 < T_s/T_m < 1$ (Zone III), the recrystallization structure resulting from diffusion at high temperature can be seen in zone III. As the substrate temperature increases, the cooling rate of the coating decreases and large columnar crystals are formed. In order to obtain the desired crystal structure, the substrate temperature should be between 800 and 900 °C during TBCs deposition.

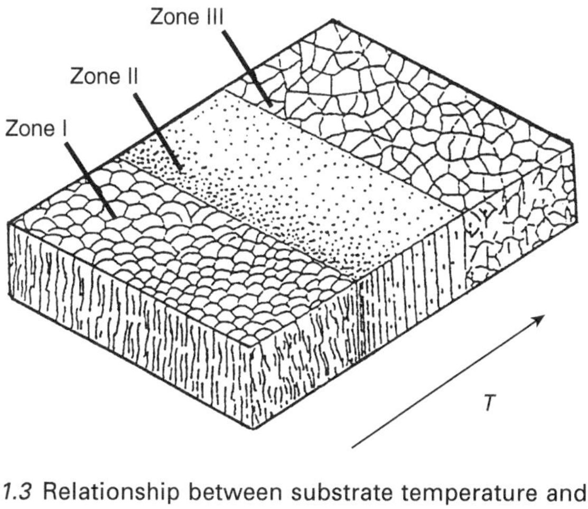

1.3 Relationship between substrate temperature and coating structure.

1.2.3 Deposition rate

For EB–PVD, both the voltage and the current are important parameters of the deposition rate. Deposition rate affects the size of the columnar crystals and the size and number of holes between the grain boundaries. As the deposition rate increases, the size of the columnar crystals and the holes between the grain boundaries also increase. For example, UE205, subject to 20 KV and 2 A, has a deposition rate of 1 μm/min, which is the optimum deposition condition needed to obtain the desired crystal structure in TBC systems.

1.3 Preparation processes of two-layered thermal barrier coatings (TBCs)

Traditional EB–PVD TBCs have a two-layered structure, comprising a 8YSZ top coat and a MCrAlY bond coat. Due to their simple structure and stability, they are currently used in turbine systems. The preparation process is described in detail below.

1. **Treatment of substrate surface**: To achieve good surface roughness and adhesion, EB–PVD requires a smooth surface to deposit coatings. The substrate surface is treated with different types of abrasive paper to ensure that Ra is about 1–3 μm before deposition. Since surface cleanliness can affect the adhesion between substrate and coatings, substrates are cleaned using acetone and alcohol in an ultrasonic bath.
2. **Preparation of MCrAlY**: To avoid oxidation, the background vacuum is about 10^{-3} Pa before deposition. Substrates should be uniformly heated to 800–900 °C. Substrate temperature plays an important role in obtaining perfect coatings. If it is not suitable for deposition, the bond coat will debond from the substrate. During deposition, sample holders should rotate to achieve uniform thickness and deposition rate. The deposition rate should be 1 μm/min and the thickness of bond coat about 60–90 μm.
3. **Anneal for diffusion**: After the bond coat is deposited, bond coat and substrate have an extremely weak adhesion. To make it stronger, samples should be annealed at high temperature for several hours. For NiCoCrAlY, the required temperature is 1050 °C and the annealing time is 2 h. At the high temperature, elements diffuse between the coating and the substrate, after which the interface between the coating and the substrate becomes a metallurgical bond. During annealing, the background vacuum should be at 10^{-3} Pa to avoid oxidation, and the rate of heating and cooling should be about 3 °C/min.
4. **Shot peening**: After annealing, samples should be shot blast in order to make the crystal structure more compact. There are two reasons for this.

First, grains grow during annealing and this decreases resistance to oxidation. To increase oxidation resistance the grain size needs to be small, and shot blast with suitable pressure (shot rate) and ball size can achieve this. Second, coatings prepared by EB–PVD have a lot of holes which allow oxygen ingress. After samples are shot blast, some of these holes are closed and hence oxidation resistance increases. The size of blast ball should be about Φ250 μm and the pressure 3 atm.

5. **Pre-oxide step**: Thermal grown oxides (TGO) comprise a compacted Al_2O_3 film formed between the top coat and bond coat which prevents oxygen from entering the bond coat and protects the bond coat from oxidizing. The film should be about 1–2 μm thick.
6. **Top coat deposition**: A top coat with ZrO_2–8wt.% Y_2O_3 is deposited on the surface of the bond coat. During depositing, the substrate is heated to 800–900 °C and the background vacuum should be at 10^{-2} Pa. Since there is oxygen in the ZrO_2–8wt.% Y_2O_3, the background vacuum is not very high. The deposition rate is 1–2 μm/min, and the top coat should be about 120 μm thick.

Thermal cycling tests are carried out to evaluate the resistance of coatings to thermal shock using the following method: the specimens are exposed to air for 0.5 h at 1323 K or 1373 K, and then cooled to room temperature within 5 min by forced air cooling.

1.3.1 Traditional TBCs

As an example, traditional two-layered structured TBCs are prepared by EB–PVD equipment comprising four electron guns heating Ni_3Al-based alloy IC6.[12] Intermetallic compound (IC) based alloys have been the subject of research for potential application in hot-section components[13–15] and a high-performance cast Ni_3Al base alloy, alloy IC6, has recently been developed for turbine blades, vanes of advanced aero-engines, and other high-temperature structural components.[16] The sample of alloy IC6 used was rectangle-shaped with dimensions about 15 mm × 10 mm × 3 mm. The TBCs top coat was 8wt.% Y_2O_3 stabilized ZrO_2 (YSZ) about 150 μm thick. Both bond coats were about 50 μm thick. Isothermal oxidation was carried out at 1373 K in air, and a thermal cycling test was performed by exposing the sample to air at 1373 K for 30 min followed by forced air-cooling to room temperature within 5 min. A TG-328A thermal analytical balance was used to measure the weight gain of the specimens during oxidation. Scanning electron microscopy (SEM) with energy disperse spectrum (EDS) was employed to study the microstructure of the coatings and the composition distribution across the bond coat and the interface between the bond coat and the alloy IC6.

Oxidation of TBCs

Figure 1.4 shows the isothermal oxidation curves of the alloy IC6 with and without coatings. Being exposed to air at 1373 K for 300 h, the weight gain is only about 0.359 mg/cm^2 for the alloy with TBCs, which is much less than that for the alloy without coatings. This indicates that the current TBC system forms an excellent oxidation protection layer. On the other hand, even though it is well known that only the bond coat in a TBCs system has a role in oxidation protection, the oxidation weight gain for the alloy with TBCs is also lower compared to that coated with only a MCrAlY layer. Therefore, the alloy with TBCs results in lower oxidation weight gain.

Figure 1.5 is a set of cross-sectional SEM images of alloy IC6 coated with MCrAlY after being isothermally oxidized at 1373 K for 0 h and 60 h. The Mo content in the coatings is 1.3wt.%. From Fig. 1.5(a) and the EDS results, it can be seen that, even in the as-deposited state, a clear interdiffusion zone can be observed between the substrate and the MCrAlY layer. Moreover, some cavity-like images were also observed, which might be brittle phase caused by the diffusion and action of Mo, Ni, and Si.[16,17] The composition analysis also shows that the Mo content in the MCrAlY layer was about 1.3wt.%. After being isothermally oxidized for 60 h, as shown in Fig. 1.5(b), the interdiffusion zone thickened to as much as ca 20 μm. However, cavity-like images observed in the as-deposited coatings disappeared. The EDS results show that Mo diffused into the MCrAlY layer not only in the interdiffusion zone, but also in the outer layer where the content was higher than 7wt.%. It is considered that such a high Mo content will be one of the causes of the spallation of the ceramic top coat.

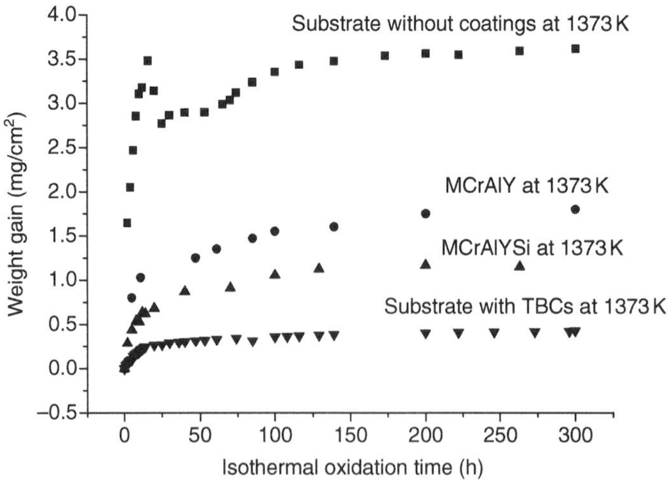

1.4 Isothermal oxidation curves of alloy IC6 with and without coatings.[12]

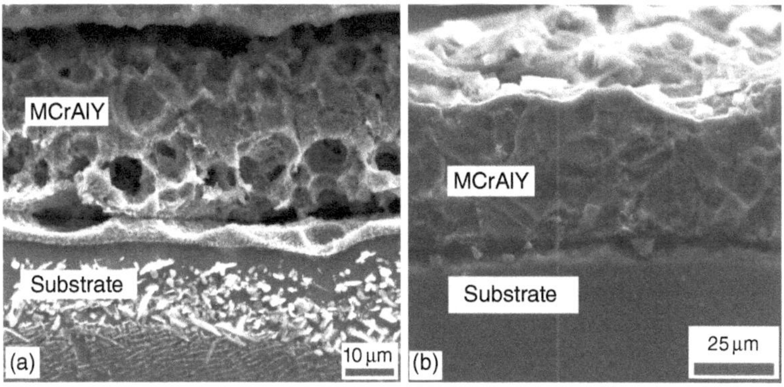

1.5 Cross-sectional SEM images of alloy IC6 coated with MCrAlY: (a) as-deposited; (b) after isothermal oxidation at 1373 K for 60 h.

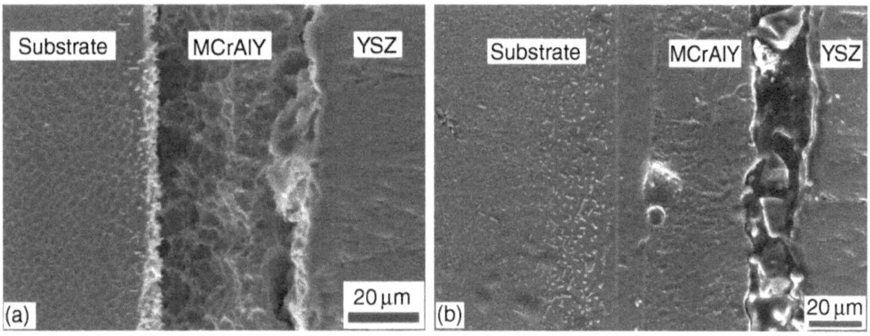

1.6 SEM cross-sectional micrographs of two-layered TBCs: (a) after thermal cycle testing at 1373 K for 30 h; (b) after thermal cycle testing at 1373 K for 300 h.

Figure 1.6 shows a cross-sectional SEM image of TBCs with MCrAlY as the bond coat after a thermal cycling test at 1373 K for 30 h and 300 h. From Fig. 1.6(a), it can be seen that a thin and uniform TGO layer formed at the interface between YSZ and MCrAlY after the thermal cycling test at 1373 K for 30 h. Figure 1.6(b) shows the cross-sectional SEM image of TBCs after the thermal cycling test at 1373 K for 300 h. It can be seen that a thick TGO layer formed at the interface between YSZ and MCrAlY, which would result in the spallation of YSZ.

1.4 Factors affecting thermal cyclic behavior of TBCs

1.4.1 Effects of pre-oxide layer thickness on thermal cyclic behavior of TBCs

Typical TBCs consist of a MCrAlY bond coat and an yttria stabilized zirconia (YSZ) ceramic top coat.[18] In such two-layered TBC systems, a TGO layer always forms at the surface of the bond coat, because the YSZ is transparent to oxygen at high temperatures. The TGO layer plays an important role in TBCs performance.[17,19–21] It is recognized that the failure of TBCs prepared by EB–PVD almost always occurs at the TGO layer with a critical TGO layer thickness of about 5 μm.[22,23] The spallation of the YSZ top coat from the bond coat is reported to be caused by stresses arising from the formation of the TGO layer and a thermal mismatch between the metallic bond coat and the YSZ top coat.[24] However, the thermal cycling test is not only a process for the accumulation of internal stress, but also a process that causes fatigue of the oxide layer and changes the interfacial adhesion, which would affect the thermal cyclic lifetime of TBCs. Conventional two-layered TBCs were prepared by EB–PVD. The substrate material in this study was a Ni-based superalloy. After vacuum annealing at 1323 K for 2 h and before ceramic deposition, the bond coat was heated to 1323 K in air for different times to obtain pre-oxide layers of about 2.5 μm, 3.0 μm, and 3.1 μm measured from SEM images. TBCs vacuum annealed for 2 h were also prepared with a pre-oxide layer about 1 μm thick. Two-layered TBCs, with pre-oxide layers of different thicknesses, were prepared by EB–PVD to investigate the effect of pre-oxide layer thickness and TGO layer microstructure on the thermal cyclic lifetime.

Figure 1.7(a) is a typical cross-sectional SEM image of the as-deposited conventional two-layered TBCs with the bond coat vacuum-annealed for 2 h. A thin pre-oxide layer about 1 μm thick can be seen between the top coat and the bond coat. Figures 1.7(b), (c), and (d) show the cross-sectional SEM images of as-deposited TBCs with pre-oxide layers about 2.5 μm, 3.0 μm, and 3.1 μm thick after exposure to air for 6 h, 14 h, and 20 h, respectively. The interface between the bond coat and the pre-oxide layer is smooth and the pre-oxide layer is intact. By comparing with Fig. 1.7(a), it can be seen that there is no distinct difference with the images except that the thickness of the pre-oxide layer is different.

Figure 1.8 is a set of cross-sectional SEM images of TBCs with different pre-heating times after they had been thermally cycled at 1323 K for 100 h. The TGO layer thickened in all specimens but at different growth rates which accelerates with increasing pre-oxide layer thickness. The uniformity and the degree of intactness of the TGO layer did not show changes in samples with a pre-oxide layer less than 3.0 μm thick. However, cracks near the interface of TGO layer and the bond coat were observed, and the TGO grains became quite coarse in the sample with pre-oxide layer about 3.1 μm thick as shown in Fig. 1.8(d), implying that the TGO layer is very brittle.

After the coatings had been thermally cycled, the thermal cycling lifetime changed with pre-oxide layers of different thickness. The thermal cyclic lifetime decreased from 730 h (1460 cycles) to 400 h (800 cycles) and the increment in TGO layer thickness for TBCs samples thermally cycled to

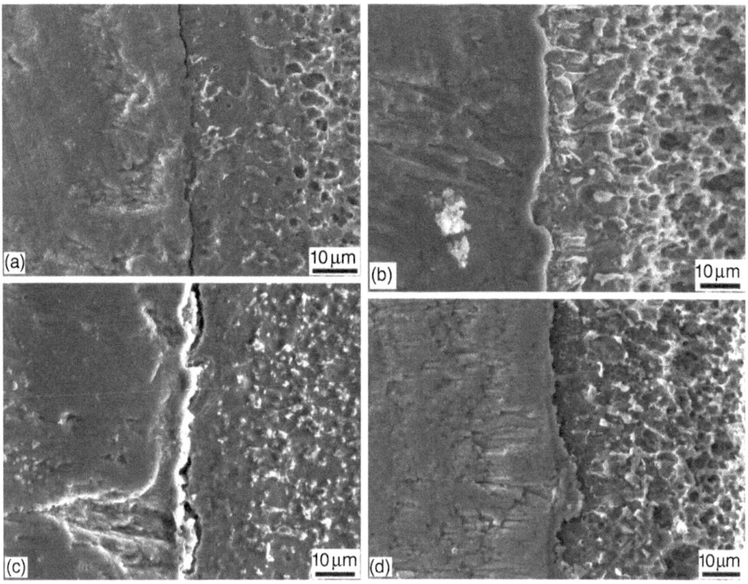

1.7 SEM cross-sectional micrographs of two-layered TBCs as-deposited: (a) a pre-oxide layer thickness of 1 μm; (b) a pre-oxide layer thickness of 2.5 μm; (c) a pre-oxide layer thickness of 3.0 μm; (d) a pre-oxide layer thickness of 3.1 μm.

failure increased from 5.6 μm to 6.7 μm with the increasing pre-heating time. The mean growth rates were about $2.1 \times 10^{-12}\ m \cdot s^{-1}$, $3.0 \times 10^{-12}\ m \cdot s^{-1}$, $3.9 \times 10^{-12}\ m \cdot s^{-1}$, and $4.7 \times 10^{-12}\ m \cdot s^{-1}$ for the TBCs with pre-oxide layers 1.0 μm, 2.5 μm, 3.0 μm, and 3.1 μm thick, respectively. The higher mean growth rate implies that cracks or fracture were easily formed in the thicker TGO layer. The total TGO layer thickness increased from 6.6 μm to 9.8 μm. The increment of oxide layer thickness results in an increase in internal stress during the thermal cycling test, and the internal stress caused by the increment of the TGO layer increased with the increase in the pre-oxide layer thickness. Given that the total oxide layer of the TBCs thermally cycled to failure also increased its thickness with the pre-heating time, at the thermal cycling to failure stage, the total internal stress would be higher for the TBCs pre-heated for a longer time compared with those pre-heated for a shorter time. If the TGO thickness were the only reason for the failure, the thermal cyclic lifetime would be much shorter than the present result (for example, 400 h). Therefore, other factors for failure have to be taken into account.

Figure 1.9 is a set of cross-sectional SEM images of Ni-based superalloy coated with TBCs with different pre-oxides after thermal cycling to spallation at 1323 K. After thermal cycling, the three kinds of TBCs had

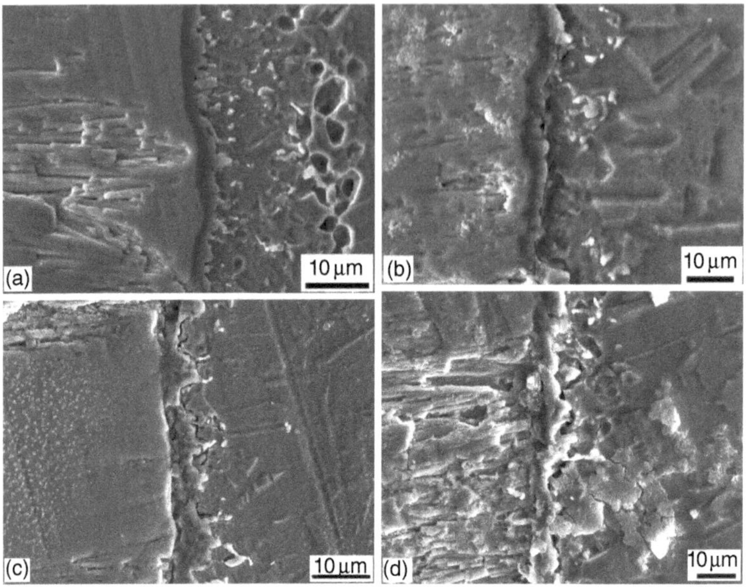

1.8 SEM cross-sectional micrographs of two-layered TBCs after thermal cycling testing at 1323 K for 100 h: (a) a pre-oxide layer thickness of 1 μm; (b) a pre-oxide layer thickness of 2.5 μm; (c) a pre-oxide layer thickness of 3.0 μm; (d) a pre-oxide layer thickness of 3.1 μm.

different lifespans. From Fig. 1.9, it can be seen that the thickness of the TGO layer increased, evidently along with the increase in thermal cycling time among the three kinds of TBCs. For the thinner pre-oxide layer, the TGO layer structure was dense, even, and intact, and, with a thicker pre-oxide layer thickness, the TGO layer had a loose, uneven, and cracked structure.

Two-layered TBCs with pre-oxides of different thickness were prepared by EB–PVD on Ni-based superalloy. As the pre-oxide thickness increased from 1 μm to 3.1 μm, the thermal cyclic lifetime of TBCs decreased from 730 h to 400 h. The growth rate of the TGO layer increased during the thermal cycling test as the pre-oxide layer thickness increased.

1.4.2 Effect of bond coat surface roughness on thermal cyclic behavior of TBCs

The TGO layer plays an important role in TBC performance.[25,26] It is well recognized that EB–PVD TBCs almost always fail at the TGO layer near the bond coat. In the present study, two-layered thermal barrier coatings were prepared by EB–PVD. The rectangle-shaped specimens of Ni-based

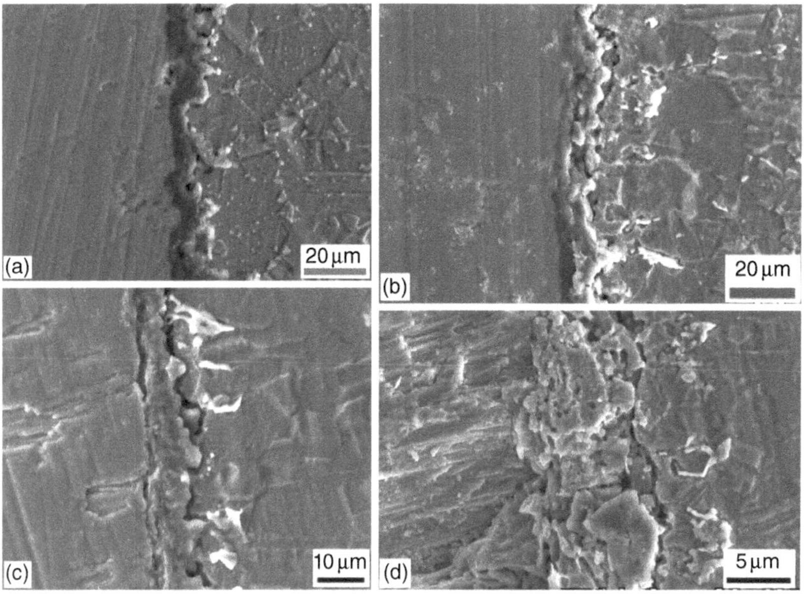

1.9 SEM cross-sectional micrographs of two-layered TBCs after thermally cycled to spallation at 1323 K: (a) a pre-oxide layer thickness of 1 μm; (b) a pre-oxide layer thickness of 2.5 μm; (c) a pre-oxide layer thickness of 3.12 μm; (d) a pre-oxide layer thickness of 4.25 μm.

superalloy DD3 with dimensions 15 mm × 10 mm × 3 mm were used as substrate materials. The YSZ (8wt.%Y_2O_3-stabilized ZrO_2) top coat and MCrAlY (Ni–24Co–15Cr–8Al–1Y, in wt.%) bond coat were about 120 μm and 60 μm thick, respectively. Before ceramic deposition, three different values of surface roughness (Ra) – 0.82 μm, 1.57 μm and 3.76 μm – were created by blasting the surface of the bond coat with different sizes of Al_2O_3 sands.

After ceramic deposition, the Ra values of the top coat were 1.51 μm, 1.71 μm, and 2.93 μm, respectively. Due to the growth of ceramic columns, the surface roughness of the top coat changed during ceramic deposition. Surface roughness of the top coat was not consistent with that of the bond coat.

Figure 1.10 shows the weight gain of TBCs vs the thermal cycling time for the specimens with surface roughness of 0.82 μm, 1.57 μm and 3.76 μm. It can be seen that the weight gain increases and the thermal cycling lifetime decreases with the increase in surface roughness. Higher roughness gives a larger surface area which results in larger weight gain during the cyclic testing. Generally, a rougher surface will increase the net area. Therefore, with the same oxidation rate, the weight gain per unit area will become

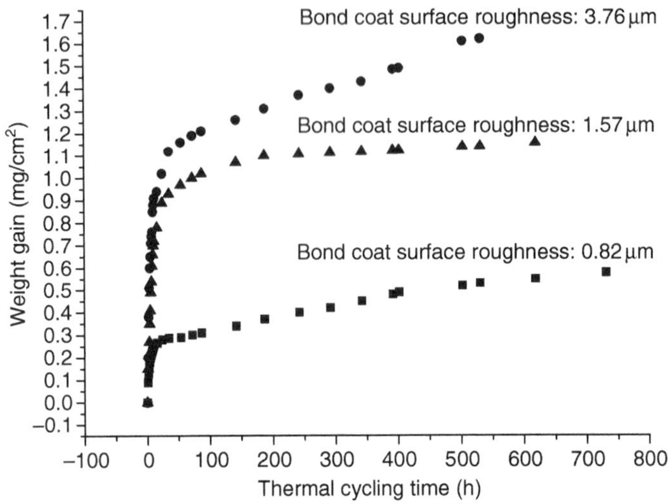

1.10 Thermal cyclic weight gain curves for three types of TBCs.[25]

larger because the area used for the calculation was fixed according to the shape of the specimen. This means that the TGO thickness does not significantly increase with surface roughness. Compared with the lifetime of 730 h (1460 cycling times) for the specimen with bond coat surface roughness of 0.82 μm, the thermal cycling lifetime for the specimen with bond coat surface roughness of 3.76 μm was only 530 h (1060 cycling times).

1.5 Preparation of graded thermal barrier coatings (GTBCs)

Traditional two-layered structured TBCs for turbine blades consist of an oxidation protective bond coat and a thermally insulating YSZ top coat.[27,28] The coating is thermally unstable under long exposure at elevated temperatures due to a TGO scale caused by the formation of alumina at the bond coat/YSZ top coat interface during selective oxidation of the bond coat.[29–32] Stresses arising from different thermal coefficients between the TGO and the bond coat lead to coating failure along the interface between the TGO and the bond coat. Compared with the traditional two-layered coating, a continuously graded coating has the advantage of forming a continuous microstructure along the cross-section of the coating and consequently avoiding stress concentrations in the coating.

There are many traditional approaches preparing GTBCs. For example, they have been prepared by the co-deposition of a tablet comprising a mixture of Al_2O_3–YSZ and YSZ onto NiCoCrAlY bond coat by means of

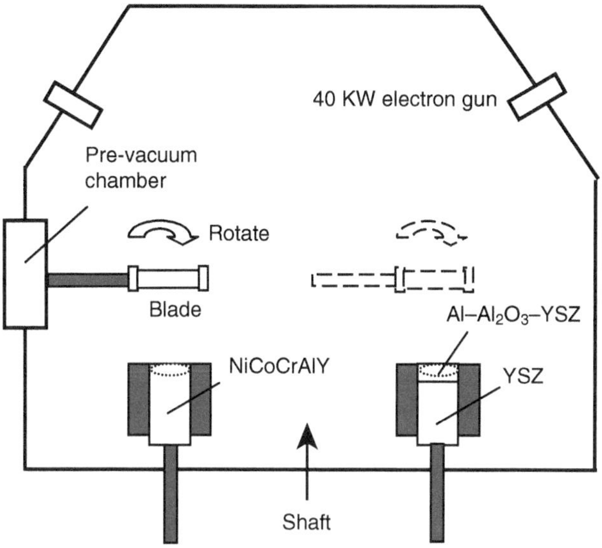

1.11 Preparation of graded TBCs.

EB–PVD. A Ni–22Co–21Cr–8Al–1Y coating about 40 μm thick was first deposited onto the substrate. Subsequently, a tablet pressed from a mixture of powders of Al_2O_3–YSZ and an ingot of ZrO_2–8wt% Y_2O_3 was evaporated by EB–PVD. As the vapor pressures of the tablet ingredients varied at the evaporation temperature, selective evaporation of the molten tablet of Al_2O_3–YSZ took place: first Al_2O_3 and then YSZ. Figure 1.11 shows the preparation diagram. As-deposited coatings were heat-treated for 4 h at 1323 K. EM with EDS was used to analyze the composition distribution across the thickness of the coating and transmission electron microscopy (TEM) to observe the microstructures. The phases in the gradient coating were analyzed by X-ray diffraction (XRD) with CuKa radiation.

Figure 1.12 shows the SEM micrograph across the thickness of the GTBC along with the EDS analysis. In the gradient coating, the left side is a NiCoCrAlY bond coat, followed by a transition layer about 10 μm thick. In the YSZ top coat, a typical columnar microstructure was observed. The EDS analysis revealed that the distribution of elements such as Ni, Al, O, and Zr changed continuously across the thickness of the coating. The transition layer consists of t–ZrO_2 phase and α-Al_2O_3 phase. In addition, $\gamma + \gamma'$ phases from Ni-based solid solution in the NiCoCrAlY bond coat can be detected. Figure 1.13 is a magnification of a cross-sectional micrograph of the transition layer. In the transition layer, a thin Al_2O_3 layer, less than 2 μm thick, and a two-phase ZrO_2/Al_2O_3 region, about 5 μm thick, are observed and the outer layer is YSZ top coat. The microstructure of the GTBC moves

1.12 Cross-section micrograph of the as-deposited gradient coating.

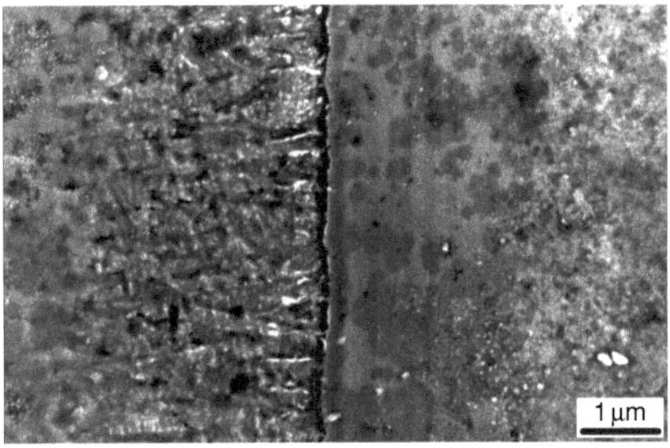

1.13 SEM cross-sectional micrograph of the transition layer in the as-deposited gradient coating.

gradually from the metallic bond coat to the YSZ top coat, due to the formation of the transition layer.

1.5.1 New investigations on bond coat

For new generation engines with a high thrust ratio, the turbine inlet temperature (TIT) could reach up to 2000 K.[33] In such a case, the traditional TBC materials, including YSZ as the ceramic top coat and MCrAlY as the metallic bond coat, might not be adequate. Recently, many efforts have been made to explore new TBC materials that can work at higher temperatures than traditional TBC materials. As a promising high-temperature material with the potential capability of working at 1473 K, NiAl could be exploited as the bond coat material in TBC systems to replace the currently used MCrAlY (M: Ni/Co or Ni+Co) bond coat.[34–38] Besides high-temperature oxidation resistance, inter-diffusion between the bond coat and the superalloy substrate should be well controlled in order to prevent the formation of the so-called secondary reaction zone (SRZ), especially in the case of Ni-based single-crystal superalloy substrate. In view of this, NiAl coatings have been modified by adding doping elements such as Pt and Hf.[35,37] Also, a double-layered structure was applied, where the NiAl was used as the outer layer, while PtAl, RuAl, and Cr formed the interlayer between the NiAl and the substrate.[39–43]

1.5.2 New investigations on top coat

6–8 wt% yttria-partially stabilized zirconia (YSZ) has been widely used as the top coat material for TBC in past decades due to its low thermal conductivity and high thermal expansion coefficient (TEC) which is compatible with that of the underlying superalloy substrate.[44,45] Recently, new ceramic candidates[46–58] for TBC application to replace the current YSZ have been increasingly studied, with the aim of developing TBCs with a higher temperature capability, a lower thermal conductivity, and longer life time. The microstructure of YSZ TBCs has also been modified to improve these properties.[59,60] The thermal conductivity of zirconia TBCs is significantly lowered by adding as dopants yttria and niobia.[61] Also, La_2O_3 has been selected as an additive to YSZ because it can significantly suppress densification of YSZ. The EB–PVD La_2O_3–YSZ coating demonstrates an extremely low thermal conductivity as well as high resistance to sintering, and the coating is characterized by feather-like sub-columns and nanopores, which contribute to limiting thermal transport.[62] It has been suggested that a thermodynamic database be established for the ZrO_2–Y_2O_3–Gd_2O_3–Al_2O_3 system in order to explore its implications for TBCs.[63]

1.6 Failure mechanism

The failure mechanism for TBCs can be considered to be the combination of internal stress accumulation as the TGO layer thickness increases and fatigue of the TGO layer during thermal cycle testing. It is believed that sintering of the ceramic top coat is also one of the reasons for failure of EB–PVD TBCs. The formation of a TGO layer and sintering of the top coat will increase residual stress and brittleness of the coating and weaken the adhesion of the top coat to the bond coat. Two-layered TBCs prepared by EB–PVD always fail between the bond coat and the TGO layer. As shown in Fig. 1.14, the SEM image indicates that the bond coat and the TGO layer debonded. When the TGO layer thickness exceeds some certain value, the interface adhesion weakens and cracks form between the bond coat and the TGO layer.

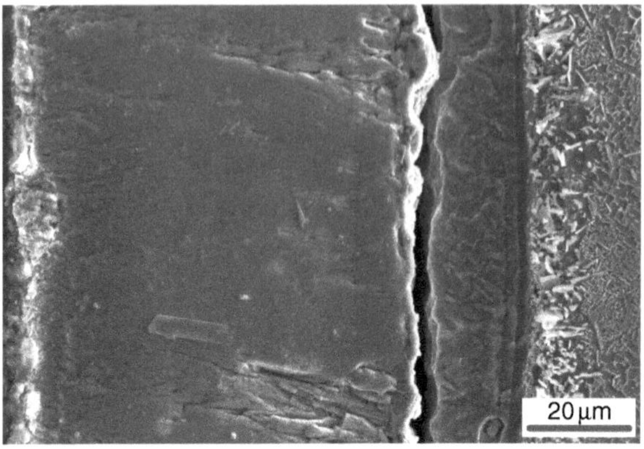

1.14 The typical cross-sectional SEM of two-layered TBCs failure.

1.7 Conclusion

Thermal barrier coatings prepared by EB–PVD have a very important role in protecting superalloys operating in high-temperature environments. Different top coat (ceramic coat) and bond coat materials will come into use in order to improve high-temperature performance. Rare earth materials with lanthanum and cerium being doped into zirconate are being investigated as elements for incorporation into EB–PVD-prepared top coat. The composition, crystal structure, and cyclic oxidation behaviour of these coatings are being studied. For the bond coat, NiAl, PtAl, and RuAl are prepared by EB–PVD and their resistance to oxidation when subjected to high temperatures is being investigated.

1.8 References

1. J.R. NICHOLLS, M.J. DEAKIN, D.S. RICKERBY, A comparison between the erosion behaviour of thermal spray and electron beam physical vapour deposition thermal barrier coatings, *Wear* **233–235** (1999) 352–361.
2. R. VAßEN, H. KAßNER, A. STUKE, F. HAULER, D. HATHIRAMANI, D. STÖVER, Advanced thermal spray technologies for applications in energy systems, *Surface and Coatings Technology* **202** (2008) 4432–4437.
3. Y. WANG, G. SAYRE, Commercial thermal barrier coatings with a double-layer bond coat on turbine vanes and the process repeatability, *Surface and Coatings Technology* **203** (2009) 2186–2192.
4. M. MOVCHAN, YU. RUDOY, Composition, structure and properties of gradient thermal barrier coatings (TBCs) produced by electron beam physical vapor deposition (EB–PVD), *Materials and Design* **19** (1998) 253–258.
5. P.S. ANDERSON, X. WANG, P. XIAO, Impedance spectroscopy study of plasma sprayed and EB-PVD thermal barrier coatings, *Surface and Coatings Technology* **185** (2004) 106–119.

6. C. VIAZZI, J.P. BONINO, F. ANSART, Synthesis by sol-gel route and characterization of yttria stabilized zirconia coatings for thermal barrier applications, *Surface and Coatings Technology* **201** (2006) 3889–3893.
7. F. TANG, J.M. SCHOENUNG, Local accumulation of thermally grown oxide in plasma-sprayed thermal barrier coatings with rough top-coat/bond-coat interfaces, *Scripta Materialia* **52** (2005) 905–909.
8. T. GOTO, Thermal barrier coatings deposited by laser CVD, *Surface and Coatings Technology* **198** (2005) 367–371.
9. L.S. PALATNIK, Y.F. KOMNIK, *Phys. Metal and Metallogr.* **10** (1960) 141–142.
10. B.A. MOVCHAN, A.V. DEMCHISHIN, Study of structure and properties of thick vacuum condensates of nickel, titanium, tungsten, aluminum oxide and zirconium dioxide, *Fizica Metalllovi Metallovedeuie* **28** (1969) 83–90.
11. R.F. BUNSHAH, Structure/property relationships in evaporated thick films and bulk coatings, *Journal of Vacuum Science and Technology* **11** (1974) 633–638.
12. D. ZHANG, S. GONG, H. XU, H. ZHANG, Y. HAN, Thermal cycling behaviors of thermal barrier coatings on intermetallic Ni_3Al based superalloy, *Surface and Coatings Technology* **168** (2003) 78–83.
13. T. NODA, Application of cast gamma TiAl for automobiles, *Intermetallics* **6** (1998) 709–713.
14. M. YAMAGUCHI, H. INUI, K. ITO, High-temperature structural intermetallics, *Acta Materialia* **48** (2000) 307–322.
15. Z.P. XING, Y.F. HAN, Z.Y. CHENG, TEM study on the microstructure of the NiCrAlYSi coating/IC6 alloy substrate, *Acta Metallurgica Sinca* **134** (1998) 1149–1152 (in Chinese).
16. Y.F. HAN, Z.P. XING, M.C. CHATURVEDI, In: *Structural Intermetallics* (1997) 713; M.V. Nathal, R. Darolia, C.T. Liu, P.L. Martin, D.B. Miracle, R. Wagner, M. Yamaguchi (eds); The Minerals, Metals and Materials Society, Warrendale, PA.
17. X.F. BI, H.B. XU, S.K. GONG, Investigation of the failure mechanism of thermal barrier coatings prepared by electron beam physical vapor deposition, *Surface and Coatings Technology* **130** (2000) 122–127.
18. D. ZHANG, S.K. GONG, H.B. XU, Effects of pre-oxide layer thickness on thermal cyclic behavior of thermal barrier coatings, *Key Engineering Materials* **326–328** (2007) 1746–1749.
19. F. JAMARANI, M. KOROTKIN, Compositionally graded thermal barrier coatings for high temperature aero gas turbine components, *Surface and Coatings Technology* **54/55** (1992) 58–63.
20. K.S. FANCEY, A. MATTHEWS, Ionisation assisted PVD of zirconia thermal barrier coatings. Proceedings of the International Conference on Metallurgical Coatings, *Journal of Vacuum Science Technology A* **4**(6) (1986) 2656–2660.
21. H.B. XU, H.B. GUO *et al.*, Development of gradient thermal barrier coatings and their hot-fatigue behavior, *Surface and Coatings Technology* **130** (2000) 133–139.
22. A.G. EVANS, D.R. MUMM, J.W. HUTCHINSON *et al.*, Mechanisms controlling the durability of thermal barrier coatings, *Progress in Materials Science* **46** (2001) 505–553.
23. A. RABIEI, A.G. EVANS, Failure mechanisms associated with the thermally grown oxide in plasma-sprayed thermal barrier coatings, *Acta Materialia* **48** (2000) 3963–3976.

24. M. KOIZUMI, K. URABE, *Funct. Gradient Compos.* **75**(6) (1989) 887–893.
25. D. ZHANG, S. GONG, H. XU, Z. WU, Effect of bond coat surface roughness on the thermal cyclic behavior of thermal barrier coatings, *Surface and Coatings Technology* **201** (2006) 649–653.
26. M.Y. ALI, S.Q. NUSIER, G.M. NEWAZ, Mechanics of damage initiation and growth in a TBC/superalloy system, *Interational Journal of Solids and Structures* **38** (2001) 3329–3340.
27. H. GUO, X. BI, S. GONG, H. XU, Microstructure investigation on porous thermal barrier coating prepared by EB–PVD, *Scripta Materiala* **44** (2001) 683–687.
28. H. GUO, S. GONG, C. ZHOU, H. XU, Investigation on hot-fatigue behaviors of gradient thermal barrier coatings by EB–PVD, *Surface and Coatings Technology* **148** (2001) 110–116.
29. B.A. MOVCHAN, I.S. MALASHENKO, K. YU. YAKOVCHUK, Two- and three-layer coatings produced by deposition in vacuum for gas turbine blade protection, *Surf. Coat. Technol.* **67** (1994) 55–63.
30. R.A. MILLER, C.E. LOWELL, Failure mechanisms of thermal barrier coatings exposed to elevated-temperatures, *Thin Solid Films* **195** (1982) 265–273.
31. S.M. MEIER, D.M. NISSLEY, K.D. SHEFFLER, T.A. CRUSE, Thermal barrier coating life prediction model development, *Trans. ASME* **114** (1992) 258–263.
32. KH. G. SCHMITT-THOMAS, Thermal barrier coatings with improved oxidation resistance, *Surf. Coat. Technol.* **68/69** (1994) 113–115.
33. H. GUO, L. SUN, H. LI, S. GONG, High temperature oxidation behavior of hafnium modified NiAl bond coat in EB–PVD thermal barrier coating system, *Thin Solid Films* **516** (2008) 5732–5735.
34. C. LEYENS, B.A. PINT, I.G. WRIGHT, Effect of composition on the oxidation and hot corrosion resistance of NiAl doped with precious metals, *Surf. Coat. Technol.* **133–134** (2000) 15–22.
35. Y. ZHANG, J.A. HAYNES, B.A. PINT, I.G. WRIGHT, W.Y. LEE, Martensitic transformation in CVD NiAl and (Ni,Pt)Al bond coatings, *Surf. Coat. Technol.* **163–164** (2003) 19–24.
36. B.A. PINT, J.R. DISTEFANO, I.G. WRIGHT, Oxidation resistance: one barrier to moving beyond Ni-base superalloys, *Mater. Sci. Eng. A* **415** (2006) 255–263.
37. B.A. PINT, K.L. MORE, I.G. WRIGHT, Effect of quaternary additions on the oxidation behavior of Hf-doped NiAl, *Oxid. Met.* **59** (3–4) (2003) 257–283.
38. B.A. PINT, The role of chemical composition on the oxidation performance of aluminide coatings, *Surf. Coat. Technol.* **188–189** (2004) 71–78.
39. D. YOSHIDA, Y. SHIBATA, S. HAYASHI, T. NARITA, Changes of an outer beta-NiAl and inner alpha-Cr coating on Ni-40 at% Cr alloy during oxidation at 1373 K in air, *Oxid. Met.* **64** (1–2) (2005) 119–130.
40. J. RIETHMUELLER, G. DEHM, E.E. AFFELDT, E. ARZT, Microstructure and mechanical behavior on of Pt-modified NiAl diffusion coatings, *J. Mater. Res.* **97** (6) (2006) 689–698.
41. B. TRYON, T.M. POLLOCK, Experimental assessment of the Ru–Al–Ni ternary phase diagram at 1000 and 1100 °C, *Mater. Sci. Eng. A*, **430** (1–2) (2006) 266–276.
42. B. TRYON, T.M. POLLOCK, M. GIGLIOTTI, *et al.*, Thermal expansion behavior of ruthenium aluminides, *Scripta Mater.* **50** (2004) 845–848.
43. Q. WEI, H. GUO, S. GONG, H. XU, Novel microstructure of EB–PVD double ceramic layered thermal barrier coatings, *Thin Solid Films* **516** (2008) 5736–5739.

44. Q. WEI, H. GUO, S. GONG, H. XU, Novel microstructure of EB–PVD double ceramic layered thermal barrier coatings, *Thin Solid Films* **516** (2008) 5736–5739.
45. D.R. CLARKE, Materials selection guidelines for low thermal conductivity thermal barrier coatings, *Surface and Coatings Technology* **163–164** (2003) 67–74.
46. J.R. NICHOLLS, K.J. LAWSON, A. JOHNSTONE, D.S. RICKERBY, Methods to reduce the thermal conductivity of EB–PVD TBCs, *Surface and Coatings Technology* **151–152** (2002) 383–391.
47. X.Q. CAO, R. VAßEN, D. STÖVER, Ceramic materials for thermal barrier coatings, *Journal of the European Ceramic Society* **24** (2004) 1–10.
48. M. GELL, L. XIE, X. MA, E.H. JORDAN, N.P. PADTURE, Highly durable thermal barrier coatings made by the solution precursor plasma spray process, *Surface and Coatings Technology* **177/178** (2004) 97–102.
49. D.L. YOUCHISON, M.A. GALLIS, R.E. NYGREN, J.M. MCDONALD, T.J. LUTZ, Effects of ion beam assisted deposition, beam sharing and pivoting in EB–PVD processing of graded thermal barrier coatings, *Surface and Coatings Technology* **177/178** (2004) 158–164.
50. S. RAGHAVAN, H. WANG, Thermal properties of zirconia co-doped with trivalent and pentavalent oxides, *Acta Materialia* **49** (2001) 169–179.
51. S. RAGHAVAN, H. WANG, Ta_2O_5/Nb_2O_5 and Y_2O_3 co-doped zirconias for thermal barrier coatings, *Journal of the American Ceramic Society* **87** (3) (2004) 431–437.
52. M. MATSUMOTO, N. YAMAGUCHI, H. MATSUBARA, Low thermal conductivity and high temperature stability of ZrO_2–Y_2O_3–La_2O_3 coatings produced by electron beam PVD, *Scripta Materialia* **50** (2004) 867–871.
53. D. ZHU, R.A. MILLER, Thermal conductivity and sintering behavior of advanced thermal barrier coatings, *Ceramic Engineering and Science Proceedings* **23** (2002) 457–468.
54. R. GADOW, M. LISCHKA, Lanthanum hexaaluminate – novel thermal barrier coatings for gas turbine applications – materials and process development, *Surface and Coatings Technology* **151–152** (2002) 392–399.
55. X. CAO *et al.*, Lanthanum-cerium oxide as a thermal barrier-coating material for high-temperature applications *Advanced Materials* **15** (2003) 1438–1442.
56. Q. XU, W. PAN *et al.*, Preparation and thermophysical properties of $Dy_2Zr_2O_7$ ceramic for thermal barrier coatings, *Materials Letters* **59** (22) (2005) 2804–2807.
57. D. MARKUS, R. VAßEN, D. STÖVER, $LaYbO_3$, a candidate for thermal barrier coating materials, *Ceramic Engineering and Science Proceedings* **24** (3) (2003) 637–643.
58. J. WU, X. WEI *et al.*, Low-thermal-conductivity rare-earth zirconates for potential thermal-barrier-coating applications, *Journal of the American Ceramic Society* **85** (12) (2002) 3031–3035.
59. S. GU, T.J. LU, D.D. HASS, H.N.G. WADLEY, Thermal conductivity of zirconia coatings with zig-zag pore microstructures, *Acta Materialia* **49** (2001) 2539–2547.
60. X. CAO, R. VAßEN, F. TIETZ, D. STÖVER, New double-ceramic-layer thermal barrier coatings based on zirconia–rare earth composite oxides, *Journal of the European Ceramic Society* **26**(3) (2006) 247–251.
61. M. MATSUMOTO, N. YAMAGUCHI, H. MATSUBARA, La_2O_3 doped Y_2O_3 stabilized ZrO_2 TBC prepared by EB–PVD, *Key Engineering Materials* **317–318** (2006) 501–504.

62. D.S. ALMEIDA, C.R.M. SILVA, M.C.A. NONO, C.A.A. CAIRO, Thermal conductivity investigation of zirconia co-doped with yttria and niobia EB–PVD TBCs, *Materials Science and Engineering, A* **443** (1–2) (2007) 60–65.
63. O. FABRICHNAYA, S. LAKIZA, C. WANG *et al.*, Thermodynamic database for the ZrO2-YO3/2-GdO3/2-AlO3/2 system and application to thermal barrier coatings, *Journal of Phase Equilibria and Diffusion* **27** (4) (2006) 343–352.

2 Ceramic thermal barrier coating materials

W. MA and H. DONG, Inner Mongolia University of Technology, China

Abstract: This chapter gives a summary of the use of ceramic materials such as the standard yttria stabilized zirconia (YSZ) and some further candidate materials, Al_2O_3, TiO_2, mullite, CaO/MgO + ZrO_2, CeO_2 + YSZ, Y_2O_3 + HfO_2, zircon, $La_2Zr_2O_7$, $La_2Ce_2O_7$, $LaMgAl_{11}O_{19}$, $SrZrO_3$, etc. as TBC materials. The thermophysical properties and corrosion-resistance characteristic of the TBC candidate materials and their thermal cycling behavior are discussed. The primary future development directions for ceramic TBC materials are also given in this chapter.

Key words: ceramic TBC material, YSZ, thermal conductivity, thermal expansion coefficient, thermal cycling.

2.1 Introduction

Some ceramic materials possessing low thermal conductivities have been used as thermal barrier coating (TBC) materials to increase turbine inlet temperatures since the 1960s. With increasing engine efficiency, they can reduce metal surface temperatures and prolong the lifetime of the hot sections of aero-engines and land-based turbines. The first application of TBCs in aerospace was developed by the National Advisory Committee for Aeronautics (NACA) and the National Bureau of Standards (NBS). The coating of calcia stabilized zirconia on the exhaust nozzle of the X-15 manned rocket plane in 1960s was believed to be the first use of TBCs in manned flight.[1] The ceramic coatings were first successfully applied to turbine section components in research engines in the mid-1970s,[2,3] and by the 1980s they were applied in the turbine section components of certain commercial gas turbine engines.[4] With further development, several ceramic materials such as YSZ (yttria stabilized zirconia), Al_2O_3, TiO_2, mullite, CaO/MgO + ZrO_2, CeO_2 + YSZ, Y_2O_3 + HfO_2, zircon, $La_2Zr_2O_7$, $La_2Ce_2O_7$, $LaMgAl_{11}O_{19}$, $SrZrO_3$, etc. have been evaluated as TBC materials. Some detailed reviews on ceramic materials for TBCs have been contributed by Cao *et al.*,[5] Levi,[6] Schulz *et al.*[7,8] and Clarke and Phillot.[9]

The selection of TBC materials is restricted by some basic requirements, such as high melting point, no phase transformation between room temperature and operating temperature, low thermal conductivity, chemical inertness, thermal expansion matching with the metallic substrate, good adherence to the metallic substrate and low sintering rate of the porous microstructure.[10,11] So far, no one material can fulfill all the requirements mentioned above; instead, only some of them can be satisfied. The properties of some ceramic materials which have been evaluated as TBC materials are listed in Table 2.1. Thermal expansion coefficient (TEC) and thermal conductivity seem to be the most important properties for TBC materials selection. These data are collected from different references and hence may not be complete. The number before YSZ represents the weight percentage of Y_2O_3 in ZrO_2. The advantages and disadvantages of some TBC materials are listed in Table 2.2.

2.2 State-of-the-art ceramic thermal barrier coating (TBC) material – yttria stabilized zirconia (YSZ)

Prior to 1975, pure zirconia (ZrO_2) was of very limited interest as a structural or engineering ceramic, and its use was restricted to refractory applications.[12] This is attributed to the displacive tetragonal (*t*) to monoclinic (*m*) phase transformation, which occurs at ~950 °C on cooling in pure ZrO_2 and is accompanied by a shear strain of ~0.16 and a volume expansion of ~4%. The change of shape with the transformed volume can result in catastrophic fracture and hence structural unreliability of fabricated components. However, ZrO_2 has other intrinsic physical and chemical properties, including high hardness, good wear resistance, low coefficient of friction, chemical inertness, ionic conductivity, low thermal conductivity and high melting point, which make it attractive as an engineering material. The focus of continued effort is to understand and improve its mechanical behavior.[13] The dramatic increase in its industrial applicability has been brought about by the discovery that the $t \rightarrow m$ phase transformation can be controlled by suitable materials processing to become the source of transformation plasticity and transformation toughening in tailored, two-phase microstructures.[14]

Pure ZrO_2 experiences the following phase transformations:

$$monoclinic(m) \xleftrightarrow[950^{\circ}C]{1170^{\circ}C} tetragonal(t) \xrightarrow{2370^{\circ}C} cubic(c) \xrightarrow{2680^{\circ}C} liquid$$

Pure ZrO_2 at ambient pressure is polymorphic, exhibiting cubic (*c*) (fluorite) structure ($Fm\bar{3}m$) at high temperatures (>2370 °C), tetragonal (*t*) structure ($P4_2/nmc$) at intermediate temperatures (1200 °C–2370 °C) and monoclinic (*m*) symmetry ($P2_1/c$) at low temperatures (<950 °C). The $t \rightarrow m$ phase transformation in pure ZrO_2 commences at ~950 °C (referred to as M_s) on cooling and is reversible at ~1150 °C on heating.[15] The *t* and *m*

Table 2.1 The properties of TBC materials[5]

Materials	T_m(K)	D_{th}(m^2s^{-1}) × 10^{-6}	λ ($Wm^{-1}K^{-1}$)	α (K^{-1}) × 10^{-6}	E (GPa)	C_p ($J_g{}^{-1}K^{-1}$)	ν
ZrO_2	2953	0.43	2.17	15.3	21		0.25
3YSZ	2973	0.58	2.12	11.5(293–1273 K)		0.64	
8YSZ (APS)				10.7(293–1273 K)	40		0.22
18YSZ				10.53			
5 wt% CaO + ZrO_2	2558 (soften)			9.91	149.3		0.28
$3Al_2O_3{\cdot}2SiO_2$	2123		3.3	5.3(293–1273 K)	30		0.25
Al_2O_3	2323	0.47	5.8	9.6	30		0.26
Al_2O_3 (TGO)				8(293–1273 K)	360		0.22
Al_2O_3 + TiO_2		0.65		5.56(1073 K)			
CeO_2	2873	0.86	2.77	13(293–1500 K)	172	0.47	0.27–0.31
$La_2Zr_2O_7$	2573	0.54	1.56	9.1(293–1273 K)	175	0.49	
$La_2Ce_2O_7$		0.19	0.6	12.3(573–1473 K)		0.43	
$BaZrO_3$	2963	1.25	3.42	8.1(293–1273 K)	181	0.45	
$SrZrO_3$	2883	1.4	2.08(1273 K)	10.9(473–1373 K)	170	0.46	0.25
YBa_2ZrO_6	1973		0.37(543 K)		120	2.8	
$CaTi_2ZrO_7$			2.1	9.04		0.7	
TiO_2	2098	0.52	3.3	9.4(293–1500 K)	283		0.28
$Y_2Ti_2O_7$	>2173			11.0(293–1293 K)			
$Y_3Al_5O_{12}$	2243		3.0	9.1			
$LaMgA_{l1}O_{19}$			1.7	10.1(298–1473 K)		0.86	
$LaPO_4$	2343		1.8	10.5	133		0.28
$CePO_4$	2318	1.4		9–11		0.43	
2MgO·NiO	2573		1.9				
$BaNd_2Ti_3O_{10}$ (BNT) (Applied Thin Film, Inc.)	>1800		0.5 (1573 K)	11.5			
NiCoCrAlY (bond coat)				17.5(293–1273 K)	86		0.3
IN737 superalloy (substrate)				16(293–1273 K)	197		0.3

Symbols have the following meanings: D_{th}, thermal diffusivity; *E*, Young's modulus; α, thermal expansion coefficient; λ, thermal conductivity; C_p, heat capacity; ν, Poisson's number; T_m, melting point; TGO, thermally grown oxide on bond coat.

Table 2.2 Some TBC materials and their characteristics[5]

Materials	Advantages	Disadvantages
7–8 YSZ	(1) high thermal expansion coefficient (2) low thermal conductivity (3) high thermal-shock resistance	(1) sintering above 1473 K (2) phase transformation (1443 K) (3) corrosion (4) oxygen-transparent
Mullite ($3Al_2O_3 \cdot 2SiO_2$)	(1) high corrosion resistance (2) low thermal conductivity (3) good thermal-shock resistance below 1273 K (4) not oxygen-transparent	(1) crystallization (1023–1273 K) (2) very low thermal expansion coefficient
Alumina (Al_2O_3)	(1) high corrosion resistance (2) high hardness (3) not oxygen-transparent	(1) phase transformation (1273 K) (2) high thermal conductivity (3) very low thermal expansion coefficient
YSZ + CeO_2	(1) high thermal expansion coefficient (2) low thermal conductivity (3) high corrosion resistance (4) less phase transformation between *m* and *t* than YSZ (5) high thermal-shock resistance	(1) increased sintering rate (2) CeO_2 precipitation (>1373 K) (3) CeO_2-loss during spraying
$La_2Zr_2O_7$	(1) very high thermal stability (2) low thermal conductivity (3) low sintering rate (4) not oxygen-transparent	(1) relatively low thermal expansion coefficient
$La_2Ce_2O_7$	(1) very high thermal stability (2) high thermal expansion coefficient (3) low thermal conductivity (4) high corrosion resistance	(1) CeO_2-loss during spraying (3) relatively high sintering rate (>1523 K)
Silicates	(1) cheap, readily available (2) high corrosion resistance	(1) decomposition into ZrO_2 and SiO_2 during thermal spraying (2) very low thermal expansion coefficient
$SrZrO_3$	(1) high thermal expansion coefficient (2) low thermal conductivity (3) low sintering rate	(1) phase transformations (2) SrO-loss during spraying

structures can be described in terms of distortions of the fluorite structure, with the major distortion associated with the $t \rightarrow m$ phase transformation corresponding to a shear of ~9° parallel to the basal plane of the *t* unit cell, to define a monoclinic β angle of ~81°. It is the apparent displacive nature of the structural change, together with the morphology of the *m* product, which has invited the application of the theory of martensite crystallography to the transformation.

The suitable addition of the stabilizing oxides such as CaO, MgO, Y_2O_3, CeO_2 and selected rare earth oxides, combined with relatively high cooling rates from sintering and solution treatment temperatures, permits metastable tetragonal ZrO_2 (t') to be retained at room temperature in the form of an equiaxed polycrystalline aggregate. Y_2O_3 stabilized ZrO_2 (YSZ) is discussed in this section while other oxides stabilized ZrO_2 are discussed in other sections. The mechanism of phase transformation toughening in ZrO_2 involves two aspects, both related to the difficulty of nucleating the martensitic transformation.[16] First, the M_s of t–ZrO_2 can be reduced to below room temperature by controlling the stabilizer concentration in this phase. More importantly, the M_s of the particle or grain size of this phase decreases with decreasing size. Zhu and Yan[17] and Basu *et al.*[18] have reported that 0.4 μm is the critical grain size for the thermally induced phase transformation of the YSZ ceramics, and that no thermally induced transformation is observed in the t–ZrO_2 materials having grain size finer than 0.4 μm.

Second, however, such fine particles undergo martensitic transformation in the stress field of propagating cracks. The volume change resulting from the stress-induced transformation in these fine t–ZrO_2 particles in a zone around the crack creates favorable tractions on the crack faces, which impedes further crack propagation and significantly increases the fracture toughness. For increasing fracture toughness, the *t*-phase transformability is the critical factor. The key is to maximize the retention of t–ZrO_2 at room temperature or application temperature with sufficient transformability in a tensile stress field. According to the above idea, partially stabilized ZrO_2, a mixture of the *c* and *m* phases, is produced by controlling the amount of stabilizer added (about 6–8 wt% Y_2O_3) with a consequence of an obvious improvement of the thermal-shock resistance compared with pure and fully stabilized zirconia materials.[19,20]

The initial YSZ TBCs contained from 12 to 20 wt% of Y_2O_3, which was added to fully stabilize the *c*-phase. Later, Stecura[21] showed that better performance could be achieved by lowering the Y_2O_3 level to between 6 and 8 wt%. So far, 7–8YSZ has been the TBC material studied and used most widely. It has a lot of good properties such as: a high melting point of ~2680 °C which makes it suitable for high-temperature applications; low thermal conductivity of ~2.1–2.2 $Wm^{-1} K^{-1}$ in bulk material at 1000 °C; high TEC of ~$11 \times 10^{-6} K^{-1}$; low density of ~6.4 g/cm^3 which can reduce the weight of the turbine components; low elastic modulus of ~40 GPa; high hardness of ~14 GPa which enables the turbine components to possess a better erosion resistance. Consequently, 7–8YSZ has been used frequently in turbine and diesel engines.[22,23] Furthermore, it has been proved that 7–8YSZ coating is more resistant against the corrosion of Na_2SO_4 and V_2O_5 compared with CaO or MgO stabilized ZrO_2 coating.[24]

Although 7–8YSZ has plenty of advantages for TBC applications, there are also some disadvantages. First, a major disadvantage of 7–8YSZ is the limited operating temperature (<1200 °C) for long-term operation. At higher temperatures, phase transformations from metastable tetragonal (t') to tetragonal and cubic ($t + c$) and then to monoclinic (m) occur, giving rise to the formation of cracks in the coating.[10,25]

Second, 7–8YSZ is sensitive to the hot corrosion.[26–28] It has a better corrosion resistance if the clean fuel is used, whereas the corrosion problem becomes more serious when using fuel containing a certain amount of S, Na and V elements. The chemical reactions between Na_2O, SO_3 or V_2O_5 in the fuel and the stabilizer of Y_2O_3 in 7–8YSZ result in the degradation of the coating. In particular, the deposit salt of V_2O_5 reacts with Y_2O_3 to get a product of YVO_4, resulting in depleting Y_2O_3 and degrading 7–8YSZ TBCs, and preventing its applications in marine gas turbine engines. The reactions may be summarized as follows:

$$Y_2O_3 \text{ (in YSZ)} + 3SO_3 \text{ (in } Na_2SO_4) \rightarrow Y_2(SO_4)_3 \text{ (in } Na_2SO_4) \quad [2.1]$$

$$Y_2O_3 \text{ (in YSZ)} + V_2O_5 \rightarrow 2YVO_4 \quad [2.2]$$

Third, the SiO_2 impurity (even as low as 1 wt%) in 7–8YSZ coating has a strong detrimental effect on the thermal cycling lifetime.[29] Firstly, in bulk ZrO_2-base ceramics, SiO_2 segregates at grain boundaries with excessive amounts collecting at triple points. SiO_2 at the grain boundaries leads to changes in the size and shape of grains and it may also dissolve Y_2O_3 from the 7–8YSZ grain boundaries leading to a localized degradation. Secondly, SiO_2 can also cause ZrO_2 polycrystal superplasticity, dramatic increase in sintering rates and decrease in electrical conductivity. Thirdly, it may also lead to increased creep rates, as what has been observed within ZrO_2-base ceramics.

Furthermore, these coatings possess a high concentration of oxygen ion vacancies, which at high temperatures assist oxygen transport and the oxidation of the bond coat at the ceramic bond/coat interface, namely the formation of thermally grown oxide (TGO) on the bond coat surface. This leads to a spallation of the ceramic, and such failure mechanism of the TBC is predominant when the coatings are thin as in gas turbines. This problem has been overcome to a large extent by providing oxidation-resistant bond coats such as alumina and mullite.[30]

2.3 Zirconia doped with one or more oxides

In addition to Y_2O_3 stabilized ZrO_2, other oxides, such as CaO, MgO, CeO_2, Nd_2O_3 or Sm_2O_3, are also investigated as the stabilizers for ZrO_2. First, CaO (3–5 wt%) stabilized ZrO_2 (CaO–ZrO_2) coating prepared by atmospheric plasma spraying (APS) has between four and three times lower thermal diffusivity compared with that of a sintered ZrO_2 at a temperature below 1027 °C.[31] This coating experiences a sintering effect above 1027 °C, resulting in a remarkable decrease of the thermal insulation effect. CaO–ZrO_2 did not prove to be a viable material for the more advanced TBC applications. The problem with this material is related to 'destabilization' from *c*–ZrO_2 that is observed in the as-sprayed coating to *m*–ZrO_2. Although the CaO is added to ZrO_2 with the intent of extending the high-temperature *c*-phase field down to room temperature, it is now known that the *c*-phase is actually not stable below 1140 °C.[32] However, it proves that the addition of CeO_2 in the CaO–ZrO_2 system conveys excellent thermal-shock resistance at 1000 °C and 1200 °C, which is superior to that of 5 wt% CaO–ZrO_2 coatings.[33] Moreover, no destabilization of *c*/*t*–ZrO_2 phase occurred during thermal shock tests.

Second, MgO (15–24 wt%) stabilized ZrO_2 (MSZ) has an ability to avoid the detrimental $t \rightarrow m$ phase transformation of ZrO_2,[34] with the equilibrium phases of *c*–ZrO_2 and MgO. MgO (24 wt%)–ZrO_2 has a relatively high TEC of 12.3×10^{-6} K^{-1}.[15] For diesel engines operating with low-quality fuels, which yield liquid salt deposits on the ceramic-coated components, the MSZ system has better properties than the YSZ system in these environments in which the corrosion resistance to vanadate attack is better, but problems with sulphidation of the MgO cannot be ruled out. The MgO reacts readily with SO_3 to form the melt-soluble $MgSO_4$, with the reaction of MgO in contact with molten $NaVO_3$ even at very low partial pressures (10^{-7} bar) of SO_3.[35] Extensive formation of $MgSO_4$ leading to the degradation of the MSZ TBCs in low-quality fuel burner rig tests has also been reported.[36] Moreover, thermal cycling to only 1000 °C/1050 °C has revealed a progressive destabilization of the MSZ coatings, apparently as the result of rapid coarsening of the tetragonal MSZ grains followed by spontaneous $t \rightarrow m$ phase transformation.[37] In addition, the MSZ system has a lower strength and erosion resistance.[38] Although the MgO is added to ZrO_2 with the intent of extending the high-temperature *c*-phase field down to room temperature, it proves that the *c*-phase is actually not stable below 1400 °C.[32] Toriz *et al.* gave 950 °C as the practical upper-use temperature for MSZ in gas turbine applications.[39] Therefore, these shortcomings combine to make MSZ an unpromising candidate for naval ship engine applications.

Third, stabilization of *t*–ZrO_2 in the CeO_2–ZrO_2 system (CeSZ) can be kept over a wide range of compositions, 16–26 wt% CeO_2, with the preferred composition being 25 wt% CeSZ. CeSZ displays considerably greater moisture stability than YSZ under similar environmental conditions.[40,41] Although the attainable fracture strength of (21 wt%) CeSZ is not as high as that of YSZ, the toughness can be considerably greater (maximum fracture toughness for YSZ is ~10 MPa·m$^{1/2}$, whereas that for CeSZ is ~17 MPa·m$^{1/2}$).[42] Some tests also indicate that CeO_2 reacts less readily than Y_2O_3 with molten vanadate–sulfate systems.[35,43,44] Miller [45] has reported that 12–20 wt% CeO_2 stabilized ZrO_2 exhibited instability, indicating they could never be realistically considered as TBC materials for gas turbines. The CeO_2(25 wt%)–ZrO_2 material offered some benefits as its thermal and mechanical properties were reasonable, but independent erosion tests showed that its erosion resistance was significantly inferior to that of the YSZ systems. However, the plasma-sprayed CeO_2(25 wt%)–ZrO_2 yields t'-phase which is retained even after 100 h annealing at 1600 °C, and does not undergo a $t \rightarrow m$ phase transformation under low-temperature cycling.[46]

Furthermore, Lee *et al.* has reported the same system with some different results.[47] CeSZ TBCs prepared by APS with 16 and 18 wt% CeO_2 consist of only t'-phase with high CeO_2 content, whereas the CeSZ coatings with 20–26 wt% CeO_2 consist of a mixture of t'-phase with high CeO_2 content and some amount of metastable cubic (c') phase with low CeO_2 content. During thermal cycling at 1130 °C, the t'-phase and the mixture of the t'-phase and the c'-phase decompose to c-phase with high CeO_2 content and t-phase with low CeO_2 content. Some of the t-phase transforms to the m-phase on cooling. More m-phase is observed to develop in the CeSZ coatings with 16 wt% CeO_2 than those with 18–26 wt% CeO_2, and the CeSZ coatings lifetimes with 18–26 wt% CeO_2 are longer than those with 16 wt% CeO_2. The tolerable amount of CeO_2 in ZrO_2 alloy required for TBCs is >16 wt%. This material is proposed for ultra-high temperature TBCs, although it may be lacking in erosion resistance as compared to YSZ. Another disadvantage for this material as a TBC material is that CeO_2 has a potential to be reduced to Ce_2O_3 with much decreased melting point.

However, it appears that even in melts such as $NaVO_3$ where the V_2O_5 activity is too low for direct reaction, CeSZ may nonetheless be destabilized by a mineralization effect. In burner rig tests with Na, S and V impurity additions, CeSZ suffered chemical attack, forming both cerium sulfates (mostly in the lower temperature tests at 700 °C) and $CeVO_4$; it was difficult to be certain, perhaps because of the high V and S levels used, that CeSZ was significantly more resistant than YSZ to hot corrosion in these particular tests.[36]

YSZ coating doped with CeO_2 is supposed to bring an improvement in thermal cycling lifetime, due to the higher TEC and the lower thermal conductivity of CeO_2 compared with that of YSZ. A remarkable improvement in thermal shock tolerability was attained by the addition of CeO_2

into YSZ.[48–50] The CeO_2-doped YSZ coating has a better thermal-shock resistance mainly because of [51]: (i) little phase transformation between the *t*-phase and *m*-phase in this coating; (ii) a smaller stress generated by the bond coat oxidation in this coating due to the better thermal insulation; (iii) a larger TEC in this coating. However, the addition of CeO_2 into YSZ has some negative effects similar to those found in the CeSZ system, such as the decrease of hardness and the stoichiometry change of the coating due to the vaporization of CeO_2,[50] the reduction of CeO_2 into Ce_2O_3 and the accelerated sintering rate of the coating.[52,53]

Fourth, some rare earth oxides have also been reported as the stabilizers of ZrO_2 besides Y_2O_3 and CeO_2, which are Dy_2O_3,[8] Nd_2O_3, Er_2O_3, Sm_2O_3,[54] Yb_2O_3,[8,55] Sc_2O_3,[56–58] Gd_2O_3,[59] Ta_2O_5, Nd_2O_5[60,61] and so on. The different stabilizers have various levels of the stabilizing ability, together with the varied stabilizing mechanisms, that are closely correlated with cation migration at high temperatures. The stabilization ability of different oxides has been summarized by Raghavan *et al.*,[60] and it has an increasing tendency with the cation radius increase from Sc^{3+} to Y^{3+}, whereas it decreases with further increase of the cation radius from Y^{3+} to La^{3+}.

Rare earth oxide stabilizers such as Dy_2O_3 and Yb_2O_3 behave similarly to Y_2O_3: a 4 mol.% addition creates a *t*′-phase while a 12 mol.% addition creates a *c*-phase. A reduction of ~40% in thermal conductivity was achieved with an optimized version of 12 mol.% DySZ. The cyclic lifetimes of these new electron beam physical vapor deposition (EB–PVD) TBCs were comparable to that of YSZ.[8] Zhang *et al.*[55] have studied the thermal cycling behavior of a ZrO_2 coating stabilized by Yb_2O_3, and proved that the cyclic lifetime of the 2.6 mol.% Yb_2O_3–ZrO_2 coating is increased by ~30% compared with that of YSZ (3.4 mol.% Y_2O_3) under the same conditions.

Khor and Yang[54,62,63] have studied coatings of ZrO_2 stabilized by Nd_2O_3 (NdSZ), Er_2O_3 (ErSZ) and Sm_2O_3 (SmSZ), respectively. Grinding of the three plasma-sprayed coatings in the 4–6 mol.% Nd_2O_3, Er_2O_3 or Sm_2O_3 revealed no $t \rightarrow m$ phase transformation. In the 2–3 mol.% Nd_2O_3, Er_2O_3 or Sm_2O_3 composition, there is *t*′-phase in different levels, respectively. Due to the fast cooling effect of the APS process, all the three coatings mainly have the *t*′-phase which would not transform into the other phases after mechanical treatment in contrast with *t*–ZrO_2. The stabilization ability of the *t*′-phase against heat treatment has the order of NdSZ < ErSZ < SmSZ, and it is supposed that this phenomenon may result from the fast diffusion speed of Nd^{3+} and the low diffusion speed of Sm^{3+} at high temperatures.[54]

Compared with the YSZ coating, the ScSZ coating (Sc_2O_3–ZrO_2) has a higher corrosion resistance against the vanadates and sulfates, and the *t*′-phase is more stable at high temperatures.[56] Jones *et al.*[57,58] have demonstrated that the Sc_2O_3, Y_2O_3-stabilized ZrO_2 (SYSZ) with 7 mol.% of total amount of stabilizer (the stabilizer contains 2.5–20 mol.% Y_2O_3) gives significantly better t'-phase stability at 1400 °C than either pure Y_2O_3- or Sc_2O_3-stabilized ZrO_2.Sc_2O_3-rich SYSZ powder shows 5–10 times better resistance to vanadate-induced destabilization and better high-temperature phase stability than YSZ powder. This suggests that SYSZ could be a possible candidate for ultra-high temperature TBC application.

Wu [59] has explored the Gd_2O_3–ZrO_2 (GdSZ) system (with chemical composition of $0.15GdO_{1.5} \cdot 0.85ZrO_2$) as a TBC candidate material. The thermal conductivity of GdSZ is ~30% lower than that of 7YSZ, and it is chemically stable at high temperatures in contact with α-Al_2O_3 (main composition in TGO). However, the thermal cycling lifetime of EB–PVD GdSZ TBC is shorter than that of 7YSZ.

Raghavan *et al.*[60,61] have investigated ZrO_2–Y_2O_3–Ta_2O_5/Nb_2O_5 as the TBC candidate materials. After choosing specific compositions, single-phase (*t*-phase) and dual-phase (*t*-phase + *c*-phase) compositions that are stable despite ageing at 1500 °C for 200 h and having a thermal conductivity as low as the 6–8YSZ were synthesized. In particular, a 16.6 wt% $YO_{1.5}$ + 16.6 wt% $TaO_{2.5}$ stabilized ZrO_2 composition was insensitive to the $t \rightarrow m$ phase transformation upon thermal cycling, and it also showed only slight evidence of corrosion in vanadate–sulfate melts after 500 h, compared with the extensive attack and spallation within much shorter time (50–100 h) for 7YSZ. Properly designed co-doping thus circumvents the problems of thermal and chemical destabilization in 7YSZ.[64]

Matsumoto *et al.* [65,66] have studied the influence of some oxides additives, namely Al_2O_3, CoO, SrO, La_2O_3, Pr_2O_3, Nd_2O_3 and Er_2O_3, on the densification of YSZ compacted powder, and found that the addition of Al_2O_3 or CoO gives significant increase in density. In terms of sintering, these compounds will be deleterious for YSZ TBC if they are used in secondary phase. In contrast, the addition of SrO, La_2O_3, Pr_2O_3 and Nd_2O_3 decreases density. Among those, La_2O_3 has the greatest effect on suppression of densification, followed by SrO, Pr_2O_3 and Nd_2O_3. The addition of Er_2O_3 has little effect on densification. The La_2O_3 doped YSZ plasma-sprayed coating has been investigated and reveals a low thermal conductivity even after high-temperature annealing. However, too much La_2O_3 addition leads to decreased t'-phase stability and cyclic lifetime of the coating; the optimum amount of La_2O_3 is about 1 mol.%.

Zhu and Miller[67] have investigated the thermal properties of ZrO_2-based TBCs stabilized by co- or multi-dopants, namely YSZ–Nd, YSZ–Yb, YSZ–Nd–Yb, YSZ–Gd–Yb and YSZ–Sm–Yb. YSZ and YSZ with one more dopant show typically higher thermal conductivities than the YSZ coatings with paired dopant additions. The total dopant concentration affects thermal conductivity as well. A low thermal conductivity regime was observed for plasma-sprayed TBCs in the range 6–15 mol.% total dopant concentration and about 10 mol.% for the EB–PVD coatings. Multi-component cluster oxide coatings were thermal cyclic tested at 1160 °C and showed longer cyclic lifetimes than YSZ at given dopant concentrations. The coating cyclic lifetime generally decreased as the total dopant concentration was increased. Ternary additions of Gd/Yb/Sc and Nd/Yb/Sc did not further improve the thermal insulation property.

2.4 Yttria stabilized hafnia and other alternative ceramic TBC materials

2.4.1 Yttria stabilized hafnia

HfO_2–Y_2O_3 (in ~19 mol.%, ~25 mol.% and ~38 mol.%) coatings prepared by EB–PVD have been studied by Singh *et al.*,[68] and the thermal conductivity of HfO_2–25 mol.% Y_2O_3 was found to be the lowest (1.1 W/mK) among the three compositions. HfO_2–25 mol.% Y_2O_3 exhibited a relatively low conductivity rate increase at high temperatures due to its good sintering resistance as compared with 8YSZ. In addition, the HfO_2–25 mol.% Y_2O_3 plasma-sprayed coating also has a better sintering resistance than YSZ coating.[69] Zhu *et al.*[70] compared the 1650 °C sintering and cyclic behavior of the HfO_2–18 mol.% (Y, Gd, Yb)$_2O_3$ coating with that of HfO_2–5 mol.% Y_2O_3 and HfO_2–15 mol.% Y_2O_3 coatings. The advanced multi-component HfO_2-based coating had a relatively slow conductivity increase during the first 20 h of steady-state testing, as well as a good water-vapor-stability. It also showed essentially no cracking and delamination during subsequent testing of 100 30-min cycles at 1650 °C, indicating excellent sintering resistance and cyclic durability.

Ibegazene *et al.*[71] have studied the influence of HfO_2-addition on the structure and mechanical properties of 8YSZ coatings prepared by APS. In the coating, only the *t*′-phase existed; this phase transformed into the mixture of $t + c$ phases during annealing at 1200 °C and the *t*-phase would further transform into the *m*-phase. Young's modulus increased with increase in the HfO_2 content.

2.4.2 Al_2O_3

Al_2O_3 did not prove to be a viable material for the more advanced TBC applications, primarily because of its relatively high thermal conductivity and low TEC compared with YSZ.[72] The plasma-sprayed Al_2O_3 coating contains mainly unstable phases such as γ- and δ-Al_2O_3. These unstable phases will transform into stable α-Al_2O_3 during thermal cycling, accompanied by a significant volume change ($\gamma \rightarrow \alpha$, ~15%) which results in microcrack formation in the coating.[73,74] The doping of transition metal oxides such as Cr_2O_3, Fe_2O_3 or TiO_2 into Al_2O_3 can only partially stabilize the α-Al_2O_3. Even though Al_2O_3 alone is not a good TBC candidate, its addition to YSZ can improve the hardness and bond strength without substantial modification of Young's modulus and toughness, as well as the oxidation resistance of the substrate.[75–77] A gradient coating of 8YSZ–Al_2O_3 has a much longer thermal cycling lifetime than that of the 8YSZ coating.[78]

2.4.3 Mullite

Mullite ($3Al_2O_3 \cdot 2SiO_2$) is an important ceramic material because of its low density, high thermal stability, high chemical stability in severe environments, low thermal conductivity and favorable strength and creep behavior.[79] Compared with YSZ, mullite has a much lower TEC and a higher thermal conductivity, and is much more oxygen-resistant than YSZ. Mullite contains a larger amount of amorphous phase resulting from the rapid cooling of molten mullite on a cold substrate during the APS process.[80] The amorphous phase in the coating crystallizes in a subsequent exposure to high temperature (750–1000 °C), accompanied by a volume contraction, causing cracking and debonding.[30] For applications such as diesel engines where the surface temperatures are lower than those encountered in gas turbines and the temperature variations across the coating are large, mullite is an excellent alternative to ZrO_2 as a TBC material. Engine tests performed with both materials show that the lifetime of the mullite coating in the engine is significantly longer than that of the YSZ coating.[81,82] Above 1000 °C, the thermal cycling lifetime of the mullite coating is much shorter than that of YSZ.[30] Furthermore, mullite is the most promising coating material for SiC substrate because their TECs are similar.[80]

Gilbert *et al.*[83,84] have investigated the thermal fracture of mullite–YSZ composite TBCs under thermal shock by both experimental and numerical studies. They demonstrated that the mullite–YSZ composite coatings reduced the driving force for coating delamination. When compared to the monolithic YSZ coatings, the composite coatings have a higher threshold temperature for both surface and horizontal crack initiation. Since the mullite–YSZ composite coating has a higher thermal resistance compared to the monolithic mullite coating, it is an excellent candidate for applications where high temperature and temperature gradients are present.

2.4.4 Pyrochlore oxides ($A_2B_2O_7$)

Pyrochlore oxides ($A_2B_2O_7$) are relatively new TBC candidate materials and have been paid much attention recently, because several zirconate pyrochlores have lower thermal conductivities than YSZ.[85] The crystal structure of pyrochlores consists of the corner-shared BO_6 octahedra forming the backbone of the network and *A* ions fill the holes which are formed by six BO_6 octahedra. It can largely tolerate vacancies at the A, B and O sites without phase transformation. Both A and B sites can be substituted by a lot of other elements with similar ionic radii when the electrical neutrality is satisfied, resulting in the potential to tailor their thermal properties. The pyrochlores are also attractive because many are refractory up to temperatures well in excess of 1500 °C and thermally stable. Of particular interest are the zirconates of the larger lanthanides (Gd → La) – an increased level of substitution relative to the co-doped compositions (from <15% to 50%$MO_{1.5}$) and the use of cations with a much higher atomic mass contrast with Zr than Y[86] are expected to yield a substantial reduction in thermal conductivity to 7YSZ.[85] Thermal conductivities ranging from ~1.1 $Wm^{-1}K^{-1}$ to ~1.7 $Wm^{-1}K^{-1}$ at temperatures between 700 °C and 1200 °C have been reported for zirconates of Gd, Eu, Sm, Nd and La.[87–92] Co-doping of pyrochlores on both A and B sites has been proposed to further reduce their thermal conductivities and also to modify their TECs. Some success has been achieved, at least up to ~800 °C, as exemplified by studies in which $La_2Zr_2O_7$ was doped with 30 at.% of Nd, Eu or Gd and the thermal conductivities were reduced from ~1.55 $Wm^{-1}K^{-1}$ to ~0.9 $Wm^{-1}K^{-1}$ for Gd doping.[92] Similar results for single- or double-oxide doped $La_2Zr_2O_7$ have also been reported by Zhu *et al.*[70] who found that $La_{1.7}Yb_{0.3}Zr_2O_7$ and $La_{1.7}Gd_{0.15}Yb_{0.15}Zr_2O_7$ had lower thermal conductivities than the undoped $La_2Zr_2O_7$ and that the Gd_2O_3 and Yb_2O_3 co-doped materials showed the lowest thermal conductivity.

Although $La_2Zr_2O_7$ has a lower thermal conductivity than YSZ, its coating did not give a longer thermal cycling lifetime than YSZ which might be explained by its relatively low TEC and poor toughness.[93] However, the double-ceramic-layer coating of YSZ/$La_2Zr_2O_7$ has increased thermal cycling lifetime and can even be used in higher temperatures compared with YSZ coating.[94,95] It has been proved that $La_2Zr_2O_7$ can be toughened by addition of $Y_3Al_5O_{12}$ or $BaTiO_3$ (both volume percent up to 20%),[96,97] which might give a positive influence for TBC applications. The addition of CeO_2 into $La_2Zr_2O_7$ ($La_2(Zr_{0.7}Ce_{0.3})_2O_7$) can improve its sintering resistance, reduce its thermal conductivity to some extent, as well as prolong its thermal cycling lifetime by applying the $La_2(Zr_{0.7}Ce_{0.3})_2O_7/La_2Zr_2O_7$ coating.[98] Unfortunately, even though $Nd_2Zr_2O_7$, $Eu_2Zr_2O_7$ and $Gd_2Zr_2O_7$ have larger TECs than $La_2Zr_2O_7$, the thermal cycling lifetimes of their single-layer coatings are very short, similar to that of the $La_2Zr_2O_7$ coating.[95,99]

The zirconates of Dy, Er and Yb with fluorite structure are also listed here, because they are very similar to the above-mentioned rare earth zirconates with pyrochlore structure. The thermal conductivities of $Dr_2Zr_2O_7$, $Er_2Zr_2O_7$ and $Yb_2Zr_2O_7$ range from ~1.3 $Wm^{-1}K^{-1}$ to ~1.6 $Wm^{-1}K^{-1}$, which are comparable to the other zirconates with pyrochlore structure, allowing them to be considered as TBC candidate materials.[100]

2.5 Lanthanum compounds, silicates and rare earth oxides

2.5.1 Lanthanum cerium oxide

Lanthanum cerium oxide ($La_2Ce_2O_7$) has been proposed as a new TBC material, which has a higher TEC and a lower thermal conductivity than YSZ.[101,102] $La_2Ce_2O_7$ is a solid solution of La_2O_3 in CeO_2 having fluorite-type structure with 1/8 O sites as vacancies. It has a good phase stability either from room temperature to 1400 °C or after annealing at 1400 °C even for 384 h. The higher TEC of $La_2Ce_2O_7$ (~12.3 × 10^{-6} K^{-1}, 300–1200 °C) compared with that of 8YSZ (10.5–11.5 × 10^{-6} K^{-1}, 30–1000 °C) [50] makes it possible to relieve thermal stress during thermal cycling. The thermal expansion improvement can be attributed to the large size of Ce^{4+} and the partial reduction of Ce^{4+} to Ce^{3+} at elevated temperatures. The minimum TEC of $La_2Ce_2O_7$ at ~250 °C is a result of the formation of oxygen vacancies. The addition of some oxides, such as Ta_2O_5 and WO_3, results in an increase in the TECs of $La_2Ce_2O_7$ below 400 °C.[103] The thermal conductivities of $La_2Ce_2O_7$ range from ~0.5 $Wm^{-1}K^{-1}$ to ~0.6 $Wm^{-1}K^{-1}$ due to the varied densities, which are obviously much lower than those of YSZ.

Thermal cycle testing of $La_2Ce_2O_7$ coating at 1250 °C has indicated that this coating has a much better thermal-shock resistance than the $La_2Zr_2O_7$ coating and even better than the 8YSZ coating.[101] Moreover, it has been proved that double-ceramic-layer coating system of $La_2Ce_2O_7$/8YSZ has a much longer thermal cycling lifetime than the $La_2Ce_2O_7$ coating alone.[104]

It is worth noting that proper control of coating composition is critical for a good thermal cycling behavior, because the coating with a composition more close to stoichiometric $La_2Ce_2O_7$ has a better thermal-shock resistance.

Another fluorite structure material, namely neodymium cerate ($Nd_2Ce_2O_7$), has also been proposed as a new TBC material.[105] It is thermally stable in the temperature range of interest for TBC applications (1200–1400 °C). The TEC of $Nd_2Ce_2O_7$ is higher than that of YSZ, and even more interesting is the TEC change as a function of temperature paralleling that of the superalloy bond coat. Moreover, the thermal conductivity of $Nd_2Ce_2O_7$ is ~30% lower than that of YSZ. The thermal-shock resistance of $Nd_2Ce_2O_7$ coating prepared by APS is inferior to that of $La_2Ce_2O_7$.

2.5.2 Silicates

There are few reports on silicates as TBC materials rather than as environmental barrier coatings. Only the spraying of zircon ($ZrSiO_4$),[106] natural zircon sand[107] and wollastonite ($CaSiO_3$)[108] have been reported. Zircon has a low TEC (4.99×10^{-6} K^{-1}, 25–1400 °C) and a high thermal conductivity (3.46 $Wm^{-1}K^{-1}$, 90–1500 °C).[15] It is chemically very stable especially at lower temperatures. With the exception of concentrated hydrofluoric acid, zircon is not generally attacked by acids but by basic materials at elevated temperatures. Zircon has a dissociation line at 1676 °C from phase diagram of the ZrO_2–SiO_2 system,[109] so that it is not possible for zircon to form from liquid phase, but via solid–solid reaction of zirconia and silica. During plasma spraying, zircon dissociates and consequently its coatings are composed of a mixture of crystalline ZrO_2 and amorphous SiO_2. When zircon is used as TBC for diesel engines, the decomposed SiO_2 in the coating may cause problems due to the evaporation of SiO and $Si(OH)_2$.[110] Ramaswamy *et al.*[107] have studied both natural zircon and a combination of zircon and Y_2O_3. The *t*–ZrO_2 phase substantially destabilized when the as-sprayed zircon coatings were annealed at 1200 °C, followed by recombination of ZrO_2 and SiO_2 to form $ZrSiO_4$. However, the *t*–ZrO_2 phase was completely stabilized in the coatings of the zircon and Y_2O_3 mixture upon annealing at 1400 °C. A greater amount of Y_2O_3 addition (up to 20 mol.%) into zircon decreases the zircon formation in the coatings, resulting in less porosity in the coating even after annealing at 1400 °C, thereby improving coating stability and adhesion.[111] Both coatings exhibited good thermal-shock resistance and thermal barrier characteristics and are supposed to be suitable for TBC applications.

As to $CaSiO_3$ coating,[108] crystalline $CaSiO_3$, glassy phase and tridymite (SiO_2) were observed in the as-sprayed coating. Tridymite (SiO_2) likely reacted with other composites such as CaO and glassy phase to form crystalline $CaSiO_3$ when the coating was heated at about 882 °C. After the first thermal cycle (from 20 and 1000 °C), the TECs and the thermal diffusivities of the coatings are $6.3–8.5 \times 10^{-6}$ K^{-1} and $3.24–3.72 \times 10^{-3}$ cm^2/s, respectively. Therefore, the APS $CaSiO_3$ coating may be used as TBC on substrate such as titanium and titanium alloys.

2.5.3 Rare earth oxides

The melting points of rare earth oxides are above 2200 °C. Except for the valence-changeable elements such as Ce, Pr, Eu and Tb, the other rare earth oxides have similar chemical, physical and mechanical properties. Coatings of rare earth oxides (La_2O_3, CeO_2, Pr_2O_3 and Nb_2O_5 as main phases) have lower thermal diffusivities and higher TECs than ZrO_2.[112] No further reports about these materials are available. Most of the rare earth oxides are polymorphic at elevated temperatures,[113] and the phase instability will certainly affect the thermal-shock resistance of their coatings to some extent. Rare earth oxides are commonly used as dopants or partial starting materials of TBC materials with the consequences of reducing their thermal conductivities, improving phase stability as well as increasing thermal cycling lifetimes.[5,6,9]

2.5.4 $Y_3Al_5O_{12}$

$Y_3Al_5O_{12}$ (YAG) has a rather complex cubic structure with 160 atoms per unit cell ($Z = 8$) with the chemical formula $A_3B_2(CO_4)_3$, where the A cation (Y^{3+}) is in a dodecahedral coordination site, the B cation (Al^{3+}) is in an octahedral site and the C cation (Al^{3+}) is in a tetrahedral site.[114] Padture and Klemens [115] were the first to propose YAG as an oxygen barrier in a TBC system due to its low oxygen diffusivity (~10^{10} lower than that in YSZ) and its comparably low thermal conductivity (3.2 $Wm^{-1}K^{-1}$, 1000 °C) at high temperatures. YAG also possesses a low creep rate, a low electrical conductivity, a phase/thermal stability (up to its melting point of 1970 °C) and an isotropic thermal expansion. Therefore, it is a promising candidate as TBCs, or as oxidation and erosion-resistant material, which can be used in gas turbine engines.[116] However, the relatively low TEC (9.1×10^{-6} K^{-1}) and the low melting point of this material seem to be the major problems. The YAG coatings prepared by different processes are reported extensively.[117–120]

2.5.5 Lanthanum aluminates

Oxides with magnetoplumbite structure of the general composition, $LnMAl_{11}O_{19}$ (Ln = La to Gd; M = Mg, Mn to Zn), have high melting points, high thermal expansion and low thermal conductivities which make them suitable for applications as high-temperature TBCs. Also, sintering in these oxides is suppressed due to the low ionic diffusion in magnetoplumbites.[121–124] In particular, the values of average TECs of $LaMgAl_{11}O_{19}$, $SmMgAl_{11}O_{19}$ and $GdMgAl_{11}O_{19}$ are about ~9.5–9.7 × 10^{-6} K^{-1} from room temperature to 1500 °C.[124] Thermal expansion behavior of all these oxides appears to be independent of the composition, indicating that the thermal expansion of these oxides is controlled by their basic magnetoplumbite crystal structure rather than having a composition based on the individual rare earth elements present. The thermal conductivities of these materials range from 2–3 $Wm^{-1}K^{-1}$ from room temperature to 1150 °C exhibiting little temperature dependence. The $SmMgAl_{11}O_{19}$ and $GdMgAl_{11}O_{19}$ compositions have lower thermal conductivities than $LaMgAl_{11}O_{19}$. Doping with Yb in the Ln site in $GdMgAl_{11}O_{19}$ can effectively reduce its thermal conductivity. Thermal conductivities of plasma-sprayed $LaMgAl_{11}O_{19}$ and $LaMnAl_{11}O_{19}$ coatings after 20 h exposure to temperatures up to 1600 °C did not change, indicating good sintering resistance. The microhardness of all $LnMAl_{11}O_{19}$ coatings is higher than that of hardened steel but is below the hardness of ZrO_2, and the elastic behavior of the coatings is superior to YSZ because of the lower indentation modulus.[122] However, the La-containing magnetoplumbites are not highly stable at elevated temperatures in water-containing atmosphere, such as in the combustion environment of turbine engines, and show significant continuous weight loss.[70]

2.5.6 $LaPO_4$

Lathanum phosphate ($LaPO_4$) has a monoclinic structure with four formula units of $LaPO_4$ in the $P2_1/n$ unit cell. This material has been proposed as a TBC candidate material by Morgan and Marshall[125] and Sudre *et al.*,[126] because it has a high-temperature stability (melting point, ~2070 °C), a high TEC ($9–10 \times 10^{-6}$ K^{-1}) and a low thermal conductivity (~1.8 $Wm^{-1}K^{-1}$ at 700 °C), as well as a low Young's modulus (133 GPa). In addition, $LaPO_4$ is expected to be relatively inert to molten salts containing sulfate or vanadate because the basic lanthanum cation is strongly bonded to the acidic phosphate group. It does not react with alumina, which is a positive attribute, but bonds poorly to it, which is a limitation on its application. In contrast, $LaPO_4$ is a line compound that melts congruently, but small deviations from stoichiometry change its solidus temperature from ~2070 °C to 1580 °C on the La-rich side or to 1050 °C on the P-rich side. From this point of view, much attention must be paid to controlling the coating process parameters and/or starting material compositions (feedstock for the plasma-spraying process and target material for the EB–PVD process) to get the desired coating composition.

Another monazite-type material $CePO_4$ has very similar properties to those of $LaPO_4$.[127] The linear TEC and thermal conductivity of the $CePO_4$ ceramics are $9–11 \times 10^{-6}$ K^{-1} (200–1300 °C) and 1.81 $Wm^{-1}K^{-1}$ (500 °C), respectively. Such properties make it suitable for consideration as a TBC candidate material. However, the $CePO_4$ ceramics are cracked or decomposed by acidic or alkaline aqueous solutions at high temperatures, which is a limitation in TBC applications.

2.6 $(Ca_{1-x}Mg_x)Zr_4(PO_4)_6$ (CMZP), perovskite oxides and metal–glass composite

2.6.1 $(Ca_{1-x}Mg_x)Zr_4(PO_4)_6$

$(Ca_{1-x}Mg_x)Zr_4(PO_4)_6$ (CMZP) is synthesized by incorporating Mg^{2+} with $CaZr_4(PO_4)_6$ and belongs to a well-known NZP family possessing an ultra-low TEC.[128] The crystal structure of NZP is a hexagonal structure with a $R\bar{3}c$ space symmetry, and is built up of PO_4 tetrahedra and ZrO_6 octahedra which are linked at corners to form a strong three-dimensional network. CMZP has a lower thermal conductivity (lower than ZrO_2 by a factor of approximately two) and its TEC is near zero.[129–131] Furthermore, this material is less dense than ZrO_2 with theoretical density of 3.2 g/cm^3. Moreover, no strength-loss after air quenching from temperatures up to 1500 °C was observed, showing a good thermal-shock resistance. Therefore, the lightweight ceramic could be used as thermal protection systems for space shuttles and for filter systems such as the filtration system for molten metals and hot gas clean-up, rather than as a TBC material in turbine engines.

2.6.2 Perovskite oxides (ABO_3)

The ideal perovskite ABO_3 is cubic symmetry, in space group $Pm\bar{3}m$. The structure is commonly visualized as a three-dimensional network of regular corner-linked BX_6 octahedra, the B cations being at the center of these octahedra and the A cations being centrally located in the spaces between them. The most commonly occurring distortion in perovskites is octahedral tilting. This means the tilting of the BX_6 octahedra about one or more of their symmetry axes, maintaining both regularity of the octahedral (approximately) and their corner connectivity (strictly). Such tilting allows greater flexibility in the coordination of the A cation, while leaving the environment of the B cation essentially unchanged. It is also a major contributor to structural variability in perovskites.[132]

Such perovskites are usually characterized by a melting point higher than 2000 °C. Furthermore, the TEC has values typically greater than 8.5–9 × 10^{-6} K^{-1} and thermal conductivity of less than 2.2 $Wm^{-1}K^{-1}$, which are advantageous for being used as a heat insulating layer.[133] Perovskites additionally offer the potential for extensive substitution of ions at A and/or B site, which enables the properties of the materials to be selectively influenced.[134]

So far, only some materials with perovskite structure have been studied as TBC candidates, such as $SrZrO_3$ and $BaZrO_3$, as well as two complex perovskites $Ba(Mg_{1/3}Ta_{2/3})O_3$ (BMT) and $La(Al_{1/4}Mg_{1/2}Ta_{1/4})O_3$ (LAMT).[89,135–137] The TECs of $SrZrO_3$, $BaZrO_3$, BMT and LAMT are 8.7–10.8 × 10^{-6} K^{-1} (200–1100 °C), ~7.9 × 10^{-6} K^{-1} (30–1000 °C), 9.6–11.3 × 10^{-6} K^{-1} (200–1200 °C) and 8.2–10.6 × 10^{-6} K^{-1} (200–1200 °C), respectively, which are comparable to that of YSZ except $BaZrO_3$. The thermal conductivities of these materials (except $BaZrO_3$) are below 2.8 $Wm^{-1}K^{-1}$ at 1000 °C, with the sequence of BMT > YSZ > $SrZrO_3$ > LAMT. The fracture toughness of $SrZrO_3$ is half of that of YSZ, whereas the values of BMT and LAMT are four times lower than that of YSZ. Finally, the Young's moduli of the above-mentioned materials are all lower than that of YSZ.

Some preliminary thermal cycling results indicate that the single-layer-coatings of these materials have shorter cycling lifetime, whereas the double-ceramic-layer coatings of $SrZrO_3$/YSZ and LAMT/YSZ have similar cycling lifetime compared with YSZ at the surface temperature of ~1250 °C. Doping Yb_2O_3 or Gd_2O_3 in $SrZrO_3$ can reduce the thermal conductivity of $SrZrO_3$,[138] and a very promising result is that the double-ceramic-layer coating of Yb_2O_3 doped $SrZrO_3$ in combination with YSZ has ~25% longer cycling lifetime than that of the optimized YSZ coating at the surface temperature of ~1350 °C.

2.6.3 Metal–glass composite

A new concept for TBC systems based on a mixture of metal and glass is called metal–glass composite (MGC).[139–141] In the MGC system, the TEC depends on the metal–glass ratio and varies from 9.5–12.3 × 10^{-6} K^{-1}. Another advantage of the gas-tight composite coatings is their ability to protect the bond coat from severe oxidation with a consequence of long lifetimes in oxidation tests. The MGC coatings have long thermal cycling lifetimes, which are supposed to be due to their high TECs, good adherence to the bond coat and the absence of open porosity. Such a coating does not have open pores, preventing the oxidation of bond coat from corrosive gases.

2.7 Future trends

The quest for new ceramic TBC candidate materials possessing low thermal conductivities is a major research area. This can be fulfilled by means of reducing phonon conduction of the materials. In real crystal structures, scattering of phonons occurs when they interact with lattice imperfections in the ideal lattice. Such imperfections include vacancies, dislocations, grain boundaries, atoms of different masses and other phonons. Ions and atoms of different ionic radius may also scatter phonons by locally distorting the bond length and thus introducing elastic strain fields into the lattice. The effects such imperfections cause can be quantified through their influence on the phonon mean free path. Moreover, materials having a large mean atomic mass, a low Young's modulus and randomly distributed point defects at a sufficiently high density favor reducing the phonon mean free path with the consequence of low thermal conductivity. In addition, materials with fine grain size in nanometer dimensions have proved to have a lower thermal conductivity resulting from much intensified grain-boundary scattering.

It is a time- and cost-consuming process to investigate the vast range of chemical compositions of all refractory oxides and minerals in order to identify promising compositions. The most rapid progress will probably only be made by using a combination of intuition about crystal structures and complementary atomic-level simulations to guide experiments. This is well practiced by Schelling *et al.*[142]

2.8 References

1. HJELM L N and BORNHORST B R (1961), 'Development of improved ceramic coatings to increase the life of XLR999 thrust chamber', in: *Research-Airplane-Committee Report on Conference on the Progress of the X-15 Project*, NASA Tech Memo X-57072, 227–253, National Aeronautics and Space Administration, Washington DC.
2. STECURA S (1976), *Two-Layer Thermal Barrier Coating for Turbine Airfoils – Furnace and Burner Rig Test Results*, NASA Tech Memo X-3425, National Aeronautics and Space Administration, Washington DC.
3. LIEBERT C H, JACOBS R E, STECURA S and MORSE C R (1976), *Durability of Zirconia Thermal-Barrier Ceramic Coatings on Air-Cooled Turbine Blades in Cyclic Jet Engine Operation*, NASA Tech Memo X-3410, National Aeronautics and Space Administration, Washington DC.
4. DEMASI J T and SGEFFKER K D (1986), *Thermal Barrier Coating Life Prediction Model Development*, NASA Contract Rep 175087, National Aeronautics and Space Administration, Washington DC.
5. CAO X, VAßEN R and STÖVER D (2004), 'Ceramic materials for thermal barrier coatings', *J Eur Ceram Soc*, **24**, 1–10.
6. LEVI C G (2004), 'Emerging materials and processes for thermal barrier systems', *Curr Opin Solid State Mater Sci*, **8**, 77–91.
7. SCHULZ U, SARUHAN B, FRITSCHER K and LEYENS C (2004), 'Review on advanced EB–PVD ceramic topcoats for TBC applications', *Int J Appl Ceram Technol*, **1**(4), 302–315.
8. SCHULZ U, RAMOND B S and LAVIGNE O (2004), 'Low thermal conductivity ceramics for turbine blade thermal barrier coating application', in: Curzio E L and Readey M J (eds), *28th International Conference on Advanced Ceramics and Composites B*, Wiley, Hoboken, NJ, **25**(4), 375–380.
9. CLARKE D R and PHILLOT S R (2005), 'Thermal barrier coating materials', *Mater Today*, **8**(6), 22–29.
10. VAßEN R, TIETZ F, KERKHOFF G and STÖVER D (1998), 'New materials for advanced thermal barrier coatings', in: Lecomte-Beckers J, Schuber F and Ennis P J (eds), *Proceedings of the 6th Liége Conference on Materials for Advanced Power Engineering*, Forshungszentrum Jülich, Jülich, Germany, 1627–1635.
11. CERNUSCHI F, BIANCHI P, LEONI M and SCARDI P (1999), 'Thermal diffusivity/microstructure relationship in Y-PSZ thermal barrier coatings', *J Therm Spray Technol*, **8**(1), 102–109.
12. RYSHKEWITCH E (1960), *Oxide Ceramics: Physical Chemistry and Technology*, Academic Press, New York.
13. HANNINK R H J, KELLY R M and MUDDLE B C (2000), 'Transformation toughening in zirconia-containing ceramics', *J Am Ceram Soc*, **83**(3), 461–487.
14. GARVIE R C, HANNINK R H J and PASCOE R T (1975), 'Ceramic steel?', *Nature*, **258**, 703–704.

15. GARVIE R C (1970), 'Zirconium dioxide and some of its binary systems', in: Alper A M (ed.), *High-temperature oxides*, Vol. 5–II, Academic Press, New York, 117–166.
16. CHEN L B (2006), 'Yttria-stabilized zirconia thermal barrier coatings – a review', *Surf Rev Lett*, **13**(5), 535–544.
17. ZHU W Z and YAN M (1997), 'Effect of grain size on the kinetics of isothermal tetragonal to monoclinic transformation in ZrO_2 (2 mol.%Y_2O_3) ceramics', *J Eur Ceram Soc*, **17**, 1729–1739.
18. BASU B, VLEUGELS J and BIEST O V D (2004), 'Transformation behaviour of tetragonal zirconia: role of dopant content and distribution', *Mater Sci Eng A*, **366**, 338–347.
19. ALVAREZ M R and TORRALVO M J (1996), 'Textural evolution of zirconia and yttria-doped zirconia on thermal treatment under different conditions', *Colloids Surf A: Physicochem Eng Aspects*, **113**, 165–174.
20. KELLY P M and ROSE L R F (2002), 'The martensitic transformation in ceramics – its role in transformation toughening', *Prog Mater Sci*, **47**, 463–557.
21. STECURA S (1978), *Effects of Compositional Changes on the Performance of a Thermal Barrier Coating System*, NASA TM-78976, National Aeronautics and Space Administration, Washington DC.
22. SLIFKA A J, FILLA B J, PHELPS J M, BANCKE G and BERNDT C C (1998), 'Thermal conductivity of a zirconia thermal barrier coating', *J Therm Spray Technol*, **7**(1), 43–46.
23. LIMA C R C and TREVISAN R DA E (1999), 'Temperature measurements and adhesion properties of plasma sprayed thermal barrier coatings', *J Therm Spray Technol*, **8**(2), 323–327.
24. KASPAR J and AMBROZ O (1988), 'Plasma spray coatings as thermal barriers based on zirconium oxide with yttrium oxide', in: Eschnauer H, Huber P, Nicoll A R and Sandmeier S (eds), *The 1st Plasma-Technik-Symposium*, Plasma-Technik AG, Wohlen, Switzerland, **2**, 155–166.
25. THORNTON J and MAJUMDAR A (1995), 'Ceria precipitation and phase stability in zirconia based thermal barrier coatings', in: Ohmori A (ed.), *Proceedings of the 14th International Thermal Spray Conference: Thermal Spray – Current Status and Future Trends*, ASM International, Materials Park, OH, 1075–1080.
26. KVERNES I, SOLBERG J K and LILLERUD K P (1979), 'Ceramic coatings on diesel engine components', in: Fairbanks J W and Stringer J (eds), *Proceedings of the 1st Conference on Advanced Materials for Alternative Fuel Capable Directly Fired Heat Engines*, Maine Maritime Academy, Castine, ME, US Department of Energy, 233–257.
27. JONES R L (1991), 'Development of hot-corrosion-resistant zirconia thermal barrier coatings', *Mater High Temp*, **9**, 228–236.
28. JONES R L (1997), 'Some aspects of the hot corrosion of thermal barrier coatings', *J Therm Spray Technol*, **6**(1), 77–84.
29. MILLER R A, BRINDLEY W J, GOEDJEN J G, TIWARI R and MESS D (1994), 'The effect of silica on the cyclic life of zirconia-yttria thermal barrier coating', in: Berndt C C and Sampath S (eds), *Proceedings of the 7th National Thermal Spray Conference: Thermal Spray in Industrial Applications*, ASM International, Materials Park, OH, 49–54.

30. RAMASWAMY P, SEETHARAMU S, VARMA K B R and RAO K J (1999), 'Thermal shock characteristics of plasma sprayed mullite coatings', *J Therm Spray Technol*, **7**(4), 497–504.
31. MANARA J, BRANDT R, KUHN J, FRICKE J, KRELL T, SCHULZ U, PETERS M and KAYSSER W A (2000), 'Emittance of Y_2O_3 stabilised ZrO_2 thermal barrier coatings prepared by electron-beam physical-vapour deposition', *High Temp-High Press*, **32**(3), 361–368.
32. STUBICAN V S (1986), 'Phase equilibria and metastabilities in the systems ZrO_2–MgO, ZrO_2–CaO and ZrO_2–Y_2O_3', *Adv Ceram*, **24**A, 71–82.
33. RAMASWAMY P, SEETHARAMU S, VARMA K B R and RAO K J (1999), 'Evaluation of CaO-CeO_2-partially stabilized zirconia thermal barrier coatings', *Ceram Inter*, **25**, 317–324.
34. BENNETT A (1986), 'Properties of thermal barrier coatings', *Mater Sci Technol*, **2**, 257–261.
35. JONES R L (1992), 'Thermogravimetric study of the 800 °C reaction of stabilizing oxides with SO_3–$NaVO_3$', *J Electrochem Soc*, **139**, 2794–2799.
36. NAGARAJ B A and WORTMAN D J (1990), 'Burner rig evaluation of ceramic coatings with vanadium-contaminated fuels', *Trans ASME*, **112**, 536–542.
37. KVERNES I (1987), 'Ceramic coatings as thermal barriers in diesel and gas turbine engine components', in: Vincenzini P (ed.), *High Tech Ceramics: Part C – Materials Science Monographs*, Elsevier, Amsterdam, 2519–2536.
38. KVERNES I (1981), 'Problems encountered in developing and using thermal barrier coatings on diesel engine components', in: Fairbanks J W and Stringer J (eds), *Proceedings of the 2nd Conference on Advanced Materials for Alternative Fuel Capable Heat Engines*, EPRI Report RD-2369-SR, Palo Alto, CA, 1–21.
39. TORIZ F C, THAKKER A B and GUPTA S K (1988), 'Thermal barrier coating for jet engines', *ASME Gas Turbine Aeroengine Congress and Exposition*, American Society of Mechanical Engineers, New York, 88-GT-279.
40. TSUKUMA K and SHIMADA M (1985), 'Strength, fracture toughness, and Vickers hardness of CeO_2-stabilized tetragonal-ZrO_2 polycrystals (Ce–TZP)', *J Mater Sci*, **20**(4), 1178–1184.
41. TSUKUMA K (1986), 'Mechanical properties and thermal stability of CeO_2-containing tetragonal-zirconia polycrystals', *Am Ceram Soc Bull*, **65**(10), 1386–1389.
42. SWAIN M V (1985), 'Inelastic deformation of Mg–PSZ and its significance for strength-toughness relationship of zirconia-toughened ceramics', *Acta Metall*, **33**(11), 2083–2091.
43. JONES R L, WILLIAMS C E and JONES S R (1986), 'Reaction of vanadium compounds with ceramic oxides', *J Electrochem Soc*, **133**, 227–230.
44. REIDY R F and JONES R L (1995), 'Thermogravimetric analysis of the reaction of CeO_2 with the $NaVO_3$–SO_3 system', *J Electrochem Soc*, **142**, 1353–1357.
45. MILLER R A (1984), 'Introduction to life modeling of thermal barrier coatings' in: *Turbine Engine Hot Section Technology Conference*, NASA CP-2339, National Aeronautics and Space Administration, Washington DC, 149–154.
46. BRANDON J B, TAYLOR R and MORRELL P (1991), 'Microstructure, composition and property relationships of plasma-sprayed thermal barrier coatings', *Surf Coat Technol*, **50**, 141–149.

47. LEE E Y, SOHN Y, JHA S K, HOLMES J W and SISSON R D (2002), 'Phase transformations of plasma-sprayed zirconia-ceria thermal barrier coatings', *J Am Ceram Soc*, **85**(8), 2065–2071.
48. HOLMES J W and PILSNER B H (1987), 'Cerium oxide stabilized thermal barrier coatings', in: Houck D L (ed.), *Proceedings of the national thermal spray conference: thermal spray – advances in coatings technology*, ASM International, Materials Park, OH, 259–270.
49. HEINTZE G N and MCPHERSON R (1987), 'Fracture toughness of plasma-sprayed zirconia coatings', in: Houck D L (ed.), *Proceedings of the National Thermal Spray Conference: Thermal Spray – Advances in Coatings Technology*, ASM International, Materials Park, OH, 271–275.
50. SODEOKA S, SUZUKI M, INOUE T, UENO K and OKI S (1996), 'Thermal and mechanical properties of plasma sprayed ZrO_2-CeO_2-Y_2O_3 coatings', in: Berndt C C (ed.), *Proceedings of The 9th National Thermal Spray Conference: Thermal Spray – Practical Solutions for Engineering Problems*, ASM International, Materials Park, OH, 295–302.
51. KIM H K, CHOI H S and LEE C H (1999), 'Failure analysis of plasma sprayed ZrO_2-CeO_2-Y_2O_3 thermal barrier coatings', in: Lugscheider E and Kamer P A (eds), *Proceedings of the United Thermal Spray Conference*, Verlag für Schweißen und Verwandte Verfahren, Düsseldorf, Germany, 740–746.
52. ROUANET A, PERAUDEAU G, SIBIEUDE F, PRIEM T, RANC R, RIGAL E and CORNU A (1995), 'Chemical and structural characterization of quenched ceria (yttria) stabilized zircoina or hafnia ceramics for plasma coatings', in: Ohmori A (ed.), *Proceedings of the 14th International Thermal Spray Conference: Thermal Spray – Current Status and Future Trends*, High Temperature Society of Japan, Osaka, Japan, 507–512.
53. THORNTON J, MAJUMDAR A and MCADAM G (1997), 'Enhanced cerium migration in ceria-stabilized zirconia', *Surf Coat Technol*, **94/95**, 112–117.
54. KHOR K A and YANG J (1997), 'Rapidly solidified neodymia-stabilised zirconia coatings prepared by DC plasma spraying', *Surf Coat Technol*, **96**, 313–322.
55. ZHANG Y H, WITHERS P J and FOX M D (1999), 'Damage mechanisms of coated systems under thermomechanical fatigue', *Mater Sci Technol*, **15**, 1031–1036.
56. LEONI M, JONES R L and SCARDI P (1998), 'Phase stability of scandia-yttria-stabilized zirconia TBCs', *Surf Coat Technol*, **108**(1–3), 107–113.
57. JONES R L, REIDY R F and MESS D (1996), 'Scandia, yttria-stabilized zirconia for thermal barrier coatings', *Surf Coat Technol*, **82**, 70–76.
58. JONES R L and MESS D (1996), 'Improved tetragonal phase stability at 1400°C with scandia, yttria-stabilized zirconia', *Surf Coat Technol*, **86/87**, 94–101.
59. WU J (2004), *Novel Low-Thermal-Conductivity Ceramics for Thermal Barrier Coating*, PhD thesis University of Connecticut, Storrs, CT.
60. RAGHAVAN S, WANG H, PORTER W D, DINWIDDIE R B and MAYO M J (2001), 'Thermal properties of zirconia co-doped with trivalent and pentavalent oxides', *Acta Mater*, **49**, 169–179.
61. RAMASWAMY P, WANG H, DINWIDDIE B, POTER W D, VAßEN R, STÖVER D and MAYO M J (2004), 'Ta_2O_5/Nb_2O_5 and Y_2O_3 co-doped zirconias for thermal barrier coatings', *J Am Ceram Soc*, **87**(3), 431–437.

62. KHOR K A and YANG J (1997), 'Lattice parameters tetragonality (c/a) and transformability of tetragonal zirconia phase in plasma sprayed ZrO_2–Er_2O_3 coatings', *Mat Lett*, **31**(1/2), 23–28.
63. KHOR K A and YANG J (1997), 'Plasma sprayed ZrO_2–Sm_2O_3 coatings: lattice parameters, tetragonality (c/a) and transformability of tetragonal zirconia phase', *J Mater Sci Lett*, **16**(12), 1002–1004.
64. PITEK F M and LEVI C G (2007), 'Opportunities for TBCs in the ZrO_2–$YO_{1.5}$–$TaO_{2.5}$ system', *Surf Coat Technol*, **21**, 6044–6050.
65. MATSUMOTO M, YAMAGUCHI N and MATSUBARA H (2004), 'Low thermal conductivity and high temperature stability of ZrO_2–Y_2O_3–La_2O_3 coatings produced by electron beam PVD', *Scripta Mater*, **50**, 867–871.
66. MATSUMOTO M, AOYAMA K, MATSUBARA H, TAKAYAMA K, BANNO T, KAGIYA Y and SUGITA Y (2005), 'Thermal conductivity and phase stability of plasma sprayed ZrO_2–Y_2O_3–La_2O_3 coatings', *Surf Coat Technol*, **194**, 31–35.
67. ZHU D and MILLER R A (2004), 'Development of advanced low thermal conductivity thermal barrier coatings', *Int J Appl Ceram Technol*, **1**(1), 86–94.
68. SINGH J, WOLFE D E, MILLER R A, ELDRIDGE J I and ZHU D (2004), 'Tailored microstructure of zirconia and hafnia-based thermal barrier coatings with low thermal conductivity and high hemispherical reflectance by EB–PVD', *J Mater Sci*, **39**, 1975–1985.
69. ZHU D and MILLER R A (1998), 'Sintering and creep behavior of plasma-sprayed zirconia- and hafnia-based thermal barrier coatings', *Surf Coat Technol*, **108/109**, 114–120.
70. ZHU D, FOX D S, BANSAL N P and MILLER R A (2004), *Advanced Oxide Material Systems for 1650 °C Thermal and Environmental Barrier Coating Applications*', NASA/TM-2004–213219, National Aeronautics and Space Administration, Washington DC.
71. IBEGAZENE H, ALPERINE S and DIOT C (1995), 'Yttria-stabilized hafnia-zirconia thermal barrier coatings: the influence of hafnia addition on TBC structure and high-temperature behaviour', *J Mater Sci*, **30**(4), 938–951.
72. WILKES K E and LAGEDROST J F (1973), *Thermophysical Properties of Plasma Sprayed Coatings*, NASA CR-121144, National Aeronautics and Space Administration, Washington DC.
73. CHRASKA P, DUBSKY J, NEUFUSS K and PISACKA J (1997), 'Alumina-base plasma-sprayed materials-part I: phase stability of alumina and alumina-chromia', *J Therm Spray Technol*, **6**(3), 320–326.
74. ILAVSKY J, BERNDT C C, HERMAN H, CHRASKA P and DUBSKY J (1997), 'Alumina-base plasma-sprayed materials-part II: phase transformation in alumina', *J Therm Spray Technol*, **6**(4), 439–444.
75. FOUJANET M C, LUMET J P and DEREP J L (1988), Structural analysis of zirconia-alumina plasma sprayed coatings, in: Eschnauer H, Huber P, Nicol A R and Sandmeier S (eds), *The 1st Plasma-Technik-Symposium*, Plasma-Technik AG, Wohlen, Switzerland, **2**, 357–366.
76. SAINTE-CATHERINE M C, DEREP J L and LUMET J P (1991), 'Zirconia-alumina plasma sprayed coatings: correlation between microstructure and properties', in: Sandmeier S, Eschnauer H, Huber P and Nicoll A R (eds), *The 2nd Plasma-Technik-Symposium*, Plasma-Technik AG, Wohlen, Switzerland, 2, 131–140.
77. KRISHNAMOORTHY P R, RAMASWAMY P and NARAYANA B H (1991), 'Evaluation of multilayer ZrO_2 and Al_2O_3 thermal barrier coatings', in: Sandmeier S, Eschnauer

H, Huber P and Nicoll A R (eds), *The 2nd Plasma-Technik-Symposium*, Plasma-Technik AG, Wohlen, Switzerland, **3**, 263–272.
78. XU H, GUO H, LIU F and GONG S (2000), 'Development of gradient thermal barrier coatings and their hot-fatigue behaviour', *Surf Coat Technol*, **130**, 133–139.
79. BRUNAUER G, FERY F, BOYSEN H and SCHNEIDER H (2001), 'High temperature thermal expansion of mullite: an in-situ neutron diffraction study up to 1600 °C', *J Eur Ceram Soc*, **21**, 2563–2567.
80. LEE K N, MILLER R A and JACOBSON N S (1995), 'New generation of plasma-sprayed mullite coatings on silicon carbide', *J Am Ceram Soc*, **78**(3), 705–710.
81. KOKINI K, TAKEUCHI Y R and CHOULES B D (1996), 'Surface thermal cracking of thermal barrier coatings owing to stress relaxation: zirconia vs mullite', *Surf Coat Technol*, **82**, 77–82.
82. YONUSHONIS T M (1997), 'Overview of thermal barrier coatings for diesel engines', *J Therm Spray Technol*, **6**(1), 50–56.
83. GILBERT A, KOKINI K and SANKARASUBRAMANIAN S (2008), 'Thermal fracture of zirconia–mullite composite thermal barrier coatings under thermal shock: an experimental study', *Surf Coat Technol*, **202**, 2152–2161.
84. GILBERT A, KOKINI K and SANKARASUBRAMANIAN S (2008), 'Thermal fracture of zirconia–mullite composite thermal barrier coatings under thermal shock: a numerical study', *Surf Coat Technol*, **203**, 91–98.
85. MALONEY M J (2000), *Thermal Barrier Coating Systems and Materials*, US Patent 6 117 560.
86. KLEMENS P G (1955), 'The scattering of low-frequency lattice waves by static imperfections', *Proc Phys Soc*, **A68**, 1113–1128.
87. SURESH G, SEENIVASAN G, KRISHNAIAH M V and MURTI P S (1997), 'Investigation of the thermal conductivity of selected compounds of gadolinium and lanthanum', *J Nucl Mater*, **249**, 259–261.
88. SURESH G, SEENIVASAN G, KRISHNAIAH M V and MURTI P S (1998), 'Investigation of the thermal conductivity of selected compounds of lanthanum, samarium and europium', *J Alloy Compd*, **269**, L9-L12.
89. VAßEN R, CAO X, TIETZ F, BASU D and STÖVER D (2000), 'Zirconates as new materials for thermal barrier coatings', *J Am Ceram Soc*, **83**(8), 2023–2028.
90. MALONEY M J (2001), *Thermal Barrier Coating Systems and Materials*, US Patent 6 1177 200B.
91. WU J, WEI X, PADTURE N P, KLEMENS P G, GELL M, GARCIA E, MIRANZO P and OSENDI I (2002), 'Low-thermal-conductivity rare-earth zirconates for potential thermal barrier coating applications', *J Am Ceram Soc*, **85**(12), 3031–3035.
92. LEHAMNN H, PITZER D, PRACHT G, VAßEN R and STÖVER D (2003), 'Thermal conductivity and thermal expansion coefficient of the lanthanum rare-earth element zirconate system', *J Am Ceram Soc*, **86**(8), 1338–1344.
93. CAO X, VAßEN R, TIETZ F, JUNGEN W and STÖVER D (2001), 'Thermal stability of lanthanum zirconate plasma-sprayed coating', *J Am Ceram Soc*, **84**(9), 2086–2090.
94. VAßEN R, DIETRICH M, LEHMANN H, CAO X, PRACHT G., TIETZ F, PITZER D and STÖVER D (2001), 'Development of oxide ceramics for an application as TBC', *Mater Sci Eng Technol*, **32**(8), 673–677.
95. VAßEN R, TRAEGER F and STÖVER D (2004), 'New thermal barrier coatings based on pyrochlore/YSZ double-layer systems', *Int J Appl Ceram Technol*, **1**(4), 351–361.

96. LI J, DAI H, ZHONG X, ZHANG Y, MA X, MENG J and CAO X (2007), 'Effect of the addition of YAG ($Y_3Al_5O_{12}$) nanopowder on the mechanical properties of lanthanum zirconate', *Mater Sci Eng*, **A460/461**, 504–508.
97. LI J, DAI H, ZHONG X, ZHANG Y, MA X, MENG J and CAO X (2008), 'Lanthanum zirconate ceramic toughened by $BaTiO_3$ secondary phase', *J Alloy Compd*, **452**, 406–409.
98. CAO X, VAßEN R, TIETZ F and STÖVER D (2006), 'New double-ceramic-layer thermal barrier coatings based on zirconia–rare earth composite oxides', *J Eur Ceram Soc*, **26**, 247–251.
99. CAO X (2007), 'Application of rare earths in thermal barrier coating materials', *J Mater Sci Technol*, **23**(1), 15–35.
100. XU Q, PAN W, WANG J, WAN C, QI L and MIAO H (2006), 'Rare-earth zirconate ceramics with fluorite structure for thermal barrier coatings', *J Am Ceram Soc*, **89**(1), 340–342.
101. CAO X, VAßEN R, FISCHER W, TIETZ F, JUNGEN W and STÖVER D (2003), 'Lanthanum-cerium oxide as a thermal barrier coating material for high-temperature applications', *Adv Mater*, **15**(17), 1438–1442.
102. MA W, GONG S, XU H and CAO X (2006), 'The thermal cycling behavior of Lanthanum–Cerium Oxide thermal barrier coating prepared by EB–PVD', *Surf Coat Technol*, **200**, 5113–5118.
103. MA W, GONG S, XU H and CAO X (2006), 'On improving the phase stability and thermal expansion coefficients of lanthanum cerium oxide solid solutions', *Scripta Mater*, **54**, 1505–1508.
104. MA W, GONG S, LI H and XU H (2008), 'Novel thermal barrier coatings based on $La_2Ce_2O_7$/8YSZ double-ceramic-layer systems deposited by electron beam physical vapor deposition', *Surf Coat Technol*, **202**, 2704–2708.
105. DAI H, ZHONG X, LI J, MENG J and CAO X (2006), 'Neodymium-cerium oxide as new thermal barrier coating material', *Surf Coat Technol*, **201**, 2527–2533.
106. CHRASKA P, NEUFUSS K, KOLMAN B and DUBSKY J (1997), 'Plasma spraying of silicates', in: Berndt C C (ed.), *Proceedings of the 1st United Thermal Spray Conference: Thermal Spray – a United Forum for Scientific and Technological Advances*, ASM International, Materials Park, OH, 477–481.
107. RAMASWAMY P, SEETHARAMU S, VARMA K B R and RAO K J (1999), 'Thermal barrier coating application of zircon sand', *J Therm Spray Technol*, **8**(3), 447–453.
108. LIU X and DING C (2002), 'Thermal properties and microstructure of a plasma sprayed wollastonite coating', *J Therm Spray Technol*, **11**(3), 375–379.
109. BUTTERMAN W C and FOSTER W R (1953), 'Zircon stability and ZrO_2-SiO_2 phase diagram', *Am Mineral*, **52**, 880–888.
110. KVERNES I, LUGSCHEIDER E and LADRU F (1998), 'Lifetime and degradation processes of TBCs for diesel engines', in: Lecomte-Beckers J, Schuber F and Ennis P J (eds), *Proceedings of the 6th Liége Conference on Materials for Advanced Power Engineering*, Forshungszentrum Jülich GmbH, Jülich, Germany, 997–1001.
111. SUZUKI M, SODEOKA S and INOUE T (2008), 'Zircon-based ceramics composite coating for environmental barrier coating', *J Therm Spray Technol*, **17**(3), 404–409.
112. DING C, XI Y, ZHANG Y, QU J and QIAO H (1991), 'Thermophysical properties of plasma sprayed rare earth oxide coatings', in: Sandmeier S, Eschnauer H,

Huber P and Nicoll A R (eds), *The 2nd Plasma-Technik-Symposium*, Plasma-Technik AG, Wohlen, Switzerland, **3**, 27–32.
113. WARSHAW I and ROY R (1965), 'Polymorphism of the rare earth sesquioxides', *J Phys Chem*, **65**, 2048–2051.
114. FLORIAN P, GERVAIS M, DOUY A, MASIOT D and COUTURES J P (2001), 'A multi-nuclear multiple-field nuclear magnetic resonance study of the Y_2O_3–Al_2O_3 phase diagram', *J Phys Chem*, **B105**, 379–391.
115. PADTURE N P and KLEMENS P G (1977), 'Low thermal conductivity in garnets', *J Am Ceram Soc*, **80**(4), 1019–1020.
116. PARTHASARATHY T A, MAH T and MATSON L E (1993), 'Deformation behavior of an Al_2O_3–$Y_3Al_5O_{12}$ eutectic composite in comparison with sapphire and YAG', *J Am Ceram Soc*, **76**(1), 29–32.
117. SU Y J, TRICE R W, FABER K T, WANG H and PORTER W D (2004), 'Thermal conductivity, phase stability, and oxidation resistance of $Y_3Al_5O_{12}$(YAG)/Y_2O_3–ZrO_2(YSZ) thermal barrier coatings', *Oxid Metals*, **61**(3/4), 253–271.
118. WU Y, DU J and CHOY K L (2006), 'Novel deposition of columnar $Y_3Al_5O_{12}$ coatings by electrostatic spray-assisted vapor deposition', *J Am Ceram Soc*, **89**(1), 385–387.
119. WEYANT C M and FABER K T (2008), 'Processing-microtructure relationship for plasma-sprayed yttrium aluminum garnet', *Surf Coat Technol*, **202**, 6081–6089.
120. SARAVANAN S, SRINIVAS G H, JAYARAM V, PAULRAJ M and ASOKAN S (2008), 'Synthesis and characterization of $Y_3Al_5O_{12}$ and ZrO_2–Y_2O_3 thermal barrier coatings by combustion spray pyrolysis', *Surf Coat Technol*, **202**, 4653–4659.
121. VAßEN R, CAO X, DIETRICH M and STÖVER D (2001), 'Improvement of new thermal barrier coating system using layered or graded structure', in: Singh M and Jessen T (eds), *The 25th Annual International Conference on Advanced Ceramics and Composites: an Advance Ceramics Odyssey*, American Ceramic Society, Westerville, OH, 435–442.
122. FRIDRICH C, GADOW R and SCHIRMER T (2001), 'Lanthanum hexaaluminate – a new material for atmospheric plasma spraying of advanced thermal barrier coatings', *J Therm Spray*, **10**(4), 592–598.
123. GADOW R and LISCHKA M (2002), 'Lanthanum hexaaluminate – novel thermal barrier coatings for gas turbine applications – materials and process development' , *Surf Coat Technol*, **151/152**, 392–399.
124. BANSAL and ZHU (2008), 'Thermal properties of oxides with magnetoplumbite structure for advanced thermal barrier coatings', *Surf Coat Technol*, **202**, 2698–2703.
125. MORGAN P, MARSHALL D B (1995), 'Ceramic composites of monazite and alumina', *J Am Ceram Soc*, **78**(6), 1553–1563.
126. SUDRE O, CHEUNG J, MARSHALL D, MORGAN P and LEVI C G (2001), 'Thermal insulation coatings of $LaPO_4$', *Ceram Eng Sci Proc*, **22**(4), 367–374.
127. HIKICHI Y, NOMURA T, TANIMURA Y, SUZUKI S and MIYAMOTO M (2005), 'Sintering and properties of monazite-type $CePO_4$', *J Am Ceram Soc*, **73**(12), 3594–3596.
128. LIU D M and BROWN J J (1993), 'Thermal expansion of porous $(Ca_{1-x}Mg_x)Zr_4(PO_4)_6$ ceramics', *Mater Chem Phys*, **33**, 43–49.

129. HIRSCHFELD D A, LIU D M and BROWN J J (1992), 'CMZP – a new high temperature thermal barrier material', in: Carlsson R, Johansson R and Kahlman L (eds), *The 4th International Symposium on Ceramic Materials and Components for Engines*, Elsevier Applied Science, London, UK, 372–380.
130. LIU D M and BROWN J J (1992), 'Lightweight $(Ca_{1-x}Mg_x)Zr_4(PO_4)_6$ ceramics', *Mater Chem Phys*, **32**, 161–167.
131. LIU D M (1994), 'Thermal diffusivity, specific heat, and thermal conductivity of $(Ca_{1-x}Mg_x)Zr_4(PO_4)_6$ ceramic', *Mater Sci Eng*, **B23**, 94–97.
132. HOWARD C J and STOKES H T (2005), 'Structures and phase transformations in perovskites – a group-theoretical approach,' *Acta Cryst*, **A61**, 93–111.
133. VAßEN R, SCHWARTZ-LUCKGE S, JUNGEN W and STÖVER D *heat-insulating layer made of complex perovskite*, US Patent 20050260435 A1.
134. HEIMBERG B, BEELE W, KEMPTER K, BAST U, HAUBOLD T, HOFFMANN M, ENDRISS A, GREIL P, HONG C, ALDINGER F and SEIFERT H, *Process for Producing a Ceramic Thermal Barrier Layer for Gas Turbine Engine Component*, US Patent 6602553 B2.
135. MA W, MACK D E, VAßEN R and STÖVER D (2008), 'Perovskite-type strontium zirconate as a new material for thermal barrier coatings ', *J Am Ceram Soc*, **91**(8), 2630–2635.
136. MA W, JARLIGO M O, MACK D E, PITZER D, MALZBENDER J, VAßEN R and STÖVER D (2008), 'New generation perovskite thermal barrier coating materials', *J Therm Spray Technol*, **17**(5/6), 831–837.
137. JARLIGO M O, MACK D E, VAßEN R and STÖVER D (2009), 'Application of plasma-sprayed complex perovskites as thermal barrier coatings', *J Therm Spray Technol*, **18**(2), 187–193.
138. MA W, MACK D, MALZBENDER J, VAßEN R and STÖVER D (2008), 'Yb_2O_3 and Gd_2O_3 doped strontium zirconate for thermal barrier coatings', *J Eur Ceram Soc*, **28**, 3071–3081.
139. DIETRICH M, VERLOTSKI V, VAßEN R and STÖVER D (2001), 'Metal-glass based composites for novel TBC-systems', *Mater Sci Eng Technol*, **32**(8), 669–672.
140. STÖVER D, PRACHT G, LEHMANN H, DIETRICH M, DÖRING J E and VAßEN R (2004), 'New material concepts for the next generation of plasma-sprayed thermal barrier coatings', *J Therm Spray Technol*, **13**(1), 76–83.
141. MACK D E, GROSS S M, VAßEN R and STÖVER D (2006), 'Metal-glass based composites for application in TBC-systems', *J Therm Spray Technol*, **15**(4), 652–656.
142. SCHELLING P K, PHILLPOT S R and GRIMES R W (2004), 'Optimum pyrochlore compositions for low thermal conductivity', *Philos Mag Lett*, **84**(2), 127–137.

3
Metallic coatings for high-temperature oxidation resistance

X. PENG, Chinese Academy of Sciences, China

Abstract: This chapter reviews metallic coatings with the ability to thermally grow a protective layer of chromia or alumina, together with the coating techniques. First generation coatings are developed using chemical vapour deposition techniques (CVD); adding small amounts of reactive elements (e.g., Ce, Y, La) or adding noble elements such as Pt can further enhance performance. Second-generation coatings of the alumina-forming M–Cr–Al–Y overlay type bring the advantages that composition can be tailored to specific environments and to match the composition of matrix alloy, in order to minimize the interdiffusion between coatings and matrix.

Key words: diffusion coating, chromized coating, aluminized coating, overlay coating, nanocrystalline coating, Pt modification, reactive element effect, high temperature oxidation.

3.1 Introduction

On being exposed to a high-temperature environment, almost all kinds of metals will spontaneously form a scale of thermally grown oxides (TGOs), which functions as a barrier between the underlying metals and environmental corrosive substances. A TGO scale with the desired degree of protectiveness will have high thermal stability, good compactness and slow growth rate, as well as strong adhesion to the metallic matrix. If a non-protective scale forms, the oxidation will cause rapid metal recession, which greatly reduces the mechanical properties and service lifetime of the metallic workpieces. Among the TGO oxides, chromia and alumina are two kinds which kinetically and thermodynamically meet the requirement of resistance to high-temperature oxidation. In general, chromia scale is ideal against oxidation below 1000 °C and Type II hot corrosion (700–800 °C), and alumina scale (particularly in its α phase form) is best against high-temperature oxidation above 1000 °C and Type I hot corrosion (800–950 °C).

Unfortunately, most conventionally coarse-grained (CG) commercial alloys cannot ensure the formation of chromia or alumina scales, because of some limitation of the Cr and Al content (e.g., high content would increase the brittleness of alloy and decrease its hot strength or creep resistance). To improve the oxidation resistance of alloys, the addition of high Cr and/or Al coatings has become a preferred choice in most cases. The first generation of these coatings comprises well-known diffusion chromizing and aluminizing coatings, which are usually manufactured by chemical vapor deposition (CVD) pack cementation.[1,2] The second-generation coatings are so-called overlay coatings, the majority of which are of M–Cr–Al type (M = Fe, Ni, Co, or their combinations) containing a small amount of yttrium. Compared to the diffusion coatings, overlay coatings have the advantage that their composition can be specifically designed and manufactured to match specific environment conditions.[3] For a M–Cr–Al coating to be resistant to high-temperature oxidation, the chromium content is normally in the range 16–24 wt% and that of aluminum in the range 13–18 wt%.[3] M–Cr–Al overlay coatings, which will be explained in more detail later, have been fabricated using various techniques.[4–10]

Recently, grain refinement, without any increase in Cr or Al, has been considered to be another approach to increasing the oxidation resistance of alloys because, in some specific cases, it can promote the formation of a protective scale of chromia or alumina.[11] Since the late 1990s, oxidation of nanocrystalline (NC) coatings has received considerable attention.[12–23] Most investigators agree that the promotion of the selective formation of chromia or alumina scales is a primary reason for the superior oxidation resistance of NC coatings. So far, the intrinsic reason for the rapid formation of chromia or alumina scales during the initial oxidation period has not been experimentally established. However, the greatly enhanced diffusion of the corresponding oxide-forming component through the numerous grain boundaries in NC coatings during oxidation is believed to be one of the main causes.

So far, the maximum temperature for practical applications of metallic coatings is generally below 1100 °C. For metals which need to be exposed to higher temperatures, such as the nickel- or cobalt-based superalloys used as structural materials in gas turbine engines, thermal barrier coatings (TBCs) are normally used.[24–27] A typical TBC is yttria-stabilized zirconia (YSZ). The application of TBCs, of which the thickness traditionally ranges from 100–500 μm, along with internal cooling of the underlying superalloy component, offers a 100–300 °C reduction in the surface temperature of the superalloy. This allows the superalloy-made workpieces to be operated at environmental temperatures above the melting temperature of the superalloy (~1300 °C), thereby improving the engine efficiency and performance. The YSZ is not applied directly onto the superalloy but deposited onto an oxidation-resistant metallic bond coat which is pre-treated on the superalloy. The bond coat is typically M–Cr–Al–Y-type overlay coating or Pt-modified aluminide diffusion coating.[24,25,27]

3.2 Oxidation-resistant metallic coatings and their fabrication techniques

3.2.1 Diffusion coatings

Diffusion coatings are fabricated using CVD. This involves the dissociation and chemical reactions of gaseous reactants in a stimulant (heat, light, plasma) environment, which is followed by the formation of a stable solid product. The deposition involves homogeneous gas phase reactions that occur in the gas phase, and heterogeneous chemical reactions which occur on a heated surface and result in the formation of powders or films, respectively.[28] Pack cementation is a variation of CVD where the volatile gaseous reactants are generated within the powder packing as compared to the usual vapor generation routes in CVD. Metals for pack cementation are usually embedded in a powder mixture of a pure metal or alloy source (e.g., Al, FeAl, Cr, Si, depending on what element will be coated), a halide salt activator (e.g., NaCl, NaF, NH_4C1) and an inert filler material (e.g., Al_2O_3). A controllable atmosphere, usually Ar or H_2/Ar, surrounds the pack as it is heated to a temperature at which the master metallic powder reacts with the halide salt activator to form volatile metal halide species of significant partial pressures. Suppose NH_4Cl is used as the activator, the reaction will be given below:

$$M(s) + xNH_4Cl\ (s,g) = MCl_x(g) + xNH_4(g) \qquad [3.1]$$

where M is the element to be coated. Several different vapor substances may be formed for aluminizing (e.g., Al, AlCl, $AlCl_2$ and $AlCl_3$) and for chromizing (e.g., Cr, $CrCl_2$, $CrCl_3$ and $CrCl_4$). Afterwards, partial pressure gradients are established between the powder and the metal for coating, which support vapor transport for each of the vapor substances to the metal matrix surface. On the surface, deposition of the desired coating element occurs via dissociation or disproportionation of the halide molecules, or by a displacement reaction with the matrix. Finally, the coating element interdiffuses and reacts with the matrix, bringing some specific phase, composition and microstructure on the surface. On this basis, the pack cementation can be summarized as four interrelated steps:[29] (i) a thermodynamic equilibrium between the activator and the powder of the master metal or alloy, which determines the vapor pressures of the active gaseous substances in the pack; (ii) the gaseous diffusion of the metal halides from the pack to the alloy matrix surface; (iii) reactions on the surface to deposit the coating elements and form the products with vapor substances; and (iv) solid-state diffusion of the coating elements into the matrix. Aluminized coatings and chromized coatings are two traditional diffusion coatings that have been extensively used for almost a century.

Aluminized coatings

Aluminized coatings are used to extend the life of metals in service in high-temperature environments by thermally growing a protective scale of alumina, a process which conventionally uses the pack-cementation method. The powder mixtures for pack aluminizing are usually composed of a pure metal or alloy (e.g., Al, FeAl), an activator (typically, a halide salt) and an inert alumina filler. The first public description of pack-cementation aluminizing was by Van Aller in a US patent filed in 1911 and in a 1914 presentation by Allison and Hawkins, all of whom are from the General Electric Research Laboratory.[24] In the 1940s, pack-cementation aluminizing low-alloy steel for a combustion chamber was reported by Franz in Germany, and pack aluminizing of cobalt-based alloy for gas turbine vane airfoils was reported around the 1950s.[24] Diffusion aluminized coatings were also fabricated by hot dip aluminizing,[30] slurry aluminizing[31] and vacuum pulse aluminizing,[32] etc. Since 1970, most vane and blade coatings have been applied by pack cementation and the more recently developed 'out-of-contact' CVD processes.[24, 32–35]

For aluminized coatings formed by pack aluminizing, it is generally accepted that their growth mechanisms can be divided into two categories:[36] (i) inward growth under the condition of low-temperature, high-activity pack; and (ii) outward growth under the condition of high-temperature, low-activity pack. The transformation temperature for the growth of aluminides from inward to outward is not easily measured, and it is affected by pack chemistry. In general, when the temperature is below 950 °C, the aluminide coating grows inwardly, forming a M_xAl_y matrix ($y \geqslant x$; M = Ni, Fe or Co depending on what metal matrix is used for coating). At a higher temperature, the aluminide coating grows outwardly, forming a MAl matrix (e.g., β-NiAl on Ni-base metals).

When exposed to high-temperature environments, the thermally grown alumina scale on the aluminized coatings always suffers stresses which are induced by thermal cycling and oxide formation and growth. These can result in loss of the adherence and spalling of the scale. However, a small concentration of a reactive element (e.g., Y, La, Hf, Ce) can improve the adherence of alumina scale. This phenomenon is referred to as the 'reactive element effect' (REE). Theories associated with the REE have been reviewed elsewhere.[37–39] It is generally accepted that the addition of RE or RE oxide particles improves the oxidation resistance of alumina formers through enhancing the alumina scale adhesion to the underlying metals by means of several mechanisms: (i) 'pegging' alumina scale to the underlying metal; (ii) preventing the coalescence of vacancies at the scale–metal interface; and (iii) eliminating dissociated sulfur at the scale–metal interface.

Different routes have been used to add reactive element into aluminized coatings. LePrince *et al.*[40] used a highly active process with a powder mixture of pure source elements to co-deposit Al and Hf onto Ni-base fibers, developing an aluminide coating (β-NiAl) with the precipitation of Ni_5Hf. The Ni_5Hf-precipitated aluminide coating showed improved oxide scale adherence during cyclic oxidation testing. Rapp *et al.*[41,42] proposed two methods which can be used for doping of a minor element (e.g., Zr or Y) onto the surface of an aluminide diffusion coating produced by a single-step pack process. One is to add a small amount of an oxide source of the desired element (e.g., ZrO_2 or Y_2O_3) into the pack powder mixture, instead of some inert filler. In this case, the oxide dopant must be converted into high halogen activity within the pack to produce additional metal halide species according to the following equation:

$$xMO_y(s) + yAlX_x(g) = xMX_y(g) + yAlO_x(s) \qquad [3.2]$$

where M is Zr or Y. The other is to add certain amounts of reactive element chlorides (e.g., $ZrCl_4$, YCl_3, or $HfCl_4$) as activator source. The chlorides can react with the master metal (or alloy) to produce $AlCl_x$ species according to the equation given by:

$$Al(s) + RECl_y(g) = AlCl_x(g) + RECl_{y-x}(g) \qquad [3.3]$$

where RE indicates a reactive element. Applying the single-step pack cementation at 1150 °C for 24 h, RE-doped aluminide coatings were developed on IN 713LC matrix. The isothermal oxidation at 1100 °C in air showed that the RE-doped aluminide coatings compared to RE-free coatings had formed α-Al_2O_3 scale with decreased growth rate but increased resistance to cracking and spalling.[43]

Tu *et al.*[44] produced a Y-modified aluminide coating by aluminizing followed by yttrizing. The coating consisted of an external aluminide layer with Y enrichment near the surface and along the grain boundaries. Yttrium was deposited in a pack powder mixture of pure Y, NH_4Cl activator and an Y_2O_3 filler which was heated at 1050 °C for 4 h. Through a 'two-step' method (Ni–La_2O_3 co-electroplating followed by aluminizing at 930 °C for 2.5 h), Peng *et al.*[45] fabricated a La_2O_3-dispersed β-NiAl coating. Compared to the La_2O_3-free β-NiAl developed on electrodeposited pure Ni, this coating exhibited decreased oxidation kinetics at 1000 °C in air. It has been proposed that during oxidation lanthanum ions from the La_2O_3 particles would incorporate the growing θ-alumina scale, where the lanthanum ions prevented the formation of microcracks in the scale by retarding the θ–α transformation, and suppressed outward aluminum diffusion for the scale growth by the segregation to the scale grain boundaries.

β-NiAl coatings developed on Ni or Ni-base alloys have low oxidation kinetics because they exclusively grow an external scale of alumina. It was found in the 1970s that addition of Pt can further improve the adhesion of the scale to the coating matrix.[46] Some recent studies have contributed an enhanced understanding of the effect of Pt addition on scale adhesion. Pt addition has been found to decrease the number of voids formed at the oxide/metal interface[47] and to minimize the segregation at the interface of sulfur (which, as a common impurity in metals, typically 1–10 ppm, usually segregates to the interface during oxidation and consequently weakens the interfacial bonding).[48,49]

Pt-modified nickel–aluminide coatings, which typically have a surface layer of either $PtAl_2$ or ($PtAl_2$ + (Ni,Pt)Al), are normally fabricated by means of a 'two-step' method (Pt electroplating followed by pack aluminizing).[50–52] A low-Al-activity CVD processing enabled a single-phase (Pt,Ni) Al coating to be developed.[53] Compared to the Pt–Al coatings, the available results[53,54] indicated that the single-phased (Pt,Ni)Al coating fabricated by pack cementation exhibited improvements in cyclic oxidation and hot corrosion resistance, thermal stability and ductility. First-principle calculations show that the presence of Pt in the B2-type NiAl intermetallic compound brings the surface electronic energies to a higher state, which results in an increased surface energy. This will increase the void nucleation energy, thus making void formation a more difficult process,[55] which may explain the beneficial effect of Pt on cyclic oxidation. Some studies also showed that the Pt-modified aluminide coatings were much more resistant to hot corrosion than the Pt-free aluminide coatings,[56–58] a result which is not yet fully understood. Some investigators have indirectly attributed it to the beneficial effect on the scale adhesion of Pt.[57,58]

Chromized coatings

Kelley first produced chromized steels by means of pack cementation in 1923.[59] Samuel and Lockington[60] and Drewett[61] published a comprehensive review of chromized coatings on steel in the early 1950s. In 1953, Gibson patented aluminizing of chromized steels with aluminum paints and heat treatment to improve high-temperature oxidation resistance.[24] In general, chromized coatings on nickel-base metals have good oxidation resistance, which can be further improved by the introduction of CeO_2 particles by the 'two-step' method (Ni–CeO_2 co-electroplating followed by pack chromizing).[62–64] However, chromized ferritic steels do not have the desired oxidation resistance, particularly in the thermal cyclic condition. One important reason is that the chromization of the steels inevitably causes the formation of a surface chromium carbide layer,[62–67] because carbon has a strong chromium affinity, which has a detrimental effect on the oxidation resistance of chromized steels.[63,64] To mitigate or exclude the carbon effect, one promising route that has been suggested but not yet fully reported is the design of a precursor film on the steels as an electrodeposited Ni–RE_xO_y film (RE = Ce, Y, La, etc.).[62–64]

M–Cr–Al type diffusion coatings

M–Cr–Al-type diffusion coatings, which have a composition similar to that of overlay coatings fabricated by means of the other techniques (see below), may be developed through pack-cementation methods such as simultaneous chromizing–aluminizing, aluminizing followed by chromizing (or vice versa). Geib and Rapp[68] simultaneously chrome–aluminized low-alloy steels and got a coating with a surface composition of 11–20 wt%Cr and 4–9 wt% Al. However, reports on the oxidation of M–Cr–Al type diffusion coatings are still rare.

Low-temperature pack cementation

The temperatures at which conventional oxidation-resistant chromized and aluminized coatings are developed by pack cementation are very high (>1000 °C for chromization and >900 °C for aluminization). Such high temperatures significantly exceed the heat treatment temperatures for ferritic steel workpieces (normally below 700 °C), resulting in noticeable grain growth and carbide precipitation which has a detrimental impact on the mechanical properties, particularly in decreasing their high-temperature strength and creep resistance. Therefore, reducing the pack-cementation temperature is generally required for widespread application of these coatings to the steels. To achieve this goal, it is essential to enhance low-temperature chromizing and aluminizing kinetics, which are mainly determined by the diffusion of the coating element in the metal matrix. Metal grain boundaries (GBs) normally act as 'short-circuit' diffusion paths because of a reduced atomic density compared to the perfect crystal lattice. NC metals consist of a large volume fraction of GBs. Even when GB diffusion is dominant in the latter,[69–71] metallic diffusion in a nano-grained polycrystalline metal would be enhanced by orders of magnitude in comparison to the diffusion in its conventionally coarse-grained (CG) counterparts with the same chemical compositions at a given temperature. The enhancement lies in a different structure which comes from an enhanced diffusivity along the triple junctions between three GBs, and a reduced concentration of impurities in comparison to the GB in CG metals. Therefore, the nanostructure of steels is likely to be a promising approach for fabricating chromized and aluminized coatings at the desired low temperatures.

Recently, Zhang *et al.*[72] developed an oxidation-resistant, low-temperature chromized coating on a mild carbon steel using the 'two-step' method: first, electrodeposition of NC Ni matrix film with dispersion of CeO_2

nanoparticles; second, pack cementation at 700 °C in a traditionally halide-activated pack mixture. The positive effects of the CeO_2 nanoparticles during chromization may be summarized as follows: (i) retarding the growth of the nickel nanograins at high temperatures by anchoring the movement of GBs; (ii) producing numerous Ni/oxide interphase boundaries, which would be additional 'short-circuit' paths other than GBs for chromium transport during chromization; and (iii) distributing the oxide more homogeneously with the particles in nano-scale instead of in micron range. Figure 3.1 shows the electron probe microanalysis (EPMA) of cross-sections of chromized coatings on a CG Ni (mean grain size: 30 μm) and CeO_2-dispersed NC Ni film (mean grain size: 56 nm). Clearly, the chromized CG Ni cannot form a continuous chromium coating on the surface. In contrast, the chromized Ni–CeO_2 film formed a surface layer 25 μm in thickness with a uniform distribution of Cr. Thus, the chromizing kinetics on the nanograined metal was greatly promoted. Moreover, CeO_2 particles are found to randomly disperse in the chromized coating, which tends to decrease the oxidation rate of the coating through the REE mentioned above.[72]

Compared to chromizing, aluminizing has greatly increased coating growth rate at low temperatures when the reaction of Al with the metal matrix such as Ni, Fe and Co increases the inward diffusion of Al. However, the temperature of aluminizing kinetics is still quite low. For example, Xiang and Datta[73] developed aluminized coatings 30 μm in thickness (η-Fe_2Al_5) on P92 ferritic steels at 650 °C for 8 h in a pack of 4Al–2$AlCl_3$–Al_2O_3 (wt%). Zhan *et al.*[74] reported that the thickness of the aluminized coating reached ~50 μm when prepared in a pack of Al–NH_4Cl–Al_2O_3 for only 30 min at 560 °C with the assumption of the steel matrix nanograined by ball milling through a severe plastic deformation mechanism. Recent studies have also shown that, for 5 h pack cementation at 600 °C in a powder mixture of 45Al–5NH_4Cl–Al_2O_3 (wt%), the aluminide (δ-Ni_2Al_3) coating formed on a ~20 μm-grained Ni 18 μm in thickness, while the thickness of the aluminide on a ~60 nm-grained Ni–CeO_2 film was increased to ~58 μm. Moreover, the low-temperature aluminized coating on the NC Ni matrix had a dispersed CeO_2 and ultrafine-grained structure,[75] which exhibited excellent spalling resistance during cyclic oxidation in air at 1000 °C.[75] It was suggested that this result occurs because the dispersed CeO_2, together with the ultrafine-grained coating structure, helped prevent the formation of cavities at the coating interface.

Pack cementation is a simple, inexpensive and reproducible method for uniformly coating metallic components of different sizes and with complex shapes. However, the method has some limitations summarized follows:[24,42] (i) easy incorporation of contaminants; (ii) labor-intensive with respect to cleaning the coated components after pack cementation; (iii) large amount of pack powder wasted together with the cost of powder disposal; and (iv) relatively long processing time.

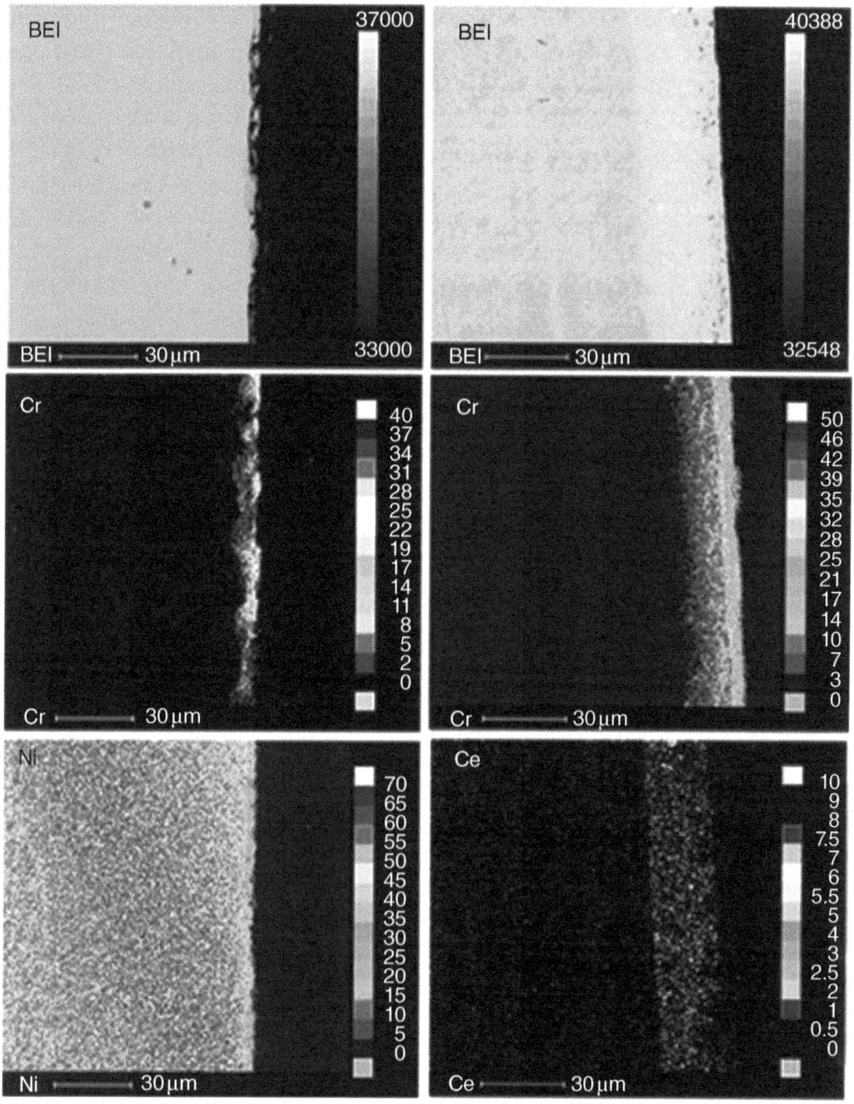

3.1 The back-scattered image, the corresponding elemental X-ray mappings of cross-sections for chromized samples on a coarse-grained Ni (left) and nanocrystalline Ni-base with dispersion of CeO_2 nanoparticles (Ni X-ray mapping not presented) (right).[72]

3.2.2 M–Cr–Al(–Y) overlay coatings

M–Cr–Al–Y overlay coatings emerged in 1960s, probably beginning with an alloy having the composition Fe–25%Cr–4%Al–1.0%Y, from G.E. nuclear programs, that exhibited exceptional alumina adherence in cyclic oxidation.[24] A model composition Fe–Cr–Al–Y, with 10–15% aluminum, was applied as a coating to Ni-base superalloys by electron beam physical vapor deposition (EB–PVD).[76] Because workpieces in industrial turbines are conventionally made of Ni- or Co-base structural superalloy, Ni-, Co- and NiCo-based M–Cr–Al–Y overlay coatings have been widely investigated, with the focus on oxidation and hot corrosion properties.[8–10,18,24,77–83] Compared to the Ni–Cr–Al–Y-type overlay coating, the Co–Cr–Al–Y-type overlay coating has generally decreased oxidation resistance but increased hot corrosion resistance. For this reason, NiCo-base overlay coatings with a compromise in oxidation and hot corrosion performance were developed. For example, the nominal composition Ni–23%Co–15%Cr–12.5%Al–0.5%Y had adequate oxidation and hot corrosion properties which proved to be satisfactory for the protection of blade airfoils in the current Pratt and Whitney commercial aircraft engines.[24] In industry, many more complex Ni–Co–Cr–Al–Y compositions containing additions of Ta, W, Ti, Nb, Re, Zr, etc., singly or in combination, have been developed for specific purposes. A comparison between oxidation and hot corrosion resistance of overlay coatings of various types is presented in Fig. 3.2.[3,84]

Oxidation and hot corrosion resistance also depend on the concentration of Cr. 16–24 wt%Cr in the overlay coatings favors not only an increase in low-temperature hot corrosion (Type II corrosion in a temperature of 650–800 °C) resistance but also oxidation and Type I hot corrosion (800–950 °C) resistance by promoting the formation of Al_2O_3 scale, through the

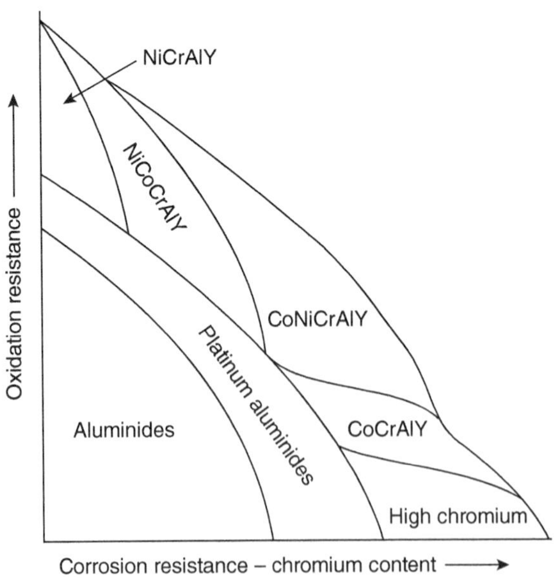

3.2 Dependence of oxidation resistance on corrosion resistance for various high-temperature coating systems.[3,84]

so-called third element effect according to Wagner theory.[85] Therefore, the overlay coatings exhibit greatly enhanced resistance to oxidation and corrosion compared with the simple and Pt-modified aluminide coating of the first generation. The overlay coatings can be specifically developed through various coating techniques, including EB–PVD,[18,76] plasma spray,[80] magnetron sputtering,[81,82] high-velocity oxyfuel (HVOF),[8,20,83] laser cladding,[10] etc. A typical Ni-base commercial overlay coating is EB–PVD ATD10 (Ni–20Cr–11Al–0.3Y).

M–Cr–Al–Y overlay coatings always contain a small amount of yttrium. The purpose of adding yttrium is to increase the resistance of the protective scale (mostly alumina) to oxidation through the so-called REEs which have been explained in detail above.[37–39] Interestingly, it was found that nanocrystallization of M–Cr–Al-based alloys without yttrium has an influence which is somewhat similar to that of yttrium additions in that it increases the adhesion of the developed alumina scale by means of several mechanisms as summarized by Peng *et al.*,[11] including forming interfacial 'oxide micropegs', minimizing the concentration of condensed cation vacancies and the segregation of sulfur at the scale/metal interface and increasing oxide diffusion creep by favoring the formation of a fine-grained scale structure. The alloy's nanocrystallization may also favor an exclusive development of a protective scale (mainly alumina), without increasing the contents of Cr and Al. This has made the synthesis of oxidation-resistant NC coatings an active field of research since the 1990s.[11–23]

Several researchers have studied the oxidation of Ni–Cr–Al-type NC coatings prepared by sputtering. Chen and Lou[86] found that a three-component NC coating containing concentrations of both Cr and Al as low as 5 wt% can develop into alumina scales. For the Ni–20Cr–xAl NC coating, only 2.4 wt% Al is needed for selective development to an α-alumina scale at 1000 °C, while the minimum content for conventionally coarse-grained three-component alloys is 6 wt% Al. Below this concentration limit, the CG alloys form an external layer of Cr_2O_3 and $NiCr_2O_4$ spinel with deep internal Al_2O_3 particles underneath.[17] An increased ability to selectively form alumina scales was also repeated for the oxidation of NiCrAl-type NC coatings prepared by HVOF using cryomilling-produced NC powders.[20]

The M–Cr–Al-type NC coatings are mostly fabricated by magnetron sputtering. The coating species (mainly atoms), which are randomly sputtered by Ar ions from the target material, are deposited onto the anode in a vacuum chamber. Due to the high-temperature state of the sputtered atoms, those arriving at the much cooler anode surface can be effectively quenched at a rate approaching 10^{15} °C/s.[87] Such a high quenching rate, together with a quick reduction of the stored energy of the sputtered atoms by collision in the flying track or on the sample surface, does not permit adequate time to rearrange the adatoms on the metal matrix through surface diffusion to form an alloy whose phase composition can be predicted according to bulk thermodynamic criteria. Therefore, the adatoms are locally accommodated to assemble a NC solid solution. In other words, the non-equilibrium characteristic of sputtering deposition fabricates a NC alloy coating with a chemical composition similar to, but phase composition dissimilar to, the target alloy. Most recently, a nanograined Ni–Cr–Al coating was fabricated by co-electrodeposition of NC Ni matrix and nanoparticles of Cr and Al.[88] Compared to ternary composites assembled with Cr and Al microparticles and conventionally coarse-grained alloys of similar composition, the nanocomposites exhibited a greatly increased ability to thermally develop chromia or alumina scales. The result constitutes a frame of reference for manufacturing novel oxidation-resistant Ni–Cr–Al-type coatings through simple electrochemical deposition, which is a sophisticated technique with the merits of simple processing, low expense, high productivity and good compositional control.

3.3 Metallic coatings as bond coats for thermal barrier coatings (TBCs)

Figure 3.3 shows a normal and practical TBC system,[25] which has three layers made of different materials with specific properties and functions; from bottom to top, they are (i) bond coat, (ii) TGO (normally Al_2O_3) and (iii) ceramic top coat (mostly YSZ). The bond coat is an oxidation-resistant

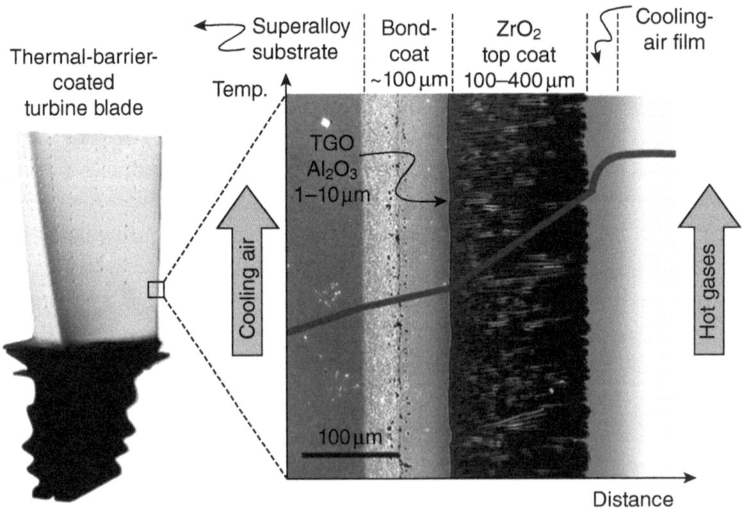

3.3 Cross-sectional morphology of an EB–PVD TBC coating onto a turbine blade, which consists of a top ceramic coat (YSZ), a TGO layer (alumina) and a bond coat. (Reprinted from reference [25] with permission from Copyright Clearance Center. INC)

metallic layer 100–200 μm in thickness. Currently, two typical bond coats are M–Cr–Al–Y alloy (such as Ni–Cr–Al–Y or Ni–Co–Cr–Al–Y), which is deposited using plasma spray or EB–PVD, and Pt-modified aluminides, which are deposited by Pt electroplating in combination with pack aluminization or CVD. In many cases, the TBC damage is correlated with the oxidation and mechanical properties of the bond coat and the TGO. Driven by compressive stresses in the TGO, brought about by oxidation-induced volume increase following the Pilling–Bedworth ratio (growth stress) or by differences in the coefficient of thermal expansion (CTE) between TGO and bond coat (thermal stress), the TGO may crack and the TGO/bond coat interface may delaminate, which is currently the key reason for spallation failure of the TBC system.[25,89]

3.3.1 Pt-modified aluminide bond coats

Contemporary commercial gas turbines usually have more advanced coating systems, e.g. platinum aluminides (or M–Cr–Al–Y, etc.), on first stage blades, and it is estimated that more than 80% of airfoils are coated by pack cementation or CVD aluminizing or chromizing processes.[24] Since the 1990s, aluminide coatings have been recognized as useful bond coats for some types of TBCs, capable of offering oxidation resistance.[90]

Pt-modified Ni–Al coatings are susceptible to the development of surface undulations (surface 'rumpling') when thermally cycled at elevated temperatures.[90–93] Figure 3.4 shows a typical evolution of surface rumpling of a Pt-modified Ni–Al coating on René N5 single crystal superalloy by platinum electroplating followed by a low-activity CVD aluminizing during cycling in air at 1200 °C.[91] The rumpling was considered to be local volume reduction of the aluminide due to the depletion of Al by oxidation.[91] This surface rumpling could cause premature coating failure during service, which normally occurs in the form of cracking and spallation of the protective oxide scale along the rumpled ridges and accelerated oxidation.

For a Pt-modified Ni–Al bond coat/EB–PVD YSZ TBC system, the growth of the TGO and the mechanical behavior of the bond coat are believed to play an extremely important role in determining the life of the system. The rumpling of the bond coat can lead to significant out-of-plane displacement of the bond coat/TGO interface, which is associated with local separations and delamination of the top ceramic coat and overall TBC failure. Moreover, other factors, such as surface roughness and defect size as well as sulfur impurity of a (Pt,Ni)Al bond coat, can influence the damage to the top YSZ.[25,89] Pt-modified Ni–Al bond coats fabricated by different routes will have different structure, chemical composition and surface morphology, which causes damage to the TGO/bond coat according to different mechanisms.

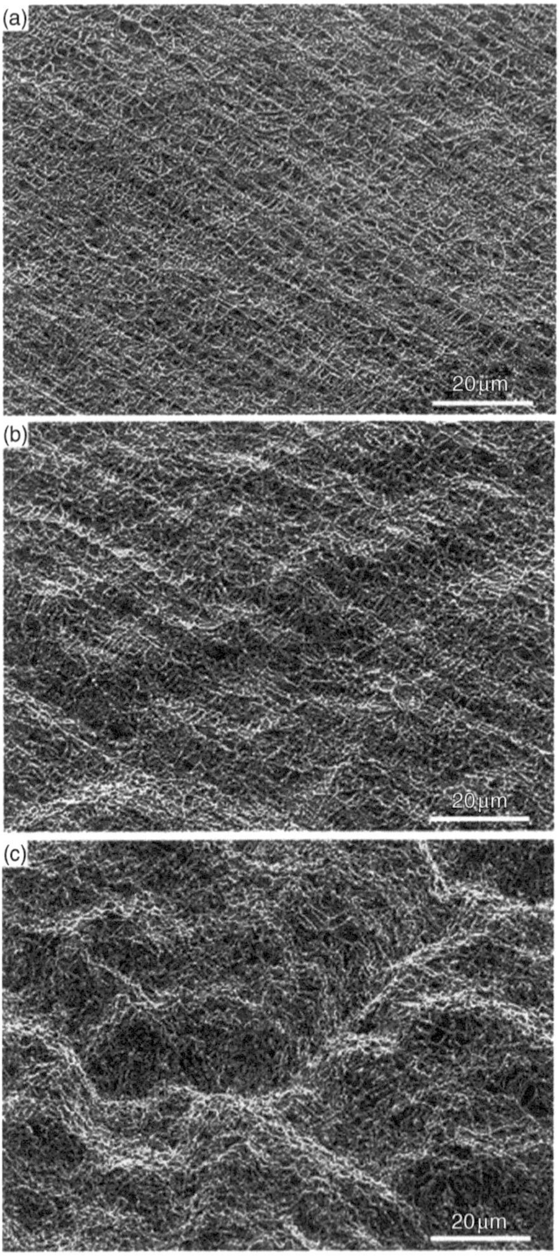

3.4 Rumpling of the bond coat surface polished with 400-grit under cyclic loads in air at 1200 °C (1 h cycles): (a) 1 cycle; (b) 10 cycles; (c) 50 cycles.[91]

3.3.2 M–Cr–Al–Y bond coats

In M–Cr–Al–Y overlay coating systems, Ni–Cr–Al–Y or Ni–Co–Cr–Al–Y are two typical coatings used as bond coats. They are normally deposited with low-pressure plasma spray (LPPS) or EB–PVD.[25,89] The bond coats usually consist of γ-Ni solid solution matrix and β-NiAl phases, although minor precipitates of γ′-Ni_3Al and σ-(Cr, Co) phases have additionally been shown to exist.[89,94–96] The concentration of the yttrium added is normally around or below 1.0 wt%. The yttrium addition can improve the adhesion of the TGO, mainly through scavenging the interface sulfur segregants, which would promote the growth of exiting interface voids[97] or weaken interface bonding.[98]

The failure mechanism of the Ni–Cr–Al–Y or Ni–Co–Cr–Al–Y bond coat/YSZ TBC system is less well understood, because the YSZ top layer often fails catastrophically by large-scale delamination.[94,99] The dominant delamination mode has been reported as separation along the TGO/bond coat interface after high-temperature exposure.[89,94,99] The interface separation depends on the synergistic effect of several factors or parameters, such as the amount of stored energy in the TGO, the interface chemistry (e.g. sulfur concentration), the interface imperfections (e.g., physical defect size, yttrium-rich oxide protrusions) and the mechanical behavior of the bond coat. The original interface defect is normally considered as one important reason for the separation of the TGO from the bond coat. When the interface defect (e.g., void and delamination) grows to a critical size, local TGO buckling occurs. The critical interface flaw ($\bar{c}$) of buckling can be given by[100,101]

$$\bar{c} = \frac{\pi h_{ox}}{2\sqrt{3}}\sqrt{\frac{E_{ox}}{(1-\nu_{ox}^2)\sigma_{ox}}} \qquad [3.4]$$

where h_{ox}, E_{ox}, ν_{ox} are the thickness, elastic modulus and Poissons' ratio of the TGO, respectively, and σ_{ox} is the stress generated in the TGO. It has also been reported that oxide intrusions (normally containing reactive element such as yttrium) with a range of size and wavelength can be the source of the initial oxide delamination.[102] In addition, it is noteworthy that a continuous γ-Ni layer can form at the TGO/bond coat interface because of the depletion of Al from the β-NiAl during oxidation. The evolution of the γ layer strongly depends on the oxidation temperature as well as the number of thermal cycles.[94] The alumina scale/γ-Ni interface has low toughness, especially when segregation or contamination is present in combination with moisture in the surroundings.[103,104] In most cases, the failure mechanism of the TGO/bond coat system follows a route of interface delamination, TGO buckling and finally the failure of the TBC system.

3.4 Conclusions

Metallic chromia- or alumina-forming diffusion coatings, which are referred to as the first-generation coatings, have been extensively used in gas turbine engines for aero-jet propulsion, land-based industrial power generation and in industrial boilers for electric generation to prevent oxidation of the structural metals in service in high-temperature environments. Compared to chromized coatings, aluminized coatings have more application advantages because they can form alumina scales that not only grow slowly but are also more thermodynamically stable at high temperatures (over >1000 °C) as well as in wet atmospheres. To further improve the oxidation resistance, composition modification of aluminized coatings by addition of specified elements has been extensively investigated. Compared to conventional NiAl coatings, platinum- and RE_xO_y (Re = Y, La, etc.)-modified nickel–aluminide coatings exhibit excellent resistance in thermal cyclic oxidation and hot corrosion. The enhanced adhesion of the alumina scales has been proposed as one of the main reasons. Grain refinement through structural tailoring may be another approach to further increase the adhesion of the alumina scales on aluminide coatings, by preventing the formation of large cavities at the oxide/metal interface, an effect for which the fundamental reason has not yet been fully explored. Fine-grained aluminide coatings can be developed on metals which are nanograined through special surface treatment or pre-deposited with a nanocrystalline metallic film such as Ni, by means of conventional pack aluminization but at much lower temperatures than normal. This processing method enables the fabrication of aluminide coatings on ferritic steels at the temperatures below the maximum temperature allowed by conventional heat treatment.

Compared to alumina-forming diffusion coatings, alumina-forming M–Cr–Al–Y overlay coatings (second-generation coatings) have the advantages that their composition can be specifically designed and manufactured to match not only the particular environment conditions but also the main compositions of matrix alloys, making it possible to minimize the inter-diffusion between coatings and matrix. Inter-diffusion can be best avoided by adding nanocrystalline M–Cr–Al-type coatings, because an alumina-forming M–Cr–Al NC coating may have concentrations of Cr and Al which are close to those in conventional M–Cr–Al-base alloys used as the matrix for coating. Oxidation of M–Cr–Al NC coatings has been widely investigated since the mid-1990s. For engineering applications, more works on the thermal stability and long-term oxidation performance of NC coatings should be conducted in the future.

TBCs (typically YSZ) are the third-generation coatings and have the most complex structure. The YSZ coat is normally deposited onto a bond coat, a metallic coating that is preferentially coated onto the matrix alloys to increase the adhesion of the ceramic during oxidation. The bond coat is normally an alumina former, because the alumina TGO has a very low growth rate at high temperatures and is phase compatible with YSZ. Two major categories of bond coats which have been developed are the Pt-modified Ni–Al and M–Cr–Al–Y alloys. The rumpling of Pt-modified Ni–Al bond coat and TGO buckling of M–Cr–Al–Y bond coat are the major mechanisms for damage to the TGO/bond coat interface, which is considered as an important reason for the mechanical failure of the TBC system at high temperatures.

3.5 Acknowledgements

We acknowledge the financial support by the National Science Foundation of China (NSFC) project under Grant No. 51071162 and 50571108.

3.6 References

1. FITZER E, MAURER H J, 'Diffusion and precipitation phenomena in aluminized and chromium-aluminized iron- and nickel-based alloys', in: *Materials and Coatings to Resist High Temperature Corrosion* (Ed. Holmes D R, Rahmel A), London, UK, Applied Science, 1978, 253–269.
2. RESTALL J E, MALIK M, SINGHEISER L, 'Metallic diffusion and overlay coatings', in: *Proceeding of High Temperature Alloys for Gas Turbines and Other Applications* (Ed. Betz W), Dordrecht, Germany, Reidel, 1986, 357.
3. NICHOLLS J R, SIMMS N J, CHAN W Y, EVANS H E, 'Smart overlay coatings – concept and practice', *Surf Coat Technol*, 2002, **149**(2–3), 236–244.
4. NICHOLLS J R, HANCOCK P, AL YASIRI L H, 'Optimising oxidation resistance of MCrAl coating systems using vapor phase alloy design', *Mater Sci Technol*, 1989, **5**, 799–805.
5. WANG F, LOU H, ZHU S, WU W, 'The mechanism of scale adhesion on sputtered microcrystallized CoCrAl films', *Oxid Met*, 1996, **45**(1–2), 39–51.
6. LIU Z, GAO W, DAHM K L, WANG F, 'Oxidation behaviour of sputter-deposited Ni–Cr–Al micro-crystalline coatings', *Acta Mater*, 1998, **46**(5), 1691–1700.
7. WAKI H, KOBAYASHI A, 'Development of lateral compression method of circular tube thin coating for mechanical properties of plasma sprayed CoNiCrAlY', *Mater Trans*, 2006, **47**(7), 1626–1630.
8. BRANDL W, TOMA D, KRUGER J, GRABKE H J, MATTHAUS G, 'The oxidation behaviour of HVOF thermal-sprayed MCrAlY coatings', *Surf Coat Technol*, 1997, **94–5**(1–3), 21–26.
9. LI C J, LI W Y, 'Effect of sprayed powder particle size on the oxidation behavior of MCrAlY materials during high velocity oxygen-fuel deposition', *Surf Coat Technol*, 2003, **162**(1), 31–41.
10. HIROSE A, KOTANI H, KOBAYASHIK K, 'Surface modification by laser cladding of Ni–Cr–Al–Y alloy', *Iron Steel Inst Jpn*, 1993, **79**(1), 105–112.
11. PENG X, WANG F, 'Oxidation resistant nanocrystalline coatings', in: *New developments in high temperature corrosion* (Ed. Gao W, Li Z), Cambridge, UK, Woodhead, 2008, 456–475.

12. WANG F, LOU H, 'Oxidation Behaviour and scale morphology of normal-grained CoCrAl alloy and its sputtered microcrystalline coating', *Mater Sci Eng*, 1990, **A43**, 279–285.
13. WANG F, 'Oxidation resistance of sputtered Ni-3(AlCr) nanocrystalline coating', *Oxid Met*, 1997, **47**(3/4), 247–258.
14. WANG F, YOUNG D J, 'Effect of nanocrystallization on the corrosion resistance of K38G superalloy in CO + CO_2 atmospheres', *Oxid Met*, 1997, **48**(5/6), 497–509.
15. CHEN G, LOU H, 'Oxidation behavior of a sputtered Ni–5Cr–5Al nanocrystalline coating', *Oxid Met*, 2000, **53**(5/6), 467–479.
16. LIU Z, GAO W, DAHM K L, WANG F, 'The effect of coating grain size on the selective oxidation behaviour of Ni–Cr–Al alloy', *Scripta Mater*, 1997, **37**(10), 1551–1558.
17. LIU Z, GAO W, DAHM K L, WANG F, 'Oxidation behaviour of sputter-deposited Ni–Cr–Al micro-crystalline coatings', *Acta Mater*, 1998, **46**(5), 1691–1700.
18. STRAUSS D, MULLER G, SCHUMACHER G, ENGELKO V, STAMM W, CLEMENS D, QUADDAKERS W J, 'Oxide scale growth on MCrAlY bond coatings after pulsed electron beam treatment and deposition of EBPVD–TBC', *Surf Coat Technol*, 2001, **135**(2–3), 196–201.
19. YANG S, WANG F, SUN Z, ZHU S, 'Influence of columnar microstructure of a sputtered NiAl coating on its oxidation behavior at 1000 °C', *Intermetallics*, 2002, **10**(5), 467–471.
20. AJDELSZTAJN L, PICAS J A, KIM G E, BASTIAN F L, SCHOENUNG J, PROVENZANO V, 'Oxidation behavior of HVOF sprayed nanocrystalline NiCrAlY powder', *Mater Sci Eng A*, 2002, **338**(1–2), 33–43.
21. MROWEC S, 'New generation of nanocrystalline coating materials resistant to high temperature corrosion', *High Temp Mater Proc*, 2003, **22**(1), 1–24.
22. PENG X, YAN J, ZHOU Y, WANG F, 'Effect of grain refinement on the resistance of 304 stainless steel to breakaway oxidation in wet air', *Acta Mater*, 2005, **53**(19), 5079–5088.
23. KHANNA A, BHAT D G, 'Nanocrystalline gamma alumina coatings by inverted cylindrical magnetron sputtering', *Surf Coat Technol*, 2006, **201**(1–2), 168–173.
24. GOWARD G W 'Progress in coatings for gas turbine airfoils', *Surf Coat Technol*, 1998, **108–109**(1–3), 73–79.
25. PADTURE N P, GELL M, JORDAN E H, 'Thermal barrier coatings for gas turbine engine applications', *Science*, 2002, **296**(5566), 280–284.
26. CLARKE D R, LEVI C G, 'Materials design for the next generation thermal barrier coatings', *Annu Rev Mater Res*, 2003, **33**, 383–417.
27. FEUERSTEIN A, KNAPP J, TAYLOR T, ASHARY A, BOLCAVAGE A, HITCHMAN N, 'Technical and economical aspects of current thermal barrier coating systems for gas turbine engines by thermal spray and EBPVD: a review', *J Therm Spray Technol*, 2008, **17**(2), 199–213.
28. CHOY K L, 'Chemical vapor deposition of coatings', *Prog Mater Sci*, 2003, **48**(2), 57–170.
29. KUNG SC, RAPP RA, 'Kinetic-study of aluminization of iron by using the pack cementation technique', *J Electrochem Soc*, 1988, **135**(3), 731–741.
30. NICHOLS E S, BURGER J A, HANINK D K, 'Protective coatings for turbine parts', *Mech. Eng.*, 1965, **87**, 52–56.

31. GALMICHE P M, 'Production of surface diffusion alloys', US Patent, 1975, No 3,900,613.
32. RESTALL J E, WOOD M I, 'Alternative processes and treatments (for gas turbine protective coating deposition)', *Mater Sci Tech*, 1986, **2**, 225–231.
33. PARZUKOWSKI R S, 'Gas-phase deposition of aluminum on nickel alloys', *Thin Solid Films*, 1977, **45**, 349–355.
34. GAUJE G, MORBIOLI R, 'Vapor phase aluminizing to protect turbine aerofoils', in: *High Temperature Protective Coatings* (Ed. Singhal SC), The Metallurgical Society of AIME, Atlanta, GA, 1983, 13–26.
35. WARNES B M, PUNOLA D C, 'Clean diffusion coatings by chemical vapor deposition', in: *24th International Conference on Metallurgical Coatings and Thin Films Surf. Coat. Technol.* (Ed. Khobaib M and Krutenat R C), 1997, Metallurgical Society, Warrendale, PA, **94–95**(1–3), 1–6.
36. GOWARD G W, BOONE D H, 'Mechanisms of formation of diffusion aluminide coatings on nickel-base superalloys', *Oxid Met*, 1971, **3**(5), 475–495.
37. WHITTLE D P, STRINGER J, 'Improvements in high temperature oxidation resistance by additions of reactive elements or oxide dispersions', *Philos Trans R Soc London Ser. A*, 1980, **295**(1413), 309–329.
38. MOON D P, 'Role of reactive elements in alloy protection', *Mater Sci Technol*, 1989, **5**(8), 754–764.
39. HOU P Y, STRINGER J, 'The effect of reactive element additions on the selective oxidation, growth and adhesion of chromia scales', *Mater Sci Eng A*, 1995, **202**(1–2), 1–10.
40. LEPRINCE G, ALPÉRINE S, VANDENBULCKE L, WALDER A, 'New high temperature-resistant NiCrAl and NiCrAl +Hf felt materials', *Mater Sci Eng*, 1989, **120–121**(Part II), 419–425.
41. RAPP R A, WANG D, WEISERT T, in: *Metallurgical Coatings* (Ed. Hobaib M and Krutenat R), TMS-AIME, Warrendale, PA, 1987, 131–141.
42. BIANCO R, RAPP R A, 'Pack cementation aluminide coatings on superalloys: codeposition of Cr and reactive elements', *J Electrochem. Soc*, 1993, **140**(4), 1181–1190.
43. BIANCO R, RAPP R A, SMIALEK J L, 'Chromium and reactive element modified aluminide diffusion coatings on superalloys: environmental testing', *J Electrochem Soc*, 1993, **140**(4), 1191–1203.
44. TU D C, LIN C C, LIAO S J, CHOU J C, 'A study of yttrium-modified aluminide coatings on IN 738 alloy', *J Vac Sci Technol*, 1986, **A4**(6), 2601–2608.
45. PENG X, LI T, PAN W P, 'Oxidation of a La_2O_3-modifed aluminide coating', *Scripta Mater*, 2001, **44**(7), 1033–1038.
46. FELTEN E J, PETTIT F S, 'Development, growth, and adhesion of Al_2O_3 on platinum-aluminum alloys', *Oxid Met*, 1976, **10**, 189–223.
47. ZHANG Y, HAYNES J A, LEE W Y, WRIGHT I G, PINT B A, COOLEY K M, LIAW P K, 'Effects of Pt incorporation on the isothermal oxidation behavior of chemical vapor deposition aluminide coatings', *Metall Mater Trans A*, 2001, **32A**, 1727–1741.
48. HOU P Y, MCCARTY K F, 'Surface and interface segregation in beta-NiAl with and without Pt addition', *Scripta Mater*, 2006, **54**, 937–941.
49. HOU P Y, TOLPYGO V K, 'Examination of the platinum effect on the oxidation behavior of nickel-aluminide coatings', *Surf Coat Technol*, 2007, **202**(4–7), 623–627.

50. LEHNARDT G, MEINHARDT H, 'Present state and trend of development of surface coating methods against oxidation and corrosion at high temperatures', *Electrodeposition Surf Treatment*, 1972, **1**(1), 71–76.
51. FARRELL M S, BOONE D H, STREIFF R, 'Oxide adhesion and growth characteristics on platinum-modified aluminide coatings', *Surf Coat Technol*, 1987, **32**, 69–84.
52. GÖBEL M, RAHMEL A, SCHÜTZE M, SCHORR M, WU W T, 'Interdiffusion between the platinum-modified aluminide coating RT22 and nickel-based single-crystal superalloys at 1000 and 1200 °C', *Mater High Temp*, 1994, **12**(4), 301–309.
53. WARNES B M, PUNOLA D C, 'Clean diffusion coatings by chemical vapor deposition', *Surf Coat Technol*, 1997, **94–95**(1–3), 1–6.
54. WARNES B M, PUNOLA D C, 'Development of diffusion coatings for gas turbines', in: *Power Generation Technology: The International Review of Preliminary Power Production* (Ed. Knox R), Sterling Publications Limited, London, UK, 1995, 207–210.
55. YU R, HOU P Y, 'First-principles calculations of the effect of Pt on NiAl surface energy and the site preference of Pt', *Appl Phys Lett*, 2007, **91**(1), 011907-1–3.
56. WU W T, RAHMEL A, SCHORR M, 'Role of platinum in the Na_2SO_4-induced hot corrosion resistance of aluminum diffusion coatings', *Oxid Met*, 1984, **22**(1–2), 59–81.
57. SCHAEFFER J, KIM G M, MEIER G H, PETTIT F S, 'The effects of precious metals on the oxidation and hot corrosion of coatings', in: *The Role of Active Elements in the Oxidation Behavior of High Temperature Metals and Alloys* (Ed. Lang E), Elsevier, Amsterdam, the Netherlands, 1989, 231–267.
58. LEYENS C, PINT B A, WRIGHT I G, 'Effect of composition on the oxidation and hot corrosion resistance of NiAl doped with precious metals', *Surf Coat Technol*, 2000, **133–134**, 15–22.
59. KELLY F C, 'Chromizing', *Trans. Electrochem. Soc.*, 1923, **43**, 351.
60. SAMUEL R L, LOCKINGTON N A, 'Metal Treatment and Drop Forging', 1951, **18**, 354–359, 407–414, 440–444, 495–506, 543–548, and 1952, **19**, 27–32, 81–85.
61. DREWETT R, 'Diffusion coatings for the protection of iron and steel', *Anti-Corrosion Methods Mater*, 1969, **16**, 543–548.
62. PENG X, YAN J, XU C, WANG F, 'Oxidation at 900 °C of the chromized coatings on A3 carbon steel with the electrodeposition pretreatment of Ni or Ni-CeO_2 film', *Metall Mater Trans*, 2008, **39A**, 119–129.
63. YAN J, PENG X, WANG F, 'Oxidation of a novel CeO_2-dispersion-strengthened chromium coating in simulated coal-combustion gases', *Mater Sci Eng A*, 2006, **426A**(1–2), 266–273.
64. ZHU L, PENG X, YAN J, WANG F, WU W, 'Oxidation of a novel chromium coating with CeO_2 dispersions', *Oxid Met*, 2004, **62**(5–6), 411–426.
65. MAZIASZ J, SWINDEMAN R W, 'Selecting and developing advanced alloys for creep-resistance for microturbine recuperator applications', *J Eng Gas Turbines Power*, 2003, **125**(1), 310–315.
66. SAUNDERS S R J, NICHOLLS J R, 'Coatings and surface treatments for high temperature oxidation resistance', *Mater Sci Technol*, 1989, **5**(8), 780–798.
67. MEIER G H, CHENG C, 'Diffusion chromizing of ferrous alloys', *Surf Coat Technol*, 1989, **39–40**, 53–64.

68. GEIB F D, RAPP R A, 'Simultaneous chromizing-aluminizing coating of low-alloy steels by a halide-activated, pack-cementation process', *Oxid Met*, 1993, **40**(3–4), 213–228.
69. ENNIS P J, ZIELINSKA-LIPIEC A, WACHTER O, CZYRSKA-FILEMONOWICZ A, 'Microstructural stability and creep rupture strength of the martensitic steel P92 for advanced power plant', *Acta Mater*, 1997, **45**(12), 4901–4907.
70. GLEITER H, 'Nanocrystalline materials', *Prog Mater Sci*, 1989, **33**(4), 223–315.
71. WÜRSCHUM R, HERTH S, BROSSMANN U, 'Diffusion in nanocrystalline metals and alloys – A status report', *Adv Eng Mater*, 2003, **5**(5), 365–372.
72. ZHANG H, PENG X, ZHAO J, WANG F, 'Prior electrodeposition of nanocrystalline Ni–CeO_2 film fabricating an oxidation-resistant chromized coating on carbon steels', *Electrochem Solid ST*, 2007, **10**(3), C12–C15.
73. XIANG Z D, DATTA, P K, 'Relationship between pack chemistry and aluminide coating formation for low-temperature aluminisation of alloy steels', *Acta Mater*, 2006, **54**, 4453–4463.
74. ZHAN Z L, HE Y D, WANG D R, GAO W, 'Low-temperature processing of Fe-Al intermetallic coatings assisted by ball milling', *Intermetallics*, 2006, **14**(1), 75–81.
75. XU C, PENG X, WANG F, 'Cyclic oxidation of an ultrafine-grained and CeO_2-dispersed δ-Ni_2Al_3 coating', *Corros Sci*, 2010, **52**(3), 740–747.
76. TALBOOM F P, GRAFWALLNER J, *Nickel or Cobalt Base with a Coating Containing Iron Chromium and Aluminum*, US Patent, 1970, No 3,542,530.
77. EVANS D J, ELAM R C, *Cobalt Base Coating for the Superalloys*, US Patent, 1972, No 3,676,085.
78. HECHT R J, GOWARD G W, ELAM R C, *High Temperature NiCoCrAlY Coating*, US Patent, 1975, No 3,928,026.
79. GIL A, SHEMET V, VASSEN R, SUBANOVIC M, TOSCANO J, NAUMENKO D, SINGHEISER L, QUADAKKERS W J, 'Effect of surface condition on the oxidation behavior of MCrAlY coatings', *Surf Coat Technol*, 2006, **201**(7), 3824–3828.
80. YANAR N M, MEIER E M, PETTIT F S, MEIER G H, 'The effect of yttrium content on the oxidation-induced degradation of NiCoCrAlY coatings and the influence of surface roughness on the oxidation of NiCoCrAlY and platinum aluminide coatings', *Mater High Temp*, 2009, **26**(3), 331–338.
81. WANG F H, LOU H Y, WU W T, 'Influence of defects on properties of sputtered CoCrAlY coating', *Acta Metall Sin*, 1992, **28**(10), B443–448.
82. TANIGUCHI S, ASANUMA N, SHIBATA T, WANG F H, LOU H Y, WU W T, 'Oxidation resistance of TiAl improved by CoCrAl abd CoCrAlY coating', *J Jpn Inst Met*, 1993, **57**, 781–789.
83. BRANDLA W, TOMAA D, GRABKE H G, 'The characteristics of alumina scales formed on HVOF-sprayed MCrAlY coatings', *Surf Coat Technol*, 1998, **108–109**(1–3), 10–15.
84. MOM A J A, *High Temperature Coatings for Gas Turbines, an Overview*, NLR Report, MP 81003U, Amsterdam. 1981.
85. WAGNER C, 'Passivity and inhibition during the oxidation of metals at elevated temperatures', *Corros Sci*, 1965, **5**(11), 751–764.
86. CHEN G F, LOU H Y, 'Predicting the oxide formation of Ni-Cr-Al alloys with nano-sized grain', *Mater Lett*, 2000, **45**(5), 286–291.

87. SHIN S M, RAY M A, RIGSBEE J M, GREENE J E, 'Growth of metastable $Cu_{1-x}Cr_x$ solid solutions by ion mixing during bias-sputter deposition', *Appl Phys Lett*, 1983, **43**(3), 249–251.
88. YANG X, PENG X, CU C, WANG F, 'Electrochemical assembly of novel Ni-xCr-yAl nanocomposites with excellent high temperature oxidation resistance', *J Electrochem Soc*, 2009, **156**, C167–C175.
89. EVANS A G, MUMM D R, HUTCHINSON J W, MEIER G H, PETTIT F S, 'Mechanisms controlling the durability of thermal barrier coatings', *Progr Mater Sci*, 2001, **46**(5), 505–553.
90. DEB P, BOONE D H, MANLEY T F, 'Surface instability of platinum modified aluminide coatings during 1100 °C cyclic testing', *J Vac Sci Technol*, 1987, **A5**(6), 3366–3372.
91. TOLPYGO V K, CLARKE D R, 'Surface rumpling of a (Ni, Pt)Al bond coat induced by cyclic oxidation', *Acta Mater*, 2000, **48**, 3283–3293.
92. TOLPYGO V K, CLARKE D R, 'On the rumpling mechanism in nickel-aluminide coatings: Part I: an experimental assessment', *Acta Mater*, 2004, **52**(17), 5115–5127.
93. TOLPYGO V K, CLARKE D R, 'Rumpling of CVD (Ni,Pt)Al diffusion coatings under intermediate temperature cycling', *Surf Coat Technol*, 2009, **203**, 3278–3285.
94. CAO F, TRYON B, TORBET C J, POLLOCK T M, 'Microstructural evolution and failure characteristics of a NiCoCrAlY bond coat in "hot spot" cyclic oxidation', *Acta Mater*, 2009, **57**, 3885–3894.
95. GOWARD G W, BOONE D H, 'Mechanisms of formation of diffusion aluminide coatings on nickel-base superalloys', *Oxid Met*, 1971, **3**(5), 475–495.
96. BAUFELD B, SCHMUCKER M, 'Microstructural evolution of a NiCoCrAlY coating on an IN100 substrate', *Surf Coat Technol*, 2005, **199**(1), 49–56.
97. GRABKE H J, WIEMER D, VIEFHAUS H, 'Segregation of sulfur during growth of oxide scales', *Appl Surf Sci*, 1991, **47**(3), 243–250.
98. HOU P Y, 'Segregation phenomena at thermally grown Al2O3/alloy interfaces', *Annu Rev Mater Res*, 2008, **38**, 275–298.
99. SERGO V, 'Observation of subcritical spall propagation of a thermal barrier coating', *J Am Ceram Soc*, 1998, **81**(12), 3237–3242.
100. HUTCHINSON J W, SUO Z, 'Mixed mode cracking in layered materials', *Adv Appl Mech*, 1992, **29**, 63–191.
101. EVANS A G, CANNON R M, 'Stresses in oxide films and relationships with cracking and spalling', *Mater Sci Forum*, 1989, **43**, 243–268.
102. HUTCHINSON J W, HE M Y, EVANS A G, 'The influence of imperfections on the nucleation and propagation of buckling driven delaminations', *J Mech Phys Solid*, 2000, **48**(4), 709–734.
103. EVANS A G, HUTCHINSON J W, WEI Y, 'Interface adhesion: effects of plasticity and segregation', *Acta Mater*, 1999, **47**(15–16), 4093–4113.
104. BONNELL D A, KIELY J, 'Plasticity at multiple length scales in metal-ceramic interface fracture', *Phys Status Solid A*, 1998, **166**(1), 7–17.

4
Nanostructured thermal barrier coatings

C. G. ZHOU and Q. H. YU, Beihang University, China

Abstract: Thermal spray zirconia–8 wt% yttria (YSZ) deposits have been employed as thermal barrier coatings (TBCs) in the hot sections of gas turbines. The use of nanostructured YSZ represents an alternative for improving the performance of these coatings. Before spraying, finely dispersed particles must be agglomerated. Air plasma-sprayed nanostructured zirconia coatings consist of only the non-transformable tetragonal phase and the microstructure is composed of nanozones (unmelted or partially melted particles), splats, microcracks and high-volume spheroidal pores. The results indicate that the microhardness, elastic modulus and roughness of the nanostructured zirconia coatings exhibit the following trends: the smoother the roughness, the higher the microhardness and elastic modulus. The decrease of thermal diffusivity of the nanostructured coating was attributed to the increase of grain boundaries, micropores and defective crystal structure. The smaller the micropores are, the more interfaces are produced, and the effect on the phonon scattering is higher, resulting in the reduction of thermal diffusivity. Meanwhile, the nanostructured TBC also exhibits the promising thermal cyclic oxidation resistance.

Key words: nanostructured YSZ coating, air plasma spraying, agglomeration, mechanical performance, thermal property.

4.1 Introduction

Since the late 1990s, nanostructured zirconia TBCs have become a hot research field due to their low thermal conductivity, high coefficient of thermal expansion and excellent mechanical properties.[1–7] These improved properties are caused by the modification to the structure.[8–16] The microstructure and the properties of plasma spray coatings depend on various process parameters, and are markedly affected by the characteristics of the feeding powder, because the nanostructured powders clearly improve the deposit efficiency, coherence strength and hardness of the coatings.[17,18] However, the individual nanoparticle is so fine that it cannot be used for plasma spray. Before spraying, finely dispersed particles must be agglomerated to a size of about 30–100 μm. The microstructure, density and grain size of the powders determine the quality of the coating.[19,20] Meanwhile, phase transformation of as-sprayed zirconia coatings in the service environment is believed to be an important factor affecting the durability of the coating.[21–23]

The spray drying process of compositing the powders, the phase stability and the mechanical and thermal properties of the nanostructured YSZ coating are discussed in this chapter. The first section is about the spray drying process of compositing the ceramic powders. In the following section, the phase composition and microstructure characterization of the nanostructured as-sprayed zirconia coatings are studied; the influence of annealing temperature and time on the phase stability and grain growth are also discussed. In the next section, the mechanical properties of nanostructured YSZ coating are discussed. In the last section, the thermophysical properties and the failure behavior of the plasma-sprayed nanostructured TBCs are illustrated.

4.2 Spray-drying process making powders

Due to its lower inertia force, the nanopowder cannot penetrate the hot gas downstream. Even if it can penetrate into the plasma plume, particles will be easily vaporized because of the low mass. This is the reason why the particle needs to be agglomerated. Spray-drying is a process by which a fluid feed material is transformed into a dry powder by spraying the feed into a hot medium.

Spray-drying is a complex process, but we now propose a simple mechanism to describe the spray-drying process of ceramics with four steps as illustrated in Fig. 4.1:[24] (i) droplet formation; (ii) evaporation and ballooning; (iii) explosion; (iv) particle formation. The drying gas is introduced through the ceiling to the drying chamber. When the suspension is pumped into the drying chamber, it is separated into small droplets immediately by the compressed feeding gas. Then, the droplet shrinks because of the fast evaporation of the solvent. During the process of liquid moving from the inside to the outside of the droplet, the solid particles are carried along. Inside the droplet, a void is produced when the evaporation rate inside the droplet is higher than the diffusion rate through the droplet surface. When the pressure inside the droplet reaches a certain level, explosion is unavoidable. In the fourth step, the particle is plasticized by the loss of solvent and the addition of organic binder. The plasticized droplet shrinks continuously due to the effect of surface tension until it is totally dried. The particle size should be mainly determined by the third step at which most of the solvent has been lost and a soft shell is formed.

Figure 4.2 shows microphotographs of some spray-dried powders.[24] SPL10 (3YSZ) (Fig. 4.2a) has a smooth surface, explaining why this powder has a good flowability. The particle is an agglomeration of a large amount of small particles which are tightly combined with each other by organic binder. Figure 4.2b shows the microstructure of SPL17 (3YSZ + 6 wt%

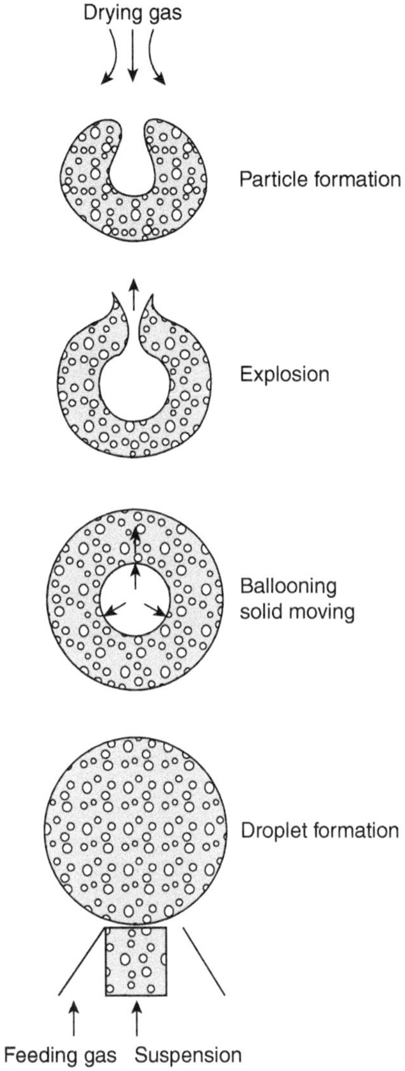

4.1 Spray-drying process of ceramics.

Al_2O_3). Particles of this powder have much larger hollow centres than those of SPL10 (3YSZ) with a low bulk density (1.56 g/cm^3) and large mean size (39.8 μm). Although its particle size is large, this powder is not suitable for plasma-spraying because of its low density. Figure 4.2c shows SPL18 (3YSZ + 6 wt% Al_2O_3) with its starting powder densified by calcination at 1200 °C for 10 h before drying. The particles are much smaller and more spherical than those of other powders. Figure 4.2d shows SPL19 (3YSZ + 1.8 wt%

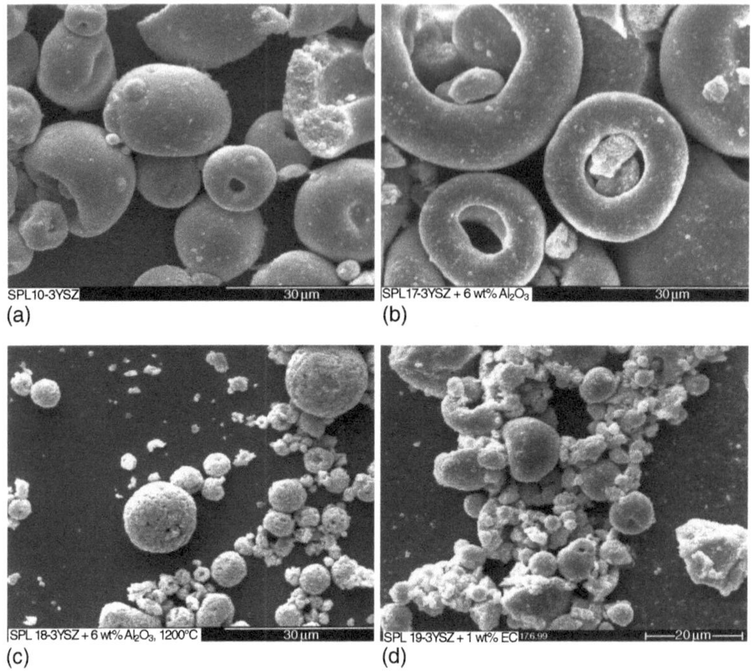

4.2 Scanning electron micrographs of spray-dried powders: (a) SPL10 (3YSZ); (b) SPL17 (3YSZ + 6 wt% Al_2O_3); (c) SPL18 (3YSZ + 6 wt%Al_2O_3, starting powder calcined at 1200 °C, 10 h); (d) SPL19 (3YSZ + 1.8 wt% PEI + 1 wt% ethylcellulose).

PEI + 1 wt% ethylcellulose). Most of the particles are not spherical having a cotton-like appearance due to the swelling or crack of cellulose. The mean particle size and bulk density of this powder are 7.3 μm and 1.27 g/cm^3, respectively. The controlling factor of the property of spray-dried powders is the suspension preparation. In order to prepare a powder with high density, large particle size and good flowability, the solid content in the suspension should be high enough, and the starting powder should be densifed at an appropriate temperature before drying if it is porous.

4.3 Phase composition and microstructure of nanostructured thermal barrier coatings (TBCs)

4.3.1 Phase composition

The mean grain sizes of the matrix t-ZrO_2 phase in heat-treated coatings were roughly estimated using the Scherrer equation[25–26]

$$B_p(2\theta) = \frac{0.9\lambda}{D\cos\theta} \qquad [4.1]$$

where D is the average dimension of crystallite, $B_p(2\theta)$ the broadening of the diffraction line measured at half maximum intensity and λ (0.154 nm) and θ denote the wavelength of the X-rays and the Bragg diffraction angle. The correction for instrumental broadening is taken into consideration in the measurement of the peak broadening by comparing the widths at half maximum intensity of X-ray reflection of the sample with the single crystalline Si standard, and then Gaussian correction is used to remove the instrumental broadening to obtain the true crystal broadening

$$B_p^2(2\theta) = B_h^2(2\theta) - B_f^2(2\theta)$$

where $B_p(2\theta)$ is the true half maximum width and $B_h(2\theta)$ and $B_f(2\theta)$ are the half maximum widths of the sample and the single crystalline Si standard.[27] The theta value of $2\theta = 30°$ is selected to calculate the crystallite size and the peaks belong to ZrO_2.

Figure 4.3 shows the X-ray diffraction patterns of the powder and as-sprayed coating.[28] It can be seen that there is no new phase in the coating and no clear difference in the peak intensities between the powder and coating. Both the powder and coating are composed of pure tetragonal zirconia, which is the stable phase in the middle temperature. However, the X-ray pattern of the coating has two more peaks than that of the powder,

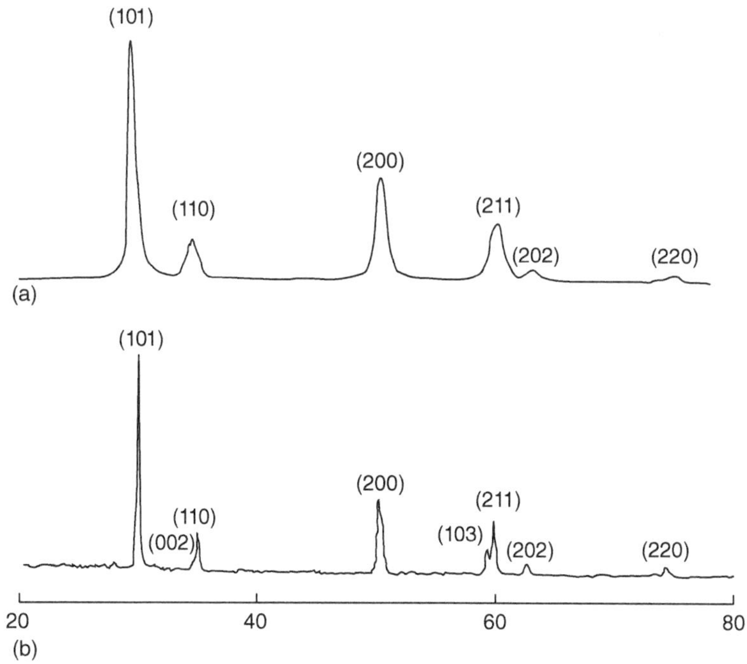

4.3 XRD patterns of (a) starting ZrO_2 powder and (b) as-sprayed ZrO_2 coating.

which indicates that the coating has a better crystal structure than the powder.

4.3.2 Microstructure

The transmission electron microscopy (TEM) image reveals the grain size of primitive particles and the as-sprayed coating, as shown in Fig. 4.4.[29] It can be seen that the non-molten phase of the coatings is composed of fine grains with sizes ranging from 30–50 nm. No smaller grains were found in the TEM image. Details of the microstructural features before and after annealing were examined through scanning electron microscopy (SEM) observation on the surface of top coat.[29] Figure 4.5 presents the surface morphology of nanostructured zirconia coatings in the as-sprayed state and after 15 h heat-treating at different temperatures. The as-sprayed microstructure is composed of the nanopowders and micropores between the nanopowders. The size of the nanopowders increases after annealing at different temperatures for 15 h. Further details of the microstructural features of the nanostructured zirconia coating after annealing were examined through fractured surfaces of coating.

The measured surface-connected porosity of the freestanding nanostructured zirconia coating is 13% and the median pore diameter (volume) is 0.3524 μm.[30] Figure 4.6a shows the cumulative porosity distribution of the coating measured by mercury intrusion porosimetry. Figure 4.6b illustrates the compositions of porosity based on the result of the porosity measurement in Fig. 4.6a. The result reveals a bimodal pore size distribution, similar

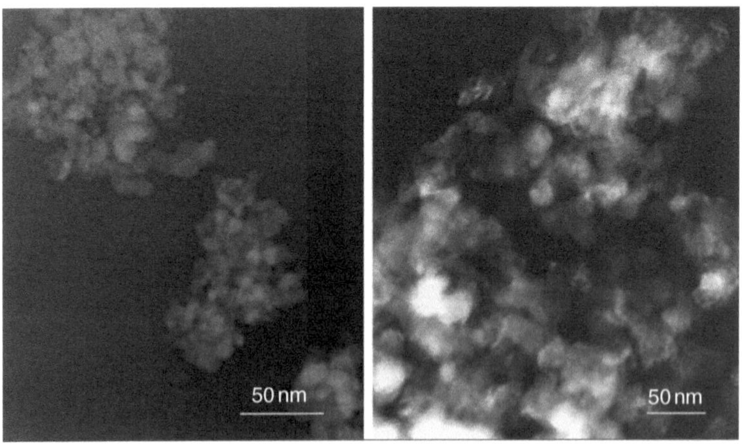

(a) (b)

4.4 TEM image of the primitive nano YSZ particles (a) and the as-sprayed coating (b).

4.5 Surface morphology of the nano YSZ coatings at different temperatures after heat treatment for 15 h: (a) as-sprayed state; (b) 600 °C; (c) 800 °C; (d) 900 °C; (e) 1000 °C; (f) 1150 °C.

to that in the conventional zirconia coatings.[31,32] Micropores contribute a larger percentage of the total porosity than macropores. The fine micropores, with diameters smaller than 1 μm, include intrasplat microcracks, intersplat gaps resulting from poor adhesion and small voids originating from the unmelted particles. A spot of macropores or large defects, with pore sizes of even up to 100 μm, are also present in the coating, which could be mainly attributed to macrocracks such as branching cracks, interlamellar pores due to improper adhesion and globular pores as a result of gas entrapment. The level of coating porosity is significantly connected with the spray conditions

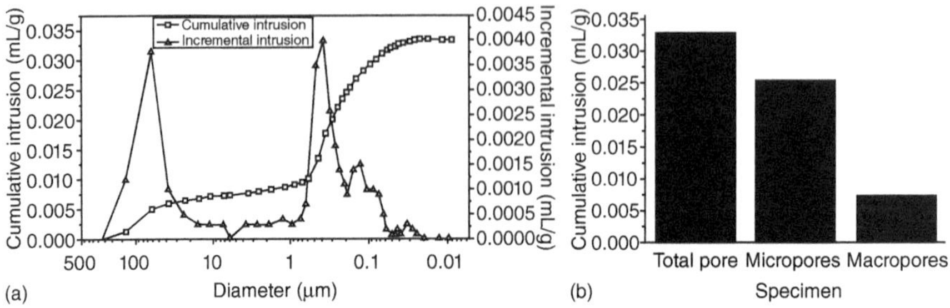

4.6 Porosity distribution of the as-sprayed nanostructured zirconia coating (a), and composition of the total porosity (b).

such as powder sizes, spray distance, spray angle, etc. To some extent, the unmelted powders in the nanostructured coating help increase the number of micropores.

4.4 Mechanical properties

4.4.1 Hardness and elastic modulus

The nanostructured PSZ (ZrO_2-7 wt% Y_2O_3) experimental powder Nanox 4007 (Inframat Corp., North Haven, CT, USA) was plasma sprayed under different parameters in air on mild steel substrates with a 40 kW, standard SG100 plasma torch (Praxair, Appleton, WI, USA). The substrates were grit-blasted with alumina just before thermal spraying (Ra = 4.1 ± 0.3 μm). The substrates were not pre-heated. Typical coating thickness was 0.9–1.1 mm. The spray parameters applied are listed in Table 4.1.[33]

Vickers microhardness measured on cross-section and top surface for 1000 g is shown in Fig. 4.7.[33] The value measured at top surface has slightly lower values than that measured at cross-section. Microhardness measurements in thermal spray coatings with respect to the planar or cross-sectional aspects of individual splats will correspond with area measurements on circles or ellipses. The planar hardness values were normalized with respect to the cross-section hardness values for indenting loads of 1000 g. Knoop microhardness values measured using 1000 g on the cross-sectional area are given in Fig. 4.8.[33] The Knoop values are lower than the Vickers values at the same load. In addition, the variation in Knoop values is higher than that in Vickers. Both phenomena are probably related to the difference in geometry of the two indenters, which provide different force fields in the coatings.

Observing Figs 4.7 and 4.8, leads to the following trend. Following the Ra axis from left to right, from spray parameter 1 to spray parameter 6, the torch power is decreasing and the spray distance is increasing. When the

Table 4.1 Spraying parameters used for PSZ coatings

Parameters	Set 1	Set 2	Set 3	Set 4	Set 5	Set 6
Power (kW)	40	40	40	40	32	24
Current (A)	800	800	800	800	800	600
Voltage (V)	50	50	50	50	40	40
Ar flow plasma (slpm)	48	38	48	38	48	48
H_2 flow plasma (slpm)	5	5	5	5	4	4
Ar flow carrier gas (slpm)	5	5	5	5	5	5
RPM-Hopper (%)	30	30	30	30	30	30
Spray distance (cm)	6	6	8	8	8	8

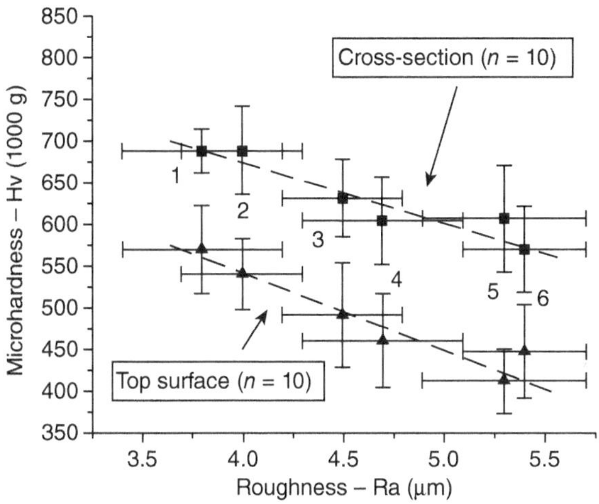

4.7 Vickers microhardness–roughness relationship (1000 g load) for different sets of spray parameters (Table 4.1).

plasma torch power is decreased, particle temperature and velocity will tend to drop, which will impede a larger spreading of the sprayed droplet. The values of elastic modulus and their relationship with roughness are shown in Fig. 4.9. Again the same trend is observed: the smoother the coating is, the higher the Knoop microhardness and elastic modulus are. A dashed line was placed in Figs 4.8 and 4.9 as a 'guideline for eye' to show this trend. The elastic modulus slightly changes from spray parameters 1 to 4, where the same plasma power was applied for different H_2/Ar ratios and a difference in spray distance of 2 cm. This experimental observation agrees with the elastic modulus measurements of PSZ coatings via the four-point bending method, which shows that the PSZ coatings did not present a significant change in elastic modulus values as varying the spray distance by 2 cm maintained the same plasma power.

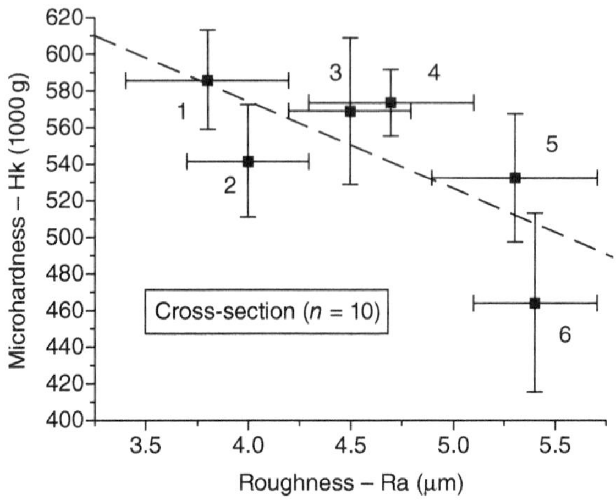

4.8 Knoop microhardness–roughness relationship (1000 g load) for different sets of spray parameters (Table 4.1).

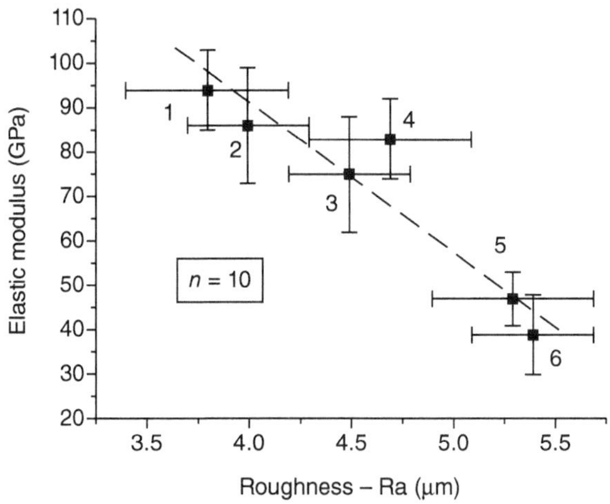

4.9 Elastic modulus–roughness relationship for different sets of spray parameters (Table 4.1).

4.4.2 Coefficients of thermal expansion (CTE)

Figure 4.10 shows the thermal expansion coefficient of nanostructured zirconia coating.[34] The average thermal expansion coefficient of the zirconia coatings at the first thermal cycle and second thermal cycle from room temperature to 1200 °C are 11.0 and 11.6 × 10^{-6} °C^{-1}, respectively. The thermal expansion coefficient of the zirconia coating increased in the begining and then decreased from room temperature to 300 °C in the first

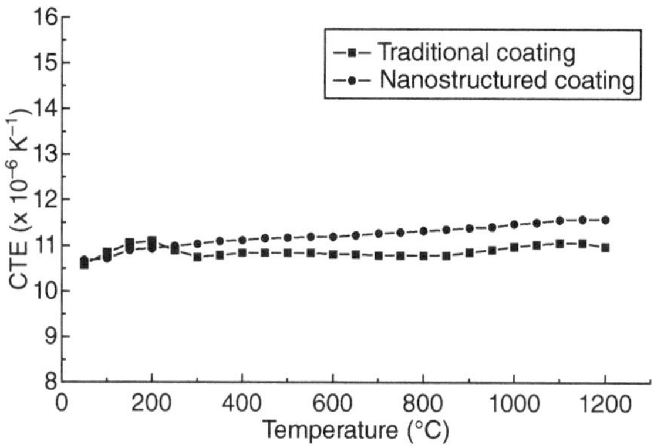

4.10 Coefficient of thermal expansion (CTE) of the nanostructured zirconia coating.

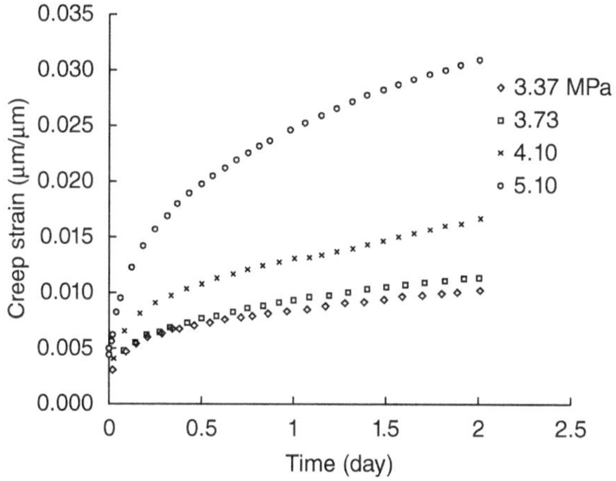

4.11 Compression creep strain of nanostructured samples under different stresses at 1000 °C.

thermal cycle. This behavior is ascribed to the oxygen loss phenomenon in ceramic coatings prepared by atmospheric plasma spraying. When the as-sprayed zirconia coating was heated in atmosphere, the oxygen atoms were added to the vacancies of the coating again. The phenomenon disappeared in the second thermal cycle. During the second thermal cycle, the coefficient of thermal expansion increased linearly according to the increasing temperature.

4.4.3 Creep

Figures 4.11 and 4.12 show the creep strain of nanostructured coatings under a range of stresses and temperatures over two days.[35] It is clear that

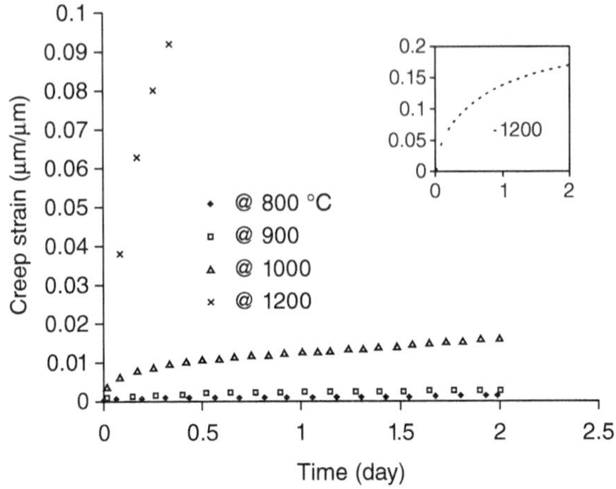

4.12 Creep strain of nanostructured samples at different temperatures under ~4 MPa compressive stress.

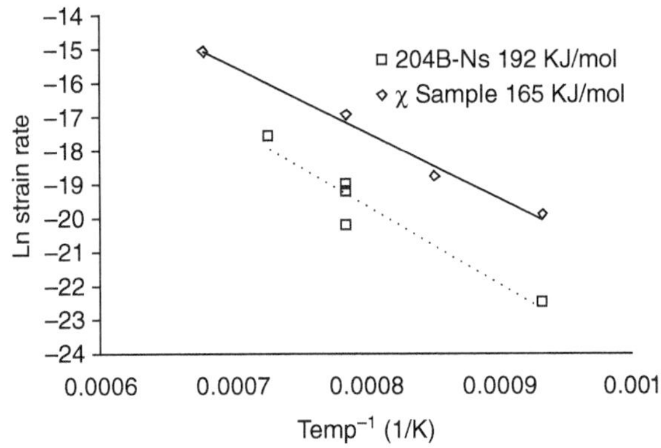

4.13 Calculation of activation energy of nanostructured and hollow spherical YSZ samples, 165 ± 40 and 192 ± 25 KJ/mol, respectively.

this nanostructured coating, in spite of applying much lower stresses, exhibits an order of magnitude higher creep strain than the hollow spherical YSZ deposits. As was observed for the hollow spherical YSZ samples, the nanostructured deposits exhibit primary and secondary creep. As load and temperature of the test conditions are increased, the amount of creep strain rises accordingly. Again, temperature has a more significant effect than load, as is clear in Fig. 4.12.

From the creep data, using the power-law creep ($\hat{O} = A\sigma^n$) and Arrhenius ($\varepsilon = A\exp - Q/RT$) equations, stress exponents and activation energies for the coatings were calculated where $\hat{O}$ is creep strain, A a constant, s applied stress in MPa, ε strain rate, Q activation energy, R gas constant and T temperature in Kelvin (Fig. 4.13).[35] The nanostructured deposit shows a slightly lower activation energy of 165 ± 40 KJ/mol compared with 192 ± 25 KJ/mol for the hollow spherical YSZ sample; however, as to the interval of values, they may not be very different.

4.5 Thermophysical properties and the failure behavior

4.5.1 Thermal diffusivity

Figure 4.14 presents the thermal diffusivity of the as-sprayed nanostructured zirconia coatings and conventional zirconia coating as the temperature rises from room temperature to 800 °C. It can be seen that the thermal diffusivity of both coatings decreases slightly as the temperature increases. The thermal diffusivity of the as-sprayed nanostructured zirconia coating was in the range $2.15–2.75 \cdot 10^{-3}$ cm^2/s during the temperature change from room temperature to 800 °C , while the thermal diffusivity of conventional zirconia coating was $2.35–2.96 \cdot 10^{-3}$ cm^2/s. The thermal diffusivity of the plasma-sprayed nanostructured zirconia coating is lower than that of conventional zirconia coating. The decrease in the thermal diffusivity is attributed to the micropores in the nanostructured zirconia and the decreased grain size of nanostructured zirconia coatings. When the temperature is below 1200 °C, phonon transport dominates the heat conduction through zirconia. For conventional materials, the grain boundary contribution to phonon scattering is thought to be small.[36,37] However, when the grain size is of the same order as the mean free path for phonon scattering, grain boundaries can have a significant effect. Kabacoff[37] investigated the influence of nanograin size on the phonon conductivity of zirconia-7 wt% yttria as a function of temperature. It was found that the thermal conductivity is

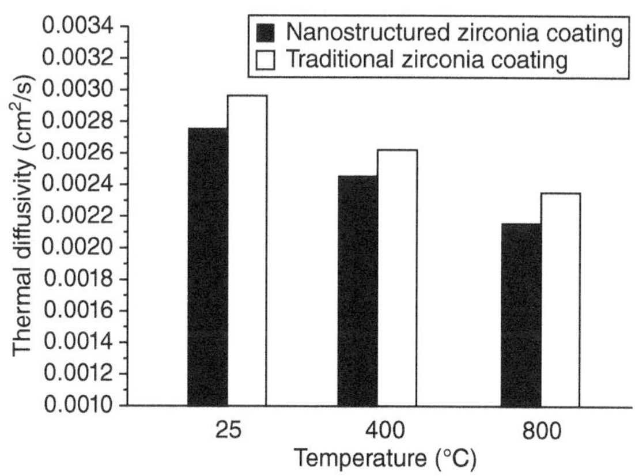

4.14 Thermal diffusivity of the as-sprayed nanostructured zirconia coating and conventional zirconia coating.

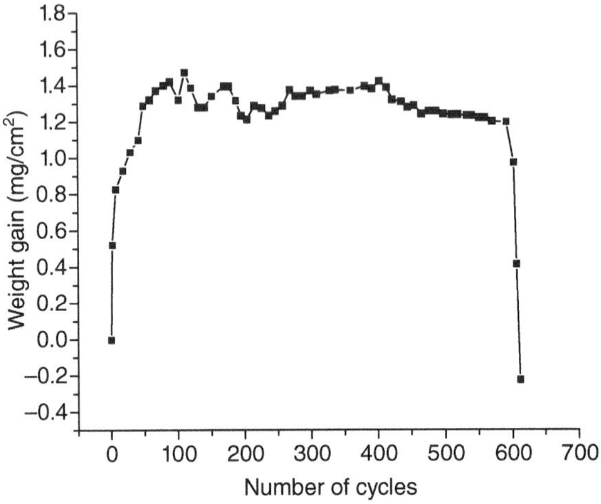

4.15 Kinetics of nanostructured TBC during 1 h cycles at 1050 °C.

drastically reduced for fine-grained materials, particularly when the grain size is of nanometer dimensions.

A cyclic oxidation test was performed to investigate the cyclic oxidation kinetics of nanostructured TBC. Figure 4.15 shows a plot of the weight change per unit area vs number of cycles for the test performed at 1050 °C in static air.[38] The nanostructured TBC had a weight gain of approximately 1.2–1.6 mg/cm^2 after 300 cycles, indicating a low oxidation rate. Moreover, the abrupt loss in weight for the nanostructured TBC did not occur until after approximately 600 cycles. The result indicates that the thermal cyclic lifetime of the nanostructured TBC is about 600 h for 1 h cycles at 1050 °C. The nanostructured TBC exhibits promising thermal cyclic oxidation resistance.

4.5.2 Failure mechanism

Table 4.2 shows the thermal cyclic life of two different TBCs under different thermal cyclings. It was found that the number of thermal shock cycles to failure at 1150 °C of the nanostructured TBCs was about two to four times larger than that of the conventional TBCs after thermal cyclic testing for 300 cycles at 1000 °C, and exhibited little increase after the isothermal oxidation for 100 h at 1000 °C.[39] These facts indicate that the influence of the low-temperature thermal cycling on the thermal cyclic lifetime is larger for the conventional TBCs than for the nanostructured TBCs. Figure 4.16 shows the cross-sectional micrograph of the failed nanostructured TBCs subjected to thermal cyclic testing at 1150 °C. The sample shown in

Table 4.2 Thermal cycles of nanostructured and conventional TBCs

Type	Nanostructured TBCs						Conventional TBCs					
Pre-treatment	1000 °C × 300 cycles			1000 °C × 100 h			1000 °C × 300 cycles			1000 °C × 100 h		
Life	19	38	38	23	36	45	7	10	10	24	29	35
Mean life	32			34			9			29		

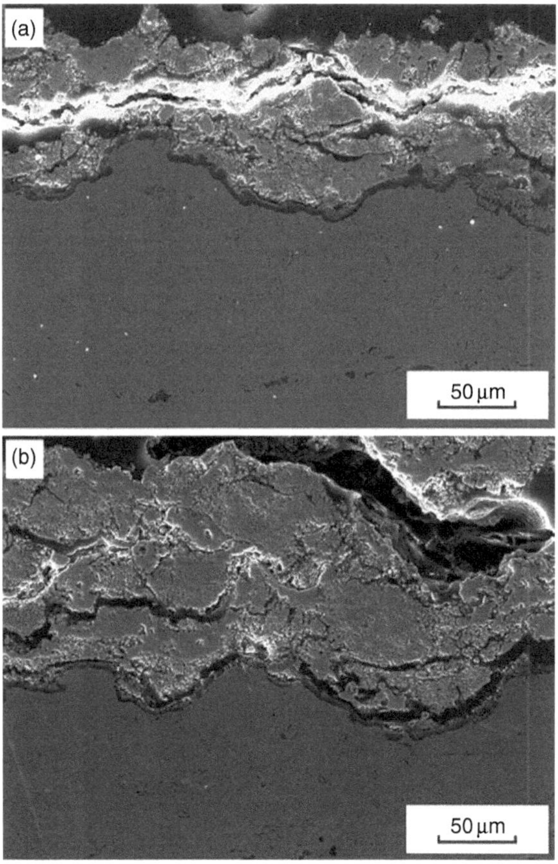

4.16 Cross-sectional micrograph of the failed nanostructured TBCs. (a) The sample experienced 38 cycles at 1150 °C after 300 cycles at 1000 °C; (b) the sample experienced 36 cycles at 1150 °C after isothermal oxidation for 100 h at 1000 °C.

Figure 4.16a has experienced thermal cyclic testing for 300 cycles at 1000 °C before the thermal cyclic test at 1150 °C, while the sample in Fig.4.16b experienced isothermal oxidation testing at 1000 °C for 100 h. The two nanostructured TBCs samples withstood a further 38 and 36 thermal cycles at 1150 °C, respectively. The failure of the coatings in two nanostructured TBC samples occurred within the top coat through spalling of the YSZ within YSZ coating close to the YSZ/thermal growth oxide interface. The failed cross-sectional microstructure of the conventional TBCs is shown in Fig. 4.17. It was noted that the failure also occurred within the top coat above thermal grown oxide (TGO), which indicates that the failure of the nanostructured TBCs is similar to the failure scenario of the conventional TBCs.[40–42] However, it was found that the TGO adheres well to the underlying bond coat. Possibly, such

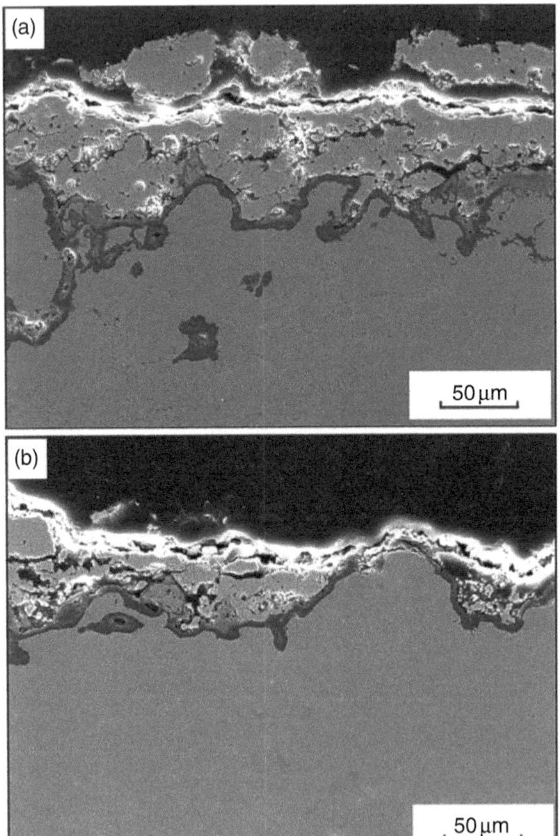

4.17 Cross-sectional micrographs of failed conventional TBCs. The sample experienced 29 cycles at 1150 °C after isothermal oxidation for 100 h at 1000 °C. (a) and (b) taken at different zones in the failed interface.

well-adhered TGO can be attributed to the nanostructure of the bond coat, which promotes the rapid formation of a uniform Al_2O_3 scale on the bond coat surface during the pre-heat treatment.[43–47]

Moreover, it can be seen in Fig. 4.18 that the buckling of the nanostructured YSZ top coat occurred before it spalled out. The buckling may suggest the possible sealing of the YSZ top coat surface through sintering. Although the failure mechanism of the TBC deposited by nanostructured YSZ in the present study is not yet clear, it is possible that the sintering of the nanosized YSZ fraction in the coating during high-temperature cyclic testing is responsible for the failure of the top YSZ coat.[48,49] This effect is possibly associated with the influence of the thermal cycling at 1000 °C on the thermal shock life of two different TBCs owing to much less sintering and possibly sustainable higher strain tolerance of the nanostructured TBCs.

4.5.3 Sintering of nanostructured Al_2O_3-yttia stabilized zirconia (YSZ) coating

Figure 4.19 shows the cumulative porosity distribution of the free-standing coating specimens which is determined by mercury porosimetry (MP). Cumulative porosity is the pore content normalized to the volume of the sample. In Fig. 4.19, it is clearly shown that the sintering of the coating leads to a lower porosity, as the annealing treatment time increased from 0 to 300 h, and the cumulative porosity decreased from 23.8 to 18%. The porosity is composed of micropores (diameter < 4 μm) and macropores (diameter > 4 μm). The presence of macropores above 4 μm in the coating reduced fractionally from 5.9 to 5.8% after heat treatment at 1100 °C over 0–300 h. However, the micropores below 4 μm reduced from 17.8 to 12.2% after annealing at 1100 °C over 0–300 h. After heat treatment at 1100 °C for 25 h and 100 h, the percentage of the small pores in the sintered sample is 14% and 17%, respectively.

The sintering resistance of the nanostructured Al_2O_3–YSZ coating can be understood by the shrinkage of the porosity.[50] During the sintering process,

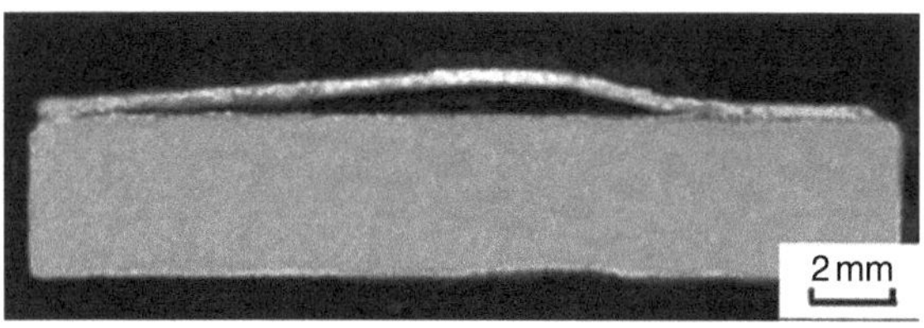

4.18 Cross-sectional micrograph of the failed nanostructured TBCs showing the overview of a buckle. The samples experienced 36 cycles at 1150 °C after isothermal oxidation of 100 h at 1000 °C.

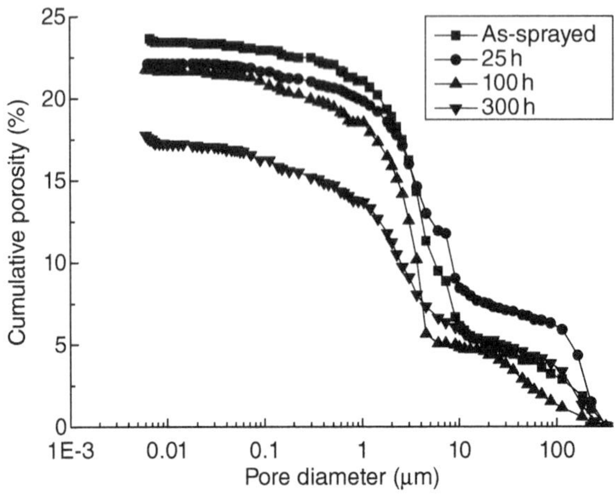

4.19 The cumulative porosity variation of the nanostructrued 13AlYSZ coating before and after heat treatment at 1100 °C for 25 h, 100 h and 300 h.

due to the presence of nanostructured particles and porosity, the porous nanozones are under a higher driving force for sintering and densification,[51] which leads to a fast sintering. Therefore, the nanozones tend to densify rapidly and the porosity of the coating decreased from 23.8 to 22.1% after annealing treatment for 25 h. With the annealing treatment ongoing, the velocity of sintering is reduced, and the porosity dropped to 21.8% after annealing treatment for 100 h. This is caused by the grain growth and the precipitation of Al_2O_3. The grain growth leads to a lower driving force for sintering and the precipitation of Al_2O_3 pins the boundaries, making sintering difficult. When the annealing time increases to 300 h, the porosity decreased to 18%. The slightly greater reduction in porosity per unit time from 100 to 300 h may simply be an experimental error since it is very difficult to measure the porosity level in a coating very accurately. Usually, an accelerated sintering occurs in the early stage, i.e. 25 h in this work, because the initial higher porosity and finer particle sizes contributed to sintering. As a result, an apparent shrinkage is observed at this stage. In this study, about 6% porosity decreased in the sintering process, similar to the tendency for the sintering of traditional 7 wt% Y_2O_3-stabilized ZrO_2 ceramic coating.[52]

4.6 Conclusion

Studies on nanostructured air plasma-sprayed YSZ coatings have shown that it is paramount to carefully control the spray parameters in order to avoid the complete melting of the nanostructured YSZ agglomerates in the plasma jet, thus preserving and embedding part of the nanostructure of the agglomerates into the coating microstructure. Those coatings were produced from microscopic porous spray-dried particles formed via the agglomeration of individual nanosized YSZ particles. The semi-molten agglomerates, once embedded in the coating microstructure, created a bimodal feature, which consisted of a structure formed by the resolidification of agglomerates that had been fully molten in the spray jet combined with zones resulting from the incorporation of semi-molten material. By controlling the amount of previously molten and porous semi-molten particles embedded in the coating microstructure, it was possible to change considerably the mechanical response of the coating. Therefore, this bimodal microstructure affected the mechanical behavior of the coating. In recent years, new classes of ceramic materials having very low thermal conductivity and better high-temperature stability have been identified as potential successors to YSZ. In order to lower thermal conductivity, YSZ coatings have been modified by doping with several additives, such as Al_2O_3, NiO, Nd_2O_3, Er_2O_3, Yb_2O_3. However, these new materials have not yet been matured for practical application because of their poor mechanical and/or chemical properties. In the future, nanostructured YSZ coatings modified by Al_2O_3, La_2O_3, HaO_2, etc. will be extensively investigated.

4.7 References

1. M. GELL, Application opportunities for nanostructured materials and coatings, *Mater. Sci. Eng. A* **204** (1995) 246–251.
2. Y.Y. ZENG, S.W. LEE, L. GAO, C.X. DING, Atmospheric plasma sprayed coatings of nanostructured zirconia, *J. Eur. Ceram. Soc.* **22** (3) (2002) 347–351.
3. R.S. LIMA, A. KUCUK, C.C. BERNDT, Evaluation of microhardness and elastic modulus of thermally sprayed nanostructured zirconia coatings, *Surf. Coat. Technol.* **135** (2001) 166–172.
4. H. CHEN, C.X. DING, Nanostructured zirconia coating prepared by atmospheric plasma spraying, *Surf. Coat. Technol.* **150** (1) (2002) 31–36.
5. C.G. ZHOU, N. WANG, S.K. GONG, H.B. XU, Thermal cycling life and thermal diffusivity of a plasma-sprayed nanostructured thermal barrier coating, *Scripta Mater.* **51** (2004) 945–948.
6. J. KARTHIKEYAN, C.C. BERNDT, J. TIKKANEN, S. REDDY, H. HERMAN, Plasma spray synthesis of nanomaterial powders and deposits, *Mater. Sci. Eng. A* **238** (1997) 275–286.
7. C.B. LIU, Z.M. ZHANG, X. L. JIANG, M. LIU, Z.H. ZHU, Comparison of thermal shock behaviors between plasma-sprayed nanostructured and conventional zirconia thermal barrier coatings, *Trans. Nonferrous Met. Soc. China* **19** (2009) 99–107.
8. R.S. LIMA, A. KUCUK, U. SENTURK, C.C. BERNDT, Integrity of nanostructured partially stabilized zirconia after plasma spray processing, *Mater. Sci. Eng. A* **313** (1/2) (2001) 75–82.

9. R.S. LIMA, A. KUCUK, U. SENTURK, C.C. BERNDT, Properties and microstructures of nanostructured partially stabilized zirconia coatings, *J. Thermal Spray Technol.* **10** (1) (2001) 150–152.
10. H. CHEN, C.X. DING, Nanostructured zirconia coatings prepared by atmospheric plasma spraying, *Surf. Coat. Technol.* **150** (1) (2002) 31–36.
11. C.G. ZHOU, N. WANG, Z.B. WANG, Thermal cycling life and thermal diffusivity of a plasma-sprayed nanostructured thermal barrier coating, *Scripta Mater.* **51** (10) (2004) 945–948.
12. B. LIANG, C.X. DING, Microstructure of nanostructure zirconia coating and its thermal shock resistance, *J. Inorg. Mater.* **21** (1) (2006) 250–256.
13. F. LIN, Y.G. YU, X.L. JIANG, K. L. ZENG, X.J. REN, Z.D. LI, Microstructures and properties of nanostructured TBCs fabricated by plasma spraying, *Chin. J. Nonferrous Met.* **16** (3) (2006) 482–487. (in Chinese).
14. M.R. GOLD, R.S. LIMA, A. KUCUK, C.C. BERNDT, Scratch testing and acoustic emission of nanostructured partially stabilized coatings, in: K. Ravi-Chandar, B.L. Karihaloo, T. Kishi, R.O. Ritchie, A.T. Yokobori Jr, T. Yokobori (Eds), *Proceedings of the 10th International Conference on Fracture*, Elsevier Science, Oxford, UK, 2001 (Paper #ICF1001074, pdf format).
15. B. LIANG, C. DING, Thermal shock resistances of nanostructured and conventional zirconia coatings deposited by atmospheric plasma spraying, *Surf. Coat. Technol.* **197** (2005) 185–192.
16. W.Q. WANG, C.K. SHA, D.Q. SUN, X.Y. GU, Microstructural feature, thermal shock resistance and isothermal oxidation resistance of nanostructured zirconia coating, *Mater. Sci. Eng. A* **424** (2006) 1–5.
17. Z. SHUI, B. WANG, A. LEVY, Erosion-corrosion of thermal spray coatings, *Surf. Coat. Technol.* **43/44** (1990) 859–874.
18. G. BARBEZAT, R. NICOLL, A. SICKINGER, Abrasion, erosion and scuffing resistance of carbide and oxide ceramic thermal sprayed coatings for different applications, *Wear* **162/164** (1993) 529–537.
19. N. BERGER-KELLER, G. BERTRAND, C. FILIATRE *et al.*, Microstructure of plasma-sprayed titania coatings deposited from spraydried powder, *Surf. Coat. Technol.* **168** (2003) 281–290.
20. R.S. LIMA, B.R. MARPLE, Thermal spray coatings engineered from nanostructured ceramic agglomerated powders for structural, thermal barrier and biomedical applications: a review, *J. Thermal Spray Technol.* **16** (1) (2007) 40–63.
21. F. TARASI, M. MEDRAJ, A. DOLATABADI, J.O. BERGHAUS, C. MOREAU, Phase formation and transformation in alumina/YSZ nanocomposite coating deposited by suspension plasma spray procees, *J. Therm. Spray. Technol.* **19** (4) (2010) 787–795.
22. W.B. GONG, C.K. SHA, D.Q. SUN, W.Q. WANG, Microstructures and thermal insulation capability of plasma-sprayed nanostructured ceria stabilized zirconia coatings, *Surf. Coat. Technol.* **201** (2006) 3109–3115.
23. B. LIANG, C.X. DING, H.L LIAO, C. CODDET, Phase composition and stability of nanostructured 4.7 wt.% yttria-stabilized zirconia coatings deposited by atmospheric plasma spraying, *Surf. Coat. Technol.* **200** (2006) 4549–4556.
24. X.Q. CAO, R. VASSEN, S. SCHWARTZ, W. JUNGEN, F. TIETZ, D. STÖVER. Spray-drying of ceramics for plasma-spray coating. *J. Eur. Ceram. Soc.* **20** (2000) 2433–2439
25. H.P. KLUG, L.E. ALEXANDER, *X-ray Diffraction Procedures for Polycrystalline and Amorphous Materials*, Wiley, London, UK, 1954.

26. L.L. SHAW, D. GOBERMAN, The dependency of microstructure and properties of nanostructured coatings on plasma spray conditions, *Surf. Coat. Technol.* **130** (2000) 1–8.
27. H. ZHOU, B. HE, J. WANG, B.D. SUN, Nanostructured yttria stabilized zirconia coatings deposited by air plasma spraying, *Trans. Nonferrous Met. Soc. China* **17** (2007) 389–393.
28. Y. ZENG, S.W. LEE, L. GAO, C.X. DING, Atmospheric plasma sprayed coatings of nanostructured zirconia, *J. Eur. Ceram. Soc.* **22** (2002) 347–351.
29. N. WANG, C.G. ZHOU, S. K. GONG, H. B. XU, Heat treatment of nanostructured thermal barrier coating, *Ceram. Int.* **33** (2007) 1075–1081.
30. H. ZHOU, F. LI, J. WANG, B.D. SUN, Microstructure analyses and thermophysical properties of nanostructured thermal barrier coatings, *J. Coat. Technol. Res.* **6** (3) (2009) 383–390.
31. H.B. GUO, R. VABEN, D. STÖVER, Thermophysical properties and thermal cycling behavior of plasma sprayed thick thermal barrier coatings, *Surf. Coat. Technol.* **192** (2005) 48–56.
32. D. CHEN, M. GELL, E.H. JORDAN, E. CAO, X. MA, Thermal stability of air plasma spray and solution precursor plasma spray thermal barrier coatings, *J. Am. Ceram. Soc.* **90** (2007) 3160–3166.
33. R.S. LIMA, A. KUCUK, C.C. BERNDT, Evaluation of microhardness and elastic modulus of thermally sprayed nanostructured zirconia coatings, *Surf. Coat. Technol.* **135** (2001) 166–172.
34. H. CHEN, X.M. ZHOU, C.X. DING, Investigation of the thermomechanical properties of a plasma-sprayed nanostructured zirconia coating, *J. Eur. Ceram. Soc.* **23** (2003) 1449–1455.
35. R. SOLTANI, E. GARCIA, T.W. COYLE, J. MOSTAGHIMI, R.S. LIMA, B.R. MARPLE, C. MOREAU, Thermomechanical behavior of nanostructured plasma sprayed zirconia coatings, *J. Thermal Spray Technol.* **15** (4) (2006) 657–662.
36. K.J. LAWSON, J.R. NICHOLLS, D.S. RICKERBY, The effect of coating thickness on thermal conductivity of CVD and PVD coatings, in: R.K. Datta, J.S. Gray (eds), *Advances in Surface Engineering*, Royal Society of Chemistry, Cambridge, UK, 1993, Vol. 1, 83–92.
37. L.T. KABACOFF, Thermal sprayed nano-structured thermal barrier coatings, *NATO Workshop on Thermal Barrier Coatings*, Aaalborg, Denmark, AGARD-R-823, 1998, paper 12.
38. C.G. ZHOU, N. WANG, Z.B. WANG, Thermal cycling life and thermal diffusivity of a plasma-sprayed nanostructured thermal barrier coating, *Scripta Mater.* **51** (2004) 945–948.
39. Q. ZHANG, C.J. LI, S.L. ZHANG, X.R. WANG, G.J. YANG, C.X. LI, Thermal failure of nanostructured thermal barrier coatings with cold-sprayed nanostructured NiCrAlY bond coat, *J. Thermal Spray Technol.* **17** (2008) 838–845.
40. K.W. SCHLICHTING, N.P. PADTURE, E.H. JORDAN, M. GELL, Failure modes in plasma-sprayed thermal barrier coatings, *Mater. Sci. Eng. A* **342** (1–2) (2003) 120–130.
41. A. RABIEI, A.G. EVANS, Failure mechanisms associated with the thermally grown oxide in plasma-sprayed thermal barrier coatings, *Acta Mater.* **48** (15) (2000) 3963–3976.
42. E.P. BUSSO, J. LIN, S. SAKURAI, A mechanistic study of oxidation-induced degradation in a plasma-sprayed thermal barrier coating system: Part II: life prediction model, *Acta Mater.* **49** (9) (2001) 1529–1536.

43. H. LAU, C. LEYENS, U. SCHULZ, C. FRIEDRICH, Influence of bond coat pre-treatment and surface topology on the lifetime of EB–PVD TBCs, *Surf. Coat. Technol.* **165** (3) (2003) 217–223.
44. M. MATSUMOTO, T. KATO, K. HAYAKAWA, N. YAMAGUCHI, S. KITAOKA, H. MATSUBARA, The effect of pre-oxidation atmosphere on the durability of EB–PVD thermal barrier coatings with CoNiCrAlY bond coats, *Surf. Coat. Technol.* **202** (12) (2008) 2743–2748.
45. R. DAROLIA, Post-deposition oxidation of a nickel-base superalloy protected by a thermal barrier coating, US Patent 6,607,611, 19 August 2003.
46. T.J. NIJDAM, L.P.H. JEURGENS, J.H. CHEN, W.G. SLOOF, On the microstructure of the initial oxide grown by controlled annealing and oxidation on a NiCoCrAlY bond coating, *Oxid. Met.* **64** (5–6) (2005) 355–377.
47. O. RACEK, C.C. BERNDT, D.N. GURU, J. HEBERLEIN, Nanostructured and conventional YSZ coatings deposited using APS and TTPR techniques, *Surf. Coat. Technol.* **201** (1–2) (2006) 338–346.
48. R.S. LIMA, B.R. MARPLE, Nanostructured YSZ thermal barrier coatings engineered to counteract sintering effects, *Mater. Sci. Eng. A* **485** (1–2) (2008) 182–193.
49. O. RACEK, C.C. BERNDT, D.N. GURU, J. HEBERLEIN, Nanostructured and conventional YSZ coatings deposited using APS and TTPR techniques, *Surf. Coat. Technol.* **201** (1–2) (2006) 338–346.
50. Q.H. YU, C.G. ZHOU, H.Y. ZHANG, F. ZHAO, Thermal stability of nanostructured 13 wt% Al_2O_3-8 wt% Y_2O_3-ZrO_2 thermal barrier coatings, *J. Eur. Ceram. Soc.* **30** (2010) 889–897.
51. R.S. LIMA, B. R. MARPLE, Nanostructured YSZ thermal barrier coatings engineered to counteract sintering effects, *Mater. Sci. Eng. A.* **485** (2008) 182–193.
52. B. SIEBERT, C. FUNKE , R. VAßEN, D. STÖVER, Changes in porosity and Young's modulus due to sintering of plasma sprayed thermal barrier coatings, *J. Mater. Process. Technol.* **92–93** (1999) 217–223.

Part II

Processing and spraying techniques

5

Plasma spraying for thermal barrier coatings: processes and applications

J. L. XU and K. A. KHOR, Nanyang Technological University, Singapore

Abstract: Plasma spraying processes are utilized for a wide range of industrial applications. This chapter reviews some fundamental aspects of plasma-spraying technology with the aim of understanding and controlling particular processes and mechanisms governing the formation and properties of plasma-sprayed coatings. The chapter mainly highlights the current status of plasma spraying in the biomedical field and the problems associated with plasma spray coating of hydroxyapatite which has shown promising effects on rapid bone remodeling and suitable functional life in orthopedic and dental applications.

Key words: plasma, plasma spraying, coating, biomaterials, hydroxyapatite, in vitro.

5.1 Introduction

Thermal plasma spraying is a process whereby a powder feedstock is injected into a high temperature plasma jet in which finely divided metallic and non-metallic materials are deposited in a molten or semi-molten state on prepared substrate.[1,2] It has been used as an effective and economical method for producing ceramic coatings on metallic substrates and production of bulk powders from spheroidisation. This has been motivated mainly by the strong drive for high-performance materials and the growing need for new material-processing technologies. Particles injected into plasma zone will experience extreme heating rates and temperatures as high as 10000 K. Since plasma spraying is a very high-temperature process, the injected particles with varying size distribution will melt to yield an aerosol of molten droplets. Particles above critical size range may experience only partial melting while small particles may have a layer of vaporised materials surrounding the liquid droplets.[3] Hence the plasma processing of a material with its stringent requirements needs to be carefully controlled to produce the desired results. Thus, the full potential of plasma technology depends on our ability to achieve the optimal match between the process requirements and the characteristics of the plasma source and reactor design used.

The plasma process has evolved over the years into a highly flexible coatings system capable of applying a wide range of materials. A major application of plasma spraying is to protect aircraft engine components by deposition of aerospace coating such as thermal barriers of zirconium/magnesium or alumina on metal substrates. Recently, it has also been used as a tool to process bio-ceramic materials such as hydroxyapatite (HA) as both coatings and powders.[4–12] In this chapter, a brief review of applications of plasma processing in biomaterials will be presented.

5.2 Basic plasma concepts

When we use the term plasma in daily life, normally plasma TV will come to mind, but what is plasma and how does it work? Apart from its application in plasma flat panel display, plasma processing has actually been used in many areas such as plastic bags, automobile bumpers, airplane turbine blades, artificial joints and semiconductor circuits. In physics and chemistry, plasma is defined as the fourth state of matter, the other three being solid, liquid and gas. Plasma is a partially ionised gas consisting of charged particles which have high kinetic energy and are normally at very high temperatures, of the order of ~10 000 K.[1,2] A better understanding of the plasma beam can be obtained from a simplistic explanation of the plasma from gas mechanics. The properties of a gas are primarily dependent on the movement of its molecules. During this movement, the molecules continually exchange energy and pulse through collisions with one another. When a certain amount of energy is supplied to a gas, the velocity of its molecules increases, the collisions become more frequent and the temperature of the gas increases. If the velocity is increased sufficiently, the molecules of the gas disintegrate into atoms upon mutual collisions, a process known as dissociation. When higher energy is applied, for example from a high electrical field, even the electrons forming the atomic shells are forced out. This process is known as ionisation. Upon ionisation, the electron released is in a higher state of energy and will soon return to its original path to minimise its energy. This stabilises the electron and releases energy in the form of kinetic or electromagnetic energy, which is the difference between the energy of the ionised electron and the energy of the electron in its original path.

The final result of the whole dissociation and ionisation process is plasma, which contains electrically charged particles. However, the plasma is outwardly neutral as it will contain the same number of electrically positive and negative charges. Also, the processes of dissociation and ionisation are thermally balanced. This means that it is dependent on temperature, where equilibrium between the dissociation and ionisation processes and the appropriate reverse reactions are achieved. Another reaction, known as the recombination process, also takes place continuously in the plasma. Depending on the way energy is released in the recombination process, several basic mechanisms of electron recombination can be categorised. Some of these are:[12]

1. Recombination with emission of radiation

$$A^+ + e \rightarrow A + h.v$$

2. Recombination with doubled excitation, when releasing energy causes displacement of two electrons with the formation of a neutral atom

$$A^+ + e \rightarrow A^{**}$$

3. Recombination at triplex, which is collision with electron, charged particle and neutral particle

$$A^+ + e + K \rightarrow A + K$$

To start the whole dissociation and ionisation process, it is first necessary to supply an external force. This can be in the form of high temperature or by the action of a high electric field.

A plasma jet can be generated by passing a gas through an electric arc. Ionisation of the gas then takes place and this gas stream is in turn converted into an extremely hot plasma jet once a process called 'electrical breakdown' takes place. The output, temperature and velocity are the main characteristics of the plasma beam. The temperature is dependent on the ionisation degree, which is then dependent on the kind of plasma gas and the parameters employed in the plasma torch. As an approximation, the mean temperature of the plasma can be calculated by dividing the mean enthalpy of the plasma by the gas volume, assuming that all heat is uniformly distributed over the gas volume.

The velocity of the plasma beam can be approximated by applying fundamental gas mechanics laws for gas flow through a cylindrical tube, during its simultaneous heating. The dependence of velocity after heating can then be approximated from the equation of constant flow rate and the state equation of the gas, provided the gas pressure is assumed constant. Due to the kinetic and thermal effects of the plasma beam, the source material entering the plasma is accelerated and, depending on the input of thermal energy, conditions arising from plasma–particle interactions such as a great variety of phase changes take place. The first phase change takes place after the solid powder is injected into the plasma torch, i.e., melting. It has been pointed out that the degree of melting is very important for the final density of deposits. When the particles leave the plasma, they quench very quickly to produce, depending on different working conditions, non–equilibrium and different amounts of both crystalline and amorphous phases.[12] From the practitioner's point of view, these phase changes can sometimes be beneficial; at other times, they are detrimental. Hence, it is important to study all of the possible phase changes occurring during plasma spraying.

5.3 Plasma spraying

There are several principal plasma generating devices currently used in material processing. These include direct current (DC) plasma torches, DC transferred arcs, radio frequency inductively-coupled plasma torches and hybrid combinations of them. DC plasma torches are the most commonly used plasma generation devices (about 99% of the market). In most cases (about 95% of the plasma spray market), DC plasma guns are working in air at atmospheric pressure. Ar, He, N_2 and their mixtures are the most commonly used gases. Figure 5.1 shows a typical DC plasma jet generated by striking a DC electric arc between two electrodes within the torch. Gas flowing between the electrodes can be heated up to temperatures of 6000–10 000 K. The electric arc partially ionises the gas to form the plasma.

In a material processing through plasma spraying, the material to be deposited (feedstock), typically in powder form but sometimes a liquid or suspension or wire, will be injected through an external feed-port or internally in the high-temperature plasma. Melting takes place almost instantaneously. The molten particles are then accelerated rapidly towards the substrate. The deposits consist of a multitude of pancake-like lamellae called 'splats', formed by flattening of the liquid droplets. As the feedstock powders typically have sizes from micrometers to above 100 μm, the lamellae have thickness in the micrometer range and lateral dimension from several to hundreds of micrometers. Due to the significant difference in the

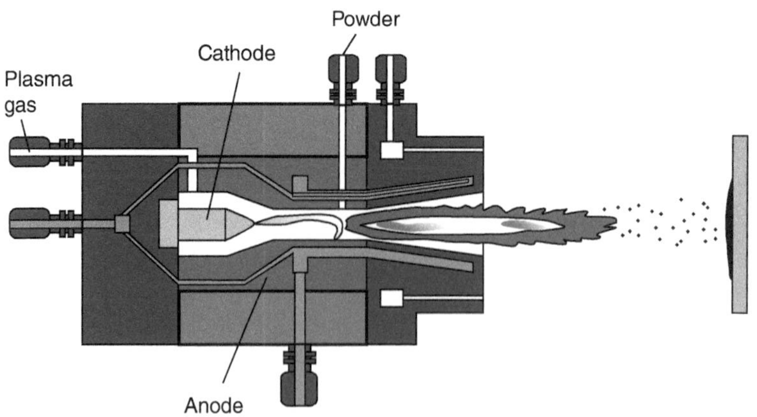

5.1 Schematic of plasma-spraying system.

temperature of the liquid droplet and the substrate, the spreading of the droplet on impact will be limited mainly by the solidification process that is taking place simultaneously. Typically, the cooling rate of the droplets is ~10^6 Ks^{-1} or greater. Adherence of the adjacent discs will likely take place through various contact points made prior to complete solidification. The microstructure of the coatings is therefore related to the nature of the interface between randomly stacked lamellae and the internal structure with each lamella produced by the rapid solidification process.

Some of the inherent defects of plasma-sprayed coating are voids, inclusions and poor adhesion of the lamellae. These may have arisen due to variation in the plasma-spraying process. Among them are fluctuations in powder feedrate, spraying cross-head speed and spraying distance. Other causes are incomplete melting of large particles and splashing of droplets on impact. In addition, due to the extreme conditions, transformations involved with feedstock related to physical and even both physical and chemical changes involving melting and rapid solidification, vaporisation, condensation and other chemical reactions might take plae. A popular application for plasma-spraying processes is to produce a coating material on a substrate, depending on its composition and desired coating performance. Figure 5.2 shows a schematic diagram of particles impinging onto a substrate during the process of building up a coating. Examples of plasma-sprayed coatings can be found in a wide range of industry types where

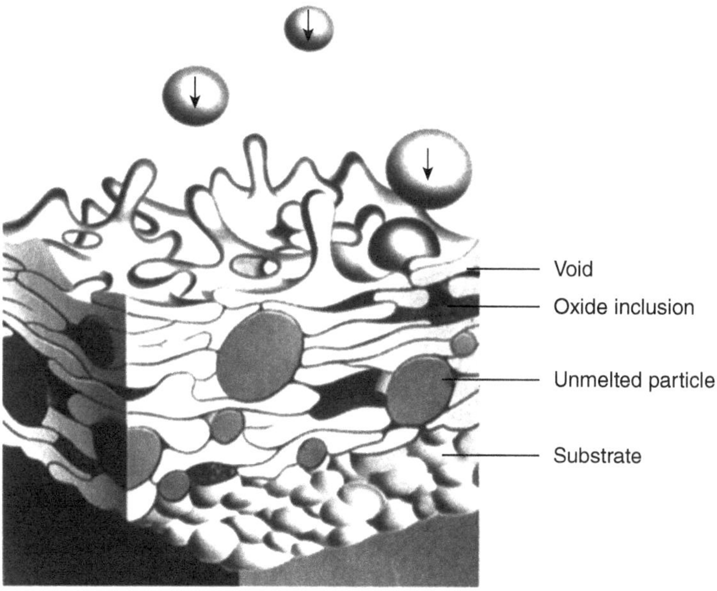

5.2 Process of building up a coating.

coatings are used to provide protection against various forms of wear and corrosion or to give thermal protection or some desired physical characteristics such as electrical insulation.

5.4 Applications of plasma spraying

5.4.1 Application in biomaterials

A biomaterial is a material in contact with fluids, cells and tissues of the living body utilised to evaluate, repair or replace any tissue or organ of the body.[13] A prerequisite for any synthetic material implanted in the body is that it should have good mechanical strength, high chemical stability, high corrosion resistance, very low toxicity and high biocompatibility.[13] Metals such as titanium, titanium alloys, stainless steel and CoCr-based alloys are popularly used as implant materials for their superior properties in terms of static and dynamic mechanical strength. In the past few years, researchers have been trying to improve on these mechanical properties and to develop new series of materials that can guarantee not only superior mechanical performance but also an excellent biological response. For this purpose, different materials have been produced to meet almost every conceivable need. Yet not all the problems connected with these implants, such as distance cell migration, breakage, stress shielding, reactivities and growth restriction, have been solved.[14] Moreover, the surface of the metal implant can become corroded and discolored and in some instances release metallic elements to the surrounding tissues when used in the corrosive environment of a human body. Ti–6Al–4V alloys show many desirable features, such as biocompatibility, high corrosion resistance, excellent strength-to-weight ratio and so on, and they have been widely used in aerospace and orthopedic applications. Their corrosion resistance can be attributed to a passivation layer on the surface. However, it can be easily worn out and wear-corrosion failure can occur because of electric cell reactions between the passivated and non-passivated layers.[15]

Biomedical coatings generally have to satisfy specific requirements such as high crystallinity, high coating adhesion and suitable porosity.[16] It is necessary to enhance biocompatibility, accelerate post-operative healing and improve adhesions. Coatings have specific functions ranging from improving fixation by establishing strong interfacial bonds, shielding the metallic implant from environmental attack or leaching effects, promoting fast tissue growth and interaction by the presence of a catalyst material and minimising adverse reaction by the provision of a biocompatible material.

Bone is mainly composed of mineral components, water and collagen fibers. A schematic anatomical view of a long bone is shown in Fig. 5.3.[17] The major subphase of the mineral consists of submicroscopic crystals of

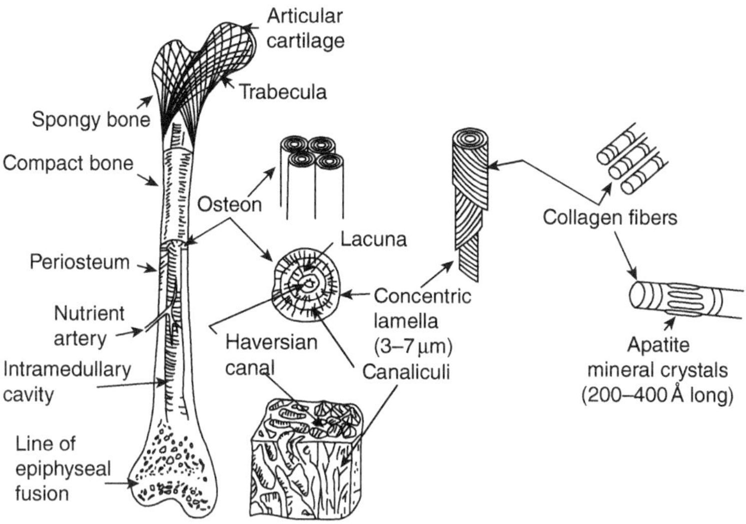

5.3 Organization of a typical bone.

an apatite of calcium and phosphate, resembling HA in its crystal structure. A typical wet cortical bone is composed of 22 wt% organic matrix, 69 wt% mineral and 9 wt% water. It has been known that HA can form strong biological bonds with bony tissue without the presence of soft fibrous tissues. The provision of a high-calcium and phosphorus-rich environment promotes rapid bone formation within the vicinity of the HA implant. HA also establishes strong interfacial bonds with titanium implants. Its excellent biointegration makes it an ideal choice for use in orthopedic and dental applications.[18] Some authors claimed that HA may act as a biological barrier to reduce toxic responses caused by the release of metallic ions from the metal substrate into the bone.[19]

The chemical formula of HA is presented as $Ca_4(I)Ca_6(II)(PO_4)_6(OH)_2$ ($Ca_{10}(PO_4)_6(OH)_2$). The Ca(I) atoms are on the fourfold symmetry 4(f) position and the Ca(II) atoms are in the sixfold symmetry 6(h) position.[19] The OH groups occupy disordered positions above and below the triangles formed by the Ca(II) atoms. The disorder of the OH groups gives rise to a 'macroscopic' space group $P6_3/m$ (as determined by X-ray diffractometry), which is lost at the level of the individual columns.[20]

As shown in Figs 5.4 and 5.5, biological HA has a hexagonal lattice with the space group of $P6_3/m$[21] with dimension $a = b = 9.432$ Å and $c = 6.881$ Å. The apatite structure of HA is represented by two sets of hexagonal cells shown in Fig. 5.5: OH atoms, represented by the solid dots, form the HA unit cell and Ca atoms, represented by the open dots, form a smaller hexagonal cell (without an indentical point in the center). The size ratio of

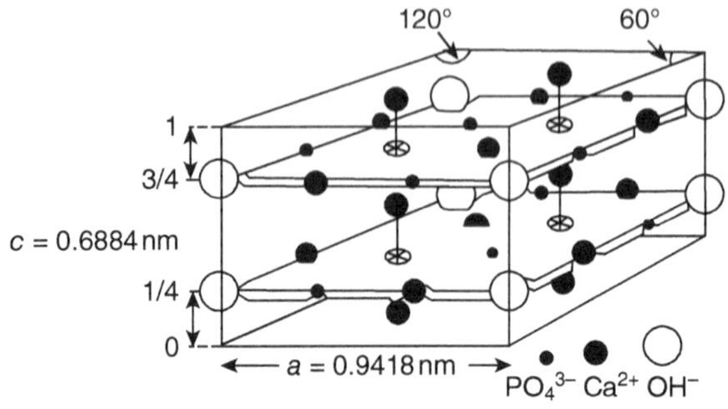

5.4 Crystal structure of HA:[21] $a = b = 9.418$ Å and $c = 6.884$ Å.

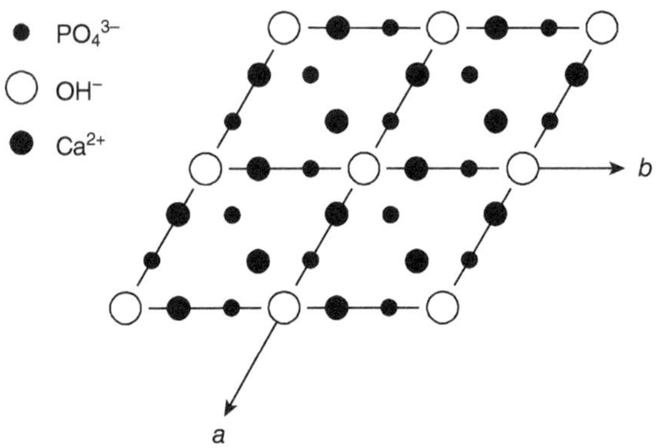

5.5 (0001)-plane of HA at $z = 3/4$.

the HA unit cell to the Ca cell is about $\sqrt{3}$. The crystal data of HA and some other physical properties are shown in Table 5.1. The ideal Ca:P ratio of HA is 10:6 and the calculated density is 3.21 g/cm.

HA promotes rapid bone growth and bonding between bony tissue and implant surfaces. Thermodynamically, HA is the most stable calcium phosphate compound at any given pH, temperature and composition of physiological fluid. However, the poor mechanical properties of bulk HA ceramics have provided impetus to research in HA coatings on metallic substrates in order to maintain biocompatibility and to improve mechanical properties. Thermal-spraying processes with their wide array of operating parameters are flexible manufacturing tools in production of bioactive HA coatings. Direct current (DC) plasma spraying of HA, in particular, since 1980s has been carried out by many researchers.[22–29] However, many challenges have been met. HA, in the light of its biomedical application, demands stringent processing conditions before a successful outcome is achieved. Consequently, several critical issues have been recognised concerning the

Table 5.1 Physical properties of HA

Molecular weight	1004.8
Space group	$P6_3/m$
Lattice constants	a = 0.9423~0.9418 nm, c = 0.6881~0.6884 nm
Chemical unit number	$Z = 1$
Theoretical density	~3.156 g/cm^3
Thermal diffusivity	~5×10^{-3} cm^2/s

plasma spraying of HA. Some have been extensively studied, thus overcoming the problem and establishing new knowledge and understanding of this class of calcium phosphates. However, many remain to be addressed. Some concerns regarding the plasma spraying of HA include:

1. induced chemical changes resulting in deviation from stoichiometry and decomposition;
2. amorphisation upon spraying;
3. induced stresses due to rapid heating and cooling;
4. full quantification of phases (including amorphous phase) after spraying.

5.4.2 Decomposition and chemical changes

Plasma spraying, being a non-equilibrium process, is bound to produce changes to the starting material after processing. When HA is introduced into the plasma, which is typically 15–20 × 10^3 K, it generally undergoes decomposition, melting and even vaporisation. The degree of these processes depends on a large number of factors, including initial size and size distribution of feedstock powders, chemical purity, pressure, feed rate, the power used for spraying and stand-off distance (for coatings). Even though the residence time of particles in the plasma is approximately 2–3 millisecond, the heat input is high enough to cause decomposition and partial melting of HA. Decomposition of HA after plasma spraying has been noted by several authors:[30–34]

$$Ca_{10}(PO_4)_6(OH)_2 \rightarrow 3Ca_3(PO_4)_2 + CaO + H_2O \quad \text{(above 1200 °C)} \qquad [5.1]$$

$$Ca_{10}(PO_4)_6(OH)_2 \rightarrow 2Ca_3(PO_4)_2 + Ca_4P_2O_9 + H_2O \quad \text{(above 1400 °C)} \qquad [5.2]$$

$$2Ca_{10}(PO_4)_6(OH)_2 \rightarrow 5Ca_3(PO_4)_2 + Ca_4P_2O_9 + CaO + 2H_2O \quad \text{(above 1400 °C)} \qquad [5.1] + [5.2]$$

$$Ca_{10}(PO_4)_6(OH)_2 \rightarrow Ca_{10}(PO_4)_6O \text{ (oxyapatite)} \quad \text{(above 1400 °C)}$$

There are also findings that the formation of the higher temperature phases such as α-tricalcium phosphate (TCP) and tetracalcium phosphate (TTCP) occurs subsequent to the formation of oxyapatite (OAp).[34] The loss of water is also thought to produce a solid solution of HA and OAp in which the chains of OH^- are replaced by O^{2-} and vacancies. The formation of CaO in the as-sprayed HA also suggests that P_2O_5 depletion occurs during the powder-spraying process.

Other than decomposed phases, the as-sprayed HA also contains amorphous material. This amorphous material has been seen as a problem by some in HA coatings, especially when the amorphous content is concentrated at the interface between implant and coating.[35,36] It is also the view of several researchers that the presence of the amorphous phase in plasma-sprayed HA coatings is detrimental not only to the interface between the coating and substrate but the coating and the implanted area in the biological environment.[37,38] Others such as Gross and Berndt, however, believe that the presence of the amorphous can be useful for absorbing mechanical mismatch and the promotion of bone remodeling.[34,39] In DC plasma spraying, the formation of the amorphous phase is thought to be related to the rapid heating and cooling experienced by material undergoing plasma spraying. It was noted too that formation of the amorphous phase is also related to the partial dehydroxylation of HA during spraying.[39] Gross and Berndt have presented a model to highlight the changes to HA after plasma spraying which gives an idea of what phases can be present in the as-sprayed powders.[39]

It is also known that the cooling rate has an important role in the formation of the various calcium phosphates. In plasma processing the cooling rate is so high that equilibrium phase transformations are a near impossibility. However, depending on the thermal treatment, different amounts of decomposition and amorphisation can take place. Other than temperature (thermal treatment), partial water vapor pressure also determines the thermal stability of HA, and hence the level of decomposition it experiences.[39,40]

It was reported that the comparative dissolution behavior of different monophasic calcium phosphates in increasing order is as shown below:

$$\text{HA} \ll \beta\text{-TCP} < \text{OHA} < \alpha\text{-TCP} < \text{CaO} < \text{TTCP} \ll \text{ACP}$$

where ACP is amorphous calcium phosphate. Since the dissolution behavior of the coatings reflects their phase distribution, the presence of metastable components (e.g., TCP, TTCP) in turn leads to an increase in instability of the coatings. Therefore the phase balance of the plasma-sprayed coating may affect its long-term behavior under *in vivo* conditions.

5.4.3 *In vitro* tests

The growth of apatite crystals on the surface determines the characteristics and performance of a number of biomaterials. Because body fluid is highly supersaturated with respect to apatite, apatite deposition may occur on the surface of any biomaterial immersed in body fluid. For biomaterials that could be used in contact with bone, it is desirable to enhance growth of apatite on the implant surface. This will subsequently result in the formation of a chemical bond between bone and biomaterial. Figure 5.6 presents the proposed bonding between bone and implant. The precipitated apatite layer will serve as a substrate for subsequent protein adsorprtion and bone cell attachment. The proteins from the body serve as cell receptors, and adsorb onto the apatite surface (cell ligand), thus enabling cells to attach onto the implant.

These biological properties of biomaterials should be tested before medical application. However, considering the inconvenience involved in testing within the living organism (*in vivo* test), many studies seeking a comprehensive understanding of the factors that control the biomechanical properties have been conducted through *in vitro* studies. Actually, *in vitro* (Latin: *within the glass*) testing refers to the technique of performing a given procedure in a controlled environment outside of a living organism.

In some *in vitro* studies, the material is immersed in a protein-free and acellular solution that has an ion concentration nearly equal to that of

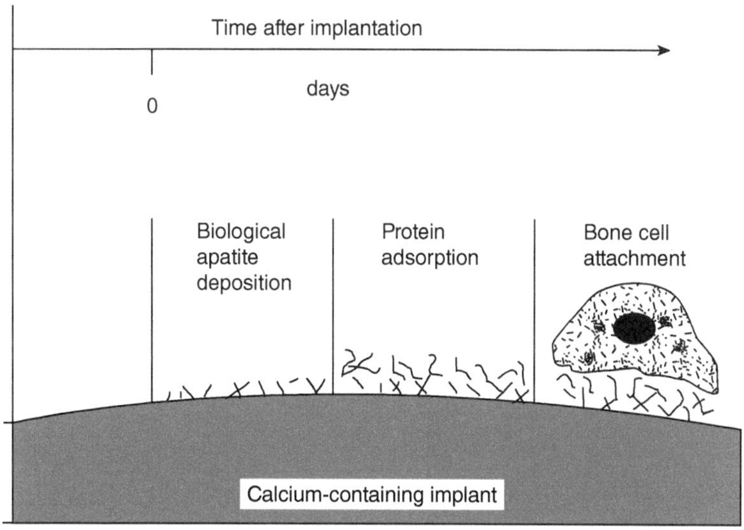

5.6 Schematic showing the proposed mechanism facilitating the bonding of bone to certain implants.[41]

human blood plasma; i.e., simulated body fluid (SBF). Because its chemical composition is close to that of human blood plasma, the bioactivity of an artificial material can be evaluated by examining the formation of apatite on its surface following immersion in SBF. It has been suggested that the efficacy of apatite nucleation is dependent not only on the composition of the biomaterial, but also on its concentration and structural arrangements. After soaking in SBF for a period, an HA sample is negatively charged immediately due to OH^- and PO_4^{3-} and the positively charged Ca^{2+} ions in the fluid. As the calcium ions accumulate, the surface imparts a positive charge, and thus combines with negatively charged phosphate ions in the fluid to form ACP. This phase is metastable, and eventually transforms into stable bone-like apatite. The apatite layer formed incorporates carbonate ions from the solution into its lattice by substituting phosphate groups. This substitution promotes the differentiation of marrow stromal stem cells into osteoblasts (bone-producing cells) that are responsible for the formation of bone. After immersion in SBF for 12 days, Yu *et al.*[42] reported the formation of a dune-like layer consisting of small granules on the as-sprayed HA coating (Fig. 5.7). With continued immersion, the layer became denser and the granules gradually grew.[43]

Cell culture is another *in vitro* technique in cellular biology that is carried out in a Petri dish. It is used not only to test material cytotoxicity (cell poisoning), but also to investigate surface-dependent responses of bone-forming cells. The quality of the bone where cell cultures are to be grown is taken into account in *in vitro* study. This kind of approach concerning *in vitro* research requires models that simulate the *in vivo* conditions: the study of primary osteoblast (bone-forming cell) behavior seems relevant to this purpose. Such a cell culture experiment provides evidence of enhanced

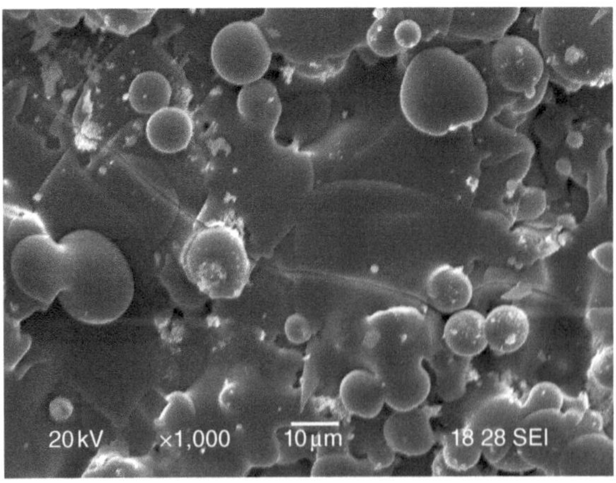

5.7 SEM micrograph of as-sprayed HA coating.

adhesion of osteoblasts on the nanophase HA. Moreover, cell culture work provides information on cell proliferation, alkaline phosphatase activity and concentration of calcium in the extracellular matrix. *In vitro* cellular models have been used to determine the efficacy of HA to serve as bone prostheses.[44,45] A study by Khor *et al.* shows that the osteoblasts proliferated well on a HA coating surface.[46] All cells spread well and grew favorably across the coating surfaces. As observed, the cells are seen to be flattened and to attach tightly onto coating surfaces with their filopodium and lamellipodium, suggesting good cell viability on HA coatings.

5.5 Conclusions

Investigation of the current state-of-the-art in plasma-spraying technology has indicated that opportunities exist to significantly improve upon modern plasma-spraying systems. One such opportunity is the tailoring of pore morphologies to improve the insulative properties of the coatings. Schematically, based on the current studies on CaP biomaterials, potential future development of new CaP biomaterials will mainly be obtained via nanotechnologies. There are several recommendations for future studies in this field:

1. Future researchers should give careful consideration to selecting the appropriate approach to synthesise nanostructured CaP materials. This judicious selection would help to transfer the technology from laboratory- to industrial-scale application.
2. In addition to the currently required characterisation, broader assessments of the new biomaterials should be conducted. For example, an evaluation of the strengths and weaknesses, short-term and long-term successes and failures through *in vivo* study would be extremely helpful with respect to medical application within the human body.

5.6 Acknowledgements

We would like to thank Nanyang Technological University for funding support. J.L. Xu is a research fellow in School of Mechanical & Aerospace Engineering, Nanyang Technological University.

5.7 References

1. X. FAN, T. ISHIGAKI, Mo5Si3-B and MoSi2 deposits fabricated by radio frequency induction plasma spraying, *Journal of Thermal Spray Technology*, **10**(4) (2001), 611–617.
2. K.A. KHOR, Y. MURAKOSH, M. TAKAHASHI, T. SANO, Plasma spraying of titanium aluminide coatings: Process parameters and microstructure, *Journal of Materials Processing Technology*, **48** (1995), 413–419.

3. S. MALMBERG, J. HEBERLEIN, Effect of plasma spray operating conditions on plasma jet characteristics and coating properties, *Journal of Thermal Spray Technology*, **2**(4) (1993), 339–344.
4. P. CHEANG, K.A. KHOR, Thermal spraying of hydroxyapatite (HA) coatings: effects of powder feedstock, *Journal of Materials Processing Technology*, **48** (1995), 429–436.
5. P. CHEANG, K.A. KHOR, Addressing processing problems associated with plasma spraying of hydroxyapatite coatings, *Biomaterials*, **17** (1996), 537–544.
6. E. CHANG, W.J. VHANG, B.C. WANG, C.Y. YANG, Plasma spraying of zirconia–reinforced hydroxyapatite composite coatings on titanium Part I: Phase, microstructure and bonding strength, *Journal of Materials Science: Materials in Medicine*, **8** (1997), 193–200.
7. Y.C. TSUI, C. DOYLE, T.W. CLYNE, Plasma sprayed hydroxyapatite coatings on titanium substrates Part 1: Mechanical properties and residual stress levels, *Biomaterials*, **19** (1998), 2015–2029.
8. H. LI, K.A. KHOR, P. CHEANG, Titanium dioxide reinforced hydroxyapatite coatings deposited by high velocity oxy-fuel (HVOF) spray, *Biomaterials*, **23** (2002), 85–91.
9. I.-S. LEE, H.-E. KIM, S.-Y. KIM, Studies on calcium phosphate coatings, *Surface and Coatings Technology*, **131** (2000), 181–186.
10. P.K. CHU, J.Y. CHEN, L.P. WANG, N. HUANG, Plasma-surface modification of biomaterials, *Materials Science and Engineering*, **R36** (2002), 143–206.
11. Y.C. TSUI, C. DOYLE, T.W. CLYNE, Plasma sprayed hydroxyapatite coatings on titanium substrates Part 2: optimization of coating properties, *Biomaterials*, **19** (1998), 2031–2043.
12. D. MATEJKA, B. BENKO, *Plasma Spraying of Metallic and Ceramic Materials*, Wiley, Chichester, UK, (1989).
13. A.M. EKTESSABI, M. HAMDI, Characterization of calcium phosphate bioceramic films using ion beam analysis technique, *Surface and Coatings Technology*, **153** (2002), 10–15.
14. M. MATTIOLI BELMONTE, A. DE BENEDITTIS, R.A.A. MUZZARELLI, P. MENGUCCI, G. BIAGINI, Bioactivity modulation of bioactive materials in view of their application in osteoporotic patients, *Journal of Materials Science: Materials in Medicine*, **9** (1998), 485–492.
15. S.Y. WANG, P.K. CHU, B.Y. TANG, X.C. ZENG, Y.B. CHEN, X.F. WANG, Radio-frequency plasma nitriding and nitrogen plasma immersion ion implantation of Ti–6Al–4V alloy, *Surface and Coating Technology*, **93** (1997), 309–313.
16. K.A. GROSS, C.C. BERNDT, Structural changes of plasma sprayed hydroxyapatite coatings during in vitro testing, in: E. Horowitz, J.E. Parr (eds), *Characterization and Performance of Calcium Phosphate Coatings for Implants*, Philadelphia, PA: ASTM, (1993): 256–268.
17. P. LUO, T.G. NIEH, Synthesis of ultrafine hydroxyapatite particles by a spray dry method, *Materials Science and Engineering*, **C3** (1995), 75–78.
18. P. CHEANG, K.A. KHOR, Addressing processing problems associated with plasma spraying of hydroxyapatite coatings, *Biomaterials*, **17** (1996), 537–544.
19. R.L. REIS, F.J. MONTEIRO, Crystallinity and structural changes in HA plasma-sprayed coatings induced by cyclic loading in physiological media, *Journal of Materials Science: Materials in Medicine*, **7** (1996), 407–411.

20. G. PENEL, G. LEROY, C. REY, B. SOMBRET, J.P. HUVENNE, E. BRES, Infrared and Raman microspectrometry study of fluor-fluor-hydroxy and hydroxy-apatite powders, *Journal of Materials Science: Materials in Medicine*, **8** (1997), 271–276.
21. T. NAKANO, T. AWAZU, Y. UMAKOSHI, Plastic deformation and operative slip system in mineral fluorapatite single crystal, *Scripta Materialia*, **44** (2001), 811–815.
22. D.P. RIVERO, J. FOX, A.R. SKIPOR, R.M. URBAN, J.O. GALANTE, Calcium phosphate coated porous titanium implants for enhanced skeletal fixation, *Journal of Biomedical Materials Research*, **22** (1988), 191–201.
23. P.S. PREVEY, X-Ray diffraction characterization of crystallinity and phase composition in plasma sprayed hydroxyapatite coatings, *Journal of Thermal Spray Technology*, **9** (2000), 369–376.
24. J.C. KNOWLES, K. GROSS, C.C. BERNDT, W. BONFIELD, Structural change of thermally sprayed hydroxyapatite investigated by Rietveld analysis, *Biomaterials*, **17** (1996), 639–645.
25. E. BOUYER, M. MULLER, R.H. HENNE, G. SCHILLER, Thermal plasma processing of nanostructured Si-based ceramic materials, *Journal of Nanoparticle Research*, **3** (2001), 373–378.
26. C.F. FENG, K.A. KHOR, Y.W. GU, P. CHEANG, Analysis of phase changes in plasma-sprayed Ti-6Al-4V/hydroxyapatite composite coating by DSC, *Materials Letters*, **51** (2001), 88–93.
27. E. PARK, R.A. CONDRATE SR, D. LEE, K. KOCIBA, P.K. GALLAGHER, Characterization of hydroxyapatite: before and after plasma spraying, *Journal of Materials Science: Materials in Medicine*, **13** (2002), 211–218.
28. L. SUN, C.C. BERNDT, K.A. KHOR, H.N. CHEANG, K.A. GROSS, Surface characteristics and dissolution behavior of plasma-sprayed hydroxyapatite coatings, *Journal of Biomedical Materials Research*, **62** (2002), 228–236.
29. J. WENG, Q. LIU, J.G.C. WOLKE, X. ZHANG, K. DE GROOT. Formation and characteristics of the apatite layer on plasma-sprayed hydroxyapatite coatings in simulated body fluid, *Biomaterials*, **18** (1997), 1027–1035.
30. A.C. TAS, F. KORKUSUZ, M. TIMUCIN, A. AKKAS, An investigation of the synthesis and high-temperature sintering behavior of calcium hydroxyapatite (HA) and tricalcium phosphate (TCP) bioceramics, *Journal of Materials Science: Materials in Medicine*, **8** (1997), 91–96.
31. E. CHANG, W.J. VHANG, B.C. WANG, C.Y. YANG, Plasma spraying of zirconia-reinforced hydroxyapatite composite coatings on titanium Part I: Phase, microstructure and bonding strength, *Journal of Materials Science: Materials in Medicine*, **8** (1997), 193–200.
32. P. CHEANG, K.A. KHOR, Addressing processing problems associated with plasma spraying of hydroxyapatite coatings, *Biomaterials*, **17** (1996), 537–544.
33. J. MOSTAGHIMI, M. PASANDIDEH-FARD, S. CHANDRA, Dynamics of splat formation in plasma spray coating process, *Plasma Chemistry and Plasma Processing*, **22** (2002), 59–84.
34. K.A. GROSS, C.C. BERNDT, In vitro testing of plasma-sprayed hydroxyapatite coatings, *Journal of Materials Science*, **5** (1994), 219–224.
35. R. MCPHERSON, N. GANE, T.J. BASTOW, Structural characterization of plasma-sprayed hydroxylapatite coatings, *Journal of Materials Science: Materials in Medicine*, **6** (1995), 327–334.

36. R. KUMAR, P. CHEANG, K.A. KHOR, Phase composition and heat of crystallization of amorphous calcium phosphate in ultra-fine radio frequency suspension plasma sprayed hydroxyapatite powders, *Acta Materialia*, **52** (2004), 1171–1181.
37. C.F. KOCH, S. JOHNSON, D. KUMAR, M. JELINEK, D.B. CHRISEY, A. DORAISWAMY, C. JIN, R.J. NARAYAN, I.N. MIHAILESCU, Pulsed laser deposition of hydroxyapatite thin films, *Materials Science and Engineering C*, **27** (2007), 484–494.
38. Z. ZYMAN, J. WENG, X. LIU, X. ZHANG, Z. MA, Amorphous phase and morphological structure of hydroxyapatite plasma coatings, *Biomaterials*, **14** (1993), 225–228.
39. K.A. GROSS, C.C. BERNDT, Thermal processing of hydroxyapatite for coating production, *Journal of Biomedical Materials Research*, **39** (1998), 580–587.
40. D.H. HARRIS, Overview of problems surrounding the plasma spraying of hydroxyapatite coatings, in: T.F. Bernacki (ed.), *Proceeding of the Third National Thermal Spray Conference*, Materials Park, OH: ASM International, 1990: 419–423.
41. M. SPECTOR, Basic principles of tissue engineering, in: S.E. Lynch, R. J. Genco, R.E. Marx (eds), *Tissue Engineering: Applications in Maxillofacial Surgery and Periodontics*, Carol Stream, IL: Quintessence, (1998): 3–16.
42. L.G. YU, K.A. KHOR, H. LI, P. CHEANG, Effect of spark plasma sintering on the microstructure and in vitro behavior of plasma sprayed HA coatings, *Biomaterials*, **24** (2003), 2695–2705.
43. X.B. ZHENG, C.X. DING, Characterization of plasma-sprayed hydroxyapatite/TiO_2 composite coatings, *Journal of Thermal Spray Technology*, **9** (4) 2000, 520–525.
44. S. ZHANG, K.E. GONSALVES, Preparation and characterization of thermally stable nanohydroxyapatite, *Journal of Materials Science: Materials in Medicine*, **8** (1997), 25–28.
45. S. SARIG, F. KAHANA, Rapid formation of nanocrystalline apatites, *Journal of Crystal Growth*, **237–239** (2002), 55–59.
46. K.A. KHOR, H. LI, P. CHEANG. Nannostructures and osteoblast behavior of thermal sprayed calcium phosphate splats, *Advances in Technology of Materials and Materials Processing*, **8**(2) (2006), 130–133.

6

Processing, microstructures and properties of thermal barrier coatings by electron beam physical vapor deposition (EB–PVD)

S. GONG and Q. WU, Beihang University, China

Abstract: Thermal barrier coatings (TBCs) produced by electron beam physical vapor deposition (EB–PVD) have exhibited some advantages over those produced by plasma spraying, being the most promising protective coatings for hot components used in rotation condition. In this work, the physical principles of EB–PVD technology are described. Processing, microstructures and properties of EB–PVD TBCs are reviewed. The failure mechanisms of EB–PVD TBCs subjected to thermal cyclic loads are also discussed.

Key words: thermal barrier coatings (TBCs), electron beam physical vapor deposition (EB–PVD), microstructure, property, processing.

6.1 Introduction

The major objectives of the turbine industry are to increase the engine combustion temperature, to double the thrust to weight ratio, and to improve the thermal efficiency of turbine components under severe environmental conditions at elevated temperatures.[1,2] The use of nickel-based superalloy materials has resulted in an increase of almost 100 °C in engine operational temperature in the past 50 years.[3] However, the limits of thermal stability have been reached. It has been suggested that a reduction in metallic temperature of 30–60 °C can produce a twofold increase in the life of turbine components. Thermal barrier coating (TBC) systems, as a means of protecting metallic substrates from high temperatures, can decrease the metallic temperature and improve the service life of turbine components. Therefore, through reducing the thermal conductivity of the TBCs by altering the microcomposition or by microstructural modifications, the life of turbine components can be prolonged.[4]

The typical TBC consists of two layers: an outer insulated ceramic top coat (TC), with its main function of reducing the heat transfer to the metallic substrate; and a metallic bond coat (BC) which mainly affects oxidation-resistance and is the actual thermal barrier. These are deposited directly onto the surface of the metallic components, BC first followed by TC.

During operational service, a thermally grown oxide (TGO) is formed between BC and TC as a result of oxidation of the BC at high temperatures. Since the top ceramic layer can endure a higher temperature than the metallic substrate, the engine operating temperatures can rise, prolonging the life of metal components when subjected to high temperature and pressure and reducing their cooling requirements, thus providing benefits in performance, efficiency and durability of the engine.[5–10]

Thermal barrier coatings deposited onto turbine components using the electron beam–physical vapor deposition (EB–PVD) process usually have unique columnar structures capable of narrowing the technological gap between insulation/life extension and reliability.[5]

The TBC materials and structures, processing techniques, properties and associated failure mechanisms have been briefly demonstrated in this chapter. Herein, we would present the processing, microstructures and properties of TBCs by EB–PVD in detail.

6.2 Description of the physical principles of electron beam physical vapor deposition (EB–PVD)

EB–PVD is widely used in the preparation of thick coatings as one of a variety of coating manufacturing methods in vacuum. In this method, a high-energy electron beam which is able to achieve rapid melting and evaporation is focused on target materials. Then the vapor is deposited onto certain substrates where it can evaporate both metals and ceramics. The high rate of deposition makes it popular in a wide range of application markets, such as optical prisms, filters, semiconductors and a high-value thermal barrier coating device, etc. The first system for the preparation of thermal barrier-coated EB–PVD in gas turbine blades was developed by the German company Leybold-Heraeus in the late 1960s.[11]

Since the 1980s, the USA, the UK, Germany and the former Soviet Union countries have been investigating the physical vapor method of deposition to prepare TBC. The early EB–PVD technology is mainly used for the preparation of MCrAlY coatings. As the equipment is expensive and the production cost is relatively high, the development of EB–PVD technology has had to cease. In the early 1980s, the American company, Airco Temescal, prepared a reproducible, high-quality TBC in the laboratory using EB–PVD technology for the first time. In the mid-1980s, P & W, GE and other companies began to utilize the EB–PVD method for the processing of TBCs in rotor blades and guide vanes in aviation turbine engines. During the same period, the Soviet Union also used the EB–PVD techniques and successfully made TBCs for rotor blades in military aircraft. In the mid-1990s, the Paton Welding Institute in Ukraine produced low-cost EB–PVD equipment, which was applied in the USA and Europe, and the EB–PVD TBC technology reached a new peak.

The electron beam is used as a heat source in EB–PVD. Since the evaporating rate is high, almost all the materials can be evaporated, and the interfacial bonding between the deposited coating and substrate is much stronger than plasma-sprayed coatings. As the processing parameters, such as electron beam power, beam spot size and location, can be controlled accurately, it is feasible to obtain a uniform coating thickness. Reactions between evaporation materials and crucible at high temperature can be avoided since the crucible is water-cooled, and it will not contaminate the coatings. The main processing procedures in the EB–PVD method are as follows: the electron beam with high energy density focuses on the target surface under magnetic or electric field and heats the target to a sufficiently high temperature to cause evaporation, then the evaporative atoms from the melted target move to the substrate surface and finally solidify to form films in a low-pressure environment.

The substrate is usually heated during the process of preparing the coating, in order to activate the surface of the substrate and then to improve the bonding between coating and substrate. When the atoms of the vapors impinge on the surface of the substrate, they may exist in several states: completely depositing onto the substrate, diffusing into the substrate and reacting with elements of the substrate, or re-evaporating from the substrate. All of these states can be controlled by changing the conditions of the substrate or the processing parameters of EB-guns.

The variable parameters include substrate temperature (T_s), shape, structure, composition and cleanliness as well as the electrical potential difference between the source of evaporation and the substrate. The substrate temperature plays a key role among these parameters. When the thickness of the films exceeds a few hundred nanometers, the crystal structure of the sediments depends on the substrate temperature. If the temperature of the substrate is $T_s < 2/3T_m$ (T_m–metal melting point, K), the metal atoms directly condense from vapor into solid phase; when $T_s > 2/3T_m$, the metal atoms change from vapor into a liquid phase (droplets); once the droplets reach a certain size they can be crystalloid.[12]

The following relationship exists between the substrate temperature and the microstructure of the coating, which is recognized as the classic EB–PVD coating-structural relationship model.

1. When $T_s/T_m < 0.3$, the coating presents as a dome of columnar structure due to a self-shadowing effect and an inadequate diffusion of the deposited atoms in the surface of the substrate. The grain boundary has more pores.

2. When $0.3 < T_s/T_m < 0.5$, a dense columnar grain structure forms and the coating structure forms owing to the condensation controlled by the surface diffusion. In this context, as T_s increases, the grain size of the columnar crystal also increases. A transition zone exists between zones I and II, which comprises close-packed fibrous grains.
3. When $0.5 < T_s/T_m < 1$, a recrystallization structure forms which is mainly controlled by volume diffusion.

6.3 Manufacturing of thermal barrier coatings (TBCs) by EB–PVD

The EB–PVD method (illustrated in Fig. 13.1) is mainly used for the preparation of gas turbine engine blades, and its industrial production system consists of a central vapor deposition chamber (main chamber) and two target storage rooms (pre-vacuum chambers), equipped with a total of 2~6 electronic guns; one gun is used to pre-heat the substrate and other guns are used for heating target materials and causing evaporation of the materials. There are three water-cooled copper crucibles in the main vacuum chamber. The target material (bar) can be fed slowly by the lower transmission components. Three guns can heat three individual crucibles, respectively. In the evaporation chamber, the substrate can rotate or tilt simultaneously according to the needs of the workpiece geometry and coating thickness distribution. There are valves between the deposition chamber and the pre-vacuum room, so that the workpiece can be fed into the deposition chamber. While one workpiece is being deposited, the other is kept in a warm condition. When the deposition is finished, another preheated workpiece passes into the deposition chamber, thereby achieving continuous production.

Multipurpose EB–PVD equipment is illustrated in Fig. 6.2. There are six electron guns and four water-cooled crucibles equipped in the main vacuum

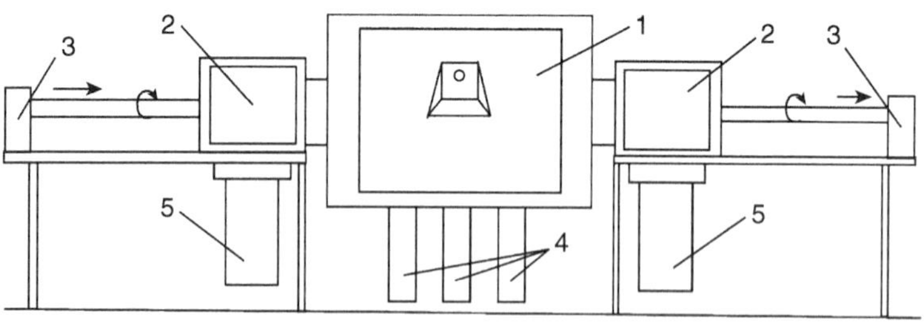

6.1 Industrial production system of EB–PVD:[13] 1. main chamber; 2. pre-vacuum chamber; 3. horizontal transmission and rotation components of workpiece; 4. transmission components for target material; 5. vacuum chamber.

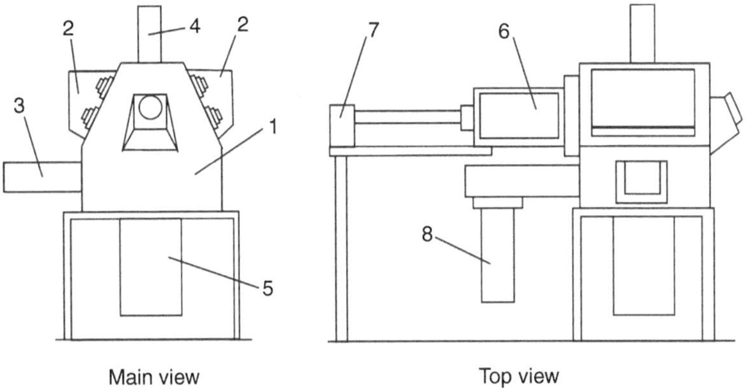

6.2 Multi-purpose EB–PVD equipment:[13] 1. main chamber; 2. electron guns and housing; 3. horizontal transmission for shielding; 4. vertical transmission for workpiece; 5. transmission for target material; 6. pre-vacuum chamber; 7. horizontal transmission for workpiece; 8. vacuum chamber.

chamber, and the device has only a pre-vacuum chamber, which is different from the industrial production system. This device also has a workpiece-supported structure which could rotate around the vertical axis and feed a frame used for vacuum melting in the horizontal direction. Therefore, it can be used not only in coating but also in preparing multi-layer film and a variety of coatings, including depositing coatings in the C/C fibers, depositing superconducting coatings in conventional wire and preparing a new alloy or melting metals and alloy with electron beam.

The coating processing is finished in the vapor deposition chamber. Because coating material vapor needs to be uniformly distributed, the speed of coating materials, the activated gas pressure and electron beam scanning on the target must be precisely controlled.

It is very important to pre-heat the vanes before the deposition of the coatings.[14] The pre-oxidation of the bond coat before deposition of the ceramic coating can not only improve the combinational strength between the ceramic coating and the bond coat, but can also decrease the oxidation rate. The metal substrate temperature strongly influences the microstructure and life of the coating. Vanes are usually heated to 1073–1323 K before deposition. Since the sizes and qualities in various parts of the blade are different, and more heat is absorbed in the thick roots, the electron guns for warming should stay in the roots longer. Pre-heating procedures are designed according to the size and quality of a blade in order to ensure that the temperature of every part is uniform, rising to 1323 K within 15 min. Pre-heating blades with electron guns could heat one part precisely and achieve a uniform temperature field. It could increase the heating speed, which boosts the rate of deposition in the coating. If the conditions are well-controlled, the blade surface temperature fluctuation will be ~10 K.

6.4 EB–PVD TBC microstructure and its advantages over plasma-sprayed coatings

The state-of-the-art TBCs consist of a ceramic top coat, typically 7 wt% yttria-stabilized zirconia (7YSZ), which has low thermal conductivity, is chemically inert in the combustion condition and has a relatively high coefficient of thermal expansion which is reasonably compatible with Ni-based superalloys.[15] There are two main categories of metallic bond coat: one is the NiCoCrAlY-based system and the other is the Pt-modified diffusion aluminide. As 7YSZ is essentially transparent to oxygen at a high temperature, a TGO layer, mainly comprising thin, dense and continuous α-Al_2O_3, is formed between the bond coat and the YSZ top coat, thus providing oxidation protection.[16] The top coats can be deposited by EB–PVD or by atmospheric plasma spraying (APS). The highly porous microstructures presented in EB–PVD and APS top coats reduce the inherent thermal conductivity. Thus, the differences in shape, orientation and pore distribution are the cause of the different values of thermal conductivity between EB–PVD and APS coatings.[17,18]

Strain-tolerant TBCs are deposited with two methods, namely EB–PVD and APS. EB–PVD evaporates the oxide and directs the vapor onto the pre-heated component from an ingot. A columnar grain structure is formed during EB–PVD deposition (Fig. 6.3a).[6] In the case of plasma spray, the raw material powders are melted and accelerated until they impinge upon the substrate at very high speed, rapidly solidifying to form a 'splat' (a flattened particle). The successive impingement of the molten particles and inter-

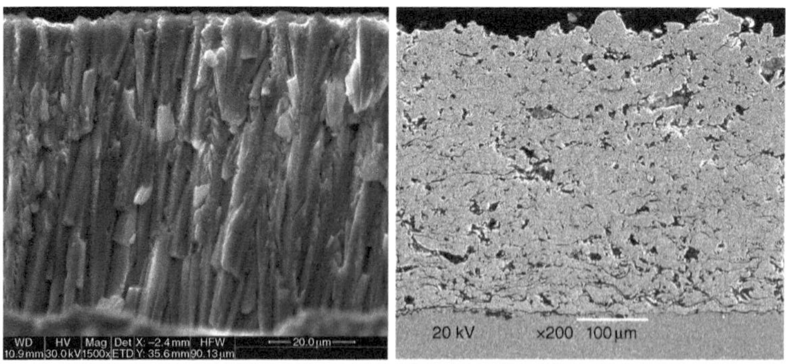

(a) (b)

6.3 SEM micrographs of cross-sections of EB–PVD coating (a) and APS coating (b).

bonding among the splats develops to deposit coatings.[19,20] The deposition is designed to incorporate inter-splat porosity with a network of crack-like voids that provide strain tolerance again, while lowering the thermal conductivity (Fig. 6.3b). Plasma spray (PS) deposition is a lower cost alternative.[6]

In the EB–PVD process, the coatings developed by vapor condensation offer superior strain tolerance and thermal shock resistance. The resistance to oxidation and corrosion is also better than APS coatings, and thus service life is prolonged.[21,22] The interface in EB–PVD is integrated mainly by chemistry bonding. Furthermore, the surface finishing is smoother, cooling hole closure is avoided and an aerodynamic design of the blades is maintained. The columnar microstructure of EB–PVD coatings has been found to offer the superior performance and durability necessary for gas turbine blade applications.[1,2] However, one disadvantage of EB–PVD coatings arises from the columnar structure, which leads to high thermal conductivity compared to that of APS coatings. APS coatings have a splat-based layered structure which offers advantages over EB–PVD in terms of insulation and cost-effectiveness.[24]

Plasma spray TBCs have been applied to commercial parts of jet engines since the 1980s because of the relatively simple and inexpensive equipment required. Plasma-sprayed coatings are still used in combustors and other hot-gas path components although they have a porous and extensive cracked structure. Studies have shown that during thermal loading, the coating integrity could be lost as a result of damage propagation within the coatings.[2]

All major gas turbine engine manufacturers have used EB–PVD to produce TBCs. With further understanding of TBC systems, we would be able to improve maintenance scheduling, allow more positive usage and design systems with improved reliability and capability. Deposition parameters such as chamber pressure, substrate temperature and rotation speed have influenced the morphology and density of the columnar microstructure of EB–PVD TBCs accordingly.[25] The low thermal conductivity and high values of strain tolerance in the ceramic top coat are due to its intrinsic columnar microstructure. The efficiency of thermal insulation is closely related to the porosity of TBC materials. In the EB–PVD process, an electron beam heats the source material, then the vapor is produced and the evaporated atoms condense onto the substrate. Crystal nuclei form on the favored sites, growing either laterally or in the through-thickness direction, so as to form individual columns, resulting in a high degree of inter-columnar porosity.[26,27] This specific coating structure provides more stress release with low in-plane Young's modulus under thermomechanical loading. Also, the structure can adapt to thermal stresses due to the coefficient of thermal expansion (CTE) mismatch. In addition, the microstructure and texture throughout the thickness of the coating are variable during deposition, since growth depends on competitive processes.[28]

6.4.1 Typical microstructure of the conventional TBCs

Due to its intrinsic columnar microstructure, the EB–PVD TBC exhibits low thermal conductivity and high values of strain tolerance. It has a unique microstructure containing individual primary columns, which grow on the substrate in a preferred crystallographic direction by adding atoms to the vapor phase. The growth of this column produces an inter-columnar open porosity which provides the TBCs with high strain tolerance. Substrates rotation during vapor deposition-processing produces an additional intra-columnar closed porosity. Furthermore, a feather-like subcolumnar structure forms at the periphery of each primary column with surfaces 45–60° inclined toward their growing axis due to over-shadowing (Fig. 6.4). The total porosity increases and therefore the thermal conductivity of the material is low. Finally, the secondary columns, so-called feather-arms, lie inside the columns due to the rotation of the specimens during the vapor deposition process, and an additional intracolumnar closed porosity inside the primary columns is produced as well (Fig. 6.4).[25,28]

The YSZ–TBC shows a typical columnar structure which is normal to the BC surface. One column is generally believed to be a single crystal that grows continuously from the substrate. A single column consists of several different misorientated fine subcolumns, with nanopores along subcolumn boundaries. The YSZ plate-like grains in each subcolumn are stacked periodically along the growing direction of the subcolumn with a constant spacing. The nanopores at the subcolumn boundaries are perpendicular to

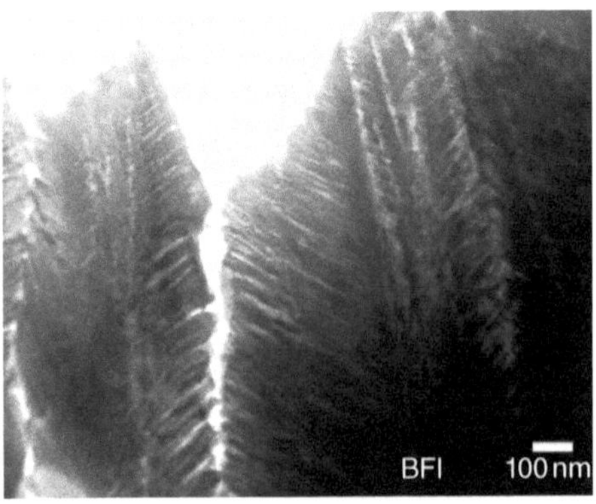

6.4 TEM showing the feather-like microstructure of EB–PVD YSZ coating.

the growth direction of the subcolumns. Such nanopores are formed due to the misorientations of YSZ plate-like grains between subcolumns.[25,29]

It is well accepted that the deposition parameters such as chamber pressure, substrate temperature and rotation speed can accordingly influence the morphology and density of the columnar microstructure of EB–PVD TBCs. Schulz and Schmucker[30] investigated microstructural properties of ZrO_2–4at% Y_2O_3 coatings deposited with different thicknesses and at different substrate rotation speeds. The 0 rpm specimen consisted of simple, straight columns. The 1 rpm specimen had a wavy microstructure. In contrast, specimens deposited at higher rotation speeds consisted of straight columns with banded structures. Nanosized pores less than 50 nm in diameter could be observed in the columnar grains. All the coatings had a columnar microstructure with gaps between columnar grains. The grain size of the columns increased with increasing coating thickness and speed of rotation. The total porosity of the coating layers increased with increasing substrate rotating speed.

6.4.2 Multi-layered and graded TBCs

To overcome the disadvantages of YSZ, candidate materials that can withstand higher gas-inlet temperatures have been investigated, and multi-layered and graded structures have been produced. It has been shown that the thermal cycling lifetimes of the multi-layered coatings are two or three times longer than those of the single ceramic layer coatings.[31] The multi-layered and graded structures are introduced below.

- **Multi-layer system:** Since thermal barrier coatings must endure arduous environments, such as corrosion, oxidation and thermal shock, it is necessary to develop a multi-layer system with several functions. Multi-layer systems generally consist of bonding layer, ceramic thermal barrier layer, anti-corrosion layer and diffusion barrier layer. The dense layer and barrier layer on the outside protect the inner layer from the diffusion of corrosive salts and other deposits into the metallic bond coat. Since there are a series of problems that remain unsolved in this multi-system, it has not yet been put into practical use. However, it is quite promising if the design of the system can be optimized.
- **Gradient-layer system:** In this system (shown in Fig. 6.5), the transition from metal layer to the ceramic layer is continuous and gradual.[32] The sharp interface between metal and ceramic disappears. As a result, the premature spalling of TBCs due to the metal/ceramic thermal expansion coefficient mismatch is avoided. This coating structure design is considered to be the best solution in prolonging the thermal cycling lifetime

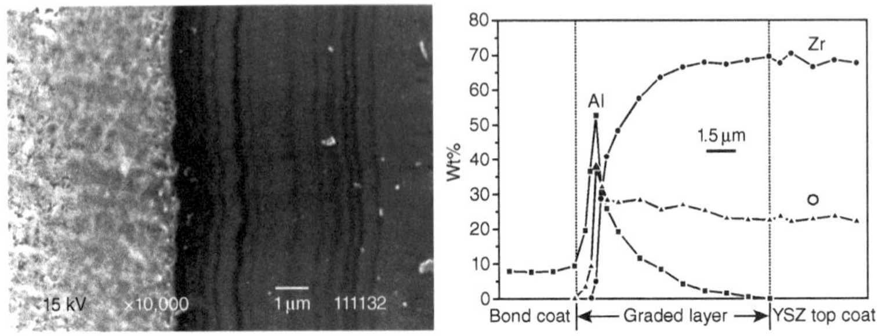

(a) (b)

6.5 SEM micrograph of cross-section of gradient thermal barrier coating (GTBC) produced by EB–PVD (a) and related element profile across the thickness of the GTBC (b).[34]

while ensuring thermal barrier insulation.[33] According to G. Zinsmeister's theory, when the electron beam evaporation power is constant, the vapor pressure of each phase in the mixture is dependent on the fraction of each phase in the mixture. Following this idea, a NiCoCrAlY/NiAl/Al_2O_3/YSZ gradient TBC was produced by co-evaporation and deposition of the mixture of Al–Al_2O_3–ZrO_2.[33–37] During the processing, since $P_{Al} > P_{Al_2O_3} > P_{ZrO_2}$, the Al phase was first evaporated from the source of the mixture. As the temperature increased, the evaporation rate of Al_2O_3 increased gradually. Finally, the evaporation of ZrO_2 phase was also finished. As a consequence, a continuous structure and the transition of the composition between the bond coat and YSZ topcoat were realized. Afterwards, the deposited coatings were annealed in vacuum. During annealing, inter-diffusion between the bond coat and the transition layer occurred, resulting in the elimination of the interface.

On thermal exposure during service, the porous coatings will sinter, caused by ageing under service conditions, bringing about changes in the distribution and morphology of the open and closed porosity. The driving force for this process is the decrease in surface energy. The gaps between the individual columns were narrowed and even disappeared. The number of gaps between the subcolumns decreased. The sharp gaps observed in the as-deposited coating became blunted, as shown in Fig. 6.6. Also, the nanopores inside each column revealed spheroid morphology.

There are two important factors which affect the performance of the entire system. The first is that the thermal conductivity as one of the key functional properties of TBC will increase. Both inter- and intra-columnar porosity decrease as a result of sintering inducing reduction of the surface area. The columns join together to form blocks separated by cracks perpendicular to the interface. The cracks originate near the substrate/coating

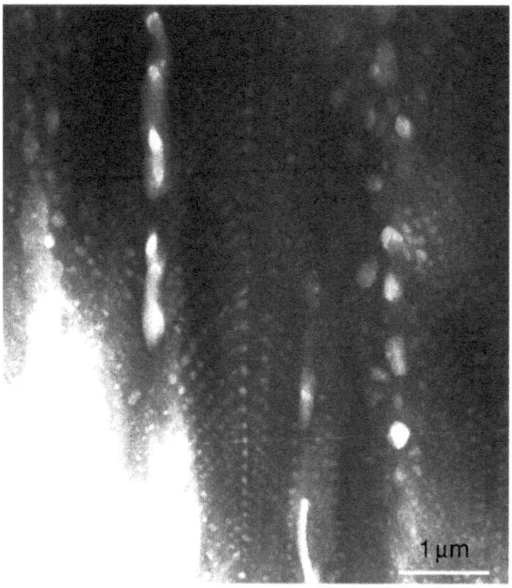

6.6 TEM image of EB–PVD YSZ coatings after 100 h heat-treatment at 1373 K.

interface and run parallel to the interface, affecting the thermal conductivity. Thermal conductivity increases due to the alteration of the porosity distribution, size and morphology and the decrease in the amount of porosity.[28] Additionally, the in-plane strain tolerance is reduced after the columns sinter together to form bridges at the contacting points. The phases also change from the initial metastable tetragonal phase into a mixture of tetragonal phase in the case of the Y–PSZ coatings at high temperature.[38]

6.5 Hot-fatigue behavior and failure mechanisms of TBCs

Thermal cycling lifetime is one of the most important properties for TBCs, which actually governs the duration of hot components in engines. However, there is lacking a reliable and precise method for evaluating damage and predicting the service life of TBCs based on thermomechanical characteristics.[39] The unique columnar microstructure of EB–PVD ceramics exhibits a higher strain tolerance, but failure still occurs by spallation.[40,41] The failure of TBC systems usually occurs near the vicinity of the TGO scale, at the TGO/bond coat interface, within the TGO and at the YSZ/TGO interface.[16,42–45] The failure mechanism is typically associated with buckling and spallation of the ceramic top coat from the TGO or at the TGO/bond coat interface.[46–49] Final spallation of the TBC involves edge-cracking resulting from oxidation and stress singularity, buckling and a 'link-up' of micro-cracks at the TGO/bond coat interface with the pre-existing microcracks at the YSZ/TGO interface.[49,50]

For EB–PVD TBCs with PtNiAl bond coats, several mechanisms leading to TBC failure have been observed, including buckling, accelerated grain boundary oxidation, excessive TGO thickness and interfacial instability such as rumpling and ratcheting.[51] There are several phenomena responsible for the spallation of YSZ after thermal cyclic oxidation, including formation and thickening of TGO resulting from oxidation of the bond coat; formation of the spinel such as Ni, Co-rich oxides; phase destabilization of YSZ coating; densification of YSZ coating by sintering; sulfur segregation; formation of voids at the TGO/bond coat interface.[49]

In general, the TGO plays a significant role in affecting the thermal cycling life of EB–PVD TBCs.[38,49] The TGO layer has two types of morphologies: regular undulation and irregular imperfections of thickness. The growth of the TGO layer follows the kinetics of parabolic law. Rumpling of the TGO/bond coat interface could cause localized cracking at the YSZ/TGO interface. Surface rumpling of the TGO occurs due to formation of voids at the TGO/bond coat interface and phase transformation of the bond coat. Formation of spinel is attributed to depletion of Al in the bond coat. The stress distribution in the direction of the thickness of the TGO layer is not uniform. The stress near the ceramic top coat is smaller than that close to the metal bond coat due to the stress in the direction of thickness.

TBCs tend to spall when they are subjected to thermal cycling due to thermal stresses generated by thermal expansion mismatching with the underneath metal substrate as well as the temperature gradients, residual stresses during deposition, phase transformations, progressive sintering of the ceramic top coat and environmental corrosion and erosion.[53] The stress in the TGO layer is significantly affected by the composition and structure of the top coat and bond coat, the creeping behavior of the bond coat and the morphologies and phase constituents of the TGO.[38]

TBCs reveal different failure modes when they suffer from different serving conditions. These serving conditions are usually combinations of temperature gradient, mechanical loads and atmospheres coupled with oxidation and hot corrosion. There are two predominant failure modes for EB–PVD TBCs on turbine airfoils according to the engine-operating conditions. TBC failure typically occurs by cracking at one of the divides by the TGO interfaces (TBC/TGO or TGO/BC) when the TGO reaches a critical thickness of more than 5 μm under long-cycle, low-temperature operating conditions (cyclic oxidation). Under short-cycle and high-temperature conditions (isothermal oxidation), the damage to the interface would accumulate and the TBC failure would finally occur when fragmentation takes place. This type of failure is named 'ratchet'. The TGO undergoes displacement instability when the TBC system is subjected to thermal cycling, being manifested as a localized penetration of the TGO into the bond coat at periodic sites along the TGO/BC interface.[53] Cracks form and extend laterally, due to the normal strains induced by the cycles by the downward displacements after being cycled. The cracks eventually coalesce, and then failure happens at one or two of the critical TGO interfaces. There is typically around 25% oxide layer debonding in length once the oxide approaches a thickness of 5~6 μm. The thermal expansion mismatching between the TGO and the substrate becomes a governing contributory factor in determining the failure under thermal cycling conditions rather than isothermal conditions.[54]

In Section 6.3 we have mentioned that the initial oxidation of the bond coat surface before deposition can improve bonding between ceramic layer and the bond coat, and decrease the oxidation rate in service. Experimental observations relating the effect of bond coat pre-oxidation to TBC life have been described.[55] It has been shown that pre-oxidation of the bond coat under a selected range of partial oxygen pressures can improve the TBC life by more than two-fold as compared to a TBC without pre-oxidation. However, in both cases, the TBCs reveal similar failure mechanism, and the final TGO thickness and the depth of TGO penetration into the bond coat are nearly the same.[54] One explanation is that pre-oxidation can reduce the rate of ratcheting, thus decreasing TGO growing rate. The grains of the oxide formed in pre-oxidation processing are finer than those formed in an uncontrolled condition. As a result, the TBC after pre-oxidation treatment exhibits lower oxidation rate and much prolonged life.[54]

It is also recognized that the failure mechanism of TBC is dependent on the type of bond coat. For TBCs with β-PtNiAl bond coats, the TGO thickens and elongates under high-temperature oxidation, which will lead to displacement instabilities. If the bond coat is relatively soft, the displacements are concessional, which would cause cracks to form in the TBC part above the TGO. Thus, the instabilities and TBC cracks propagate on a cycle-by-cycle basis, by a ratcheting mechanism. Another common system comprises a two-phase ($\beta + \gamma'$) NiCoCrAlY bond coat, which reveals a different TGO-based failure mechanism.[56] More than five metal elements exist in this bond coat. As a consequence, the TGO develops thickness heterogeneities which are a main feature of this system. The TGO contains various oxides rather than pure α-Al_2O_3, such as yttrium aluminum garnet (YAG) and other spinels. Both the magnitude of the 'pegs' and related geometric imperfections are sensitive to the composition of the bond coat, the deposition condition and the porosity, as well as the surface condition. The infusion of Pt into the surface can ameliorate these imperfections. In some cases, spinel oxides occur within the 'pegs', often accompanied by localized TGO cracks. These cracks seem to be confined to the region around the imperfections. However, whether these localized cracks can be extended beyond the imperfections to cause delamination still remains to be ascertained. The PtNiAl system delaminates primarily along the interface between the TGO and the bond coat. In the NiCoCrAlY bond coat system, the cracks extend through the TGO only at the locations of the 'pegs', which do not enter into the TBC layer. A dominant delamination-cracking not only extends along the interface between the TGO and the bond coat but also traverses the 'pegs', leaving the TGO islands embedded in the bond coat.[57] The delamination cracks nucleate at occasional large 'edge' imperfections. Once the imperfections are formed, they will extend along the interface and finally result in failure of the TBC during the subsequent thermal cycles.[57]

6.6 References

1. A. MARICOCCHI, D. BARTZ, D. WORTMAN, PVD TBC experience on GE aircraft engines, in W.J. Brindley (ed.), *Thermal Barrier Coating Workshop*, NASA-CP-3312, 1995, 79–83, NASA, Washington DC.
2. R.A. MILLER, Thermal barrier coatings for aircraft engines – history and directions, in W.J. Brindley (ed.), *Thermal Barrier Coating Workshop*, NASA-CP-3312, 1995, 17–34, NASA, Washington DC.
3. P. SCARDI, M. LEONI, L. BERTINI, L. BERTAMINI, Residual stress in partially-stabilised-zirconia TBCs: experimental measurement and modelling, *Surf. Coat. Technol.*, **94–95** (1997) 82–88.
4. Data file for X-ray diffraction, JCPDS-ICDD 27-997, International Center for Diffraction Data, Swarthmore, PA.
5. A. KULKARNI, A. GOLAND, H. HERMAN, *et al.*, Advanced neutron and X-ray techniques for insights into the microstructure of EB–PVD thermal barrier coatings, *Mater. Sci. Eng. A*, **426** (2006) 43–52.
6. M. MARTENA, D. BOTTO, P. FINO, *et al.*, Modeling of TBC system failure: stress distribution as a function of TGO thickness and thermal expansion mismatch, *Eng. Fail. Analy.*, **13** (2006) 409–426.
7. H.B. GUO, S.K. GONG, K.A. KHOR, H.B. XU, Effect of thermal exposure on microstructure and properties of EB–PVD gradient thermal barrier coatings, *Surf. Coat. Technol.*, **168**(1) (2003) 23–29.
8. H.B. GUO, H.B. XU, S.K. GONG, A study on EB–PVD gradient thermal barrier coatings under high-temperature hot-corrosion, *J. Mater. Sci.*, **37**(24) (2002) 5333–5337.
9. H.B. GUO, S.K. GONG, C.G. ZHOU, H.B. XU, Evaluation of hot-fatigue behaviors of gradient thermal barrier coatings by EB–PVD, *Mater. Sci. Eng. A*, **325** (2002) 261–269.
10. H.B. GUO, H.B. XU, S.K. GONG, Preparation of Al_2O_3–YSZ composite coatings by EB–PVD, *Mater. Sci. Eng. A*, **325** (2002) 389–393.
11. H.B. GUO, *Thermal fatigue behavior and failure mechanism of gradient thermal barrier coatings by electron beam physical vapor deposition*, Beijing University of Aeronautics and Astronautics, PhD Thesis, 2001.
12. X.Q. CAO, *Materials in Thermal Barrier Coatings*, Beijing, China: Science Press, 2007.

13. H.B. XU, S.K. GONG, F.S. LIU, Electron beam physical vapor deposition technology in Barton Welding Institute of Ukraine, *Aviation Engineering & Maintenance*, **6** (1997) 6–8.
14. E. REINHOLD, C. NEUS, P. BOTZLER, B.D. WENZEL, *et al.*, EB-preheating of turbine blades – the completion of EB-technology for thermal barrier coating, *Surf. Coat. Technol.*, **111** (1999) 10–15.
15. C. LEYENS, U. SCHULZ, B.A. PINT, *et al.*, Influence of electron beam physical vapor deposited thermal barrier coating microstructure on thermal barrier coating system performance under cyclic oxidation conditions, *Surf. Coat. Technol.*, **120–121** (1999) 68–76.
16. R.A. MILLER, Oxidation-based model for thermal barrier coating life, *J. Am. Ceram. Soc.*, **67**(8) (1984) 517–521.
17. D.D. HASS, A.J. SLIFKA, H.N.G. WADLEY, Low thermal conductivity vapor deposited zirconia microstructures, *Acta Mater.*, **49** (2001) 973–983.
18. J.R. NICHOLLS, K.J. LAWSON, A. JOHNSTONE, D.S. RICKERBY, Methods to reduce the thermal conductivity of EB–PVD TBCs, *Surf. Coat. Technol.*, **151–152** (2002) 383–391.
19. S.H. HILLMAN, R.M. GERMAN, Constant heating rate analysis of simultaneous sintering mechanisms in alumina, *J. Mater. Sci.*, **27** (1992) 2641–2648.
20. D. SEN, *et al.*, Pore morphology in sintered ZrO_2–8% mol Y_2O_3 ceramic: a small-angle neutron scattering investigation, *J. Alloys Compd.*, **340** (2002) 236–241.
21. R.A. MILLER, Current status of thermal barrier coatings – an overview, *Surf. Coat. Technol.*, **30** (1987) 1–11.
22. M. GELL, K. VAIDYANATHAN, B. BARBER, *et al.*, Mechanism of spallation in platinum aluminide/electron beam physical vapour deposited thermal barrier coatings, *Metall. Trans.*, **30A** (1999) 427–435.
23. U. SCHULZ, K. FRITSCHER, C. LEYENS, Two-source jumping beam evaporation for advanced EB–PVD TBC systems, *Surf. Coat. Technol.*, **133–134** (2000) 40–48.
24. D.J. WORTMAN, B.A. NAGARAJ, E.C. DUDERSTADT, Thermal barrier coatings for gas turbine use, *Mater. Sci. Eng. A*, **121** (1989) 433–440.
25. F. RENTERIA, B. SARUHAN, U. SCHULZ, *et al.*, Effect of morphology on thermal conductivity of EB–PVD PYSZ TBCs, *Surf. Coat. Technol.*, **201** (2006) 2611–2620.
26. A.J. ALLEN, S. KRUEGER, G.G. LONG, *et al.*, Small-angle neutron scattering studies of ceramic nanophase materials, *Nanostruct. Mater.*, **7**(1–2) (1996) 113–126.
27. A.J. ALLEN, J. ILAVSKY, G.G. LONG, Microstructural characterization of yttria-stabilized zirconia plasma-sprayed deposits using multiple small-angle neutron scattering, *Acta Mater.*, **49**(9) (2001) 1661–1675.
28. F. RENTERIA, B. SARUHAN, Effect of ageing on microstructure changes in EB–PVD manufactured standard PYSZ top coat of thermal barrier coatings, *J. Eur. Ceram. Soc.*, **26** (2006) 2249–2255.
29. Y. JUNG, T. SASAKI, T. TOMIMATSU, *et al.*, Distribution and structures of nanopores in YSZ-TBC deposited by EB–PVD, *Sci. Technol. Adv. Mater.*, **4** (2003) 571–574.
30. U. SCHULZ, M. SCHMUCKER, Microstructure of ZrO_2 thermal barrier coatings applied by EB–PVD, *Mater. Sci. Eng. A*, **276** (2000) 1–8.
31. J. ILAVSKY, G.G. LONG, A.J. ALLEN, C.C. BERNDT, Evolution of the void structure in plasma-sprayed YSZ deposits during heating, *Mater. Sci. Eng. A*, **A272** (1999) 215–221.

32. W.J. BRINDLEY, R.A. MILLER, TBCs for better engine efficiency, *Adv. Mater. Process.*, **8** (1989) 29–33.
33. H.B. GUO, H.B. XU, X.F. BI, S.K. GONG, Preparation of Al_2O_3–YSZ composite coating by EB–PVD, *Mater. Sci. Eng. A*, **325** (2002) 389–393.
34. H.B. XU, H.B. GUO, S.K. GONG, Development of gradient thermal barrier coatings by EB–PVD and their hot-fatigue behaviors, *Surf. Coat. Technol.*, **130** (2000) 133–139.
35. H.B. GUO, X.F. BI, S.K. GONG, H.B. XU, Microstructure investigation on gradient porous thermal barrier coatings prepared by EB–PVD, *Scripta Mater.*, **44** (2001) 683–687.
36. H.B. GUO, H.B. XU, S.K. GONG, Investigation on hot-corrosion of functionally gradient thermal barrier coatings, *J. Mater. Sci. Technol.*, **18**(1) (2002) 27–30.
37. H.B. GUO, S.K. GONG, C.G. ZHOU, H.B. XU, Investigation on hot-fatigue behaviors of gradient thermal barrier coatings by EB–PVD, *Surf. Coat. Technol.*, **148** (2001) 110–116.
38. A. AZZOPARDI, R. MEVREL, B. SAINT-RAMOND, *et al.*, Influence of aging on structure and thermal conductivity of Y–PSZ and Y–FSZ EB–PVD coatings, *Surf. Coat. Technol.*, **177–178** (2004) 131–139.
39. T. TOMIMATSU, S. ZHU, Y. KAGAWA, Effect of thermal exposure on stress distribution in TGO layer of EB–PVD TBC, *Acta Mater.*, **51** (2003) 2397–2405.
40. K. GREEN, T.M. POLLOCK, R.C. REED, *et al.*, *Superalloys 2004*, Warrendale, PA: The Minerals, Metals and Materials Society, 2004.
41. C.G. LEVI, Emerging materials and processes for thermal barrier systems, *Curr. Opin. Solid State Mater. Sci.*, **8** (2004) 77–91.
42. T.E. STRANGMAN, Thermal barrier coatings for turbine airfoils, *Thin Solid Films*, **127** (1985) 93–105.
43. M.J. STIGER, N.M. YANAR, M.G. TOPPINGS, F.S. PETTIT, G.H. MEIER, Thermal barrier coatings for the 21st century, *Z. Met.kd.*, **90** (1999) 1069–1078.
44. A.G. EVANS, D.R. MUMMS, J.W. HUTCHINSON, G.H. MEIER, F.S. PETTIT, G.H. MEIER, Mechanisms controlling the durability of thermal barrier coatings, *Prog. Mater. Sci.*, **46** (2001) 505–553.
45. P.K. WRIGHT, A.G. EVANS, Mechanisms governing the performance of thermal barrier coatings, *Curr. Opin. Solid State Mater. Sci.*, **4** (1999) 255–265.
46. S.M. MEIER, D.K. GUPTA, The evolution of thermal barrier coatings in gas turbine engine applications, *J. Eng. Gas Turbines Power*, **116** (1994) 250–257.
47. D. ZHU, R.A. MILLER, Determination of creep behavior of thermal barrier coatings under laser imposed temperature and stress gradients, *J. Mater. Res.*, **14** (1999) 146–161.
48. A.G. EVANS, M.Y. HE, J.W. HUTCHINSON, Mechanics-based scaling laws for the durability of thermal barrier coatings, *Prog. Mater. Sci.*, **46** (2001) 249–271.
49. S. DAS, S. DATTA, D. BASU, *et al.*, Thermal cyclic behavior of glass-ceramic bonded thermal barrier coating on nimonic alloy substrate, *Ceram. Int.*, **35** (2009) 2123–2129.
50. E.W. DOUGLAS, S. JOGENDER, M.A. ROBERT, Tailored microstructure of EB–PVD 8YSZ thermal barrier coatings with low thermal conductivity and high thermal reflectivity for turbine application, *Surf. Coat. Technol.*, **190** (2005) 132–149.
51. B.W. KEMPSHALL, Y.H. SOHN, S.K. JHA, *et al.*, A microstructural observation of near-failure thermal barrier coating: a study by photostimulated luminescence spec-

troscopy and transmission electron microscopy, *Thin Solid Films*, **466** (2004) 128–136.

52. E. MARK WALTER, B. EIGENMANN, The mechanical response of three EB–PVD thermal barrier coating microstructures, *Mater. Sci. Eng. A*, **282** (2000) 49–58.
53. J.A. RUUD, A. BARTZ, M.P. BOROM, C.A. JOHNSON, Strength degradation and failure mechanisms of electron-beam physical-vapor-deposited thermal barrier coatings, *J. Am. Ceram. Soc.*, **84**(7) (2001) 1545–1552.
54. I. SPITSBERG, K. MORE, Effect of thermally grown oxide (TGO) microstructure on the durability of TBCs with PtNiAl diffusion bond coats, *Mater. Sci. Eng. A*, **417** (2006) 322–333.
55. I.T. SPITSBERG, D.R. MUMM, A.G. EVANS, On the failure mechanisms of thermal barrier coatings with diffusion aluminide bond coatings, *Mater. Sci. Eng. A*, **394** (2005) 176–191.
56. G.M. KIM, N.M. YANAR, E.N. HEWITT, F.S. PETTIT, G.H. MEIER, The effect of the type of thermal exposure on the durability of thermal barrier coatings, *Scripta Mater.*, **46** (2002) 489–495.
57. T. XU, S. FAULHABER, C. MERCER, Observations and analyses of failure mechanisms in thermal barrier systems with two phase bond coats based on NiCoCrAlY, *Acta Mater.*, **52** (2004) 1439–1450.

7

Processing, microstructures and properties of thermal barrier coatings (TBCs) by plasma spraying (PS)

L. CHEN, China University of Petroleum, China

Abstract: A review is given of the process, microstructure and properties of plasma-sprayed thermal barrier coatings (PS–TBCs). The chapter first summarizes the main process methods, parameters and model of PS–TBCs before introducing the basic microstructural characteristics – splat, porosity and microcrack – and main properties – modulus, bonding strength and thermal conductivity – of PS–TBCs. The chapter goes on to discuss the relationship between the microstructure and mechanical properties of PS–TBCs and plasma-spraying parameters. In conclusion, some likely future trends in the field of PS–TBCs are described.

Key words: plasma spray, thermal barrier coatings, process methods, process parameters, process model, splat, porosity, microcrack, modulus, bonding strength, thermal conductivity.

7.1 Introduction

Plasma-sprayed thermal barrier coatings (PS–TBC) were invented in the 1960s to reduce the surface temperature of hot components by as much as 100–200 °C. Since then, they have been widely used in the fields of diesel, aerospace and land-based gas turbines and jet engines.[1,2] Typical PS–TBCs usually consist of three layers, namely top coat (partially stabilized zirconia), an overlay of bond coat (MCrAlY) and the substrate. Partially stabilized zirconia (PSZ) containing 6–8 wt% of Y_2O_3 coatings (YSZ) demonstrates excellent thermal shock resistance, low thermal conductivity and relatively high coefficient of thermal expansion (CTE). The bond coat (MCrAlY) provides oxidation resistance.

The thermomechanical properties of TBCs are the key features making these materials useful. Notably, the porous, defected, layered microstructure of the plasma-sprayed coatings can substantially reduce the already low intrinsic thermal conductivity of YSZ by as much as 60% (typical thermal conductivities of plasma-sprayed YSZ are about 1 W/mK compared to the bulk value of 2.5 W/mK).[3,4] Furthermore, the porosity, cracks and myriad arrays of interfaces also confer mechanical compliance on this system, enabling thermomechanical compatibility during cyclic thermal exposure to operating temperatures between room temperature and over 1000 °C. Finally, the YSZ system is also noted to have superior fracture toughness, adequate erosion resistance and thermochemical compatibility with the underlying bond coats and the dynamically evolving alumina scale at the interface.

By the 1990s, a second type of deposition process, electron beam physical vapor deposition (EB–PVD) TBC, had been successfully brought into commercial use. TBCs deposited by EB–PVD have been characterized as having a higher strain tolerance, smoother surface coating and better ability to control the microstructure of the deposited ceramics component to those deposited by plasma spraying. However, the PS–TBCs are currently the main preparation method in the fields of aerospace and power generation gas turbines because of their low cost, high production efficiency, adjustable and large range of coating thickness, easy control of ingredients and so on.[5] The microstructure of a typical plasma-sprayed thermal barrier is illustrated in Fig. 7.1 and compared with that produced by EB-PVD, shown in Fig. 7.2.

Evolution of the complex microstructure and properties of plasma-sprayed coatings is related to a number of processing variables related to device, feedstock, process, deposition, etc. Despite scientific and technical progress on many fronts, the extraordinary complexity of the process and material variables has stymied both scientific understanding of the process–structure–property relationships and industry/application-related attributes, such as coating design, property characterization and, perhaps most importantly, reliability. In this chapter, the process, microstructure and properties and their relationship to PS–TBC will be mainly discussed.

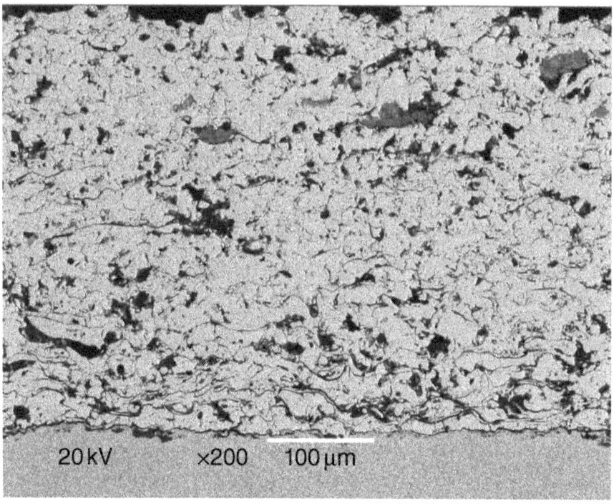

7.1 Cross-sectional micrograph of a current state-of-the-art plasma-sprayed zirconia–yttria/MCrAlX TBC showing a porous and microcracked ceramic layer over a relatively dense metallic bond coat.

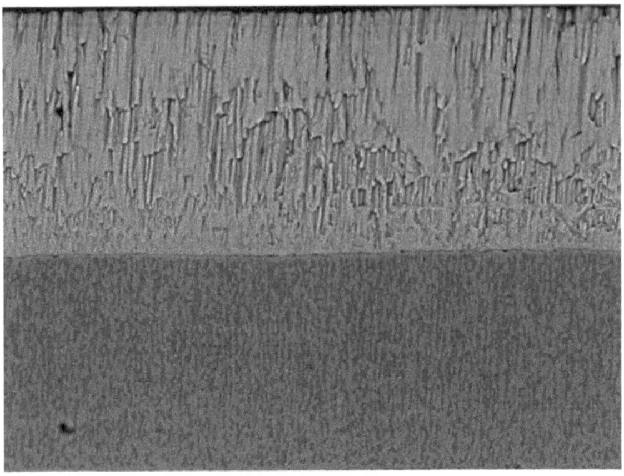

7.2 Cross-sectional photomicrograph of a current state-of-the-art EB–PVD YSZ/MCrAlX TBC showing a columnar ceramic over a relatively dense bond coat.

7.2 Processing of thermal barrier coatings (TBCs) by plasma spraying (PS)

Plasma-spray processing utilizes electrically generated plasma to heat and melt the feedstock material, a very intensive process requiring significant electrical power. It offers a method of depositing a feedstock material which is plasma-sprayed as a solid coating over an underlying target material. Deposits having a thickness from just a few micrometers up to several millimeters can be produced using a variety of feedstock materials, including metals and ceramics. In the plasma-spraying process, the material to be deposited, typically as a powder but sometimes as a liquid suspension, is introduced into the plasma jet emanating from a plasma torch. In the jet, where the temperature is about 10000 K, the material is melted and propelled towards a substrate. Then, the molten droplets flatten, rapidly solidify and form a deposit. Usually, the deposits remain adherent to the substrate as coatings. There are a large number of technological parameters influencing the interaction of the particles with the plasma jet and the substrate and therefore the deposit properties. These parameters include feedstock type, plasma gas composition and flow rate, energy input; torch offset distance, substrate cooling, etc. A typical plasma spray system consists of the following:

- spray torch (or spray gun) – the core device performing the melting and acceleration of the particles to be deposited;
- feeder – for supplying the powder or liquid to the torch;
- media supply – gases for generating the flame or plasma jet and carrying the powder, etc.

- robot – for manipulating the torch or the substrates to be coated;
- power supply – often for the torch only;
- control console(s) – either integrated or individual for all of the above.

7.2.1 Process method

The durability and failure mechanisms of PS–TBC are strongly dependent on the interfacial and microstructural features of the TBC system which are, in turn, linked to the method employed to deposit the individual layers. At present, there are several main processes for plasma spray preparation of TBC, such as atmospheric plasma spraying (APS), low-pressure plasma spraying (LPPS), laser cladding plus plasma spraying and solution precursor plasma spraying (SPPS), etc.

Atmospheric plasma spraying (APS) is a coating process carried out at normal pressure, which is usually applied to the ceramic coating on metallic structural components. In plasma-sprayed coating deposition, particles melt in the process of being deposited and when they impinge on the substrate, they flatten, undergo rapid solidification and form deposits known as splats, as shown schematically in Fig. 7.3.[6] The high energy associated with the plasma-sprayed torches, and hence the impinging particles, gives rise to a

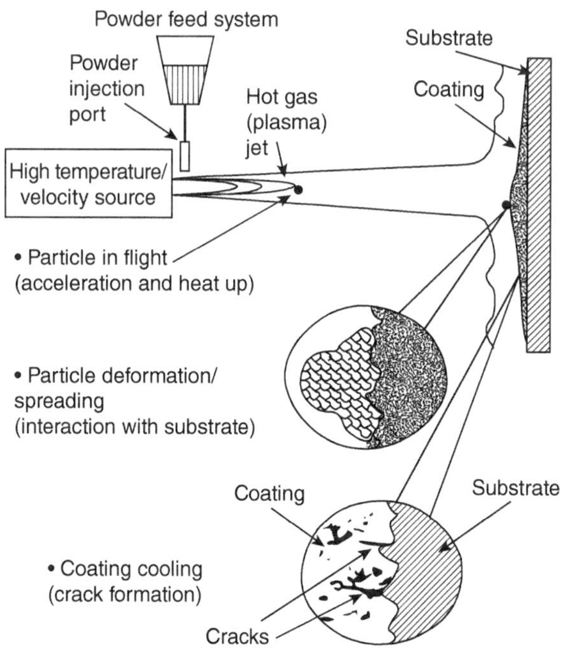

7.3 Schematic of air plasma spray process.[6]

rather rough and wavy interface which is fundamentally different from that produced by other methods such as EB–PVD. Furthermore, the presence of microstructural defects in the ceramics, i.e. porosities, generally introduces a long-term reliability problem as they tend to sinter after prolonged exposure at high temperatures leading to an increase in the coating elastic stiffness.

In order to overcome the limitations of APS, LPPS and vacuum plasma spray technologies are being developed. In the LPPS coating process, plasma spray gun, work piece and running machine are put in an airtight room with a low-vacuum environment. A low-pressure environment decreases the rate of oxidation of the feedstock powder as it melts in the plasma plume, and it also provides increased acceleration of the melted powder due to the decreased air resistance. The lower pressure also extends the length of the plasma arc, so that spray distances can be increased from 3–5 inches for normal plasma spray to 11–13 inches for LPPS. Compared with the ordinary APS coating, the LPPS coating is characterized by uniform and compact microstructure, low porosity, high purity, without oxide inclusion and smooth surface. The coating oxidation resistance performance is also distinctly enhanced.

At present, laser re-melting of the superficial layer of TBC prepared by PS may further remove the pores, inclusions, microcracks and other coating defects and enhance the coating quality and service life. Roughness results demonstrate a significant decrease after laser glazing.[7]

It is widely recognized that deposition of small, melted particles achieves a fine microstructure, which in turn leads to improvements in certain desirable mechanical properties such as strength and hardness. Unfortunately, it is generally impossible to feed powders finer than 5–10 μm due to the effects of surface forces on powder flow. Recently, the suspension plasma spray (SPS) process has been developed, partially to overcome this limitation. In this process, nano-sized particles are suspended in a liquid before being injected into the plasma plume, thus circumventing normal feeding problems.

Recently, a relatively new spraying technique, solution precursor plasma spray (SPPS), has been developed to deposit highly durable thermal barrier coatings. Similar to the existing plasma deposition methods, SPPS is based on the injection of liquid chemical precursors into a plasma jet and forming of a ceramic by pyrolysis in the plasma jet. In this process, an aqueous solution of the constituent metal salts (Zr and Y) is atomized and injected into a direct current (DC) arc plasma jet. The ceramic is formed by pyrolysis in the plasma jet before reaching the target material or, in some cases, after reaching the surface. This spray method is identical to APS deposition with the substitution of a solution atomizer for the solid powder feeder. The atomization characteristics can be varied, but it involves initial droplet sizes between 20 and 50 μm at various velocities up to 100 m/s. Plasma torch operating power and gas flow rates are in the same range as used for depositing APS coatings.[8]

However, the primary difference attribute between SPPS and other methods is that the feedstock material is in the form of a liquid solution rather than a powder or a powder suspension. As a result, the SPPS process is able to deliver fine splats without the difficulties and limitations of both fabricating and feeding fine solid powders. Instead, SPPS has the challenges of formulating high-molarity solutions and engineering their chemical and physical properties.

7.2.2 Process parameters

The properties of TBC systems prepared by plasma-sprayed technology are dependent not only on the material being used and its microstructure but also on the process method and its parameters. There are many process parameters that influence plasma spray coating microstructure and properties. These parameters (summarized in Table 7.1.[9]) can be categorized into four groups: (i) plasma characteristics, such as plasma power, plasma gas flow rate, and plasma gas species, and geometry of the electrodes; (ii) powder feed parameters, such as size and shape of the powder, flow rate of the powder and carrier gas and powder injection site; (iii) plasma jet parameters, such as jet velocity and temperature, particle velocity and temperature, particle trajectories; (iv) particles at impact, such as particle velocity

Table 7.1 Processing parameters in plasma spraying[9]

Plasma characteristics	Power supply Plasma gas Mass flow rate plasma gas Cooling Mass flow rate cooling gas Nozzle geometry
Powder feed	Powder fraction and shape Thermal properties of powder material Carrier gas flow rate Injection geometry
Plasma jet	Jet velocity and temperature Particle velocity and temperature Particle trajectories
Particle impact	Particle velocity at impact Particle impact angle Molten state of particle at impact Substrate type Substrate temperature and roughness

at impact, particle impact angle, molten state of particle at impact, substrate type, roughness and temperature. Those parameters are not independent and they interact with one another.

Research has indicated that power input, substrate temperature, coating thickness and spray distance are closely correlated with the coating microstructure. The coating porosity may be reduced through enhancing power input, pre-heating substrate, reducing spray coating distance and adjusting appropriate spray angle. In order to reduce the hot corrosion attack of TBCs by oxidizing environments, some studies have tried to seal open pores on the TBC surface with organic materials, while others have used hot isostatic pressing (HIP) and laser remelting to post-treat the TBCs. These methods attempted to create a dense layer on the coating surface or a dense coating to prevent the oxidizing environment entering into the coating. On the other hand, studies have shown that inserting a graded coating between the ceramic and bond coat or changing the chemical composition of the ceramic coating can reduce the thermal stresses building up within the coating.

The average particle temperature and velocity distribution determine coating bond strength, porosity and coating inclusions such as oxides. It is generally accepted that higher velocities result in higher particle/droplet impact energy. This high kinetic energy leads to higher degree of particle deformation, producing better coating bonding strength and higher coating density. The temperature, velocity and environmental interactions of the particles are related to the particular jet velocity, temperature, resultant particle dwell time, all of which determine the subsequent coating microstructure and properties.

The dwell time of particles is controlled by gas velocity and powder particle characteristics. Gas velocity in turn is determined by the total gas flow through the nozzle and the gas characteristics. Particle velocity is then a function of jet velocity coupled with particle characteristics – that is, size, morphology and mass. The temperature of a given particle is thus a function of enthalpy, velocity, trajectory and its own physical and thermal properties.

The degree of melting particles in the plasma jet depends on their trajectories. The temperature and velocity vary considerably in the plasma jet, the most significant factor being the difference in temperature between the hot core of the jet and its relatively cold surrounding environment. The temperature drop from the jet core to the boundary is typically several thousand degrees over a few millimeters. This drop in temperature causes a corresponding drop in gas velocity and an increase in viscosity toward the outer boundaries. Temperature and velocity also decay as the jet exits the nozzle, so there are both radial and axial gradients of gas temperature, gas velocity and gas viscosity. Distributions of particles traveling in these thermal jets are accelerated by the gas-velocity distributions through a distribution of temperature. Therefore, velocities and temperature of the particles in the plasma jet are dependent on the trajectories. Those particles injected into the core of the jet will melt and form splats to impact the substrate, while those with trajectories along the periphery of the jet will not melt or only partially melt. The unmelted or partially melted particles may be incorporated into the deposit causing lower deposit densities.

7.2.3 Process modeling

Due to the complexity of the phenomena, the process remains highly empirical, even after the practical use of plasma spraying for many years. The production of coatings is still based on trial and error approaches, depending on the experience and the instinctive feelings of the researchers. The irreproducibility of a high-quality plasma coating therefore continues to be a major problem for the rapid expansion of this technology. The main reason for this is the lack of understanding of the underlying physical processes during plasma spraying.

In order to create a higher reliability of coating production, it is necessary to gain a scientific understanding of different processes, namely the fluid dynamics of the plasma jet, the plasma–particle interactions and the particle impacts onto the surface. During the last years, many efforts by different authors have been made to gain knowledge of the interdependencies of plasma-spraying parameters and coating characteristics. However, the process is rather complex and only some of the phenomena involved were described sufficiently in the past. Reasonable tools to close this lack of understanding are simulation models as well as measuring and diagnostic techniques. Some comprehensive models that involve major physical mechanisms for plasma spray, i.e. transport phenomena of plasma flame, particle heating, acceleration, evaporation and oxidation, have been developed for predicting coating characteristics independent of selected spray parameters.[9]

7.2.4 Plasma jet modeling

A range of models has been proposed to calculate plasma jet temperature and velocity distributions. These models start with the results of the calculations described in the previous section or calculate temperature and velocity distributions at the nozzle exit consistent with the plasma gas flow rate and enthalpy. The codes consist of momentum and thermal energy equations for the multicomponent fluid mixture, continuity equations for each component of the mixture, taking into account the mixing with the surrounding atmosphere but usually neglecting chemical reactions, and state relations for an ideal gas mixture with temperature-dependent specific heat, enthalpy and transport properties.[10–12]

The large velocity difference between the plasma jet and its surrounding atmosphere results in a large-scale engulfment of the outside gas which has been corroborated by shadowgraph, CARS spectroscopy measurements and probe sampling of the plasma jet. A multiphase model for turbulent flow could reproduce these effects. Such a model has been developed for an argon plasma jet flowing into an argon stagnant environment. The plasma jet is considered as a two-fluid mixture consisting of hot, out-moving fragments and cold, in-moving fragments. The results obtained for the axial distributions of temperature and velocity are in good agreement with measurements.[12]

7.2.5 Particle in flight modeling

The modeling of particle acceleration and heating in the flow uses the conventional equations for droplet motion in flows and heat transfer. However, some effects specific to the thermal plasma environment must be taken into account. The main effects are due to the high temperature gradient prevailing in the boundary layer surrounding the particles, strongly varying plasma properties, non-continuum conditions, thermophoresis, turbulent dispersion, evaporation and possible chemical reactions on the particle surface. The models based on the heat conduction equation are generally one-dimensional and allow the tracking of the melting, evaporation and possible resolidification front in a particle. However, recent work has shown that the large difference in velocity between the flow and particles can induce convective movements inside the particles.[12]

In the modeling of the plasma spray process, the injection conditions of the particles in the flow govern the particle trajectory distribution and therefore their treatment. Thus, these conditions must be as realistic as possible and involve the particle size, injection velocity and direction distribution at the injector exit. The description of powder injection that implicates a stochastic aspect is still a challenging problem.[12]

7.2.6 Splat formation modeling

With regard to splat formation, analytical models have been proposed for droplets impacting at 90° onto a smooth surface. They express the flattening degree as a function of dimensionless groups characteristic of the impact and spreading process: the Reynolds number that quantifies the viscous dissipation of the inertia forces, the Weber number that expresses the conversion of the kinetic energy into surface energy and a heat transfer parameter, the Peclet number, used to characterize the solidification rate. The most popular analytical model provided by Madejski attempted to include viscous force, surface tension and crystallization kinetics. It can be simplified by assuming that the spreading is complete before solidification becomes significant and that the surface tension effects can be neglected. Surface tension effects are generally acknowledged to be minor under thermal spray conditions, except perhaps for droplet size <10 μm and at the end of the flattening process. Similar expressions with different constants have been established from analysis analogous to that of Madejski or from the fit of the results of numerical modeling.[12]

Over the last years, various numerical studies have dealt with the impact of a thermal-sprayed particle on a smooth surface. The majority assume a two-dimensional geometry and the most advanced solve the flow equations taking into account convective, viscous and surface tension effects. They track the material deformation during impact and involve the solidification process and heat interaction of the droplet with the underlying surface through an interfacial thermal contact. Such models enable the evaluation of the effect of particle parameters at impact on splat formation.[12,13]

7.3 Microstructures of TBCs processed by PS

At present, the TBC topcoat is typically sprayed either by APS or LPPS. The plasma spraying process results in unique TBC microstructures which are greatly different from those of bulk materials and, consequently, many coating properties differ hugely from those of the corresponding bulk materials. The inherent complexity of the plasma spray process and the TBC microstructures represents a great challenge to the establishment of process–microstructure–property relations. A fundamental understanding of the microstructure properties is essential for process optimization and microstructural control.

The coating microstructures can be summarized as complex porous lamella structures consisting of a broad range of features, as shown Fig. 7.4. The pancake-shaped splat, about 1–50 μm in thickness and 10–50 μm in diameter, is the basic structural unit of a coating.[14] A few layers of splats form a lamella which extends horizontally inside the coating, and distinct boundaries exist between lamellae. Inside a lamella, columnar grain structures which grow through splats can be observed. This indicates that the gradient direction of heat conduction is perpendicular to the spraying surface. An integral coating is made up of numerous lamellae.

Besides the layered structure, another prominent microstructural feature of a thermal spray coating is the 'random' distribution of a huge number of ubiquitous pores and voids of various sizes and shapes throughout the coating, which makes the coating appear sponge-like under scanning electron microcopy (SEM). To be precise, the pore distribution is actually

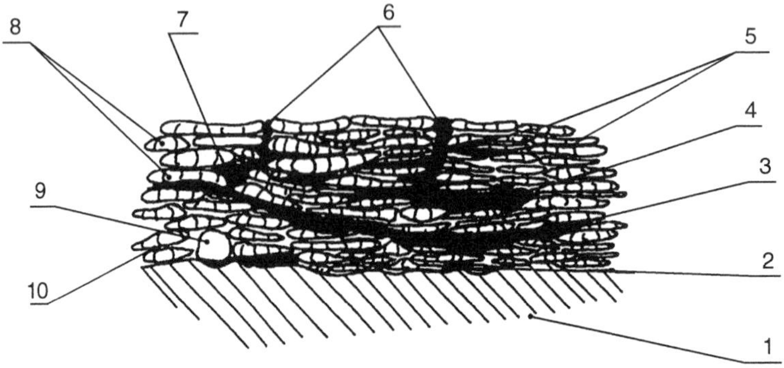

7.4 Schematics of typical plasma-sprayed coating: 1. Substrate; 2. Adherence defect; 3. Coating stratification (coherence defect) due to torch passes; 4. Coating stratification (coherence defect) due to relaxation of the residual stresses; 5. Cracks inside lamellae; 6. Vertical cracking resulting from residual stresses and open pores; 7. Closed pores; 8. Fully molten particles; 9. Unmelted particles; 10. Bad contact between lamellae.[14]

pseudo-random because the pores are more likely to be oriented close to the direction parallel or perpendicular to the deposition surface than to be at a much slanted angle. A truly random pore distribution will result in an isotropic coating with reduced stiffness.

Besides round, oval and elliptical pores, numerous long sharp crack-like horizontal delamination and vertical microcracks are visible. The delamination may lie either along the lamella boundaries (inter-lamella) or on the boundaries of splats in a lamella (infra-lamella). The vertical microcracks are caused by weak adhesion between lamellae or between splats. They usually initiate from the low-adhesive grain boundaries or some initial defects in the coatings and are believed to be caused by quenching and subsequent stress relaxation during the coating cool-down process when high biaxial tensile residual stresses are induced in the deposition surface.

This whole gamut of features encompasses a range of macro-, micro- and even nano-length scales. In this section, microstructural characterization across length scales of PS–TBCs will be discussed.

7.3.1 Splat formation

The basic characteristic of a PS–TBCs lies in its layered nature. As such, it is quite different from a bulk material. Since the deposit is a stack of layers, there is an inherent porosity associated with the structure. Also, this type of stacked structure introduces anisotropy in terms of microstructure and properties. The interfaces between layers play an important role in modifying any property along the cross-sectional direction. Within a single spraying process, the differences in parameters generate a variety of particle temperature–velocity histories which in turn introduce a great amount of inhomogeneity within a coating. This layered coating structure is thus built up of a splat structure in which it is important to study an individual splat which is the building block.

Splats represent unit cells of plasma-sprayed coatings and their morphological and microstructural characteristics affect the properties of the coating. Many properties of the sprayed coatings, especially the adhesive strength to the substrate, depend remarkably on the splat behavior of the individual particles; hence splat morphologies become an important aspect. Splats are formed upon impact and flattening of molten droplets. The droplet size and impact velocity play important roles in the kinetics of the flattening and splat morphologies obtained. Flattening and cooling begin immediately after, and in some cases simultaneously with, droplet impingement. This flattening process includes spreading and solidification.

When a particle hits the substrate, the sudden deceleration causes a pressure build-up at the contact area. The high pressure forces melted material to jet laterally away from the point of impact. This causes fragmentation of splats and generates different morphologies. The liquid spreads outward from the point of impact and forms a splat. A very low kinetic energy means poor contact with the substrate and causes the splat to bounce off whereas a very high kinetic energy will cause it to splash. The arresting of spreading results from the conversion of particle kinetic energy into work of viscous deformation and surface energy, and it may also be effected through solidification. The spreading kinetics of the impacting droplet governs the splat shape and thickness.

Splat formation is also subject to the surface topography, wettability and reactivity of the substrate. Wetting affects the contact thermal resistance at the splat–substrate interface.[12] Under the same spray conditions, porosity increases if wetting is poor.[15] If the surface of the substrate layer is rough, the flattening is also affected and the splats formed are thicker. The ability of the sprayed material to fill existing voids affects the quality of contact between the splat and the underlying layer, thereby controlling local cooling rate and overall heat transfer. Solidification and mechanical constraints can interfere with the flattening process.[15]

The study by Fukumoto *et al.* described the mechanisms of formation of fragmented splats and disk splats on room-temperature and high-temperature substrates.[16] For a splash splat at room temperature, the bottom part of the splat solidifies rapidly after impacting onto the substrate and a porous microstructure is formed. This microstructure decreases the ability to transfer heat and acts as a heat insulator. Then the inside of the splat solidifies slowly and the viscosity increase is slow: therefore, the droplet flattens rapidly with low viscosity. According to Raleigh–Taylor instability, the periphery of the splat becomes unstable and a splash splat forms. For a disk splat at high temperature, the contact condition is good owing to the good wettability and because the formation of the porous solidification layer is suppressed. Following this, the inside of the splat solidifies quickly. Therefore the viscosity of the droplet increases and most of the kinetic energy effectively is transferred to the viscous energy. The droplet thus flattens at lower velocity, resulting in the formation of a disk splat.

Splat morphology can be quantified in terms of properties such as splat flattening ratio and splat fragmentation index. Fragmentation degree in a way defines the degree of splashing and is given by $P^2/4\pi A$ where P is the perimeter of the splat and A is the splat area. Flattening degree describes how well the splat has undergone spreading and this is given by a ratio of diameter D of the splat over diameter d of the droplet.

Another important aspect of a single splat is the residual stresses that are generated upon its deposition and rapid solidification. Quenching of the impinging splat and restricted contraction lead to tensile stresses. Plasticity upon impact causes compressive stresses. Thermal mismatch between the coating and the substrate can be either tensile or compressive. Relaxation of these tensile stresses leads to microcracking of splats, a phenomenon more characteristic of ceramic splats than metallic splats.

An assemblage of millions of such splats consolidates to form the deposit. Therefore the coating structure is dependent on the details of solidification and resulting microstructure and morphology of splats. The physical phenomena involved in this process include heat and mass transfer, melt undercooling, selective nucleation of stable and metastable phases, non-equilibrium kinetics of crystalline growth, solidification, thermal resistance at the splats/substrate and splat/splat interfaces and substrate melting and resolidification.

In most cases, heterogeneous nucleation takes place at the contact between the molten splat and the colder substrate. The crystalline nuclei then grow into the undercooled melt in all directions and quickly form a solid layer covering the substrate surface followed by columnar growth normal to the substrate surface. A large temperature gradient exists in the solidified parts because of the high rate of heat transfer from the splat to the substrate. This rate can vary with substrate surface roughness, substrate temperature and substrate thermal properties, and also varying thickness regions of the substrate. Cooling rates in excess of 10^6 °C/sec result in solidification front velocities greater than 10 cm/sec with interface controlled heat transfer. This results in a stable solidification front leading to columnar solidification.[17] Another study verifies that a stable planar interface does exist when the interfacial heat transfer coefficient across the melt and the substrate interface is large.[18]

It has been shown before that spreading of a splat generates a core and a rim region. In the core region, where the particle droplet first comes in contact with the substrate, heat is extracted through the substrate and the solid liquid interface moves essentially upwards resulting in a columnar grain structure. In the case of the peripheral areas, the heat is extracted through the core region leading to radially elongated grains.[19]

Chraska and King studied single splats of YSZ using transmission electron microscopy (TEM), and a rapid solidification model was proposed for the observed narrow columnar grain structure: the heterogeneous nucleation and epitaxal growth mechanisms involved in the formation of first two splat layers are described for the YSZ ceramic coating. Nucleation in the second splat was seen to start at a significantly smaller undercooling than in the first splat.[20,21]

7.3.2 Porosity

In addition to a 'brick-wall' structure, the thermal-sprayed coating also exhibits defect features such as globular pores, interlamellar pores, cracks, etc. Incomplete contact between successive splats and the presence of unmelted particles lead to the formation of voids. Rapid solidification and high-velocity impact of the molten particles give rise to inter-lamellar pores, while cracks are formed due to solidification thermal stresses and relaxation of tensile quenching stresses. Knowledge of the extent of these imperfections is critical since they may influence a wide range of coating properties and behaviors in service. Globular pores introduce porosity compliance which might be the origin of failure. Inter-lamellar pores enhance the thermal barrier effect but are also a source of delamination. The vertical microcracks in the splat relieve residual stresses, which are the first to sinter during thermal cycling. The sintering effects significantly change the pore architecture, with healing of intra-splat microcracks and enhanced inter-splat bonding. These are accompanied by pronounced grain growth. These changes lead to significant increases in both in-plane stiffness and through-thickness thermal conductivity. The latter is sensitive to increased inter-splat area, while the former is affected by both microcrack healing and improved inter-splat bonding.[22] An illustration of how both inter-splat voids and intra-splat microcracks can coarsen and partially heal via sintering phenomena at elevated temperature is provided by the pair of micrographs shown in Fig. 7.5. While the temperature used for the heat treatment to which the coating in Fig. 7.5b was subjected was relatively high (1400 °C), the period concerned (10 h) was short compared to those common under service conditions.

Cipitria *et al.* have used a model to predict changes in pore architecture during sintering, and also the build-up and relaxation of global elastic

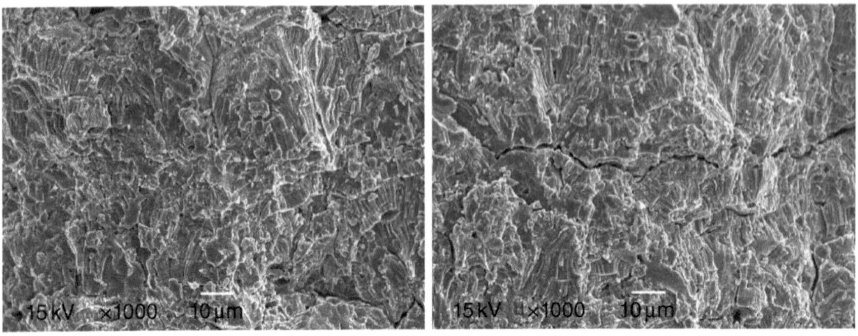

7.5 SEM micrographs of fracture surface of a YSZ coating produced by APS; (a) as-sprayed; (b) after heat treatment for 10 h at 1400 °C.

strains and stresses. While being stored, elastic strain energy levels in these coatings arising from (in-plane) inhibition of sintering shrinkage tend to remain low, while those from differential thermal expansion can be high immediately after a temperature change, particularly after being cooled to a temperature at which stress relaxation rates are slow. It has been shown that such strain energies, when converted to strain energy release rates per unit area for typical coating thicknesses, could exceed the fracture energy of the interface, and thus make spallation energetically favorable, particularly if the coating stiffness has been raised by sintering.[23]

The amount of porosity, pore shape and distribution in TBC play an important role in its performance. Porosity is engineered into TBC to provide strain tolerance for the coating and to optimize thermal cycle life. If a TBC does not have adequate porosity then it may spall off because it cannot accommodate strains induced upon thermal cycling or by residue stress. If a TBC has too much porosity (by which the thermal resistivity is increased, but the mechanical bonding with the bond coat is decreased), then it has the potential to spall off due to oxidation of the bond coat. The pore shape (such as circular, oval, cylinder or cubic) and distribution (elongated transversely or longitudinally) also strongly affect the properties and performance of TBC. For example, if elongated pores are transversely distributed in TBC, the fracture energy is higher, thus TBC has higher resistance to spallation. However, TBCs with this kind of pore shape and distribution have higher thermal conductivity. Vice versa, if elongated pores are longitudinally distributed in the top layer, the TBC would be easier to delaminate due to lower fracture toughness, but may decrease the damage of the bond coat incurred by oxidation or hot corrosion and possess a lower thermal conductivity.[24]

Many efforts have been made to classify the origin of different pores and estimate the pore size and shape distribution quantitatively.[25] The porosity rate of a coating is usually from 1 to 20%, varying largely with different coating methods. The porosity of APS ceramic coatings is, in general, between 3 and 20%. Generally, industry-standard TBC with 10–15% porosity is accepted. The existence of so many pores greatly reduces the elastic modulus and thermal conductivity of the coating. For example, the elastic modulus of a ceramic coating is only about one fourth or less of that of bulk ceramic, i.e. the ceramic coating is much more compliant.[26] Besides the reduced stiffness, many coatings also exhibit a certain degree of anisotropy or, to be precise, transverse isotropy. Five independent elastic constants exist for a transversely isotropic coating. The anisotropy is partially induced by the afore-mentioned pseudo-random distribution. Reduction of the coating stiffness in the direction perpendicular to the deposition surface is usually more significant, i.e. the effective modulus measured in that direction is often smaller than the effective modulus measured in the direction parallel to the deposition surface.

Microstructural characterization of thermal spray coatings involves quantitative measurements of geometrical features such as porosity (the form of voids, cracks and other defects) and analysis of material aspects in the coatings such as splat structure, interfaces, phases, etc. Various methods are employed for measurement of porosity, a prevalent feature in the microstructure.[27] Lavigne *et al.* and others have presented a reliable and reproducible procedure for characterization of porosity in PS–TBC using SEM and image acquisition (IA).[28] They have enlisted the methods used for metallographic preparation and SEM IA. Certain image analysis techniques have been employed to divide the net porosity into classes, such as globular pores and crack network.

7.3.3 Crack and fragment

As a key microstructure feature of PS–TBC, microcracks determine the performance and service lifetimes of the coatings by influencing their thermal conductivity and mechanical properties. In this section, crack formation in plasma-sprayed microstructures and its effects on mechanical and thermal properties are discussed.

Thin bonded layers on rigid surfaces show the familiar mud cracking pattern of meandering cracks intersecting to fragment the layer into independent pieces under biaxial stress. It is often seen on ceramic glazes, dry earth and tectonic plates. Three processes are involved in the formation of these features: vertical through-thickness cracking of the film, lateral elongation of this vertical crack and cracking parallel to the interface within the film, interface or substrate causing delamination.[29]

The residual stresses induced in the fabrication process of PS–TBCs are associated with crack and fragment. In addition, they will strongly influence the coating adhesion strength, resistance to thermal shock, life under thermal cycling and erosion resistance.[30] Residual stresses are believed by many researchers to affect TBC integrity in high-temperature applications.[31] Some experiments conducted show that the thermal fatigue lifetime of some TBCs is directly related to the magnitude of the residual stresses. As another example, the compressive residual stresses were sometimes thought to be the cause of coating spallation from the ceramic bond interface. The mechanisms which cause residual stresses in thermal spray coatings have been discussed extensively.[32] Generally, three distinct processes have been singled out as the most conspicuous mechanisms active in inducing residual stresses in the coatings. These are (i) phase transformation of ceramic in the cooling process; (ii) primary cooling of the hotly melted spray particles from their impinging temperature to the elevated substrate temperature (deposition temperature), namely, the solidification process of splats; (iii) secondary cooling of the coating system, including substrate, from the elevated deposition temperature to room temperature, i.e. the thermal mismatch effect.

The first mechanism concerns phase transformations. For example, unstabilized pure zirconia which cools down from an impinging temperature well above 2500 °C to room temperature will transform from tetragonal phase to monoclinic phase. This phase transformation will bring about an internal volume increase of about 8% inside the zirconia and vice versa. Consequently, very large residual stresses will appear. However, in real applications, pure zirconia is rarely used. Instead, PSZ, for example, ZrO_2–6~8 wt% Y_2O_3 where yttria is added as stabilizer, is used.[33] The added stabilizer can minimize the phase transformation effect and keep the highly disruptive volume change at bay. Moreover, a small amount of phase transformation in the coating is thought to actually 'toughen' the coating and render major cracks less likely to form. This phenomenon is called the 'transformation toughening effect'.[34] The benefit of transformation toughening is to render the PSZ coating superior to the fully stabilized zirconia coating, but there have been few theoretical analyses to explain the physics of such a phenomenon.

The second mechanism is the solidification process of the molten coating powder on top of its substrate after exiting from the nozzle of the spray gun. This process is usually called quenching or primary cooling. The temperature (T) of the molten powder is usually between 2500 and 3000 °C. Since the cooling rate of the spraying particle is very high, typically about 10^6 °C/sec, it takes only about 1 μs for a very thin splat (a few micrometers in thickness) to contract and condense on the underlying material which is extremely thick relative to the splat. Therefore, the underlying substrate can be regarded as rigid compared to the impinging and condensing thin layer of splat. Consequently, the ensuing residual stresses (quenching stress) in the condensing splats due to solidification are tensile stresses. The magnitude of the tensile residual stress from the above scenario can be estimated from $\sigma = E * \alpha (T - T_{dep})$, where E and α are the elastic modulus and the coefficient of thermal expansion (CTE) of the splat, respectively, and T_{dep} is the deposition temperature (200–400 °C). According to such a formula, the stress can reach as high as 200 MPa.

The above estimation is purely theoretical. In practice, residual stresses due to solidification are not so significant since other ongoing processes reduce the tensile residual stress. First, the tensile stress inside coatings will inevitably induce microcracking and other structural degradation which in turn can reduce the tension inside the coatings. This process is similar to stress relaxation and the stress relaxation of a NiCrAIY coating has been studied.[31] On the other hand, the continual deposition of material on top of the previously deposited laminar coating will induce compressive stress underneath the current layer of deposition. As the coating thickness builds up, much of the tensile residual stress induced during solidification might be compensated by the compressive stress. These facts imply that the actual residual stress due to primary cooling is much smaller than that predicted theoretically. Some authors have even argued that solidification is quite insignificant in terms of generating residual stresses in coatings.[35]

The third mechanism is secondary cooling where the coated substrate cools down from the deposition temperature to room temperature after thermal spraying is accomplished. The mismatch of the CTEs of different materials is the major reason that large residual stresses ensue after cooling. In fact, the thermal mismatch effect can be so great that it may constitute the major source of the residual stress in the coatings. Consider a common ceramic/bond coat/substrate system. The residual stress due to thermal mismatch will be largely compressive in the ceramic coating under ambient conditions because of the relatively low CTE possessed by the ceramic coating. Such compressive stress is in contrast with the tensile residual stress generated by the solidification/quenching process. It should be noted that the residual stress in the bond coat due to secondary cooling is also tensile. Secondary cooling is the only mechanism of residual stress generation.

Residual stresses develop in TBCs during processing and thermal cycling as a result of the thermal expansion mismatch between the thermal grown oxide (TGO) and the underlying metal and because of bond coat oxidation. These stresses provide the driving force for the failure of TBCs.

The residual stresses in TBCs can be distinguished according to their origin: stresses developing during oxidation at high temperature are called growth stresses; stresses which develop during cooling due to the difference of the thermal expansion coefficient of the metallic substrate and the oxide are called mismatch stresses. The ambient stress status in each layer of a TBC will change with the variation of TGO thickness, TGO/bond coat interface morphology and material properties.[36,37]

For APS TBCs, the coating usually fails within the ceramic top coat and near the YSZ/TGO interface.[38] Therefore, the stress state, especially the stress normal to the YSZ/TGO interface, in the APS YSZ coating is of great interest in the understanding of coating failure mechanisms and life prediction modeling. The stress distribution within the YSZ depends on TGO thickness, material properties of YSZ and YSZ/TGO/bond coat interfaces morphologies.[39–43] When the TGO is thin, the YSZ on top of the peaks of the wavy interface is in tension, whilst the YSZ above the trough is in compression. With the thickening of TGO, the stress of YSZ above the peaks transforms from tension to compression, while the stress status of YSZ on top of the trough (about the same altitude of the peak) transfers from compression to tension. It was found that the maximum tensile stress occurred around the shoulder of the peak.[43,50]

The stress in the TGO has an important influence on TBC failure. There are two main sources of stress: one from the thermal expansion misfit upon cooling and the other from TGO growth. Ambient temperature measured by the photo-stimulated laser piezo-spectroscopic technique indicates that the TGO is in compression (3–6 GPa).[44–47] The stress in TGO may be alleviated by TGO creep or TGO/bond coat interface rumpling and redistributed in the vicinity of imperfections.[48,49]

The stress normal to the YSZ/TGO and TGO/bond coat interfaces is related to TGO thickness and interface morphology. It has been shown that for a thin TGO layer, the stress at the YSZ/TGO interface is tensile at convex locations and compressive at concave locations.[50] With the thickening of TGO, the normal stress at the YSZ/TGO interface is usually compressive. For the TGO/bond coat interface, the convex locations are always in tension while the concave places are in compression.[50]

Residual stresses can be measured by monitoring the bending curvatures of free-standing as-sprayed coatings.[51] A more accurate method is X-ray diffraction. However, conventional X-ray diffraction is only able to measure the residual stresses at a depth within 10 μm of free surface. Any measurement of deep residual stresses requires chemical or mechanical removal of layers.

Bond strength and thermal fatigue life are strongly related to the thermal mismatch strains within the TBC and between the TBC and the substrate materials. One way of reducing the thermal mismatch strain is to form functionally graded materials (FGMs). The gradient layer can be continuous on a microscopic level. FGM coatings can be laminates comprising gradients of metals, ceramics or just a variation in porosity/density, etc. They have superior properties to duplex coatings. The bond strengths for FGM and duplex coatings are about 18 MPa and 9 MPa, respectively. The thermal cycling resistance of the FGM coating is six times that of the duplex coating.[52]

7.4 Properties of TBCs processed by PS

There are a number of mechanisms identified for the failure of TBC in service. Coating-delamination failure is closely related to the thermal stresses (including the internal stresses of the top coat, stresses inside TGO and mismatch in the coefficients of thermal expansion between top coat and bond coat) in the coating systems as well as the oxidation of the bond coat and substrate. Coating failure may be further accelerated by shrinkage-cracking and elastic-modulus increase from ceramic sintering and creep with an associated density increase at high temperatures.

Ceramic coating sintering and creep at high temperatures are among the most important issues for the development of advanced TBC. Sintering and creep not only can result in considerable coating thermal conductivity and elastic-modulus increase that will lead to reduced thermal insulation and increased thermal stresses, but they can also cause shrinkage-cracking, eventually leading to spallation of the coating. The densification and deformation of porous, microcracked ceramic TBC due to sintering and creep involve stress and thermally-activated diffusion and mechanical compaction processes. Faster shrinkage rates are observed initially, followed by slower but relatively constant rates for longer sintering times. Higher temperatures and compressive stresses will lead to a higher sintering/creep rate.

7.4.1 Modulus

The sintering densification process also results in an increase in the elastic modulus of the porous ceramic coating at high temperatures. Depending upon the actual temperatures and stresses experienced by the coating, a significant elastic modulus gradient will develop and evolve with time under high heat flux conditions. The surface of exposed TBC ceramic coating showed very fast modulus increase rate while much longer time is required for the inner layers of the ceramic coating to attain the maximum modulus value. The coating modulus increase results from the coating sintering and densification.

Measuring the properties as well as internal stresses of a thermal spray coating always represents a challenge since the coating itself is usually very thin and attached to a substrate. Any effort to separate the coating from its substrate may alter the internal stresses and even some properties of the coating. Only a brief introduction to various experimental methods for modulus and yield strength is presented here.

There are several methods for measuring Young's modulus of plasma-sprayed TBGs. These include indentation, four-point bending and ultrasonic techniques. Subsized and nano-indentation techniques are often used to measure mechanical properties at the surface of thin films. The nano-indentation technique was originally developed to measure Young's modulus and hardness of materials by interpreting the load–penetration response of the sample. It has subsequently been widely used for measuring properties of bulk materials and thin films.

In combination with analytical and/or numerical models, a load–displacement curve formed by the load and penetration depth in a sub-sized indentation test can be used to estimate mechanical properties. As the area tested by nano-indentation only samples a very small region of material with a typical size of less than one grain, mechanical properties determined by this technique are in general in a probability distribution, affected by a variety of factors such as defects. To describe the consistency of materials, a Weibull distribution based on the weakest-link principle is usually adopted to interpret the nano-indentation results.

Using micro-/nano-hardness indentation to determine mechanical properties of materials such as modulus and yield strength is largely an empirical technique based on some well-derived data from known materials. In most cases, the correlation is realized in a form of $H \sim C * \sigma$, where H is some micro-/nano-hardness measure, C is a constant or function of material property and σ is the material property of interest. When applied to an 'unknown' material, it becomes invalid. Moreover, as the indentation size is typically much less than the material grain size, the experimental results are sensitive to the microstructural characteristics in the vicinity of the tested area, which may cause a scattering of test results.

Although they are widely used to reveal materials properties, subsized and nano-indentation techniques are not suitable for characterizing the full range of desired mechanical properties of bond coats as a function of temperature. The Vickers microindentation technique has been used to estimate strengths of different bond coats at room temperature.

Four-point bending techniques have been used extensively in the characterization of various mechanical properties of metallic and ceramic materials, including Young's modulus, strength, fracture toughness, stress relaxation interfacial fracture resistance and fracture strain. Figure 7.6 is a schematic of four-point bend testers. Load and displacement of the cross-head are recorded for measurement of properties. A typical load–displacement curve obtained in a four-point bending test is shown in Fig. 7.7. Young's modulus and yield stress were derived from the following equations:

$$E = 2D^3 \, \Delta P/(3 \, I \, \Delta y) \qquad [7.1]$$

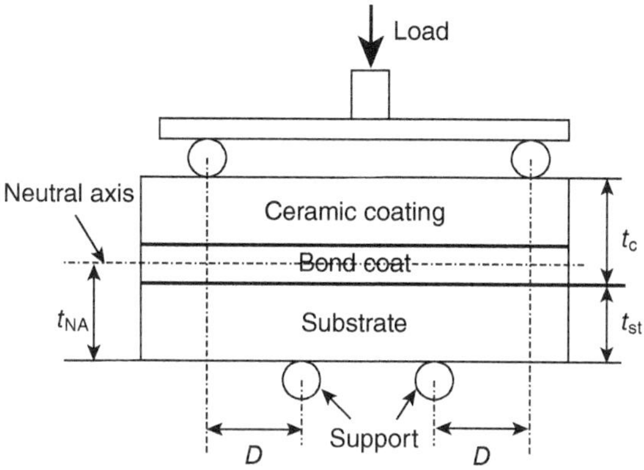

7.6 Schematic of four-point bend testers.

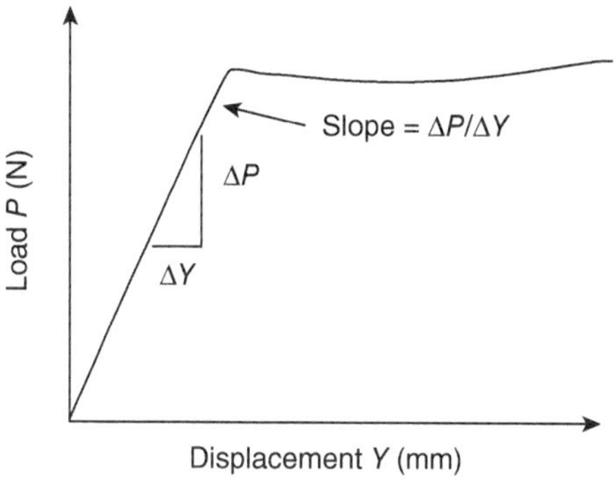

7.7 Typical load–displacement curve obtained in a four-point bending test.

$$\sigma_y = P_y \, D \, t/(4 \, I) \qquad [7.2]$$

where E is modulus, σ_y is the yield stress, D is the distance between the loading and supporting point, I is moment of inertia, $\Delta P/\Delta y$ is the slope of 'elastic' portion of the load–displacement curve, P_y is the yielding load and t is the thickness of the specimen.

Although four-point bending techniques are convenient to measure a variety of mechanical properties of different materials, their application on thin-foil layers (<250 μm) generally needs an addition of a much thicker substrate whose properties are well known. Moreover, the statistically determined properties are impinged on by various parameters such as residual stress, layer thickness and substrate temperature. All these factors mean that four-point bending techniques are not appropriate for accurate measurements of the properties of (Ni,Pt)Al bond coats, especially at elevated temperatures.

7.4.2 Bonding strength

Typically, grit-blasted, rough substrate surfaces are used for thermal spray coating deposition because of the established relationship between substrate surface roughness and coating adhesion.[53] The surface roughness effect on mechanical interlocking by calculations which show that a higher maximum pressure is generated at particle impact after increased substrate surface roughening. Alternatively, other researchers have demonstrated the importance of chemical bonding by showing comparable adhesion of thermal spray coatings onto smooth surfaces that were laser surface prepared, during deposition, to remove native oxides.

In general, the inter-atomic forces for particle bonding after impact during thermal spray deposition can range from weak van der Waals forces to strong chemical bonding forces, i.e., ionic, covalent or metallic bonding. For the case of air plasma spray deposited coatings, where complete particle melting leads to malleable particles impacting the substrate, mechanical interlocking must also be considered for long-range (1–10 μm) bonding.

Mechanical interlocking as a theory for thermal spray coating adhesion arises from the early observation that coatings adhered best to roughened substrates. Mechanical interlocking is evident as the penetration of impacted particles into valleys of the substrate or previously deposited layers. This is an especially important bonding mechanism for molten particles, which easily deform upon impact, even at the high strain rates occurring during thermal spray deposition.

While mechanical interlocking may be a primary bonding mechanism for molten powders, it does not adequately explain the adhesion of partially molten or non-molten ceramic YSZ particles. When unbonded surface zirconium, yttrium and oxygen sites on Y_2O_3-stabilized ZrO_2 particles meet with other unbonded atoms within a 0.15 nm distance at elevated temperatures, a fraction of the unbonded sites will chemically bond. The close inter-atomic distance required for chemical bonding is considered to be traversed by solid-state and liquid phase diffusional transport of atoms.[54] The kinetics of diffusional transport must be evaluated in order to fully analyze the bonding mechanism.

Although there is no agreement on how to explain all observed coating failures, it is agreed that choosing an appropriate measurement to gauge the bonding strength of a thermal spray coating is of utmost importance in order to quantitatively calibrate aspects of coating integrity such as failure strength and fatigue lifetime. Such a bonding strength measurement will be essential for designing a thermal spray coating. So far, the widely accepted measurement of coating bonding strength has been the average debonding stress from the standard tensile adhesion test stipulated in ASTM C633-01.[55,56]

The average debonding stress is without doubt simple to obtain, but it ignores the fact that a thermal spray coating is full of defects such as pores and cracks. It was found that there was a large variability in the bonding strength determined by the tensile adhesion test.[57] It is well known that existence of a crack will reduce the debonding stress, and the larger the crack the lower the debonding stress. Obviously, the difficulty of applying the debonding stress according to ASTM C633-01 is that it does not account for the structural defects inherent in the coating. Therefore, the debonding stress may not be an appropriate measurement of coating bonding strength. Instead, the fracture toughness which accounts for the interior coating defects is recommended as the measurement of coating bonding strength.

The fracture toughness of a YSZ coating has also been studied by Heintze and McPherson.[58] Coating toughness was found to be related to the composition of transformable tetragonal phase, i.e., addition of stabilizer such as yttria changed the toughness of zirconia coating.

The term 'fracture toughness' is sometimes ambiguous in context. It is often referred to as the critical stress intensity factor (SIF) since the SIF is an old and more understood fracture mechanics concept. However, by this reckoning, the fracture toughness implies the critical energy release rate which is more appropriate than the critical SIF because of the inherent mixed-mode situation encountered in the multi-layered thin coatings.

7.4.3 Thermal conductivity

The temperature- and time-dependent change in coating thermal conductivity is one of the most important parameters required for coating design and life prediction. The increase in thermal conductivity due to ceramic sintering can result in reduced coating thermal insulation and increased bond-coat/substrate oxidation.

There are three possible sources of heat conduction: lattice vibrations (phonons), photons and electrons. In dielectric materials, phonons dominate thermal transfer. They can be scattered by other phonons, crystal imperfections and surfaces. Phonon conductivity is the proportionality constant that relates the rate of heat flow per unit area and the temperature gradient. Thermal conductivity depends on the phonon mean path and the velocity as well as the lattice heat capacity. The phonon mean free path is the average distance traveled between scattering sites. Porosity, cracks and lattice defects all serve as phonon scattering centers, decreasing the phonon mean path and its conductivity. As discussed above, porosity and microcracking are typical microstructural features found in plasma-sprayed coating, which tend to lower thermal conductivity.

Pore shape and orientation, as well as percentage of porosity, affect coating properties such as thermal conductivity and elastic response.

Inter-splat porosity can be thought of as parallel cracks oriented perpendicular to the heat flow. Cracks oriented parallel to the direction of heat flow do not affect a material's thermal resistance, whereas those oriented perpendicular to the heat flow direction provide the maximum reduction in conductivity. It is clear that a low thermal conductivity coating should possess porosity oriented perpendicular to the temperature gradient, i.e., parallel to the substrate. Thus plasma-sprayed coatings are quite suitable for thermal barrier applications due to the lamellae structure.[59]

There are two mechanisms by which thermal resistance can be enhanced via interfaces in multi-layer structures: (i) process of phonon reversal and (ii) interfacial resistance. The dominating operating mechanism depends on the individual layer's thickness. The process is only likely to occur given a layer thickness on the order of nanometers. For example, a TiO film on a fused silica substrate with a moderate interfacial resistance of 10^{-7} ($m^2 \cdot K$)/W was experimentally found to decrease the thermal conductivity in a 1 EuN film by 50% relative to a bulk material.[60] There was further reduction in conductivity with thinner films on the order of 10 nm thick. On the other hand, interfacial resistance can contribute in systems with micron thick layers. Since boundaries serve as phonon scattering sites, the degree of scattering will theoretically increase with the number of interfaces in a given volume of material. This enhanced phonon scattering is the underlying source of interfacial resistance. The general theory of improving thermal resistance by layering has been studied using plasma-sprayed interfacial resistance. The theory is commonly determined experimentally by measuring the thermal conductivity of layered materials with varying numbers of interfaces.[61]

This interfacial phenomenon has also been observed for internal boundaries within a given material. Thermal conductivity of heat-treated plasma-sprayed YSZ coatings has been found to be higher than that of as-sprayed coatings.[61] Microstructural evaluation indicates that sintering during heat treatment eliminates many inter-splat interfaces. The increase in thermal conductivity has been caused by a decrease in the number of boundaries between splats and pores.

7.5 Conclusion

Since the 1960s, the development of PS–TBC technology has made reasonable progress and achieved considerable economic efficiency. However, it is far from being able to meet the demand for technological progress in terms of spraying equipment conceptualization, spray craft optimization, coating evaluation and coating life forecast as well as functional coating. In the future, research in PS–TBCs technology should concentrate on the following aspects: (i) Online monitoring technology should be used to measure the hydrodynamics characteristic of powder in plasma flame – this will assist in achieving a thorough understanding of the forming process of TBC and the mechanism by which various spray parameters influence TBC structure and performance. Multi-scale simulation and advanced test methods should be used to study the physical, chemical and mechanical behaviour of PS–TBC and to consolidate understanding of correlation failure theory for coating material and structure in a high-temperature environment. (ii) An established method and standard for PS–TBC is required along with development of testing technology for physical and chemical properties. (iii) The direction of progress in plasma-spraying equipment is likely to be in the development of high-energy and high-speed plasma spray along with powder feeding techniques, multi-purpose integrated technologies and real-time control technology. (iv) Establishment of a spray parameter database and model will allow the design and optimization of PS–TBC to continue. Functional gradient coating will develop and will enable controllability of coating performance. This new generation of TBC has evolved from an understanding of improvements to the structure, properties, processing and failure modes of earlier TBC systems. Such understanding will continue to improve TBC performance.

7.6 References

1. HODGE P.E., MILLER R.A., GEDWILL M.A. (1980), Evaluation of the hot corrosion behavior of thermal barrier coatings, *Thin Solid Films*, **73**, 447–453.
2. BENNET A., TORIZ F.C., THAKLEER A.B. (1987), A philosophy for thermal barrier coating design and its corroboration by 10000 h service experience on RB211 nozzle guide vanes, *Surf. Coat. Technol.*, **32**, 359–375.
3. CAO X.Q., VASSEN R., STÖVER D. (2004), New double-ceramic-layer thermal barrier coatings based on zirconia–rare earth composite oxides, *J. Eur. Ceram. Soc.*, **24** (1), 1–10.
4. CLARKE D.R., LEVI C.G. (2003), Materials design for the next generation thermal barrier coatings, *Annu. Rev. Mater. Res.*, **33**, 383–417.
5. MILLER R.A. (1997), Thermal barrier coatings for aircraft engines: history and directions, *J. Therm. Spray Technol.*, **6**, 35–42.
6. SEMENOV S.Y. (2002), Experimental diagnostics and modeling of thermal spray deposition processes of nano-structured wear resistant and thermal barrier coatings, UMI Number: 307671, 15.
7. BATISTA C., PORTINHA A., RIBEIRO R.M., TEIXEIRA V., COSTA M.F., OLIVEIRA C.R. (2005), Surface laser-glazing of plasma-sprayed thermal barrier coatings, *Appl. Surf. Sci.*, **247**, 313–319.
8. JORDAN E.H., XIE L., MA X., GELL M., PADTURE N.P., CETEGEN B., OZTURK A., ROTH J., XIAO T.D., BRYANT P.E.C. (2004), Superior thermal barrier coatings using solution precursor plasma spray, *J. Therm. Spray Technol.*, **13**(1), 57–65.
9. LUGSCHEIDER E., BARIMANI C., ECKERT P., ERITT U. (1996), Modeling of the APS plasma spray process, *Comput. Mater. Sci.*, **7**(1–2), 109–114.

10. DUSSOUBS B., FAUCHAIS P., VARDELLE A., VARDELLE M., THEMELIS N.J. (1997), Computational analysis of a threedimensional plasma spray jet, in: Berndt C.C. (ed.), *ThermalSpray: A United Forum for Scientific and Technological Advances*, ASM International, Materials Park, OH, 557–565.
11. PFENDER E., CHANG C.H. (1998), Plasma spray jets and plasma particulate interactions: modeling and experiment, in: Coddet C. (ed.), *Proceedings of ITSC 98*, ASTM International, Materials Park, OH, 315–327.
12. FAUCHAIS P., VARDELLE A. (2000), Heat, mass and momentum transfer in coating formation by plasma spraying, *Int. J. Therm. Sci.*, **39**, 852–870.
13. BUSSMANN M., AZIZ S.D., CHANDRA S., MOSTHAGIMI J. (1998), 3-D modelling of thermal spray dnoplet splashing, in: Coddet C. (ed.) *Proceedings of ITSC 98*, ASM International, Materials Park, OH, USA, 413–418.
14. MONTAVON G., SAMPATH S., BERNDT C.C., HERMAN H., CODDET C. (1995), Effects of vacuum plasma spray processing parameters on splat morphology, *J. Spray Technol.*, **4**, 67–74.
15. FAUCHAIS P., MONTAVON G., VARDELLE M., CEDELLE J. (2006), Developments in direct current plasma spraying, *Surf. Coat. Technol.*, **201**, 1908–1921.
16. FUKUMOTO M., NISHIOKA E., MATSUBARA T. (1999), Flattening and solidification behavior of a metal droplet on a flat substrate surface held at various temperatures, *Surf. Coat. Technol.*, **139**, 131–137.
17. SAMPATH S., HERMAN H. (1996), Rapid solidification and microstructure development during plasma spray deposition, *J. Therm. Spray Technol.*, **5**(4), 445–456.
18. WANG G.X., SAMPATH S., PRASAD V., HERMAN H. (1997), The stability of rapid planar solidification during melt substrate quenching, *Mater. Sci. Eng.*, **A226–A228**, 1035–1038.
19. HERMAN H. (1988), Plasma sprayed coatings, *Scientific American*, **259**(3), 112–117.
20. CHRASKA T., KING A.H. (2001), Transmission electron microscopy study of rapid solidification of plasma sprayed zirconia: part I First splat solidification, *Thin Solid Films*, **397**, 30–39.
21. CHRASKA T., KING A.H. (2001), Transmission electron microscopy study of rapid solidification of plasma sprayed zirconia: part II Interfaces and subsequent splat solidification, *Thin Solid Films*, **397**, 40–48.
22. PAUL S., CIPITRIA A., GOLOSNOY I.O., XIE L., DORFMAN M.R., CLYNE T.W. (2007), Effects of impurity content on the sintering characteristics of plasma-sprayed zirconia, *J. Therm. Spray Technol.*, **16**(5–6), 798–803.
23. CIPITRIA A., GOLOSNOY I.O., CLYNE T.W. (2009), A sintering model for plasma-sprayed zirconia thermal barrier coatings. Part II: Coatings bonded to a rigid substrate, *Acta Mater.*, **57**, 993–1003.
24. LEGER A.C., WIGREN J., HANSSON M.O. (1998), Development of a process window for a NiCoCrAlY plasma-sprayed coating, *Surf. Coat. Technol.*, **108–109**, 86–92.
25. HARDING J.H., MULHERAN P.A., CIROLINI S., MARCHESE M., JACUCCI G. (1997), Modeling deposition process of thermal barrier coating, *J. Therm. Spray Technol.*, **4**, 34–40.
26. MCPHERSON R. (1989), A review of microstructure and properties of plasma sprayed ceramic coatings, *Surf Coat. Technol.*, **39**, 173–181.
27. ANDREOLA F., LEONELLI C., ROMAGNOLI M., MISELLI P. (2000), Techniques used to determine porosity, *Am. Ceram. Soc. Bull.*, **79**, 49–52.

28. LAVIGNE O., RENOLLET Y., POULAIN M., RIO C., MORETTO P., BRANNVALL P., WIGREN J. (1999), Microstructural characterization of plasma sprayed thermal barrier coatings by quantitative image analysis, in: Strang A., Cawley J. (eds), *Quantitative Microscopy of High Temperature Materials Conference*, Institute of Materials, London, UK, 22–25.
29. KULKARNI A., WANG Z., NAKAMURA T., SAMPATH S., GOLAND A., HERMAN H., ALLEN A.J., ILAVSKY J., LONG G., FRAHM J., STEINBRECH R.W. (2003), Comprehensive microstructural and predictive property modeling of plasma-sprayed zirconia coatings, *Acta. Mater.*, **51**(9), 2457–2475.
30. KURODA S., CLYNE T.W. (1991), The quenching stress in thermally sprayed coatings, *Thin Solid Film*, **200**, 49–66.
31. BRINDLEY W.J., WHITTENBERGER J.D. (1993), Stress relaxation of low pressure plasma-sprayed NiCrAIY alloys, *Mater. Sci. Eng.*, **A163**, 33–41.
32. HOBBS M.K., REITER H. (1988), Residual stresses in Zr02–8%Y203 plasma-sprayed thermal barrier coatings, *Surf. Coat. Technol.*, **34**, 33–42.
33. HARMSWORTH, P.D., STEVENS R. (1992), Phase transformation and properties of plasma-sprayed zirconia thermal barrier coatings, *J. Mater. Sci.*, **27**, 611–615.
34. HERMAN H., SHANKAR N.R. (1987), Survivability of thermal barrier coatings, *Mater. Sci. Eng.*, **88**, 69–74.
35. TAKEUCHI S.M., ITO M., TAKEDA K. (1990), Modeling of residual stress in plasma-sprayed coatings: effect of substrate temperature, *Surf. Coat. Technol.*, **43/44**, 426–436.
36. BUSSO E.P., LIN J., SAKURAI S., NAKAYAMA M. (2001), A mechanistic study of oxidation-induced degradation in a plasma-sprayed thermal barrier coating system. Part I: model formulation, *Acta Mater.*, **49**, 132–140.
37. CHENG J., JORDAN E.H., BARBER B., GELL M. (1998), Thermalresidual stress in an electron beam physical vapor deposited thermal barrier coating system, *Acta. Mater.*, **46**, 5839–5842.
38. MILLER R.A., LOWELL C.E. (1982), Failure mechanisms of thermal barrier coatings exposed to elevated temperatures, *Thin Solid Films*, **95**, 265–270.
39. ROSIER J., BAKER M., VOLGMANN M. (2001), Stress state and failure mechanisms of thermal barrier coatings: role of creep in thermally grown oxide, *Acta Mater.*, **49**, 3659–3662.
40. BUSSO E.P., LIN J., SAKURAI S. (2001), A mechanistic study of oxidation-induced degradation in a plasma-sprayed thermal barrier coating system. Part II: Life prediction model, *Acta Mater.*, **49**, 1529–1534.
41. TZIMAS E., MULLEJANS H., PETEVES S.D., BRESSERS J., STAMEN W. (2000), Failure of thermal barrier coating systems under cyclic thermomechanical loading, *Acta Mater.*, **48**, 18–20.
42. HSUEH C.H., FULLER E.R. (2000), Residual stresses in thermal barrier coatings: effects of interface asperity curvature/height and oxide thickness, *Mater. Sci. Eng.*, **A283**, 46–49.
43. BUSSO E.P., LIN J., SAKURAI S. (2001), A mechanistic study of oxidation-induced degradation in a plasma-sprayed thermal barrier coating system. Part I model formation, *Acta Mater.*, **49**, 1515–1528.
44. SCHLICHTING K.W., VAIDYANATHAN K., SOHN Y.H., JORDAN E.H., GELL M., PADTURE N.P. (2000), Application of Cr^{3+} photoluminescence piezo-spectroscopy to

plasma-sprayed thermal barrier coatings for residual stress measurement, *Mater. Sci. Eng.*, **A291**, 68–72.

45. JORDAN E.H., SOHN Y.H., XIE W., GELL M., XIE L., TU F., PATTIPATI K.R., WILLET P. (2001), Residual stress measurement of thermal barrier coatings using laser fluorescence technique and their life prediction, in Autotestcon, Valley Forge, PA.
46. GELL M., JORDAN E., VAIDYANATHAN K., MCCARRON K., BARBER B., SOHN Y.-H., TOLPYGO V.K. (1999), Bond strength, bond stress and spallation mechanisms of thermal barrier coatings, *Surf. Coat. Technol.*, **53**, 120–121.
47. CHRISTENSEN R.J., LIPKIN D.M., CLARKE D.R., MURPHY K. (1996), Nondestructive evaluation of the oxidation stresses through thermal barrier coatings using Cr3+piezospectroscopy, *Appl. Phys. Lett.*, **69**, 3754–3760.
48. TOLPYGO V.K., CLARKE D.R. (2000), Surface rumpling of a (Ni, Pt)Al bond coat induced by cyclic oxidation, *Acta Mater.*, **48**, 3283–3288.
49. HAYNES J.A., RIGNEY E.D., FERBER M.N., PORTER W.D. (1996), Oxidation and degradation of a plasma-sprayed thermal barrier coating system, *Surf. Coat. Technol.*, **86–87**, 102–108.
50. BUSSO E.P., LIN J., SAKURAI S. (2001), A mechanistic study of oxidation-induced degradation in a plasma-sprayed thermal barrier coating system. Part II: Life prediction model, *Acta Mater.*, **49**, 1529–1532.
51. TSUI Y.C., HOWARD S.J., CLYNE T.W. (1994), The effect of residual stresses on the debonding of coatings – II. An experimental study of a thermally sprayed system, *Acta Mater.*, **42**, 2837–2844.
52. KHOR K.A., DONG Z.L., GU Y.W. (1999), Plasma sprayed functionally graded thermal barrier, *Mater. Lett.*, **38**, 437–444.
53. STÖVER D., FUNKE C. (1999), Directions of the development of thermal barrier coatings in energy applications, *J. Mater. Process. Technol.*, **92–93**, 195–202.
54. SUN J.H., CHANG E., WU B.C. (1993), Performance of CVD A1203 intermediate layer between zirconia top coat and NiCrAIY bond coat, *Mater. Trans., JIM*, **34**(7), 614–621.
55. COBLE R.L. (1958), Initial sintering of alumina and hematite, *J. Am. Ceram. Soc.*, **41**(2), 55–62.
56. HAN W., RYBICKI E.F., SHADLEY J.R. (1992), Bond strength testing of thermal spray coatings using ASTM C633-79: Effect of specimen size on test results, Proceedings of the International Thermal Spray Conference, Orlando, FL, 28 May–5 June, 911–914.
57. BERNDT C.C., LIN C.K. (1993), Measurement of adhesion for thermally sprayed materials, *Adhesion Sci. Tech.*, **7**, 1235–1264.
58. HEINTZE G.N., MCPHERSON R. (1988), Fracture toughness of plasma-sprayed zirconia coatings, *Surf. Coat. Technol.*, **34**, 15–23.
59. JENNIFER SU Y. (2001), *Processing and Performance of YSZ-Based Multilayer Thermal Barrier Coatings*, UMI Number: 3033564, 26–31.
60. WU Z.L., KUO P.K., WRI L., GU L.S., THOMAS R.L. (1993), Photothermal characterization of optical thin films, *Thin Solid Film*, **236**, 191–198.
61. RAVICHANDRAN K.S., AN K. (1999), Thermal conductivity of plasma-sprayed monolithic and multilayer coatings of alumina and yttria-stabilized zirconia, *J. Am Ceram. Soc.*, **82** (3), 673–821.

8

Plasma-sprayed thermal barrier coatings with segmentation cracks

H. GUO, R. YAO and L. ZHOU, Beihang University, China

Abstract: Plasma-sprayed thermal barrier coatings (TBCs) have been widely used on stationary components in turbine engines. The thermal shock resistance of such coatings is usually limited by their typical lamellar structure. In this work, a segmented structure was induced in the plasma-sprayed yttria stabilized zirconia (YSZ) coatings by using a 'hot' spraying condition. The thermophysical properties, mechanical properties and thermal cycling behaviors and associated failure mechanism of the segmented TBCs were addressed. The segmented TBCs exhibited very promising potential in improving the thermal cycling lifetime as compared to traditional non-segmented TBCs.

Key words: plasma spray, thermal barrier coatings (TBCs), segmentation cracks, thermal cycling, thermophysical property.

8.1 Introduction

Plasma-sprayed thermal barrier coatings (TBCs) have been extensively used to protect the components of gas turbines including combustion chambers, vanes and transition pieces from hot gases. The utilization of TBCs, accompanied by internal cooling, can achieve a temperature reduction of up to 300 K on the surface of the superalloy, which greatly extends the duration of its components.[1,2] Moreover, this enables engines to operate at temperatures above the melting point of the superalloy, and thus their efficiency and energy performance are tremendously improved.

An increasing demand for engine efficiency and performance in a context of higher operating temperatures and lower emissions is propelling the rapid development of TBC technology. Current interests focus on plasma-sprayed TBCs that comprise an yttria stabilized zirconia (YSZ) top coat and a NiCoCrAlY bond coat. These are usually used in crucial hot sections of gas turbines to increase the turbine inlet temperature (TIT), hence enhancing the efficiency and performance of the turbines. However, sintering and progressive destabilization of the YSZ top coat caused by mismatched thermal stresses and the formation of thermally grown oxide (TGO) may lead to spallation failure of the plasma-sprayed TBCs.[3–7]

Although TBCs thicker than 1 mm have better thermal barrier capability than thinner ones, such as those with a thickness of only a few tenths of 1 mm, some critical problems remain to be solved in the present work.[8–16] One serious problem is that thick TBCs present low thermal shock resistance when exposed to a high-temperature environment. An increase in coating thickness causes greater thermal gradients in the transient phase across the coating, and consequently leads to higher thermal stresses in the coating. Also, thick coatings will store more elastic strain energy, hence increasing the rate of energy release in a crack. These two factors have caused spalls of the thick TBCs, tensile cracks or chippings in the bulk of the coating materials. This failure mechanism is different from the conventional interface delamination failure that is mainly caused by thermal stresses resulting from differences in the thermomechanical properties between the coating and the substrate. Another factor that influences the performance of the thick coatings is the edge effect, defined as stress singularities resulting from moments and shear stresses at the coating edge. The stress will approach a singularity at the edge of the TBCs, especially those having a thickness of more than 500 μm. However, the edge effect might be avoided by optimization of the sample edge.

To improve the thermal shock performance of thick TBCs, some methods such as fabricating coatings with graded porosity or gradient layer structure have been attempted.[17–19] Recently, macrocracks or segmentation cracks in TBCs that are perpendicular to the substrate have been reported to be conducive to an increase in the thermal shock resistance of plasma-sprayed TBCs, mainly because they can accommodate the stresses resulting from the different thermal expansions between the coating and the substrate.[14,20–24] These macrocracks can be induced through post treatment such as laser glazing, or directly introduced during the deposition by applying large heat input to the substrate and using a high lamellar thickness.[11,12]

Segmented TBCs have a higher amount of microcracks and lower Young's modulus and can tolerate greater strains.[14,23] Those features that determine the thermal shocking behaviors of segmentation cracks TBCs are greatly ameliorated compared to the traditional non-segmented TBCs. This chapter addresses the manufacturing, thermophysical properties, mechanical properties and thermal cycling behaviors and associated failure mechanism of segmented TBCs.

8.2 Manufacturing of segmented thermal barrier coatings (TBCs)

When the coating is subjected to biaxial tensile stress, segmentation cracks may start and develop in the deposition phase because of stress relaxation.[13,25,26] During spraying, the molten droplets impinge on the surface of the substrate and quickly begin to solidify as a splat. When the solidified splat cools, shrinkage, usually associated with the decrease in temperature, is inhibited because the splat is attached closely to the substrate, and thus in-plane stresses are generated in the splat as a result. However, these stresses may be released immediately by the initiation of microcracks in the first splat.[25–27]

It is well accepted that the temperature of a substrate is the primary factor determining the development of segmentation cracks.[13,20] When spraying is conducted on high-temperature substrate, the difference in temperature between the first splat and the following splats will be large enough to remelt the surface of the first splat, thereby creating good connections among the splats. Then, the tensile stresses developing in the second splat will be released by the proliferation of the microcracks pre-existing in the first splat. This circle continues until the spraying process is finished. In other words, the microcracks propagate in the first lamellae throughout the thickness. In the deposition of the second lamellae, contacting areas still exist between the first splat in the second lamellae and the final splat in the first lamellae, even if the heat input is not high enough to remelt the underlying splat. Because of these connections, some of the microcracks in the first lamellae propagate and penetrate into the next lamellae, and consequently initiate segmentation cracks. Such cracks remain all along the following processes and eventually develop into segmentation cracks.[20]

Nevertheless, if the spraying is performed with low substrate or particle temperatures due to a long spraying distance, the heat from the new liquid splat is not high enough to create a good bonding to the underlying splat. Therefore, the real area of the connection between the splats is rather limited and estimated to be approximately 20%.[28] If so, a large number of horizontal cracks will present in the form of inter-splat an interlamellar delaminations, leading to obvious stress relaxation. Additionally, microcracks will form continuously within each splat and cause further tensile stresses, rather than propagating from splats to splats because of the poor bonding between the splats.[13,20]

Passage thickness is another significant parameter affecting the density of the segmentation crack.[20,27] The density is found to increase along with the increase in passage thickness. In one study, passage thickness is controlled by the plasma gun speed and the powder feed rate. It has been shown that segmentation cracks are much denser when the passage is fabricated thick by using a low plasma gun speed. It is generally accepted that when the plasma gun passes across the substrate materials at a low speed, more heat is transferred to the substrate resulting in a higher substrate temperature. Moreover, to a given coating thickness, the number of interlamellar delaminations decreases accordingly, owing to the increased lamellar thickness resulting from the low plasma gun speed. However, if controlled by a powder feed rate, the segmentation crack density is not as sensitive to the passage thickness as that controlled by plasma gun speed. As shown in Fig. 8.1, the difference in segmentation crack density is not as much for the coatings sprayed with different passage thickness. A high feeding rate usually leads to thicker lamellae when hundreds of particles are deposited during one pass. These particles are probably still rather hot, so the cooling of the substrate leads to reduced tensile strain difference among the adjacent lamellae, thereby reducing the possibility of microcrack formation. However, it has been reported by Guo *et al.*[29] that an overhigh powdering rate tends to decrease the density of the microcrack slightly compared to the low powder feeding rate, which may be attributed to incomplete melt of the powders at such a high feed rate.

The particle sizes of the powder feedstock have critical effects on porosity and growth of segmentation cracks in the coatings.[13,14] As discussed above, the coatings sprayed at a high plasma power and short spraying distance tend to have a high segmentation cracks density. It has been shown that since 20~45 μm powders have higher depositing efficiency and better capability of producing segmented coatings than commercially available, wide size distribution powders, they are considered to be preferable for processing segmented TBCs. Diagnostics of the in-flight particles demonstrates that the 20~25 μm powders yielded higher heat transmission because of their greater specific surface area compared to larger powders. This leads to more heat input to the substrate and a higher top coat temperature, as well as a higher velocity of the fine powders. Thus, the coatings sprayed from the fine powders exhibit a lower porosity, primarily due to the decrease of the volume of the inter-splat gaps and voids.

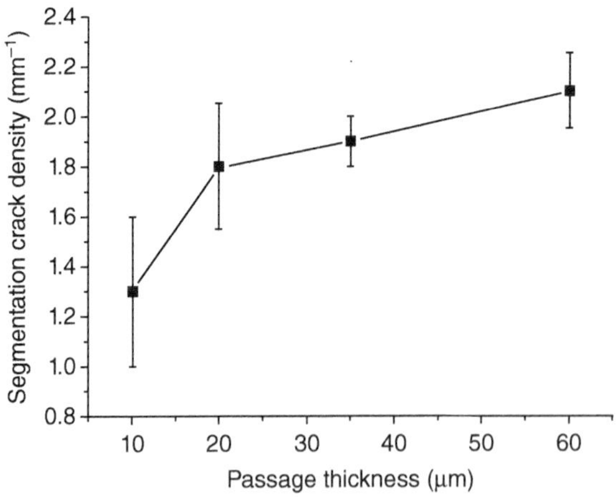

8.1 Segmentation crack densities of YSZ coatings sprayed with different feeding rates.

The type of feedstock tremendously influences the properties of the sprayed YSZ coatings because different powders interact diversely in a plasma flame, thus affecting the development of the microstructure.[13,14,29] Generally, the coatings sprayed from hollow powders contain more unmelted particles because their particle size is broadly distributed. As a result, the sprayed coatings had a relatively higher segmentation crack density than the coating with high porosity because the origin and propagation of segmentation cracks are mainly hindered by delamination cracks and large defects. Therefore, the fused and crashed powder has greater capability of producing segmentation cracks in the coatings compared with the hollow powder.

Cooling rate during deposition is also one of the main parameters determining the development of segmentation cracks. High cooling rate is conducive to the generation of segmentation cracks.

8.3 Microstructure of segmented TBCs

Figures 8.2a and b are the scanning electron microscope (SEM) micrographs of the cross-section of the YSZ coatings sprayed from the HOSP feedstocks. The number of segmentation cracks in the coating sprayed with a longer spraying distance and lower plasma power is rather limited (Fig. 8.2a). In contrast to this, the number of the segmentation cracks of the coating sprayed with a higher substrate temperature (Fig. 8.2b) increased enormously, with most passing throughout the coating thickness.

In Figures 8.3a and b, the polished surfaces of the above YSZ coatings are compared. From the surface appearance, the non-segmented coating contains a large amount of micrometer-sized pores, while the segmented coating has a fine segmentation crack network with a diameter in the order of 0.5 mm^{-1}.

Figures 8.4a and b show the fracture surfaces of the cross-sections of the YSZ coatings. Some pores larger than 1 μm are presented in the non-

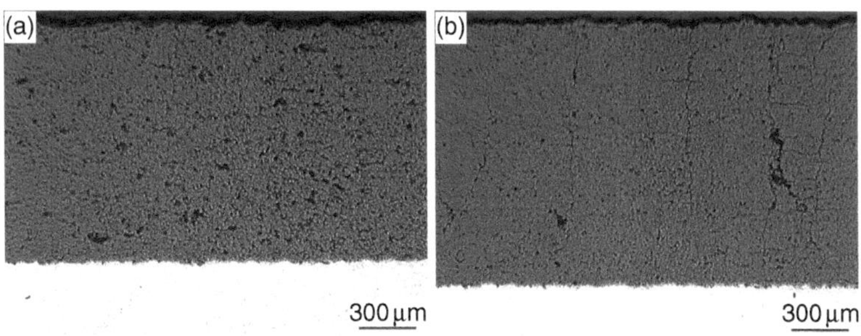

8.2 SEM micrographs of cross-sections of YSZ coatings: (a) traditional non-segmented; (b) highly segmented.

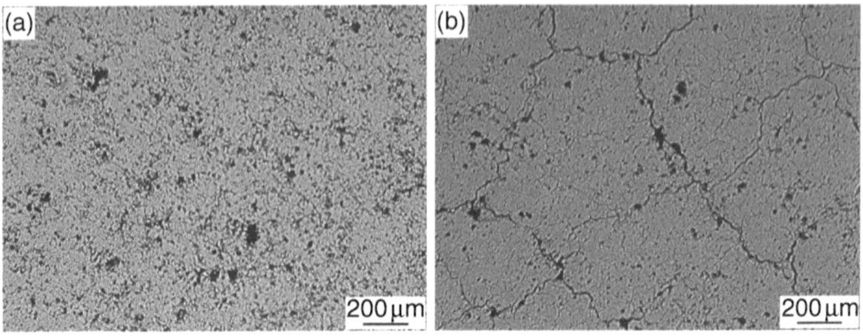

8.3 Surface morphologies of YSZ coatings: (a) traditional non-segmented; (b) highly segmented.

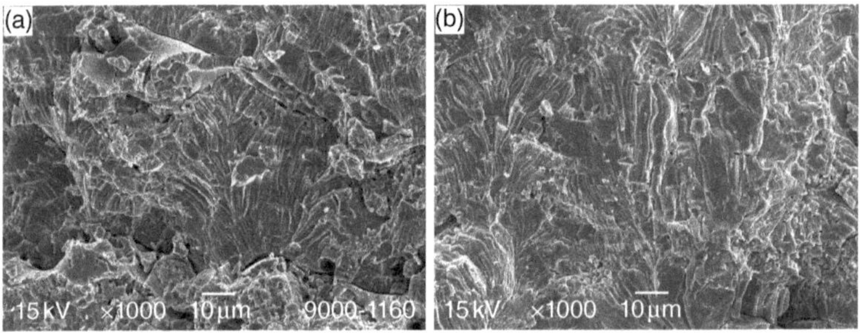

8.4 SEM micrographs of fracture cross-sections of YSZ coatings: (a) traditional non-segmented; (b) highly segmented.

segmented coating, as shown in Fig. 8.4a. These pores usually result from unmolten particles and gas entrapment during spraying. The segmented coating was apparently denser than the non-segmented coating, because fewer inter-splat gaps and voids are visible in the coating. Furthermore, the columnar grains go through several lamellae. Moreover, the contact between the lamellae is significantly improved, as shown in Fig. 8.4b.

8.4 Thermophysical and mechanical properties of segmented TBCs

The thermal diffusivities of the YSZ coatings were measured by laser flash apparatus. The thermal diffusivity values for the non-segmented coating range from 0.002–0.006 $cm^2 s^{-1}$ between room temperature and 1200 °C, while the segmented coating exhibits a perceptible increase in thermal diffusivity compared to the non-segmented coating. The difference indicates that the values of the thermal diffusivity largely depend on the microstructure of the coatings.

Thermal conductivity is calculated using the equation: $K(T) = \alpha(T)\cdot C_P\cdot\rho$ where $K(T)$ is the thermal conductivity, $\alpha(T)$ is the thermal diffusivity, C_P is the specific heat and ρ is the bulk density of the coating. The thermal conductivity of the segmented coating, with an increase of nearly 20% compared to that of the non-segmented coating,[14,23] is approximately 1.75 W/mK.

Plasma-sprayed TBCs reveal a two-dimensional microcrack network: one is defined as a horizontal microcrack, resulting from the flawed contacts between the inter-splats of the coatings; the other is oriented vertically, primarily resulting from the microcracking of the splats during cooling. The former is particularly effective in reducing the thermal conductivity of the sprayed coatings, because the interfaces formed by such cracks are perpendicular to the main heat flux. Among the coatings mentioned above, the segmented coating has a lower porosity than the non-segmented coating, and thus has a better connection between the splats, which contributes to reducing the amount of horizontal microcracks. This explains why the segmented coating tends to have a slightly higher thermal conductivity compared to the non-segmented coating. Moreover, dispersed spheroid microporosity is also helpful in lowering the thermal conductivity of the YSZ coatings. It is worth noting that branching cracks are also introduced when the spraying is performed with thick lamellae, accompanying the generation of segmentation cracks. The branching cracks play a role in lowering the thermal conductivity of the segmented TBC.[20]

Both the non-segmented and segmented coatings reveal very similar thermal expansion when heating to 1200 °C from room temperature, despite the different processing parameters and microstructure between the coatings. The average values of the thermal expansion coefficient (TEC) are in the range $10\sim11.2 \times 10^{-6}$ K^{-1}.

8.5 Thermal shock resistance and associate failure mechanism

Thermal shock testing was conducted using a gas burner rig with natural gas and oxygen to evaluate the thermal shock resistance of the coatings (see Fig. 8.5). The temperatures of the coating surface and the back side (substrate) were measured using a pyrometer and a thermocouple fixed at the back of the substrate, respectively. During the test, the gas burner has a broad flame in order to produce a homogeneous temperature distribution in the center of the sample. The sample was heated to a desired temperature within 20 s, and then maintained at the temperature for 5 min. The back side of the sample was cooled by compressed air in order to keep a controlled temperature gradient across the thickness of the sample and, in this case, the temperature gradient was nearly ~500 °C. The burner was removed

8.5 Burner rig set-up for thermal shock testing.

automatically from the coating during the cooling, and the sample was cooled for 2 min using compressed air at a rate of more than 100 °C/s. The lifetime of the coating is defined as the number of thermal cycles when the spallation of the coating is clearly visible to the naked eye.

The segmented specimen had a thermal cycling lifetime of more than 2000 cycles at 1200 ± 50 °C (surface)/1000 ± 50 °C (substrate), while the non-segmented one had a lifetime of less than 500 cycles. The segmented coating exhibited a much improved thermal shock resistance. Figure 8.6 gives a photograph showing the surface morphology of the segmented TBC after 3200 cycles of thermal shock testing. Spallation of the YSZ coating occurred starting from the edge of the specimen, which is a typical failure mechanism caused by oxidation of the TBC, as shown in Fig. 8.7.

The traditional non-segmented specimen is usually spalled off from the substrate by cracking at the interface between the YSZ top coat and the bond coat.[30–36] This is a typical failure mechanism of plasma-sprayed TBCs caused by thermal stresses deriving from the mismatch of thermal expansions between the ceramic top coat and the superalloy substrate. In the segmented specimen, some chippings on the surface of the top coat were observed before bulk spallation. The failure of the segmented coating then started from the edge of the specimen and propagated to the center. Finally, partial spallation of some segments rather than spallation of the whole coating occurred, which is quite different from the failure mechanism observed in traditional TBCs.[13,20,31,37,38]

Segmented TBCs have a much longer thermal cycling lifetime than traditional non-segmented TBCs and a distinct failure mechanism, to which the contribution of the segmentation crack network should be attributed.

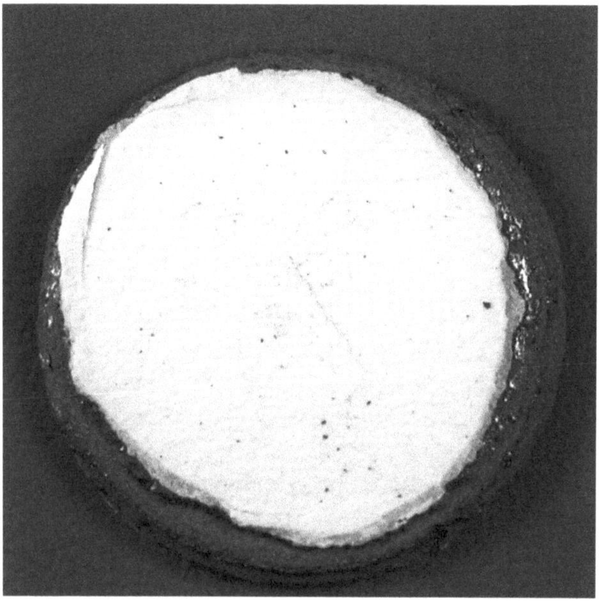

8.6 Photograph of the segmented TBC after thermal shock resistance of 3200 cycles.

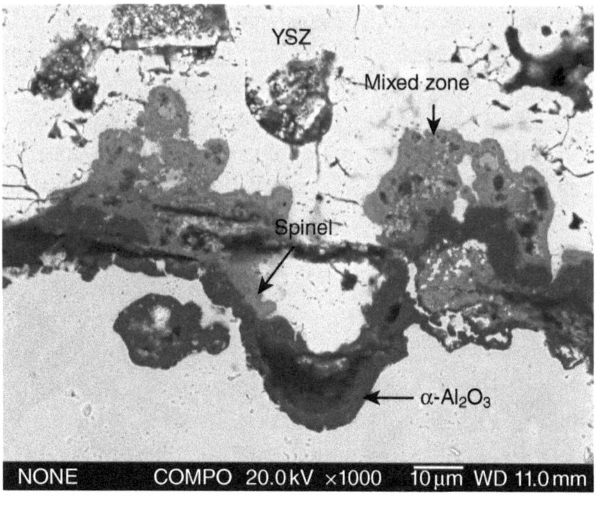

8.7 SEM micrograph of cross-section of the segmented TBC after thermal shock testing, showing the morphology of TGO.

These cracks can improve the straining tolerance of the coatings by opening during tensile loading and closing during compressive loading, just like the behavior of the inter-column gaps of electron beam physical vapor deposition (EB–PVD) coatings. The straining tolerance may be seen as a reduction of the stress concentration that existed at various edges of a TBC specimen. In particular, at free edges, such as at the rim of a disk specimen or around the periphery of each segment in the segmented coating, shear stresses at the interface between the coating and the substrate will develop because of different thermal expansions, hence placing TBCs in lateral tension or compression depending on the instantaneous temperature as well as the specimen's thermal history. The thickness and the area of each coating are the geometrical factors that can determine the magnitude of the maximum shear stresses at the edges. To a constant coating thickness, a smaller area will lead to smaller stresses. Therefore, it is reasonable to anticipate that a highly segmented specimen can generate relatively small shear stresses compared to a non-segmented specimen, and hence the force to produce an interface crack is also significantly reduced.

Several factors, such as porosity and other horizontal cracks, may affect the thermal properties of TBCs, but it is found in the present study that segmentation cracks dominate the lifetimes of TBCs. Some other factors that are related to the changes of materials microstructure still need further investigation.

Plasma-sprayed 8YSZ coatings mostly consist of a non-equilibrium tetragonal phase (denoted as t') resulting from the rapid cooling during the spraying process.[7] Nevertheless, exposure to high temperature leads to the decomposition of t'-phase. During the high-temperature stage, the t'-phase first transforms to the stable tetragonal phase (t) with low yttria content, and then to the monoclinic phase (m) during the cooling stage. The phase transformation of t' to m is associated with a volume expansion of up to 5%, which will deteriorate the integrity of the YSZ top coat and can hence cause the failure of TBCs.

For the segmented coating sprayed under 'hot' condition, the issue of concern was the phase stability of the YSZ coating. Only a small amount of monoclinic phase was detected in both the as-sprayed coating and the coating after thermal shock testing.[3,4,7,37] The presence of the monoclinic phase in the as-sprayed coating could be ascribed to the special processing condition for spraying highly segmented coatings, in which the substrate temperature could reach as high as 1273 K. It can be inferred that the coating surface temperature was even higher than the substrate temperature during the spraying. Therefore, it is possible that the phase transformation of t to m occurred just in the as-sprayed coating. As for the TBC specimen after thermal shock testing, the maximum temperature that the specimen was exposed to is lower than 1523 K. It is well recognized that the decomposition of the *t*′-phase in 8YSZ TBCs is very slow at temperatures below 1523 K. Therefore, it can be predicted that decomposition of TBC phase could not cause severe damage to the integrity of segmented TBCs.

As shown in Fig. 8.7, a TGO layer of about 5 μm thickness was formed on the bond coat and some horizontal cracks were developed in the top coat close to the TGO. The TGO consists of three zones: the α-Al_2O_3 zone close to the NiCoCrAlY bond coat; the spinel zone above the α-Al_2O_3 zone; and the mixed zone of spinel oxide and ZrO_2. The formation of spinel oxides is related to depletion of Al in the bond coat.[7,39–41] In the case of the segmented TBC, it usually survives hundreds of hours under a hot gas environment. As a result, severe oxidation inevitably occurs and a thick TGO is formed between the YSZ top coat and the bond coat. With the qualitative analysis given above, we currently believe that the geometrical effects of reducing stresses by segmentation cracks do indeed play a dominant role in prolonging the thermal cycling life of TBCs, but further quantitative investigations should be carried out in the near future.

Thick TBCs with different levels of segmentation crack densities can be produced by controlling the heat input to the substrate. The thermal cycling lifetimes of the sprayed TBC are strongly dependent on the segmentation crack density. Highly segmented TBCs achieve a significant improvement in thermal cycling lifetime as compared to TBCs without segmentation cracks. It has been shown that the segmentation crack network is quite stable during thermal cycling testing. No segmentation cracks disappear during sintering of YSZ coatings.

Segmentation cracks can be also applied to quite thin coatings. As shown in Fig. 8.8, a thin YSZ TBC with segmentation cracks was produced and exhibited a very satisfactory thermal shock resistance as compared to the traditional TBC. The failure of the thin coating was governed by the oxida-

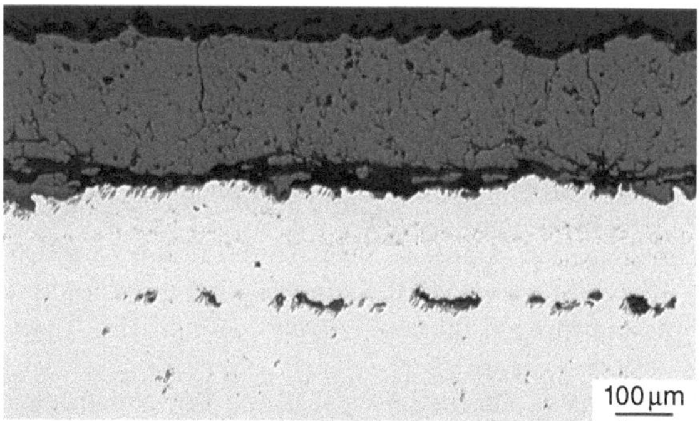

8.8 Optical micrograph of cross-section of the thin TBC with segmentation cracks after 2320 cycles thermal shock testing.

tion of the bond coat because the temperature of the bond coat in the case of the thin TBC was almost 50~100 °C higher than that in the case of the thick TBC.

8.6 Future trends

With the development of technology, the production of highly segmented TBCs with high segmentation crack densities has become possible by using advanced processing methods. Even higher segmentation crack densities, as well as a high porosity level and hence low thermal conductivities, can be obtained by using suspensions as feedstock for the plasma-spraying process. One problem associated with the plasma-spraying process is the blocking of the cooling holes during deposition, and the search for solutions to overcome this problem is still in progress.

8.7 References

1. N.P. PADTURE, M. GELL, E.H. JORDAN, Thermal barrier coatings for gas-turbine engine applications, *Science*, **296** (2002): 280–284.
2. R.L. JONES, Thermal barrier coatings, in K.H. Stern (Ed.), *Metallurgical and Ceramic Coatings*, Chapman and Hall, London, UK, 1996, 194–253.
3. R. VASSEN, F. TIETZ, G. KERKHOFF, *et al.*, New materials for advanced thermal barrier coatings, in J. Lecomte-Becker, F. Schubert and P.J. Ennis (eds), *Materials for Power Engineering*, (1998), Forshungszentrum Jülich, Jülich, Germany, 1627–1635.
4. R.A. MILLER, Thermal barrier coatings for aircraft engines: history and directions, *J. Therm. Spray Technol.*, **6**(1) (1997): 35–42.
5. A. RABIEI, A.G. EVANS, Failure mechanisms associated with the thermally grown oxide in plasma-sprayed thermal barrier coatings, *Acta Mater.*, **48** (2000): 3963–3976.
6. K.W. SCHLICHTING, N.P. PADTURE, E.H. JORDAN, M. GELL, Failure modes in plasma-sprayed thermal barrier coatings, *J. Mater. Sci. Eng. A*, **342** (2003): 120–130.
7. H. GUO, H. MURAKAMI, S. KURODA, Thermal cycling behavior of plasma sprayed segmented thermal barrier coatings, *Mater. Trans.*, **47**(2) (2006): 306–309.
8. G. GUALCO, S. CORCORUTO, A. CAMPORA, Highly porous thick thermal barrier coatings produced by air plasma spraying of a plastic-ceramic mixed powder, in: C.C. Berndt (Ed.), *Thermal Spray: A United Forum for Scientific and Technological Advances, Proceedings of the 1st National Thermal Spray Conference*, ASM International, Materials Park, OH, 305–315.
9. H.-D. STEFFENS, Z. BABIAK, M. GRAMLICH, Some aspects of thick thermal barrier coating lifetime prolongation, *J. Therm. Spray Technol.*, **8**(4) (1999): 517–522.
10. R. TAYLOR, J.R. BRANDON, P. MORRELL, Microstructure composition and property relationships of plasma-sprayed thermal barrier coatings, *Surf. Coat. Technol.*, **50**(2) (1992): 141–149.
11. H. GUO, R. VAßEN, D. STÖVER, Thermophysical properties and thermal cycling-behavior of plasma sprayed thick thermal barrier coatings, *Surf. Coat. Technol.*, **192** (2005): 48–56.

12. P. BENGTSSON, T. ERICSSON, J. WIGREN, Thermal shock testing of burner cans coated with a thick thermal barrier coating, *J. Therm. Spray Technol.*, **7**(3) (1998): 340–348.
13. H. GUO, H. MURAKAMI, S. KURODA, Effect of hollow spherical powder size distribution on porosity and segmentation cracks in thermal barrier coatings, *J. Am. Ceram. Soc.*, **89**(12) (2006): 3797–3804.
14. H. GUO, S. KURODA, H. MURAKAMI, Microstructures and properties of plasma-sprayed segmented thermal barrier coatings, *J. Am. Ceram. Soc.*, **89**(4) (2006): 1432–1439.
15. E.F. REJDA, D.F. SOCIE, T. ITOH, Deformation behavior of plasma-sprayed thick thermal barrier coatings, *Surf. Coat. Technol.*, **113** (1999): 218–226.
16. D. SCHWINGEL, R. TAYLOR, T. HAUBOLD, J. WIGREN, C. GUALCO, Mechanical and thermophysical properties of thick PYSZ thermal barrier coatings: correlation with microstructure and spraying parameters, *Surf. Coat. Technol.*, **108–109** (1998): 99–106.
17. M. TAMURA, M. TAKAHASHI, J. ISHII, K. SUZUKI, M. SATO, K. SHIMOMURA, Multilayered thermal barrier coating for land-based gas turbines, *J. Therm. Spray Technol.*, **8**(1) (1999): 68–72.
18. W.Y. LEE, D.P. STINTON, C.C. BERNDT, F. ERDOGAN, Y.D. LEE, Z. MUTA-SIN, Concept of functionally graded materials for advanced thermal barrier coating applications, *J. Am. Ceram. Soc.*, **79**(12) (1996): 3003–3012.
19. C. GUALCO, E. CORDANO, F. FIGNINO, C. GAMBARO, S. AHMANIEMI, S. TUURNA, T. MÄNTYLÄ, P. VUORISTO, An improved deposition process for very thick porous thermal barrier coatings, in: E. Lugscheider, C.C. Berndt (eds), *Tagungsband Conference Proceedings, Proceedings of the International Thermal Spray Conference*, DVS, Dusseldorf, Germany, 2002, 196–201.
20. H.B. GUO, R. VAßEN, D. STÖVER, Atmospheric plasma sprayed thick thermal barrier coatings with high segmentation crack density, *Surf. Coat. Technol.*, **186** (2004): 353–363.
21. H.L. TSAI, P.C. TSAI, Performance of laser-glazed plasma-sprayed (ZrO_2-12wt.%Y_2O_3)/(Ni-22wt.%Cr-10wt.%Al-1wt.%Y) thermal barrier coatings in cyclic oxidation tests, *Surf. Coat. Technol.*, **71** (1995): 53–59.
22. S. AHMANIEMI, P. VUORISTO, T. MÄNTYLÄ, C. GUALCO, A. BONADEI and R. DIMAGGIO, Thermal cycling resistance of modified thick thermal barrier coatings, *Surf. Coat. Technol.*, **190** (2005): 378–387.
23. H. GUO, H. MURAKAMI, S. KURODA, Materials, effects of heat treatment on microstructures and physical properties of segmented thermal barrier coatings, *Mater. Trans.*, **46**(8) (2005): 1775–1778.
24. H. GUO, H. MURAKAMI, S. KURODA, Segmentation cracks in plasma sprayed thin thermal barrier coatings, *Advanced Ceramic Coatings and Interfaces: Ceramic Engineering and Science Proceedings*, **27**(3) (2007): 17–27.
25. S. KURODA, T.W. CLYNE, The quenching stress in thermally sprayed coatings, *Thin Solid Films*, **200** (1991): 49–66.
26. J.W. HUTCHINSON, Z. SUO, Mixed mode cracking in layered materials, in: J.W. Hutchinson, T.Y. Wu (eds), Advances in Applied Mechanics, Vol. 29(1992): Academic Press, New York, 1289–1298.
27. P. BENGTSSON, J. WIGREN, Segmentation cracks in plasma sprayed thick thermal barrier coatings, in: P.J. Maziasz, I.G. Wright *et al.* (eds), *Gas Turbine Materials*

Technology, Proceedings of ASM Materials Solutions, ASM International, Materials Park, OH, 1999, 92–101.
28. R. MCPHERSON, The relationship between the mechanism of formation, micro structure and properties of plasma sprayed coatings, *Thin Solid Films*, **83** (1981): 297–310.
29. H. GUO, S. KURODA, H. MURAKAMI, Segmented thermal barrier coatings produced by atmospheric plasma spraying hollow powders, *Thin Solid Films*, **506** (2006): 136–139.
30. Z.Z. MUTASIM, Y.L. NAVA, Development and performance evaluation of thick air plasma sprayed thermal barrier coatings, in: C.C. Berndt (Ed.), *Surface Engineering via Applied Research, Proceedings of the 1st International Thermal Spray Conference*, ASM International, Materials Park, OH, 2000, 1325–1330.
31. T.A. TAYLOR, D.L. APPLEBY, A.E. WEATHERILL, J. GRIFFITHS, Plasma-sprayed yttria-stabilized zirconia coatings: structure-property relationships, *Surf. Coat. Technol.*, **43/44** (1990): 470–480.
32. Y.W. GU, K.A. KHOR, Y.Q. FU, Y. WANG, Functionally graded ZrO_2-NiCrAlY coatings prepared by plasma spraying using pre-mixed, spheroidized powders, *Surf. Coat. Technol.*, **96**(2–3) (1997): 305–312.
33. K. KOKINI, J. DEJONGE, S. RANGARAJ, B. BEARDSLEY, Thermal shock of functionally graded thermal barrier coatings with similar thermal resistance, *Surf. Coat. Technol.*, **154** (2002): 223–231.
34. H.A. NIED, Edge crack stress intensity factors in layered ceramic-metal composites due to thermal mismatch, *Proceedings of the International Conference and Exhibits on Failure Analysis*, ASM International, Materials Park, OH, 1991, 59–68.
35. D. ZHU, R.A. MILLER, Investigation of thermal high cycle and low cycle fatigue mechanisms of thick thermal barrier coatings, *Mater. Sci. Eng. A*, **245** (1998): 212–223.
36. R.E. TAYLOR, X. WANG, X. XU, Thermophysical properties of thermal barrier coatings, *Surf. Coat. Technol.*, **120–121** (1999): 89–95.
37. Y. WANG, H. GUO, S. GONG, Thermal shock resistance and mechanical properties of $La_2Ce_2O_7$ thermal barrier coatings with segmented structure, *Ceram. Int.*, **35** (2009): 2639–2644.
38. R. VAßEN, F. TRAEGER, D. STÖVER, Correlation between spraying conditions and micro crack density and their influence on thermal cycling life of thermal barrier coatings, in: B.R. Marple, C. Moreau (eds), *Advancing the Science and Applying the Technology, Proceedings of International Thermal Spray Conference*, ASM International, Materials Park, OH, 2003, 1455–1462.
39. T.J. NIJDAM, L.P.H. JEURGENS, W.G. SLOOF, Modeling the thermal oxidation of ternary alloys – compositional changes in the alloy and the development of oxide phases, *Acta Mater.*, **51** (2003): 5295–5307.
40. T.J. NIJDAM, L.P.H. JEURGENS, W.G. SLOOF, Promoting exclusive α-Al_2O_3 growth upon high-temperature oxidation of NiCrAl alloys: experiment versus model predictions, *Acta Mater.*, **53** (2005): 1643–1653.
41. H. GUO, H. MURAKAMI, S. KURODA, Thermal cycling and hot-corrosion behaviors of plasma sprayed segmented thermal barrier coatings, *Mater. Sci. Forum*, **546–549** (2007): 1713–1716.

9
Detonation gun sprayed thermal barrier coatings

P. L. KE, Y. N. WU and C. SUN, Chinese Academy of Sciences, China

Abstract: Detonation gun spraying (D-gun spraying) is believed to be a promising new spraying technology in thermal barrier coatings (TBCs) for application to aircraft turbines. This chapter focuses mainly on two issues of aircraft turbine applications. It discusses first the low thermal conductivity of the new type of D-gun sprayed YSZ top coating. The failure mechanism of thermal shock cycling aimed at improving the lifetime of D-gun sprayed TBCs is then described.

Key words: thermal barrier coatings (TBCs), detonation gun spraying (D-gun spraying), thermal conductivity, thermal shock cycling, failure mechanism.

9.1 Introduction

Gas detonation gun (D-gun) technology and equipment were originally developed and patented in 1955 by Union Carbide (now Praxair).[1,2] These approaches and capabilities, have not, however, been opened to the public until recently. Even so, in the 1960s, the Institute of Materials Science and Welding Institute of Ukraine, SSR managed to utilize D-gun technology, and eventually developed D-gun equipment. With the disintegration of the Soviet Union, D-gun spraying technology and equipment were offered for sale and thus made known to the research field. Since then, D-gun-sprayed coatings have found wide application, especially in the aircraft industries of the USA, Japan and the former Soviet Union.[3–5] In light of this promising processing method, Beijing Institute of Aviation Materials, China, began to launch independent innovations in D-gun technology and equipment in 1985, but there remains a large gap compared with Ukraine. At the time of writing, there are about ten D-gun facilities in use in China, most of which were imported from Ukraine, and the technology has been studied systematically since the 1990s.[6–11] D-gun technology has now gained an ever-extending application in various engineering fields, such as the medical, aircraft, agricultural and machinery industries, especially for surface strengthening and modifying of machinery components.

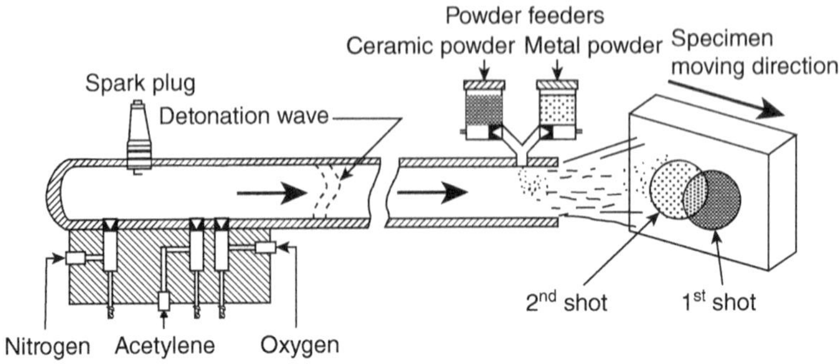

9.1 Schematic illustration showing the principle of the D-gun spraying process.[16]

A schematic representation of a D-gun spraying device is given in Fig. 9.1. The main feature of D-gun spraying is that high-velocity particles, typically 800–1200 m/s, are accelerated by the detonation wave during spraying; therefore, higher bond strength, higher hardness and lower porosity of the coating can be achieved. This method allows the deposition of quality coatings for various purposes using various powder materials. Alloy(s), oxides, hard metal, carbide cermets as well as composite materials can be sprayed to form coatings with properties of wear-resistance, friction-reduction, thermal-resistance, insulation, conduction and thermal shock resistance. These are extensively used as wear-resistance coatings or seal coatings. The duration of parts can be extended several dozen fold by application of D-gun-sprayed coatings on the top coat.

The two processes in commercial use for thermal barrier coating (TBCs) are air plasma spraying (APS) and electron beam physical vapor deposition (EB–PVD). APS TBCs have lower cost and thermal conductivity compared with EB–PVD TBCs, but at the expense of durability.[12–14] One means of improving the quality of TBCs, is to try to modify the current APS and EB–PVD techniques. The alternative is to develop a new generation of deposition technology, such as D-gun spraying. This might be a promising technology with the potential to solve some technical difficulties, such as the low torch temperature and very short heating time for ZrO_2–Y_2O_3 (YSZ) ceramic particles during the spraying process. Until 1999, Kim *et al.* had utilized D-gun spraying equipment to obtain TBCs successfully for the first time through optimizing the spray parameters, especially the ratio of C_2H_2 to O_2.[15–16]

9.2 Detonation gun (D-gun) sprayed thermal barrier coatings (TBCs)

9.2.1 Microstructure of as-sprayed D-gun TBCs

It is known that the D-gun-sprayed TBCs have a typical layer structure similar to that of coatings deposited using other thermal-spraying processes. D-gun spraying offers the highest particle velocity amongst all the thermal spraying technologies which means that it is able to achieve satisfactory lamellar bonding at the interfaces between the flattened particles. Based on the work of Wu and Ke,[17–19] it is clear that as-sprayed D-gun YSZ top coats have a unique microstructure with vertical cracks, splats, unmelted particles and a rough bond coat surface with protrusions on the top coat. The unmelted particles, the porosity and especially the through-thickness cracks are all beneficial in imparting strain-tolerance to the TBC, while the porosity helps reduce the thermal conductivity. Moreover, it is generally agreed that a rough bond coat surface can increase the duration of a typical plasma-sprayed TBC system during thermal cycling. It has been assumed that the roughness of the bond coating provides mechanical interlocking at the interface between the top and bond coat; however, a recent theoretical analysis showed that periodic undulations of the surface of the bond coat can provide oscillatory energy-release rates, which would cause cracking in the top coat to be exposed locally, which, in turn, prevents or postpones TBC spallation. Inter-splat porosities and splat boundaries can also be seen under the protrusions in the bond coat around the interface. These can possibly provide a pathway for oxygen ingress into the bond coat when connecting to the interface of a relatively porous structure typical of the YSZ top coat.

9.2.2 Thermal conductivity

When designing a TBC, a thermally protective TBC layer with a low thermal conductivity is introduced to maximize the thermal drop across the thickness of the coating in order to protect the substrate alloy. Much research work focuses on how to further decrease the thermal conductance, especially from the initial ZrO_2 (between 2.5 and 4.0) to YSZ (2.3–2.6 in the dense form). Generally, there are three approaches to lowering the thermal conductivity of TBCs: (i) to lower the thermal conductivity of the coating material; (ii) to lower the thermal conductivity by increasing the porosity of the coating; and (iii) to increase the thickness of the coating. In this section, we will deal with two YSZ powder materials – normal solid ZrO_2–8%Y_2O_3 (NSP–YSZ) and hollow spherical ZrO_2–8%Y_2O_3 (HSP–YSZ) – sprayed by D-gun with a low thermal conductivity.

The thermal conductivity of TBCs is related to the microstructure and density of the YSZ. A relation which has been derived for calculation of the thermal conductivity (λ) values of YSZ coatings is[20]

$$\lambda = \alpha \cdot \rho \cdot C_p \tag{9.1}$$

where α, C_p, ρ are thermal diffusivity, specific heat and density, respectively. With regard to the normal widely used solid powder of YSZ (NSP–YSZ), it has been reported that D-gun-sprayed YSZ free coatings had a thermal conductivity (1.0–1.4 W/mK) approaching that of the plasma-sprayed coatings (1.0–1.5 W/mK[17]), and much lower than their EB–PVD counterparts (>2.0 W/mK[18]).

In order to obtain a top coat with even better thermal insulation Ke *et al.*[21] introduced HSP–YSZ as a new top coat powder material. Different from the NSP–YSZ, as shown in Fig. 9.2, HSP–YSZ has a high closed porosity (>10 vol. %), which is advantageous in increasing the effectiveness of thermal protection and enhancing thermal shock resistance. Correspondingly, the thermal conductivities of the freestanding NSP–YSZ and HSP–YSZ ceramic coatings were 1.0–1.4 W/mK and 0.8–1.2 W/mK from 200 °C to 1200 °C, respectively. In short, HSP–YSZ ceramic coating by D-gun spraying has lower thermal conductivity. Reducing thermal conductivity will help (i) to improve TBC durability by reducing the metal temperature and retarding the thermally-activated processes responsible for failure and/or (ii) to improve engine efficiency at higher temperatures.

Hasselman analyzed the effect of cracks of various orientations on thermal conductivity and showed that the maximum reduction in thermal conductivity could be achieved when all cracks were oriented perpendicular to the direction of heat flow.[22] Moreover, cracks oriented parallel to the direction of heat flow have negligible effects on thermal conductivity. Similar to a plasma-sprayed coating, a D-gun sprayed coating is formed by

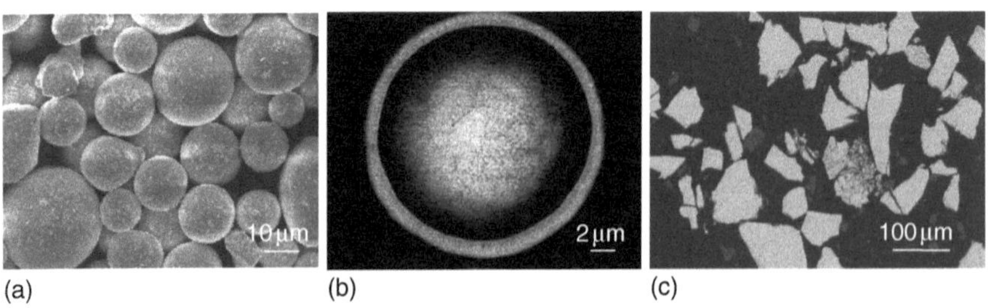

9.2 SEM images of YSZ powders; (a) surface view of HSP; (b) cross-sectional view of HSP; (c) cross-sectional view of NSP.

a succession of molten and partly-molten droplets impacting on a substrate followed by lateral flattening, rapid solidifying and cooling, resulting in a ceramic top coat with a regular and thin lamellar microstructure. The lamellar intervals are applicable to the cracks oriented perpendicular to the direction of heat flow. As a result, the thermal conductivity of D-gun YSZ coatings is lower than that of bulk materials and EB–PVD YSZ coatings which have a columnar microstructure parallel to the direction of heat flow.

9.2.3 Failure mechanism of D-gun TBCs during thermal shock cycling

For plasma-sprayed TBCs with a rough bond coat, failure occurs during crack evolution in the TBC itself, which occurs on planes nominally parallel to the substrate. Conversely, EB–PVD coatings are deposited on smooth bond coats and failed, typically, by loss of adherence at the thermal grown oxide (TGO)/bond coat interface. In both cases, failure is initiated by the energy density in the TGO and exacerbated by sintering of the TBC. This occurs because a thin, highly stressed TGO develops beneath the TBC. The specific mechanisms that contribute to failure of D-gun TBCs are described in this section, with a view to enabling the design of TBCs with greater reliability and durability.

Stress within the TGO exerts a crucial influence on TBC failure.[23] Understanding the stress distributions is important to any failure mechanism model, and curvature and X-ray residual stress evaluation are the methods most commonly adopted.[24] However, curvature methods are not usually appropriate for stresses between alternate layers and thus may not lead to any bending of the multi-layer as a whole. Whether X-ray diffraction provides a measure of the stress in the top layer only, or an average over many layers, curvature methods depend largely on the layer thickness. Moreover, problems include peak broadening for very fine grained coatings, loss of peak intensity and a depth of penetration (some tens of micrometers) which can be on a scale equal to the surface roughness for thick coatings. Other methods such as Raman spectroscopy have also been used. One particularly promising method is photo-stimulated luminescence spectroscopy (PSLS)[25] since it is non-destructive and it produces a direct measure for local elastic strain energy in the TGO.

There are two main sources of the stresses.[26,27] (i) thermal expansion mismatching upon cooling; and (ii) TGO growth. Studies have indicated that growth stresses range from near zero to about 1 GPa, and laser piezo-spectroscopic measurement shows that thermal expansion misfit results in compressive stresses ranging from 3–4 GPa. The stresses within the TGO layer are non-uniform, but all compressive stresses use Eq. [9.2] which is similar to that given in reference[25] but modified to reflect one-axis stress within the cross-sectional TGO layer

$$\Delta\upsilon = \frac{1}{3}\Pi_{ii}\sigma \qquad [9.2]$$

where Π_{ii} is the trace of the piezo-spectroscopic coefficient tensor with a value of 7.6 cm^{-1}/GPa, $\Delta\upsilon$ is the frequency shift (cm^{-1}) of a Ruby (R2) line at about 14430 cm^{-1} and σ is the stresses in the TGO scale. Most of the stress values (thermal stresses together with growth stresses) ranged from 1–3 GPa. A low stress in these coatings provides little strain energy to drive failure and can be seen as the primary reason for their long spallation life under excessively severe thermal cycling conditions applied in D-gun-sprayed TBCs.

It should be noted that relatively higher stresses (4–5 GPa) were also found when measurements were taken where Ni, Cr-rich oxides were overlaid on the Al_2O_3 layer while 2–3 GPa was found in regions consisting of complete Al_2O_3. This higher stress in the TGO with Ni, Cr-rich oxides is not well understood. A possible reason is that the measured stresses may cover not only the thermal stress, caused by thermal expansion mismatches between metals and oxides, and growth stresses but also phase stresses due to thermal expansion misfit between Al_2O_3 and Ni, Cr-rich oxides.

The combination of stress distributions in the TGO layer and damage which occurred during thermal shock cycling together with the proposed failure mechanism for the damage to D-gun TBCs, which occurs strictly in the ceramic layer and at the TGO/YSZ interface close to the Ni and Cr-rich oxides, is illustrated in Fig. 9.3.

- Undulated TGO formed at the interface during thermal shock cycling results in microcracking at the YSZ/TGO interface and the formation of voids, decreasing toughness between TGO and YSZ (Fig. 9.3a).
- As a result of depletion of Al due to severe internal oxidation within the bond coat, voluminous breakaway oxides containing Cr and Ni formed at the interface, especially around protrusions into the top coat (Fig. 9.3b). Strain caused by thermal expansion misfit during cooling and tensile stresses in the oxides resulted in formation of cracks and voids within the non-alumina oxides, leading to delamination of the outer ceramic layer and failure of the coating at the regions where chemical failure has occurred.
- The nucleation and propagation of the cracks in the ceramic coat along the splat boundary due to the D-gun intermittent spraying process as well as cracks normal to the interface will induce wedge-shaped spallation areas, even in the entire ceramic coat. (Fig. 9.3c)

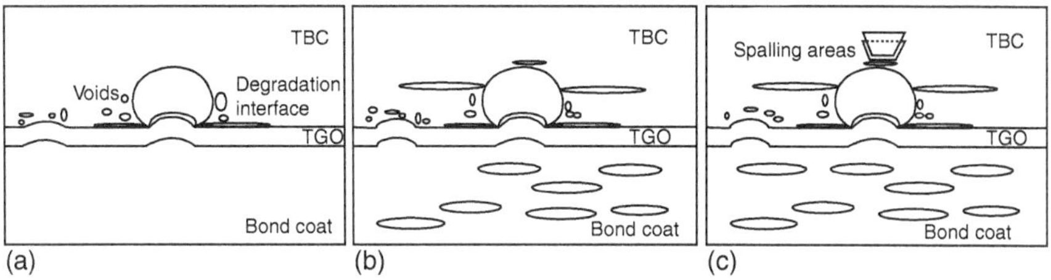

9.3 Schematic illustrations showing progressive microstructural evolution and damage accumulation during thermal shock cycling leading to spallation failure of the TBC: (a) interface delamination due to undulation development-induced cracking at the YSZ/TGO interface and formation of voids; (b) internal oxidation in bond coat and formation of Ni/Cr-rich oxides due to Al depletion – crack initiation and voids formation within the non-alumina oxides caused by strain evolving from thermal expansion misfit and tensile stress in non-alumina oxides; (c) wedge-shaped spalling areas caused by the propagation and linkage of the cracks along the splat boundary and vertical cracks in the ceramic coat.

9.3 TBCs deposited through arc ion plating (AIP)/D-gun two-step technology

As noted above, the primary failure mechanism of D-gun TBCs involves the depletion of Al in the bond coat and formation of non-alumina break-down oxides within the TGO layer at the bond coat and ceramic top coat interface. With regard to durability and reliability, the challenge is to improve the oxidation resistance of the bond coat at high temperature. Many attempts have been made to deposit metallic coatings other than MCrAlY in order to improve substrate properties. Here we will give one example, an arc ion plated (AIP) NiCoCrAlYSiB alloy coating, designed to confer a better oxidation resistance.

Arc ion plating (AIP), also referred to arc plasma deposition (APD), is a well recognized method of producing protective coatings due to its high ionization efficiency and excellent film adhesion. It is a promising method for depositing oxidation-resistant coatings on gas turbine components. This novel composite two-step AIP/D-gun method has been designed to produce TBCs with better oxidation resistance and thus better thermal shock resistance.[28]

9.3.1 Bond coatings obtained by AIP

TGO growth is the dominant factor in controlling durability; therefore, the chemistry and microstructure of the bond coat are crucial, because of their influence on the structure and morphology of the TGO. In this case, alloy coatings acting as bond coat deposited using AIP will be described below.

The provided NiCoCrAlYSiB bond coat has a relatively rough surface. Different sizes of particles were dispersed on the surface with smaller particles nucleating on larger ones. The larger particles were of different morphologies while smaller ones were globular. According to the X-ray diffusion (XRD) results, the as-deposited bond coat was composed mainly of γ'-Ni_3Al and γ-Ni phase. Within the as-deposited TBCs obtained by the AIP/D-gun two-step method, there was a uniform and dense NiCoCrAlY-SiB bond coat (generally 35–40 μm thick) with good adhesion to the substrate and the top coat.

As expected, the TGO scale is only Al_2O_3 phase during the thermal exposure of AIP/D-gun TBCs. Raman luminescence spectra on the TGO scale reveal additional peaks, together with α-Al_2O_3 peaks, after 1 h thermal exposure. These additional peaks are arrowed in Fig. 9.4, and reported to coincide with that of θ-Al_2O_3, indicating the co-existence of θ-Al_2O_3 and α-Al_2O_3 in the TGO at the early stage of oxidation. XRD analysis also confirmed this result. At 1100 °C, θ-Al_2O_3 and α-Al_2O_3 co-exist after 1 h oxidation. However, after 5 h oxidation only a weak θ-Al_2O_3 peak was identified, and α-Al_2O_3 was the dominant phase present in the oxide scale; extended oxidation causes no further change in phase composition of the TGO scale, only increasing the intensity of the α-Al_2O_3 peaks. This is borne out by Brumm and Grabke[29] who identified a region of phase co-existence at about 850–1200 °C.

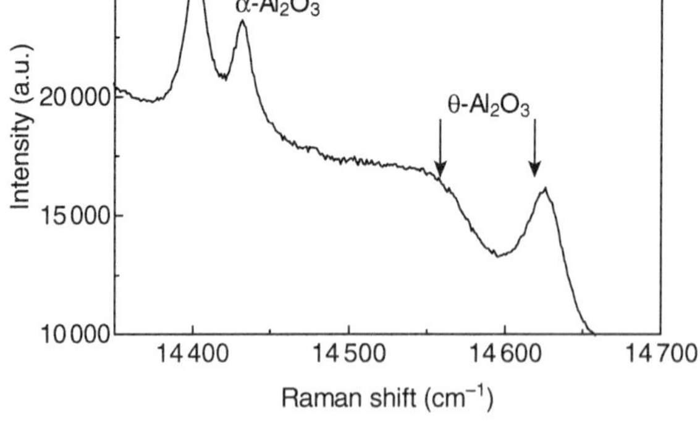

9.4 Typical R1 and R2 fluorescence spectra for α-Al_2O_3, and an additional peak coinciding with θ-Al_2O_3.

9.3.2 Stress distributions of thermal grown oxide (TGO)

Equation [9.3][25] is used to estimate internal stresses in the TGO when the TGO is assumed to be untextured and polycrystalline under biaxial-stresses:

$$\Delta\upsilon = \frac{2}{3}\Pi_{ii}\sigma \qquad [9.3]$$

Spectra were recorded from samples that were thermally exposed for different times at 1100 °C. On the basis of the observed frequency shifts of the R2 line, the corresponding biaxial compressive stresses can be produced by inverting Eq. [9.3]. The measured stresses were varied. It should be noted that the transformation of metastable θ-Al_2O_3 to α-Al_2O_3 is generally accompanied by volume shrinkage of about 13%, creating a tensile stress (or decreasing compression). Therefore, compressive stresses of TGO during the transient oxidation period from the metastable θ-Al_2O_3 to the stable α-Al_2O_3 (~5 h) were at a lower level. After the transient period, the average stresses ranging from 0.5–1.5 GPa were almost independent of oxidation time. However, at long oxidation times (thermally exposed for 500 h), the stresses appeared to decrease due to the spallation failure.

9.3.3 Microstructure evolution during thermal exposure

The changes in coatings microstructure with increasing oxidation time, especially at the TGO/TBC and TGO/bond coat interfaces shown in Fig. 9.5a–f have four features. (i) a more undulated and thickened TGO layer; (ii) separation between the bond coat and the TGO; (iii) cracks running across the TGO; and (iv) cracking within the top coat and along the TGO/bond coat interface. Initial cracking across the TGO was commonly observed around all undulation crests in those TBCs thermally exposed after 200 h, which can be seen from Fig. 9.5e. Meanwhile, the cracks within the top coat increased in length over time. It has been suggested that their eventual linkage would lead to the spallation of the lamellar TBC in the ceramic coat, which may account for the thinner top coat, with thickness of 70 μm for the sample after 400 h thermal exposure (as shown in Fig. 9.6a) compared with ~200 μm for the as-received samples. AIP/D-gun TBCs after 500 h decayed by instant peeling of the ceramic top coat from the substrate on cooling to room temperature (Fig. 9.6b). The ceramic top coat split into thin pieces and peeled off, although some YSZ still adhered to the substrate. Energy dispersive X-ray (EDX) composition analysis of the spallation area in the plan view (Fig. 9.6c) implied that the light region consisted of some remaining YSZ, while the area where TBC had been spalled off during cooling to room temperature was Al_2O_3 TGO layer (the grey black region)

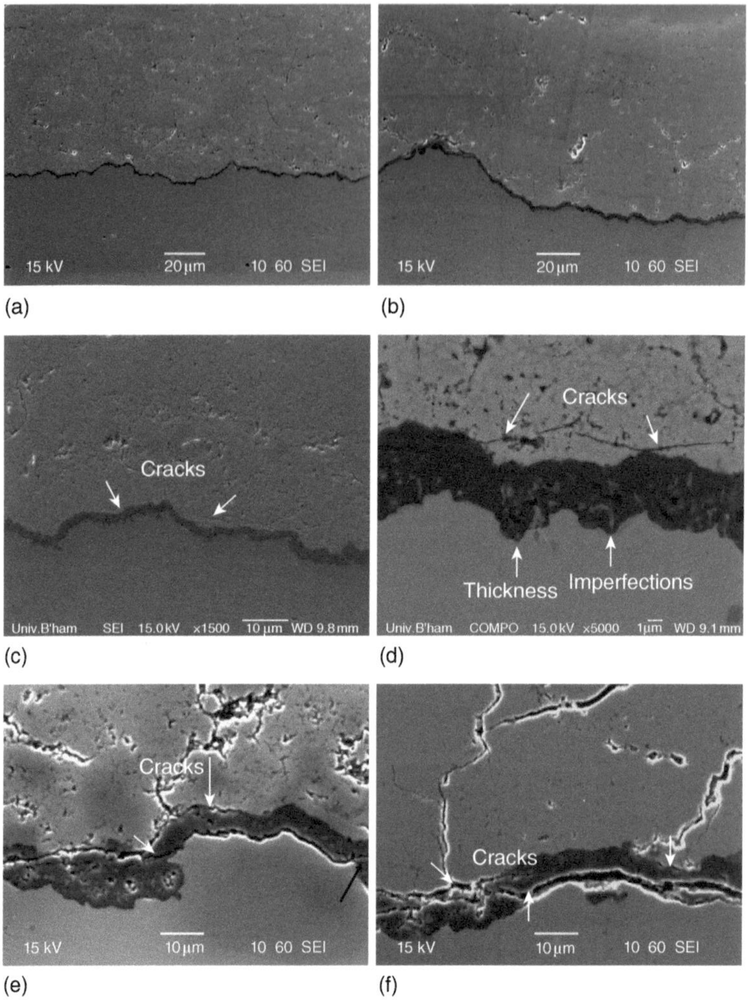

9.5 Change of coating microstructure with thermal exposure at 1100 °C for (a) 1 h; (b) 5 h; (c) 10 h; (d) 25 h; (e) 200 h; (f) 300 h, showing the TGO thickening and becoming more undulated, cracks running across the TGO and debonding damage along the bond coat/TGO interface.

and bare bond coat (Ni/Cr rich indicated by arrows). The fracture surface from the sample suggested that the failure of the AIP/D-gun TBCs occurred at the TGO/bond coat interface, leading to lamellar spallation of TBC and even peeling of the entire ceramic top coat.

9.6 (a) Cracking sequence caused by growth misfit, followed by cooling to ambient and the radial crack penetrating through the TGO at its inner edge. (b) Surface images of sample after 500 h cooling to ambient from macro-view. (c) Back scattered electron image of spallation area.

9.3.4 Failure mechanism of AIP/D-gun TBCs during thermal exposure

Compared with the thermal mismatch stresses (generally 3–4 GPa), the stresses in AIP/D-gun TGO were much lower. The exact cause for this was not well understood. Possible reasons could include different instrumentation as well as different methods of analyzing the recorded spectra. However, a reasonable deduction might be linked to greater strain tolerance of the HSP–YSZ top coat and better oxidation resistance in AIP/D-gun TBCs compared to D-gun TBCs with a NSP–YSZ top coat.

The fact that the stress values after the transient oxidation period did not change much with thermal exposure time, together with a slight decrease for the sample after 500 h thermal exposure, and the occurrence of catastrophic spallation failure at ambient temperature after cooling, suggests that spallation failure might be determined by an environmental-driven sub-critical crack along with the increase in the growth along the bond coat/TGO scale interface. Therefore, we could find that the decrease in biaxial stress at long oxidation times was a direct result of the growth of debonding damage at the TGO/bond coat interface which would result in stress relief as the materials expanded.

9.4 Future trends

Better thermal insulation of the hot path components is needed in state-of-the-art gas turbines and in diesel engines because of the increasing demands created by high process temperatures. As previously described, taking advantage of the higher closed porosity for HSP–YSZ, HSP–YSZ D-gun TBCs with HSP–YSZ have a lower thermal conductivity than those TBCs with NSP–YSZ obtained by D-gun, PS and EB–PVD. Thus the next step in the development of D-gun TBCs should be aimed at lowering thermal conductivity and increasing coating thickness, thick TBCs being an attractive option. One problem in the application of thick coatings which needs addressing is the tensile stresses within a coating system. It is important to note that D-gun TBCs have been found to be a compressive stress state within the coatings system, which has potential for thick (TBCs).

It is a fact that the failure of the conventional duplex TBC systems is mainly caused by the thermal expansion mismatch between the ceramic top coat and metal bond coat of the systems. One method of solving this problem is to introduce the concept of functionally gradient material (FGM) into TBCs, which are referred to as 'FGM–TBCs'; however, the fabrication of a fine mixture of ceramics and metals with a compositional gradient is quite difficult. Several processing techniques have been explored, such as plasma spraying, powder metallurgy, and *in situ* synthesis, but finding the optimum process for the fabrication of FGM–TBCs still remains a challenging task. Despite the difficulty, a new approach (a 'shot control method') in D-gun spraying with the potential to achieve an excellent FGM–TBC was successfully developed by the work of Kim *et al.*[16] The FGM layer of the FGM–TBCs prepared using this method exhibited a well-mixed microstructure of metals and ceramics with a desired compositional gradient in the thickness direction. Microhardness measurements and XRD analyses revealed that the ceramics and metals mixed in the FGM layer maintained their individual properties without severe oxidation or phase transformation. The expected improvement in the thermal shock resistance due to the realization of a functionally graded layer between the ceramic and the metal coating layers in TBC systems was also proved, as shown by thermal shock tests using a burner rig tester.

Another viable method is to produce nano-structured coatings because of their extraordinary properties. Novel properties of nano-structures have been widely reported. For example, nano-sized powder materials, nanocomposite or nano-scale multi-player thin films frequently exhibit novel properties that cannot be implemented in materials of conventional microstructures, demonstrating enhanced properties, including hardness, strength, ductility and toughness, and nano-structured YSZ coatings have low thermal conductivity, high coefficient of thermal expansion and excellent mechanical properties.[30–32] Therefore, use of nano-structure materials is expected to improve the performance of TBCs.

One of the biggest challenges in thermal spraying of nano-materials is to retain the pre-existing nano-structure of the feedstock. During the plasma spraying of ceramics, it is necessary to partially melt the powder particles in order to achieve the necessary physical conditions for cohesion and adhesion. If nano-structured powder particles are fully melted during spraying, then the traditional behavior of thermal spray particles, such as solidification, nucleation and growth, will take place. Such processes will destroy the original nano-structured features of the feedstock. D-gun spraying technology has advantages over both plasma spraying and high-velocity oxygen fuel (HVOF), since it has a moderate flame temperature for melting nano-materials instead of a high velocity to produce high-performance coatings.

9.5 Conclusion

In aircraft turbine applications, the spallation resistance and the thermal conductivity of the coating system are the main properties. This chapter focused on these two issues. The results showed that D-gun-sprayed TBCs with a lamellar structure had a low thermal conductivity, close to the thermal conductivity of the plasma-sprayed YSZ and much lower than their EB–PVD counterparts, due to microstructural difference.

Different process methods lead to different microstructure features which are associated with different failure mechanisms. Unlike PS where failure occurred due to crack evolution in the TBC itself and EB–PVD where failure, typically, occurred at the TGO/BC interface due to loss of adherence, the events that lead to the failure of D-gun sprayed TBCs are motivated by chemical failure and occur strictly in the ceramic layer and at the TGO/YSZ interface near Ni and Cr-rich oxides. Regarding TBCs modified by a two-step AIP/D-gun process, spallation failure might be determined by an environmental-assisted sub-critical crack with growth along the bond coat/TGO scale interface.

9.6 References

1. U. S. PATENT. 2714563, 1955.
2. ENGLISH PATENT. 742387, 1952.
3. R. C. TUCKER, JR. Structure property relationships in deposits produced by plasma spray and detonation gun techniques. *J. Vac. Sci. Technol.*, 1974 **11**(4):725–734.
4. Y. A. KHARLAMOV. Detonation spraying of protective coatings. *Mater. Sci. Eng.*, 1987 **93**:1–37.
5. X. Q. MA, Y. M. RHYIM, H. W. JIN, C. G. PARK and M. C. KIM. Application of fractal dimension to optimum deposition of NiCrAlY coating by D-gun spray. *Mater. Manu. Proc.*, 1999, **14**(2):195–204.
6. Y. N. WU. *Study on the High Temperature Properties and Failure Mechanisms of Protective Coatings for Blades of Gas Turbine*, Postdoctoral Research report, IMR, CAS. Shenyang, China, 2003.
7. P. L. KE. *Study on Thermal Shock Cycling Behaviors and Failure Mechanisms of Thermal Barrier Coatings Obtained by Detonation Gun Spraying*. Dissertation for degree of Doctor of Philosophy, IMR, CAS, Shenyang China,. 2006.

8. Q. T. MI, J. P. CHEN and Y. K. FENG, Detonation gun spraying, *J. Mater. Eng.*, 1990, **6**:1–5.
9. S. LIU, Application of detonation gun spraying, *Welding and Joining*, 1997, **9**:2–5.
10. J. X. TANG, A. B. ZHANG, J. P. CHEN and Q. T. MI, Analysis on processing principle of detonation gun spraying, *Mater. Prot.* 2000, **9**(32):33–35.
11. G. H. LU, P. ZENG, H. P. HUANG, G, T, WANG, G. R. XIE, Z. P. PAN and S. J. HU, Microstructure and properties of Al_2O_3 ceramic gradient coatings deposited by detonation gun spraying, *J. Mater. Eng.*, 2000, **32**(4):30–33.
12. A. G. EVANS, D. R. MUMM, J. W. HUTCHINSON, G. H. MEIER and F. S. PETTIT, Mechanisms controlling the durability of thermal barrier coatings, *Prog. Mater. Sci.*, 2001, **46**(5):505–553.
13. N. P. PADTURE, M. GELL, and E. H. JORDAN, Thermal barrier coatings for gas-turbine engine applications, *Science*, 2002, **296**:280–284.
14. M. GELL, L. D. XIE, X. Q. MA, E. H. JORDAN and N. P PADTURE, Highly durable thermal barrier coatings made by the solution precursor plasma spray process, *Surf. Coat. Technol*, 2004, **177–178**:97–102.
15. X. Q. MA, Y. M. RHYIMN, H. W. JIN, C. G. PARK and M. C. KIM. Effect of aluminum oxide diffusion barrier layer on the oxidation behavior of detonation gun-sprayed thermal barrier coating at 1200 °C, *Corrosion*, 1999, **55**:805–813.
16. J. H. KIM, M. C. KIM and C. G. PARK. Evaluation of functionally graded thermal barrier coatings fabricated by detonation gun spray technique, *Surf. Coat. Technol.*, 2003, **168**:275–280.
17. Y. N. WU, F. H. WANG, W. G. HUA, J. GONG, C. SUN and L. S. WEN, Thermal barrier coatings obtained by detonation spraying, *Surf. Coat. Technol.*, 2003, **166**:189–194.
18. P. L. KE, Y. N. WU, Q. M. WANG, J. GONG, C. SUN and L. S. WEN. Study on thermal barrier coatings deposited by detonation gun spraying, *Sur. Coat. Technol.*, 2005, **200**:2271–2276.
19. P. L. KE, Q. M. WANG, J. GONG, C. SUN and Y. C. ZHOU. Progressive damage during thermal shock cycling of D-gun sprayed thermal barrier coatings with hollow spherical ZrO_2–$8Y_2O_3$, *Mater. Sci. Eng. A*, 2006 **435–436**:228–236.
20. R. E. TAYLOR. Thermal conductivity determinations of thermal barrier coatings, *Mater. Sci. Eng. A*, 1998, **245**:160–167.
21. P. L. KE, Q. M. WANG, M. H. GUO, J. GONG, C. SUN and Y. C. ZHOU. Failure mechanism of D-gun sprayed thermal barrier coatings subjected to thermal shock cycling, *Mater. Sci. Forum*, 2005, **475–479**:3977–3980.
22. D. P. H. HASSELMAN. Effect of cracks on thermal conductivity, *J. Compos. Mater.*, 1978, **12**:403–407.
23. C. H. HSUEH and E. R. FULLER, Analytical modeling of oxide thickness effects on residual stresses in thermal barrier coatings, *Scripta. Mater.*, 2000, **42**: 781–787.
24. P. J. WITHERS and H. K. D. H. BHADESHIA, Overview: residual stress Part 2 – Nature and origins, *Mater. Sci. and Technol.*, 2001, **17**:366–375.
25. R. J. CHRISTENSEN, D. M. LIPKIN and D. R. CLARKE. Nondestructive evaluation of the oxidation stresses through thermal barrier coatings using Cr^{3+} piezospectroscopy. *Appl. Phys. Lett.*, 1996, **69**:3754–3756.
26. D. M. LIPKIN and D. R. CLARKE. Measurements of the stress in oxide scales formed by oxidation of alumina-forming alloys, *Oxid. Met.*, 1996, **45**:267–280.

27. J. R. CHRISTENSEN, V. K. TOLPYGO and D. R. CLARKE, The influence of the reactive element yttrium on the stress in alumina scales formed by oxidation, *Acta Mater*, 1997, **45**:1761–1766.
28. C. SUN, P. L. KE, Y. N. WU, W. G. HUA, J. GONG, L. S. WEN, Q. M. WANG, J. Q. XIAO, A. Y. WANG, A. L. JI, J. D. ZHENG and H. T. CAO. *A Method of Producing Thermal Barrier Coatings with Thermal Shock Resistance*, Chinese Patent No. 03133344.3, 2006.
29. H. J. GRABKE, M. W. BRUMM, and B. WAGEMANN, The oxidation of NiAl, *Mater. Corros.* 1996, **47**:675–677.
30. B. H. KEAR and G. SKAUDAN, Thermal spray processing of nanoscale materials, *Nanostruct. Mater.*, 1997, **8**:765–769.
31. M. GELL, Application opportunities for nanostructured materials and coatings, *Mater. Sci. Eng. A*, 1995, **204**:246–251.
32. G. SOYEZ, J. A. EASTMAN, L. J. THOMPSON, R. J. DIMELFI, G. R. BAI and P. M. BALDO, Grain-size-dependent thermal conductivity of nanocrystalline yttria-stabilized zirconia films grown by metal-organic chemical vapor deposition, *Appl. Phys. Lett.*, 2000, **77**:1155–1157.

Part III
Performance of thermal barrier coatings

10
Oxidation and hot corrosion of thermal barrier coatings (TBCs)

C. G. ZHOU and Y. X. SONG, Beihang University, China

Abstract: Thermal barrier coatings (TBCs) with a two-layered structure consisting of an oxidation protective bond coat and a thermally insulating yttria stabilized zirconia (YSZ) top coat are used to provide thermal insulation and oxidation resistance to metallic components from the hot gas stream in gas turbine engines used for aircraft propulsion, power generation and marine propulsion. The oxidation behavior of the bond coat, including both MCrAlY and Pt aluminide coating, has a major impact on the life of the TBCs. The oxidation behavior and failure mechanisms of TBC systems are summarized. In addition, the degradation mechanisms experienced by TBC systems exposed to deposits are also discussed.

Key words: thermal barrier coatings (TBCs), oxidation behavior, failure mechanism, degradation mechanism.

10.1 Introduction

Oxidation is a phenomenon whereby metallic elements of materials exposed to oxygen or oxygen-containing atmospheres at elevated temperatures are converted into their oxides.[1] The successful application of many materials requires excellent high-temperature properties. Surface stability of the candidate material is necessary in order to prevent rapid material consumption at high temperatures. In general, this indicates that the exposed material should be able to form a slow-growing, stable, adherent surface scale, which would protect the underlying substrate material. The most important and effective scales are oxides; they are easily produced through reaction with oxygen in the air or with oxygen-containing molecules such as H_2O, CO_2 or CO in other gaseous environments. The most sought-after oxide scales are Al_2O_3, Cr_2O_3 and SiO_2 because of their outstanding protective qualities. Many of the high-temperature components used in aircraft and ground power gas turbines rely on these scales for protection.[2]

When materials are exposed simultaneously to oxygen and some other environmental constituents such as CO_2 and SO_2, fused or molten salts and solid particles, the interaction between these environmental constituents and the materials results in corrosion and erosion.[1] Hot corrosion is a serious problem in power generation equipment, gas turbines, internal combustion engines, fluidized bed combustion, industrial waste incinerators and the paper and pulp industries. This phenomenon has been noticed in boilers, internal combustion engines, gas turbines, fluidized bed combustion and industrial waste incinerators since the 1940s. However, it was not until the late 1960s, when gas turbine engines in military aircraft suffered severe corrosion while operating over sea water during the Vietnam conflict, that it became a topic of importance and general interest.[3,4] Although there are some alloy compositions that require a long initiation time, during which the hot corrosion process moves from the initiation stage to the propagation stage, there is no alloy which is immune to hot corrosion attack indefinitely. Superalloys have been developed for high-temperature applications. However, these alloys are not always able to meet the requirements of both high-temperature strength and high-temperature corrosion resistance simultaneously. To achieve both strength and resistance to environmental degradation, these two functions should be separated. The load capability is provided by the substrate alloy, while the oxidation and corrosion resistance is achieved by the application of coatings. The high-temperature protection system has to meet several criteria, such as providing adequate environment resistance, being chemically and mechanically compatible with the substrate and being applicable in practice, reliable and economically viable.[5]

10.2 Oxidation of thermal barrier coatings

Typical TBCs have a two-layered structure consisting of an oxidation protective bond coat and a thermally insulating yttria-stabilized zirconia (YSZ) top coat.[6] The metallic bond coat not only provides a surface for the top coat to bond to, but also prevents oxidation and corrosion of superalloy substrates as YSZ coating is essentially a porous layer that is transparent to oxygen penetration.

It has been determined that the oxidation behavior of the bond coat has a major impact on the life of the TBCs. Several factors can affect oxidation resistance of the bond coat, including: (i) bond coat and substrate materials; (ii) bond coat and substrate interdiffusion; (iii) characteristics and morphologies of the thermal grown oxides (TGO); (iv) phase transformation and cracking of the coating.[7–12] There are two different types of bond coats currently used in TBC systems, namely the MCrAlY bond coat, where M is a metal such as Ni and/or Co, and the platinum-modified nickel aluminides. The purpose of both coating types is to provide oxidation and corrosion resistance as metallic coatings. Thus, a brief overview of oxidation will be presented.

10.2.1 Oxidation mechanisms of metals and alloys

There are two means of oxidation of metals and alloys: metal atoms diffuse outwards from the metal and react with oxygen at the outer surface of the oxide, or oxygen diffuses and reacts with metal atoms at the metal oxide interface. Additionally, oxides can form subsurface precipitates instead of a discrete layer; this type of oxidation is especially common for high-temperature alloys exposed to an oxidation environment for extended time periods.[13] In some cases, the diffusion of both metals and oxygen is significant and new oxides can form within the oxide layer.[14] The growth of oxides thus forms a continuous, slow-growing scale.

In a binary alloy A–B, the more reactive element may be preferentially oxidized, either internally or externally, depending on content. It is possible that the oxidized element will form a protective scale on the surface of the alloy, so that the other elements will not be oxidized. The growth of a protective scale requires a sufficient supply of the metal that is oxidizing. In order to keep the protective scale growing, the rate of metal consumption due to oxide scale growth has to be slower or equal to the rate at which the metal atoms can diffuse to the scale.[13]

Stress can be generated in a metal/oxide system due to either the volume the oxide takes up being different from that of the metal it replaces, or differences in thermal expansion coefficients between the metal and the oxide, or more complicated mechanisms taking place, especially when alloys are considered. For an oxide that grows at the oxide/metal interface, the compressive stress slows down the flux of anion vacancies; thus the diffusion of oxygen ions to that interface[15] can be slowed down. An additional oxidation mechanism may cause stress due to oxygen going into solution in the substrate. Some metals have a very high solubility for oxygen, causing a large dilational strain that can lead to tensile failure of the oxide.[15]

10.2.2 Oxidation behavior of thermal barrier coatings (TBCs)

When a MCrAlY bond coat is used, the TGO frequently exhibits an inhomogeneous morphology and/or composition, especially in the early stages of oxidation.[16] This is caused by the multi-phase character of the MCrAlY coating, which generally consists of γ-Ni and β-NiAl; depending on composition and temperature, further phases, such as γ'-Ni_3Al, α-Cr, Ni–Y and/or σ-CoCr, may appear.[17] By first exposing to a high temperature, i.e., during heat treatment, coating manufacturing and/or actual service, every individual phase prevailing in the coating will tend to form a different type of oxide, e.g., α-Al_2O_3, θ/γ-Al_2O_3, Cr_2O_3, NiO, $(Ni,Co)(Cr,Al)_3O_4$, etc.[16] The transition of this heterogeneous scale into a stable α-Al_2O_3-based layer can lead to stress initiation and/or result in formation of cracks, which may be the starting points for failures during high-temperature service, especially during thermal cycling.

The addition of precious metals, such as Pt, Ru or Ir, to aluminide coating allows the formation of an adherent α-alumina scale and improves resistance to oxidation and hot corrosion. Since the early 1970s, Pt–aluminide diffusion layers have been successfully deposited as protective layers on high-temperature components.[17] However, initially, these coatings served exclusively for protecting metals from high-temperature oxidation and hot gas corrosion, and therefore did not consist of a ceramic top coat for insulation. The application of Pt–aluminide coatings as a bond coat was first mentioned in 1993 in the patent of Duderstadt and Bangalore.[18] These layers are normally produced by initial electrodeposition of a Pt layer 8–10 mm thick followed by a diffusion treatment, either as a separate thermal treatment or combined with the aluminizing step. The aluminization is performed using a pack cementation or a chemical vapor deposition (CVD) process.[19] The outer layer of the modified aluminides comprises one of the intermetallic phases $PtAl_2$, Pt_2Al_3, PtAl[20] or Pt in solid solution, depending on the deposition parameters such as thickness of the Pt layer, duration and temperature of the thermal treatment or the activity of Al during the pack cementation process. According to Tawancy *et al.*,[9] Pt excludes refractory metal strengthening elements out of the layer. This promotes the selective oxidation of Al.

Pt-modified aluminide bond coats have shown substantially improved oxidation resistance for TBCs, but they are prone to 'rumpling', during which TGO can initiate cracks, leading to spallation of the top coat.[20–24] A rumpled bond coat has a very negative effect on TBC adhesion as shown in Fig. 10.1.[25] In addition, Pt lacks adequate strength at high temperature,[26] and the cost of Pt is much higher than that of other noble metals. As

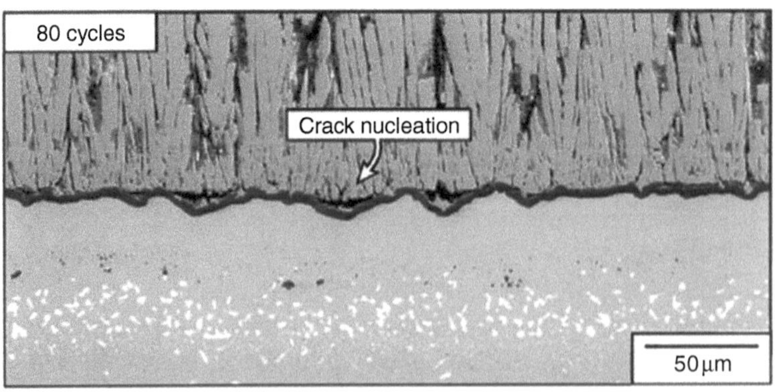

10.1 Rumpling on a NiAl bond coat after thermal cycling.[25]

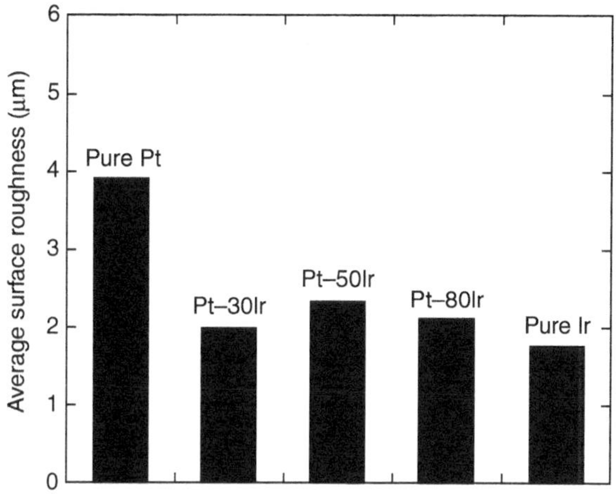

10.2 Average surface roughness of specimens measured by a surface profiler after cyclic oxidation.[27]

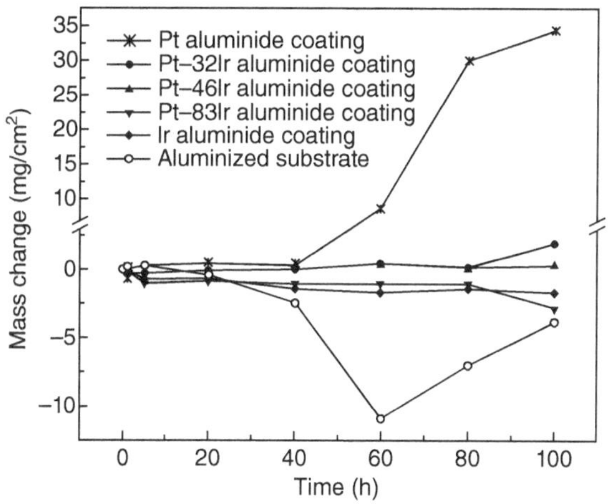

10.3 Corrosion kinetics for Pt–Ir aluminide coatings and aluminized substrate at 1173K coated with 90% Na_2SO_4 + NaCl salts.[28]

Fig. 10.2 illustrates, the addition of Ir to Pt-modified aluminide coating can be one means of retarding surface rumpling.[27] In addition, it was reported that PtIr-modified aluminide coating could enhance type I hot corrosion resistance, as shown in Fig. 10.3.[28] On the other hand, adding Ru to Pt-modified aluminide coating is of interest also because the addition of Ru to Ni-base superalloys has been proved to have the capacity to improve creep strength.[29]

It has also been proposed that Pt group metals, such as Ru, can potentially act as diffusion barriers preventing the migration of detrimental elements toward the BC/TGO interface, thus possibly minimizing the interdiffusion effects.[30–32] Recently, the replacement of Pt with Ru in the bond coat system has been investigated to fabricate a creep-resistant bond coat.[33] These experiments showed that Ru-modified aluminide bond coats offered less favorable oxidation resistance and oxide adherence in comparison with conventional Pt-modified aluminides.[29,33]

In order to utilize the beneficial creep properties of RuAl[34,35] without impairing the oxidation resistance of the bond coats, Tryon *et al.* investigated the hybrid Ru/Pt-modified bond coatings fabricated by electron beam physical vapor deposition (EB–PVD). It was found that these hybrid coatings exhibited comparable cyclic oxidation resistance and better creep properties compared with traditional Pt-modified coatings.[36] Three-layered Pt/Ru-modified aluminide coatings were developed by electrodeposition followed by the conventional aluminizing process. The Pt/Ru-modified aluminide coating showed better thermal cyclic oxidation resistance than Pt-modified aluminide coating since Ru provided an increase in the creep strength of the bond coat, with the result of retarding the rumpling phenomenon. The addition of Pt and Ru to aluminide has also been proved capable of increasing the surface concentration of Al and retarding the interdiffusion of the alloying elements.[37]

10.3 Failure mechanisms of TBCs

Experimental observations have shown that the failure of TBC systems is very complex and mostly affected by a superposition of different mechanisms. Previous experiments[38–42] have demonstrated that oxidation of the bond coat and thermal mismatch stress[43,44] contribute to the rapid degradation of TBCs. The major cause of TBC failure is delamination of the outermost coating layer from the bond coat to which it is applied.

The spallation of the TBC top coat is a very destructive event, as it will generally result in failure of the substrate in a relatively short time due to a large increase in the surface temperature of the metal. TBC spallation is initiated by cracking either at the TGO/bond coat interface or at or near the TGO/TBC interface.[45] It is widely accepted that once enough cracks form, they eventually coalesce into a crack large enough for the TBC to buckle and spall. In most cases, the cracking is attributed to the mechanical forces that result from the changing from a planar to a non-planar interface. During thermal cycling, progressive roughening of the bond coat/TGO/top coat interfaces occurs due to cyclic creep of the bond coat.[46] Such roughening, often termed 'ratcheting,' occurs only under thermal cycling, but not during isothermal exposures. It also requires an initial surface-geometry imperfection of some minimum dimension and the progressive lengthening of the TGO caused by TGO cracking or in-plane growth during oxidation.[47,48] For a nominally flat bond coat/TGO interface (EB–PVD TBC), this roughening is manifested in the form of TGO penetrating into the bond coat.[25,49] In an already undulated interface, the undulation amplitude is amplified – atmospheric plasma spraying (APS) APS TBC.[20] All of these geometrical factors result in out-of-plane stresses normal to the metal/ceramic interface, the severity of which increases with thermal cycling. These stresses, in combination with the interfacial imperfections, are primarily responsible for the TBC failure. The relative locations of the tensile stresses that form around a perturbation are shown in Fig. 10.4.[50]

Damage initiation and progression in the form of microcracks can occur in several different ways in TBCs, depending on the particularity of TBC, the portion of spallation life already consumed and the thermal cycling environment. The coalescence of these cracks results in the ultimate spallation failure of the TBC.[51] The initiation of cracks close to the interface between the TBC and the bond coat seems to be fully explained by the residual stress calculations made by Chang *et al.*[52] and others.[25,53–55] Meanwhile, numerous models based on different conceptions have been proposed, which might qualitatively explain how cracks propagate.[20,25,38,56] All models consider oxidation at the rough interface between TBC and bond coat as well as the phenomena resulting from reasons such as the redistribution of residual stresses induced by volume growth or morphological instability of the TGO layer[15,56–60] as the driving force of crack propagation. A fracture mechanics approach was developed,[61] by which cracks are modeled in the critical areas of the TBC system and assessed by using the modified crack closure integral method to determine the mode-dependent crack loading. The crack propagation capability is then predicted by using a recently-implemented mixed mode failure criterion and proper fracture toughness data.

The growth of the TGO during engine operation is the most important phenomenon responsible for the spallation failure of TBCs. TGO growth sometimes results in a constrained volume expansion that leads to compressive 'growth' stresses[59] persisting at all temperatures. Upon cooling, the thermal-expansion mismatch between the TGO and the bond coat leads

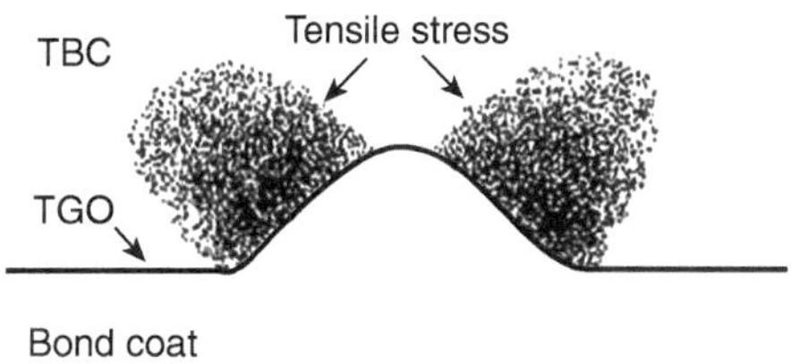

10.4 Tensile stresses in the TBC resulting from a surface imperfection.[50]

to very high 'thermal' compressive residual stresses in the TGO which reach a maximum at ambient temperature.[62,63] The strain energy in the TGO scales linearly with the TGO thickness and quadratically with the TGO stress, and drives fracture. In addition, a coefficient of thermal expansion mismatch between bond coat and ceramic topcoat can exacerbate a coating failure during thermal cycles.[64,65] It is seen that bond coat exhibits a higher coefficient of thermal expansion than ceramic top coat.

Evans *et al.*[66] provided an in-depth analysis of the various failure mechanisms of TBC systems. The mechanical failure mechanisms are the result of heterogeneities of TGO thickness in an otherwise planar interface, the formation of interface undulations or foreign object damage. The chemical failure mechanisms which result in the formation of Ni-rich oxides such as spinels are attributed to either Al-depletion, where the spinel forms below the alumina layer, or Ni diffusing through the alumina to form spinel on top of the alumina layer. In either case, the failure occurs within the spinel layer. The formation and growth of the α-Al_2O_3 TGO result in the depletion of Al in the bond coat. The Al depletion, if severe, results in the formation of other oxides, such as Ni- and Co-containing spinels, $Y_3Al_5O_{12}$ and Y_2O_3.[67,68] The formation of these phases compromises the structural integrity of the TGO and accelerates localized oxidation by providing fast oxygen-diffusion paths.

The detailed failure mechanisms in the two classes of TBCs (APS and EB–PVD) are discussed separately.[51] There are at least four primary failure mechanisms in APS TBCs driven by the out-of-plane stresses, which are shown schematically in Fig. 10.5a.[38,69,70] As the TGO thickens, the tensile stress at the bond coat/TGO interface increases, which causes cracking at the bond coat/TGO interface at crests (mechanism I in Figs. 10.5 a and b).[71,72] The thermal expansion mismatch between the top coat and the metal (bond coat/superalloy) puts the top coat in overall compression at room temperature. Because of the highly undulating nature of the metal/ceramic interface, out-of-plane stresses result in the vicinity of the TGO/top coat interface: tension at the crests and compression at the troughs.[73] The tension causes fracture along the TGO/top coat interface at the crests (mechanism II in Figs. 10.5 a and b) and cracking within the highly brittle top coat in the vicinity of the crests (mechanism III in Figs. 10.5 a and b).[69,70] Beyond a certain TGO thickness, the thermal expansion coefficient of the bond coat/TGO composite becomes lower than that of both the top coat and the bond coat, which converts the compression force in the top coat undulation troughs into the tension force.[70,74] This reversal causes cracking within the top coat in the 'valleys' between the crests (mechanism IV in Figs. 10.5 a and b).[70]

There are three main types of failure mechanisms in EB–PVD TBCs, two of which are schematically shown in Fig. 10.6a. Mechanism I is the separa-

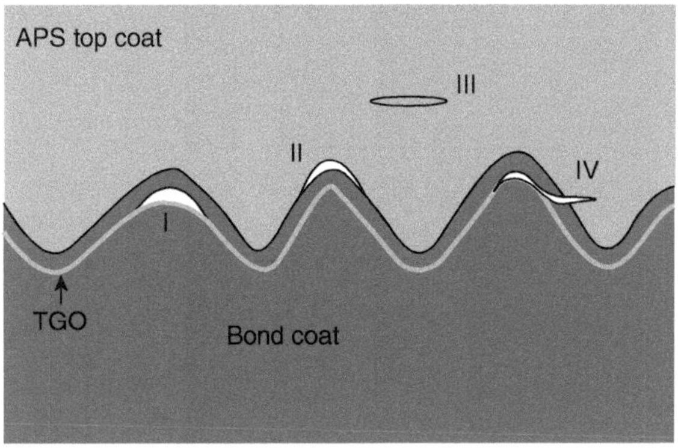

(a)

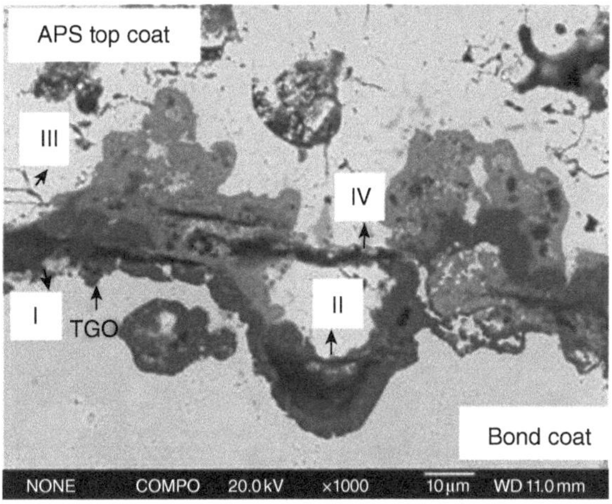

(b)

10.5 (a) Schematic diagram showing the four cracking mechanisms in APS TBC. (b) Cross-sectional SEM of a failed APS TBC showing the various cracking mechanisms illustrated in (a).[51]

tion of the bond coat/TGO interface, which is the same as that described in the APS TBC case (mechanism I in Fig. 10.5). The crests in the case of EB–PVD are 'ridges' present on the bond coat surface before top coat deposition (Fig. 10.6b).[75] Mechanism II in Fig. 10.6a is separation of the TGO/top coat interface and penetration of the TGO into the bond coat resulting from one or more of the following mechanisms: (i) progressive TGO roughening caused by bond coat cyclic creep (Fig. 10.6b);[25,49] (ii) accelerated growth of embedded oxides due to localized TGO cracking;[46,68,76] and (iii) cavity formation in the bond coat.[46] For EB–PVD TBCs with relatively flat and defect-free interfaces, the compression in the TGO causes large-scale buckling.[75,77]

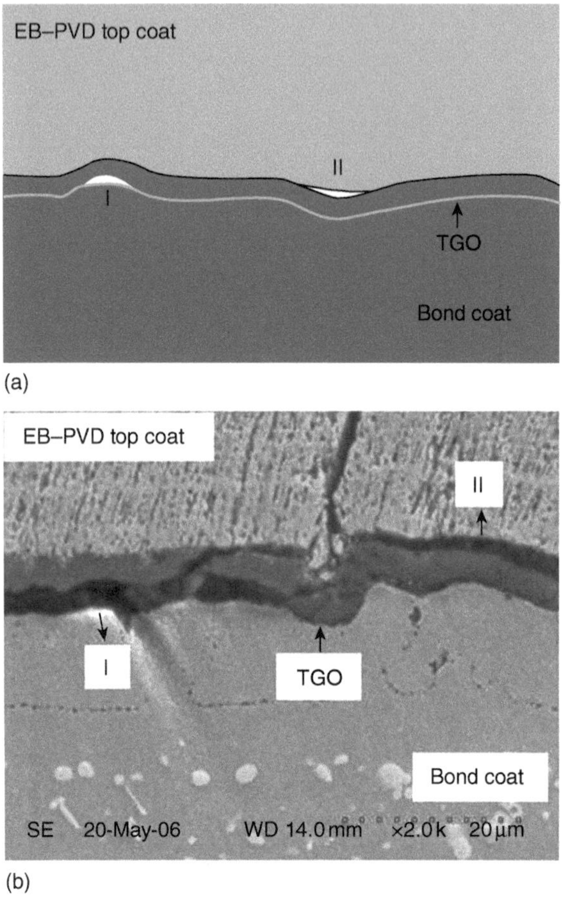

10.6 (a) Schematic diagram showing two of the three cracking mechanisms in EB–PVD TBC. (b) Cross-sectional SEM of a failed EB–PVD TBC showing the cracking mechanisms illustrated in (a).[51]

10.4 The degradation mechanisms experienced by TBC systems exposed to deposits

Corrosion induced by molten salts in an oxidizing gas at elevated temperatures is called 'hot corrosion'.[78–81] It occurs when metals and alloys are covered with salt films at a temperature typically between 700 and 925 °C. According to Hancock[82] and Eliaz *et al.*,[83] hot corrosion is an accelerated form of oxidation that occurs when metals are heated in the temperature range 700–900 °C in the presence of sulphate deposits, which are formed as a result of the reaction between sodium chloride and sulphur compounds in the gas phase surrounding the metal.

Corrosion which occurs above the melting point of the salt is called 'type I hot corrosion', while the corrosion at the lower end of the temperature range is called 'type II hot corrosion'.[1] The range of the two types of hot corrosion measured in terms of metal loss under the impact of temperature is schematically illustrated in Fig. 10.7.[1] In both types of hot corrosion, fluxing with corroding salts defeats the protective oxide scale that formed on superalloys and coatings. Once the protective scale is rendered ineffective, the substrate alloy or coating becomes vulnerable to internal oxidation and sulphidation. There are generally two stages of hot corrosion. The first stage is the initiation stage that involves the breakdown of protective oxide scale. The second stage is known as the propagation stage, in which the salts have access to the unprotected metal and corrosion continues deteriorating at exceedingly high rates.

The degradation mechanisms experienced by TBC systems exposed to deposits during service may be manifold. Deposits may affect both the zirconia top coat and the bond coat, and their effect may be predominantly either chemical or mechanical in nature. Molten salt can penetrate into the YSZ coatings along pores and cracks in YSZ and it reacts with the metallic bond coat.[84–88] The formation of monoclinic zirconia and/or the accelerated sintering of zirconia in the presence of oxides resulting from the chemical reaction of the deposits with the stabilizing phases in the zirconia top coat lead to the degradation of TBCs.[89–93] Basically, both the formation of monoclinic zirconia and enhanced sintering contribute to the increased stress build-up within the zirconia layer. Transformation of metastable tetragonal zirconia into the monoclinic phase results in a volume expansion of approximately 4% which, given a significant occurrence in a coating, can lead to significant degradation of the structural integrity of the ceramic coating. Sintering of the zirconia coating increases its Young's modulus, and thus adversely affects its mechanical properties.[94] In addition, infiltration of

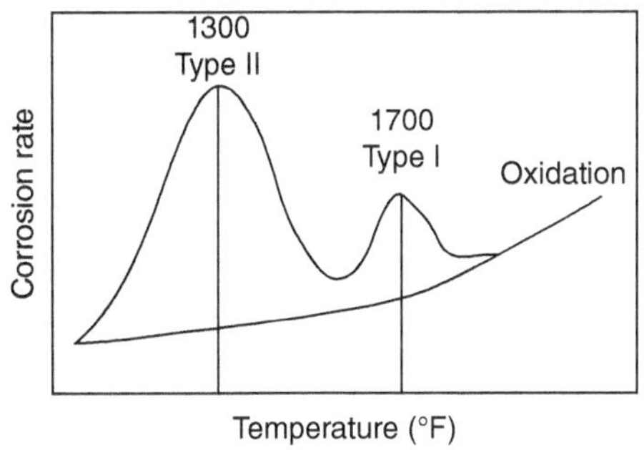

10.7 Schematic illustration of temperature effect on rate of damage to superalloys based on type I and II hot corrosion superimposed on contribution due to oxidation.[1]

molten deposits into the microstructure of a YSZ porous coating would cause additional stresses because of a thermal expansion mismatch between the deposits and the zirconia.[95,96] It has been suggested that TBC damage from the deposits will only be significant if the deposits melt and infiltrate the ceramic coating. Non-infiltrating deposits were considered benign as they do not impose additional stress on the TBC. However, it is possible for a chemical reaction of yttria from the YSZ coating to give rise to a tetragonal-to-monoclinic phase transformation. Borom *et al.*[96] proposed a damage map consisting of four in-service failure regimes for APS YSZ TBCs (Fig. 10.8). They categorized the regimes as infant mortality, particle erosion, infiltration of molten particles and thermochemical phenomena, such as, sintering, phase changes and bond coat oxidation, etc. Notably, for APS TBCs, hot corrosion attack of the bond coat seemed to play a significant role only after a very long operating period. In comparison, infiltration could affect the integrity of the ceramic coating in the range of several hundred to a few thousand hours, depending on TBC surface temperature and extent of deposit load.

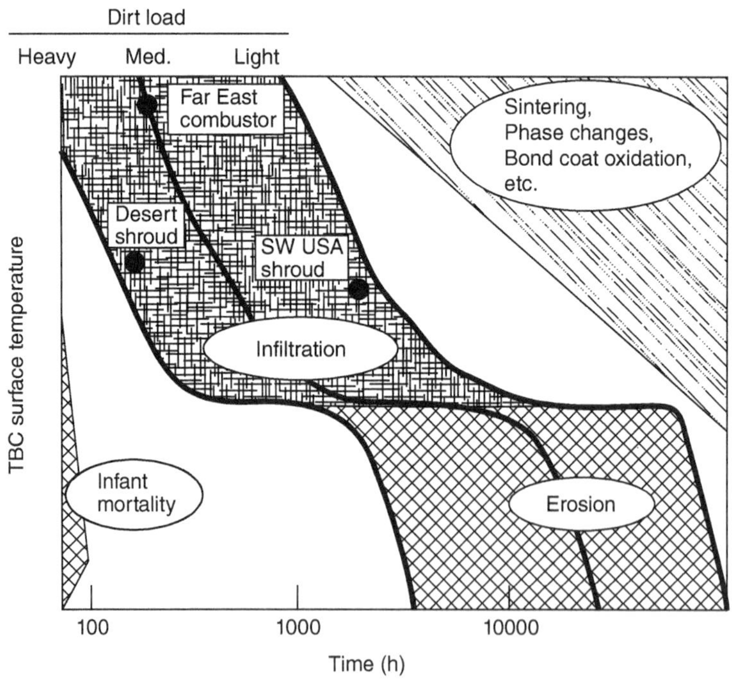

10.8 Hypothetical in-service failure regimes for air plasma-sprayed YSZ TBCs shown schematically on a plot of TBC surface temperature vs log of hours to failure.[96]

For sustained hot corrosion, a model that requires a negative solubility gradient has been proposed by Rapp and Goto,[97] i.e.,

$$\{d(\text{oxide solubility})/dx\}_{\text{at the oxide–salt interface}} < 0$$

where x is the distance from the oxide–salt interface. This mechanism does not consume the molten salt and therefore will not be sustained until either the melt becomes more basic and oxide ions are not produced or the oxide scale is completely removed and the metal becomes accessible to the salt. The self-sustaining corrosion process can prevent the formation of continuous and protective Al_2O_3.

The salts involved in hot corrosion are typically alkali and alkaline earth sulphates. The exact composition of the salts depends on the particular industrial process, impurities involved and the fuel, air and coolant compositions. Khanna and Jha[98] reported on the chemistry of formation of salt during the combustion of coal/fuel. According to them, the sulphur present in coal and fuel oils yields SO_2 on combustion, which can be partially oxidized to SO_3. NaCl (either impurity in the fuel or in the air) reacts with SO_3 and water vapor at the combustion temperature to yield Na_2SO_4, which deposits on the metal/alloy surfaces and would be present as liquid at a sufficiently high temperature. Small amounts of vanadium present in the fuel, on combustion, form V_2O_5, which can react with Na_2SO_4 to form low melting sodium vanadates which are highly corrosive. At high temperatures, deposits of Na_2SO_4 are molten (m.p. 884 °C) and can cause accelerated attack on the Ni and Co-base superalloys.

Accelerated corrosion can also be caused by other salts, viz. vanadates or sulphates–vanadate mixtures or in the presence of solid or gaseous salts, such as chlorides.[99] Vanadium and sodium are common impurities in low-grade petroleum fuels. Molten sulphate–vanadate deposits resulting from the condensation of combustion products of such fuels are extremely corrosive to high-temperature materials in combustion systems.[100] DeCrescente and Bornstein[86] were the first to demonstrate that a sodium sulphate deposit must be fused rather than solid in order to cause a significant attack. With regard to experiments conducted in air, Goebel *et al.*[101] found that basic fluxing in combination with sulphidation could cause the hot corrosion of Ni. The cyclic oxidation behavior of TBC exposed to NaCl vapor shows that the failure of the TBC occurs within the top coat and close to the YSZ/TGO interface. The formation of voluminous and non-protective oxide scales and the increased TGO thickness may be responsible for the accelerated failure of TBC.[102]

A finite element analysis was employed to analyze the stress distribution in the coatings.[103] The computed result showed that: (i) maximum stresses occurred at the bond coat/TGO interface near the edge of the sample; (ii) the increased TGO thickness caused an increase in stress in the TGO/

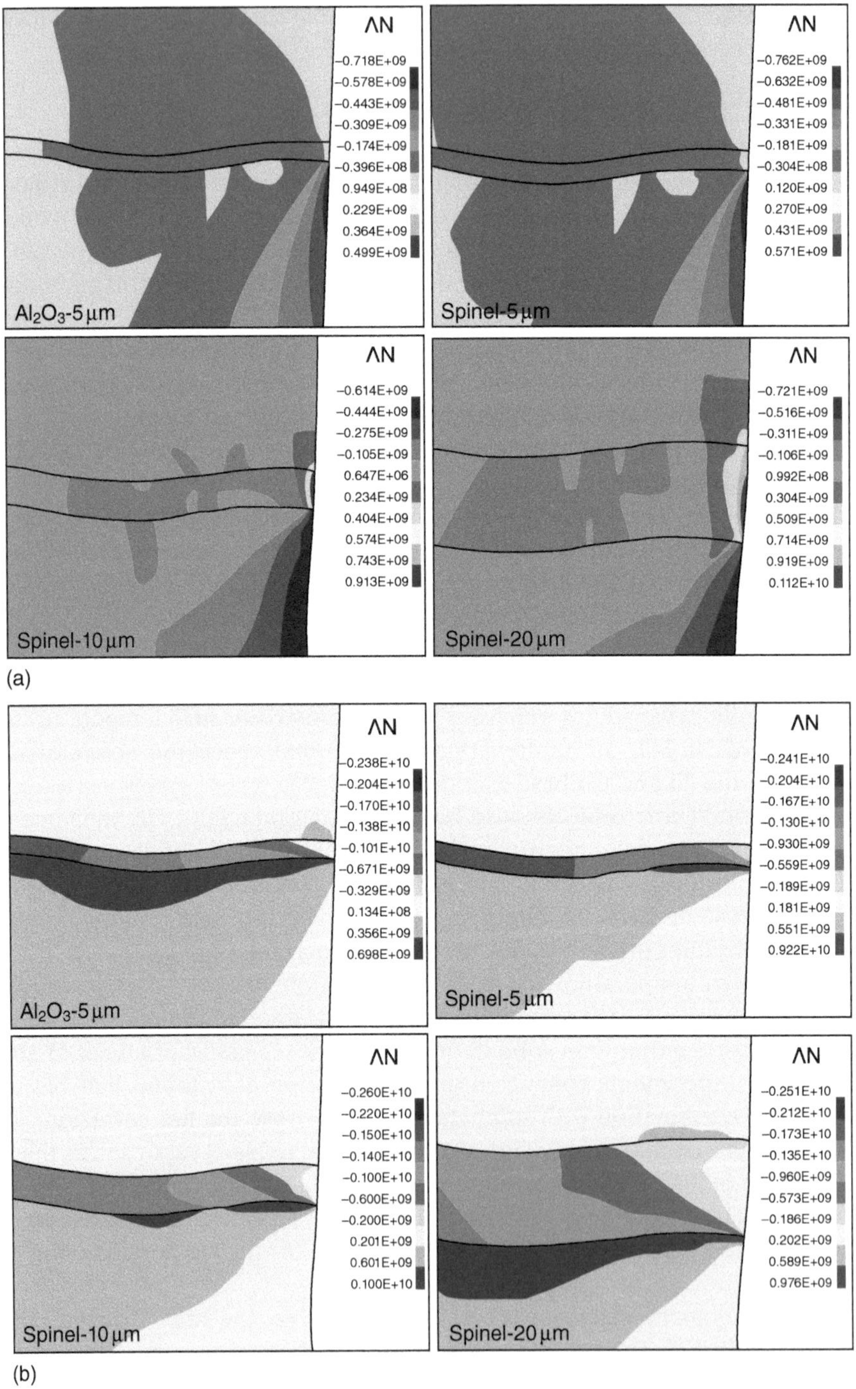

10.9 Typical contour plot of (a) axial and (b) radial stress distribution of TBC with TGO consisting of Al_2O_3 or spinel.[103]

YSZ interface, as shown in Fig. 10.9, and (iii) the spinel TGO resulted in significantly higher stresses than Al_2O_3 TGO, which implies that the formation of spinel plays a dominant role in shortening the coating cycling lifetime. Reese and Grabke[104] investigated the effect of NaCl on the oxidation of high-Cr and Ni–Cr steels. It has been found that the presence of NaCl(s) leads to the formation of voluminous, non-protective layers. The oxides grow according to the mechanism of active oxidation in which Cl plays a catalytic role.[104] Similar results have been found for Fe–Cr–Ni and Fe–Mn–Al alloys with NaCl deposits.[105–107] After several cycles, depletion of Al in the coating surface occurs. Due to the depletion of Al, Ni and Cr react with oxygen to form NiO and Cr_2O_3. The resultant NiO and Cr_2O_3 react with each other to form $NiCr_2O_4$ spinel. The oxide scale consisted of $NiCr_2O_4$ spinel, NiO and Al_2O_3. Moreover, the TGO thickness is higher in the presence of NaCl (g) in dried air.[108] Therefore, the decrease in thermal cycling resistance for the TBC can be attributed to the increased TGO thickness and the significant formation of voluminous spinel, which result in higher residual stresses in the coating.

The synergistic effect of NaCl and water vapor shortened the cyclic oxidation life of NiCrAlY+YSZ coating at 1050 °C as illustrated in Fig. 10.10.[109] The failure of the coating mainly occurred near the interface between bond coat and YSZ. There was no phase transformation of ZrO_2 in NaCl plus water vapor. The thickness of the TGO became larger in the presence of water vapor and the amount of Cr and Ni in the TGO increased. The presence of water vapor accelerated the failure of TBC, which is in accordance with Wang *et al.*'s research into the corrosion behavior of Ti60 alloy,[110] 1Cr11Ni2W2MoV steel,[111] pure Cr,[112] Fe–Cr alloy[113] and K38G superalloy[114] in O_2 plus water vapor.

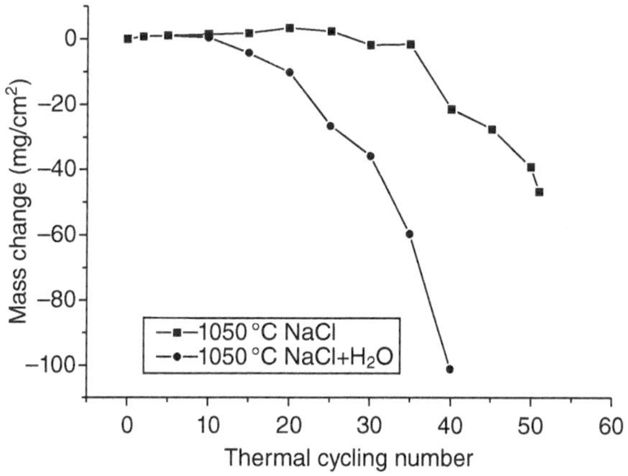

10.10 Corrosion kinetics of TBC at 1050 °C in NaCl vapor and NaCl plus water vapor.[109]

10.5 Conclusions

Oxidation is a common phenomenon when metallic elements are exposed to oxygen or oxygen-containing atmospheres at elevated temperatures. Thermal barrier coatings (TBCs) are used to provide thermal insulation and oxidation resistance to metallic components from the hot gas stream in gas turbine engines used for aircraft propulsion, power generation and marine propulsion. The oxidation behavior of the bond coat, including both MCrAlY and Pt-aluminide coating, has a major impact on the life of the TBCs. The failure of TBC systems is very complex and mostly affected by different mechanisms, wherein the oxidation of the bond coat and thermal mismatch stress contribute to the rapid degradation of TBCs. When metals and alloys are covered with salt films at elevated temperatures in an oxidizing gas, hot corrosion occurs. The degradation mechanisms experienced by TBC systems exposed to deposits were discussed.

10.6 References

1. B. SUDHANGSHU, *High Temperature Coatings*, Burlington, MA, Elsevier Science & Technology Books (2007).
2. J.L. SMIALEK, Advances in the oxidation resistance of high-temperature turbine materials, *Surf. Interface Anal.* **31** (7) (2001) 582–592.
3. R.A. RAPP, Chemistry and electrochemistry of the hot corrosion of metals, *Corrosion* **42** (10) (1986) 568–577.
4. R.A. RAPP, Hot corrosion of materials: a fluxing mechanism, *Corros. Sci.* **44** (2) (2002) 209–221.
5. T.S SIDHU, R.D. AGRAWAL, S. PRAKASH, Hot corrosion of some superalloys and role of high-velocity oxy-fuel spray coatings – a review, *Surf. Coat. Technol.* **198** (1–3) (2005) 441–446.
6. D.M. ZHU, R.A. MILLER, Thermal conductivity and elastic modulus evolution of thermal barrier coatings under high heat flux conditions, *J. Therm. Spray Technol.* **9** (2) (2000) 175–180.
7. B.A. PINT, I.G. WRIGHT, W.Y. LEE, Y. ZHANG, K. PRÜßNER, K.B. ALEXANDER, Substrate and bond coat compositions: factors affecting alumina scale adhesion, *Mater. Sci. Eng. A* **245** (1998) 201–211.
8. B. LOWRIE, D.H. BOONE, Composite coatings of CoCrAlY plus platinum, *Thin Solid Flims* **45** (3) (1977) 491–498.
9. H.M. TAWANCY, N. SRIDHAR, B.S. TAWABINI, N.M. ABBAS, T.N. RHYS-JONES, Thermal stability of a platinum aluminide coating on nickel-based superalloys, *J. Mater. Sci.* **27** (1992) 6463–6474.
10. H.M. TAWANCY, N.M. ABBAS, T.N. RHYS-JONES, Role of platinum in aluminide coatings, *Surf. Coat. Technol.* **49** (1–3) (1991) 1–7.
11. H.J. KIM, M.E. WALTER, Characterization of the degraded microstructures of a platinum aluminide coating, *Mater. Sci. Eng. A* **360** (1–2) (2003) 7–17.
12. G. NAVES, L. VILORIA, Laboratory and field corrosion behavior of coatings for turbine blades, *Surf. Coat. Technol.* **94–95** (1997) 161–167.

13. B. GLEESON, High-temperature corrosion of metallic alloys and coatings, in: M. Schutze (ed.), *Corrosion and Environmental Degradation, Vol. II*: Vol. 19 of the Materials Science and Technology Series, Weinheim, Germany Wiley-VCH (2000) 173–228.
14. C. WAGNER, Theoretical analysis of the diffusion processes determining the oxidation rate of alloys, *J. Electrochem. Soc.* **99** (1952) 369–380.
15. H.E. EVANS, Stress effects in high temperature oxidation of metals, *Int. Mat. Rev.* **40** (1995) 1–40.
16. D. CLEMENS, F. TIETZ, R. MUNOZ-ARROYO, R. ANTON, L. SINGHEISER, J. QUADAKKERS, Influence of composition and phase distribution on the oxidation behaviour of NiCoCrAIY Alloys, *Mat. Sci. Forum* **369–372** (2001) 165–172.
17. K. BUNGARDT, G. LEHNERT, H. MEINHARDT, German Patent 1 796 175, 1 July 1971.
18. E.C. DUDERSTADT, A.N. BANGALORE, U.S. Patent 5 238 752, 24 August, 1993.
19. Z. MUTASIM, W. BRENTNALL, Characterzation and Performance evaluation of pack cementation and chemical vapor deposition platinum aluminide coatings, *J. Am. Soc. Mech. Eng.* 96-GT-436, 1996.
20. A.G. EVANS, D.R. MUMM, J.W. HUTCHINSON, G.H. MEIER, F.S. PETTIT, Mechanisms controlling the durability of thermal barrier coatings, *Prog. Mater. Sci.* **46** (5) (2001) 505–553.
21. V.K. TOLPYGO, D.R. CLARKE, Surface rumpling of a (Ni, Pt)Al bond coat induced by cyclic oxidation, *Acta Mater.* **48** (13) (2000) 3283–3293.
22. V.K. TOLPYGO, D.R. CLARKE, On the rumpling mechanism in nickel-aluminide coatings: Part I: an experimental assessment, *Acta Mater.* **52** (17) (2004) 5115–5127.
23. V.K. TOLPYGO, D.R. CLARKE, On the rumpling mechanism in nickel-aluminide coatings: Part II: characterization of surface undulations and bond coat swelling, *Acta Mater.* **52** (17) (2004) 5129–5141.
24. D.R. MUMM, A.G. EVANS, I.T. SPITSBERG, Characterization of a cyclic displacement instability for a thermally grown oxide in a thermal barrier system, *Acta Mater.* **49** (12) (2001) 2329–2340.
25. A.G. EVANS, M.Y. HE, J.W. HUTCHINSON, Mechanics-based scaling laws for the durability of thermal barrier coatings, *Prog. Mater. Sci.* **46** (2001) 249–271.
26. P.J. HILL, P. ELLIS, L.A. CORNISH, L.A. WITCOMB, I.M. WOLFF, The oxidation behaviour of Pt–Al–X alloys at temperatures between 1473 and 1623 K, *Proceedings of High Temperature Corrosion And Protection 2000*, Japan, (2000) 185–190.
27. A. SUZUKI, Y.N. WU, A. YAMAGUCHI, H. MURAKAMI, C.M.F. RAE, Oxidation behavior of Pt–Ir modified aluminized coatings on Ni-base single crystal superalloy TMS-82+, *Oxid Met* **68** (1–2) (2007) 53–64.
28. Y.N. WU, A. YAMAGUCHI, H. MURAKAMI, S. KURODA, Hot corrosion behavior of Pt-Ir modified aluminide coatings on the nickel-base single crystal superalloy TMS-82+, *J. Mater. Res.*, **22** (1) (2007) 206–216.
29. B. TRYON, F. CAO, K.S. MURPHY, C.G. LEVI, T.M. POLLOCK, Ruthenium-containing bond coats for thermal barrier coating systems, *JOM*, **58** (1) (2006) 53–59.
30. I.T. SPITSBERG, R. DAROLIA, M.R. JACKSON, J. ZHAO, J.C. SCHAEFFER, U.S. Patent No. 6,306,524 B1, 2001.
31. P.K. DATTA, J.S. BURNELL-GRAY, K. NATESAN, Coating technology, in: J.H. Westbrook and R.L. Fleischer (eds), *Intermetallic Compounds*, Wiley, New York, **3** (2002) 561–588.

32. B. TRYON, Q. FENG, T. POLLOCK, Intermetallic phases formed by ruthenium–nickel alloy interdiffusion, *Intermetallics*, **12** (7–9) (2004) 957–962.
33. B. TRYON, Q. FENG, R.W. WELLMAN, K.S. MURPHY, J. YANG, C.G. LEVI, J.R. NICHOLLS, T.M. POLLOCK, Multilayered ruthenium-modified bond coats for thermal barrier coatings, *Metall. Mater. Trans. A* **37** (2006) 3347–3358.
34. T.K. NANDY, Q. FENG, T.M. POLLOCK, Elevated temperature deformation and dynamic strain aging in polycrystalline RuAl alloys, *Intermetallics*, **11** (10) (2003) 1029–1038.
35. T.K. NANDY, Q. FENG, D. BANERJEE, M.F.X. GIGLIOTTI, T.M. POLLOCK, Mechanical behavior of ternary and quaternary RuAl alloys, *Mater. Res. Soc. Symp. Proc.* **753** (2003) BB2.11.11.
36. B. TRYON, K.S. MURPHY, J.Y. YANG, C.G. LEVI, T.M. POLLOCK, Hybrid intermetallic Ru/Pt-modified bond coatings for thermal barrier systems, *Surf. Coat. Technol.* **202** (2007) 349–361.
37. Y. X. SONG, H. MURAKAMI, C. ZHOU, submitted to *J. Mater. Sci. Technol.*
38. R.A. MILLER, C.E. LOWELL, Failure mechanisms of thermal barrier coatings exposed to elevated temperatures, *Thin Solid Films*, **95** (3) (1982) 265–273.
39. C.C. BERNDT, H. HERMAN, Failure during thermal cycling of plasma-sprayed thermal barrier coatings, *Thin Solid Films*, **108** (4) (1983) 427–437.
40. B.C. WU, E. CHANG, D.C. TU, S.L. WANG, Microstructures, properties and failure analysis of (ZrO_2 – 8wt.%Y_2O_3)/((Co, Ni)-Cr-Al-Y) thermal barrier coatings, *Mater. Sci. Eng. A* **111** (1989) 201–210.
41. J.H. SUN, E. CHANG, B.C. WU, C.H. TSAI, The properties and performance of (ZrO_2 – 8 wt.% Y_2O_3) / (chemically vapour-deposited Al_2O_3) / (Ni – 22 wt.% Cr – 10 wt.% Al – 1 wt.% Y) thermal barrier coatings, *Surf. Coat. Technol.*, **58** (2) (1993) 93–99.
42. H.L. TSAI, P.C. TSAI, Microstructures and properties of laser-glazed plasma-sprayed ZrO_2-$YO_{1.5}$/Ni-22Cr-10AI-1Y thermal barrier, *J. Mater. Eng. Perform.* **4** (1995) 689–696.
43. H.L. TSAI, P.C. TSAI, Thermal cyclic response of laser-glazed plasma-sprayed ZrO_2 – 19.5 wt.% Y_2O_3 / Ni – 22 Cr – 10 Al – 1 Y thermal barrier coatings, *Mater. Sci. Eng. A* **177** (1–2) (1994) 227–232.
44. H.L. TSAI, P.C. TSAI, Performance of laser-glazed plasma-sprayed (ZrO_2 – 12 wt.% Y_2O_3) / (Ni – 22 wt.% Cr – 10 wt.% Al – 1 wt.% Y) thermal barrier coatings in cyclic oxidation tests, *Surf. Coat. Technol.*, **71** (1) (1995) 53–59.
45. H.E. EVANS, M.P. TAYLOR, Delamination processes in thermal barrier coating systems, *J. Corr. Sci. Eng.* **6** (Paper H011) (2003).
46. M. GELL *et al.*, Bond strength, bond stress and spallation mechanisms of thermal barrier coatings, *Surf. Coat. Technol.* **120–121** (1999) 53–60.
47. J.M. AMBRICO, M.R. BEGLEY, E.H. JORDAN, Stress and shape evolution of irregularities in oxide films on elastic–plastic substrates due to thermal cycling and film growth, *Acta Mater.* **49** (9) (2001) 1577–1588.
48. A.M. KARLSSON, A.G. EVANS, A numerical model for the cyclic instability of thermally grown oxides in thermal barrier systems, *Acta Mater.* **49** (10) (2001) 1793–1804.
49. Y.H. SOHN, J.H. KIM, E.H. JORDAN, M. GELL, Thermal cycling of EB–PVD/MCrAlY thermal barrier coatings: I. Microstructural development and spallation mechanisms, *Surf. Coat. Technol.* **146–147** (2001) 70–78.

50. B.J. ZIMMERMAN, *Rumpling phenomenon in platinum modified Ni-Al alloys*, Master Degree Thesis, Iowa State University, 2005.
51. N.P. PADTURE, M. GELL, E.H. JORDAN, Thermal barrier coatings for gas-turbine engine applications, *Science* **296** (2002) 280–284.
52. G.C. CHANG, W. PHUCHAROEN, R.A. MILLER, Behaviour of thermal barrier coatings for advanced gas turbine blades, *Surf. Coat. Technol.* **30** (1987) 13–28.
53. A.M. FREBORG, B.L. FERGUSON, W.J. BRINDLEY, G.J. PETRUS, Modeling oxidation induced stresses in thermal barrier coatings, *Mater. Sci. Eng. A* **A245** (1998) 182–190.
54. J. RÖSSLER, M. BÄKER, M. VOLGMANN, Stress state and failure mechanisms of thermal barrier coatings: role of creep in thermally grown oxide, *Acta Mater* **49** (2001) 3659–3670.
55. S.R. CHOI, J.W. HUTCHINSON, A.G. EVANS, Delamination of multilayer thermal barrier coatings, *Mech. Mater.* **31** (1999) 431–447.
56. A.G. EVANS, G.B. CRUMLEY, R.E. DEMARAY, On the mechanical behavior of brittle coatings and layers, *Oxid. Met.* **20** (1983) 193–216.
57. W.J. QUADAKKERS, A.K. TYAGI, D. CLEMENS, R. ANTON, L. SINGHEISER, The significance of bond coat oxidation for the life of TBC coatings, in: Hampikian JM, Dahotre NB (eds), *Elevated Temperature Coatings: Science and Technology III*, Warrendale, PA, The Minerals, Metals & Materials Society (1999) 119–130.
58. D.M. LIPKIN, D.R. CLARKE, M. HOLLATZ, M. BOBETH, W. POMPE, Stress development in alumina scales formed upon oxidation of (111) NiAl single crystals, *Corros. Sci.* **39** (2) (1997) 231–242.
59. E. SCHUMANN, C. SARIOGLU, J.R. BLACHERE, F.S. PETTIT, G.H. MEIER, High temperature stress measurements during the oxidation of NiAl. *Oxid. Met.* **53** (3–4) (2000) 259–272.
60. V.K. TOLPYGO, D.R. CLARKE, Competition between stress generation and relaxation during oxidation of an Fe–Cr–Al–Y alloy, *Oxid Met.* **49** (1–2) (1998) 187–212.
61. J. AKTAA, K. SFAR, D. MUNZ, Assessment of TBC systems failure mechanisms using a fracture mechanics approach, *Acta Mater.* **53** (16) (2005) 4399–4413.
62. D.M. LIPKIN, D.R. CLARKE, Measurement of the stress in oxide scales formed by oxidation of alumina-forming alloys, *Oxid. Met.* **45** (3–4) (1996) 267–280.
63. K.W. SCHLICHTING *et al.*, Application of Cr^{3+} photoluminescence piezo-spectroscopy to plasma-sprayed thermal barrier coatings for residual stress measurement, *Mater. Sci. Eng. A* **291** (1–2) (2000) 68–77.
64. S. KURODA, T.W. CLYNE, The quenching stress in thermally sprayed coatings, *Thin Solid Films* **200** (1) (1991) 49–66.
65. J.D. LEE, H.Y. RA, K.T. HONG, S.K. HUR, Analysis of deposition phenomena and residual stress in plasma spray coatings, *Surf. Coat. Technol.* **56** (1) (1992) 27–37.
66. A.G. EVANS, D.R. MUMM, J.W. HUTCHINSON, G.H. MEIER, F.S. PETTIT, Mechanisms controlling the durability of thermal barrier coatings, *Prog. Mat. Sci.* **46** (2001) 505–553.
67. E.Y. LEE, R.R. BIEDERMAN, R.D. SISSON, Diffusional interactions and reactions between a partially stabilized zirconia thermal barrier coating and the NiCrAlY bond coat, *Mater. Sci. Eng. A* **120–121** (1989) 467–473.

68. D.R. MUMM, A.G. EVANS, On the role of imperfections in the failure of a thermal barrier coating made by electron beam deposition, *Acta Mater.* **48** (8) (2000) 1815–1827.
69. A. RABIEI, A.G. EVANS, Failure mechanisms associated with the thermally grown oxide in plasma-sprayed thermal barrier coatings, *Acta Mater.* **48** (18) (2000) 3963–3976.
70. K.W. SCHLICHTING, N.P. PADTURE, E.H. JORDAN, M. GELL, Failure modes in plasma-sprayed thermal barrier coatings, *Mater. Sci. Eng. A* **342** (1–2) (2003) 120–130.
71. X.Y. GONG, D.R. CLARKE, On the measurement of strain in coatings formed on a wrinkled elastic substrate, *Oxid. Met.* **50** (5–6) (1998) 355–376
72. D.R. CLARKE, W. POMPE, Critical radius for interface separation of a compressively stressed film from a rough surface, *Acta Mater.* **47** (6) (1999) 1749–1756.
73. C.H. HSUEH, P.F. BECHER, E.R. FULLER, S.A. LANGER, W.C. CARTER, Surface-roughness induced residual stresses in thermal barrier coatings: computer simulations, *Mater. Sci. Forum* **308–311** (1999) 442–449.
74. C.H. HSUEH, E.R. FULLER, Analytical modeling of oxide thickness effects on residual stresses in thermal barrier coatings, *Scripta Mater.* **42** (8) (2000) 781–787.
75. K. VAIDYANATHAN, M. GELL, E.H. JORDAN, Mechanisms of spallation of electron beam physical vapor deposited thermal barrier coatings with and without platinum aluminide bond coat ridges, *Surf. Coat. Technol.* **133–134** (2000) 28–34.
76. J. CHENG, E.H. JORDAN, B. BARBER, M. GELL, Thermal/residual stress in an electron beam physical vapor deposited thermal barrier coating system, *Acta Mater.* **46** (16) (1998) 5839–5850.
77. A.G. EVANS, J.W. HUTCHINSON, On the mechanics of delamination and spalling in compressed films, *Int. J. Solids Struct.* **20** (5) (1984) 455–466.
78. F.S. PETTIT, C.S. GIGGINS, Hot corrosion, in: C.T. Sims, N.S. Stoloff, W.C. Hagel (eds), *Superalloys II*, Wiley, New York (1987) 327–358.
79. G.W. GOWARD, Low-temperature hot corrosion in gas turbines: A review of causes and coatings therefor, *J. Eng. Gas Turbine Power, Tran ASME*, **108** (1986) 421–425.
80. R.A. RAPP, Y.S. ZHANG, Hot corrosion of materials: fundamental studies, *JOM*, **46** 47–55.
81. T.S. SIDHU, S. PRAKASH, R.D. AGRAWAL, Hot corrosion and performance of nickel-based coatings, *Curr. Sci.* **90** (1) (2006) 41–47.
82. P. HANCOCK, Vanadic and chloride attack of superalloys, *Mater. Sci. Technol.* **3** (1987) 536–544.
83. N. ELIAZ, G. SHEMESH, R.M. LATANISION, Hot corrosion in gas turbine components, *Eng. Fail. Anal.* **9** (1) (2002) 31–43.
84. J. STRINGER, *Hot Corrosion in Gas Turbines*, New York, Metals and Ceramics Information Center (1972) MCIC Report 72–08.
85. D.A. SHORES, K.L. LUTHRA, *Fundamentals of High-Temperature Corrosion*, New York, Academic Press (1992).
86. M.A. DECRESCENTE, N.S. BORNSTEIN, Formation and reactivity thermodynamics of sodium sulfate with gas turbine alloys, *Corrosion* **24** (1968) 127–133.

87. J.A. GOEBEL, F.S. PETTIT, G.W. GOWARD, Mechanisms for the hot corrosion of nickel-base alloys, *Metall. Trans.* **4** (1973) 261–278.
88. C. LEYENS, I.G. WRIGHT, B.A. PINT, Hot corrosion of an EB–PVD thermal-barrier coating system at 950 °C, *Oxid. Met.* **54** (5–6) (2000) 401–424.
89. I. GURRAPPA, Thermal barrier coatings for hot corrosion resistance of CM 247 LC superalloy, *J. Mater. Sci. Lett.* **17** (1998) 1267–1269.
90. S.Y. PARK, J.H. KIM, M.C. KIM, Microscopic observation of degradation behavior in yttria and ceria stabilized zirconia thermal barrier coatings under hot corrosion, *Surf. Coat. Technol.* **190** (2–3) (2005) 357–365.
91. R.L. JONES, The development of hot-corrosion-resistant zirconia thermal barrier coatings, *Mater. at High Temp.* **9** (1991) 228–236.
92. H.E. EATON, N.S. BORNSTEIN, J.T. DEMASI-MARCIN, The effects of environmental contaminants on industrial gas turbine thermal barrier coatings, *Proc. Int. Gas Turbine Aeroengine Congress and Exhibition*, ASME, New York (1996) Paper 96-GT-283.
93. M.F. TRUBELJA, D.M. NISSLEY, N.S. BORNSTEIN, J.T.D. MARCIN, Pratt & Whitney thermal barrier coating development, *Proc. Advan. Turbine Systems Annu. Program Rev. Meet*, U.S. Department of Energy, Washington, DC (1997).
94. C. LEYENS, U. SCHULZ, B.A. PINT, I.G. WRIGHT, Influence of electron beam physical vapor deposited thermal barrier coating microstructure on thermal barrier coating system performance under cyclic oxidation conditions, *Surf. Coat. Technol.* **120–121** (1999) 68–76.
95. F.C. TORIZ, A.B. THAKKER, S.K. GUPTA, Thermal barrier coating for jet engines, *Proceedings of the International Gas Turbine Aeroengine Congress*, ASME, Paper 88-GT-279 (1988).
96. M.P. BOROM, C.A. JOHNSON, L.A. PELUSO, Role of environment deposits and operating surface temperature in spallation of air plasma sprayed thermal barrier coatings, *Surf. Coat. Technol.* **86–87** (1996) 116–126.
97. R.A. RAPP, K.S. GOTO, Hot corrosion of metals by molten salts, in: J. Braustein, J.R. Selman (eds), *Molten Salts I*, Pennington, NJ, Electrochemical Society (1981) 159–177.
98. A.S. KHANNA, S.K. JHA, Degradation of materials under hot corrosion conditions, *Trans. Indian Inst. Met.* **51** (5) (1988) 279–290.
99. N.S. BORNSTEIN, M.A. DECRESCENTE, H.A. ROTH, The relationship between relative oxide ion content of Na_2SO_4, the presence of liquid metal oxides and sulfidation attack, *Metall. Trans.* **4** (1973) 1799–1810.
100. Y.S. HWANG, R.A. RAPP, Thermochemistry and solubilities of oxides in sodium sulfate-vanadate solutions, *Corrosion* **45** (11) (1989) 933–937.
101. J.A. GOEBEL, F.S. PETTIT, G.W. GOWARD, Mechanisms for the hot corrosion of nickel-base alloys, *Metall. Trans.* **4** (1973) 261–278.
102. C.G. ZHOU, Y.X. SONG, C.L. WANG, H.B. XU, Cyclic-oxidation behavior of thermal-barrier coatings exposed to NaCl vapor, *Oxid. Met.* **69** (1–2) (2008) 119–130.
103. C.G. ZHOU, C.L. WANG, Y.X. SONG, Evaluation of cyclic oxidation of thermal barrier coatings exposed to NaCl vapor by finite element method, *Mater. Sci. Eng. A* **490** (2008) 351–358.
104. E. REESE, H.J. GRABKE, Einfluß von natriumchlorid auf die oxidation von hochlegierten chrom- und chrom-nickel-stählen, *Mater. Corr.* **44** (1993) 41–47.

105. C.J. WANG, Y.C. CHANG, Y.H. SU, The hot corrosion of Fe–Mn–Al–C alloy with NaCl/Na_2SO_4 coating mixtures at 750 °C, *Oxid. Met.* **59** (2003) 115–133.
106. B.P. MOHANTY, D.A. SHORES, Role of chlorides in hot corrosion of a cast Fe–Cr–Ni alloy. Part I: Experimental studies, *Corr. Sci.* **46** (12) (2004) 2893–2907.
107. D.A. SHORES, B.P. MOHANTY, Role of chlorides in hot corrosion of a cast Fe–Cr–Ni alloy. Part II: thermochemical model studies, *Corr. Sci.* **46** (12) (2004) 2909–2924.
108. C.L. WANG, *High temperature Corrosion Behavior of Thermal Barrier Coating Exposed to NaCl Vapor*, Master Degree Thesis, Beijing University of Aeronautics and Astronautics, 2006.
109. Y.X. SONG, C.G. ZHOU, H.B. XU, Corrosion behavior of thermal barrier coatings exposed to NaCl plus water vapor at 1050 °C, *Thin Solid Films* **516** (16) (2008) 5686–5689.
110. Y. SHU, F. WANG, W. WU, Corrosion behavior of Ti60 alloy coated with a solid NaCl deposit in O_2 plus water vapor at 500–700 °C, *Oxid. Met.* **52** (1999) 463–473.
111. Y. SHU, F. WANG, W. WU, Synergistic effect of NaCl and water vapor on the corrosion of 1Cr-11Ni-2W-2Mo-V steel at 500–700 °C, *Oxid. Met.* **51** (1999) 97–110.
112. Y. SHU, F. WANG, W. WU, Corrosion behavior of pure Cr with a solid NaCl deposit in O_2 plus water vapor, *Oxid. Met.* **54** (2000) 457–471.
113. F. WANG, Y. SHU, Influence of Cr content on the corrosion of Fe–Cr alloys: the synergistic effect of NaCl and water vapor, *Oxid. Met.* **59** (2003) 201–214.
114. F. WANG, S. GENG, S. ZHU, Corrosion behavior of a sputtered K38G nanocrystalline coating with a solid NaCl deposit in wet oxygen at 600 to 700 °C, *Oxid. Met.* **58** (2002) 185–195.

11
Failure mechanism of thermal barrier coatings by electron beam physical vapor deposition (EB–PVD) under thermomechanical coupled loads

C. ZHANG, Baosteel Group Corporation, China and
C. CHEN, AVIC Commercial Aircraft Engine Co. Ltd, China

Abstract: This chapter discusses the failure mechanism of thermal barrier coatings (TBCs) in service in gas turbine engines. A service environment simulation system was established in order to examine the failure behaviour of TBCs subjected to a gradient temperature field coupled with complex mechanical loadings. The experiment took the form of a finite element analysis accompanied by non-destructive evaluation (NDE) in order to analyse the TBCs failure mechanism.

Key words: thermal barrier coatings, failure mechanism, thermomechanical loadings, non-destructive evaluation, service environment simulation.

11.1 Introduction

Ceramic coatings have emerged as vital constituents of hot section components in aero-propulsion systems. These act as both thermal (TBC) and environmental (EBC) barriers.[1,2] Most coated components are actively cooled from the back side, resulting in temperature gradients along the through-thickness direction during operation.[3–9] Moreover, during engine shut down, the surface of coatings can cool quite rapidly, which causes additional gradients. At the same time, the coating is also subject to temperature gradient along the surface direction due to the non-uniform heating of the combustion chamber. Therefore, TBCs in the service environment are subject to temperature gradient in both the through-thickness and the surface direction. Accompanied by non-uniform thermal loading, TBCs coated gas turbines also underwent complicated mechanical loading, including tension, compression, fatigue and so on.

However, most studies are focused on the degradation of TBCs subjected to either uniform temperature field or mechanical loading. Few investigations have been applied to the evaluation of TBCs subjected to coupled thermomechanical loading. Recent observations can be made only on engine components which have been removed from service.[10–12] In order to observe the failure mechanism of TBCs in service, an effective service environment simulation system is required and the corresponding failure mechanism of TBCs subjected to coupled thermomechanical loading must be studied.

Non-destructive evaluation (NDE) techniques, including acoustic emission (AE), photo-simulated luminescence spectroscopy (PSLS) and impedance spectroscopy (IS), are widely used to study the degradation of TBCs. IS, in particular, is an efficient, quick and powerful testing method for ceramic materials.[13–15] Voltage and current responses, on application of an alternating current (a.c.) with varying frequency, relating to yttrium stabilized zirconia (YSZ), thermal grown oxides (TGO) and mixed oxide layers have been observed in the impedance spectra.[13,15] The influence of changes in the thickness and conductivity of the TGO layer on the impedance spectra has been identified by both experimental and modelling results.[16] In this investigation, TGO and interfacial microcracks were studied *in situ* using IS.

Due to the complexity of the thermomechanical coupled effect and non-uniform temperature field, finite element (FE) analysis was also introduced to simulate the stress fields in the whole experiment, including the heating, temperature dwell and cooling procedures. The simulation results are in good agreement with the NDE and scanning electron microscopy (SEM) observations during the degradation of TBCs. By this means, the failure mechanism of TBCs subjected to thermomechanical loading is attained.

11.1.1 Commentary

Thermal barrier coatings are widely used as the protective coating in gas turbine engines and their reliability is a matter of much concern. In order to predict the lifetime of TBCs during in-service operation, the first requirement is a thorough investigation of the failure mechanism of the coating. In the service environment, TBCs are subject to high temperature, non-uniform heating along both the through-thickness and the surface direction, fast heating and cooling, hot corrosion and complicated mechanical loadings. Due to the lack of effective service condition simulation, most investigations are focused on the failure mechanism of TBCs either in a uniform temperature field or subjected mechanical loading. Therefore, simulation of the coating and gas turbine engines in service and the corresponding failure mechanism is considered to be one of the most important directions for future research.

Some research has been carried out on failure mechanisms of TBC under a temperature gradient field.[9–11] It has been reported that the failure mode in this circumstance is different from that which is found under isothermal heat treatment.[10,17] Subjected to a temperature gradient field, delamination occurred in the ceramic coating instead of the TBC bond coat interface. Subjected to a thermal gradient, isolated cracks parallel to the interface are envisaged, that experience an energy release rate and exfoliate.[17] However, under conditions of isothermal oxidation, most failure occurred at the bond coat/TGO or TGO/TBC interface.[8] A situation comprising three-dimensional temperature gradient fields is more close to the real service conditions of TBCs in gas turbine engines.[10,12,17,18] This suggests that further study of the failure mechanism of TBCs under a temperature gradient field is required.

In order to obtain the temperature gradient field, some simulation systems have been reported, such as furnace cycling and burner rig testing.[18–20] Using such equipment, a temperature gradient along the through-thickness direction can be obtained. However, very few reports are concerned with the 3D temperature gradient field, which encompasses a temperature gradient along both the through-thickness direction and the in-plane direction (parallel to the surface of the coating). In this investigation, hollow tube samples coated with EB–PVD TBC were employed to simulate turbine blades. Infrared light was focused on a very small area of the TBC surface, accompanied by compressed air through the tube to cool the substrate, by this means, thermal gradients along both the axial and radial directions were obtained.

The usual procedure for heating furnaces is not capable of satisfying the high heating and cooling rates, which are considered to occur during engine start-up and shut-down, respectively.[20] In order to simulate the service temperature gradient field and high heating rate of a gas turbine, an infrared thermal shock testing system was employed in this investigation. Using the infrared heating system, an average heating rate of 55 °C/s was attained.

In this chapter, the results obtained from this gas turbine service environment simulation system are analyzed, accompanied by finite element (FE) modeling. Hence, the method of studying the failure mechanism of TBCs and the corresponding laws governing failure are established.

11.2 Establishment of gas turbine service environment simulation system

11.2.1 Thermomechanical loading test system

In the investigation, a thermomechanical loading system was employed in order to simulate the service condition of a gas turbine blade. The basis is a mechanical loading system coupled with a thermal shock system using focused infrared light. The photo and schematic of the system are shown in Fig. 11.1.

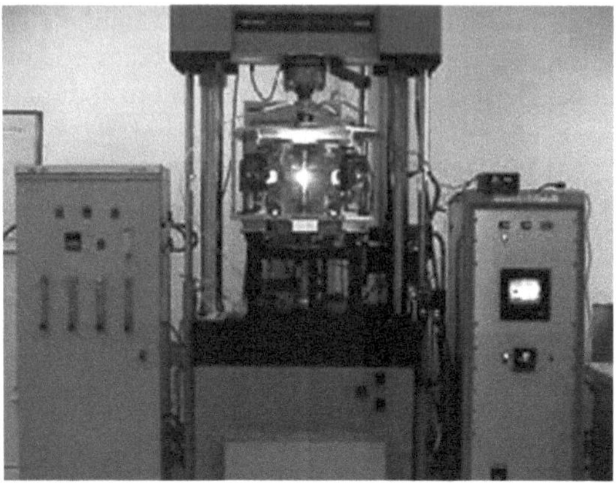

(a)

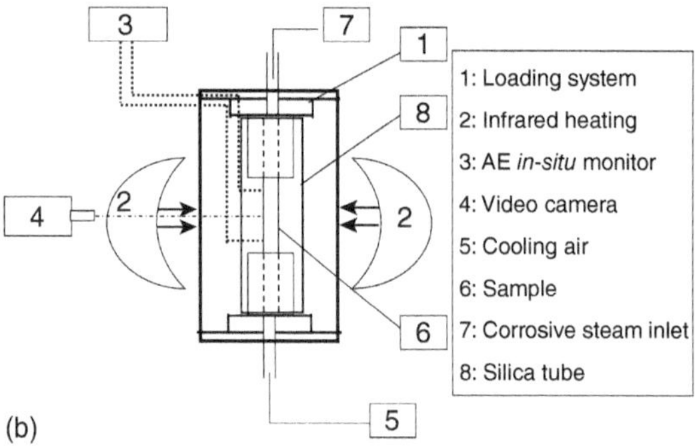

(b)

11.1 TBCs service environmental performance simulation test system: (a) photograph; (b) schematic diagram.

An Instron 8803 was employed in the mechanical loading system. The maximum load was 250 kN, while the maximum alternate loading frequency was 40 Hz. WaveMaker^{TM} software was used to programme and apply the loading wave pattern in the experiment.

In the thermal shock system, the infrared light was generated by four 4 kW bulbs, and focused on the sample using concave mirrors. Therefore, the heated length of the sample which could be controlled was only 20 mm. Taking advantage of such a small heated area, the temperature gradient along the axial direction was obtained.

Heating and cooling took place in three stages:

- **Heating procedure**: the surface temperature is increased to 1100 °C from room temperature – the high heating rate of the system (55 °C/s on average) enables this to be done in less than 20 s:
- **Temperature dwell procedure**: the temperature is kept at 1100 °C for 5 min;

- **Cooling procedure**: the infrared lights are switched off and the sample is cooled to room temperature using compressed air.

In the experiment, the surface temperature was controlled by the power of the infrared light, which was calibrated by thermocouple.

11.2.2 Sample description

In this investigation, Ni-based superalloys were used as substrate materials. The specimens were fabricated into cylinders with both ends screwed to fit the grips of the mechanical loading machine, as shown in Fig. 11.2. The diameter of the specimen was 8 mm, and the length of the coated part 50 mm.

The samples were heated using an infrared heating system with a controllable heating rate. The heated length along the axial direction was 1.5 mm. In the specimen, the centre of the heated zone is indicated as No. 3, as marked in Fig. 11.2. The edges of the heated zone are indicated as Nos. 2 and No. 4, and the area far from the heated zone as Nos. 1 and 5. The desired temperature was reached in 20 s. In this case, the surface temperature of the specimen was around 1373 K. After holding at this temperature for 5 min, the heating power was reduced to 0 in 20 s, and then the specimen was cooled down to room temperature in 3 min by compressed air.

It can be seen from Fig. 11.2 that standard tensile samples have been used in order to facilitate fitting to the mechanical loading system. It should be noted that an acoustic probe can be also embedded at the end of the sample. Hence crack propagation during the experiment can be successfully monitored *in situ*. However, this is not included in this investigation.

11.2.3 Experimental parameters

Different mechanical loadings were applied to the sample, and the parameters and loading types are illustrated in Table 11.1. In this investigation, samples S1 and S2 denote the solid cylinder samples, while H1 and H2 denote the tube-shaped samples.

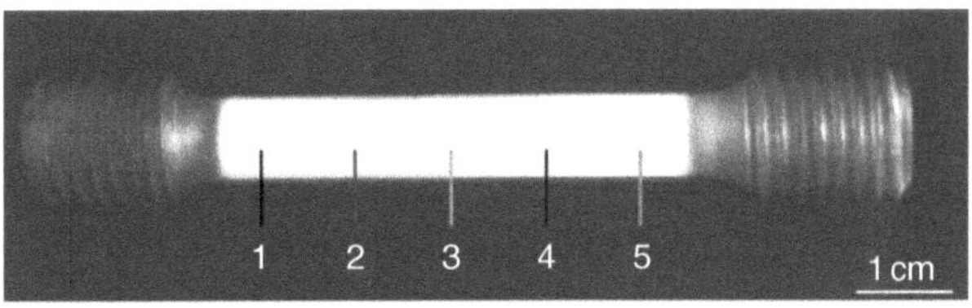

11.2 Photograph of the sample. In the thermal shock test, infrared light is focused on the area marked as No. 3.

Table 11.1 Experimental parameters for thermomechanical loading test

Sample code	S1	S2	H1	H2
Mechanical loading (MPa)*	99.5	119.4	143.3	132.3
Mechanical loading description	Constant tensile loading	Fatigue loading		
Temperature/power	45%	50%	50%	50%
Cycling	1670	1670	660	180
in Section	11.3.1	11.3.2	11.4.1	11.4.1

*The mechanical loading value in fatigue loading denotes the peak value in the loading curve.

Table 11.2 Parameters for super alloy*[19]

T(K)	293	473	573	673	773	873	973	1073	1173	1273	1373
k(W·m^{-1}·K^{-1})		9.67	11.47	13.44	14.99	16.79	17.96	19.63	19.51	19.43	19.00
c(J·Kg^{-1}·K^{-1})		385	427	456	481	498	506	506	489	473	443
α(×10^{-6} K^{-1})			12.36	12.91	13.15	13.41	13.98	14.49	15.01	15.93	16.94
E(GPa)	127.6	124.0	121.0	118.0	114.0	110.0	105.0	100.0	93.5	84.0	
ν	0.33		0.33		0.34	0.34	0.36	0.36	0.37	0.38	

* T = testing temperature, k = thermal conductivity, α = thermal expansion coefficient, E = Young's modulus, ν = Poisson's ratio.

During mechanical loading, the samples were heated by focused infrared light. The highest thermal loading and mechanical loading were achieved at the same time in order to simulate the whole engine operation procedure, from engine start-up to temperature dwell and then shut-down.

In this investigation, FE analysis was performed using the ANSYS 9.0 software developed by ANSYS Inc., Canonsburgh, PA, USA. In the simulation, PLANE 13 (four-node quadrilateral element) was selected for coupled-field thermal structure analysis.

Material parameters employed in the simulation were assumed to be isotropic and temperature-dependent and are illustrated in Tables 11.2–11.5. It should be noted that in the simulation, only elastic deformation was considered. Material evaluation during thermal cycling was not included in the simulation.

Table 11.3 Parameters for NiCoCrAlY bond coat[19]

T(K)	293	473	673	873	1073	1273	1373
k($W \cdot m^{-1} \cdot K^{-1}$)	5.8	7.5	9.5	12.0	14.5	16.2	17.0
α($\times 10^{-6}$ K^{-1})	13.6	14.2	14.6	15.2	16.1	17.2	17.6
E(GPa)	200	190	175	160	145	120	110
ν	0.30	0.30	0.31	0.31	0.32	0.33	0.33

Table 11.4 Parameters for TGO layer[19]

T(K)	293	473	673	873	1073	1273	1373
k($W \cdot m^{-1} \cdot K^{-1}$)	10	7.8	6.0	5.1	4.4	4.4	4.0
α($\times 10^{-6}$ K^{-1})	8.0	8.2	8.4	8.7	9.0	9.3	9.6
E(GPa)	400	390	380	370	355	325	320
ν	0.23	0.23	0.24	0.24	0.25	0.25	0.25

Table 11.5 Parameters for YSZ top coat[19]

T(K^{-1})	293	473	673	873	1073	1273	1373
k($W \cdot m^{-1} \cdot K^{-1}$)	1.9	1.8	1.7	1.6	1.6	1.7	1.7
α($\times 10^{-6}$ K^{-1})	9.0	10.0	9.7	9.8	10.0	10.3	10.3
E(GPa)	48	47	44	40	34	26	22
ν	0.10	0.10	0.10	0.11	0.11	0.12	0.12

11.3 Failure mechanism of EB–PVD TBC under in-plane thermal gradient coupled with mechanical loading

11.3.1 Failure mechanism of EB–PVD TBC under in-plane thermal gradient coupled with constant tensile loading

In this experiment, a constant load of 99.5 MPa was applied to the sample during testing. The test was carried out using INSTRON 8803. After 1670 cycles of thermal shock testing, the specimens were cut into five pieces corresponding to those parts marked in Fig. 11.2, respectively. The impedance spectroscopy of the cycled specimens was measured at 673 K using an impedance analyzer (Solatron ® 1260 coupled with 1296). Zview software was used for impedance spectroscopy analysis. Platinum wire was fixed to the superalloy substrate as an electrode, and the silver paint on the top coat functions as another electrode.

Failure modes during thermal shock testing

Figure 11.3 shows optical micrographs of the specimen. Surface morphology varies whith the position of the sample along the axial direction. Few cracks can be observed on the coatings at the two ends of the specimens (Nos. 1 and 5). Crack networks are observed on the surfaces of the coatings at the edge of the heated zone and in the hottest zone. It should be noted that the coating surface in the hottest zone is denser and smoother than those at the edge of the heated zone, possibly due to sintering of YSZ coating at high temperature.

Figure 11.4 shows SEM micrographs of cross-sections from different parts of the specimen. It can be seen that in Parts 1 and 5 which are far away from the heated zone the YSZ top coats are still intact and well bonded to the bond coat. However, a large number of vertical cracks running perpendicular to the coating surface can be seen at the edge of the heated zone (Nos. 2 and 4). Moreover, some horizontal cracks propagated along the interface between YSZ top coat and bond coat. Cracks along the interface are not observed in the hottest zone (No. 3), although the coating was exposed to the highest temperature. This suggests that the surface temperature is not the dominant reason for the failure of TBC. The largest temperature gradient along the axis direction is attained at the edge of the hottest zone (Nos. 2 and 4), which causes higher thermal stresses. Consequently, cracks of the TBCs are initiated due to the thermal gradient coupled with the constant tensile mechanical loading.

The formation of vertical cracks in the coatings can be attributed to the temperature gradient through the coating thickness. Transient tensile stresses were generated at the surface of the coatings on cooling from high temperature to room temperature. Generation of the horizontal cracks was partly caused by the temperature gradient along the axial direction.

Evaluation of impedance spectroscopy

Figure 11.5 shows the impedance spectroscopy results for the specimen after thermal shock. Figure 11.5a shows the lg|Z|–lgf curve of the Bode plot, Fig. 11.5b the θ–lgf curve of the Bode plot and Fig. 11.5c is the Nyquist plot.

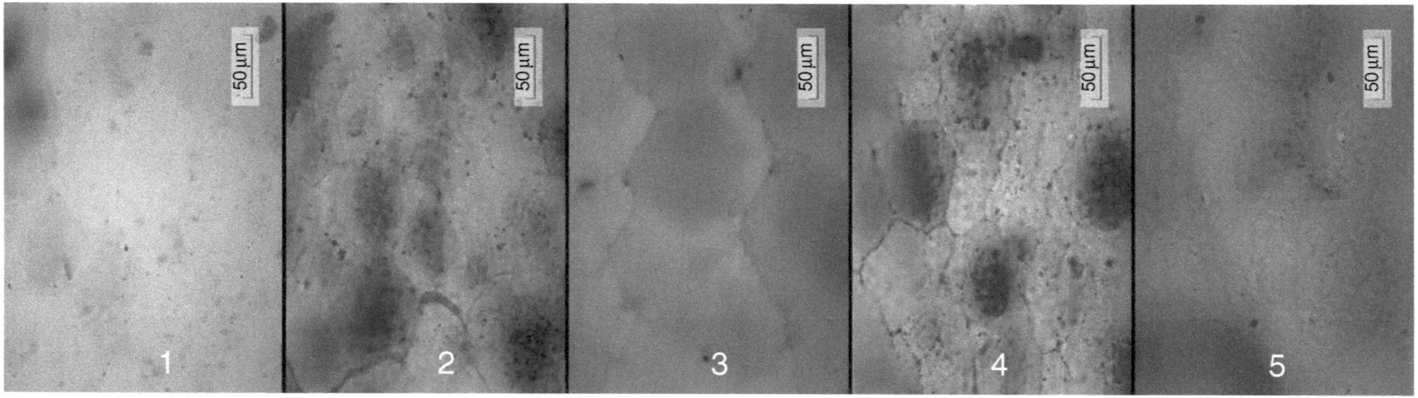

11.3 Surface morphology of the sample after thermomechanical loading test at different positions.[21]

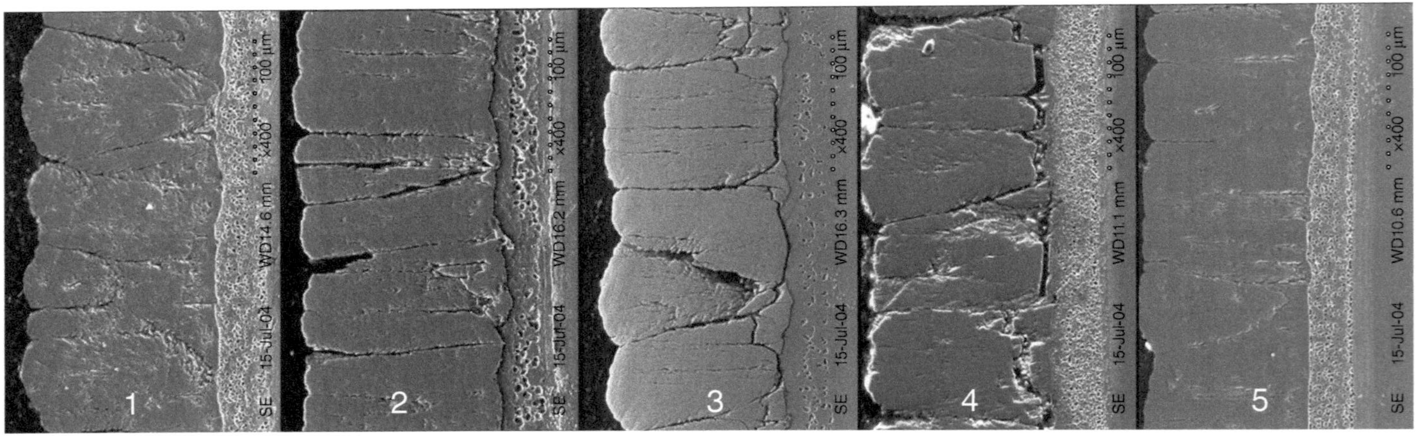

11.4 Cross-sectional microstructure at different positions of the sample. The markers of different positions correspond with those shown in Fig. 11.3.[21]

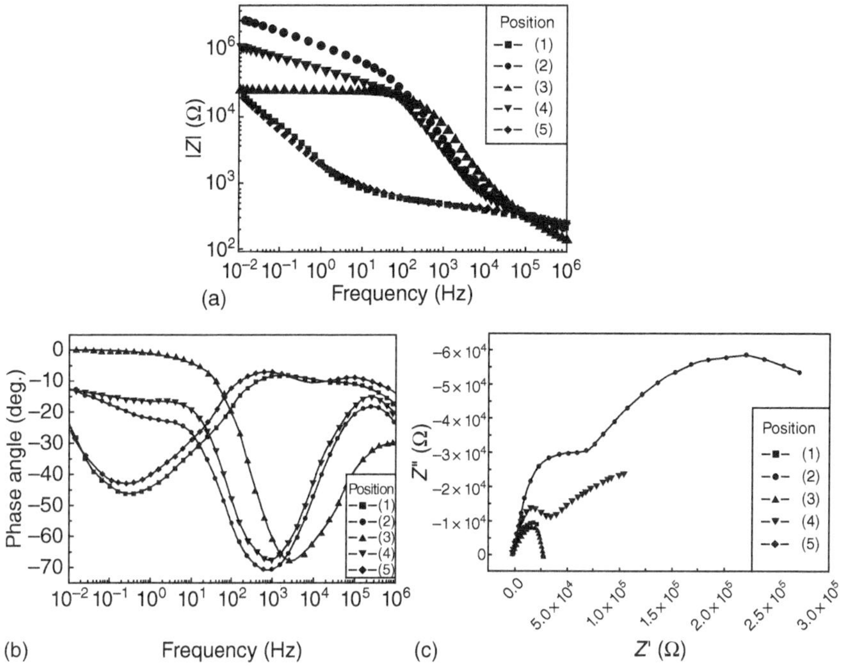

11.5 Impedance spectra of different parts of the sample after thermomechanical loading:[21] (a) |Z|–f; (b) phase angle–f; (c) Z″–Z′.

There are three kinds of zones in the specimen, as mentioned in Figs. 11.2 and 11.4. The impedance spectroscopies are rather different from each other, depending on the zones measured. In the low-frequency range, the resistance values obtained at the edge of the heated zone are the highest, while those from the far ends are the lowest. The resistance value in the low-frequency range (0.01 Hz–100 Hz) detected from the hottest zone is almost constant. Also, in the θ–lgf plot, the curve corresponding to the coating from the hottest zone is kept at zero degree. This means that the electrical characteristics make the largest contribution to the resistance value at the low end of the frequency range.

It can be seen from the SEM micrographs that the coating in the hottest zone shows the thickest TGO layer. Assuming all the coating structures to be the same, the resistance value of the coating corresponding to No. 3 should be the largest. However, the actual experimental results are different from those expected. There are three reasons for this:

1. Nos. 2 and 3 are at the edge of the heated zone. The surface temperature in these zones is around 1200 K. At this temperature, the bond coat was

oxidized. The TGO in these zones is only a little thinner than that in the hottest zone.
2. Sintering of the ceramic coating occurred at the centre of the heated zone. This gave rise to increased coating conductivity, with consequent decrease in the value.
3. Nos. 2 and 4 experienced the largest thermal gradient along the axial direction. As a result, more cracks and defects, such as horizontal cracks, were generated in the coating, which contributed to the increased resistance value.

It can be concluded that the failure mechanism of the TBC-coated cylinder specimens subjected to thermal load coupled with mechanical load was investigated. The failure modes of the TBC after thermal shock were evaluated by impedance spectroscopy together with SEM. Thermal gradients across the coating thickness and parallel to the coating plane, rather than coating surface temperatures, are the most important factors responsible for the formation of vertical and horizontal cracks which eventually led to the failure of the TBCs.

11.3.2 Failure mechanism of EB–PVD TBC under in-plane thermal gradient coupled with cycling fatigue loading

Failure modes

Cross-sectional microstructures of the sample after thermomechanical loading tests at different positions are shown in Fig. 11.6. The morphology varies greatly with different testing positions and temperatures. In the low-temperature zone (position 5 in Fig. 11.2, also position 1 in Figs 11.2 and 11.6) few cracks and little failure can be seen either in the ceramic coating or at the interface between the ceramic coating and the metallic coating. At the mid-temperature zone (position 2 or 4 in Figs 11.2 and 11.6), which is at the edge of the heated zone, vertical cracks are present in the ceramic coating and these emerged as horizontal cracks. It should be noted that, in this area, the vertical cracks emerged in the ceramic coating, rather than at the TBC/TGO interface, which has always been considered to be the weakest part in the TBC system. Compared to other positions, the highest crack density in the ceramic coating occurs in this area. At the same time, it can be seen in Fig. 11.6 that interfacial failure occurred at the highest temperature zone, with both delamination of the ceramic coating and TGO growth shown. At the low temperature zone, however, TGO is much thinner (hardly visible even) and no apparent interfacial failure occurs. It can be easily seen that TGO growth is correlated with testing temperature. TGO tends to attain a higher growth rate at the heated area, which results in

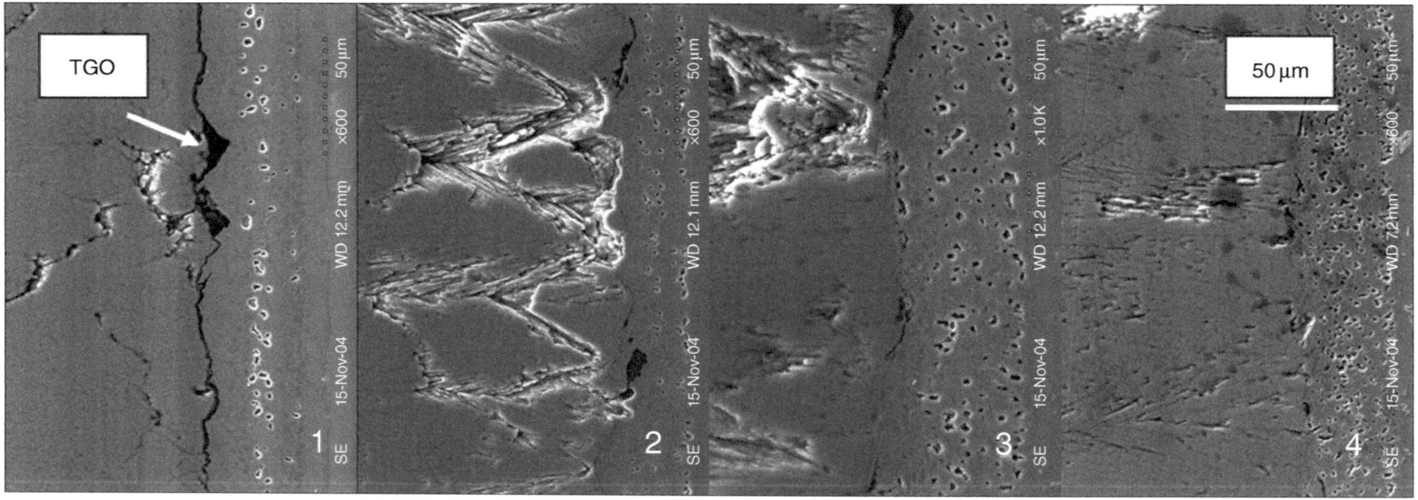

11.6 SEM images for the sample at different positions. The observation position is marked in the schematic. The position markers correspond with those shown in Fig.11.2.

thicker TGO in the higher temperature region. TGO growth coupled with mechanical loading caused dislocation and slipping of grains at the interface, which is considered to result in stress concentration.

Impedance evaluation

Interfacial failure after thermomechanical testing at different positions of the sample was evaluated using impedance spectroscopy. Figure 11.7b shows a Bode plot (phase angle vs frequency) and Figure 11.7c shows a Nyquist plot (imaginary impedance (Z'') vs real impedance (Z')) from impedance measurements of sample S1 at different positions subjected to thermomechanical loading. It was reported that the relaxation process is correlated with the electrical resistance and capacitance of each equivalent element, which is considered to represent the electrical properties of each layer.[20,21] In the TBC system, the TGO relaxation process is presented as a Bode plot (phase angle vs frequency) at a frequency of around 10^3 Hz.[20]

Fig. 11.7a illustrates the impedance resistance moduli at different positions, a clear trend of the resistance modulus increasing with testing temperature can be seen. A higher electrical resistance modulus is obtained in the area tested at higher temperature. In Fig. 11.7b, different electrical responses are exhibited at different positions. From positions 2–4, which were subjected to higher temperature, a relaxation process at a frequency of around 10^3 Hz is observed, while in the positions at the low-temperature zone, position 5 in Fig. 11.7, no apparent relaxation process is observed. This suggests that the TGO at the low temperature zone is so thin that it makes little contribution to the electrical relaxation.

From Fig. 11.7c, it can also be seen that the electrical resistance measured at the high temperature zone is much higher compared to the low-temperature zone. From the low-temperature to the high-temperature zone, the increasing electrical resistance of the TGO denotes increasing TGO thickness. It is considered that thicker TGO tends to form at the high the temperature zone rather than at the low-temperature zone.

TGO growth

As the infrared heating light was focused on a small area of the sample, the temperature varied according to position during the thermomechanical loading test, resulting in the TGO thickness varying according to distance from the heated area. These phenomena can be seen from the cross-sectional microstructure of samples after the thermomechanical loading test – a result confirmed by the impedance measurements at different positions. The simulated temperature profiles along the axial direction at the TGO/TBC interface and the TBC layer are illustrated in Fig. 11.8. Comparing the

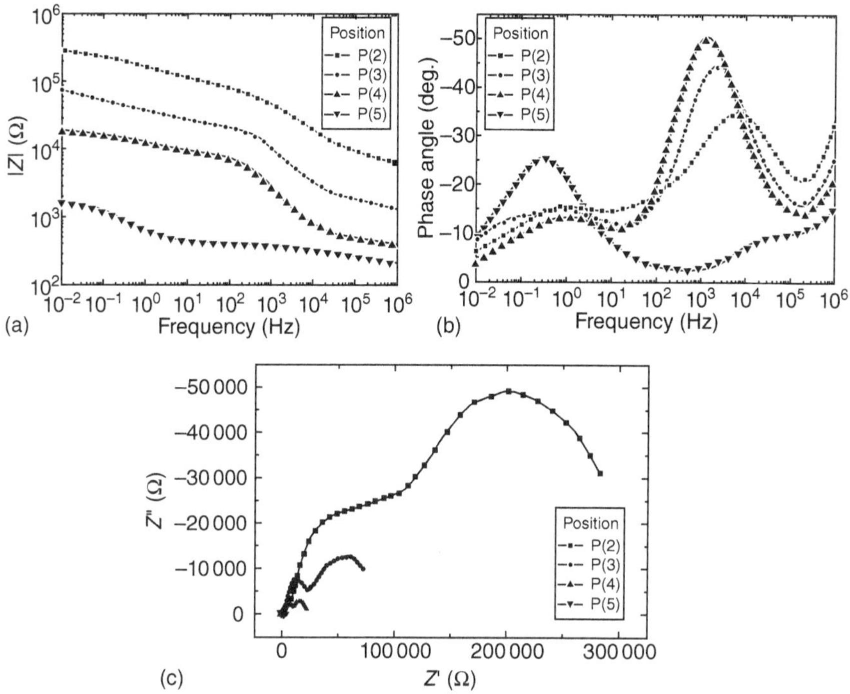

11.7 Impedance spectra of sample S2 obtained at different positions.

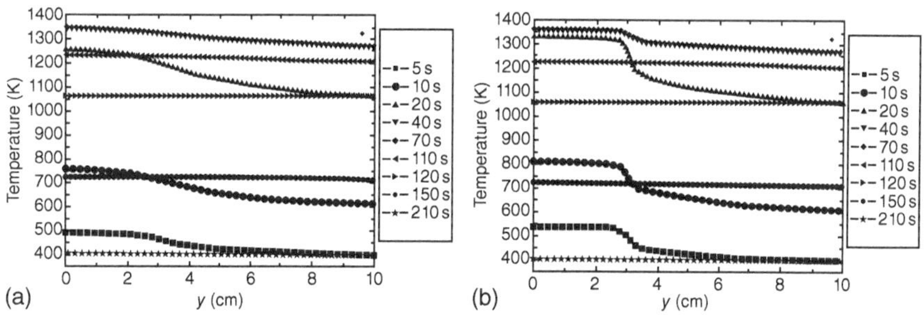

11.8 Simulated temperature profile at the TGO/TBC interface (a) and the TBC layer (b) along the axial direction during thermomechanical loading experiment. *y* on the *x*-axis denotes the distance from heated centre; the heated length in the model is 2.5 mm. Temperature profiles for each experimental stage are shown.

temperature profiles of both TGO/TBC interface and TBC layer at 70 s, which is also the temperature dwell procedure in the test, no apparent difference is obtained. This implies that the temperature gradient along the radius direction can be ignored, compared to that along the axial direction. At the same time, an obvious temperature drop takes place along the axial direction in the range 70–200 °C. This is considered to be the dominant reason for TGO at the heated area being much thicker than that at the cold end. The higher growth rate of TGO at the heated zone is considered to be one of the reasons responsible for the failure of TBCs.

Stress analysis

The simulated equivalent stress distribution profile, as a function of distance from heated centre along the axial direction obtained from interfaces and TBC layer, is illustrated in Fig. 11.9. It can be seen from the figure that, except for the substrate/bond coat interface, the highest stress level is present at 70 s, which is the temperature dwelling procedure. No apparent stress drop along the axial direction is observed in each interface and metallic layer, which is considered to be a consequence the high thermal conductivity the metallic layers. However, in the TBC layer, a clear temperature drop is present at the edge of the heated centre, particularly during the heating procedure (at 20 s).

Stress profiles along the radial direction in each interface and ceramic coating are illustrated in Fig. 11.10. In each figure, a stress peak is present at the edge of the heated zone, which changes from tensile stress to com-

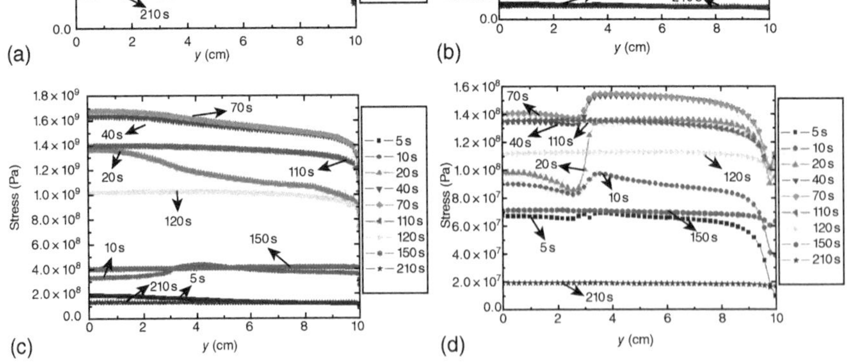

11.9 Stress spatial distribution at interfaces and in ceramic coating vs axial position at different periods of cycling: (a) substrate/bond coat; (b) bond coat/TGO; (c) TGO/YSZ; (d) YSZ layer.

pressive stress. This implies that a large shear stress along the radial direction is present in this area. The shear stress profiles along the axial direction are shown in Fig. 11.11. It can be seen from Fig. 11.6 that, at the edge of the heated zone (position 3), a large number of cracks are present, particularly vertical cracks. This is considered to be induced by the shear stress along the radial direction in this area. With the emerging of vertical cracks, horizontal cracks are formed, which will finally result in delamination of the ceramic coating.

This stress trend is also illustrated in a normal stress profile along the axial direction in Fig. 11.12. Looking at the lowest testing temperature at the non-heated area, no failure or cracks should occur. However, it can be observed from the microstructure of the sample that a large number of vertical cracks are present in the area. This can be attributed to the large tensile stress applied on the coating, making it very likely that cracks will propagate along the through-thickness direction following mode I fracture.

As is the case for the stress profile along the radial direction, a stress peak at the edge of the heated zone can also be seen in the shear stress profile along the axial direction. This might be responsible for the formation of horizontal cracks and the delamination of the coating.

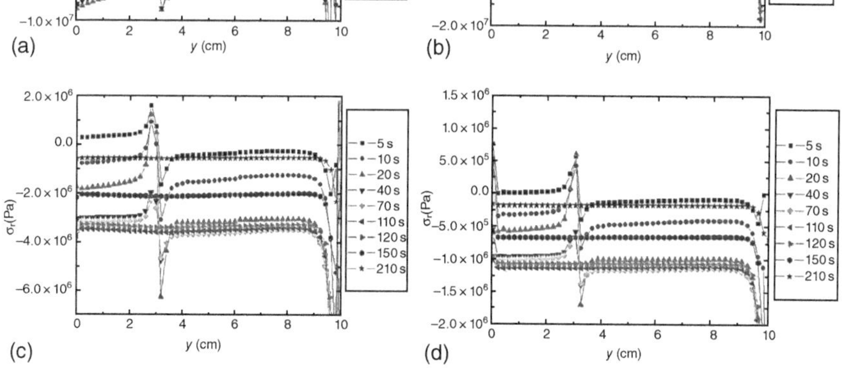

11.10 σ_r spatial distribution of interfaces vs axial position at different times for cylinder specimen with loading: (a) substrate/bond coat; (b) bond coat/TGO; (c) TGO/YSZ; (d) YSZ layer.

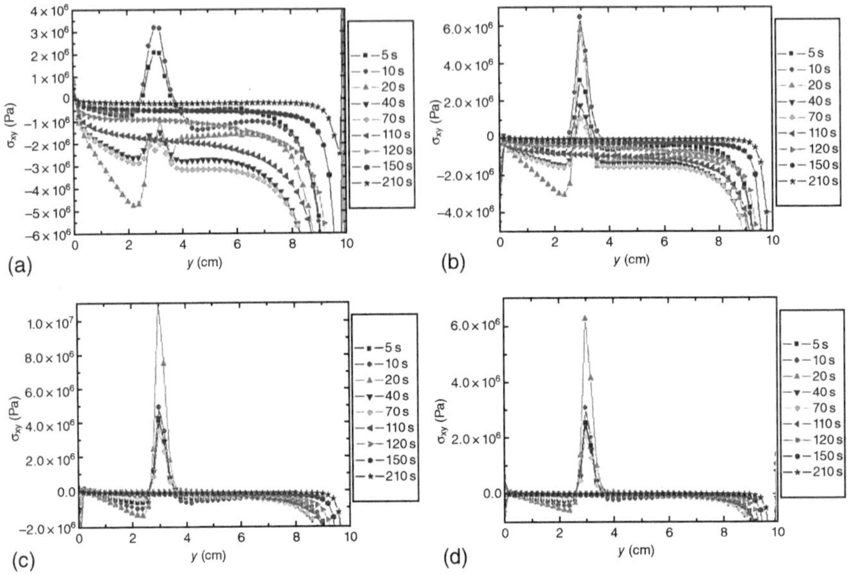

11.11 σ_{xy} spatial distribution of interfaces vs axial position at different times for cylinder specimen with loading: (a) substrate/bond coat; (b) bond coat/TGO; (c) TGO/YSZ; (d) YSZ layer.

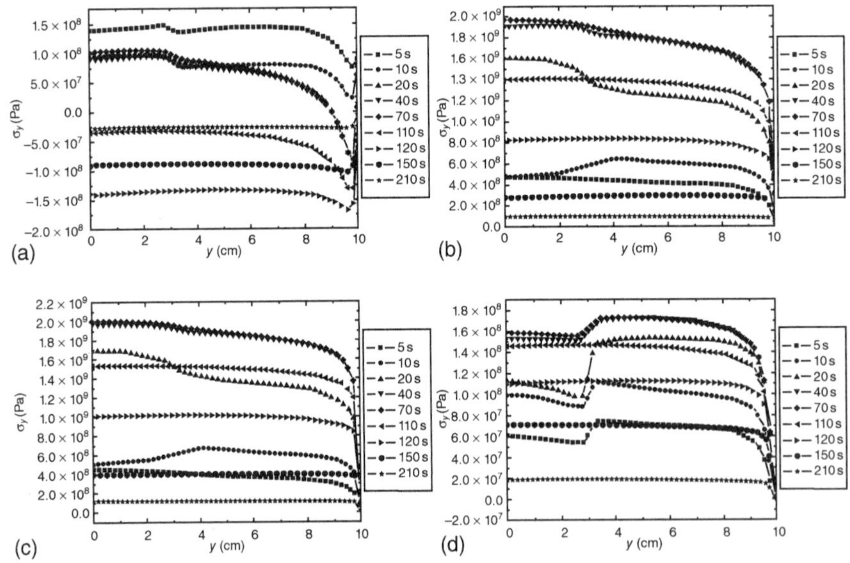

11.12 σ_y spatial distribution of interfaces vs axial position at different times for cylinder specimen with loading: (a) substrate/bond coat; (b) bond coat/TGO; (c) TGO/YSZ; (d) YSZ layer.

The shape of the stress profile in the TBC system suggested a possible failure mechanism of the coating. The shrinkage of the substrate is inevitable when it is subjected to tensile stress along the axial direction σ_y coupled with compressive stress along the radial direction, σ_r. It can be seen from the simulated stress profiles that the highest values of both σ_y and σ_r are obtained in substrate at the heated centre. Therefore, the shrinkage of the substrate in the radial direction will result in a defect between bond coat and substrate, which is also considered to be de-cohesion between bond coat and substrate.

Subjected to applied mechanical loading, due to the large thermal mismatch between TBC and bond coat, the stress concentration in the TGO layer is enormous. It can be seen from Fig. 11.11 that the shear stress level in TGO-related interfaces is much larger than in other layers. This enables the crack to propagate along the interface following mode II fracture. The interfacial failure is observed in the heated centre, where TGO is thickest in all positions due to high temperature during the thermomechanical test.

Failure mechanism

Subjected to in-plane thermal loading with mechanical loading, the substrate is under compression caused by the coupled effect of normal axial tensile stress σ_y and radial compressive stress σ_r. The highest values for the both are presented at the heated zone, which enhances the contraction of the substrate. Hence, defects occurred at the interface between substrate and bond coat, which is a possible reason for the de-cohension of the interface.

Due to high temperature and mechanical fatigue loading, plastic deformation and slip bands occurred in the metallic bond coat. These defects in the ceramic coating gave rise to the formation of the oxidation penetration paths. Consequently, abnormal oxidation in the bond coat and oxidation at the surface of the bond coat are observed.

Stress concentration in the TGO layer is one of the most important reasons for the failure of TBCs. An order of 1 GPa is shown at both bond coat/TGO interfaces and TGO/TBC interfaces. Delamination and crash of the TGO are also observed.

11.4 Failure mechanism of EB–PVD TBC under 3-D thermal gradient coupled with mechanical loading

In this investigation, a tube-shaped sample was employed. Subjected to infrared heating at a local area of the sample surface coupled with cooling air through the hollow in the sample, a temperature gradient field along both radius and axial directions was obtained.

11.4.1 Failure modes

The cross-sectional microstructure of the sample at different positions is illustrated in Fig. 11.13. Failure mode varies according to the position of the sample along the axial direction.

It can be seen from Fig. 11.13 that at the area far away from the heated zone, no apparent damage can be observed in the coating. No obvious cracks are present in the ceramic coating and the interfaces. It can also be seen that the coating is well attached to the substrate by means of the bond coat. As the testing temperature at this area is relatively low, accompanied by a low stress level, failure hardly occurred in this area. The corresponding stress level details are discussed in Section 11.4.2.

As the test area moves towards the heated centre, failures present in the TBC layer at the edge of the heated zone. The arrows in Fig. 11.13 denote the vertical cracks present in the ceramic coating in this area. The mechanism of nucleation and propagation of the vertical cracks will be discussed in detail using simulated stress analysis in Section 11.4.2. However, the figure clearly shows that vertical cracks emerged in both the ceramic coating (mark 3) and the TBC/baud coat interface (mark 2), and these are considered to result in the local delamination of TBC. In the low-temperature area and in the TGO layer (mark 1 in Fig. 11.13), very few interfacial failures can be observed. This suggests that in this situation, the ceramic coating has good adhesion with the metallic coating and the stress concentration in the TBC layer is of more concern.

At the heated centre, three kinds of failure mode are observed. At the edge of the heated centre, both vertical cracks (arrows denoted) and local spallation (mark 3) caused by vertical cracks emerging can be observed. At the same time, severe interface delamination occurrs at this area (mark 2). Compared to the interfacial failure shown at the edge of the heated zone, the interface delamination shown in this area is induced by the decrease in interface adhesion rather than by the emerging of vertical cracks. The decrease in adhesion is considered to be induced by the growth of TGO during heat treatment. It should be noticed that, in the heated area, failure in the bond coat occurrs (mark 4), which is different from that in other positions. The same phenomena are widely observed in this area and shown in Fig. 11.14.

It is well known that the fracture toughness of the bond coat is much higher than that of the ceramic coating. As a large number of slip bands are present in the metallic coating, plastic deformation occurrs at the tip of the cracks, leading to a compressive stress field at the crack tip, and finally

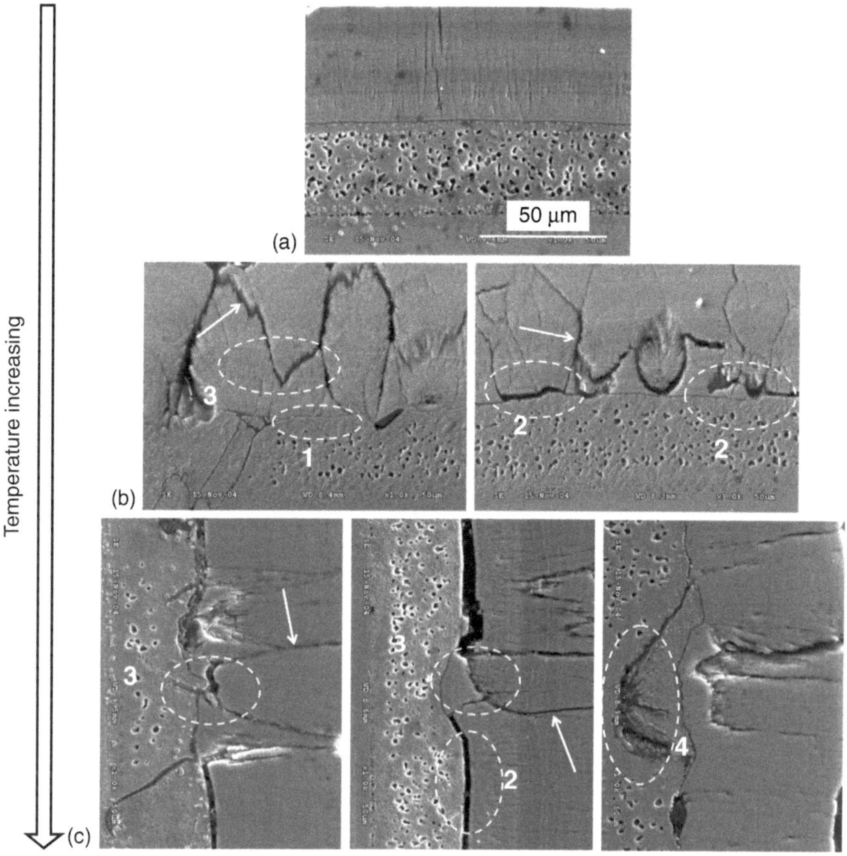

11.13 Cross-sectional images for sample H2 at different positions along the radius direction.

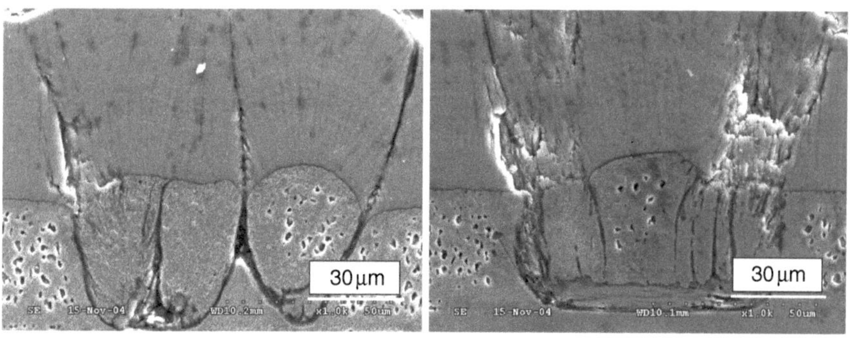

11.14 Failure modes for sample H1 along the direction perpendicular to loading.

resulting in the inhibition of crack propagation. It is considered that when subjected to thermomechanical coupled loading, creep of the bond coat is accelerated. Therefore, the plastic deformation of the metallic coating is completed at a relatively early stage of the test. Consequently, during the following cycling test, there are insufficient slip bands in the metallic coating, which presents as a decrease in fracture toughness.

Compared to the failure mechanism of the solid cylinder sample subjected to the thermomechanical loading test, some differences are shown in the tube sample. In the tube sample, the outer surface is locally heated, while internal cooling is done using cooling air. Therefore, a thermal gradient along both the radial and the axial direction is obtained, which is considered to be responsible for the failure of TBCs. However, in the solid cylinder sample, the thermal gradient along the radius direction is ignored. Therefore, in the solid cylinder sample, only the thermal gradient along the axial direction is considered.

Consequently, different failure modes are present in the solid cylinder sample and the tube sample. In the solid cylinder sample, vertical cracks in the ceramic coating and the emerging of vertical cracks near the TBC/TGO interface are the dominant phenomena. However, in the tube sample, crack propagation along both the through-thickness direction and the interface direction are observed. This is considered to be responsible for the delamination of the coating.

11.4.2 Stress analysis

In the stress analysis, a tensile stress profile ranging from 0–160 MPa was applied to the sample, in accordance with the experimental parameters. The periodicity of the stress profile corresponds to that of the infrared heating. Equivalent stress profiles along the axial direction of each layer and interface are shown in Fig. 11.15. In each part of the figure, the heating procedure takes place over 0–20 s and the temperature dwell procedure over 20–70 s. Cooling to room temperature is reached at 120 s.

It can be seen from the stress profile in the bond coat/substrate interface that the highest stress occurred in the heated area and that stress increased with temperature. At the same time, the equivalent stress shows a large drop at the edge of the heated area. This can be attributed to the coupled effect of applied tensile stress and the temperature-induced thermal expansion.

The highest stress level is shown in the TGO-related interfaces due to the high Young's modulus of the TGO layer, and the stress trend along the axial direction is similar to that which occurs in the substrate/bond coat interface. However, it can be seen in Fig. 11.15b that the stress level at the edge of the heated zone is the lowest in the whole profile. This explains why few interfacial failures are observed in this area (Fig. 11.13).

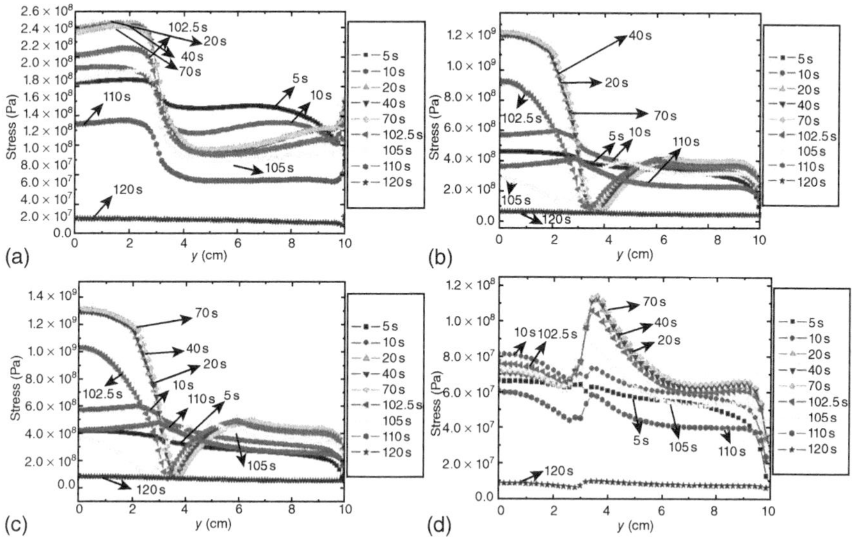

11.15 Stress spatial distribution of interfaces vs axial position at different times for hollow cylinder specimen with loading: (a) substrate/bond coat; (b) bond coat/TGO; (c) TGO/YSZ; (d) YSZ layer.

A different phenomenon is shown in the stress profile in the YSZ layer, where the highest stress peak value is present at the edge of the heated zone. In this layer in the heated area, the stress value is slightly decreased towards the edge of the heated area, and no obvious stress gradient is shown in the non-heated area along the axial direction in the whole experimental procedure. Comparing the stress value in the heated area with that in the non-heated area, no obvious difference can be seen. It should also be noted that in the whole experimental procedure, the stress level change is not so much as in other layers. This suggests that fatigue failure is not the dominant reason for the failure in the ceramic coating.

The stress distributions along the radial direction in each layer and the interface are illustrated in Fig. 11.16. A stress value peak is present at the edge of the heated zone, whereas a relatively low stress level is shown in other positions along the axial direction. It should be noted that the stress status changed from compression to tension at the edge of the heated zone, where the lattice distortion is much larger than at other positions, and hence large lattice strain energy was stored. Therefore, crack formation and subsequent propagation occurred releasing the induced strain energy. At the same time, it can be seen from the stress profiles along the axial direction that, except at the edge of the heated centre, the stress value is not high enough to cause any failure.

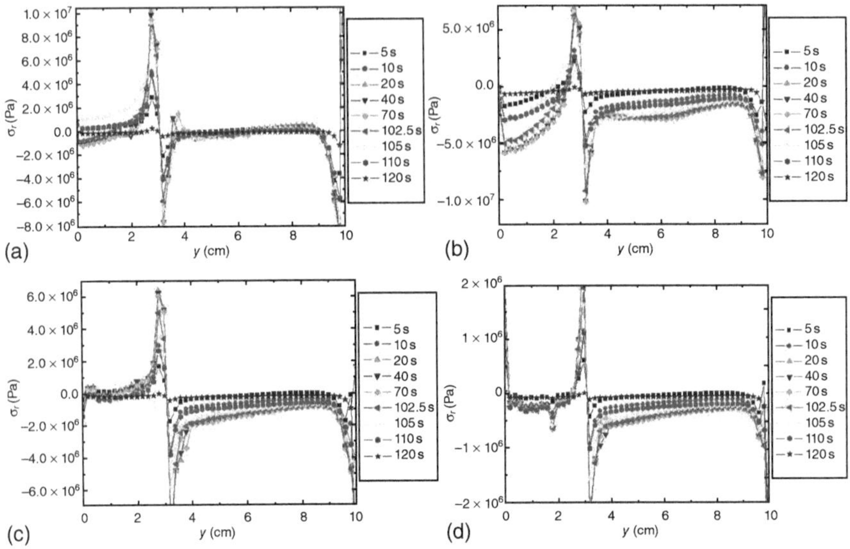

11.16 σ_r spatial distribution of interfaces vs axial position at different times for hollow cylinder specimen with loading: (a) substrate/bond coat; (b) bond coat/TGO; (c) TGO/YSZ; (d) YSZ layer.

The in-plane normal stress distribution along the axial direction is illustrated in Fig. 11.17, and the stress is as the same order as equivalent stress. This implies that σ_y is the dominant stress component in equivalent stress. Regarding the texture of EB–PVD TBC, the in-plane fracture toughness of the ceramic coating is lower than that in the through-thickness direction. Hence σ_y is of great concern as the dominant cause for cracks along the through-thickness direction. The presence of vertical cracks causes a moment to be applied on the ceramic column due to applied in-plane stress which acts as the driving force behind the propagation of horizontal cracks.

The shear stress profiles are illustrated in Fig. 11.18 which shows a trend similar to that as the stress distribution along the radius direction. The stress level is relatively low in most positions except at the edge of the heated zone, where a stress peak value is presented. This suggests that occurrence of fracture is highly possible at this area.

11.4.3 Failure mechanism

Subjected to high temperature coupled with fatigue load cycling , substrate and bond coat are under tension, σ_y, with the highest stress level being 150 MPa. At the edge of the heated area, the substrate and bond coat underwent shear stress over 10 MPa caused by non-uniform thermal

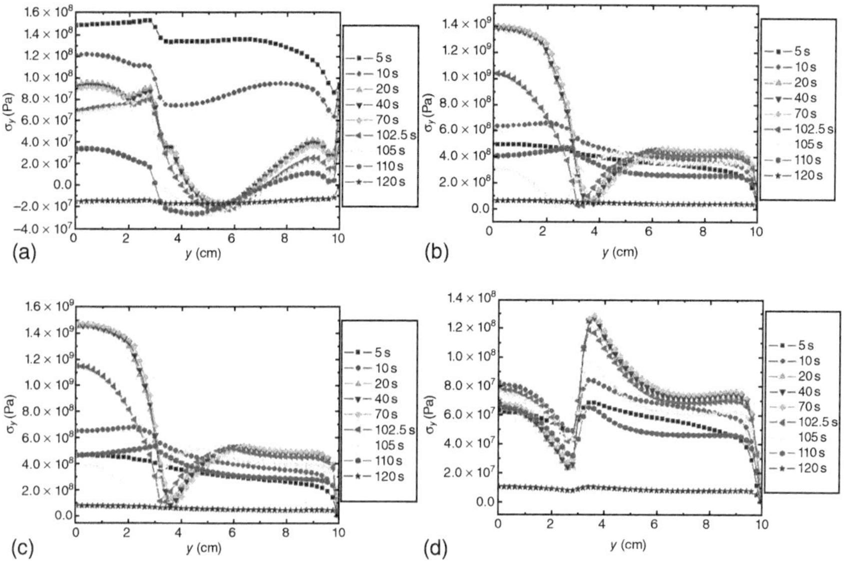

11.17 σ_y spatial distribution of interfaces vs axial position at different times for hollow cylinder specimen with loading: (a) substrate/bond coat; (b) bond coat/TGO; (c) TGO/YSZ; (d) YSZ layer.

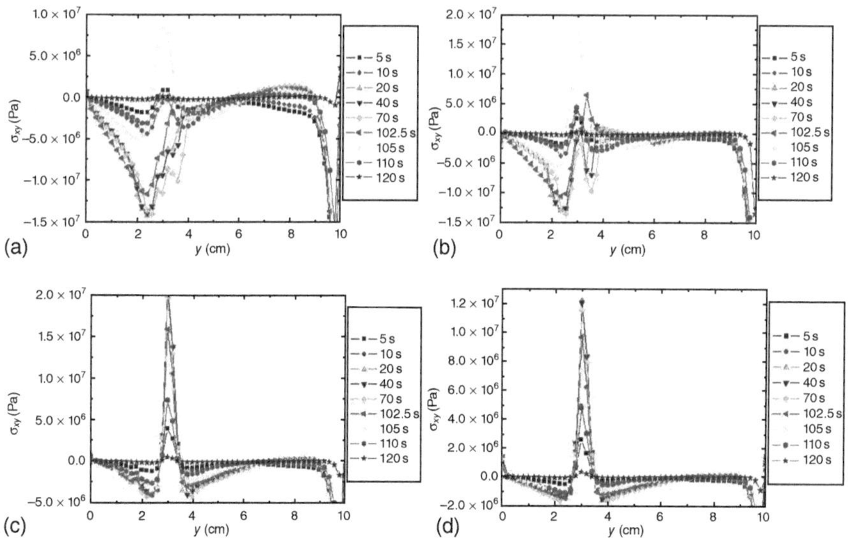

11.18 σ_{xy} spatial distribution of interfaces vs axial position at different times for hollow cylinder specimen with loading:[22] (a) substrate/bond coat; (b) bond coat/TGO; (c)TGO/YSZ; (d)YSZ layer.

loading. Therefore, slip bands formed in the bond coat at an angle of 45° from the interface. Cracks induced at these defects tend to propagate in the bond coat, since the slip bands will be blocked at the interface. Hence, the bond coat is internal oxidized.

Stress concentration in the TGO layer is one of the most important reasons for the failure of TBCs. An order of 1 GPa is shown at both bond coat/TGO interfaces and TGO/TBC interfaces. Delamination and crash of TGO is also observed.

Subjected to mechanical loading, normal stress, σ_y is at a relatively high level, accompanied by radius stress and shear stress, and residual stress in the ceramic coating increases. During the experiment, the coating is exposed to high temperature, so the strength of the ceramic coating decreases, particularly at the area with the highest temperature gradient. Therefore, vertical cracks and delamination in the ceramic coating are considered to be the most important reason for the failure of TBCs.

11.5 Conclusions

The failure mechanism of the sample subjected to thermomechanical loading was obtained using microstructure analysis and impedance spectroscopy accompanied by stress simulation. Some conclusions are drawn as follows.

Cracks initiated and propagated in ceramic coating are considered to be the dominant reason for the failure of TBCs. This is entirely different from the failure mechanism in a uniform temperature field without mechanical loading. It is also proved by FEM stress analysis that the degradation of the ceramic coating is accelerated under applied mechanical loading.

The highest crack density in ceramic coatings occurred at the edge of the heated area, which is also considered to be the area with the highest temperature gradient. According to the results obtained by impedance spectroscopy, the highest electrical resistance was presented at the area with the highest temperature gradient rather than the highest temperature. This is induced by the vertical crack density.

In the tube sample, the temperature gradient along both the radius and the axial directions was obtained. In this situation, vertical cracks occurred in the ceramic coating. At the same time, delamination of the ceramic coating was also observed. The emerging of vertical cracks and delamination of the coatings are considered to be responsible for the failure of TBCs.

In the solid cylinder sample, only the temperature along the axial direction was considered. Subjected to applied mechanical loadings, accompanied by thermal stress, cracks along the through-thickness direction are the dominant reason for the failure TBCs.

11.6 References

1. ANDERSON, P.S., X. WANG and P. XIAO, Impedance spectroscopy study of plasma sprayed and EB-PVD thermal barrier coatings. *Surface & Coatings Technology*, 2004. **185**(1): 106–119.
2. WANG, X. and P. XIAO, Characterisation of clay sintering process using impedance spectroscopy. *Journal of The European Ceramic Society*, 2002. **22**(4): 471–478.
3. MILLER, R.A., Oxidation-based model for thermal barrier coating life. *Journal of the American Ceramic Society*, 1984. **67**(8): 517–521.
4. STRANGMAN, T.E., Thermal barrier coatings for turbine airfoils. *Thin Solid Films*, 1985. **127**(1–2): 93–106.
5. EVANS, A.G., M.Y. HE and J.W. HUTCHINSON, Mechanics-based scaling laws for the durability of thermal barrier coatings. *Progress in Materials Science*, 2001. **46**(3–4): 249–271.
6. PADTURE, N.P., M. GELL and E.H. JORDAN, Thermal barrier coatings for gas-turbine engine applications. *Science*, 2002. **296**(5566): 280–284.
7. EVANS, A.G., *et al.*, Mechanisms controlling the durability of thermal barrier coatings. *Progress in Materials Science*, 2001. **46**(5): 505–553.
8. WRIGHT, P.K. and A. EVANS, Mechanisms governing the performance of thermal barrier coatings. *Current Opinion in Solid State and Materials Science*, 1999. **4**: 255–265.
9. BOROM, M.P., C.A. JOHNSON and L.A. PELUSO, Role of environmental deposits and operating surface temperature in spallation of air plasma sprayed thermal barrier coatings. *Surface and Coatings Technology*, 1996. **86–87**: 116–126.
10. EVANS, A.G. and J.W. HUTCHINSON, The mechanics of coating delamination in thermal gradients. *Surface & Coatings Technology*, 2007. **201**(18): 7905–7916.
11. MERCER, C., *et al.*, A delamination mechanism for thermal barrier coatings subject to calcium-magnesium-alumino-silicate (CMAS) infiltration. *ACTA Materialia*, 2005. **53**(4): 1029–1039.
12. KRAMER, S., *et al.*, Mechanisms of cracking and delamination within thick thermal barrier systems in aero-engines subject to calcium-magnesium-alumino-silicate (CMAS) penetration. *Materials Science and Engineering*, 2008. **490**(1–2): 26–35.
13. WANG, X., J.F. MEI and P. XIAO, Non-destructive evaluation of thermal barrier coatings using impedance spectroscopy. *Journal of The European Ceramic Society*, 2001. **21**(7): 855–859.
14. SONG, S.H. and P. XIAO, Relation of microstructural and compositional features to the electrical properties in degraded thermal barrier coating systems. *Journal of Materials Science*, 2003. **38**(8): 1661–1665.
15. DENG, L.F. and P. XIAO, Characterisation of alumina scales on FeCr alloy using impedance spectroscopy. *Thin Solid Films*, 2008. **516**(15): 5027–5031.
16. DENG, L.F., Y.S. XIONG and P. XIAO, Modelling and experimental study of impedance spectra of electron beam physical vapour deposition thermal barrier coatings. *Surface & Coatings Technology*, 2007. **201**(18): 7755–7763.
17. HUTCHINSON, J.W. and A.G. EVANS, On the delamination of thermal barrier coatings in a thermal gradient. *Surface & Coatings Technology*, 2002. **149**(2–3): 179–184.
18. HE, M.Y., D.R. MUMM and A.G. EVANS, Criteria for the delamination of thermal barrier coatings: with application to thermal gradients. *Surface & Coatings Technology*, 2004. **185**(2–3): 184–193.

19. ZHANG, D.B., *et al.*, Measurements of the thermal gradient over EB–PVD thermal barrier coatings. *Vacuum*, 2003. **70**(1): 11–16.
20. VASSEN, R., *et al.*, Recent activities in the field of thermal barrier coatings including burner rig testing in the European Union. *Advanced Engineering Materials*, 2008. **10**(10): 907–921.
21. CHEN, C., *et al.*, Failure of EB–PVD thermal barrier coatings subjected to thermo-mechanical loading. *Chinese Journal of Aeronautics*, 2006. **19**: s82–s85.
22. PENG, H., *et al.*, Numerical Analysis of EB–PVD Thermal barrier coatings under thermal-mechanical coupled environment. *Materials Science Forum*, 2007. **546–549**: 1795–1799.
23. ZHOU, Y.C. and T. HASHIDA, Coupled effects of temperature gradient and oxidation on thermal stress in thermal barrier coating system. *International Journal of Solids And Structures*, 2001. **38**(24–25): 4235–4264.

12

Non-destructive evaluation (NDE) of the failure of thermal barrier coatings

G. CHEN, Northwestern Polytechnical University, China

Abstract: This chapter discusses the development of non-destructive evaluation (NDE) methods used for the detection of premature failure of thermal barrier coatings (TBCs) systems. The fundamentals and application of different NDE methods are reviewed, with particular interest focused on the latest results with regard to the promising emission spectroscopy NDE method used for *in situ* TBCs failure inspection.

Key words: thermal barrier coatings (TBCs), non-destructive evaluation (NDE), high-temperature oxidation, gas turbine.

12.1 Introduction

A large number of coatings have been developed and are widely used to protect a variety of structural engineering materials from corrosion, oxidation, wear and erosion. Of them all, thermal barrier coatings (TBCs) have the most complex structure. TBCs are applied to the hot-section components of gas turbines and they are required to operate under the most demanding conditions encountered by the structural components. The demand for improved performance has led to the increasing use of TBCs in gas turbines, and they are the most important materials on the technology roadmap in that field. The function of TBCs is to provide thermal insulation and oxidation protection simultaneously, and TBCs have become the crucial element in the pursuit of improved efficiency, longer lifetime and reduced environmental impact.[1–9]

The state-of-the-art TBCs consist of an yttria stabilized zirconia (YSZ) top coat, a thin thermally grown oxide (TGO) reaction layer and a metallic bond coat on a superalloy substrate. The thermal insulation is provided by an outer YSZ layer deposited by either electron beam physical vapor deposition (EB–PVD) or plasma-spaying (PS) techniques. The oxidation protection is provided by a thin TGO layer. Since superalloys are not generally optimized for oxidation resistance, a bond coat (BC) is usually used to develop a durable oxidation barrier, and it must be designed to maintain a stable TGO over the life of the system. The schematic representations of TBCs prepared by EB–PVD and PS are illustrated in Fig. 12.1, which shows

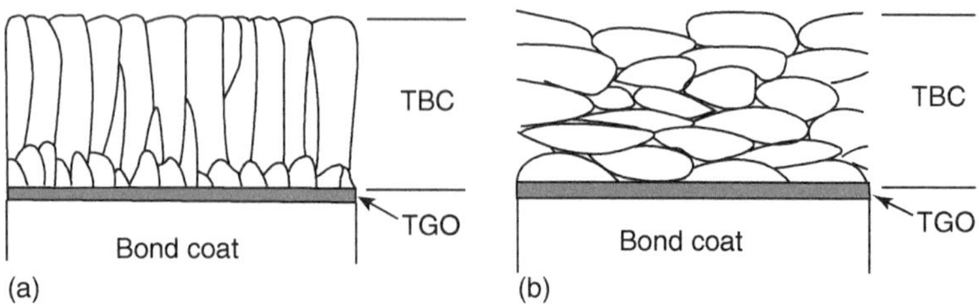

12.1 Schematic representation of TBCs obtained by EB–PVD (a) and PS (b).

the different structures obtained by EB–PVD and PS. In the modern TBCs system, the TGO layer is the source of the major strain incompatibility that leads to failure.[2,6] Evolutionary processes in the surrounding layers have deleterious effects that converge on the TGO, further aggravating the problem. The adhesion and growth kinetics of a TGO layer and hence the system life can be strongly influenced by the composition and/or morphology of the bond coat.

The application of TBCs in conjunction with internal cooling of the underlying superalloy component can provide a surface temperature reduction of 50–150 °C, with potential of over 170 °C.[9] This has enabled the operation of modern gas turbines having combustion gas temperatures in excess of 250 °C above the melting temperature of the protected superalloy components, thereby improving turbine efficiency and performance. Alternatively, TBCs help reduce metal temperature and make gas turbine components more durable at lower operating temperatures. Therefore, the growing reliance on TBCs to increase the gas inlet temperature of gas turbines requires that TBCs must be 'prime reliant', and must not fail over the life of the components. However, lack of reliability and concern over premature failure have slowed the general application of TBCs in gas turbines and, even where used, the benefits offered by TBCs, such as the temperature advantage, and hence energy efficiency and operating lifetime, have not yet been fully realized.[4] The exact failure mechanisms of TBCs are still not clear due to the many contributing factors. The type and fabrication procedures of the bond coats, as well as the TBCs, may result in different behaviors that could complicate understanding of TBCs failure even more.[2]

Due to the severity of TBCs failure, early detection of TBCs degradation and failure is becoming critical to prevent catastrophic failure as more advanced gas turbines depend on TBCs for their high-temperature operation. TBCs on turbine hot-section components, therefore, require regular maintenance. A typical periodic maintenance is a visual inspection of the TBCs for cracks, spallation, erosion or other damage that will indicate the remaining TBCs life. These visual inspections require the gas turbine to be taken out of service and to be partially disassembled to provide access to the TBCs by boroscope. This type of maintenance is costly because it requires skilled labor and the interruption of service. Moreover, in most cases only the regions with very large spalled area can be seen visually, and this is the basis for boroscope inspection today. Therefore, for the reliable application of TBCs, it is highly desirable to have non-destructive evaluation (NDE) methods that can detect the presence of small defects in TBCs at an early stage, monitor their growth and predict the remaining life of TBCs.

The present chapter summarizes the current status of understanding of TBCs failures and the development of failure detection methods; particular interest will be concentrated on the NDE methods and the latest development of *in situ* NDE of TBCs failure during gas turbine operation.

12.2 Failure of thermal barrier coatings (TBCs)

As mentioned above, the actual TBCs failure mechanisms are complex and difficult to understand due to the large number of contributing factors. Even so, some general failure models have been produced based on a large number of results.[2,6,9,10] The present topic of NDE development for TBCs failure evaluation/detection does not allow the author to give a detailed introduction to each failure mechanism of TBCs, therefore, a general summary of TBCs failure will be presented in terms of TBCs materials, structure and application, in order to facilitate a basic understanding of the different NDE methods used in this chapter. For more information readers should refer to Chapters 8, 11 and 13.

12.2.1 Materials

Bond coat

Bond coat is recognized as the most crucial component in TBCs and a critical factor affecting the stability and durability of TBCs.[4,6,9] Its chemistry and microstructure are dependent on the compositions and preparation processes, and then have significant influence on the durability through the structure and morphology of TGO formed in service. TBCs failure is often observed either at the interface between the bond coat and the TGO or between the TGO and the YSZ, or within the TGO or within the YSZ near the TGO, and the failure is dependent on the bond coat used. Thus, increasing the adhesion and integrity of the interfacial TGO layer may greatly contribute to the reliability of TBCs. Generally, the onset of TBCs spallation is triggered by some form of microstructural imperfections at or close to the TGO layer. These imperfections in or around the TGO convert the misfit stress, which results from the different mechanical properties of different layers in the TBCs, into a driving force for failure by crack initiation, propagation at the interface and final large-scale buckling of TBCs. Two principal failure models have been proposed, even though the relative importance of these two mechanisms has not been established for different bond coats. One is the development of undulations with thermal cycling, in accordance with wrinkling and ratcheting mechanisms.[11,12] Such imperfections develop primarily in bond coats most susceptible to visco-plastic deformation. They induce cracks in the superposed TBCs that extend laterally and link together to cause failure. The other is the extent of oxidation, which leads to heterogeneity of the TGO layer thickness.[13] These imperfections give rise to failure by decohesion of the interface between the TGO and the bond coat, which may be accelerated by impurity segregation.[14,15]

At present, bond coat materials are in two categories, one of which is based on NiCoCrAlY and the other is Pt-modified diffusion aluminide. Both of them result in distinct TGO features as well as differing tendencies to plastic deformation. Accordingly, the failure mechanisms are often different.[16] Both TGO growth and inter-diffusion with the substrate contribute to the bond coat change in terms of chemistry and phase structures and have a similar effect. Undulations can be promoted by softening or local volume changes arising from phase transformations in the bond coat.[11,17] Impurities migrating from the substrate can embrittle the interface or produce local oxide penetrations. In the extreme case, Al depletion from bond coat/substrate inter-diffusion can reduce the ability of bond coat to sustain protective Al_2O_3 growth and destabilize the Al_2O_3 layer by introducing unwanted elements in favor of the formation of inter-phases with lower toughness or adherence.[18] Under some circumstances, thermal stress relaxation by plastic flow of the bond coat can lead to cracking of the TGO upon reheating, giving rise to local oxide penetrations.[19]

Yttria stabilized zirconia (YSZ) insulation layer

Durability and stability concerns in the YSZ layer arise primarily from the anticipated increases in surface temperature in advanced gas turbine technologies. The consequences are increased susceptibility to morphological changes of the pore structure as well as densification of the compliance-enabling features, i.e. the columnar structure YSZ produced by EB–PVD, with attendant degradation of the insulation efficiency and strain tolerance. Moreover, current YSZ layers are becoming more susceptible to destabilization of the 'non-transformable' tetragonal ZrO_2 structure (t') as the higher operating temperature reduces the diffusional hindrance to partitioning,[20] while the effect of the subsequent monoclinic transformation is generally considered undesirable. The problem can be further aggravated by chemical reaction with molten salt deposits from the combustion environment, mostly vanadates, sulfates and silicates.[21–23]

12.2.2 Structure

As seen in Fig. 12.1, two different TBCs microstructures can be produced by EB–PVD and PS methods. A columnar TBCs grain structure with multi-scale porosity produced by EB–PVD has much more strain tolerance capability than the splat structure produced by PS, while the inter-splat/intra-splat porosity and a network of crack-like voids in the PS TBCs can provide much lower thermal conductivity compared to that of EB–PVD TBCs. Both of them make different contribution to the failure of TBCs in the initiation or propagation of the defects/imperfections. A more detailed description of the failure from EB–PVD and PS TBCs can be found in Reference 6.

12.2.3 Application

TBCs applied in aero gas turbines for propulsion will experience numerous thermal cycles whereas TBCs in industrial gas turbines for power generation will have a long isothermal history. For the latter, much longer service lifetime will be demonstrated with the same TBCs, and the failure process model is also different. Another issue occasionally encountered is the mechanical degradation of TBCs by foreign object impact and/or erosion.[24] Usually this effect will contribute along with other factors to the failure.

12.3 Development of failure inspection methods

With the application of TBCs in the early 1980s, a method of evaluating TBCs failure has been sought, proposed and developed. The major impetus of this thrust for TBCs failure inspection was to improve the integrity and performance during service of the coatings. Acoustic emission (AE) was the first method applied to qualitative and quantitative evaluation of the cracking process within the YSZ in TBC systems.[25–27] The AE method can be used for destructive and non-destructive testing. Indentation, tensile adhesion tests and four-point bending test coupled with AE can be taken as the destructive way. *In situ* AE evaluation can also be applied to the inspection of micro- and macrocracking initiation and propagation in TBCs. This method is focused only on the detection of the defects in the form of cracks due to the limited lifetime exhibited by the TBCs at that time.

However, there are numerous defects present in well-developed TBCs, such as delamination, spallation and TGO development. Thus, infrared (IR) thermography was developed to evaluate effectively the delamination defects in TBCs.[28,29] Besides the delamination defects, IR thermography was also used to inspect the degradation of YSZ in TBCs, because there existed a correlation between YSZ degradation and TBCs lifetime.[30,31] For this technique, the sample is usually heated from front or back surfaces by a flash of radiation, and the thermal image is captured using an IR camera. Delaminations, large internal cracks or changes in thermal diffusivity all result in a non-uniform temperature distribution and hence image contrast.

With the development of TBCs materials and preparation processes, more and more results suggested that the TBCs lifetime has a close relation with the reaction layer of TGO. The evaluation of the formation of TGO is important and non-destructive evaluation and detection of these coating failures at the TGO layer is necessary. All the previously mentioned NDE methods, however, did not take the TGO layer into account. Therefore, some new NDE methods were developed, e.g. electrochemical impedance spectroscopy (EIS)[32–35] and photo-stimulated luminescence spectroscopy (PSLS).[35] These methods consider the role of TGO in the determination of defects evolution and eventual failure of TBCs. EIS can determine the evolution of YSZ as well as TGO,[33] whereas PSLS focuses only on the state of TGO.[35] Recently, a new kind of NDE method has been proposed by doping marker ions in the YSZ layer in TBCs system; *in situ* evaluation of TBCs failure during engine operation could be made, demonstrating its key and promising role in determination of the failure of TBCs.[36,37]

Based on the above-mentioned methods, the techniques used in NDE of failure of TBCs can be categorized into four groups based on different aspects of TBCs discussed in detail as follows:

1. crack or delamination development in TBCs;
2. thermal property change of TBCs;
3. TGO development and its interface with YSZ and bond coat;
4. doping ions in YSZ.

12.3.1 Inspection of crack or delamination in TBCs

In the early stage of TBCs development, crack development plays a major role in determining the lifetime of TBCs.[26,27] Failure-related cracking in the YSZ is of great concern; its initiation and evolution can give rise to noise and so can be monitored by the AE technique, whereby the noise response can be correlated with the evaluation of the coating and its lifetime predicted. Acoustic emission (AE) is a term describing a class of phenomena whereby transient elastic waves are generated by the rapid release of energy from localized sources within a material.[38] Studies on AE responses during the thermal cycling of plasma-sprayed TBCs have shown that the type of cracking and the AE responses may be influenced by a number of factors, such as the porosity, the presence of metastable phases, the thermal expansion mismatch between substrate and coating and the anisotropy of thermal expansion among the phases. A stable microcracking network can be generated for TBCs with good adhesion strength. These coatings remain effective even after the formation of a microcrack network. In contrast, catastrophic failure occurred because of unstable crack growth. A 'crack density function' (CDF),[39] considering both the number and size of cracks, has been proposed, and it is found that macrocracking events tend to occur at low values of the CDF. One of the shortcomings is that this method provides temporal information of cracking in the system, but no spatial information.

With the development of TBCs including material for TBCs as well as the process methods used in the preparation of TBCs and their increasingly wide application, crack detection no longer effectively predicts the failure of TBCs. Interest has moved to the delamination of the top YSZ coat at the interface or in itself which would lead to the spallation of the top layer partly or totally, eventually causing the TBCs failure. Harada *et al.*[28] and Saitoh *et al.*[29] proposed an evaluation method for delamination effects using infrared thermography, and further work has since been carried out to study the use of this technique to predict TBCs failure.[30,31,40] The detection of the delamination in TBCs with the infrared thermography method is always coupled with measurement of the thermal properties of TBCs components; detailed information will be given in the following section.

12.3.2 Inspection of thermal property change of YSZ in TBCs

Changes in all kinds of coating properties after long-term operation may degrade the performance of the coatings. The thermal barrier performance of TBCs (in other words, the thermal resistance of TBCs) can be degraded by means of decreases in TBCs thickness, rises in TBCs density and peeling of TBCs. Thermal properties are usually treated as just basic material parameters and have not been utilized widely in TBCs characterization and failure analysis. This ageing process can alter the thermal properties of the coating (through phase transformation and microstructure change) and introduce defects like cracks, delaminations and spallations. Thermal transport properties such as thermal conductivity and thermal diffusivity are closely related to the microstructural changes of the TBCs. Therefore, it is highly desirable to have an evaluation technique sensitive to the coating properties as well as the thermal contact with the substrate.

A variety of measurement techniques have been developed to determine the thermal properties of coatings. Measurements are distinguished by a wide variety of methods employed for heating and temperature sensing. Several photothermal methods utilize front surface temperature sensing, and these include photoacoustic measurements,[41] IR thermography,[31] deflection method,[42] etc. in which IR thermography can be used to non-destructively inspect the TBCs failure.[43–46] In this method, a laser heats the part to be inspected and the surface temperature of the heated part is measured using a radiation thermometer or infrared camera. The thermal resistance of the TBCs is evaluated based on the measured temperature. An important requirement in this non-destructive inspection method is to divide the heating wavelength and measuring wavelength in order to remove the influence of the reflected light. Another requirement is to perform the heating and measurement of the target part continuously and at a fixed speed. This method is believed to be effective for complex-shaped metal parts of actual gas turbines coated with TBCs. Also, the TBCs thickness can be evaluated by the IR method.[31,43–46] Further related information about optical NDE techniques for TBCs can be found in Ellingson *et al.*[47]

Recently, a new generation of NDE IR method, phase of thermal emission spectroscopy, has been developed for TBCs failure analysis.[48] Phase of thermal emission spectroscopy is a technique whereby harmonic heating is induced in the coating by laser excitation and the resulting temperature field is detected through its thermal emission. The thermal emission comes from the volume of the coating experiencing transient temperature variation and carries information regarding heat transport through the TBCs related to the coating's thermal properties, thickness, the thermal contact (delamination) with the substrate and the substrate thermal properties. By changing the laser modulation frequency, the thermal penetration depth relative to the coating thickness is altered. Therefore, phase of thermal emission spectroscopy can be used for characterization of the coating properties over a large range of thermal penetration depths. A combination of features makes phase of thermal emission spectroscopy unique and potentially superior to existing IR techniques for coating property measurements. This technique does not require any special sample preparation and requires access only to the top surface of the coating, permitting non-destructive measurements to be made on coatings in almost any condition with arbitrary substrate shape and dimension. The method can simultaneously provide thermal conductivity and volumetric heat capacity of the coating as well as the thermal contact resistance of the interface. This method is not sensitive to environmental temperature, and therefore can be utilized at the high operating temperatures of the coating.

Impedance spectroscopy (IS) has also been used for NDE of TBCs.[32,33,49] Impedance spectroscopy is a method for the evaluation of defects and degradation of materials using the AC frequency response of the electrical impedance of a material. The impedance characteristic of a material depends on temperature, thickness or area of the coating, formation of reaction layer or delamination. These degradations and defects are evaluated by measuring the variation of impedance caused by these microstructural changes. Therefore, the IS method enables the thickness reduction of TBCs, initiation of defects, formation of oxide film and coating delamination to be detected by these resistance and capacitance data.[32] Moreover, defects such as delamination, spalling and cracking and other forms of material damage can be detected using voltage and current response when alternating current with varying frequency is applied. Changes of both resistance and capacitance associated with the degradation can be measured. Thickness reduction of TBCs has already been successfully evaluated by IS.[33] Impedance spectroscopy is a cheap, quick and powerful testing method, which is particularly useful for ceramic materials. Voltage and current responses, on application of an alternating current with varying frequency, relating to YSZ, TGO and mixed oxide layers were observed in the impedance spectra.[34] These responses can be modeled using equivalent circuits software. In all, IS is a powerful NDE method capable of detecting different types of defects mentioned above; changes in other physical properties and even the formation, growth and thickness of the TGO can also be monitored by IS.[49,50]

Compared to the 'dry' IS technique (tested at high temperatures of 200–350 °C, EIS was developed as a NDE technique for the life-remaining prediction and quality control of TBCs.[51–53] Electrochemical impedance response was acquired at room temperature and analyzed with an AC equivalent circuit. EIS is an NDE technique that measures electrochemical impedance of materials using an applied voltage with varying frequency. The electrochemical impedance characteristics of TBCs may depend on the thickness and microstructure of the YSZ coating, the thickness and microstructure of the TGO and, potentially, the damage at critical interfaces in the vicinity of TGO.[35]

12.3.3 Inspection of thermal grown oxide (TGO) and its interface with YSZ and bond coat

Although these above-mentioned non-destructive inspection techniques can characterize the degradation of TBCs, studies on NDE of the detrimental influence of the TGO reaction layer are limited. Moreover, only a few attempts have been made at NDE of the TBCs lifespan. The formation of the TGO layer in TBCs and its evolution and damage play an important role in determining the life of TBCs in service. There is a real need for non-destructive tools to evaluate whether TBCs are damaged and, on the basis of the damage detected, determine how much life remains before failure will occur. Once the TBCs have buckled or spalled they can generally be seen optically, but it is essential to develop a tool capable of detecting damage evolution before it reaches the critical buckling size. Thus, it is a prerequisite that the non-destructive tool is able to probe through the thickness of the TBCs to detect any damage related to the TGO. The evaluation of the TGO reaction layer can be realized by using the IS method.[33,35,50] The physical properties, the formation and the thickness of the reaction layer can be evaluated using this method together with sensitivity analysis. The other promising method is PSLS since it is non-destructive and produces a direct measure of local elastic strain energy in the TGO.[54–68] This process also provides a basis for the quantitative determination of the extent of local damage prior to the spalling from an analysis of the shape of the luminescence spectra.

The application of PSLS as a NDE technique for TBCs was pioneered by Clarke and refined by some other researchers.[54,55] This technique is concentrated on the stress state of the TGO. Adhesion of the TBCs to the bond coat has been shown to be dependent upon a good bond being maintained between TGO and bond coat. The thermal expansion mismatch between the metallic bond coat and the alumina TGO is such that the TGO is under a large compressive stress (about 4 GPa) at room temperature. Laser excitation of small amounts of Cr ions in the TGO causes fluorescence to occur and the spectral response is pressure (stress) dependent. If the TGO becomes detached, the stress would be reduced and this in turn results in a peak shift in the fluorescent spectra, which is illustrated in Fig. 12.2.

In α-alumina the R-line fluorescence is split into a doublet (R1, R2) by crystal field effects. Of the two lines, it has been found that the R2 line has a more linear pressure dependence and is also more isotropic for different

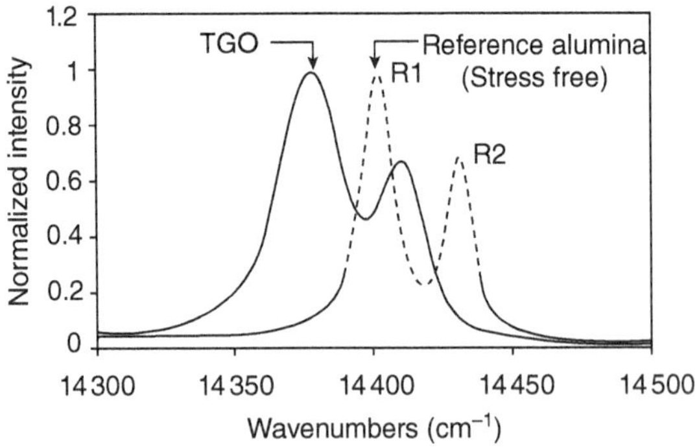

12.2 The typical R1/R2 fluorescence spectra for chromium containing stress-free and stressed α-Al_2O_3.

crystallographic orientations. Hence, the stress measurements reported have been primarily based on measurements of R2. For polycrystalline materials where the sample volume probed is large compared with the grain size, the frequency shift, $\Delta\nu$, of the R2 line is given by:

$$\Delta\nu = 2.54\ (\text{cm}^{-1}\ \text{GPa}^{-1})\sigma \qquad [12.1]$$

and for the case of a scale which we take to be stressed in only two dimensions, Eq. (12.1) becomes:

$$\Delta\nu = 5.07\ (\text{cm}^{-1}\ \text{GPa}^{-1})\sigma_{2D} \qquad [12.2]$$

where the biaxial stress, σ_{2D}, is taken to be homogeneous.

However, the spectra observed from aged TBCs were not usually similar to the clear doublet observed in single-crystal material. The shape and position of the R1R2 doublet for the measured TGO luminescence spectra changes with stress, temperature, oxidation time and even Cr^{3+} ion concentration,[69–71] and therefore allows quantitative information on these conditions to be extracted from analysis of the fluorescence spectrum.[67] Some spectra can be described as being due to two stress states in the material – physically interpreted as being due to bonded (stressed) and de-bonded (unstressed) material. More often, although there may be dominant contributions from two stress states, there is evidence that there may be a distribution of stressed states present. This makes the extraction of quantitative data complicated.

A program was developed to separate the contribution from the different stress states.[70] Each spectrum is considered to consist of an underlying linear baseline (due to background signal and instrumental effects) together with multiple R1R2 curves. The R1 peak is used to obtain an initial estimate of the parameters for fitting to a single doublet, and the optimization routine uses a sequential quadratic programming method combined with a line search to find the best-fit parameters subject to constraints on the intensity ratios, half-width ratios, R1 and R2 shapes and R1–R2 separation. If the fit to a single doublet is deemed unsatisfactory, a fit to two doublets is implemented, with the parameters from the fit to a single doublet utilized to obtain an initial estimate of the parameters for the optimization routine – the typical de-convoluted spectrum is demonstrated in Fig. 12.3. The appropriateness of the fitting program was compared in an international round-robin[67] and the present program was demonstrated to be in agreement with that used by other leading workers in the field.

Figure 12.4 shows the stress map for EB–PVD/Pt-aluminide after 400 1 h cycles at 1100 °C in the fast cool ageing test, EB–PVD/MCrAlY after 4000 h at 1000 °C in the slow cool ageing test and after 1175 1 h cycles at 1000 °C in the fast cool ageing test. Based on the stress analysis, the detachment of the top TBCs coat can be extrapolated. For example, the highly stressed

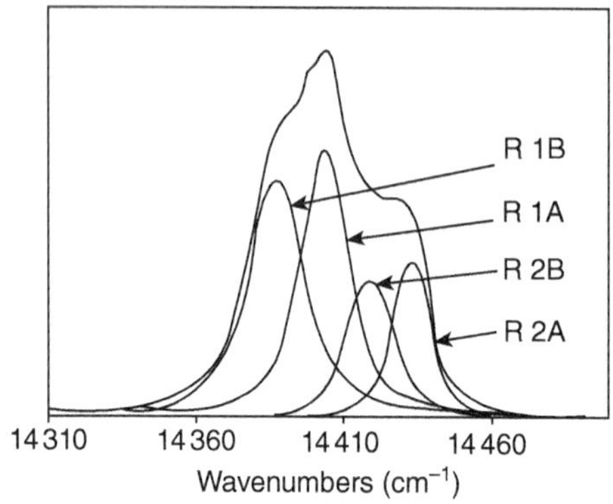

12.3 The de-convolution of the fluorescence spectrum.

regions with dark area represent the attached TBCs and low stressed regions are detached TBCs.

12.3.4 In situ inspection by doping in YSZ

Substantial savings in gas turbine maintenance could be obtained if the TBCs materials could be made to emit signals that contain information about their deterioration (cracking/spalling) and instrumentation could be developed to detect these signals *in situ* during gas turbine operation. An exciting development in rocket engine health monitoring shows great promise for application in *in situ*, real-time TBCs health monitoring.[72] The technique consists of optical detection of species in the vapor phase inside the chemically reacting or expanding flow. Species originating from the deteriorating TBCs would appear as atomic vapor, oxides and/or hydroxide species in the engine flow path. During flow expansion through the turbine and nozzle, these species equilibrate, typically by collisional de-excitation or emission of radiation. This radiation can be monitored anywhere in the engine where optical access can be designed into the engine components. The use of fiber optics as light guides facilitates the optical access and could allow the design of low-cost, *in situ* TBCs health monitoring equipment.

An ideal engine monitoring system would correctly identify the engine part before it fails. It is desirable to detect and track the TBCs failure throughout the life of the component. Therefore, TBCs can be doped with signature marker materials that can be monitored optically in the combustion gas or the exhaust. The marker materials can be incorporated in the area where the failure is generally initiated, which is the inner YSZ at the TGO/YSZ interface in the case of the plasma-sprayed TBCs. From the presence of the marker species in the combustion gas or the exhaust, deter-

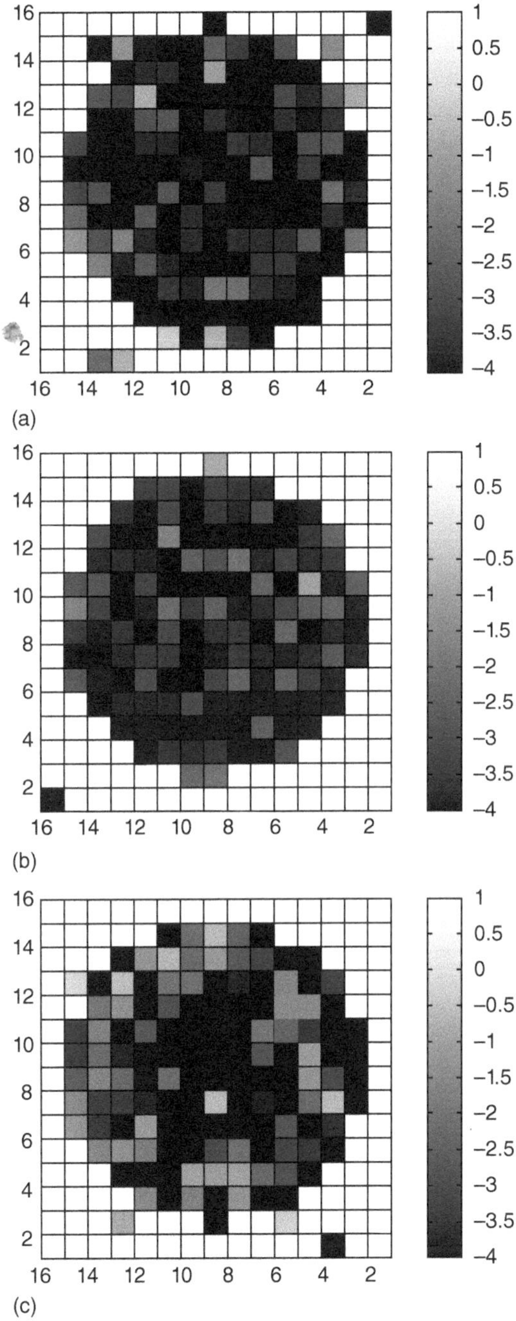

12.4 The stress map for EBPVD/Pt aluminide after 400 1-h cycles at 1100 °C in the fast cool ageing test (a), EBPVD/MCrAlY after 4000 h at 1000 °C in the slow cool ageing test (b) and after 1175 1-h cycles at 1000 °C in the fast cool ageing test (c) in their high stress component (R2B).

mined by a spectroscopic interpretation of the observed optical emission, the mechanical condition and deterioration of TBCs can be inferred.

Our preliminary research demonstrated the feasibility of using spectroscopic *in situ* NDE to monitor the deterioration of Li_2O-doped TBCs.[36] The intensity of characteristic luminescence response of lithium ions at its wavelength of 670.8 nm can be used to extrapolate the degradation degree of TBCs, which is demonstrated in Figs 12.5 and 12.6. In these figures, the top YSZ layer was drilled out in the samples in order to expose the bottom Li_2O-doped layer and simulate coating spallation. The figures show a linear

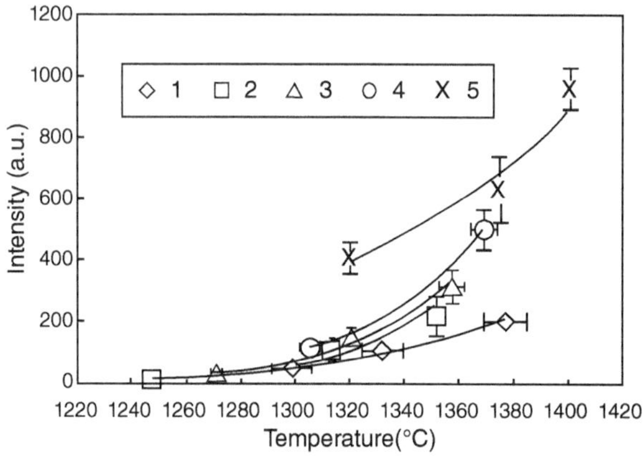

12.5 Effect of flame temperature and number of 1 mm diameter holes on the intensity of slurry dip-processed 1 wt% Li_2O-doped YSZ/YSZ TBCs.

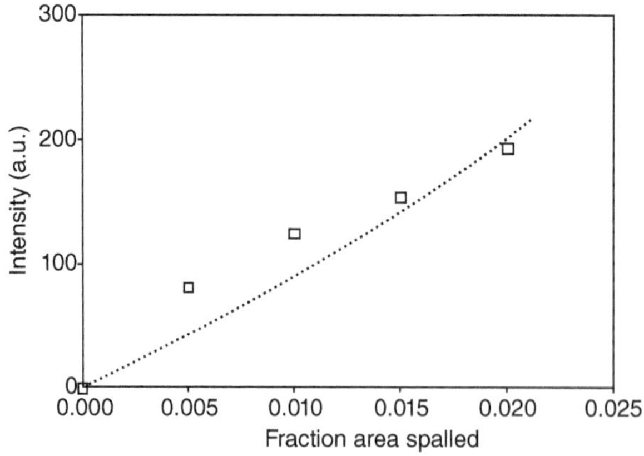

12.6 Emission intensity measured at 1320 °C vs fraction area spalled (drilled out) for the slurry-dip processed 1 wt% Li_2O-doped TBCs.

increase in the emission intensity with increasing extent of spallation under the condition that there is a certain amount of Li dopant. It demonstrates the viability of using Li emission spectroscopy as a tool to monitor the health of TBCs.

However, this technique requires that the addition of Li_2O in the YSZ layer would not impair the performance of TBCs compared with the widely accepted normal plasma-sprayed TBCs during application, particularly regarding the stability and durability of TBCs in service. Therefore, the actual configuration design for the plasma-sprayed TBCs was also carried out. Plasma-sprayed TBCs with a Li_2O-doped YSZ inner layer and a regular YSZ outer layer configuration could be an option for the modified TBCs used in emission spectroscopic failure analysis. Thermal cycling tests at 1125 and 1150 °C indicated that the modified plasma-sprayed TBCs with a 1/9 (the thickness of the inner and outer YSZ layer in *mil*, 1 mil = 25.4 μm) configuration doped with medium Li_2O concentration of 1 and 3 wt% exhibited longer durability,[37] which can be seen in the Tables 12.1 and 12.2.

Another *in situ* doping NDE method has been proposed, but it is still in its infancy.[73] It consists of an inner layer of the coating doped with a chromophore adjacent to the bond coat. Where the coating is intact, the doped region is buried and so can only be probed with a laser wavelength to which the coating is transparent. Where the coating has spalled away or has been eroded, the doped layer is exposed and can be illuminated with a laser in the UV as well as in the visible. As the TBC erodes away, successive doped

Table 12.1 Thermal cycling results for the plasma-sprayed TBCs tested at 1115 °C

Configuration	0.3 wt% Li_2O	1 wt% Li_2O	3 wt% Li_2O	5 wt% Li_2O
5/5	2	2	Failed in annealing	Failed in annealing
1/9	NA	35	>75	NA

Note: Standard no Li_2O doped sample failed after 75 cycles and annealing was performed at 1150 °C in Ar + $5H_2$ atmosphere for 4 h, all the samples were subject to annealing before thermal cycling.

Table 12.2 Thermal cycling results for the plasma-sprayed TBCs tested at 1150 °C

Configuration	1 wt% Li_2O	3 wt% Li_2O
1/9	22	>22
2/8	20	19
3/7	20	3

Note: Standard no Li_2O doped sample failed after 22 cycles.

layers will be removed and so the visible luminescence will characterize the remaining layers, enabling the remaining thickness of the coating to be assessed.

12.4 Future trends

With the development of TBCs, new kinds of materials will be applied in TBC systems to make the next generation of TBCs more reliable and durable. These include a more stable TBCs insulation layer with much lower thermal conductivity and, higher temperature capability as a substitute for the current YSZ[8,74] and more reliable bond coat materials.[9] In the mean time, a deeper and more comprehensive understanding of the failure of the new TBCs will be generated and the life of TBCs will be easier to predict. Concurrently, new kinds of NDE methods for assessing TBCs degradation will be developed and matured, for example, femtosecond laser-induced breakdown spectroscopy[75] and microwave[76] to detect thickness loss in TBCs. Even so, more issues will need to be addressed relating to the full development of currently proposed NDE methods and considerable attention should also be paid to NDE instrumentation and application for field inspection of the TBCs coated gas turbine vanes/blades.

12.5 References

1. STIGER M J, YANAR N M, TOPPING M G, PETTIT F S and MEIER G H (1999), 'Thermal barrier coatings for the 21st century', *Z Metallkd*, **90**, 1069–1078.
2. EVANS A G, MUMM D R, HUTCHINSON J W, MEIER G H and PETTIT F S (2001), 'Mechanisms controlling the durability of thermal barrier coatings', *Prog Mater Sci*, **46**, 505–555.
3. PETERS M, LEYENS C, SCHULZ U and KAYSSER W A (2001), 'EB-PVD TBCs for aeroengines and gas turbines', *Adv Eng Mater*, **3**, 193–204.
4. PADTURE N P, GELLM and JORDAN E H (2002), 'Thermal barrier coatings for gas-turbine engine applications', *Science*, **296**, 280–284.
5. SCHULZ U, LEYENS C, FRITSCHER K, PETERS M, SARUHAN-BRINGS B and LAVIGNE O (2003), 'Some recent trends in research and technology of advanced thermal barrier coatings', *Aerospace Sci Technol*, **7**, 73–80.
6. NICHOLLS J R (2003), 'Advances in coating design for high-performance gas turbines', *MRS Bull*, **28**, 659–670.
7. CLARKE D R and LEVI C G (2003), 'Materials design for the next generation thermal barrier coatings', *Ann Rev Mater Res*, **33**, 383–417.
8. LEVI C G (2004), 'Emerging materials and processes for thermal barrier systems', *Curr Opin Solid State Mater Sci*, **8**, 77–91.
9. GLEESON B (2006), 'Thermal barrier coatings for aeroengine applications', *J Propul Power*, **22**, 375–383.
10. SPITSBERG I T, MUMM D R and EVANS A G (2005), 'On the failure mechanisms of thermal barrier coatings with diffusion aluminide bond coatings', *Mater Sci Eng A*, **394**, 176–191.

11. HE M Y, EVANS A G and HUTCHINSON J W (2000), 'The ratcheting of compressed thermally grown thin films on ductile substrates', *Acta Mater*, **48**, 2593–2601.
12. EVANS A G, HE M Y and HUTCHINSON J W (1997), 'Convergent debonding of films and fibers', *Acta Mater*, **45**, 3543–3554.
13. MUMM D R and EVANS A G (2000), 'On the role of imperfections in the failure of a thermal barrier coating made by electron beam deposition', *Acta Mater*, **48**, 1815–1827.
14. HUTCHINSON J W, HE M Y and EVANS A G (2000), 'The influence of imperfections on the nucleation and propagation of buckling driven delaminations', *J Mech Phys Solids*, **48**, 709–734.
15. GAUDETTE F A, SURESH S, EVANS A G, DEHM G and RUHLE M (1997), 'The influence of chromium addition on the toughness of γ-Ni/α-Al_2O_3 interfaces', *Acta Mater*, **45**, 3503–3513.
16. YANAR N M, PETTIT F S and MEIER G H (2006), 'Failure characteristics during cyclic oxidation of yttria stabilized zirconia thermal barrier coatings deposited via electron beam physical vapor deposition on platinum aluminide and on NiCoCrAlY bond coats with processing modifications for improved performances', *Metal Mater Trans*, **37A**, 1563–1580.
17. TOLPYGO V K and CLARKE D R (2000), 'Surface rumpling of a (Ni, Pt)Al bond coat induced by cyclic oxidation', *Acta Mater*, **48**, 3283–3293.
18. SHILLINGTON E A G and CLARKE D R (1999), 'Spalling failure of a thermal barrier coating associated with aluminum depletion in the bond-coat', *Acta Mater*, **47**, 1297–1305.
19. TOLPYGO V K and CLARKE D R (1999), 'Tensile cracking during thermal cycling of alumina films formed by high-temperature oxidation', *Acta Mater*, **47**, 3589–3605.
20. JONES R L (1996), 'Thermal barrier coatings', in: Stern K H (ed.), *Metallurgical and Ceramic Protective Coatings*, Chapman & Hall, London, UK, 195–235.
21. JONES R L and WILLIAMS C E (1987), 'Hot corrosion studies of zirconia ceramics', *Surf Coat Technol*, **32**, 349–358.
22. STOTT F H, DE WET D J and TAYLOR R (1994), 'Degradation of thermal barrier coating at very high temperatures', *MRS Bull*, **19**, 46–49.
23. BOROM M P, JOHNSON C A and PELUSO L A (1996), 'Role of environment deposits and operating surface temperature in spallation of air plasma sprayed thermal barrier coatings', *Surf Coat Technol*, **86–87**, 116–126.
24. CHEN X, WAND R, YAO N, EVANS A G , HUTCHINSON J W and BRUCE R W (2003), 'Foreign object damage in a thermal barrier system: mechanisms and simulations', *Mater Sci Eng A*, **352**, 221–231.
25. SHANKAR R N, BERNDT C C, HERMAN H and RANGASWAMY S (1983), 'Acoustic emission from thermally cycled plasma-sprayed oxides', *Am Ceram Soc Bull*, **62**, 614–619.
26. BERNDT C C and HERMAN H (1983), 'Failure during thermal cycling of plasma-sprayed thermal barrier coatings', *Thin Solid Films*, **108**, 427–437.
27. BERNDT C C and MILLER R A (1984), 'Failure analysis of plasma sprayed thermal barrier coatings', *Thin Solid Films*, **119**, 173–184.
28. HARADA S, IKEDA Y, MIZUTA Y, SUGITA Y and ITO A (1995), 'The inspection of delaminated defects in ZrO 2-coated material by infrared thermography techniques', in: Sekine K (ed.), *Proceedings of the 1st Symposium on Thermographic NDT and E Techniques*, Tokyo, Japan, 27–28 Nov, 47–52.

29. SAITOH M, ITOH Y and ISHII J (1995), 'Evaluation of thermal barrier coating using infrared thermography', in: Sekine K (ed.), *Proceedings of the 1st Symposium on Thermographic NDT and E Techniques*, Tokyo, Japan 27–28 Nov, 53–58.
30. WANG H and DINWIDDIE R B (2001), 'Characterization of thermal barrier coatings using thermal methods', *Adv Eng Mater*, **3**, 465–468.
31. MANDELIS A (2001), 'Diffusion-wave laser radiometric diagnostic quality-control technologies for materials NDE/NDT', *NDT&E Int*, **34**, 277–287.
32. NAKAJIMA M, SHOJI T and HASUDA H (1995), 'Nondestructive characterization of silicon carbide by impedance spectroscopy method', *Trans Japan Soc Mech Eng*, **61**, 940–945.
33. OGAWA K, MINKOV D, SHOJI T, SATO M and HASHIMOTO H (1999), 'NDE of degradation of thermal barrier coating by means of impedance spectroscopy', *NDT&E Int*, **32**, 177–185.
34. ANDERSON P S, WANG X, XIAO P (2004), 'Impedance spectroscopy study of plasma sprayed and EB-PVD thermal barrier coatings', *Surf Coat Technol*, **185**, 106–119.
35. SOHN Y H, JAYARAJ B, LAXMAN S, FRANKE B, BYEON J W, KARLSSON A M (2004), 'The non-destructive and nano-microstructural characterization of thermal-barrier coatings', *JOM*, **56**, 53–57.
36. CHEN G F, LEE K N and TEWARI S N (2006), 'Emission spectroscopy analysis for the non-destructive evaluation of the health of thermal barrier coatings', *J Mater Sci*, **41**, 6855–6860.
37. CHEN G F, LEE K N and TEWARI S N (2007), 'The concept and development of emission spectroscopy for the non-destructive evaluation (NDE) of the failure of thermal barrier coatings', *NDT&E Int*, **40**, 265–270.
38. LIN C K and BERNDT C C (1998), 'Acoustic emission studies on thermal spray materials', *Surf Coat Technol*, **102**, 1–7.
39. BERNDT C C (1989), 'Failure processes within ceramic coatings at high temperature', *J Mater Sci*, **24**, 3511–3514.
40. IBARRA-CASTANEDO C, GOMZÁLEZ D, KLEIN M, PILLA M, VALLERAND S and MALDAGUE X (2004), 'Infrared image processing and data analysis', *Infrared Phys Technol*, **46**, 75–83.
41. ROSENCWAIG A and GERSHO A (1976), 'Theory of the photoacoustic effect with solids', *J Appl Phys*, **47**, 64–69.
42. BERTOLOTTI M, LIAKHOU G, VOTI R L, MICHELOTTI F and SIBILIA C (1993), 'Method for thermal diffusivity measurements based on photothermal deflection', *J Appl Phys*, **74**, 7078–7084.
43. FUJII T and TAKAHASHI T (1998), *Estimation of Thermophysical Properties of Coating Layers for Gas Turbine Hot Parts – Part 1: Measurement of Thermophysical Properties of Coating Layer and Superalloy, and Comparison Between Virgin and Aged Material*, CRIEPI Report W97017, Central Research Institute of Electric Power Industry, Tokyo, Japan.
44. FUJII T and TAKAHASHI T (1999), *Estimation of Thermophysical Properties of Coating Layers for Gas Turbine Hot Parts – Part 2: Comparison Between Virgin and Aged Thermal Barrier Coatings in Terms of Plasma Spray Powder Size And Shape*, CRIEPI Report W98017, Central Research Institute of Electric Power Industry, Tokyo, Japan.

45. MORINAGA M, TAKAHASHI T and FUJII T (1998), *Development of Non-Destructive Inspection For The Performance Of Thermal Barrier Coating – Part 1 Numerical Analysis For TBC Thermal Resistance Detecting Method With Transient Heating*, CRIEPI Report W97021, Central Research Institute of Electric Power Industry, Tokyo, Japan.
46. MORINAGA M and TAKAHASHI T (1999), *Development of Non-destructive Inspection for the Performance of Thermal Barrier Coating – Part 2 Experimental Study for TBC Thermal Resistance Detecting Method with Transient Heating*, CRIEPI Report W99005, Central Research Institute of Electric Power Industry, Tokyo, Japan.
47. ELLINGSON W A, VISHER R J, LIPANOVICH R S, DEEMER C M (2006), 'Optical NDT techniques for ceramic thermal barrier coating: NDT of ceramic: an introduction to the technical focus issue', *Mater Eval*, **64**, 45–51.
48. YU F L and BENNETT T D (2005), 'Phase of thermal emission spectroscopy for properties measurements of delaminating thermal barrier coatings', *J Appl Phys*, **98**, 103501–103505.
49. WANG X, MEI J and XIAO P (2001), 'Non-destructive evaluation of thermal barrier coatings using impedance spectroscopy', *J Euro Ceram Soc*, **21**, 855–859.
50. WANG X, MEI J and XIAO P (2001), 'Determining oxide growth in thermal barrier coatings (TBCs) non-destructively using impedance spectroscopy', *J Mater Sci Lett*, **20**, 47–49.
51. JAYARAJ B, VISHWESWARAIAH S, DESAI V H and SOHN Y H (2004), 'Electrochemical impedance spectroscopy of thermal barrier coatings as a function of isothermal and cyclic thermal exposure', *Surf Coat Technol*, **177–178**, 140–151.
52. JAYARAJ B, DESAI V H, LEE C K and SOHN Y H (2004), 'Electrochemical impedance spectroscopy of porous ZrO_2 8 wt% Y_2O_3 and thermally grown oxide on nickel aluminide', *Mater Sci Eng A*, **372**, 278–286.
53. BYEON J W, JAYARAJ B, VISHWESWARAIAH S, RHEE S, DESAI V H and SOHN Y H (2005), 'Non-destructive evaluation of degradation in multi-layered thermal barrier coatings by electrochemical impedance spectroscopy', *Mater Sci Eng A*, **407**, 213–225.
54. MA Q and CLARKE D R (1993), 'Stress measurement in single-crystal and polycrystalline ceramics using their optical fluorescence', *J Am Ceram Soc*, **76**, 1443–1448.
55. CHRISTENSEN R J, LIPKIN D M, CLARKE D R and MURPHY K (1996), 'Non-destructive evaluation of oxidation stresses through thermal barrier coatings using Cr^{3+} piezospectroscopy', *Appl Phys Lett*, **69**, 3754–3756.
56. TOLPYGO V K, DRYDEN J R and CLARKE D R (1998), 'Determination of the growth stresses generated during the oxidation of Fe–22Cr–4.8Al–0.3Y alloy', *Acta Mater*, **46**, 927–937.
57. TOMIMATSU T, ZHU S and KAGAWA Y (2003), 'Effect of thermal exposure on stress distribution in TGO layer of EB–PVD TBC', *Acta Mater*, **51**, 2397–2405.
58. SERGO V and CLARKE D R (1998), 'Observation of sub-critical spall propagation of a thermal barrier coating', *J Am Ceram Soc*, **81**, 3237–3242.
59. SOHN Y H, KIM J H, JORDAN E H and GELL M (2001), 'Thermal cycling of EB–PVD/MCrAlY thermal barrier coatings: II. Evolution of photo-stimulated luminescence', *Surf Coat Technol*, **146–147**, 102–109.

60. XIE L, SOHN Y H, JORDAN E H and GELL M (2003), 'The effect of bond coat grit blasting on the durability and thermally grown oxide stress in an electron beam physical vapor deposited thermal barrier coating', *Surf Coat Technol*, **176**, 57–66.
61. GONG X Y and CLARKE D R (1998), 'On the measurement of strain in coatings formed on a wrinkled elastic substrate', *Oxid Met*, **50**, 355–376.
62. ATKINSON A, SELCUK A and WEBB S J (2000), 'Variability of stress in alumina corrosion layers formed in thermal barrier coatings', *Oxid Met*, **54**, 371–384.
63. SELCUK A and ATKINSON A (2002), 'Analysis of the Cr^{3+} luminescence spectra from thermally grown oxide in thermal barrier coatings', *Mater Sci Eng A*, **335**, 147–156.
64. SELCUK A and ATKINSON A (2003), 'The evolution of residual stress in the thermally grown oxide on Pt diffusion bond coats in TBCs', *Acta Mater*, **51**, 535–549.
65. WEN M, JORDAN E H and GELL M, 'Evolution of photo-stimulated luminescence of EB–PVD/(Ni, Pt)Al thermal barrier coatings', *Mater Sci Eng A*, **398**, 99–107.
66. PENG X and CLARKE D R (2000), 'Piezospectroscopic analysis of interface debonding in thermal barrier coatings', *J Am Ceram Soc*, **83**, 1165–1169.
67. MURPHY S (2003), 'NDE assessment of TBCs: an interim report of a photo-stimulated luminescence "round-robin" test', *Surf Coat Technol*, **163**, 87–94.
68. SRIDHARAN S, XIE L, JORDAN E H and GELL M (2004), 'Stress variation with thermal cycling in the thermally grown oxide of an EB–PVD thermal barrier coating', *Surf Coat Technol*, **179**, 286–296.
69. CHEN G F (2004), 'Cr-related luminescence in α-alumina prepared by sintering and oxidizing', *Mater Sci Eng B*, **113**, 106–109.
70. SAUNDERS S R J, BANKS J P, CHEN G F and CHUNNILALL C J (2004), 'Measurement of residual stress in thermally grown oxide layers in thermal barrier coating systems – development of non-destructive test methods', *Mater Sci Forum*, **383**, 461–464.
71. CHEN G F and RUHLE M (2005), 'Transient alumina transformation on a sputtered K38G nanocrystalline coating', *Surf Coat Technol*, **191**, 263–266.
72. MADZSAR G C, BICKFORD R L and DUNCAN D B (1992), 'An overview of in-flight plume diagnostics for rocket engines', in: *Proceedings of the 28th AIAA/ASME/SAE/ASEE Joint Propulsion Conference*, Nashville, TN, 6–8 July, 1–23.
73. GENTLEMAN M M and CLARKE D R (2004), 'Concepts for luminescence sensing of thermal barrier coatings', *Surf Coat Technol*, **188–189**, 93–100.
74. CLARKE D R, LEVI C G and EVANS A G (2006), 'Enhanced zirconia thermal barrier coating systems', *Proc IMechE Part A: J Power and Energy*, **220**, 85–92.
75. DAS D K, MCDONALD J P, LEVI C G, YALISOVE S M and POLLOCK T M (2008), 'Detection of a marker layer in a 7YSZ thermal barrier coating by femtosecond laser-induced breakdown spectroscopy', *Surf Coat Technol*, **202**, 3940–3946.
76. SAYAR M, SEO D and OGAWA K (2009), 'Non-destructive microwave detection of layer thickness in degraded thermal barrier coatings using K- and W-band frequency range', *NDT&E Int*, **42**, 398–403.

13
Substrate and bond coat related failure of thermal barrier coatings

R. T. WU, National Institute for Materials Science (NIMS), Japan

Abstract: The current chapter describes research aimed at improving understanding of the high-temperature degradation mechanisms of thermal barrier coating (TBC) systems on different bond coats and single-crystal superalloys. The first part presents a systematic study aimed at demonstrating that the compatibility of modern nickel-base single-crystal superalloys with TBC systems is influenced strongly by the content of alloying element additions in the superalloy substrate. This is followed by a detailed study of the bond coat effect on TBC degradation, based on an electron beam physical vapour deposition (EB-PVD) yttrium stabilized zirconia (YSZ) top coat and a substrate material of CMSX-4 superalloy.

Key words: nickel-based single crystal superalloys, thermal barrier coatings, compatibility, oxidation, interface, residual stress, rumpling, interfacial adhesion, fracture toughness.

13.1 Introduction

The success of turbine technology can be largely attributed to the development and utilization of Ni-base superalloys as the material for hot-section turbine components. Following the Kyoto Protocol aiming at mitigating CO_2 emission to prevent global warming, there has been a considerable international effort aimed at improving the efficiency of the gas turbines used for jet propulsion and electricity generation, due to the price of fossil fuels and widespread environmental concerns about the effects of CO_2 emissions. As is widely known, high-temperature materials are important in this respect, since fuel economy and rate of emissions scale directly with the temperature of the hot gas stream exiting the combustor regions; this effect explains the incentive to raise the operating temperatures experienced by the turbomachinery in these engines. In fact, a substantial enhancement of the engine is based on the increase of the maximum turbine entry temperature over the years as shown in Fig. 13.1.[1] It is not surprising therefore that all the major original equipment manufacturers (OEMs) possess research programmes which seek to develop new grades of alloy for use in their engines.

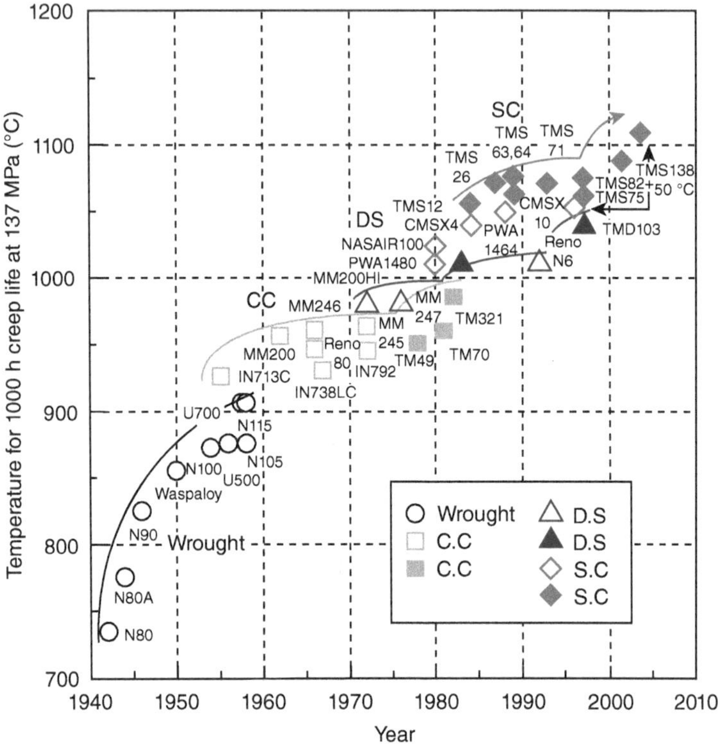

13.1 Evolution of the high-temperature capability of the superalloys over a period of 60 years since their emergence in the 1940s.[1]

Ni-base superalloys[1–4] play a vital role in this, since they have emerged as the materials of choice for the turbine blades, nozzle guide vanes and hot-section seals in the very hottest parts of these engines. These are amongst the most complex of the alloys produced by man, owing to the addition of many different alloying elements such as Co, Cr, Mo, W, Al, Ta, Re, Ru, which are added to provide a balance of properties, e.g. creep resistance, tensile strength, fatigue capability, oxidation resistance, corrosion resistance and castability. While the alloy designer must consider all of these properties in order to meet the turbine design specifications, advanced 4th and 5th generation Ni-base single crystal superalloys tend to exhibit improved creep resistance at a cost of sacrificing high-temperature oxidation resistance. Thus, advanced thermal barrier coatings (TBCs) – usually based on yttria stabilised zirconia (YSZ) – must now be applied to the surface of alloy components as a functional material to further enhance the temperature capability of the turbine blade aerofoils.

With the introduction of TBCs onto the surface of the alloy substrate being exposed to the high-temperature environment, the composite structure results in a complex interplay between the inter-diffusion and internal stresses due to mismatches in coefficients of thermal expansion (CTE). Thus, TBC-coated components are subjected to several potentially inter-related failure mechanisms due to their structural complexity.[5–10]

In the first experimental section of this chapter, TBC systems on different single-crystal superalloys will be presented. This section will focus on substrate-related failure of TBCs, aiming at demonstrating that the compatibility of modern Ni-base single-crystal superalloys with TBC systems is influenced strongly by the content of alloying element additions in the superalloy substrate. The results can be explained by postulating that the fracture toughness parameters controlling de-cohesion are influenced strongly by small changes in composition arising from inter-diffusion with the bond coat, which itself inherits elemental changes from the substrate.

The second experimental section of this chapter involves a comparative study of different bond coats (two β-structured Pt–Al types and a γ/γ' Pt diffusion type) in TBC systems based on an electron beam physical vapour deposition (EB–PVD) YSZ top coat and a substrate material of CMSX-4 superalloy. Generation of stress in the thermally grown oxide (TGO) on thermal cycling, and its relief by plastic deformation and fracture, have been the subject of detailed experimental investigation.

13.2 Substrate related failure of thermal barrier coatings (TBCs)

Single-crystal (SC) superalloys[1] have traditionally been designed with the priority of optimising the creep, fatigue, oxidation and corrosion-resistant properties. However, as the demand for more fuel-efficient turbine systems continues, the substrate materials must now be coated with TBCs[11,12] to enhance the high-temperature capability of critical turbine components such as turbine blades and guide vanes. Yet, it is known that TBC-coated Ni-base superalloys are prone to spallation;[13] for this reason, several bond coat technologies have been developed as a means to enhance the mutual compatibility of the two by improving the oxidation resistance of the coated specimens.

Unfortunately, little work[14,15] has been reported to elucidate the compatibility issues of superalloys with the TBCs. More importantly, the influence of the substrate composition on the TBC spallation life is not well understood. This is despite much progress which has been made to determine the micromechanics of the failure mode of TBCs.[16] In particular, the approach in modeling TBC failure[17–20] generally relies on the treatment of the oxidation-induced stresses that drive TBC spallation, despite the fact that it might reasonably be assumed that the TBC life and the modes of failure (i.e.

location of interfacial failure) are also influenced significantly by the inherited chemistry of the underlying superalloys.

In this section, the compatibility of a number of Ni-base SC superalloys with TBC systems will be discussed to demonstrate that the compositions of these alloys play an important role in determining TBC spallation resistance.

13.2.1 Experimental details and background information

Fully heat-treated (solution and primary-aged) SC alloys of SRR99, CMSX-4, TMS-82+, PWA-1484 and TMS-138A processed using conventional investment casting methods into cylindrical rods (10 mm in diameter), with the long axis of each aligned close (±5°) to the ⟨001⟩ axis, were used as substrate materials for this research. The chemical compositions of the alloys considered are given in Table 13.1 (major elemental additions) and Table 13.2 (trace elements). These chemical analyses were conducted by Cannon Muskegon Corporation using LECO™ Carbon & Sulfur Determinator, X-ray fluorescence (XRF) and inductively coupled plasma mass spectrometry (ICPMS) methods, to industry leading levels of precision (i.e. ppm levels). Three different bond coats, a high-temperature low-activity Pt–Al (HT Pt–Al), a low-temperature high-activity Pt–Al (LT Pt–Al) and

Table 13.1 Nominal chemical composition, wt%, of the Ni-based superalloys considered

Element	Co	Cr	Mo	W	Al	Ti	Ta	Hf	Re	Ru	C	Ni
SRR99	5.0	8.0	–	9.5	5.5	2.2	2.8	–	–	–	0.011	Bal.
TMS-82+	7.8	4.9	1.9	8.7	5.3	0.5	6.0	0.1	2.4	–	–	Bal.
PWA1484	10.0	5.0	2.0	6.0	5.6	–	8.7	0.1	3.0	–	–	Bal.
CMSX-4	9.6	6.5	0.6	6.4	5.6	1.0	6.5	0.1	3.0	–	–	Bal.
TMS-138A	5.8	3.2	2.8	5.6	5.7	–	5.6	0.1	5.8	3.6	–	Bal.

Note: – means that the element is not included in the chemical analysis.

Table 13.2 Levels of trace elements in the superalloy substrates examined, as determined by XRF, LECO and ICPMS methods (in ppm by weight)

Element	S	B	O	N	P	Zr
SRR99	2	2	2	4	2	4
TMS-82+	2	1	2	2	11	13
PWA1484	3	2	2	1	6	19
CMSX-4	2	8	2	2	4	7
TMS-138A	2	3	2	1	18	3

a Pt-diffusion bond coat, were subsequently deposited onto the flat surfaces of all substrate materials.

The HT Pt–Al and the LT Pt–Al bond coat specimens were prepared by first electrodepositing a thin layer of platinum of 5 and 7 μm, respectively, followed by a vacuum heat treatment applied at 1100 °C for 1 h. Next, 5 h vapour phase and 20 h pack-aluminisation processes were applied at 1080 °C and 870 °C, respectively. Both the vapour phase and pack processes involve the generation of vapours containing Al, which are usually halides due to their high vapour pressures. The vapours flow and react with the substrate surface upon contact forming the aluminide coating. In the aluminisation process, specimens are either embedded in a pack mix (i.e. LT Pt–Al) or suspended in a vapour-filled environment (i.e. HT Pt–Al) in a heated chamber.

The Pt-diffusion bond coat specimen was first electroplated with a 10 μm layer of platinum and then a vacuum heat treatment at 1150 °C for 1 h was subsequently carried out. No aluminisation process was applied in this case. All specimens were further heat-treated for 1 h at 1100 °C in an argon atmosphere and a ZrO_2/7 wt% Y_2O_3 (YSZ) top coat 175 μm in thickness was then deposited by EB–PVD. A further vacuum heat treatment (1100 °C, 1 h) and ageing (870 °C, 16 h) were applied to all specimens before testing.

The TBC coated disc-shape specimens were then subjected to cyclic oxidation testing in a purpose-built rig (see Fig. 13.2). Each thermal cycle involved 1 h at a furnace temperature of 1135 °C (approximately 10 min to reach the peak temperature); after which, the specimens were removed automatically from the rig and fan-cooled by laboratory air for 1 h. Five individual specimens of each TBC system were cycled to failure to determine the average TBC spallation lifetime, in order to evaluate the scatter in the experimental results. In addition to the cycled to failure test, additional specimens were removed from the rig at specific intervals for

(a) (b) (c)

13.2 Pictures showing the use of the thermal cycling furnace (a), the placement of specimens on the sample stage (b) and the heating/cooling conditions (c).

luminescence measurements (see below) and then returned for further thermal cycling. Some specimens were withdrawn from cycling, mounted in epoxy, cut through by an abrasive alumina blade to reveal the cross-section and polished to mirror-finish for microstructural characterisation. An additional three coated disc-shape specimens were isothermally heated at 1135 °C in the same rig for 150 h.

Microstructure characterisation was carried out in a field emission gun scanning electron microscope (FEG–SEM) equipped with energy dispersive X-ray (EDX) using the secondary electron detection mode at a working distance of 15 mm and an accelerating voltage of 20 kV. For high-resolution quantitative elemental analysis, a prototype field emission electron probe microprobe analyser (FE–EPMA) was used for analysis by wavelength dispersive X-ray spectrometry (WDS). A probe current of 5.0×10^{-8} A was applied within a probe diameter of about 0.25 μm, at an accelerating voltage of 20 kV.

13.2.2 Characterisation of coating cross-sections – as received condition

Scanning electron micrographs illustrating the microstructures of the three coatings in the as-received condition are given in Fig. 13.3. In all cases, the YSZ top coat layer had almost the same thickness. Discernible differences can be seen between the microstructures and the thicknesses of the bond coats. Pt–Al bond coat systems produce a single-phase β-(Pt,Ni)Al microstructure (i.e. NiAl with Pt in solid solution) next to the YSZ. Below the grit-line (residual Al_2O_3 particles remaining from the grit-blasting process prior to Pt plating) is the inner multi-phase layer of the coating consisting of precipitates rich in refractory metals (Ta, W, Mo) in the β phase matrix. The Pt-diffusion bond coat consists of a two-phase γ and γ′ microstructure both above and below the grit-line; the β-NiAl phase is completely absent since the aluminisation step was absent. Al and Pt EDX concentration line-

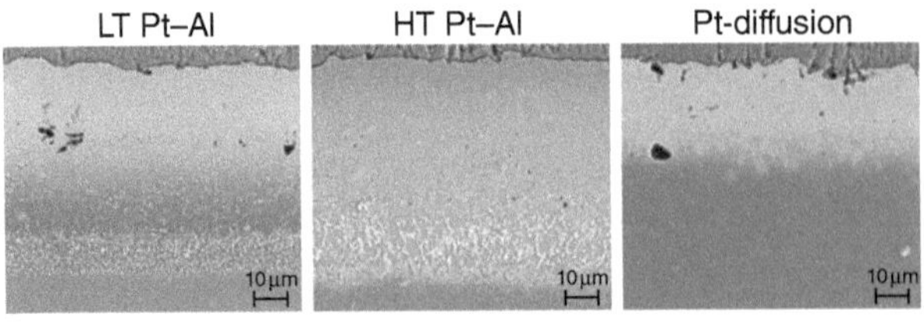

13.3 The cross-section microstructures of the bond coats (LT Pt–Al, HT Pt–Al and Pt-diffusion) in the as-received coatings on CMSX-4.

profiles of these coatings are shown in Fig. 13.4. As can be seen, the aluminisation process significantly increases both the Al concentration and the bond coat thickness in the Pt–Al systems. The main difference between the LT Pt–Al and HT Pt–Al bond coats is in the Pt profile. The LT Pt–Al has a higher Pt concentration within the first 20 μm, whereas the HT Pt–Al bond coat has a higher Pt concentration at 30–60 μm. The Pt-diffusion bond coat, on the other hand, inherits the original Al content from the superalloy substrate and has higher Pt concentration than the Pt–Al systems due to the thicker electroplated Pt prior to the diffusion-heat treatment process. In all cases, the post-processing heat treatment resulted in the formation of a thin thermally grown oxide (TGO) layer, which shows as a thin dark line between the bond coat and YSZ, even before any thermal cycling had taken place. The microstructures of Pt-diffusion coatings on superalloy substrates are also shown in Fig. 13.5.[21] Again, the TBC top coats were prepared to similar thickness.

Based on the results of the thermal cycling experiment, it was found that coating life varied significantly with the composition of the superalloy substrate as illustrated in Fig. 13.6[21] with the error bars indicating ± one

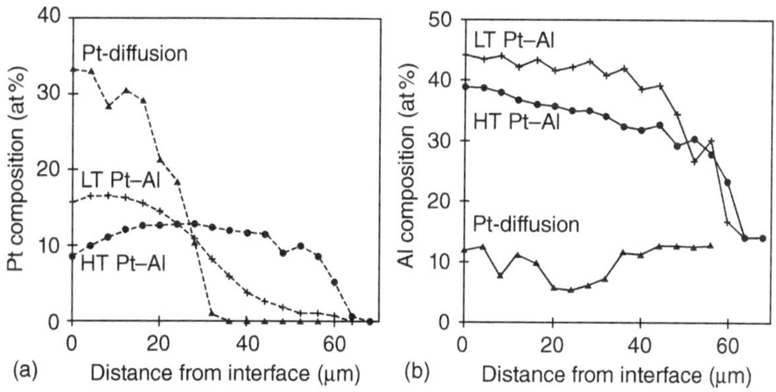

13.4 EDX concentration line-profile of Pt (a) and Al (b) in the as-received bond coats.

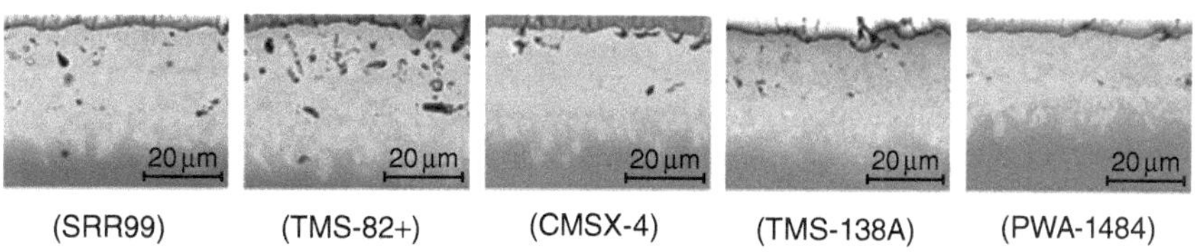

13.5 The microstructures of Pt-diffusion coatings on the superalloy substrates in the as-received condition.

standard deviation in the lifetimes of each set of five specimens. For instance, TBC-coated PWA1484 exhibited spallation lifetimes at least three times superior to those of first-generation superalloy SRR99 regardless of the type of bond coat being applied, suggesting that strong chemical effects were inherited from the substrate. In addition, the spallation resistance of TBC-coated fourth generation TMS-138A behaved significantly better than that of SRR99 and comparable to that the second-generation superalloy CMSX-4, suggesting that no detrimental effect is inherited from ruthenium addition to the substrate, at least for the Pt-diffusion coating. It should be noted here that the thermal exposure of PWA-1484 specimens coated with Pt-diffusion bond coat was interrupted at the 1000th cycle as the purpose of demonstrating TBC spallation resistance had been fulfilled. These findings confirm a dependence of the TBC spallation life on the superalloy composition, and can be explained only by the different degradation mechanisms taking place in the bond coat or near the TGO interfaces (since the bond coat and the ceramic top coat were prepared in the same way on all different substrates).

In terms of the bond coat technology, it was demonstrated in general that the type of bond coat being applied had less influence on the TBC spallation life than did the substrate. It was shown as well that Pt-diffusion bond coats outperformed both Pt–Al systems.

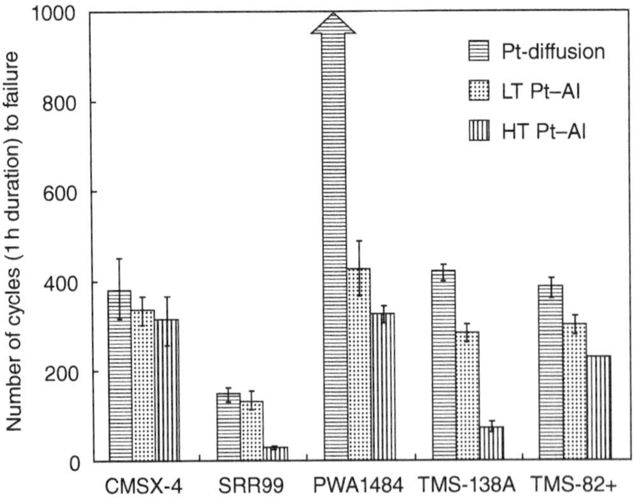

13.6 TBC spallation life for cyclic oxidation testing (1 h thermal cycling to peak temperature) of TBC coated superalloys. The error bars indicate ± one standard deviation in the lifetimes of each set of five specimens.

13.2.3 Mechanism of TBC failure

As previously demonstrated, the composition of the superalloy substrates has a larger influence on the TBC spallation life than that of the bond coat materials. For this reason, attention is focused in this section on how the substrate affects the mechanism of TBC failure, at a fixed TBC preparation condition. Pt-diffusion bond coat was chosen for this investigation as it showed the largest substrate effect on the coating spallation life.

SEM images of the fracture interface for each of the five substrates are shown in Fig. 13.7.[21] It was found that de-cohesion occurred at a position which depended on the substrate composition. In particular, it was observed that specimens with shorter TBC spallation life tended to fail at the TGO/bond coat interface; while more spallation-resistant coatings failed within the TGO. In addition, by importing higher magnification files of these SEM images into imaging software, the thickness of the TGO at failure for each of the specimens could be quantised by making boundary selections (i.e. defining the top and bottom TGO interfaces) and calculated with reference to the scale bar of these images. These TGO thicknesses were plotted along with the TBC life for comparison as illustrated in Fig. 13.8.[21]

Based on these results, it can be seen that the TGO thickness at failure in TMS-138A is considerably thicker than that of the TMS-82+, even though TBC-coated TMS-138A offered slightly longer TBC spallation life than TMS-82+. Moreover, despite the fact that the TBC lifetimes of CMSX-4, TMS-138A and TMS-82+ were within a few cycles of each other, their TGO growth kinetics are very different. Thus, it seems that there is little correlation between the oxide growth kinetics and the TBC spallation life.

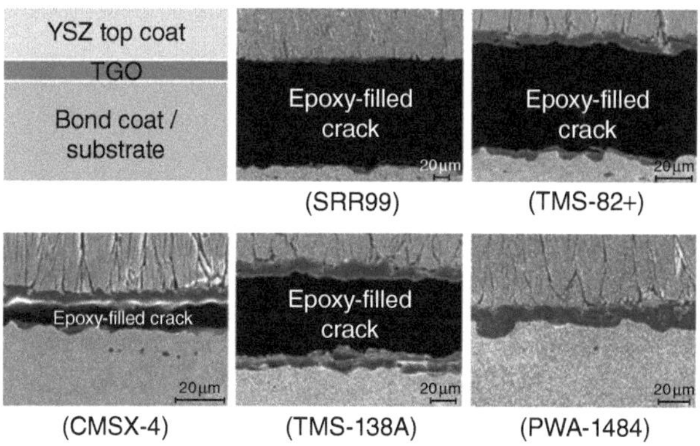

13.7 SEM micrographs illustrating the location of interfacial spallation failure of TBC (Pt-diffusion bond coat) coated superalloys.

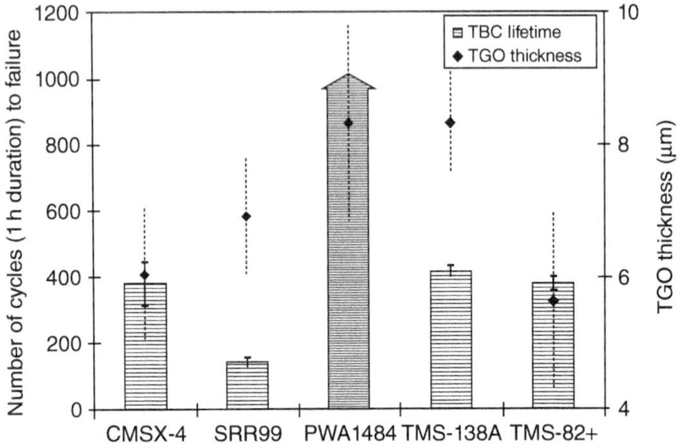

13.8 TBC spallation life vs the thickness of TGO upon failure; illustrating no clear correlation between the two.

13.2.4 High-resolution analysis and mapping by electron probe microprobe analyser/wavelength dispersive spectroscopy (EPMA/WDX)

Elemental compositions of the TGO layers on the three bond coats were characterized using WDS analysis (see Table 13.3). The results are quoted only for specimens with TGO thickness greater than the spatial resolution of the FE–EPMA/WDX technique (~1 μm). Chemical mapping of the bond coat near the TGO interfaces was also carried out to determine the elemental chemical distributions on specimens after 100 cycles thermal exposure as shown in Figs 13.9–13.13. The results suggest that for all specimens, the oxide layers are primarily of Al_2O_3. However, it has also been observed for some specimens that either islands or continuous layers of Ni, Cr and or Co-rich oxides are present above the primarily Al_2O_3 layer. For example, CMSX-4, TMS82+, TMS-138A and PWA1484 all contain islands which are rich in Cr; while TMS82+, TMS138A and PWA1484 further contain islands rich in Ni. In addition, the presence of Co-rich regions has also been noticed on TMS-82+ and TMS-138A. For SRR99 specimens, fairly continuous Ni, Cr and Co-rich oxides are present. However, it should be recognised that the number of spinel phases in the TGO is very small in any of the TBC-coated systems; the TGO thickness upon failure varies between different alloys and is essentially a measure of the amount of alumina.

Table 13.3 Elemental compositions (wt%) of the TGO layers measured by WDS FE–EPMA analysis

Element	Pt	Zr	Y	Co	Re	Ta	Mo	Al	Ni	Ru	W	Ti	Cr	O
SRR99	0.3	0.8	0.1	0.1	–	0.0	–	51.6	0.5	–	0.1	0.0	0.2	46.2
TMS-82+	0.5	0.7	0.1	0.1	0.1	0.1	0.0	51.6	0.6	–	0.1	0.0	0.1	45.9
PWA1484	0.5	0.9	0.1	0.1	0.1	0.1	0.0	51.6	0.7	–	0.1	–	0.1	45.6
CMSX-4	0.3	0.9	0.1	0.1	0.1	0.1	0.0	48.8	0.7	–	0.0	0.0	0.1	48.7
TMS-138A	0.4	0.8	0.1	0.1	0.1	0.0	0.0	51.7	0.6	0.0	0.1	–	0.0	46.0

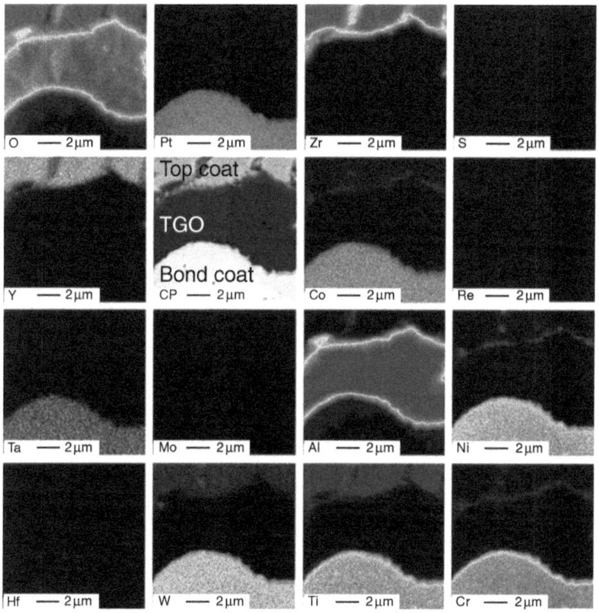

13.9 Quantitative WDX maps of the TBC coated SRR99–Pt-diffusion system following 100 cycle exposure at 1135 °C.

At the TGO/bond coat interface, an enriched layer (i.e. a bright line) of Ni, Cr, Al and Ti was observed in the SRR99 specimen. TMS-82+ and CMSX-4 also have enriched layers of Ni and Cr near the interface. Both TMS-138A and PWA1484, in contrast, had no noticeable enrichment of alloying constituents.

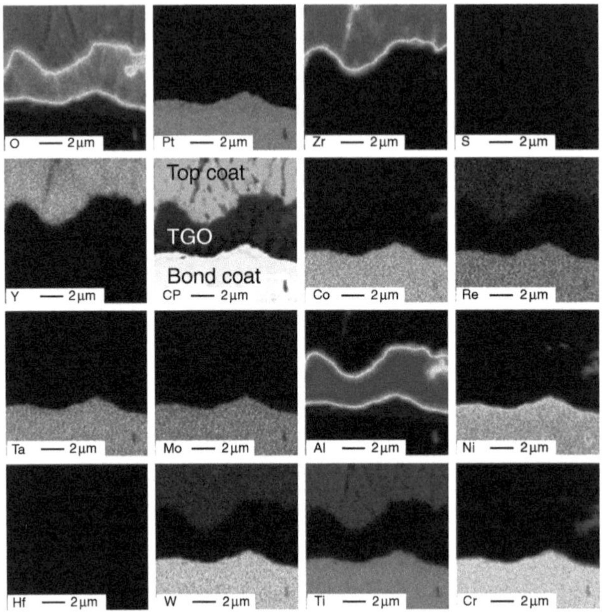

13.10 Quantitative WDX maps of the TBC coated TMS-82+–Pt-diffusion system following 100 cycle exposure at 1135 °C.

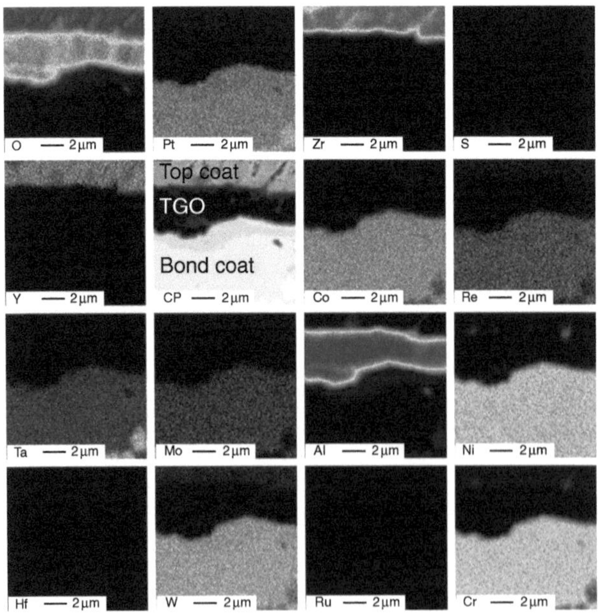

13.11 Quantitative WDX maps of the TBC coated PWA1484–Pt-diffusion system following 100 cycle exposure at 1135 °C.

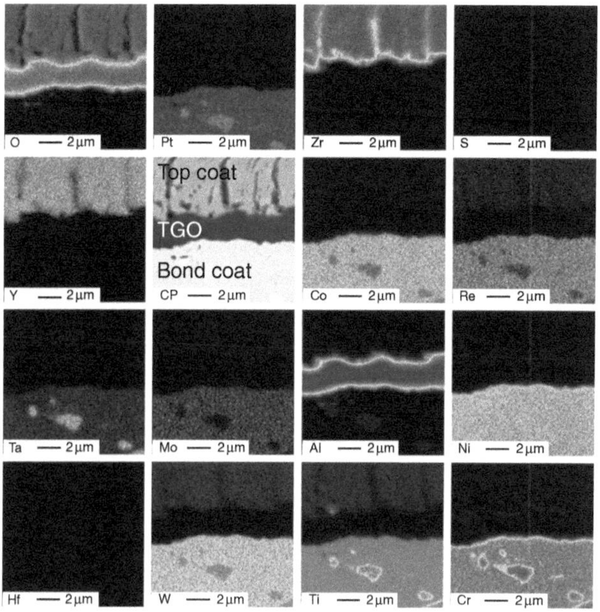

13.12 Quantitative WDX maps of the TBC coated CMSX-4–Pt-diffusion system following 100 cycle exposure at 1135 °C.

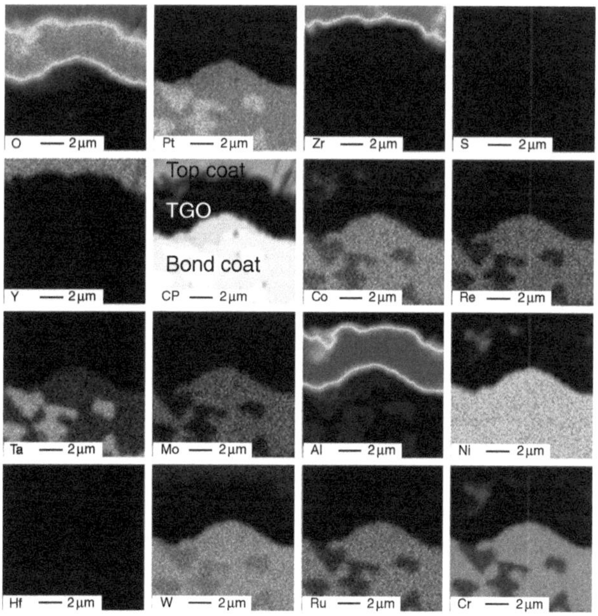

13.13 Quantitative WDX maps of the TBC coated TMS138A–Pt-diffusion system following 100 cycle exposure at 1135 °C.

13.3 Compatibility issues of nickel-based single-crystal superalloys with thermal barrier coating systems

The studies presented here provide new insights into the current understanding of TBC systems. Traditionally, it is considered that TBC spallation occurs by the accumulation of local damage to form a weakly bonded area that can be susceptible to buckling.[22–24] Choi *et al.*[22] have studied the competing failure processes of TBCs, such as edge delamination and small/large-scale buckling. Their results show that the failure mechanism depends strongly on both the thickness and in-plane modulus of the YSZ top coat. Assuming the in-plane elastic modulus for the EB–PVD top coat is in the range 0.05–0.25 times the bulk modulus of dense YSZ[24] and taking the top coat to be approximately 175 μm in thickness as is the case in the current study, substantial suppression of TGO buckling is believed to be present. Therefore, the final failure mode is more likely to be associated with edge delamination or wedging[25] irrespective of the initiation process.

Consider that TBC spallation occurs as the stored elastic energy of the TGO and top coat, which stiffens with time due to sintering, exceeds the interfacial adhesion between the TGO and bond coat. In the case of the steady-state energy release rate, a driving force G exists for the delamination of the multilayer TBC as:[24]

$$G = \frac{\left(\sum_i \sigma_0^i h_i\right)^2}{2\sum_i \frac{E_i}{(1-\nu_i^2)} h_i}$$

where E_i, h_i and ν_i are the elastic modulus, thickness and Poisson's ratio of each layer i, respectively and σ_0 is the equi-biaxial residual plane stress. The general argument is that the driving force for spallation G increases as the TGO thickens; delamination occurs when G reaches the critical interfacial fracture toughness G_c. In fact, as the EPMA mapping has revealed, the segregation of Ni, Cr and Co-rich oxides above the Al_2O_3 increased the overall TGO thickness and, thus, further increased this misfit strain energy.

However, the findings have conclusively confirmed that the resistance of TBC to spallation during thermal cycling is strongly dependent upon the composition of the superalloy substrate upon which the TBC system is placed. Since there is little correlation between the oxide growth kinetics and the TBC spallation life and the levels of trace elements present (e.g. sulphur) do not vary substantially from alloy to alloy, any differences in TBC lives will be attributable to differences in major elemental additions rather than trace elements.

The presence of Ti as an alloying element in SRR99, TMS-82+ and CMSX-4 is suggested to have shortened the TBC life. EPMA mapping of

Ti shows a concentration-enriched layer (i.e. a bright layer) near the TGO/bond coat interface in the coated SRR99 system. In fact, previous studies[26,27] have suggested that fast diffusion of Ti to the surface leads to the formation of Ti-rich oxides, potentially weakening the adhesive strength at the TGO/bond coat interface. In addition, Ti^{4+} ions, reported[28] to substitute for Al^{3+} ones to create Al vacancies within the alumina scale, may also degrade the properties of the TGO/bond coat interface. Thus, it is not difficult to rationalise the reason why SRR99, among the alloys considered, exhibits the fastest oxidation rate and the shortest TBC life. TMS-138A and PWA1484, in contrast, have no detected enrichment of alloy constituents at the TGO/bond coat interface and represent the two most spallation-resistant systems.

Regarding the spallation mechanisms as mentioned previously, it appears that the more spallation-resistant coatings failed at the top coat/TGO interface or tended to fail within the TGO; while specimens with shorter TBC spallation life preferentially failed at the TGO/bond coat interface, which again is the interface most likely to be affected by diffusion of harmful elements from the superalloys. This leads to a proposed argument that the TGO/bond coat interfacial adhesive strength of TBCs, which controls the TBC lifetime, is a dynamic materials property dependent on and influenced considerably by the composition of the substrate materials (i.e. superalloys). This situation can be qualitatively represented by a schematic drawing (Fig. 13.14) which illustrates the possible variation of the driving force G and the interfacial fracture toughness G_c with the time of thermal exposure.

Based upon the steady-state energy release rate equation for the delamination of the multi-layer TBC, the driving force G increases with TGO thickness, which is assumed to grow parabolically with time t, then thermal cycling at higher temperatures will accelerate the kinetics. The possible variation of G_c is also shown. At the beginning, for any choice of bond coat, G_c is assumed to be roughly the same for all alloy systems consistent with the identical processing conditions employed. However, upon thermal exposure, as the harmful elements such as Ti diffuse to the TGO interface, the value of G_c is believed to decrease – rapidly for a substrate system such as the SRR99 and modestly for PWA1484 with other alloys under consideration falling somewhere between the two.

It should be recognised that studies published so far have tended to consider just a single substrate composition with a TBC placed upon it; thus, any influence of the substrate chemistry is then factored out from the experiment. It would appear that the superalloy composition has a major influence on the lifetime of the systems, potentially due to the sensitivity of the interfacial fracture toughness to substitutional elements diffusing through the bond coat system from the superalloy substrate beneath it. These findings have implications for the design of TBC systems for the protection of the turbine blade aerofoils used for high-temperature applica-

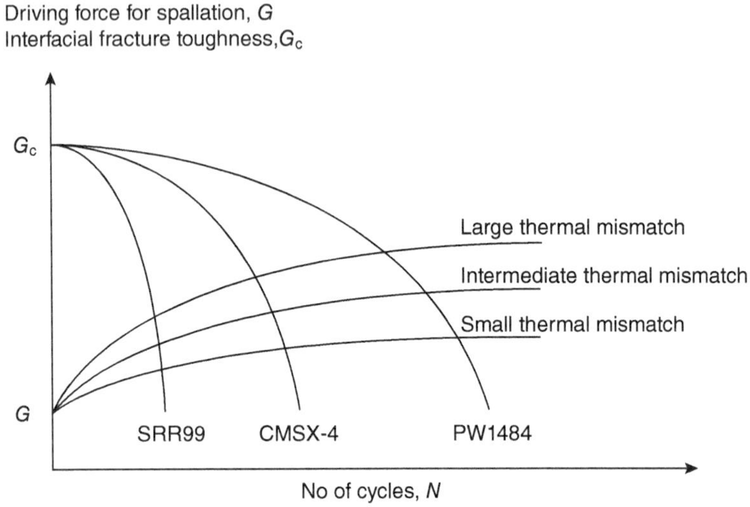

13.14 Schematic illustration of the proposed variation of driving force G and interfacial fracture toughness G_c during thermal cycling.

tions. Traditionally, Ni-base SC superalloys have been designed with their mechanical properties – particularly in creep and fatigue – in mind. However, as the operating conditions of modern gas turbines continue to become more aggressive and use of TBCs for the provision of thermal insulation is increasing, a further property of the superalloy substrate is then required: that of compatibility with the TBC system which it is required to support. In particular, the influence of different alloying elements present in the superalloy on the TBC's interfacial fracture toughness needs to be better understood.

13.4 Bond coat related failure of TBCs

Traditionally, the study of TBCs has been focused on the investigation of individual bond coat systems.[29–32] However, as the operating temperature of advanced gas turbines increases, new bond coats based on existing ones are required to meet this demand and suit different applications.[33,34] Thus, it becomes necessary not only to understand how a coating system performs but also to know how and why existing coating systems behave differently.

Little work[35] has been reported in which the modes of degradation responsible for TBC spallation are compared for different bond coats. This is despite much progress that has been made on the micromechanics of the failure of aluminide bond coat systems.[30,31] In particular, the usual approach to modelling TBC failure[36–41] concentrates on the treatment of the oxidation-induced stresses that drive TBC spallation-despite the fact that it is likely that the modes of failure[33,42,43] and evolution of the residual stress are influenced considerably by the bond coat systems.

The work reported here was carried out with these factors in mind. Three different representative bond coat systems were deposited on SC CMSX-4 substrates and then subjected to identical (industry standard) EB–PVD YSZ coating processes. It has been shown that both the evolution of the residual stress in the TGO and the mode of TBC failure depend on the bond coat composition. FEG-SEM and EDX analysis were used to study the composition and microstructure changes occurring near the TGO/bond coat interface.

13.4.1 Experimental details and background information

The TBC spallation life (defined as detachment of top coat by approximately 20% in area) in thermal cycling conditions was found to be slightly dependent on the type of bond coat system used (see Fig. 13.15). Pt-diffusion bond coat specimens achieved a mean TBC life of 384 cycles; 15% and 23% longer than for LT and HT Pt–Al bond coats. It should be pointed out that, although the three bond coats differed significantly in composition and were prepared by different coating manufacturing processes, all three systems showed remarkably similar spallation life. However, it was found that depending on the type of bond coat, de-cohesion of the TBC occurred at different interfaces. Since the substrate and the ceramic top coat were prepared to be identical, this can only be explained by the different behaviours of the bond coats, as will be discussed later.

13.4.2 Residual stress measurement by luminescence spectroscopy

Residual stress in the TGO can be measured by piezospectroscopy which acquires Cr^{3+} luminescence spectra[30–33] emitted from α-Al_2O_3 as a result of excitation by a laser beam. The details for luminescence experiments are given in Levi.[33] Luminescence spectra were acquired using a Renishaw Raman optical microprobe (model 2000) fitted with a motorised mapping stage. The luminescence from the TGO is due to the photo-stimulation of trace Cr^{3+} impurities present within the alumina. The spectrum has a featured intense doublet, the R1 and R2 lines, where R2 represents the peak with the higher wave number and lower integral peak area.[34] The characteristic R-line luminescence is a phenomenon resulting from the non-phonon radiative decay from the Cr^{3+} first excited state to the ground state within the *d* orbital.[35] The laser source was a green Ar+ laser with a

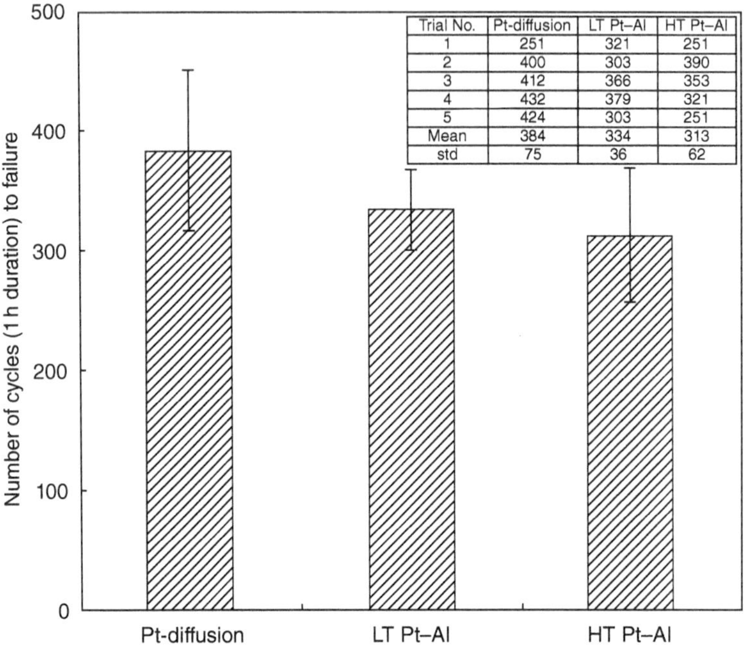

13.15 TBC spallation life for cyclic oxidation testing (1 h at 1135 °C) of TBC systems. The error bars indicate ± one standard deviation in the lifetimes of each set of five specimens.

wavelength of 514.5 nm, which is focused on the specimen surface through an objective lens. The focused spot size can be adjusted by using different objective lens, with a minimum spot size of approximately 2 μm. However, upon incident on the specimen surface, the laser beam scatters within the YSZ top coat and consequently, results in an effective spot size at the TGO of approximately 20 μm.[35,36] The light scattered back from the specimen was then collected by the objective lens, dispersed by a 1800 lines/mm grating and then received by a charge coupled device (CCD) detector.

An air conditioning unit was used to maintain a stable room temperature at 22 ± 0.3 °C. An acquisition time of 1 s was used for each spectrum. Measurements (121 per specimen) were taken on a square grid of 200 × 200 μm with a pitch of 20 μm located in the centre of the flat face of the disk specimen. Before and after each measurement, the spectrometer was calibrated by taking a spectrum from a strain-free SC sapphire sample. The residual stress in the TGO was estimated from the R2 peak shift assuming a planar equi-biaxial stress: $\sigma(\text{GPa}) = 5.07\Delta\nu\ (\text{cm}^{-1})$. Transition phases of alumina (e.g., θ-Al_2O_3) produced luminescence lines at 14260 cm^{-1}, 14330 cm^{-1},

14546 cm^{-1} and 14626 cm^{-1}, which are distinguishable from the R1 and R2 lines of α-Al_2O_3.[44] However, no significant transition phases were seen in the TGO for all the cases in this study. Five individual specimens of each TBC system were used for the above analyses to ensure consistency of the experimental results.

Luminescence measurements illustrating the evolution of residual stress in the TGO layer are given in Fig. 13.16. It was found that that the evolution of this stress was substantially different between the Pt–diffusion and the Pt–Al bond coat systems, despite the fact that they all had similar initial stress level in the as-coated condition. The Pt-diffusion bond coat systems began with a compressive stress of 2.8 GPa that gradually increased to 3.5 GPa over the coating lifetime, while those of the Pt–Al coatings decreased from an initial compressive stress of about 2.3 GPa to between 1 and 1.5 GPa during the first 75–100 thermal cycles, before maintaining a steady level to near the end of life. The error bars in Fig. 13.16 represent ± one standard deviation of the 121 measurements made at each timestep. It is therefore notable that the standard deviation increases significantly for the Pt-diffusion bond coats just before failure.

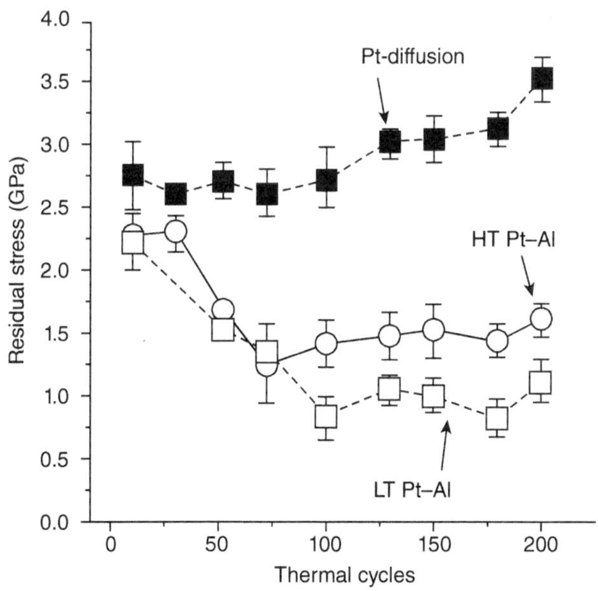

13.16 Compressive residual stress in the TGO shift vs number of cyclic oxidation tests of the bond coat systems.

13.4.3 Microstructure imaging of thermal cycled specimens using a field emission gun scanning electron microscope/energy dispersive X-ray (FE-SEM/EDX)

Cross-section scanning electron micrographs illustrating the microstructures of the three coatings after 10, 30 and 200 thermal cycles are shown in Fig. 13.17. Based on these SEM images, the oxidation kinetic (TGO thickness vs time) curve is plotted in Fig. 13.18. A few observations can be made with regard to the thermal-cycling-induced degradation. First, it can be seen that the TGO morphology evolved quite differently for the three bond coats. For example, the TGO interfaces in the Pt–diffusion system remained intact on thermal cycling for most of its spallation life. In contrast, local interfacial separation of the TGO in the Pt–Al systems occurred at less than 30% of their lifetimes. Furthermore, final spallation occurred at the TGO/YSZ interface for the LT Pt–Al bond coat but at the TGO/bond coat interface for the Pt-diffusion system. The HT Pt–Al bond coat system showed a mixture of the two failure modes. Finally, the waviness of the TGO layer in Pt–Al systems appears to amplify significantly upon thermal cycling.

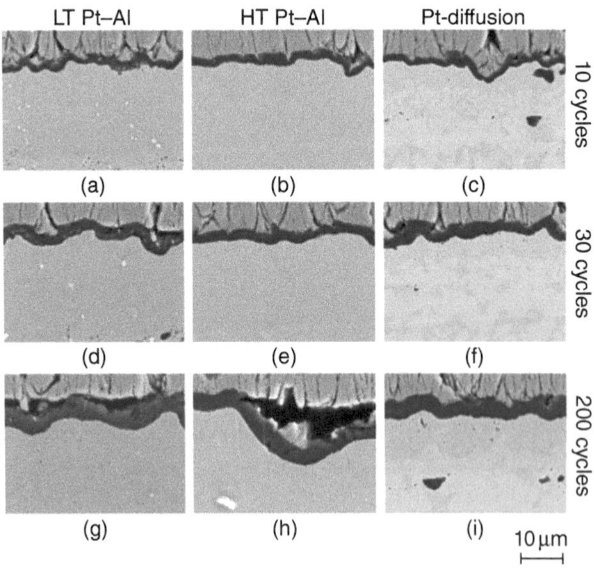

13.17 A series of SEM micrographs illustrating the near-TGO microstructure of the coatings after 10, 30 and 200 thermal cycles.

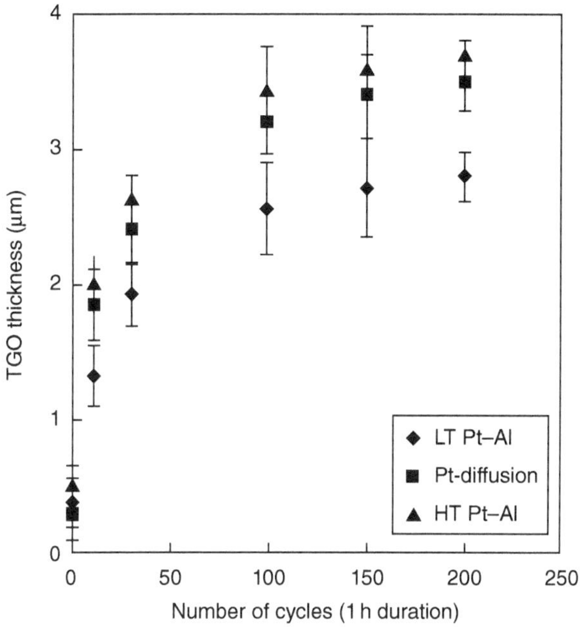

13.18 TGO thickness vs number of thermal cycles (1 h at 1135 °C) for the three bond coat systems.

13.4.4 Quantification of rumpling at the thermally grown oxide (TGO)/bond coat interface

Measurement of the interfacial waviness was carried out by tracking the normal displacement along the TGO/bond coat interface in a cross-section SEM image over a distance of approximately 640 μm with a step size of 1 μm. Each position on the interface was represented by a pair of X (distance parallel to the specimen surface) and Y (distance normal to the specimen surface) co-ordinates. Figure 13.19 shows the evolution of the TGO/bond coat interface morphology for each of the three bond coat systems. The amplitude of waviness was quantified by determining the standard deviation (STD) of each profile. Figure 13.20 shows the standard deviation from the mean for each profile as a function of the number of thermal cycles. Results of isothermal tests were also plotted for reference. As can be seen in both Figs 13.19 and 13.20, the LT Pt–Al system exhibited a marked increase in TGO/bond coat interfacial waviness with thermal cycling. The HT Pt–Al bond coat also demonstrated a tendency to roughen with thermal cycling, but at a considerably slower rate. In contrast to the

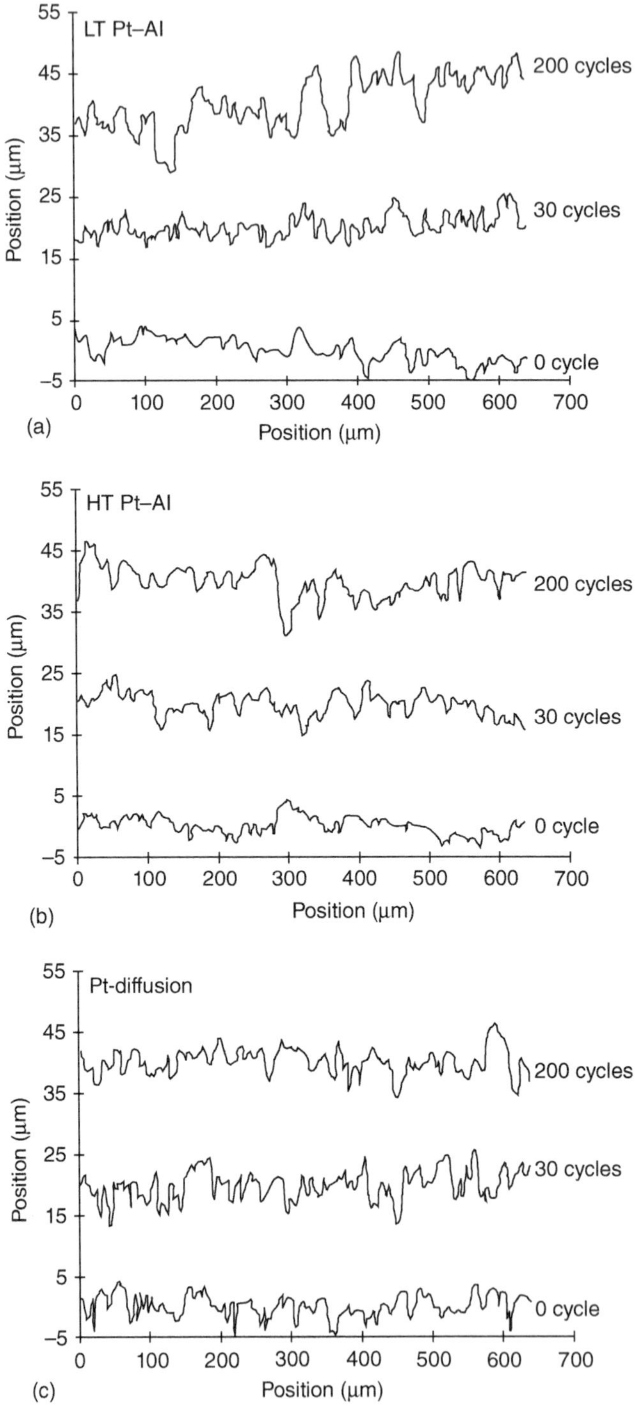

13.19 TGO/bond coat interface profiles of: (a) LT Pt–Al; (b) HT Pt–Al; (c) Pt-diffusion bond coat at stages of the thermal cycling history.

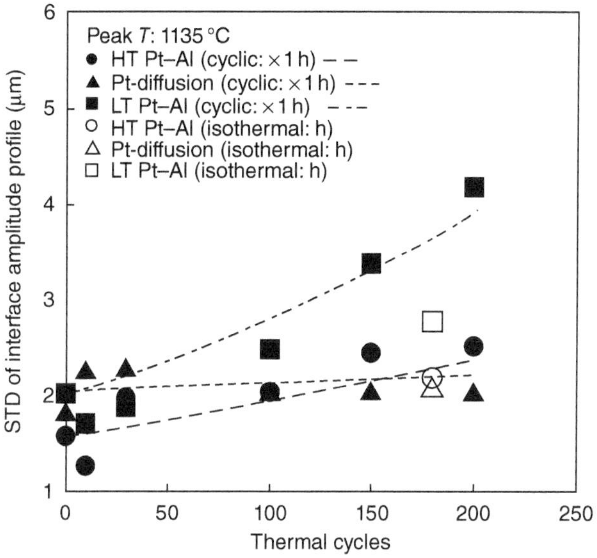

13.20 Standard deviation of interface amplitude profile (i.e. magnitude of rumpling) as a function of the number of thermal cycles for the three different bond coat systems.

Table 13.4 Elemental compositions (at. %) taken using WSD FE–EPMA analysis of the TGO layer from specimens at 200 thermal cycles

Bond coat	Pt	Zr	Y	Co	Re	Ta	Al	Ni	Ti	Cr	O
LT Pt–Al	0.04	0.21	0.01	0.04	0.01	0.01	37.05	0.23	0.01	0.04	62.35
HT Pt–Al	0.02	0.20	0.02	0.04	0.00	0.02	38.60	0.29	0.02	0.04	60.75
Pt-diffusion	0.04	0.23	0.02	0.02	0.01	0.01	37.75	0.21	0.01	0.02	61.68

two Pt–Al systems, the Pt-diffusion bond coat showed no detectable roughening.

13.4.5 Chemical analyses of TGO layers and interfaces

Elemental compositions of the TGO layers on the three bond coats were characterised using WDS analysis after 200 thermal cycles (see Table 13.4). The results are quoted only for locations having TGO thickness greater than the spatial resolution of the FE–EPMA/WDS technique. The results show that for all specimens, the oxide layers consisted of more than 99 at% Al_2O_3. Less than a total of 1 at% of other elements originating from the top coat, bond coat and substrate was detected in the oxide layer, possibly due to solid solution and particle segregation within the alumina scale. In

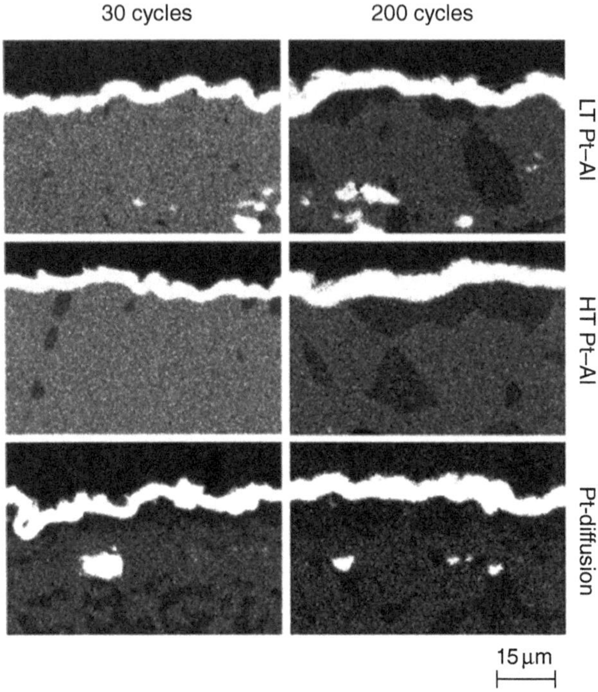

13.21 A series of Al EDX maps illustrating the microstructural evolution of the bond coat systems after 30 and 200 thermal cycles.

addition, high-resolution Al-mapping (Fig. 13.21) of the bond coat near the TGO interfaces using FE–SEM/EDX revealed how the different bond coats respond to the Al loss resulting from TGO growth and inter-diffusion with the substrate. Both Pt–Al bond coats showed a progressive phase transformation from the as-deposited single-phase β-(Pt,Ni)Al to a two-phase β-(Pt,Ni)Al and γ′-$(Pt,Ni)_3Al$ microstructure. These images reveal that the phase transformation in the Pt–Al systems was detectable within the first 30 thermal cycles and the rate of transformation in the HT Pt–Al bond coat appeared to be faster. The Pt-diffusion bond coat, on the other hand, showed a distinct γ′ to γ phase transformation near the TGO/bond coat interface. Upon thermal cycling, a continuous γ layer was eventually formed directly below the Al_2O_3 layer and thickened as the Al depletion continued.

13.5 Effect of bond coat on the TBC degradation mechanisms

The results presented here provide a detailed comparison of the progressive degradation of TBCs with three different bond coat systems upon thermal exposure. The results show that the three TBC-coated bond coat systems failed with quite different characteristics, despite having similar lifetimes. In particular, the location of the TBC spallation interface varied strongly with type of bond coat employed. In addition, both the bond coat rumpling and the TGO residual stress measurements showed significantly different behaviour for the different bond coats. Since the substrate and the ceramic top coat were identical for all three systems, these differences can only be explained by the different processes taking place in the bond coats.

Quantification of the bond coat/TGO interfacial roughness indicated that rumpling was detectable within the first 30 cycles of thermal exposure in the Pt–Al systems. Upon further thermal cycling, the undulation in the TGO amplified in the LT Pt–Al system and resulted in localised vertical separation between the YSZ (which remained at its initial roughness) and the TGO, as can be clearly seen in Figs. 13.17g and k. TGO rumpling is capable of relieving the residual stress in the TGO by increasing its waviness,[45] which explains the reduction of the TGO residual stress in the Pt–Al coatings. In contrast, the absence of rumpling in the Pt-diffusion bond coat system helped in ensuring that the YSZ/TGO and the TGO/bond coat interfaces remained intact over the majority of the coating lifetime. Consequently, when there was significant rumpling (i.e. the LT Pt–Al system), coating de-cohesion took place primarily in the YSZ/TGO interface. In the absence of rumpling as in the case of the Pt-diffusion system, spallation eventually occurred at the bond coat/TGO interface. The HT Pt–Al bond coat system showed a mixture of these two failure modes, presumably due to the fact that its rate of rumpling was considerably less than that of the LT Pt–Al system.

In addition to the observed consequences of rumpling, the mechanism by which rumpling occurs is also of importance. In particular, the dependence of rumpling on the properties of the bond coat materials must be considered. Upon depletion of Al, a phase transformation from the as-deposited single-phase β-(Pt,Ni)Al to a two-phase β-(Pt,Ni)Al and γ'-$(Pt,Ni)_3Al$ microstructure takes place in Pt–Al bond coats. It has been proposed[46] that this phase transformation induces non-uniform volume changes of the bond coat and thus leads to rumpling.[30,32,44,47–49] However, this mechanism does not appear to play a major role in inducing bond coat rumpling, at least in the current study, for two reasons. First, according to the proposed mechanism, the LT Pt–Al bond coat, which exhibited strongest rumpling behaviour in the current study, should correspondingly have the fastest oxidation kinetics (i.e. fastest Al-depletion rate). However, the present results showed that it had the slowest oxidation rate among the three systems studied. Second, the lower-Al containing Pt–Al bond coat (HT Pt–Al), according to the proposed mechanism, should be more sensitive to Al depletion and, thus, prone to the phase transformation-induced rumpling than its high-Al Pt–Al counterpart (LT Pt–Al). However, results showed the opposite, in that the rate of rumpling in the HT Pt–Al system was considerably lower than that of the LT Pt–Al system.

Based on these considerations, it is suggested that rumpling is not primarily induced by the phase transformation associated with the Al-depletion, but depends on the high-temperature mechanical properties of the bond coats. Thus, it is hypothesised that the rumpling behaviour observed in this study may be controlled by the high-temperature resistance to plastic deformation of these bond coat materials. The fact that the magnitude of rumpling was lower in the isothermal conditions suggested that the rumpling mechanism may be mainly due to micromechanical interactions of the TBC systems and strongly related to the resistance of the bond coat materials to plastic deformation at elevated temperatures; but not the temperature and time-dependent phase transformation process. The Pt-diffusion bond coat, which inherited the two-phase γ and γ' microstructure of the superalloy, should be more resistant to plastic deformation at elevated temperatures than the single-phase β-(Pt,Ni)Al Pt–Al bond coats. Similarly in the Pt–Al system, a previous study[50] has suggested that a strengthening effect (i.e. high value creep index and activation energy) can be attributed to the precipitation of finely dispersed γ' within the β matrix. The extent of γ' phase precipitation depends not only on the Al depletion by inter-diffusion or oxidation, but primarily on how close the composition is to the $\beta/\beta + \gamma'$ phase boundary (Fig. 13.22)[50] at elevated temperatures. The HT Pt–Al system, leaner in both Pt and Al near its rumpling interface, is situated closer to the $\beta/\beta+\gamma'$ phase boundary than the LT Pt–Al system. Thus, one would expect the HT Pt–Al system to be stronger in creep than the LT Pt–Al system, due to a higher volume fraction of precipitated γ'. This explanation is consistent with the observation that the Al-rich Pt–Al (LT Pt–Al) bond coat exhibited a considerably faster rumpling rate than the low Al-containing one (HT Pt–Al). Clearly, further studies need to be done to clarify the role of high-temperature plasticity in the rumpling phenomenon.

13.6 Conclusions and future trends

The importance of superalloy substrate composition in determining the oxidation kinetics and the cycling life of TBC systems has been demonstrated here. The TBC spallation life was found to vary by a factor of up to three, depending upon the chemical composition of the superalloy substrate even though the bond coat and top coat were nominally the same. This effect implies that considerable chemical effects are at play. By comparison, the choice of bond coat among common types has a smaller effect in determining TBC spallation life. The experimental results can be explained only if the fracture toughness parameters controlling de-cohesion (i.e. the frac-

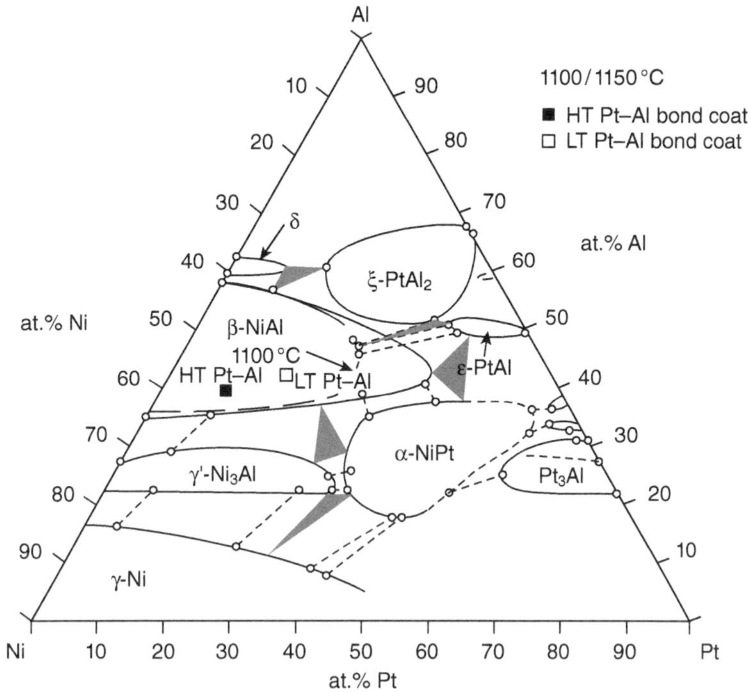

13.22 Ternary Ni–Al–Pt phase diagram at 1100 and 1150 °C[21] showing the two Pt–Al bond coat systems considered in the current study.

ture toughness of the TGO and the fracture toughness of the interfaces bounding it) are influenced strongly by small changes in composition arising from inter-diffusion with the bond coat, such that they are degraded during thermal cycling. Chemical analysis shows that trace elements such as sulphur (on the basis of bulk chemical analysis) in the superalloy substrate are not responsible for the effects reported, indicating that major elemental additions, such as Ti in particular, are the cause.

Although results obtained here provide some insight to assist the design of advanced superalloys or oxidation-resistant coatings, further work needs to be carried out to elucidate more detailed mechanisms associated with the degradation of TBCs.

The importance of substrate composition on the lifetime of TBCs was highlighted. The relationship between substrate composition and the composition of the Pt-diffused layer is central to the beneficial effect of Pt-modification in the context of diffusion coating technology. The Pt-diffusion bond coat is resistant to rumpling and results in an enhanced TBC life, but this effect is diminished in alloys containing elements that have an adverse effect on adhesion at the bond coat/YSZ interface. In view of this, future use of the Pt-diffusion bond coat requires a more considered approach towards the chemistry of the underlying superalloy and the composition of the Pt-diffused substrate prior to top coat application. For optimum turbine blade aerofoil characteristics in service, it may now be necessary to balance the mechanical behaviour of the substrate and its compatibility with the TBC systems in order to improve the performance of the system as a whole.

It is believed that with the development of more sophisticated models, the important phase transformations and oxidation behaviour which govern the degradation of coated superalloys will be modeled, with the goal of developing a predictive capability which is presently lacking; this activity will require the identification of the physical factors which dictate substrate/coating interactions – it will also provide the basis for a chemistry-based approach to coating life prediction.

In terms of the work on the study of bond coat rumpling, the current study showed the different rumpling behaviours between the Pt-diffusion and Pt–Al bond coats. Both Pt–Al systems exhibited rumpling behaviour, but the LT Pt–Al bond coat system showed a significantly faster rumpling rate than the HT Pt–Al bond coat system. The Pt-diffusion system, in contrast, showed no tendency to rumple. It would be of interest as a continuation of the current work to study the effect of the substrate composition on bond coat rumpling. If such an effect is apparent, it would be interesting to correlate the TBC spallation life with the magnitude of rumpling.

13.7 References

1. R.C. REED, *Superalloys: Fundamentals and Applications* (Cambridge, UK: Cambridge University Press 2006).
2. C.T. SIMS, N.S. STOLOFF and W.C. HAGEL (eds), *Superalloys II: High-Temperature Materials for Aerospace and Industrial Power* (New York: Wiley 1987).
3. K.A. GREEN, T.M. POLLOCK, H. HARADA *et al.* (eds), *Superalloys 2004, Proceedings of the Tenth International Symposium on the Superalloys* (Warrendale, PA: The Minerals, Metals and Materials Society (TMS), 2004).
4. W. BETTERIDGE and S.W.S. SHAW, 'Development of superalloys,' *Materials Science and Technology*, **3**, 1987, 682–694.
5. A.M. KARLSSON and A.G. EVANS, 'A numerical model for the cyclic instability of thermally grown oxides in thermal barrier systems,' *Acta Materialia*, **49**, 2001, 1793–1804.
6. N. CZECH, H. FIETZEK, M. JUEZ-LORENZO, V. KOLARIK and W. STAMM, 'Studies of the bond-coat oxidation and phase structure of TBCs,' *Surface and Coatings Technology*, **113**, 1999, 157–164.
7. V. LUGHI, V.K. TOLPYGO, and D.R. CLARKE, 'Microstructural aspects of the sintering of thermal barrier coatings,' *Materials Science & Engineering A*, **368**, 2004, 212–221.

8. K.T. VOISEV and T.W. CLYNE, 'Laser drilling of cooling holes through plasma sprayed thermal barrier coatings,' *Surface and Coatings Technology*, **176**, 2004, 296–306.
9. A. NUSAIR KHAN, J. LU and H. LIAO, 'Heat treatment of thermal barrier coatings,' *Materials Science & Engineering A*, **359**, 2004, 129–136.
10. M. ANDRITSCHKY, P. ALPUIM, D. STÖVER and C. FUNKE, 'Study of the mechanics of the delamination of ceramic functional coatings,' *Materials Science and Engineering A*, **271**, 1999, 62–69.
11. C.G. LEVI, 'Emerging materials and processes for thermal barrier systems,' *Current Opinion in Solid State and Materials Science*, **8**, 2004, 77–91.
12. N.P. PADTURE, M. GELL and E.H. JORDAN, 'Thermal barrier coatings for gas-turbine applications,' *Science*, **296**, 2002, 280–284.
13. V.K. TOLPYGO and D.R. CLARKE, 'On the rumpling mechanism in nickel-aluminide coatings, part I: an experimental assessment,' *Acta Materialia*, **52**, 2004, 5115–5127.
14. B.A. PINT, I.G. WRIGHT, W.Y. LEE, Y. ZHANG, K. PRUSSNER and K.B. ALEXANDER, 'Substrate and bond coat compositions: factors affeccting alumina scale adhesion,' *Materials Science and Engineering A*, **245**, 1998, 201–211.
15. B.A. PINT, 'The role of chemical composition on the oxidation performance of aluminide coatings,' *Surface and Coatings Technology*, **188–189**, 2004, 71–78.
16. B.A. PINT, J.A. HAYNES, K.L. MORE and I.G. WRIGHT, 'The use of model alloys to understand and improve the performance of Pt-modified aluminide coatings,' in: *Superalloys 2004*, eds K.A. Green, T.M. Pollock, H. Harada, T.E. Howson, R.C.Reed, J.J. Schirra and S. Walston, (Warrendale, PA, USA: The Minerals, Metals and Materials Society (TMS), 2004), 597–606.
17. E.P. BUSSO, J. LIN and S. SAKURAI, 'A mechanistic study of oxidation-induced degradation in a plasma-sprayed thermal barrier coating system: Part II: Life prediction model,' *Acta Materialia*, **49**, 2001, 1529–1536.
18. F. TRAEGER, M. AHRENS, R. VAßEN and D. STÖVER, 'A life time model for ceramic thermal barrier coatings,' *Materials Science and Engineering A*, **358**, 2003, 255–265.
19. E.P. BUSSO, L. WRIGHT, H.E. EVANS, L.N. MCCARTNEY, S.R.J. SAUNDERS, S. OSGERBY and J. NUNN, 'A physics-based life prediction methodology for thermal barrier coating systems,' *Acta Materialia*, **55**, 2007, 1491–1503.
20. M. BAKER, J. ROSLER and G. HEINZE, 'A parametric study of the stress state of thermal barrier coatings Part II: cooling stresses,' *Acta Materialia*, **53**, 2005, 469–476.
21. R.T. WU, K. KAWAGISHI, H. HARADA and R.C. REED, The retention of thermal barrier coating systems on single-crystal superalloys: effects of substrate composition, *Acta Materialia*, **56**, 2008, 3622–3629.
22. S.R. CHOI, J.W. HUTCHINSON and A.G. EVANS, 'Delamination of multilayer thermal barrier coatings,' *Mechanics of Materials*, **31**, 1999, 431–447.
23. M.Y. HE, D.R. MUMM and A.G. EVANS, 'Criteria for the delamination of thermal barrier coatings; with application to thermal gradients,' *Surface and Coatings Technology*, **185**, 2004, 184–193.
24. A. SELCUK and A. ATKINSON, 'The evolution of residual stress in the thermally grown oxide on Pt diffusion bond coats in TBCs,' *Acta Materialia*, **51**, 2003, 535–549.

25. A. STRAWBRIDGE and H.E. EVANS, 'Mechanical failure of thin brittle coatings,' *Engineering Failure Analysis*, **2**, 1995, 85–103.
26. B.A. PINT, J.A. HAYNES, K.L. MORE, I.G. WRIGHT and C. LEYENS, 'Compositional effects on aluminide oxidation performance: objectives for improved bond coats,' in: *Superalloys 2000*, eds T.M. Pollock, R.D. Kissinger, R.R. Bowman, K.A. Green, M. McLean, S.L. Olson and J.J. Schirra, (Warrendale, PA, USA: The Minerals, Metals and Materials Society (TMS), 2000), 629–638.
27. A.L. PURVIS and B.M. WARNES, 'The effects of platinum concentration on the oxidation resistance of superalloys coated with single-phase platinum aluminide,' *Surface and Coatings Technology*, **146–147**, 2001, 1–6.
28. P. FOX and G.J. TATLOCK, 'Effect of tantalum additions on oxidation of overlay coated superalloys,' *Materials Science and Tecchnology*, **5**, 1989, 816–827.
29. A.G. EVANS, D.R. MUMM, J.W. HUTCHINSON, G.H. MEIER and F.S. PETTIT, 'Mechanisms controlling the durability of thermal barrier coatings,' *Progress in Materials Science*, **46**, 2001, 505–553.
30. V.K. TOLPYGO and D.R. CLARKE, 'On the rumpling mechanism in nickel-aluminide coatings, part I: an experimental assessment,' *Acta Materialia*, **52**, 2004, 5115–5127.
31. V.K. TOLPYGO and D.R. CLARKE, 'On the rumpling mechanism in nickel-aluminide coatings, part II: characterisation of surface undulations and bond coat swelling,' *Acta Materialia*, **52**, 2004, 5129–5141.
32. M. GELL, K. VAIDYANATHAN, B. BARBER, J. CHENG and E. JORDAN, 'Mechanism of spallation in platinum aluminide/electron beam physical vapor-deposited thermal barrier coatings,' *Metallurgical and Materials Transactions A*, **30**, 1999, 427–435.
33. C.G. LEVI, 'Emerging materials and processes for thermal barrier systems,' *Current Opinion in Solid State and Materials Science*, **8**, 2004, 77–91.
34. N.P. PADTURE, M. GELL and E.H. JORDAN, 'Thermal barrier coatings for gas-turbine applications,' *Science*, **296**, 2002, 280–284.
35. R.C. PENNEFATHER, and D.H. BOONE, 'Mechanical degradation of coating systems in high-temperature cyclic oxidation,' *Surface and Coatings Technology*, **47**, 1995, 76–77.
36. E.P. BUSSO, J. LIN and S. SAKURAI, 'A mechanistic study of oxidation-induced degradation in a plasma-sprayed thermal barrier coating system: Part II: Life prediction model,' *Acta Materialia*, **49**, 2001, 1529–1536.
37. F. TRAEGER, M. AHRENS, R. VAßEN and D. STÖVER, 'A life time model for ceramic thermal barrier coatings,' *Materials Science and Engineering A*, **358**, 2003, 255–265.
38. E.P. BUSSO, L. WRIGHT, H.E. EVANS, L.N. MCCARTNEY, S.R.J. SAUNDERS, S. OSGERBY and J. NUNN, 'A physics-based life prediction methodology for thermal barrier coating systems,' *Acta Materialia*, **55**, 2007, 1491–1503.
39. M. BAKER, J. ROSLER and G. HEINZE, 'A parametric study of the stress state of thermal barrier coatings Part II: cooling stresses,' *Acta Materialia*, **53**, 2005, 469–476.
40. M. JINNESTRAND and S. SJOSTROM, 'Investigation by 3D FE simulations of delamination crack initiation in TBC caused by alumina growth,' *Surface and Coatings Technology*, **135**, 2001, 188–195.

41. A.M. KARLSSON and A.G. EVANS, 'A numerical model for the cyclic instability of thermally grown oxides in thermal barrier systems,' *Acta Materialia*, **49**, 2001, 1793–1804.
42. A.G. EVANS, M.Y. HE and J.W. HUTCHINSON, 'Mechanics-based scaling laws for the durability of thermal barrier coatings,' *Progress in Materials Science*, **46**, 2001, 249–271.
43. J.S. WANG and A.G. EVANS, 'Measurement and analysis of buckling and buckle propagation in compressed oxide layers on superalloy substrates,' *Acta Materialia*, **46**, 1998, 4993–5005.
44. D.R. MUMM and A.G. EVANS, 'On the role of imperfections in the failure of a thermal barrier coating made by electron beam deposition,' *Acta Materialia*, **48**, 2000, 1815–1827.
45. G. LEE, A. ATKINSON and A. SELCUK, 'Development of residual stress and damage in thermal barrier coatings,' *Surface and Coatings Technology*, **201**, 2006, 3931–3936.
46. V.K. TOLPYGO and D.R. CLARKE, 'Surface rumpling of a (Ni, Pt)Al bond coat induced by cyclic oxidation,' *Acta Materialia*, **48**, 2000, 3283–3293.
47. D.R. MUMM, A.G. EVANS and I.T. SPITSBERG, 'Characterisation of a cyclic displacement instability for a thermally grown oxide in a thermal barrier coating system,' *Acta Materialia*, **49**, 2001, 2329–2340.
48. V.K. TOLPYGO and D.R. CLARKE, 'Morphological evolution of thermal barrier coatings induced by cyclic oxidation,' *Surface and Coatings Technology*, **163–164**, 2003, 81–86.
49. B. GLEESON, W. WANG, S. HAYASHI and D. SORDELET, 'Effects of platinum on the interdiffusion and oxidation behavior of Ni–Al-based alloy,' *Materials Science Forum*, **213**, 2004, 461–464.
50. M.P. TAYLOR, H.E. EVANS, E.P. BUSSO and Z.Q. QIAN, 'Creep properties of a Pt-aluminide coating,' *Acta Materialia*, **54**, 2006, 3241–3252.

14
Life prediction of thermal barrier coatings

H. Y. QI and X. G. YANG Beihang University, China

Abstract: Three viscoplastic constitutive models were developed for top coats and substrate. Some important material properties were tested and evaluated under high temperature, and thermal fatigue experiments for atmospheric plasma spraying (APS) and electron beam physical vapor deposition (EB–PVD) were designed and carried out. A fatigue life model was set up and the lifetime of a vane with thermal barrier coating (TBC) was predicted.

Key words: APS, EB–PVD, viscoplastic constitutive model, thermal fatigue, life prediction, vane.

14.1 Introduction

At time of writing, mainstream turbine manufacturers provide an initial gas temperature level of 960–1100 °C for basic service and 1065–1300 °C for peak service. In military aircraft turbines, the gas temperature rises to 1600 °C.[1] The application of thermal barrier coating (TBC) is an effective means of protecting aero-engine blades from hot gas. Currently, there are two methods used for depositing TBC on substrate, namely plasma spray (PS) and electron beam physical vapor deposition (EB–PVD). A typical TBC system consists of two thin coatings: a ceramic coating and a metallic bond coat (BC).

Since the 1980s, the failure mechanisms of atmospheric plasma spraying (APS) and EB–PVD have been analyzed and spallation and local buckling found to be major failure modes. TBC failure results in a combination of oxidation and thermal fatigue.[2–4] Most life prediction models are based on NASA's initial work on phenomenological and macroscopic approaches.[5] However, the potential of TBC cannot be fully utilized due to the lack of a reliable life prediction method and accurate stress–strain results.

In much of the literature, methods based on elastic–plastic theory using an axisymmetric model have been used to analyze failure mechanism and life prediction. Additionally, by adopting finite element method (FEM) theory and simulative technology, many models such as an axisymmetric and elastic–plastic model with realistic oxide growth and an elastic–plastic model with creep of the bond coating, have been simulated and analyzed.[6] However, these 'traditional' models were not able to simulate exactly the complex mechanical behaviours of TBC in a high-temperature environment. Some viscoplastic constitutive models, such as 'Walker' and 'Willam–Warnke', have been developed for TBC systems. Compared to the classical model, the viscoplastic constitutive models allow both cyclic plasticity and creep deformation to be modeled within the framework of continuum damage mechanics. Viscoplastic constitutive models cover all aspects of inelastic deformation, such as monotonic and cyclic plasticity, strain-rate sensitivity, creep and creep relaxation.

Ability to operate in an increasingly high-temperature environment is a key performance indicator in the development of jet engines. With the extensive application of advanced technology, such as directionally solidified superalloys on hot sections, investigation of methods of analyzing the structures of complex heterogeneous materials as well as their failure mechanisms are becoming increasingly urgent. Spallation of TBC in service, deformation analysis and thermal fatigue life prediction methods for a vane with TBC have been intensively investigated and discussed.

14.2 The mechanical behavior of thermal barrier coating (TBC) systems under elevated temperatures

14.2.1 Introduction

Modern methods for life prediction of structures are based on inelastic analyses, which lead to large numbers of cycles of computation, especially under high temperature and cyclic loadings. In order to achieve a higher level of accuracy without increasing the computing time required, better constitutive equations need to be employed to describe material behaviour.[7]

An advanced constitutive model based on the Willam–Warnke flow surface has been developed for concreteness.[8] The model has ability to describe non-equal yield stresses in compression and tension and rate-independent features. Based on this model and the experimental data, Xie developed a genetic algorithm to optimize material constants. The FEM model for PWA 264 TBC with thermal grown oxide (TGO) has been simulated and analyzed successfully.[9]

Using experimental data from NASA's Hot Section Technology (HOST) program, and based on the Walker viscoplastic material model, an alternative unified viscoplasticity constitutive model was adopted for EB–PVD coating. The substrate used is DZ40M, a kind of directionally solidified superalloy. Based on the Chaboche constitutive model, an advanced viscoplasticity constitutive model for orthotropic substrate was developed[10,11] to describe behavior under high temperature.

14.2.2 Material constitutive laws

TBC generally comprises a multi-layer structure with a top coat of brittle material while the bond coat and substrate are ductile material. Thus, when evaluating material behaviour with regard to temperature, each layer has its own specific constitutive law. The Willam-Warnke model was used for the APS ceramic top coat while the Walker model was adopted for EB–PVD. The Chaboche constitutive law was used for the metal substrate.

Willam–Warnke model for the top coat by APS

A viscoplastic constitutive model based on the theory of finite deformation for PS TBCs able to describe characteristics of tension/compression asymmetry, strain rate sensitivity and hydrostatic stress dependency has been introduced. The APS TBC exhibits non-equal yield stresses in compression and tension and non-linear rate-dependent behaviour under high temperature. In order to simulate realistic deformation characteristics, the viscoplastic Willam–Warnke model was adopted for the top coat. The constitutive equations are as follows:[8,9]

$$f = K^2\left[\left(\frac{1}{2\mu}\right)\int F^n \mathrm{d}F + \left(\frac{R}{H}\right)\int G^m \mathrm{d}G\right] \qquad [14.1]$$

$$\dot{\varepsilon}_{ij}^I = c_0\left[c_1\delta_{ij} + c_2(S_{ij} - a_{ij}) + c_3\left((S_{iq} - a_{iq})(S_{qj} - a_{qj}) - \frac{2J_2\delta_{ij}}{3}\right)\right] \qquad [14.2]$$

$$\dot{a}_{ij} = h\left[\dot{\varepsilon}_{ij}^I - c_4\left(c_1\delta_{ij} + c_5 a_{ij} + c_6\left(a_{iq}a_{qj} - \frac{2J_2\delta_{ij}}{3}\right)\right)\right] \qquad [14.3]$$

where f is the viscoplastic potential function, F is the three-parameter Willam–Warnke yield criterion, G is the recovery function, σ_{ij} is a stress component; δ_{ij} is Kronecker delta, S_{ij} and S_{iq} are the deviatoric stress component and J_2 is the second stress invariant. Because the second and third invariants of the deviatoric part of the Cauchy stress tensor are adopted, this Willam–Warnke model features the non-equal yield stresses in compression and tension, rate and hydrostatic independence. $\dot{\varepsilon}_{ij}^I$ indicates non-linear strain rate tensor, a_{ij} indicates internal variables of non-linear kinematic hardening, C_0~C_6 indicate functions of the stress and back stress tensors and their invariants and K, R, H, m, n, h, and μ are material constants.

Walker model for the top coat by electron beam physical vapor deposition (EB–PVD)

Within the framework of unified viscoplasticity, the complete set of equations is given by Walker. This model offers a great deal of scope for the description of temperature-dependent phenomena. The Walker model contains one scalar internal state variable K, called drag strength, and an internal variable Ω called back stress. The viscoplastic potential was as follows:[12,13]

$$f = \frac{K}{n+1}\left[\frac{J(\sigma - X)}{k(\dot{p})}\right]^n \quad [14.4]$$

where K and n are material parameters and p is internal state variable. In this formula $J(\sigma - X)$ is the internal stress invariant. The viscoplastic strain rate was written as

$$\dot{\varepsilon}_{ij}^I = \left[\frac{\sqrt{\frac{2}{3}\left(\frac{3}{2}S_{ij} - \Omega_{ij}\right):\left(\frac{3}{2}S_{ij} - \Omega_{ij}\right)}}{K_d}\right]^n \frac{\left(\frac{3}{2}S_{ij} - \Omega_{ij}\right)}{\sqrt{\frac{2}{3}\left(\frac{3}{2}S_{ij} - \Omega_{ij}\right):\left(\frac{3}{2}S_{ij} - \Omega_{ij}\right)}} \quad [14.5]$$

The drag stress was written as

$$K = K_1 - K_2 e^{(-n_7 \dot{p})} \quad [14.6]$$

where $\dot{p}$ is the rate of the accumulated inelastic strain:

$$\dot{p} = \sqrt{\frac{2}{3}\dot{\varepsilon}_{ij}^I : \dot{\varepsilon}_{ij}^I} \quad [14.7]$$

The evolution equation for the back stress tensor was shown as

$$\dot{\Omega}_{ij} = (n_1 + n_2)\dot{\varepsilon}_{ij}^I - \left(\Omega_{ij} - \dot{\Omega}_{ij} - n_1 \varepsilon_{ij}^I\right)\dot{G} \quad [14.8]$$

where K_1, K_2, n_1, n_2, and n_7 are temperature-dependent material parameters.

Chaboche model for the substrate

A modified Chaboche constitutive model for directionally solidified superalloy substrate able to describe characteristics of loading direction dependency, creep, cyclic hardening/softening and mean stress relaxation has been developed. A fourth-order tensor M_{ijkl} was induced for the viscoplastic constitutive law, which laid a foundation for achieving accurate stress and strain analysis on a turbine vane with PS TBC.

The substrate used for specimens and a vane in our work is DZ40M, a kind of directionally solidified superalloy. Based on the Chaboche constitutive model, an advanced constitutive model for orthotropic substrate has been developed.[10,11] The constitutive relation is as follows:

$$f = \sqrt{\frac{3}{2}M_{ijkl}(\sigma'_{ij} - X'_{ij})(\sigma'_{kl} - X'_{kl})} - R - K \quad [14.9]$$

$$\dot{\varepsilon}_{ij}^{I} = \frac{\partial f}{\partial \sigma_{ij}} = \frac{3}{2}\left\langle \frac{f}{K} \right\rangle^{n} \frac{M_{ijkl}(\sigma'_{kl} - X'_{kl})}{\sqrt{\frac{3}{2} M_{ijkl}(\sigma'_{ij} - X'_{ij})(\sigma'_{kl} - X'_{kl})}} \quad [14.10]$$

$$\dot{X}_{ij} = \frac{2}{3} a c \dot{\varepsilon}_{ij}^{I} - c\left(\frac{J(X_{ij})}{a}\right)^{m} \dot{p} X_{ij} - \beta |J(X_{ij})|^{r-1} X_{ij} \quad [14.11]$$

$$\dot{R} = b(Q - R)\dot{p} \quad [14.12]$$

where f is the yield function, X_{ij}, X_{kl} and R indicate internal stresses for kinematic hardening and isotonic hardening, respectively, and $J(x_{ij})$ is the internal stress invariant. In the general case of an initially anisotropic material, a fourth-order tensor M_{ijkl} was introduced. To simplify M_{ijkl}, there are only three parameters, namely M_{11}, M_{33} and M_{55}. $\dot{\varepsilon}_{ij}^{I}$ indicates the viscoplastic strain rate. The formula $\dot{p} = \sqrt{3/2\, \dot{\varepsilon}_{ij}^{I} M_{ijkl}^{-1} \dot{\varepsilon}_{kl}^{I}}$ is called the accumulated plastic strain rate. K, n, k, a_1, c_1, a_2, c_2, β, m, r_1, b and Q in the equations are temperature-dependent parameters. This model is able to describe direction-dependent characteristics, hardening properties, relaxation of the mean stress and cyclic creep.

14.2.3 Material parameters of TBCs under high temperature

Measurement of Young's modulus for the top coat

While the thermal-sprayed coatings bring many advantages, there are still some aspects requiring further investigation and better understanding in order to improve performance and increase the scope of applications. One important property of thermal-sprayed coatings, which is difficult to determine, is Young's modulus. An accurate Young's modulus value is necessary to evaluate residual stresses, bond strength, fatigue crack growth rates and coating stress under in-service loading conditions.[14]

Because all coatings are used while they are bonded to a substrate, it is important to use a procedure capable of evaluating the Young's modulus *in situ*. One of the problems encountered is that the ceramic top coat is usually thin, and therefore it is difficult to obtain a specimen entirely made of coating material to conduct a modulus test. Nowadays, by using warm hydrochloric acid, the top coat specimen of TBCs can be detached. However, it is difficult to conduct the measurement of the applied load and displacement under high temperature.[15]

APS top coat was selected to measure the Young's modulus using the resonant frequency method. The Young's modulus of top coat under high temperature measured using the bending resonance method was the particular focus. The testing system included both hardware and software components.

Specimens were produced using a METCO-7M APS of a CoNiCrAlY bond coat and a top-coat of ZrO_2–6–8 wt% Y_2O_3. The bond coat thickness was about 120 μm, while that of the top coat was about 240 μm. Some specimens for the resonant frequency method were deposited onto 1Cr18 Ni9Ti stainless steel substrate. With approximate dimensions 120 × 5 × 1.5 mm. The resonant flexural frequencies were tested by suspending the rectangular beam specimens from hanging string at the node positions for the fundamental flexural resonance.[16,17] The modulus of the coating can be determined in a straightforward manner from the stiffness of the composite beam, which is equal to the sum of the stiffnesses of the components

$$E_{c+s}I_{c+s} = E_s I_s + E_c I_c \qquad [14.13]$$

where E is the Young's modulus, I is the moment of inertia and subscripts c + s, c and s refer to the quantity for composite beam with both coating and substrate, for coating only and for substrate only, respectively. Young's modulus for a homogeneous rectangular beam is related to the resonant frequency by

$$E = 0.94642\frac{TL^3}{h^3}\frac{mf^2}{t} \qquad [14.14]$$

where L, h and t are the length, height and thickness of a homogenous specimen, respectively, T indicates a dimensionless shape factor (dependent on the specimen length, height and Poisson's ratio), m indicates the mass and f indicates the resonant frequency of the fundamental mode of vibration. The combination of the two equations leads to the following expression for the modulus of the coating, E_c:

$$E_c = E_s\left[\frac{m_{c+s}}{m_s}\left(\frac{f_{c+s}}{f_s}\right)^2 - 1\right]\frac{t_s}{t_{c+s} - t_s} \qquad [14.15]$$

E_s is determined by the resonant frequency of the bar after the coating is removed and E_c could be deduced by the change in resonant frequency and sample dimensions between the composite beam with and without the coating.

A testing system was set up based on the resonant frequency and composite beam theory. The system, shown in Fig. 14.1, consists of a high-temperature furnace, a signal generator and a pair of hanging strings, together with DASP2005 signal processing software and so on. This system is able to measure the Young's modulus of the film of composite beam and laminate at room temperature or at high temperature. The highest temperature bearable for the furnace is up to 1300 °C.

14.1 Young's modulus testing system.

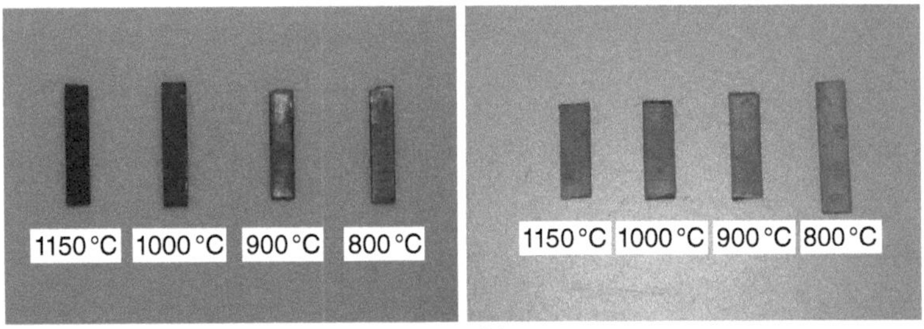

(a) (b)

14.2 Specimens for oxidation experiment: (a) 1Cr18Ni9Ti substrate with no TBC; (b) specimens with TBC.

Antioxidative properties are a key material parameter. If the substrate is oxidized, the resonance frequencies values will not be accurate. The substrate used for the TBC system was 1Cr18Ni9Ti stainless steel, which is a kind of austenitic refractory steel extensively applied in many fields.[18] A 1Cr18Ni9Ti stainless steel plate was cut using a wire cutting machine, polished by abrasive paper for degreasing and derusting and treated by sand blasting so as to reach a specified roughness. These samples were shorter than those for measuring Young's modulus.

The oxidation of the substrate was designed and carried out at a temperature of 200°–1150 °C. The oxidation weight of the specimens was close to zero. The substrates after the oxidation experiment and without TBCs are shown in Fig. 14.2a. The experiment shows that, the 1Cr18Ni9Ti stainless steel could be manufactured for use as the TBC substrate material. Specimens with TBCs are shown in Fig. 14.2b. After 2 h of testing using resonance frequencies, the ceramic top coat did not detach from the substrate and the experiment could be considered a success.

Table 14.1 Young's modulus under different temperatures

T (°C)	20	100	200	300	400
E (GPa)	62.5	59.6	55.7	50.3	44.9
T (°C)	600	800	900	1050	1150
E (GPa)	29.8	22.2	21.5	19.8	18.6

Three specimens were fabricated with ceramic top coat. The resonance frequencies of the specimens were obtained at room temperature up to 1150 °C, and every specimen was tested three times at each temperature. The average Young's modulus of the ceramic top coat is shown in Table 14.1, in which the Young's modulus decreases as the temperature increases. From room temperature to 800 °C, the Young's modulus decreases rapidly. However, at temperatures above 800 °C, it decreases slowly.

Evaluation of thermal conductivity for the top coat

Thermal spraying is a process whereby a stream of molten droplets impacts on the substrate followed by flattening, rapid solidification and cooling processes. The individual molten droplets spread to thin lamellae, the stacking of which constitutes the deposit.[19] Thermally-sprayed coatings generally have a lamellar structure. The properties of a thermally sprayed coating depend on its microstructure which, in turn, depends on the spraying process used. The relationship between the properties and structure of a coating is of key importance in determining whether a desirable coating can be produced. For the top-coat of a TBC system, the fundamental microstructural property may be the porosity. Currently, evaluation of porosity volume is the major analytical method for TBC microstructure.[20,21]

The TBC system under evaluation was a two-layer material system consisting of a 7–8% Y_2O_3–ZrO_2 top coating insulator over a CoNiCrAlY bond coat. The substrate material was 1Cr18Ni9Ti stainless steel. The Mcpherson model was used to study the relationship between the deposit structure and thermal conductivity properties in this work. The relative thermal conductivities can be derived by the following equation:[22]

$$\frac{\lambda_c}{\lambda} = \frac{2\alpha}{\pi}\frac{\delta}{a}\left(1+\frac{2\alpha\delta}{a\pi}\right)^{-1} \qquad [14.16]$$

where λ_c and λ are thermal conductivities of deposit and splat materials, α indicates the bonding ratio between lamellae, δ indicates the mean lamellae thickness and a indicates the size of bonding region. Obviously, the thermal conductivity also depends on the geometric dimensions of bond region.

Table 14.2 Thermal conductivity of topcoat under high temperature

T (°C)	20	800	900	1000	1100
λ_c (W/mK)	0.59	0.84	0.91	0.99	1.18

The relationship between the microstructural parameters and the thermal conductivity was demonstrated experimentally. The mean lamellar thickness $\delta = 9.12$ μm, can be obtained by microscopic image test using a scanning electron microscope. Some parameters of equation [14.16] are $\alpha = 0.32$ and $a = 3.62$. The thermal conductivity of 6–8% Y_2O_3–ZrO_2 is given in Table 14.2. The thermal conductivity at high temperature is higher than that at room temperature, the maximum relative value being 50%.

14.2.4 Finite element method (FEM) calculation and analysis

Although ABAQUS is a complete material library, the advanced constitutive equation cannot be modified in the ABAQUS/CAE interface. However, CAE software provides a development function allowing users to add a user-defined material model (UMAT) in order to find accurate solutions to their practical problems. Three constitutive equations were developed in FORTRAN for APS, EB–PVD and metallic substrate, respectively.

Simulation for the top coat

Regarding the APS top coat, the Willam–Warnke model has the ability to describe the non-equal yield stresses in compression and tension for the top coat while incorporating strain rate dependency under high temperature. A FORTRAN program was implemented for the UMAT subroutine in ABAQUS, and the resulting developed UMAT code was successfully applied to describe the complex mechanical behaviors of material under high temperature. The material constants were from references.[8] Some simulation results are shown in Figs. 14.3, 14.4 and 14.5.

Simulation for DZ40M superalloy

Because the Chaboche viscoplastic constitutive model is able to simulate the monotonic and cyclic behaviour of DZ40M relatively well, an ABAQUS UMAT subroutine was developed to implement it. Some comparisons are given in Figs. 14.6 and 14.7, which show advanced non-linear and anisotropic behaviour at 500 °C and cyclic behavior at 900 °C.

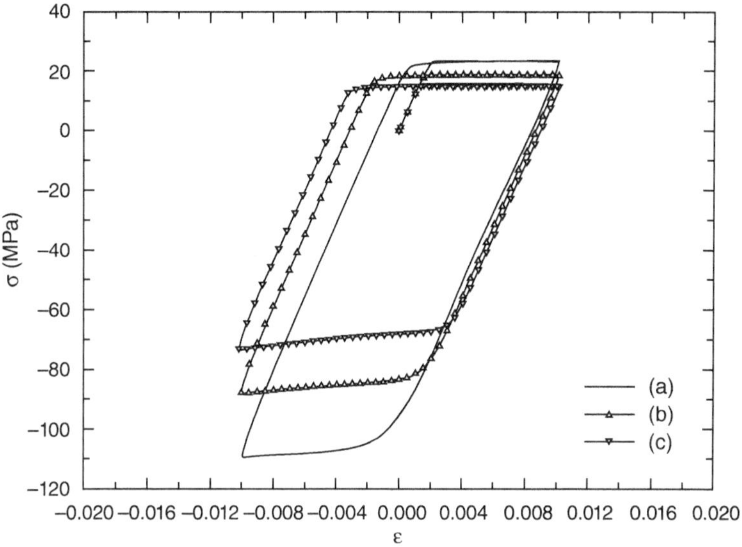

14.3 Stress–strain curves under incorporating strain rate dependency: (a) strain rate 1.0×10^{-3}; (b) strain rate 1.0×10^{-4}; (c) strain rate 1.0×10^{-5}.

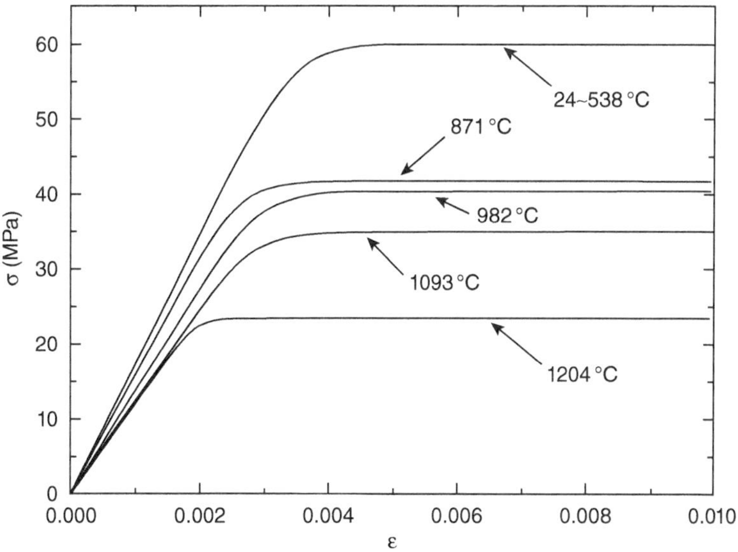

14.4 Monotonic tensile curves under strain rate 1.0×10^{-3}.

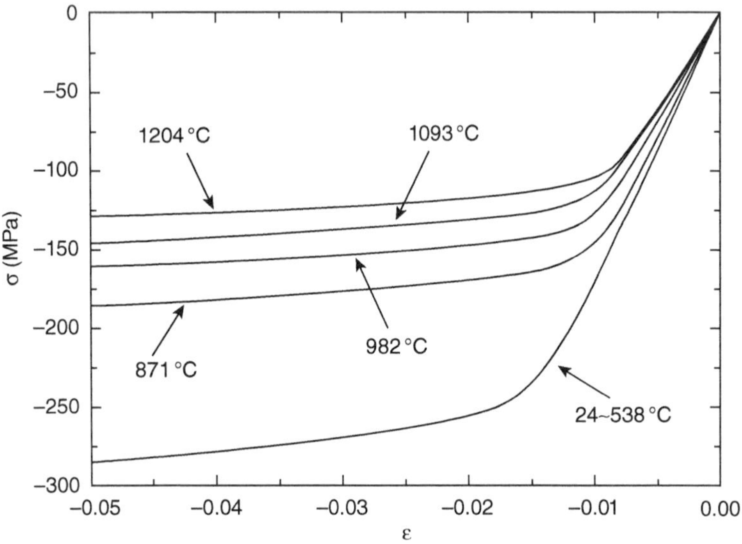

14.5 Monotonic compressive curves under strain rate 1.0×10^{-3}.

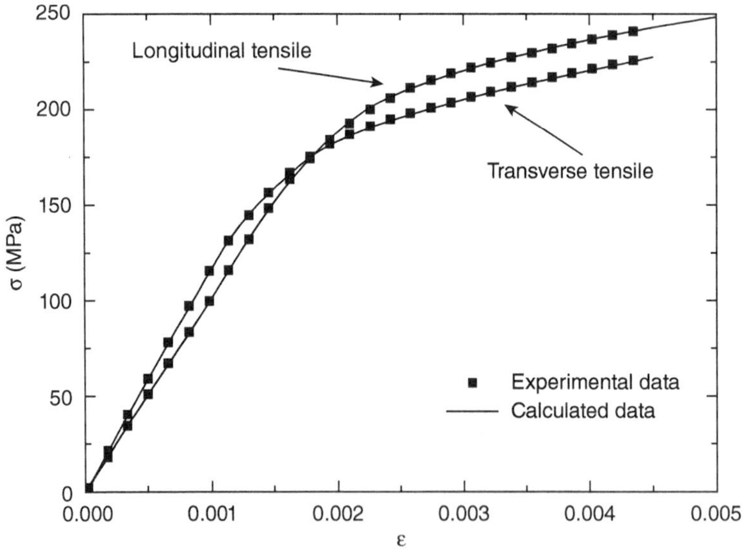

14.6 Monotonic tensile stress–strain curves under 500 °C.

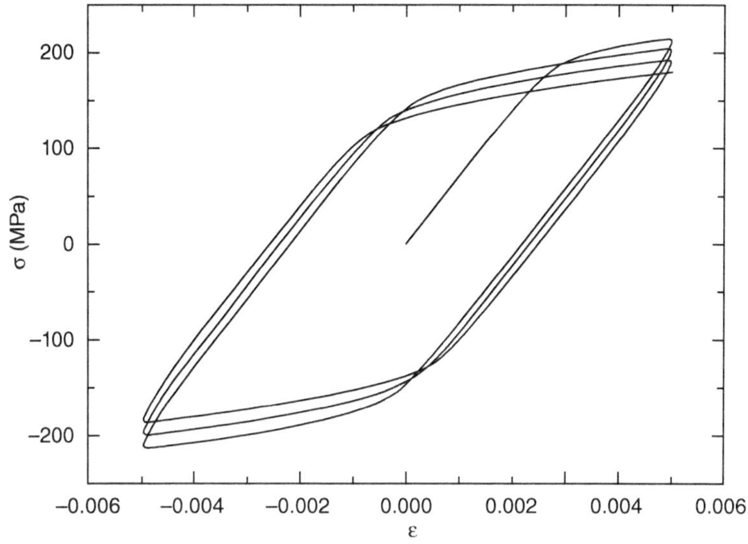

14.7 Cyclic behaviour under 900 °C.

14.3 Life prediction for TBCs

14.3.1 Life prediction model

Phenomenological fatigue life model

Based on the Manson–Coffin model, the prediction model is as follows:

$$N_f = (\Delta\varepsilon/\Delta\varepsilon_f)^{-b} \quad [14.17]$$

where N_f is number of cycles to failure, $\Delta\varepsilon$ is strain range, $\Delta\varepsilon_f$ is critical strain range and b is a material parameter.

Oxidation will be considered in this model due to its key significance. When bond coating was oxidized for different lengths of time, $\Delta\varepsilon_f$ decreased with the increase in TGO thickness and this value changed from $\Delta\varepsilon_{f0}$ to $\Delta\varepsilon_f$. The equation is as follows:[23]

$$\Delta\varepsilon_f = \Delta\varepsilon_{f0}\left(1-\frac{\delta}{\delta_c}\right)^c + \Delta\varepsilon\left(\frac{\delta}{\delta_c}\right)^c \quad [14.18]$$

where δ is TGO thickness, δ_c is critical TGO thickness when TBC cracks are caused by oxidation only and c is a material parameter. Therefore, the life prediction model for TBCs is as follows:

$$N_f = \left[(\Delta\varepsilon_{f0}/\Delta\varepsilon)\left(1-\frac{\delta}{\delta_c}\right)^c + \left(\frac{\delta}{\delta_c}\right)^c\right]^b \quad [14.19]$$

For Eq. [14.19], the parameters with the exception $\Delta\varepsilon_f$ and δ_c, are determined by thermal testing and $\Delta\varepsilon$ is obtained by finite element analysis. The thermal fatigue life could be obtained by using Miner linear cumulative damage theory.

This approach, used by General Electric Company, employed time-dependent, non-linear finite element modeling of the stresses and strains present in the APS TBC. The modified model is expressed as:[24]

$$\Delta\varepsilon_{RZ} + 0.4\Delta\varepsilon_R = 0.121 N_f^{-0.486} \quad [14.20]$$

where $\Delta\varepsilon_{RZ}$ is the shear strain range, $\Delta\varepsilon_R$ is the normal strain range and N_f is the number of cycles to failure. The subscripts R and Z represent two directions for the cylindrical coordinate system. The model is the only one to consider failure induced by edges.

An EB–PVD ceramic has a columnar structure as a result of the vapor deposition process. Failure in an EB–PVD TBC occurs in the TGO layer adjacent to the metallic bond coating. Compared to the failure mechanism of APS TBC, this model has extended the life prediction of the EB–PVD coating. On the other hand, the strain range of TGO will be adopted as an important parameter in the fatigue model.

Fatigue model based on imperfections

With respect to the imperfections in TBC system, the driving stress used for predicting cracking and failure is written as follows:[25]

$$\sigma_c = \frac{E(\varepsilon_g - 1)}{2(1-\nu)\varepsilon_g}\left(\frac{h}{R}\right) \quad [14.21]$$

where E is Young's modulus, h is the TGO thickness and ε_g is growth strain at highest temperature. By equating the crack diameter $2R$ to the spacing $2d$ between imperfections, a critical TGO thickness, h_c yields

$$h_c = \frac{2\sqrt{\pi}(1-\nu^2)\varepsilon_g d^{3/2} K_{Ic}}{(\varepsilon_g - 1)RE} \quad [14.22]$$

where K_{Ic} is the stress intensity factor. According to the kinetic oxidation model, the failure time is given by

$$t_f = h_c^2/2k_p \quad [14.23]$$

where k_p is the parabolic rate constant, which is a strong function of temperature and optimized using static oxidation experimentation. If the cyclic behavior $d\Delta_R/dN$ is specified, the fatigue model is as follows:

$$N_f \approx \frac{2\sqrt{\pi}(1-\nu^2)d^{3/2} K_{Ic}}{E(d\Delta_R/dN)L} \quad [14.24]$$

In this formula, E is the Young's modulus of the TBC with wavelength $2L$ imperfection and ν is Poisson's ratio. The fatigue model has two features: (i) the model is based on microstructural imperfection and renders the physical meaning of failure mechanism more obvious; (ii) the parameters such as $d\Delta_R / dN$ which are required for TBC thermal fatigue experimentation are difficult to obtain. Further study is required for thorough verification of the model.

Some other models based on fracture mechanics utilize such concepts as time-dependent interface toughness and strain energy release rate, with the emphasis placed within the TGO or at the TGO/bond coat interface. The durability of the TBC is controlled by the energy density in the TGO and the formation or presence of imperfections in the vicinity of the TGO.

Fatigue model based on damage mechanics

Busso quantified the effects of different microstructural variables and service conditions on the local stresses induced by oxidation, sintering processes and thermal cycling in a typical PS TBC system. The driving force σ_{22max} was combined with a damage mechanics approach. Two boundary conditions for this fatigue model were set as $D = 0$ at the beginning of thermal cycling and $D = 1$ at thermal fatigue failure. The continuous evolution of microscopic damage with thermal cycles is given by[26,27]

$$dD = \hat{D}\{\sigma_{22\max}, \bar{\sigma}_{22\max}, D\}dn \tag{14.25}$$

where n is the number of cycle under a load case, and can be expressed in functional form in terms of the maximum cycle stress, σ_{22max}, the mean cycle stress, $\bar{\sigma}_{22max}$, and the current microscopic damage state, D. This model was suitable only for measuring the thermal fatigue load.

14.3.2 Life prediction for plasma sprayed PS thermal barrier coatings

In order to set up a fatigue model for APS TBC, thermal fatigue experiments were designed. The load case was designed using rapid heating (120 s) and rapid cooling (300 s), simulating the shut down of a turbine engine. By using induction heating and air compressor technology, the APS TBC structure was heated to 1050 °C and cooled to 100 °C.

Instead of a vane, hollow cylindrical specimens were designed with internal radius 11 mm, external radius 15 mm and length 80 mm. The specimens were APS coated with the zirconia–yttria (6%~8% Y_2O_3–ZrO_2) on a nickel–chromium–aluminum (Ni22Cr10Al1.0Y) bond coat. The metal substrate was Ni-based superalloy (DZ40M). The ceramic layer was 0.250 mm thick while the bond coat was 0.125 mm thick.

Using macroscopic images and FEM simulation, the damage evolution of the TBC could be summarized as follows. Under thermal fatigue load, cracks nucleated in the TBC, which were initially in-plane Mode I and Mode II or mixed-mode tangential cracks. The crack in the ceramic layer was driven primarily by the normal stress or the in-plane shear stress. If the compressive stress along the axial direction occurs at the same time, then these cracks link together and protrusion occurs as shown in Fig. 14.8a. Due to induction heating, the length of the heated specimen is finite. The part far away from the heated section will provide displacement constraint similar to plane-strain constraint. Meanwhile, a temperature gradient exists along the axial direction. An axial compressive stress results from the temperature gradient along the axis of the specimen and a mismatch in the structure. Based on the protrusion, the ceramic layer will buckle as shown in Fig. 14.8b. On the other hand, it should be noted that the circular stress in the specimen could be compressive stress. However, the temperature gradient along the circular direction is close to zero, so it is impossible for these specimens to be buckling. Simultaneously, the axial direction of the specimens is sufficiently long compared to the diameter (circular direction). The buckling phenomenon is not only a function of force but also of distance; thus the buckling does not occur along the circular direction. Finally, TBC spallation occurs along the axial direction as shown in Fig. 14.8c.

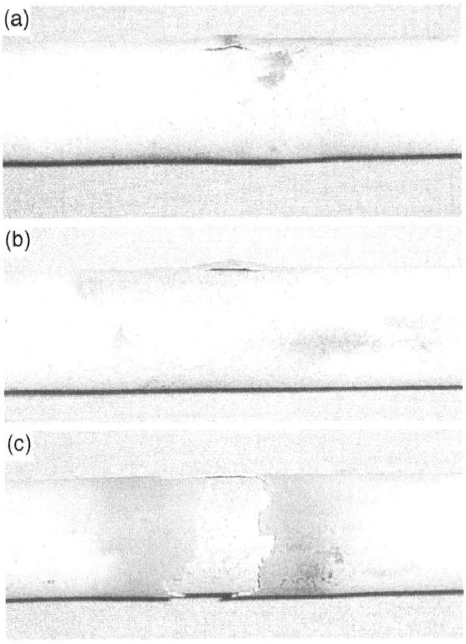

14.8 Failure mode of APS TBC under thermal fatigue: (a) protrusion; (b) buckling; (c) spallation.

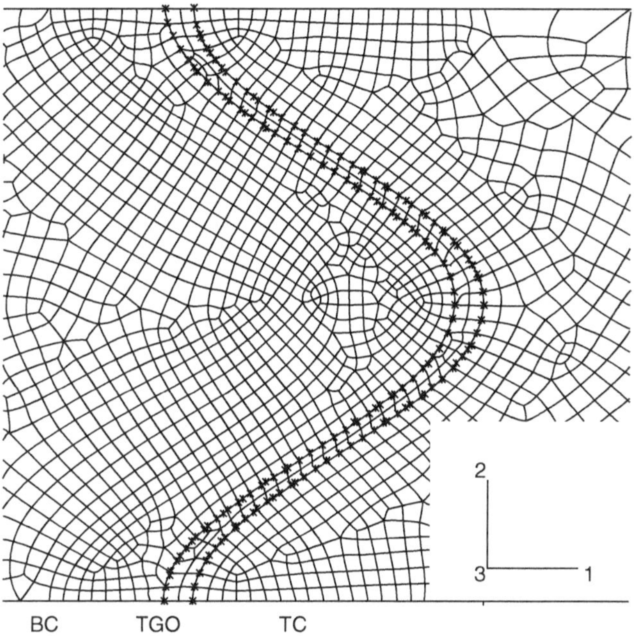

14.9 FEM model for TBC.

In order to consider the TGO thickness and interface, the interface morphology is conceptualized as a periodic array of wavy-type segments characterized by co-sinusoidal interface in a FEM model. This interface between the bond coat and the ceramic top coat is introduced, with a period as well as peak-to-valley amplitude of 0.01 mm and peak-to-peak dimensions up to 0.04 mm. An axisymmetric model of this specimen was set up in Fig. 14.9. By introducing a co-ordinate system, an orthonormal basis {1, 2, 3} was defined, where directions 1, 2 and 3 denote the normal, axial and circular direction, respectively.

Many TBC life models were based on the assumption that oxidation and thermal strain range are important factors limiting the lifetime of the coating. The strain range was always defined as the equivalent strain range, such as axial strain or shear strain range. According to the failure mechanism and coupling different strain ranges, a life model for TBC was set up in the present work:

$$N = \left[\left({}^{\Delta\varepsilon_{f0}} \Big/ {\Delta\varepsilon_{RZ} + a\Delta\varepsilon_Z} \right) \left(1 - \frac{\delta}{\delta_c} \right) + \left(\frac{\delta}{\delta_c} \right) \right]^b \qquad [14.26]$$

where $\Delta\varepsilon_{RZ}$ and $\Delta\varepsilon_Z$ are the shear and axial strain range at critical location and δ is TGO thickness at any Nth cycles. a, b, $\Delta\varepsilon_{f0}$ and δ_c are parameters of this model. The fatigue life was predicted by the linear cumulative damage hypothesis. According to this rule, the fatigue damage was $D_m = 1/N_m$ after one cycle. Applying n_m cycles with thermal fatigue load k and corresponding fatigue life endurance N_m is equivalent to overall damage $D = \sum_{m=1,k} n_m \cdot D_m = \sum_{m=1,k} n_m / N_m$. When the criterion $D \geq 1$, spallation failure of the top coat occurs. The parameters of the fatigue predictive model were optimized and are listed in Table 14.3 and the predicted lives for APS are listed in Table 14. 4.

14.3.3 Life prediction for EB–PVD thermal barrier coatings

EB–PVD specimens with the same geometry size as APS samples were coated with zirconia–yttria (6~8 % Y_2O_3–ZrO_2) on a nickel–chromium–aluminum (Ni22Cr10Al1.0Y) bond coat. The metal substrate was Ni-based superalloy (DZ40M). The ceramic layer was 0.125 mm thick while the bond coat was 0.07~0.08 mm thick. The load-case consist of a 2.5 min heating from room temperature to 1150 °C, a holding time of 0.5 min and a cooling time of 4.5 min.

As a result, the EB–PVD coating was reported to be more strain-tolerant in the direction perpendicular to the columns (in-plane coating direction) and thus more resistant to crack growth compared with the PS coatings. Another difference was the morphology of the bond coat/top coat interface, mechanical adhesion for PS and chemical adhesion for EB–PVD. The top coat/bond coat interface has been modified by EB–PVD processing.

The failure location of EB–PVD TBCs was within the TGO (usually Al_2O_3) that formed between the ceramic coating and the bond coating. In

Table 14.3 Parameters for the fatigue life model

a	b	$\Delta\varepsilon_{f0}$	δ_c
2.202	3.751	0.087	0.058

Table 14.4 Thermal fatigue data of APS

Specimen	Pre-oxidation time (h)	Lifecycle	Predicted lifecycle
PL100a	100	430	398
PL100b	100	350	398
PL200a	200	280	306
PL200b	200	380	306

other literature, the failure location was in the TGO/bond coat interface. The failure mode of EB–PVD TBCs under cyclic thermal load is shown in Fig. 14.10. Compared with APS TBC, this difference in failure mechanism can primarily be attributed to the difference in the microstructure of the TBC. The EB–PVD TBC interface is relatively smooth, so the stress concentration is correspondingly small. Therefore, the stress or the strain state in the TGO is important and should be focused on. Because the TGO is very thin (less than 0.1 mm) and usually remains elastic during the load case, it is difficult to obtain the strain range from finite element analysis (FEA). It is assumed that the TGO has a negligible influence on mean stress/strain of the bond coating. The TGO strain will be determined based on the bond coating. The mean stress/strain can be obtained from FEA analysis. The formula for TGO strain calculation can be obtained as follows. First, the TGO expansion strain is given by:

$$\varepsilon = \frac{\delta L(T)}{L} = \int_{T_{sf}}^{T} \alpha(T)\mathrm{d}T \tag{14.27}$$

where T is temperature and T_{sf} is the stress-free temperature. Assuming that the expansion parameters of Al_2O_3 change linearly with temperature, the TGO strain is:

$$\varepsilon_{(T)} = \varepsilon_{ss(T)} - \varepsilon\frac{\delta L(T)}{L} \tag{14.28}$$

Where $\varepsilon_{ss(T)}$ is strain of bond coating. For a cyclic load, the evolution of bond coating strain at given temperature can be calculated using FEA analysis. Then by using Eqs. [14.27] and [14.28], the evolution of TGO strain and the strain range can be obtained. The predicted lives are listed in Table 14.5.

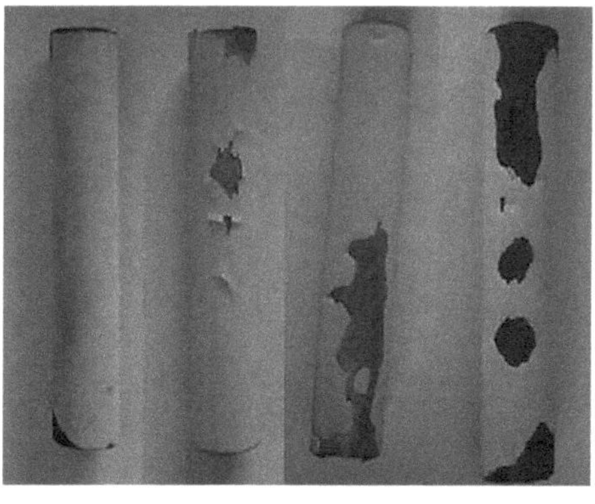

14.10 Failure mode of EB–PVD TBC under thermal fatigue.

Table 14.5 Thermal fatigue data of EB–PVD

Specimen	Pre-oxidation time (h)	Lifecycle	Predicted lifecycle
EB50a	50	202	226
EB50b	50	250	226
EB100a	100	145	173
EB100b	100	200	173

14.3.4 Life prediction for the vane with TBCs

In order to analyze the local stress–strain field, a sub-model method, which is included in commercial FEM code, is always used for large and complicated structures such as a vane which has a twist in its shape from the root to the end and a body which is thin and uneven at different positions. However the sub-model is not suitable for a vane with TBC because the top coat and bond coat are meso-scale. The sub-model will comprise a great many meshes for a part of a vane with TBC. So it is necessary to establish an equivalent way in which macro-scale stress state and meso-scale will link each other.

First, a cylinder-shaped sample with different inner radius was simulated in meso-scale. The results showed that different radius could not affect the stress–strain fields close to the interface. In other words, the sections at the leading and trailing edge could be assumed as samples with different radius. Thus the stress–strain field at the leading and trailing edge could be neglected, which results in different radius.

Secondly two interfaces (such as straight and cosine interface) between top coat and bond coat were simulated for TBC. The local field close to the interface and that far away from the interface were defined as 'near-field' and 'far-field', respectively. Compared to the results of macro-scale and different interfaces, there is a proportional factor to the strain range in near-field and far-field.

The vane TBC was modelled and meshed using solid elements. The ceramic top coat was meshed into two through-thickness layers with 1100 elements. The bond coat was meshed into one layer with 550 elements. Because the TGO thickness is very small, the interface between top coat and bond coat was neglected.

The initial state was set as 400 °C, a temperature at which the vane is in a stress-free state. The boundary conditions were applied to fix the inner root of the vane. The heating and cooling times were 155 s and 109 s, respectively. The two cyclic thermal mechanical behaviors of the vane were simulated by the constitutive equation mentioned above.

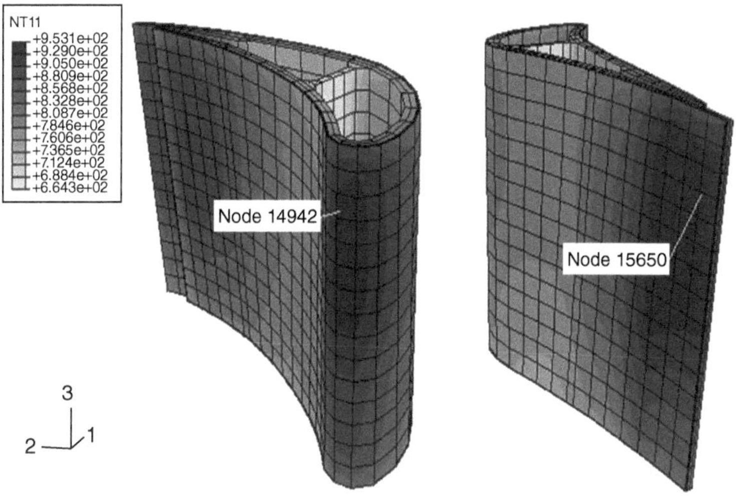

14.11 Critical locations at leading and trailing edge.

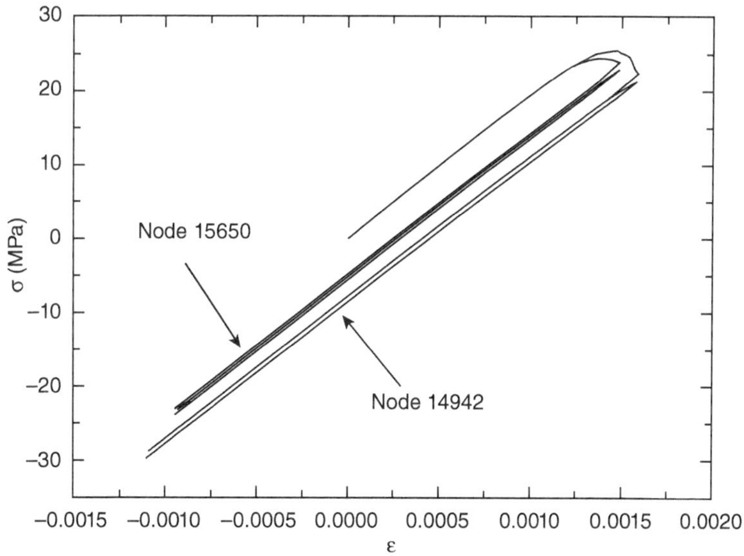

14.12 Cyclic behavior of two critical locations.

The temperature field of the vane with TBC is shown in Fig. 14.11. Generally, failure will occur at the top coat. Nodes 14942 and 15650 at the leading and trailing edge were selected as the critical locations for the vane with TBC. The ceramic stress–strain behavior was calculated for two cycles as shown in Fig. 14.12. After the second cycle, the cyclic behavior is stable.

According to the fatigue model and Miner's law, the predicted life at the leading and trailing edge was 334 h and 255 h, respectively.

14.4 Future trends

1. Life prediction methodology for the failure of the PS TBC under thermal fatigue loading condition has been studied in the previous work. There are two primary goals in developing semi-empirical models: (i) to include correct variables such as mechanical strain range; (ii) to quantify models for some processes, especially with regard to oxidation behavior. The weight gained and thicknesses at high temperature were adopted by the NASA life model to describe oxidative damage. However, this describes only the diffusion of Al in bond coat towards the bond coat surface. In order to quantify oxidation damage totally, it is necessary to include the diffusion of Al towards the substrate. Our results confirm that microstructural evolution is controlled by the diffusion of Al towards the substrate as well as towards the TBC. A mathematical model of Al depletion has now been presented, which can provide a basis to develop a more accurate life prediction model in further work.
2. For most thermal fatigue experiments, inductive heating was selected for specimens with TBC. During the heating of the specimens, it is very difficult to detect the damage evolution online because the temperature gradient decreases from the outer to the inner edges of the specimens. Although acoustic emission (AE) technology has been used for online damage evaluation and measurement, operating environments are often very noisy and the acoustic emission signals are usually very weak. Thus, signal discrimination can be very difficult, yet it is extremely important for successful applications. The next step is likely to be the design of flame heating methods to reduce acoustic noise.
3. Based on the failure mechanism, the radial strain range was defined as an important parameter in the NASA life model. Presently, effective strain range is defined as axial strain, coupled shear and radial strain, respectively. In the future, more effective use will need to be made of strain range in life prediction.
4. For blades and vanes with APS or EB–PVD coating, different scales means that it is very difficult to measure the strain range at the critical location. Because sub-model technology comprises a great many meshes for a part of a vane and blade with TBC, it should be further studied to find how to calculate the stress–strain field at a critical location.

14.5 Conclusion

The TBC generally consists of a multi-layered structure. Every layer has different constitutive laws because the top coat is brittle material and the bond coat and substrate are ductile material. Viscoplasticity law for top coat and substrate has been developed and integrated in FEM code. The important parameters of the top coat, such as Young's modulus and thermal conductivity, have been tested and obtained at high temperature.

Generally, it has been shown in our work that the failure mechanism of APS TBC is mode I, mode II or mixed. The shear and axial strain range were selected as the parameters for the life model. Compared with the experimental data, the predicted life is considered adequate. For EB–PVD coating, the result is better. However, using those models for blade and vane, many important factors need to be studied further.

14.6 References

1. ZHANG H., SUN K., XU Q. (2009), 'Preparation and thermal conductivity of $Sm_2(Zr_{0.6}Ce_{0.4})_2O_7$ ceramic', *Journal of Materials Engineering and Performance*, Published online.
2. SHEFFLER K. (1988), 'Current status and future trends in turbine application of thermal barrier coatings', *ASME Journal of Engine for Gas Turbine and Power*, **110**, 605–609.
3. LI W., THOMAS S. (2004), 'An analysis of the indentation test to determine the interface toughness in a weakly bonded thin film coating–substrate system', *Acta Materialia*, **52**, 2989–2999.
4. GOWARD G. (1988), 'Progress in coatings for gas turbine airfoils', *Surface Coating Technology*, **156**, 73–79.
5. TRAEGER F., AHRENS M. (2003), 'A life time model for ceramic thermal barrier coatings', *Materials Science and Engineering*, **A358**, 255–265.
6. RÖSLER J., BÄKER M., VOLGMANN M. (2001), 'Stress state and failure mechanisms of thermal barrier coatings: role of creep in thermally grown oxide', *Acta Materialia*, **49**(18), 3659–3670.
7. CHABOCHE J.L. (1989), 'Constitutive equations for cyclic plasticity and cyclic viscoplasticity', *International Journal of Plasticity*, **5**, 247–302.
8. XIE W., WALKER K.P., JORDAN E.H. (2003), 'Implementation of a viscoplastic model for a plasma sprayed ceramic thermal barrier coating', *Journal of Engineering Materials and Technology*, **125**, 200–207.
9. JANOSIK L.A., DUFFY S.F. (1998), 'A viscoplastic constitutive theory for monolithic ceramics-I', *ASME Journal of Engineering for Gas Turbines and Power*, **120**, 155–161.
10. QI H., MA H. (2008), 'Thermal fatigue of thermal barrier coatings by atmospheric plasma spraying', *Journal of Key Engineering Materials*, **385–387**, 405–408.
11. HAN S.L., LI S.X., SMITH D.J. (2001), 'Comparison of phenomenological and crystallographic models for single crystal nickel base superalloys I analytical identification', *Mechanics of Materials*, **33**, 251–266.

12. GUNTHER H. (1990), 'Comparison of the uniaxial behaviour of the inelastic constitutive models of Miller and Walker by numerical experiments', *International Journal of Plasticity*, **6**, 189–206.
13. FREED A.D., WALKER K.P. (1993), 'Viscoplasticity with creep and plasticity bounds', *International Journal of Plasticity*, **9**, 213–242.
14. RYBICKI E.F., SHADLEY J.R. (1995), 'In situ evaluations of Young's modulus and Poisson's ratio using a cantilever beam specimen', in: Berndt C.C. and Sampath S. (eds), *Proceedings of the 8th National Thermal Spray Conference*, ASM, Materials Park, OH. 409–414.
15. POMEROY M.J. (2005), 'Coatings for gas turbine materials and long term stability issues', *Materials and Design*, **26**, 223–231.
16. JING B.-K., HIDEAKI M. (2005), 'Influence of porosity on hardness and Young's modulus of nanoporous EB–PVD TBCs by nanoindentation', *Materials Letters*, **59**, 3462–3466.
17. TAO L., CHEN F., DING H. (2005), 'Properties of plasma sprayed NiCrAlY + (ZrO_2 + Y_2O_3) coating on refractory steel surface', *Journal Central South University of Technology*, **12**, 246–249.
18. QI H., LI R., YANG X. (2007), 'Thermal fatigue life of thermal barrier coatings', *Journal of Rare Earths*, **25**, 370–374.
19. LI C. (2002), 'Relationship between the microstructure and properties of thermally sprayed deposits', *Journal of Thermal Spray Technology*, **11**, 365–374.
20. STRANGMAN T., RAYBOULD D. (2007), 'Damage mechanisms, life prediction, and development of EB-PVD thermal barrier coatings for turbine airfoils', *Surface and Coatings Technology*, **202**, 658–664.
21. DU Q., LIU X. (2005), 'Thermal barrier coating – thermal conductivity measurement and methods to reduce the thermal conductivity', *Journal of Functional Materials*, **36**, 1100–1106.
22. MCPHERSON R. (1984), 'A model for the thermal conductivity of plasma-sprayed ceramic coatings', *Thin Solid Films*, **112**, 89–95.
23. BOSE S. (2007), *High Temperature Coatings*, Oxford, UK, Butterworth-Heinemann.
24. MILLER R.A. (1988), *Life Modeling of Thermal Barrier Coatings for Aircraft Gas Turbine Engines*, NASA Technical Memorandum 100283, National Aeronautics and Spare Administration, Washington DC.
25. EVANS A.G., MUMM D.R., HUTCHINSON J.W. (2001), 'Mechanisms controlling the durability of thermal barrier coatings', *Progress in Materials Science*, **46**, 505–553.
26. BUSSO E.P., LIN J., SAKURAI S. (2001), 'A mechanistic study of oxidation-induced degradation in a plasma-sprayed thermal barrier coating system. Part I: model formulation', *Acta Materialia*, **49**, 1515–1528.
27. BUSSO E.P., LIN J., SAKURAI S. (2001), 'A mechanistic study of oxidation-induced degradation in a plasma-sprayed thermal barrier coating system, Part II: life prediction model', *Acta Materialia*, **49**, 1529–1536.

15

New materials, technologies and processes in thermal barrier coatings

H. XU and J. WU, Beihang University, China

Abstract: To meet the demand for advanced turbine engines, new materials and new manufacturing technologies for thermal barrier coatings (TBCs) applications heve been sought. Ceramic materials with high operating temperature capabilities and other excellent thermo-physical and mechanical properties have been investigated by materials researchers, with the purpose of replacing the currently used yttria stabilized zirconia (YSZ). Also, several novel processing methods have been proposed recently by researchers to develop more reliable and durable TBCs.

Key words: thermal barrier coatings (TBCs), turbine engines, temperature capability, thermo physical property.

15.1 Introduction

Thermal barrier coatings (TBCs) are widely used in gas turbines to increase the turbine inlet temperature (TIT) or to reduce cooling requirements, thus achieving higher engine efficiency, lowering emissions and increasing performance goals.[1–5] Currently, commercial standard TBCs are typically two-layered structures, consisting of a ceramic top coat and a metallic bond coat deposited onto a superalloy substrate. The actual thermal barrier is the ceramic top coat with the prime function of reducing the transfer of heat to the metallic substrate. Owing to its excellent mechanical, chemical and thermal properties, 7–8 wt% (4–4.5 mol%) yttria stabilized zirconia (YSZ) is the primary industrially employed TBC, usually applied either by plasma spraying (PS) or electron beam physical vapor deposition (EB–PVD), producing characteristic microstructures with certain desirable attributes.[6–10] While PS leads to the pores between the splats being oriented parallel to the substrate surface, revealing lower thermal conductivity (0.8–1.1 $W \cdot m^{-1}K^{-1}$), the columnar grain microstructure created by the EB–PVD process contains channels between the columns and pores within the grains which are oriented perpendicular to the substrate surface, exhibiting a higher thermal conductivity (1.5–1.9 $W \cdot m^{-1} \cdot K^{-1}$). Conversely, the columnar distribution of the grains allows an increase strain tolerance and, hence, EB–PVD TBCs present greater durability compared to PS TBCs.

Nevertheless, the YSZ coatings, together with a considerable increase in these effects at 1200 °C and above, which affects the reliability and service life of the TBCs, suffer from both phase transformation and sintering, even at temperatures as low as 1100 °C.[11–13] Therefore, different strategies are being investigated to replace the current YSZ coatings in order to develop more durable and reliable TBCs.[14–17] The approaches can be broadly classified as chemically modified YSZ,[18–30] novel alternate low thermal conductivity (κ) materials and microstructurul modifications. Additionally, advanced processing technologies to produce TBCs with tailored microstructure are also essential. They will be addressed in more detail below.

15.2 Chemically modified yttria stabilized zirconia (YSZ)

A key physical property of the ceramic top coat is its thermal conductivity (κ). Use of an appropriate dopant introduces vacancies due to the different valences of the host atoms. It also strains the lattice because of different atomic sizes. Both of these phenomena reduce the mean free path of the phonon scattering. Instead of one (yttria), the addition of two or more dopants to stabilize zirconia, i.e. chemically modified YSZ, is then expected to incorporate a substantial amount of vacancies, providing an efficient source of scattering phonons with lower κ.

Schulz *et al.* prepared ZrO_2–25 wt% CeO_2–2.5 wt% Y_2O_3 (CeSZ) TBCs by EB–PVD. The CeSZ has a lower thermal conductivity than YSZ and, at the same time, presents good corrosion resistance and superior phase stability at high temperature. Cyclic tests carried out in a furnace at 1100 °C showed a higher thermocyclic resistance and a longer lifetime of CeSZ TBC compared to YSZ TBC.[24–25]

Nicholls *et al.* evaluated dopants including nickel oxide (NiO), neodymia (Nd_2O_3), gadolinia (Gd_2O_3), erbia (Er_2O_3) and ytterbia (Yb_2O_3) as both ternary and quaternary additions. It was found that Gd_2O_3, Nd_2O_3, Er_2O_3 and Yb_2O_3 lowered κ of zirconia-4.5 mol% yttria (8 wt% yttria) when added at a 4 mol% level. NiO, the only divalent oxide addition used, produced only a small decrease. Er_2O_3 (4 mol%) reduced κ by 25% and colored the ceramic pink; Nd_2O_3 (4 mol%) lowered κ by 42% and colored the ceramic blue. The best system, Gd_2O_3, resulted in a thermal conductivity of 0.88 $W{\cdot}m^{-1}{\cdot}K^{-1}$ (a 47% reduction), achieving thermal conductivities comparable to PS TBCs in an EB–PVD system. In all cases, the ceramic crystal structure remained t prime phase, and the modified EB–PVD TBCs had a columnar microstructure.[26,28]

Zhu and Miller developed the multiple co-doping of TBCs, being designated as 'defect cluster systems' and typically comprising Y plus two other cations, one smaller (Yb, Sc) and one larger (Sm, Nd, Gd). They found that

YSZ and YSZ with one or more dopant show typically higher κ than the coatings of ZrO_2–Y_2O_3 with paired dopant additions. The total dopant concentration affects κ as well. A low κ regime was observed for the PS TBCs in the range 6–15 mol% total dopant concentration and about 10 mol% for the EB–PVD coaings. The multicomponent oxide clustering coatings were thermally cycled at 1160 °C and showed better cyclic lives than YSZ at the given dopant concentrations. The coating cyclic life generally decreases with increase in the total concentration. It was proposed that these added dopant oxides effectively promote the creation of thermodynamically stable, highly defective lattice structures with essentially immobile defect clusters and/or nanoscale ordered phases, thus reducing κ and improving coating sintering resistance.[29,30]

15.3 Alternate low thermal conductivity (κ) materials

15.3.1 Fluorite oxides

Fluorite oxides as candidates of TBCs mainly include HfO_2, CeO_2 and ThO_2. Although CeO_2 is not a practical choice because of its volatilization, reduction of CeO_2 into Ce_2O_3 and the accelerated sintering rate of the coating at elevated temperatures,[31–33] it has higher coefficient of thermal expansion (CTE) and lower (κ) than YSZ. Fluorite lanthanum cerate ($La_2Ce_2O_7$), which is a solid solution of La_2O_3 in CeO_2, does not present phase transformation after long-term annealing at 1400 °C and, at the same time, has a large TEC that is close to that of the bond-coat layer even at high temperature. Thermal cycling testing indicated that the single $La_2Ce_2O_7$ (LC) coating has a thermal cycling life comparable to the conventional YSZ coating and in addition the $La_2Ce_2O_7$/8YSZ double-ceramic-layer (DCL) systems showed improved thermal cycling life.[34–36]

HfO_2-containing compositions have been widely investigated, both ZrO_2-based and ZrO_2-free materials.[37–39] It has been found that HfO_2–Y_2O_3 TBCs display thermal cycling resistance comparable to 7YSZ, but at higher Y_2O_3 compositions of up to 27 wt% when the crystal structure is fully cubic phase.[38] The lowest thermal conductivity determined at 1316 °C has been obtained for ZrO_2-free 25 wt% Y_2O_3-stabilized HfO_2, with a value of 1.1 $W \cdot m^{-1} \cdot K^{-1}$, which also has a low rate of conductivity increase, and good sintering resistance, as compared with YSZ.[39]

15.3.2 Zirconate pyrochlore oxides ($RE_2Zr_2O_7$)

Rare earth (RE) zirconate pyrochlores ($RE_2Zr_2O_7$) are of interest for use as TBCs due to their lower thermal conductivities than YSZ and thermal stability at high temperatures.[16,40–43] It has been found that the thermal conductivities at temperatures between 700 °C and 1200 °C for zirconates of Gd, Eu, Sm, Nd and La are in a range 1.1~1.7 $W \cdot m^{-1} \cdot K^{-1}$, which is significantly below that of 7YSZ.[16,42,44–47] However, their CTEs are much lower than those of the metallic bond coat materials, which induced a reduction in the TBC thermal cycling properties. A double and graded layered TBC structure of $La_2Zr_2O_7$/YSZ has been developed to overcome that problem, offering the potential for TBCs at temperatures higher than 1250 °C.[16,48,49]

15.3.3 Lanthanum hexaluminate ($LaMgAl_{11}O_{19}$, LHA)

Lanthanum hexaluminate ($LaMgAl_{11}O_{19}$, LHA), which displays the magnetoplumbite structure, crystallizes in the habit of platelets. The magnetoplumbite structure is characterized by the highly charged La^{3+} cation located in an oxygen position in the hexagonal close-packed structure of oxygen ions. Ion diffusion is strongly suppressed vertical to the crystallographic *c*-axis, thus hindering sintering densification. The low thermal conductivity of LHA is caused by its microstructure, i.e. a random arrangement of LHA platelets that build up a well-balanced microporous coating and the insulating properties of the material with its crystallographic feature. LHA is considered to be a promising competitor to YSZ for operations above 1300 °C since the crystalline phase is stable to at least 1300 °C and its platelike structure enables excellent ageing behavior at 1300 °C (eight-fold less shrinkage than that of YSZ).[50–52]

15.3.4 Nanocrystalline materials

Nanocrystalline materials offer the potential of limiting thermal conductivity by incorporating grain boundary scattering as an extrinsic phonon-scattering phenomenon. Additional benefits are provided for TBC applications because of the possibility of improving toughness and ductility compared to that of coarser-grained ceramics.[53–55]

Soyez *et al.* evaluated the thermal conductivity of nanocrystalline YSZ films as prepared by metal-organic chemical vapour deposition (MOCVD). A significant reduction in thermal conductivity in these films compared to that of bulk YSZ is observed for grain sizes smaller than the phonon mean free path, and the observed changes in thermal conductivity are more likely to be attributable to the effects of grain size than the porosity.[56]

Nanocrystalline YSZ TBCs were subsequently investigated using standard plasma-spraying technology.[57–73] It was found that nanostructured YSZ TBCs possessed higher hardness, higher CTE, lower thermal diffusivity and better thermal shock resistance compared to conventional TBCs, although the grain size tended to increase during annealing at elevated temperatures, which could result in substantial degree of sintered density. These results demonstrated potential benefits of grain refinement to the nanometer scale in TBC applications, but also indicated that extremely small grain sizes will need to be produced and stabilized against grain growth in order to obtain significant improvements in thermal property.

15.4 Microstructure modification

Considerable efforts have been made to develop EB–PVD TBCs with lower thermal conductivity and PS TBCs with improved strain tolerance by microstructural modification through advanced processing technologies. They are summarized below.

15.4.1 Electron beam physical vapor deposited thermal barrier coatings (EB–PVD TBCs)

EB–PVD TBCs have three main porosity features.[74] Columns and inter-columnar gaps (type I) lead to an in-plane anisotropy which has a significant impact on the strain tolerance of the EB–PVD TBC system.[75–79] The other two arrangements of intra-columnar pores, aligned spherical pores (type II) and random spherical pores (type III), are much more effective at impeding heat flow through the coating as they are generally inclined to the heat flow direction.[74] According to this, the changes in the parameters of the deposition process lower the thermal conductivity of TBCs by modifying the pore microstructure in the top coat. For instance, a decrease of 15% in YSZ thermal conductivity has been obtained by increasing the chamber pressure and decreasing the substrate temperature.[80] More remarkable is the 40% reduction reached with a modified microstructure based on inclined columnar growth, known as the 'zig-zag' or 'herringbone' structures, varying the vapour incidence from the normal direction.[81,82] Nevertheless, the erosion resistance decreases with the column angle and, therefore, depending on the specific application of the component, a compromise between the thermal conductivity and the erosion resistance will be necessary.

Recently, a new tailored microstructure of EB–PVD 8YSZ TBC has been proposed.[83] By periodically interrupting the condensing vapor during the deposition process to create strain fields, the phonon and photon scattering processes can be enhanced, leading to a lower thermal conductivity and a higher reflectance, respectively. The changes in the microstructure, mainly in the porosity and in the grain boundaries, lead to a reduction in thermal conductivity after 2 h testing at 1316 °C from 1.8 to 1.2 $W \cdot m^{-1} \cdot K^{-1}$ (~30% reduction), and also to an increase in the reflectance of up to 56%. Furthermore, an improvement of 100% in oxidation cyclic life as well as a better strain tolerance than the non-modified 8YSZ TBC structure has been attained. The new tailored YSZ is expected to offer the potential within the turbine industry of longer turbine component life and reduced fuel consumption.

15.4.2 Plasma sprayed (PS) TBCs

In air plasma-sprayed (APS) coatings, inter-splat pores are roughly aligned parallel to the substrate surface and are accompanied by microcracks and a fine grain size.[84] This pore orientation is highly effective for impeding the flow of heat through the coating. Therefore, APS coatings have low thermal conductivities which are usually in the range 0.8–1.1 W/mK.[14] However, this coating has less in-plane strain tolerance to accommodate the differences of strain between the metallic bond coat and the ceramic top layer during thermal cycling.[84,85] This constrains their use in low thermal cyclic environments such as land-based power generation turbines. Modifications to the APS deposition conditions have enabled dense, vertically cracked (DVC) coatings to be deposited.[86–93] The roughly equally spaced vertical cracks enable the coating to expand with the metal during heating, and they may become important for aero gas turbine applications in the years ahead. A new, potentially lower-cost, solution precursor plasma spray (SPPS) deposition process has also been recently proposed.[94–96] It is claimed to have the potential for creating more durable TBCs with reduced strain compatibility limitations compared to conventional APS TBCs.

15.5 Advanced processing technologies

In order to develop 'tailored' microstructure, a processing approach must be identified or developed for which coating architecture can be manipulated and controlled. Several alternative methods have been proposed to develop more reliable and durable TBCs.

15.5.1 Solution precursor plasma spray (SPPS)

The SPPS process was first reported and applied in the synthesis and deposition of nano-powders and deposited by Herman *et al.*[97–99] The process was then developed successfully in a joint effort by the researchers at the Institute of Materials Science and Mechanical Engineering Department at the University of Connecticut and the Inframat Corporation, making high-durability, low-thermal conductivity 7YSZ thermal barrier coatings needed for marine, aircraft and industrial engine applications.[94–96,100–105]

The SPPS process, which is schematically illustrated in Fig 15.1, is fundamentally different in that, instead of ceramic powder, an aqueous solution of zirconium and yttrium salts is injected into the plasma gas environment.[94–96,100–105] Upon entrainment into the plasma, the subsequent vaporiza-

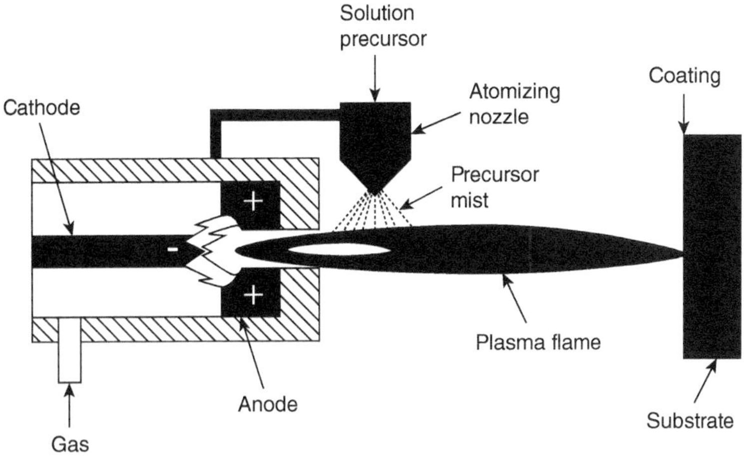

15.1 Schematic illustration of the SPPS process.[96]

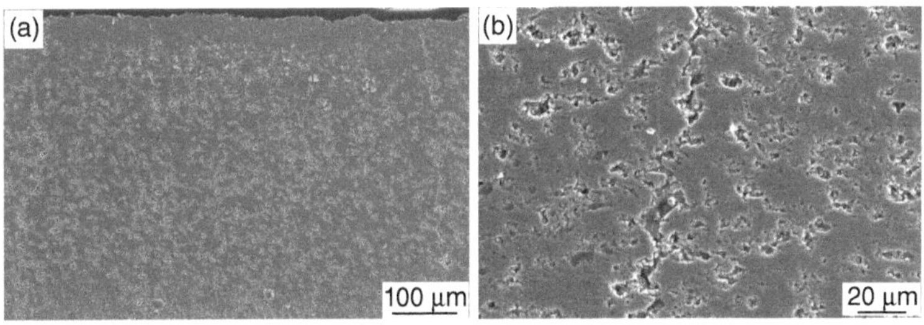

15.2 SEM micrographs of polished cross-sections of the SPPS TBC: (a) low-magnification; (b) higher-magnification.

tion of the water in the droplets and decomposition of the acetate and nitrite groups result in the formation of ceramic material, which contains the desired nano-structure in the final coating structure.

The SPSS process enables the creation of some unique features and so makes it possible to manufacture high-strength TBCs, as shown in Fig. 15.2. The basic characteristics of the SPPS TBCs are: uniformly dispersed porosity, absence of unwanted coarse splat boundaries and straining tolerance providing evenly spaced vertical cracks. Thermal cyclic durability tests of TBCs deposited by APS, SPPS and EB–PVD demonstrated that the spallation times of SPPS coatings were 2.5 times longer than those by EB–PVD coatings and 1.5 times longer than the APS coatings under the same conditions. During these tests, vertical cracks and ultra-fine splats in the SPPS coatings were found to remain stable.

15.5.2 Electron beam directed vapor deposition (DVD)

Some of the limitations of the traditional EB–PVD process conducted in vacuum include low coating rates, difficulties in controlling and modifying microstructure and deposition limited to the line of sight.[106,107] The last limitation is alleviated by rotation of the substrate so that every point on the surface to be coated has a chance to receive the deposit. Many engineering components, however, have design features that give rise to shadowing of some areas of the components, resulting in vapor molecules being unable to arrive at those sites. Either the shadowed areas get no coating, or the deposited coating is thin with poor column orientation.

Researchers at the University of Virginia in the USA and the Fraunhofer Institute for Electron Beam and Plasma Technology in Germany have devised a modification of the EB–PVD process called electron beam directed vapor deposition (DVD), which increases the deposition rate significantly by improved mixing of vapors from multiple ingot sources. Additionally, the process can modify the microstructure, vary porosity and direct some of the vapor to the other shadowed areas of the depositing surface.

In the DVD process, material is vaporized by using a continuous high-voltage/medium-power (60 kV/10kW) e-beam gun (modified to function in a low vacuum environment by incorporating differential pumping of the gun column and the use of a very small (~3 mm) e-beam exit opening). The vapor is entrained in a carrier gas stream and deposited onto a substrate at a high rate (>10 $\mu m \cdot min^{-1}$). Atoms in the carrier gas flow generate collisions with the vapor molecules. The collisions keep the vapor stream directed at the surface, including the shadowed regions to be coated. The collimating effect of the carrier gas has also been found to increase deposition rate. During processing of DVD, the carrier gas stream is created by a rarefied, inert gas supersonic expansion through a nozzle. The speed and flux of the gas atoms entering the chamber, the nozzle parameters and the operating chamber pressure can all be significantly varied to lead to a wide range of processing conditions. These changes are expected to affect the morphology of porous coating structures.

The supersonic carrier gas stream, which is maintained by achieving a high upstream pressure (i.e. the gas pressure prior to its entrance into the processing chamber) and a lower downstream pressure (or chamber pressure), is critical to the process. The ratio of the pressure from upstream to downstream along with the size and shape of the nozzle opening control the speed of the gas entering the chamber. The molecular weight of the carrier gas (compared to that of the vapor) and the speed of the carrier gas control its effectiveness in redirecting the vapor atoms via binary collisions towards the substrate. High molecular weight carrier gases are preferred, but they have a high electron scattering cross-section. Helium is therefore used for entrainment. High gas speeds are thus required to redirect the vapor toward the substrate with a minimal amount of gas flow. High gas speeds also facilitate the formation of a wall jet (i.e. a component of the gas speed generally aligned parallel to the substrate surface) through the interaction of the carrier gas jet with the substrate.

Compared to conventional EB–PVD processing, two major differences exist in DVD conditions. The first is the use of an elevated chamber pressure. Typical chamber pressures in DVD are between 0.1 and 1 Torr which have higher order of magnitude than those found in EB–PVD. The second is the high-speed carrier gas stream used to alter the direction and speed of the vapor species. The presence of the carrier gas strongly affects the incident angle of the vapor species during deposition and can thus modify the growing process. Recently, e-beam multisource evaporation, a plasma activation system to ionize the vapor and the gas stream, together with electrical biasing of the substrate were incorporated into the DVD process, which opened new possibilities for the creation of more desirable film structures and compositions.

15.6 Future trends

To exploit the full potential of TBCs, further studies are required for substantial improvement of the TBC system. Up to now many of the newly developed EB–PVD TBCs have been introduced into service or ground testing with only an underlying standard YSZ ceramic layer. Future research must generate an increasing understanding of the processing–property interplay of these systems to provide reliable data for improved servce life.

15.7 References

1. R.A. MILLER, *Surf. Coat. Technol.*, **30** (1987) 1.
2. R.A. MILLER, *J. Therm. Spray Technol.*, **6**(1) (1997) 35.
3. R.L. JONES, in *Metallurgical and Ceramic Coatings*, K. H. Stern, Ed., Chapman and Hall, London UK, 1996, 194.
4. N.P. PADTURE, M. GELL, E.H. JORDAN, *Science*, **296** (2002) 280.
5. S. BOSE, in *High Temperature Coatings*, S. Bose, Ed., Butterworth-Heinemann, Elsevier, Oxford, UK, 2007, 155.
6. S. STECURA, *Adv. Ceram. Mater.*, **1**(1) (1986) 68.
7. D.P.H. HASSELMAN, L.F. JOHNSON, L.D. BENTSEN, R. SYED, H.L. LEE, M.V. SWAIN, *Am. Ceram. Soc. Bull.*, **66** (1987) 799.
8. S. BOSE, J. DEMASI, *J. Therm. Spray Technol.*, **6**(1) (1997) 99.
9. W.A. NELSON, R.M. ORENSTEIN, *J. Therm. Spray Technol.*, **6**(2) (1997) 176.
10. J. THORNTON, *Mater. Sci. Forum*, **22** (1998) 159.
11. J. ILAVSKY, J.K. STALICK, *Surf. Coat. Technol.*, **127**(2–3) (2000) 120.
12. U. SCHULZ, *J. Am. Ceram. Soc.*, **83** (2000) 904.
13. D. R. CLARKE, C. G. LEVI, *Annu. Rev. Mater. Res.*, **33** (2003) 383.

14. J.R. NICHOLLS, *Mater. Res. Soc. Bull.*, **28**(9) (2003) 659.
15. C.G. LEVI, *Curr. Opin. Solid State Mater. Sci.* **8**(1) (2004) 77.
16. R. VASSEN, X. CAO, F. TIETZ, D. BASU, D. STÖVER, *J. Am. Ceram. Soc.* **83**(8) (1999) 2023.
17. D. STÖVER, C. FUNKE, *J. Mater. Process. Technol.*, **92–93** (1999) 195.
18. R.L. JONES, R.F. REIDY, D. MESS, *Surf. Coat. Technol.*, **82** (1996) 70.
19. R.L. JONES, D. MESS, *Surf. Coat. Technol.*, **86–87** (1996) 94.
20. P.D. HARMSWORTH, R. STEVENS, *J. Mater. Sci.*, **26** (1991) 3991.
21. D.J. KIM, *J. Am. Ceram. Soc.*, **73**(1) (1990) 115.
22. S. RAGHAVAN, H. WANG, W.D. PORTER, *et al.*, *Acta Mater.*, **49** (2001) 169.
23. R. VAßEN, N. CZECH, W. MALLÉNER, *et al.*, *Surf. Coat. Technol.*, **141** (2001) 135.
24. U. SCHULZ, K. FRITSCHER, M. PETERS, *Surf. Coat. Technol.*, **82** (1996) 259.
25. C. LEYENS, U. SCHULZ, K. FRITSCHER, *Mater. High Temp.*, **20** (2003) 475.
26. Y.A. TAMARIN, E.B. KACHANOV, S.V. ZHERZDEV, *Mater. Sci. Forum*, **251–254** (1997) 949.
27. J.R. NICHOLLS, K.J. LAWSON, A. JOHNSTONE, D.S. RICKERBY, *Mater. Sci. Forum*, **369** (2001) 595.
28. J.R. NICHOLLS, K.J. LAWSON, A. JOHNSTONE, D.S. RICKERBY, *Surf. Coat. Technol.*, **151–152** (2002) 383.
29. D.M. ZHU, R.A. MILLER, *Ceram. Eng. Sci. Proc.*, **23** (2002) 457.
30. D.M. ZHU, Y.L. CHEN, R.A. MILLER, *Ceram. Eng. Sci. Proc.*, **24** (2003) 525.
31. S. SODEOKA, M. SUZUKI, T. INOUE, *et al.*, in: *Proceedings of the 9th National Thermal Spray Conference: Thermal Spray-Practical Solutions for Engineering Problems* C. C. Berndt Ed., ASM International, Materials Park, OH, 1996, 295.
32. J. THORNTON, A. MAJUMDAR, G. MCADAM, *Surf. Coat. Technol.*, **94/95** (1997) 112.
33. A. ROUANET, G. PERAUDEAU, F. SIBIEUDE, *et al.*, in: *The 14th international thermal spray conference: Thermal Spray-Current Status and Future Trends* A. Ohmori Ed., High Temperature Society of Japan, Osaka, 1995, 507.
34. X. CAO, R. VASSEN, W. FISCHER, *et al.*, *Adv. Mater.*, **15** (17) (2003) 1438.
35. W. MA, S.K. GONG, H.B. XU, X.Q. CAO, *Scripta Mater.*, **54** (2006) 1505.
36. W. MA, S.K. GONG, H.B. XU, X.Q. CAO, *Surf. Coat. Technol.*, **200** (2006) 5113.
37. D.M. ZHU, R.A. MILLER, *Surf. Coat. Technol.*, **108–109** (1998) 114.
38. K. MATSUMOTO, Y. ITOH, T. KAMEDA, *Sci. Technol. Adv. Mater.*, **4** (2003) 153.
39. J. SINGH, D. E. WOLFE, R. A. MILLER, *et al.*, *J. Mater. Sci.*, **39** (2004) 1975.
40. M.J. MALONEY, US Patent 6117560, 2000.
41. M.J. MALONEY, European Patent EP 0 992 603 A1, 2000.
42. J .WU, X. WEI, N.P. PADTURE, *et al.*, *J. Am. Ceram. Soc.*, **85** (2002) 3031.
43. M.J. MALONEY, US Patent 20030049470 A1, 2003.
44. G. SURESH, G. SEENIVASAN, M.V. KRISHNAIAH, P.S. MURTI, *J. Nucl. Mater.* **249** (1997) 259.
45. G. SURESH, G. SEENIVASAN, M.V. KRISHNAIAH, P.S. MURTI, *J. Alloys Compd.* **269** (1998) L9.
46. LEHMANN, H., *et al.*, *J. Am. Ceram. Soc.*, **86** (2003) 1338.
47. B. SARUHAN, P. FRANCOIS, K. KRITSCHER, U. SCHULZ, *Surf. Coat. Technol.*, **182** (2004) 175.
48. D. STÖVER, G. PRACHT, H. LEHMANN, *et al.*, *J. Therm. Spray Technol.*, **13** (2004) 76.
49. R. VASSEN, F. TRAEGER, D. STÖVER, *Int. J. Appl. Ceram. Technol.*, **1** (2004) 351.
50. R. GADOW, M. LISCHKA, *Surf. Coat. Technol.*, **151** (2002) 392.
51. X.Q. CAO, R. VASSEN, D. STÖVER, *J. Euro. Ceram. Soc.*, **24** (2004) 1.

52. X.Q. CAO, Y.F. ZHANG, J.F. ZHANG, *et al.*, *J. Euro. Ceram. Soc.*, **28** (2008) 1979.
53. M. GELL, *Nanostruct. Mater.*, **6** (1995) 997.
54. M. GELL, *Mater. Sci.Eng. A*, **204** (1995) 246.
55. U. BETZ, H. HAHN, *Nanostruct. Mater.*, **12** (1999) 911.
56. G. SOYEZ,. A. EASTMAN, L.J. THOMPSON, G.R. BAI, P.M. BALDO, A.W. MCCORMICK, R.J. DIMELFI, A.A. ELMUSTAFA, M.F. TAMBWE, D.S. STONE, *Appl. Phys. Lett.*, **77**(8) (2000) 1155.
57. R.S. LIMA, A. KUCUK, C.C. BERNDT, *Surf. Coat. Technol.*, **135** (2001) 166.
58. R.S. LIMA, A. KUCUK, C.C. BERNDT, *Mater. Sci. Eng. A*, **313** (2001) 75.
59. R.S. LIMA, A. KUCUK, U. SENTURK, C.C. BERNDT, *J. Therm. Spray Technal.*, **15**(1) (2001) 150.
60. H. CHEN and C. X. DING, *Surf. Coat. Technol.* **150**(1) (2002) 31.
61. R.S. LIMA, A. KUCUK, C.C. BERNDT, *Mater. Sci. Eng. A*, **327** (2002) 224.
62. H. CHEN, X.M. ZHOU, C.X. DING, *J. Euro. Ceramal. Soc.*, **23**(9) (2003) 1449.
63. H. CHEN, C.X. DING, S. LEE, *Mater. Sci. Eng. A*, **361**(1–2) (2003) 58.
64. C.G. ZHOU, N. WANG, Z.B. WANG, *Scripta Materi.*, **51**(9) (2004) 945.
65. R.S. LIMA, S.E. KRUGER, G. LAMOUCHE, B.R. MARPLE, *J. Therm. Spray Technol.* **14**(1) (2005) 520.
66. B. LIANG, C.X. DING, *Surf. Coat. Technol.*, **191** (2005) 267.
67. B. LIANG, C.X. DING, *Surf. Coat. Technol.*, **197**(2–3) (2005) 185.
68. B. LIANG, C. DING, H. LIAO, C. CODDET, *Surf. Coat. Tech.*, **200** (2006) 4549.
69. R.S. LIMA, B.R. MARPLE, *J. Therm. Spray Technol.*, **16**(1) (2007) 40.
70. C.G. ZHOU, N. WANG, H.B. XU, *Mater. Sci. Eng. A*, **452–453** (2007) 569.
71. N. WANG, C.G. ZHOU, S.K. GONG, H.B. XU, *Ceramics International*, **33** (2007) 1075.
72. R.S. LIMA, B.R. MARPLE, *Mater. Sci. Eng. A*, **485** (2008) 182.
73. J. WU, H.B. GUO, L. ZHOU, L. WANG, S.K. GONG, *Asian Thermal Spray Conference*, Xi'an China, 22–24 Oct. 2009.
74. T.J. LU, C.G. LEVI, H.N.G. WADLEY, A.G.EVANS, *J. Am. Ceram. Soc.*, **84** (2001) 2937.
75. S.G. TERRY, C.G. LEVI, *J. Am. Ceram. Soc.*, **84** (2001) 2937.
76. U. SCHULZ, K. FRITSCHER, C. LEYENS, M. PETERS, *Ceram. Eng. Sci. Proc.*, **22(**41) (2001) 347–356.
77. K. FRITSCHER, F. SZIICS, U. SCHULZ, *et al.*, *Ceram. Eng. Sci. Proc.*, **23**(4) (2002) 341–352.
78. C.A. JOHNSON. J.A. RUUD, R. BRUCE, D. WORTMAN, *Surf. Coat. Technol.*, **108–109** (1998) 80.
79. U. SCHULZ, K. FRITSCHER, C. LEYENS, M. PETERS, *J. Eng. for Gas Turbines Power*, **124**(2) (2002) 229–234.
80. D.V. RIGNEY, A.F. MARICOCCHI, A.J. WORTMANN, R.W. BRUCE, J.D. RIGNY, U.S. Patent 6 447 854, 2002.
81. U. SCHULZ, B. SARUHAN, K. FRITSCHER, C. LEYENS, *Int. J. Appl. Ceram. Technol.* **1** (2004) 302.
82. W. BEELE, G. MARIJNISSEN, E. VERGELDT, E. HURON, R. DAROLIA, in *Turbine Forum, Advanced Coatings for High Temp enraptures*, Nice, 17–19 Apr. 2002, 1.
83. D.E. WOLFE, J. SINGH, R.A. MILLER, J.I. ELDRIDGE, D.M.ZHU, *Surf. Coat. Technol.* **190** (2005) 132.
84. J.R .NICHOLLS, M.J. DEAKIN, D.S. RICKERBY, *Wear*, **233–235** (1999) 352.
85. T.E. STRANGMAN, *Thin Solid Films* **127** (1985) 93.
86. T.A. TAYLOR, U.S. Patent 5073433, 1991.

87. T.A. TAYLOR, D.L. APPLEBY, A.E. WEATHERILL, J. GRIFFITHS, *Surf. Coat. Technol.*, **43/44** (1990) 470.
88. T.A. TAYLOR, in: N. B. Dahotre, R. J. Gaster, R. A. Hill and O. O. Popoola (Eds), *Heat Treating and Surface Engineering: Proceedings of the 22nd International Surface Engineering Conference*, ASM International, Materials Park, OH, 2003, 520.
89. R. VASSEN, F. TRAEGER, D. STOVER, *J. Therm. Spray. Technol.*, **13**(3) (2004) 396.
90. H.B. GUO, S. KURODA, H. MURAKAMI, *J. Am. Ceram. Soc.*, **89**(4) (2006) 1432.
91. H.B. GUO, S. KURODA, H. MURAKAMI, *Thin Solid Films*, **506–507** (2006) 136.
92. P. BENGTSSON, J. WIGREN, in: P.J. Maziasz, I.G. Wright *et al.* (Eds), *Gas Turbine Materials Technology: Conference Proceeding from ASM Materials Solutions*, ASM International, Materials Park, OH, 1999, 92.
93. H.B. GUO, R. VASSEN, D. STÖVER, *Surf. Coat. Technol.*, **186** (2004) 353.
94. L. XIE, X. MA, E.H. JORDAN, N.P. PADTURE, T.D. XIAO, M. GELL, *J. Mater. Sci.*, **39** (2004) 1639.
95. L. XIE, X. MA, E.H. JORDAN, N.P. PADTURE, T.D. XIAO, M. GELL, *Mater. Sci. Eng. A.*, **362** (2003) 204.
96. L. XIE, X. MA, E.H. JORDAN, N.P. PADTURE, T.D. XIAO, M. GELL, *Surf. Coat. Technol*, 177–178 (2004) 97.
97. J. KARTHIKEYAN, C.C. BERNDT, J. TIKKANEN, J.Y. WANG, A.H. KING, H. HERMAN, *Nanostruct. Mater.* **9**(1) (1997) 137.
98. J. KARTHIKEYAN, C.C. BERNDT, S. REDDY, J-Y. WANG, A.H. KING, H. HERMAN, *J. Am. Ceram. Soc.*, **81** (1998) 121.
99. H. HERMAN, S. SAMPATH, R. MCCUNE, *MRS Bull.*, **7** (2000) 17.
100. N.P. PADTURE, K.W. SCHLICHTING, T. BHATIA, A. OZTURK, B. CETEGEN, E.H. JORDAN, M. GELL, S. JIANG, T.D. XIAO, P.R. STRUTT, E. GARCIA, P. MIRANZO, M.I. OSENDI, *Acta Mater.*, **49** (2001) 2251.
101. E.H. JORDAN, L. XIE, C. MA ,M. GELL, N. PADTURE, B. CETEGEN, J. ROTH, T. D. XIAO, P. E. C. BRYANT, *J. Therm. Spray*, **13**(1) (2004) 57.
102. A. JADHAV, N. PADTURE, F. WU, E. JORDAN, M. GELL, *Materi. Sci. Engi. A*, **405** (2005) 313.
103. X. MA, F. WU, J. ROTH, M. GELL, E. JORDAN, *Surf. Coat. Technol.*, **201** (2006) 3343.
104. X.Q. MA, T.D. XIAO, J. ROTH, L.D. XIE, E.H. JORDAN, N.P. PADTURE, M. GELL, X.Q. CHEN, J.R. PRICE, in: D. von Hofe (Ed.), *Thermal Spray 2004: Advances in Technology and Application*, ASM International, Materials Park, OH, 2004, 1103.
105. M. GELL, F. WU, E.H. JORDAN, N.P. PADTURE, B.M. CETEGEN, L. ZIE, A. OZTURK, E. CAO, A. JADHAV, D. CHEN, X. MA, in: *Proceedings of 2005, ASME Turbo Expo 2005*, ASME New York, 905.
106. J.F. GROVES, G. MATTAUSCH, H. MORGNER, D.D. HAAS, H.N.G. WADLEY, *Surf. Eng.*, **16**(6) (2000) 461.
107. Z. YU, D.D. HASS, H.N.G. WADLEY, *Mater. Sci. Eng. A*, **394**(1–2) (2005) 43.

Index

Kolibri 360 GmbH
Ferdinand-Jühlke-Straße 7
99095 Erfurt, Deutschland
produktsicherheit@kolibri360.de

Global Logistics Network Modelling and Policy

Quantification and Analysis for International Freight

Edited by

Ryuichi Shibasaki
Hironori Kato
Cesar Ducruet

ELSEVIER

Elsevier
Radarweg 29, PO Box 211, 1000 AE Amsterdam, Netherlands
The Boulevard, Langford Lane, Kidlington, Oxford OX5 1GB, United Kingdom
50 Hampshire Street, 5th Floor, Cambridge, MA 02139, United States

Library of Congress Cataloging-in-Publication Data
A catalog record for this book is available from the Library of Congress

British Library Cataloguing-in-Publication Data
A catalogue record for this book is available from the British Library

ISBN: 978-0-12-814060-4

Publisher: Joe Hayton
Acquisitions Editor: Brian Romer
Editorial Project Manager: Megan Ashdown
Production Project Manager: Swapna Srinivasan
Cover Designer: Matthew Limbert

Typeset by SPi Global, India

Contents

Contributors

César Ducruet Centre National de la Recherche Scientifique

Masahiko Furuichi International Association of Ports and Harbors (IAPH), The University of Tokyo

David Guerrero AME-SPLOTT, Univ Gustave Eiffel, IFSTTAR

Hidekazu Itoh Kwansei Gakuin University

Takashi Kadono NEWJEC Inc.

Hironori Kato The University of Tokyo

Taiji Kawakami Ministry of Land, Infrastructure, Transport and Tourism, Japan

Tomoya Kawasaki Tokyo Institute of Technology

Masaya Kobayashi Nippon Express Co., Ltd.

Kentaro Nishimura The University of Tokyo

Takashi Riku The University of Tokyo

Ryuichi Shibasaki The University of Tokyo

Takashi Shimada The Overseas Coastal Area Development Institute of Japan (OCDI)

Masaru Suzuki Nikken Kogaku Co., Ltd.

Satoshi Tanabe The University of Tokyo

Preface: Globalisation and global logistics

In our daily lives, we are surrounded by many goods from all over the world. My shirt, for example, is from Vietnam, my trousers are made in China, and my jacket comes from Italy. My watch is Swiss made, and my glasses are imported from the United States. The coffee cup on my desk is produced in France, and the coffee in it is made from blended beans from Brazil and Guatemala. Perhaps my laptop is assembled in my own country, but its parts are imported. In the same way, many goods are produced in various regions, transported across locations, and finally consumed at other places. This complicated production–consumption system has been facilitated by the globalised logistics system wherein products are transported internationally, in addition to the international movements of people, currency, and information.

The establishment of the global logistics system is the result of three major factors: technological innovation, infrastructure investment, and evolution of the institutional system. Containerisation was an epoch-making event that transformed the logistics industry from a labour-intensive to a capital-intensive industry (Levinson, 2006). Containerisation has facilitated complex cargo-handling tasks such as loading/discharging and trans-shipment of cargo to and from vessels at ports, upgraded safety standards, reduced damage to cargos, and also enabled an intermodal transport network connecting ships, railways, and trucks. Vessel design technology has also evolved, resulting in larger vessels over time (Rodrigue, 2017). The upward trend in the sizes of containerships is mainly motivated by economies of scale, which has led to a significant reduction in average cargo transport costs. International cargo traffic flows are also supported by sophisticated transactions of commercial information and currency. Notably, recent revolutions in information and communication technology (ICT) have enhanced the efficiency and safety of global logistics operations. These developments include sophisticatedly digitalised operating systems such as electronic data interchange (EDI) processing, radio-frequency identification (RFID) processing, and optimisation of cargo handling at automated container terminals (Saragiotis, 2019; Al-Fuqaha et al., 2015; Steenken et al., 2004). Regarding the institutional framework, trade obstacles due to traditional manual transactions at cross-border points have been gradually removed under the guidance of regional strategies. The liberalisation of international trade can be realised by trade facilitation and implementation of cross-border paperless trade, which includes simplifying required paperwork, modernising procedures, harmonising customs requirements, and introducing a single-window system (Tijan et al., 2019). Reductions in time and costs of cross-border point transactions enable nations to seamlessly connect with others, which can facilitate the process of evolution into an integrated global production chain.

Rapid globalisation with reduced transport costs has motivated global firms to optimise their manufacturing clusters located in regions with the cheapest labour and material costs. This significantly diminishes production and operating costs. This trend is further accelerated by a business model of horizontal specialisation at a global scale (Bloch, 1995). Firms that specialise horizontally identify a specific market to which it can offer a complete business solution, which may involve offering a wide range of components, products, and services to a narrow range of customer types (Williams and Aaron, 2018). Additionally, this business model has promoted the rapid development of manufacturing industries located in the markets of emerging economies. To keep updated, major logistics industry players have adapted to rapid globalisation. Shipping liner companies have established a global hub-and-spoke shipping network (Farahani et al., 2013). Tough competition amongst shipping companies has also encouraged mergers and acquisitions under horizontal integration since the 1990s (Notteboom et al., 2017). This has led to the establishment of giant shipping liners and promoted global strategic alliances amongst shipping companies (Crotti et al., 2019). Furthermore, the increased sizes of vessels require massive investment in port and/or canal facilities. Many governments have participated in the global competition (Parola et al., 2017) to construct large-scale hub ports to lead the global supply chain and earn benefits from saved transport costs by facilitating the movement of direct shipping services to and from hub ports. An interregional intermodal transport network has been formulated (Reis et al., 2013) under the international development strategies of regional bodies. Such efforts to improve efficiency in the global logistics system have accelerated international business activities.

In summary, the globalisation of the supply chain, in line with innovation in logistics and institutional systems and massive investment in freight transport infrastructure, has enabled many firms to diversify their procurement sources, which has led to lower supply costs. End users now have more options in consumption goods, whilst prices have also significantly reduced. This has improved the quality of life for people whilst increasing tax revenues for governments through the revitalisation of economic activities. The above-mentioned causality is supported by much empirical evidence, particularly on the significant associations between international trade and global GDP growth (Alcalá and Ciccone, 2004; Frankel and Romer, 1999) and positive impacts of the liberalisation of international trade on economic efficiency (Pavcnik, 2002; Bloom et al., 2016). Many studies have indicated that even in the least-developed countries, export growth could stimulate economic growth (e.g. Ghirmay et al., 2001) following two paths: increasing investments (capital accumulation) and enhancing efficiency. This has contributed to addressing poverty and other global issues, which are targeted by the sustainable development goals. The rapid development of the global logistics network has upgraded accessibility to and from landlocked regions where no seaport is available as well as remote areas located far from major markets (Faye et al., 2004). These improvements in accessibility have encouraged global firms to invest in such landlocked and remote areas whilst also promoting exports from those areas that could create more jobs, generate better salaries, and improve the quality of life of the people in these places. Similarly, enhancement of regional connectivity amongst nations enables the development of an integrated economic market, fosters regional competitions, stimulates international trade and leads to better economic growth even in less-developed regions.

Nonetheless, the negative aspects of globalisation also exist. A commonly discussed issue is its damaging effect on local economies and domestic jobs. Developed countries that outsource manufacturing to other regions to exploit cheaper labour costs could suffer from employment insecurity; developing countries could also be affected in their domestic employment although the impacts are still inconclusive (Lee and Vivarelli, 2006). This may lead to political movements favouring protectionism and isolationism (Stiglitz, 2017). Another issue relating to globalisation is the income inequality compromised for the sake of countries' economic growth. This typically indicates a 'core-periphery' structure, wherein the core contains the major wealthy and powerful countries, with countries that cannot reap the benefits of global wealth located at the periphery (Hartmann et al., 2019). Core countries settle on a diverse set of knowledge-intensive and value-added products, and peripheral developing countries specialise in exports of simple resources and labour-intensive products to higher blocks of the hierarchy (Kostoska et al., 2020). This could be regarded as neocolonialism—the dark side of globalisation (Rao, 2000). Additionally, the world system may be more vulnerable under globalisation, as supply chain disruptions could transfer to and significantly affect international trade. Such system disturbances may be caused by natural disasters, trade embargoes, or disruptive demand change (Sprecher et al., 2015). Many researchers in recent years have highlighted the resilience and vulnerability of the supply chain (Elleuch et al., 2016). The tremendous impacts of the COVID-19 pandemic in 2020 are still ongoing.

This book attempts to provide a reference for discussions on the above-mentioned issues from a global logistics system perspective by presenting a technical tool to investigate the international freight transport network and its related policies. Part 1 contains four chapters which cover introductory topics regarding the container shipping market, hinterlands, cross-border logistics, and container demand forecast. Part 2 presents the model and data, which are complemented with quantitative simulations and later applied to case studies in many regions. A macroscopic network modelling technique is employed to analyse cargo flows in international freight networks. The practical modelling approach enables us to examine the expected impacts of the logistics system changes in a realistic manner. As explained earlier, these logistics changes include technological innovation, infrastructure investment, and evolution of the institutional system. The proposed simulation model can assist in evaluating the policies intended to change the logistics system in terms with their impacts on accessibility, traffic volume, trade patterns, or share of transport modes. Their expected influences on regional/local economic performance could also be discussed if the logistics network model is integrated with another international economic simulation model. The logistics transport network assumed in this book's model covers multiple transport modes, including maritime shipping, railway, trucks, and inland waterway transport, although maritime shipping should be highlighted since its role is dominant in international trade. Part 3 demonstrates a series of cases wherein the proposed model is customised and applied to seven regions across the world. They highlight the logistics network in developing countries since the major problems in logistics systems are observed in less-developed regions. These case studies are expected to contribute to the policy debates in each region.

The unique contribution of this book is in providing a useful tool and verifying its application in various regions for the decision-making of stakeholders in logistics industries, related government authorities, and international donors concerned with global issues. We hope that our modelling approach can assist various individuals, such as supply chain and logistics professionals, university students interested in logistics and freight transport, and experts in logistics and transport planning/policy in exploring novel directions in logistics research.

References

Alcalá, F., Ciccone, A., 2004. Trade and productivity. Q. J. Econ. 119 (2), 613–646.

Al-Fuqaha, A., Guizani, M., Mohammadi, M., Aledhari, M., Ayyash, M., 2015. Internet of things: a survey on enabling technologies, protocols, and applications. IEEE Commun. Surv. Tutor. 17 (4), 2347–2376.

Bloch, B., 1995. Specialization and its critical role in business. Manag. Decis. 33 (6), 51–58.

Bloom, N., Draca, M., Van Reenen, J., 2016. Trade induced technical change? The impact of Chinese imports on innovation, IT and productivity. Rev. Econ. Stud. 83 (1), 87–117.

Crotti, D., Ferrari, C., Tei, A., 2019. Merger waves and alliance stability in container shipping. Marit. Econ. Log. https://doi.org/10.1057/s41278-019-00118-6.

Elleuch, H., Dafaoui, E., Elmhamedi, A., Chabchoub, H., 2016. Resilience and vulnerability in supply chain: literature review. IFAC-PapersOnLine 49-12, 1448–1453.

Farahani, R.Z., Hekmatfar, M., Arabano, A.B., Nikbakhsh, E., 2013. Hub location problems: a review of models, classification, solution techniques, and applications. Comput. Ind. Eng. 64 (4), 1096–1109.

Faye, M.L., McArthur, J.W., Sachs, J.D., Snow, T., 2004. The challenges facing landlocked developing countries. J. Hum. Dev. 5 (1), 31–68.

Frankel, J.A., Romer, D.H., 1999. Does trade cause growth? Am. Econ. Rev. 89 (3), 379–399.

Ghirmay, T., Grabowski, R., Sharma, S.C., 2001. Exports, investment, efficiency and economic growth in LDC: an empirical investigation. Appl. Econ. 33 (6), 689–700.

Hartmann, D., Bezerra, M., Lodolo, B., Pinheiro, F.L., 2019. International trade, development traps, and the core-periphery structure of income inequality. Economia. https://doi.org/10.1016/j.econ.2019.09.001.

Kostoska, O., Mitikj, S., Jovanovski, P., Kocarev, L., 2020. Core-periphery structure in sectoral international trade networks: a new approach to an old theory. PLoS ONE 15 (4), e0229547.

Lee, E., Vivarelli, M., 2006. The social impact of globalization in the developing countries. Int. Labour Rev. 145 (3), 167–184.

Levinson, M., 2006. The Box: How the Shipping Container Made the World Smaller and the World Economy Bigger. Princeton University Press.

Notteboom, T.E., Parola, F., Satta, G., Pallis, A.A., 2017. The relationship between port choice and terminal involvement of alliance members in container shipping. J. Transp. Geogr. 64, 158–173.

Parola, F., Risitano, M., Ferretti, M., Panetti, E., 2017. The drivers of port competitiveness: a critical review. Transp. Rev. 37 (1), 116–138.
Pavcnik, N., 2002. Trade liberalization, exit, and productivity improvements: evidence from Chilean plants. Rev. Econ. Stud. 69 (1), 245–276.
Rao, N., 2000. "Neocolonialism" or "globalization"?: postcolonial theory and the demands of political economy. Interdiscip. Lit. Stud. 1 (2), 165–184.
Reis, V., Meier, J.F., Pace, G., Palacin, R., 2013. Rail and multi-modal transport. Res. Transp. Econ. 41, 17–30.
Rodrigue, J.P., 2017. The Geography of Transport Systems, fourth ed. Routledge, New York.
Saragiotis, P., 2019. Business process management in the port sector: a literature review. Marit. Bus. Rev. 4 (1), 49–70.
Sprecher, B., Daigo, I., Murakami, S., Kleijn, R., Vos, M., Kramer, G.J., 2015. Framework for resilience in material supply chains, with a case study from the 2010 rare earth crisis. Environ. Sci. Technol. 49 (11), 6740–6750.
Steenken, D., Voß, S., Stahlbock, R., 2004. Container terminal operation and operations research: a classification and literature review. OR Spectr. 26, 3–49.
Stiglitz, J.E., 2017. The overselling of globalization. Bus. Econ. 52, 129–137.
Tijan, E., Agatic, A., Jovic, M., Aksentijevic, S., 2019. Maritime national single window: a prerequisite for sustainable seaport business. Sustainability 11, 4570. https://doi.org/10.3390/su11174570.
Williams, R., Aaron, J., 2018. Specialization as a small business strategic approach. Small Bus. Inst. J. 14 (2), 1–15.

Part One

General introduction

Introduction to global container shipping market

1

César Ducruet[a] *and Hidekazu Itoh*[b]
[a]Centre National de la Recherche Scientifique, [b]Kwansei Gakuin University

Introduction: Containerisation and global logistics

In 26 April 1956, an American land transporter named Malcom McLean started competing with freight railway companies on interstate long distance transport in the United States. He first navigated a hopped-up container ship from Newark, New Jersey, to Houston, Texas, along the US East Coast by his shipping company (later named Sea-Land). Maritime containers were acquired for two main purposes: (1) to reduce port handling costs by the unitisation (container 'box') of cargo and (2) to reduce truck transport cost on long-distance delivery. Indeed, the container ships are permitted to deliver cargo through intermodal transport on land and sea (Levinson, 2006).

The strongest advantage of containerisation is that cargo handling on the docks could be managed in a more efficient way. At this time, a container was mounted on a wagon for land transport, or current roll-on/roll-off (Ro-Ro) shipping. However, because the system was initially inefficient due to the weight and space of wagons, container ships used cranes to handle the box between the ship and the yard. Finally, the Sea-Land company launched a modern full-container ship without crane onboard in 1966 to cross the Atlantic, as European ports such as Antwerp became able to handle containers in the late 1960s (Morel and Ducruet, 2015). After certain technological progress, gantry cranes were placed on berths to carry containers between the ship and the terminal, whilst chassis and trailer moved containers inside the container terminal.

Containerisation helped reducing handling time on both sea and land sides. At the time of early containerisation, the total duration of a round trip in the Pacific Ocean between East Asia (e.g. Kobe, Japan) and North America (NA) (e.g. San Francisco, US East Coast) by conventional ship (general cargo) was about 80 days (35 days on sea and 45 days on land) in 1956. However, in 1968, after full-container ships were launched, and the total duration of a round trip decreased to 30 days (23 days on sea and 7 days on land) between Tokyo and Los Angeles (Hoshino, 1995). Containerisation had contributed to minimising the temporal gaps between origin and destination along supply chains, whilst accelerating global trade and horizontal division of production. In 2017, because of slow steaming and multistops at hub-ports, most of round-trip durations were 35 days (5 weeks) or 42 days (6 weeks) on the route (International Transport Handbook, 2017), as explained in the following.

Global Logistics Network Modelling and Policy. https://doi.org/10.1016/B978-0-12-814060-4.00001-0

Finally, maritime transport business had changed from labour-intensive to capital-intensive industry. For example, global major ports have invested heavily in new gantry cranes for faster handling operations. In addition to container terminal development, container ships grown in size to achieve economies of scale. Indeed, after the introduction of over-Panamax[a] vessels in 1988, shipping lines built ever-larger container ships (Table 1.3). Such vessels again needed investments on the terminal side to accommodate ship calls all over the world. Deeper container berths, mega-gantry cranes, and larger container yards became the norm for terminal operations. For instance, such cranes must be cover 24 lines for the beam of 18,000 TEUs class container ships today.

Such operational and technological changes are both causes and consequences of wider global economic (e.g. manufacturing shifts) dynamics affecting the global port hierarchy, as presented in Table 1.1 for the period 1975–2016. In 1975, most of the top ranked container ports were North American, European, and Japanese ports due to the provision, in the 'Triade' (Ohmae, 1985), of container berths with gantry and terminal cranes that were still lacking at developing countries. However, in 2016, 7 of the top 10 ports were Chinese (including Hong Kong), following high economic growth and rapid port development since the 2000s. The other three major ports are also in Asia, like Singapore, Busan, and Dubai.

Despite their initial domination within Asia, Japanese ports, especially the port of Kobe, had been taken over by other East Asian ports, especially by the port of Busan due to network effects and the 1995 Hanshin Earthquake (Xu and Itoh, 2018). In a similar vein, and after playing a crucial role as a gateway and hub for mainland China due to its pre-1997 status as an independent city-state with Western trade practices, Hong Kong lost cargo in the last decade to Shenzhen. By contrast, Singapore maintained its port growth as the transit point between Pacific Ocean and Indian Ocean connecting Asia with Europe, and by providing highly frequent feeder services to neighbouring Southeast Asian countries. Yet, competitors started to emerge such as Tanjung Pelepas in Malaysia (2000), Cai-Mep Thi-Vai in Vietnam (1996), and Jakarta/New Priok in Indonesia (under construction) to provide alternative transit points and enhance their respective local economies.

In this chapter, we discuss the changes in maritime and port logistics caused by containerisation in the last 50 years. In "Economic growth and container cargo movements" section, we show the impact of containerisation on the world economy and global maritime networks including supply chain. "Port development and terminal operations" section discusses the function of container terminal enhancing maritime transport and connecting the land and sea transports, especially the constraints and challenges of port development for larger ships, and the port management and terminal operation. In "Global maritime container shipping" section, we present operational logics of shipping lines and alliances whilst providing concrete empirical evidences on changing patterns of global container flows. As reference, Chapter 2 discusses port hinterland which is the connection to port on the land side with shippers.

[a] Based on the ship size which can navigate on Panama Canal; the ship which cannot navigate on the canal is defined as post-Panamax (or over-Panamax) vessel.

Table 1.1 The container handling ranking changes at ports (unit: thousand TEUs).

	1975		1980		1985		1990		1995	
1	NY/NJ	1730	NY/NJ	1947	Rotterdam	2655	Singapore	5220	Hong Kong	12,550
2	Rotterdam	1079	Rotterdam	1901	NY/NJ	2367	Hong Kong	5100	Singapore	11,846
3	Kobe	905	Hong Kong	1465	Hong Kong	2289	Rotterdam	3670	Kaohsiung	5232
4	San Juan	877	Kobe	1456	Kaohsiung	1901	Kaohsiung	3490	Rotterdam	4787
5	Hong Kong	802	Kaohsiung	979	Kobe	1857	Kobe	2600	Busan	4503
6	Oakland	522	Singapore	917	Singapore	1699	Busan	2350	Hamburg	2890
7	Seattle	481	San Juan	852	Yokohama	1327	Los Angeles	2120	Yokohama	2757
8	Baltimore	421	Long Beach	825	Antwerp	1243	Hamburg	1970	Los Angeles	2555
9	Bremen	410	Hamburg	783	Long Beach	1172	NY/NJ	1900	Long Beach	2390
10	Long Beach	391	Oakland	782	Hamburg	1159	Keelung	1810	Antwerp	2329
11	Tokyo	369	Seattle	782	Keelung	1158	Yokohama	1650	NY/NJ	2276
12	Melbourne	365	Antwerp	724	Busan	1115	Long Beach	1600	Tokyo	2177
13	Keelung	246	Yokohama	722	Los Angeles	1104	Tokyo	1560	Keelung	2170
14	Hamburg	326	Bremen	703	Tokyo	1004	Antwerp	1550	Dubai/Jebel Ali	2073
15	Antwerp	297	Keelung	660	Bremen	986	Felixstowe	1420	Felixstowe	1898
16	Virginia	292	Busan	634	San Juan	882	San Juan	1380	Manila	1668
17	Sydney	262	Los Angeles	633	Oakland	856	Seattle	1170	San Juan	1593
18	London	260	Tokyo	632	Seattle	845	Bremen	1160	Oakland	1550
19	Yokohama	329	Jeddah	563	Felixstowe	726	Oakland	1120	Shanghai	1527
20	Le Havre	232	Baltimore	523	Baltimore	706	Manila	1039	Bremen	1526

Continued

Table 1.1 Continued

	2000		2005		2010		2015		2016	
1	Hong Kong	18,100	Singapore	23,192	Shanghai	29,069	Shanghai	36,537	Shanghai	37,130
2	Singapore	17,040	Hong Kong	22,427	Singapore	28,431	Singapore	30,922	Singapore	30,900
3	Busan	7540	Shanghai	18,084	Hong Kong	23,699	Shenzhen	24,204	Shenzhen	23,979
4	Kaohsiung	7426	Shenzhen	16,197	Shenzhen	22,510	Ningbo	20,620	Ningbo	21,560
5	Rotterdam	6280	Busan	11,843	Busan	14,194	Hong Kong	20,114	Busan	19,850
6	Shanghai	5613	Kaohsiung	9471	Ningbo	13,144	Busan	19,469	Hong Kong	19,580
7	Los Angeles	4879	Rotterdam	9300	Guangzhou	12,550	Guangzhou	17,625	Guangzhou	18,885
8	Long Beach	4601	Hamburg	8088	Qingdao/Tsingta	12,012	Qingdao/Tsingta	17,510	Qingdao/Tsingta	18,000
9	Hamburg	4248	Dubai/Jebel Ali	7619	Dubai/Jebel Ali	11,600	Dubai/Jebel Ali	15,592	Dubai/Jebel Ali	14,772
10	Antwerp	4082	Los Angeles	7485	Rotterdam	11,146	Tianjin	14,100	Tianjin	14,500
11	Shenzhen	3994	Long Beach	6710	Tianjin	10,080	Rotterdam	12,235	Port Kelang	13,183
12	Port Kelang	3207	Antwerp	6482	Kaohsiung	9181	Port Kelang	11,890	Rotterdam	12,385
13	Dubai/Jebel Ali	3059	Qingdao/Tsingta	6307	Port Kelang	8870	Kaohsiung	10,264	Kaohsiung	10,465
14	NY/NJ	3050	Port Kelang	5716	Antwerp	8468	Antwerp	9654	Antwerp	10,037
15	Tokyo	2899	Ningbo	5208	Hamburg	7900	Dalian	9450	Dalian	9614
16	Felixstowe	2853	Tianjin	4801	Tanjung Perapus	6530	Xiamen	9183	Xiamen	9414
17	Bremen	2712	NY/NJ	4793	Long Beach	6263	Tanjung Perapus	9120	Hamburg	8910
18	Gioia Tauro	2653	Guangzhou	4685	Xiamen	5820	Hamburg	8821	Los Angeles	8857
19	Tanjung Priok/Ja	2476	Tanjung Pelepas	4177	NY/NJ	5292	Los Angeles	8160	Tanjung Perapus	8029
20	Yokohama	2317	Laem Chabang	3766	Dalian	5242	Long Beach	7192	Laem Chabang	7227

From Containerization International, 2009. Containerization International Yearbook. Informa UK Ltd, and United Nations, 2016. Review of maritime transport. In: United Nations Conference on Trade and Development, Geneva.

Economic growth and container cargo movements

The innovation of containerisation in maritime trade cause a rapid expansion of global trade (see also Bernhofen et al., 2013). Fig. 1.1 compares the evolution of different maritime trades in the last four decades (the handling level of 1990 is the baseline). Most of the goods shipped in containers being general cargo, or intermediate and finished goods, container traffic had expanded much quickly than general cargo, especially after 1998 (see also Fig. 1.4). Indeed, a growing share of general cargo had become containerised in the last two decades. In addition, the impacts of global recession for maritime trade were much bigger on container than on general cargoes.

The relationship between maritime trade, especially containerisation, and economic activities, Fig. 1.2, shows the increasing rate changes of container handling volumes (TEUs) and dry cargo (ton base) compared to Global Domestic Product (GDP) in the last four decades or so. Except for 1998 and 2009, the growth rate of container handling volumes has been higher than that of dry cargo and GDP. The average growth rate of container handling volume is about 9.6% as compared to 4.1% for dry cargo and 3.0% for GDP. On the other hand, the standard deviation of container handling growth rate is 5.8, as compared with 5.0 and 1.3 for dry cargo and GDP, respectively. Several factors explain such a result. First, handling items in containers are mainly high value-added goods (i.e. consumption and intermediate goods), so the demand for container transport is less stable than that for general cargo and natural resources (i.e. bulks). Second, container handling is highly connected with economic circulation. For example, the correlation coefficients of the growth rate with GDP are 0.66

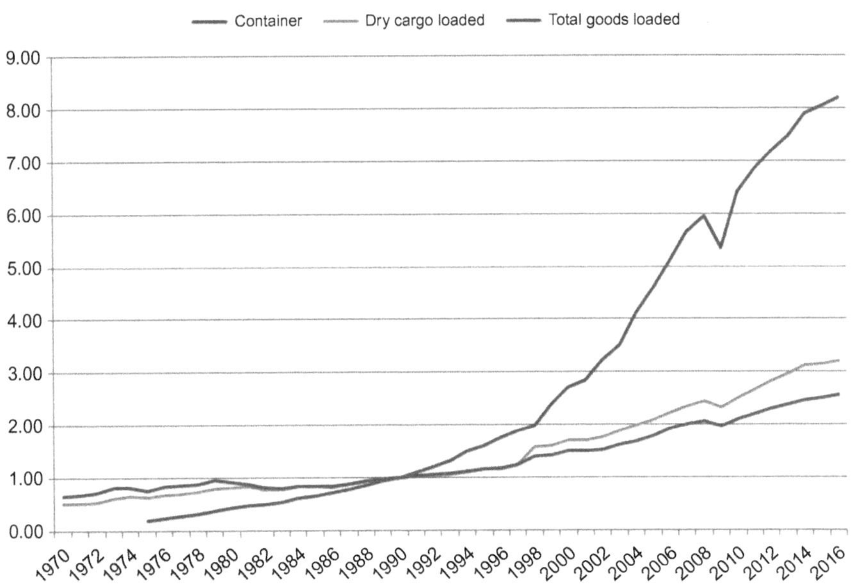

Fig. 1.1 The relative expanding speeds of maritime trade. World Bank Open Data, and UNCTAD Stat.

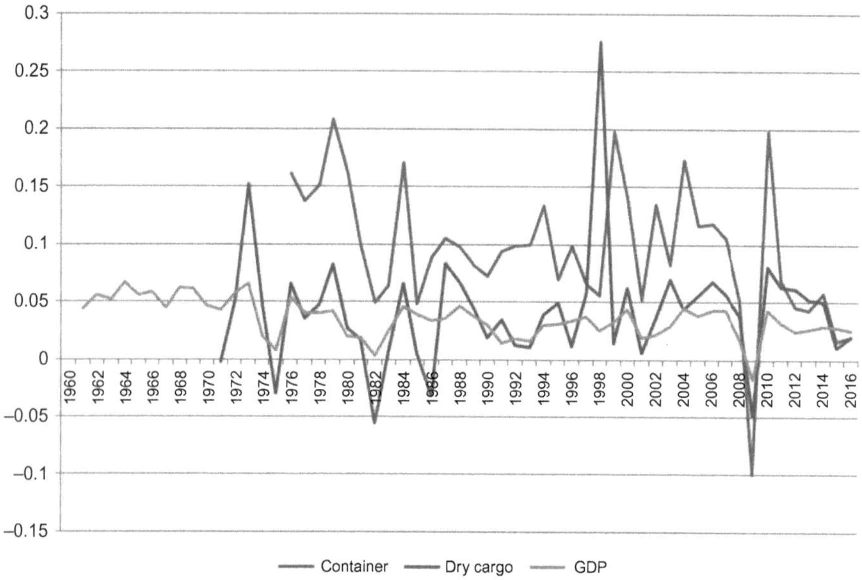

Fig. 1.2 The increasing rate changes of container, dry cargo, and GDP. World Bank Open Data, and UNCTAD Stat.

with container and 0.47 with dry cargo. Third, slowing of trade, or increase in global GDP, is higher than that for global trade, as the Lehman shock ended in 2017 (CPB Netherlands Bureau for Economics Policy Analysis, 24 November 2017).

When observing the growth rate of container handling volumes and GDP by countries (see Appendix, Table 1.A1), again, the growth rate of container handling was found to be higher than that of GDP. However, the current Chinese growth rate of container handling remains rather moderate, whilst that in Hong Kong has been negative in the last 5 years. The centre of gravity of economic expansion had, indeed, shifted toward South Asia, like Indonesia and Vietnam (Itoh, 2012).

Fig. 1.3A and B show the relative scales of container handling volumes and value-added goods (GDP) by regions/countries (see Appendix, Table 1.A2). Until the mid-1990s, most of the container traffic was handled in advanced economies and regions as mentioned earlier, until Chinese ports increased their share after 1995, and especially in 2001 [entry of China in the World Trade Organisation (WTO)]. Currently, the total Chinese share including Hong Kong is more than 30%, whilst European ports witnessed a decrease from 30% in 1975 to 12% in 2015. Although Hong Kong had increased its global share until the middle of 1990s, its share was taken by the mainland Chinese ports by the container terminal developments, turning it into a global financial and value-added centre instead of a cargo handling hub (see Wang and Chen, 2010).

On the other hand, the relative shares of GDP have been changing more smoothly than that for container handling. For example, although the Chinese economy including Hong Kong occupied about 12% in 2015, the advanced economies, like NA, Europe (Germany, United Kingdom, and France), and Japan still take their position

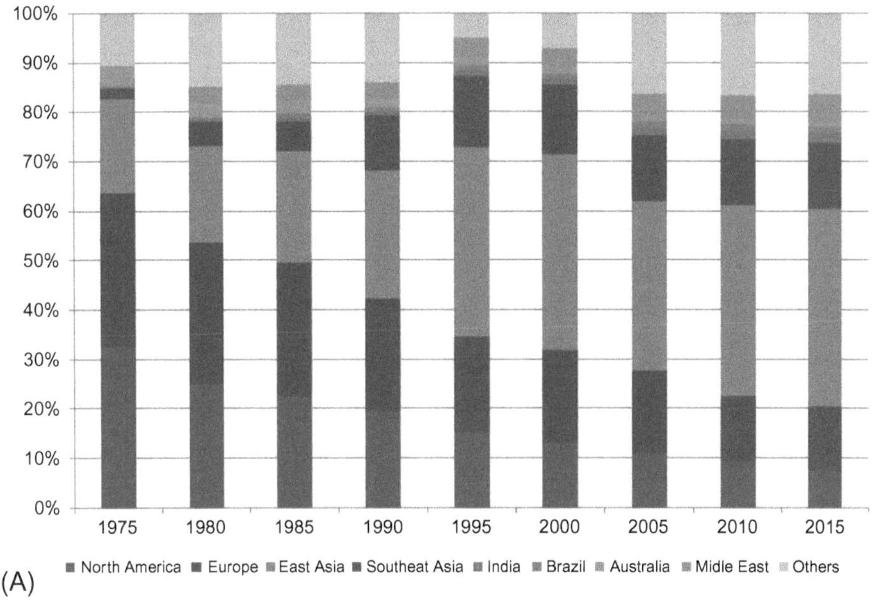

(A)

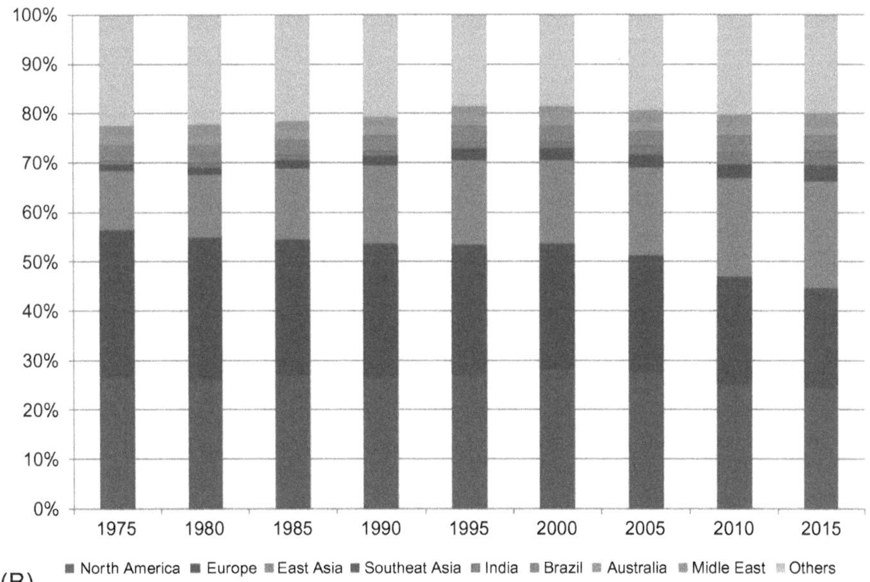

(B)

Fig. 1.3 The shares for global total by countries/regions. (A) Container handling volumes. Based on the data from Table 1.A2A; (B) Gross domestics products (GDP). Based on the data from Table 1.A2B.

to some extent. This result is partly due to the fact that container handling volumes are sometimes inflated by official statistics because of large transshipment volumes, leading to double counts of each container move.

Until the end of 1990s, the relative changes in economic activities (GDP) and cargo movements (containers) had maintained relatively tight linkages. In Fig. 1.4, the correlation coefficients between the relative shares of GDP and container handling on countries (see Appendix, Table 1.A2) had been increasing until 1999. The decrease after 2000, including China's entry in the WTO and the global financial crisis effects, can be explained by the rapid progress of supply chain development in emerging economies (i.e. BRICS countries), especially in Asia, and a growing imbalanced international horizontal division of production (see Table 1.2).

Table 1.2A and B present the inter- and intraregional container movements in 1998 and 2016. As discussed earlier, the distribution of container and economic activities had been closely connected until 1999. However, they are less connected in the 2000s. The centre of gravity of container movements have been shifting to intraregional activities in Asia in a context of increased regional integration, thereby concentrating more than 25% of global container movements. Currently, the impact of economic growth on container movements is amplified and imbalanced on routes, or highly weighted inside Asia. Containerisation has increased the speed of economic growth at emerging economies and expanded the imbalance of cargo movements on routes and regions throughout the world.

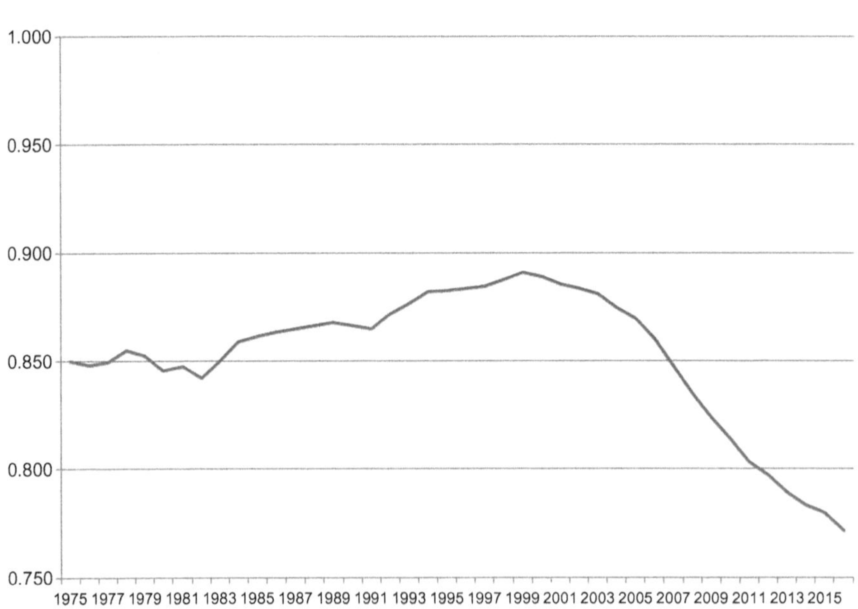

Fig. 1.4 The correlation coefficient between GDP and container shares. This figure is based on the data sets on Table 1.A1A and B.

Table 1.2 World container movements (unit: thousand TEUs).

(A) 1998			
Origin/destination	**North America**	**Europe**	**East/Southeast Asia**
North America	–	2036 (5.7%)	3338 (9.4%)
Europe	1509 (4.2%)	–	2296 (6.5%)
East/Southeast Asia	5938 (16.7%)	4246 (12.0%)	5873 (16.5%)
(B) 2016			
Origin/destination	**North America**	**Europe**	**East/Southeast Asia**
North America	482 (0.3%)	2048 (1.3%)	7252 (4.7%)
Europe	3913 (2.6%)	6928 (4.5%)	7022 (4.6%)
East/Southeast Asia	16,708 (10.9%)	15,049 (9.8%)	39,214 (25.6%)

(A) The estimated total container handling volumes are 35,528,000 TEUs.
(B) The estimated total container handling volumes are 153,270,049 TEUs.
(A) From MOL Research Institute, 1999. Current Status of Liner Shipping Service 1998–1999 (in Japanese); (B) From Japan Maritime Public Relations Center, 2017. Shipping Now 2017–2018 (in Japanese). see https://www.kaijipr.or.jp/shipping_now/.

Port development and terminal operations

Terminal development and new port opening

The first ship in 1956 by McLean delivered 58 boxes. Containerships expanded through economies of scale (see Table 1.3) as underlined by Cullinane and Khanna (2000). For instance, although the ship capacity was less than 1000 in the 1960s, currently the world's largest containership 'MOL Triumph' of the shipping line MOL built in 2017 reached more than 20,000 twenty-foot equivalent unit (20,170 TEU), 400 m in length, and 59 m in width. As a consequence, such 'mega-ships' require high-standard container terminals in terms of both berth length and depth (see Ducruet and Berli, 2018 for an empirical analysis of the global distribution of mega-ship traffic). These facilities affected port operational efficiencies (Tongzon, 2001; Itoh, 2002) and led to a debate on whether ports and nation-states should follow such a trend. As discussed previously, the so-called 'Post-Panamax II' ships carrying about 8000 TEU needed container berth with 16–18 m water depth and 18,000 TEU capacity on the terminal (ULCS).

Large-size container ship causes the expansion of container terminals. Except the ports of Marseilles and Barcelona in Europe, Los Angeles, and Long Beach in NA, numerous ports are located upstream rivers and estuaries as a reflection of their historical background of port development, so their expansion faces important limitations, with Antwerp and Hamburg being exceptional cases (Notteboom, 2016). This explains why larger ships increasingly call at deepwater ports, the exemplary case being London, mainly served by Felixstowe (operated by Hong Kong-based Hutchinson Whampoa), 250 km away, although recently the global operator Dubai Ports World

Table 1.3 The changes of container ship sizes.

	1	2	3	4	5	6	7	8	9
Category	Early	Fully cellular	Panamax	Panamax Max	Post Panamax	Post Panamax II	New-Panamax	VLCS (very large)	ULCS (ultra large)
Year	1956–	1970s	1980	1985–	1988–	2000–	2014–	2006–	2013–
Capacity	500–800	1000–2500	3000–3400	3400–4500	4000–6000	6000–8500	12,500	11,000–15,000	18,000–21,000
Length (m)	200	215	250	290	300	340	366	397	400
Beam (m)	20	20	32	32	40	43	49	56	59
Draught (m)	9	10	12.5	12.5	13	14.5	15.2	15.5	16
Across	6	10	13	13	15	17	19–20	22	23
High[a]	4 (4)	5 (4)	6 (5)	8 (6)	9 (5)	9 (6)	10 (6)	10 (8)	10 (8)

[a] High is the number of container high on deck, the number in parentheses is below deck.

Based on Rodrigue, J.P., 2017. The Geography of Transport Systems, fourth ed. Routledge, New York.

(DPW) developed the new London Gateway container terminal with the slogan 'ship closer, save money'. Rotterdam is a special case as it has been relocating its container terminals by sea reclamation through the Maasvlakte projects. Most of the ports and container terminals now have to invest in mega gantry cranes, or longer arm of crane, for lifting on and off the containers inside a width ship.

In East Asia, the ports of Shanghai and Busan also have constructed new container terminals through sea reclamation (i.e. Yangshan deepwater port in 2005, Busan New port in 2006). Elsewhere, ports of Laem Chabang in Thailand (1991) and Cai-Mep Thi-Vai in Vietnam (1996) were newly constructed for containerisation away from their old river ports (i.e. Bangkok and Ho Chi Minh, respectively). Lastly, the new port of Tanjung Pelepas in Malaysia (2000) was constructed for competing with Singapore (see Table 1.1). At its opening, APM Terminals, the main terminal operator of Maersk, shifted from Singapore to Tanjung Pelepas, reducing Singapore's traffic by 30%, for the sake of competition and also as a way to bargain port costs. Following Maersk, Evergreen also shifted their hub-port to Tanjung Pelepas. Although Maersk returned to Singapore less than a year later, Tanjung Pelepas has been increasing its container throughput in the last few years.

Port management and terminal operations

Although the port logistics industry before containerisation was more labour-intensive, container terminal activity remains a highly capital-intensive industry. As discussed earlier, large-size container ships need high-standard container terminals, handling facilities, and equipments. Most of the global major container terminals are operated by (public and private) port operating companies and terminal operators instead of (central and local governmental) port managers (Fig. 1.5), because the operation and management of container terminal needs massive investments and management technology, and has been making profits with the growth of container transport.

In general, port management is classified into the following four types: (1) central government management, (2) local government management, (3) public enterprise management, and (4) private company management. Type (3) and (4) port managements are consolidated as port operating company as shown in Fig. 1.5. Type (1) port management, central government management, due to the lack of reliability, agility, and efficient management, is not so much common, except in the case of Hong Kong before the handover to China (1997). However, central governments are concerned by the development of basic facilities on the sea and port side (waterways, breakwater, and infrastructure). Most ports are managed by type (2), local governments (e.g.

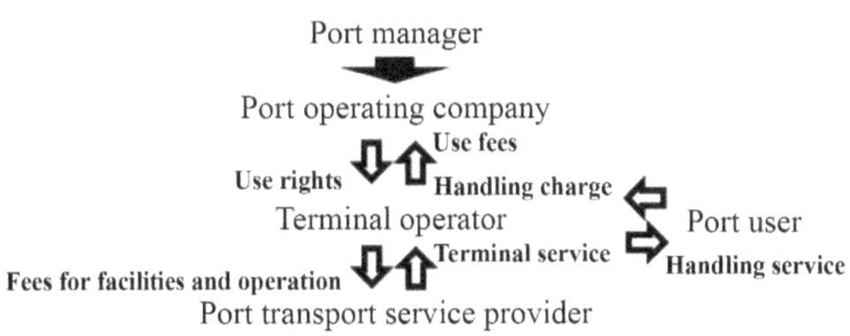

Fig. 1.5 Container port management system.

European major ports, the ports of Los Angeles, Long Beach, Miami, and Everglade, Japanese major ports), and type (3) port managements, public firms, because of the large local economic impacts of port activities. Type (3) port management, public businesses, is divided based on the independence of budget and decision making on port business and planning. For example, local governments established public firms (i.e. government-affiliated firms, such as Korea Container Terminal Authority and Singapore Port Authority—PSA, before privatisation), which manage port operation. They often establish public corporations under specific activity framework at the region level (e.g. Kaohsiung, Keelung, Seattle, and Tacoma). Lastly, a firm, which has public elements and constraints, is established under (general) company law. As the example of type (4) port management, private company, the ports of United Kingdom and New Zealand are managed by fully private companies (Doi, 2003).

Port operations are thus classified into four types based on the construction, ownership, and operation of infra- and superstructure between port manager and port (terminal) operator. In addition, leased home is divided into two types for port facilities (infrastructure and superstructure) (Table 1.4). About term-leased land, port operator constructs not only superstructure (gantry and yard cranes, container yard, shed, and warehouse) but also infrastructure (berthing facilities). For example, most Western ports (i.e. Europe and NA) are leased home without any type of superstructure. However, the port of Hong Kong is term-leased land without no-profit type infrastructure (or, waterway and breakwater). The terminal operator at Hong Kong pays concession money of terminal development to Hong Kong government, and develops new terminal including landfill, terminal facilities' construction, and their equipment procurements. Although new terminal developed by the operator belongs to Hong Kong government, the operator borrows terminals and operates the terminal. The port of Singapore was also managed by the central government. However, all its terminals were transferred to PSA by privatisation in 1997. UK ports were also privatised in the second half of the 1980s (Kurihara, 2014).

Terminal operators are classified into two types: (1) global stevedores' terminal operators and (2) global carrier's terminal operators (Table 1.5). The first type of operator is a preplay company focusing on port service business. Most of stevedores mainly operate at their home base and neighbouring areas focusing on competitive and efficient operations. For example, HPH had 52 terminals in 26 countries in 2017, as well as 9 terminals in China and Hong Kong. The second type of operator is divided into two subtypes, namely cost centre and profit centre. On the one hand, terminal operation focuses on profit centre inside their group company working for other shipping lines. On the other hand, the cost centre focuses on more efficient operations inside their group company's shipping network. In the past, major carriers had been making their global terminal operation supporting maritime business of their mother companies (or, cost centre). However, under difficult situations in a sensitive shipping market (cf. Fig. 1.2), port service business' profit has become more stable than maritime business. Therefore, some shipping lines changed their terminal service not only for their own shipping lines but also for other (competitive) shipping lines and group at their own/dedicated terminals for getting lower handling charges. Currently, about 80% of total container throughput at terminals are handled by global operators, and this share has been increasing (78.8% in 2015, cf. Table 1.6).

Table 1.4 The classification of port management organisation and operational form.

			1. Government	2. Leased home: With equipment	2. Leased home: Without equipment	3. Term-leased land	3. Term-leased land	4. Ownership
Development plan and permission			Port manager	Port manager	Port manager	Port manager	Port manager	Operator
Construction	Infra-	Nonprofit	Port manager	Port manager	Port manager	Port manager	Port manager	Operator
		Profit				Operator	Operator	
Ownership	Super-				Operator			
	Infra-nonprofit		Port manager	Port manager	Port manager	Port manager	Port manager	Operator
	Infra-profit						Operator	
	Super-				Operator	Operator	Operator	
Operating			Port manager	Operator	Operator	Operator	Operator	Operator
Example			Tanjung Priok (Indonesia), Laem Chabang (Thailand), Durban (South Africa), Haifa (Israel)	Kaohsiung, (Taiwan) Busan (Korea), Japanese major ports, Seattle (United States), Chinese ports, Dubai (United Arab Emirates)	European major ports (Rotterdam, Humbug), Los Angeles, Long Beach, NY/NJ (United States)	Hong Kong	Kaohsiung (Taiwan), Busan (Korea), Leam Chabang (Thailand), Jawaharlal Nehru Port (India)	Singapore, United Kingdom, New Zealand

Based on Kurihara, Y., 2014. The Global Trend of Port Service Industry, Mitsui & Co. Global Strategic Studies Institute. (in Japanese).

Table 1.5 The classification of global/international terminal operators.

	Classification	Management type	Ex.
Stevedores	Profit centre	Public company	PSA International, DP World, HHLA
		Private company	Hutchinson Port Handling (HPH), Eurogate, SSA Marine, Dragados, Crup TCB, ICTSI
Carriers	Cost centre	Public company	Terminal Investment Limited (TIL)
		Private company	CMA/CGM, Evergreen, APL, Hanjin, K-line, MISC, MOL, Yang Ming, Hyundai (HMM)
	Profit centre	Public company	COSCO Pacific
		Private company	APMT, NYK

From Mori, T., 2018. Introduction to Modern Logistics, third ed. Dobunkan Publisher, Tokyo (in Japanese).

Table 1.6 Global/International terminal operators' throughput.

	Operator	Million TEU	Share (%)
1	China COSCO Shipping	85.5	12.2
2	Hutchinson Ports	79.1	11.3
3	APM Terminals	71.4	10.2
4	PSA International	67.3	9.6
5	DP World	62.4	8.9
6	Terminal Investment Limited (TIL)	37.7	5.4
7	China Merchants Port Holdings	28.5	4.1
8	CMA CGM	16.6	2.4
9	Hanjin	14.0	2.0
10	Eurogate	11.9	1.7
11	SSA Marine/Carrix	10.6	1.5
12	NYK	9.6	1.4
13	Evergreen	9.4	1.3
14	ICTSI	8.7	1.2
15	OOCL	6.7	1.0
16	China Shipping Terminal Development	6.4	0.9
17	MOL	5.9	0.8
18	Yildirim/Yilport	5.6	0.8
19	Yang Ming	4.4	0.6
20	Bollore	4.3	0.6
Global/international operators total		555.1	79.4

Data from Global Container Terminal Operators: Annual Review and Forecast, 2017. Drewry, London.

Global maritime container shipping

It is now well known that horizontal integration pushed shipping lines to deploy global networks by the principle of merger and acquisition, such as Maersk (Frémont, 2007) and CMA-CGM (Frémont, 2015), although there are variations in the way this process had taken place amongst companies and amongst regions (Slack and Frémont, 2009). Global maritime container flows are currently transported by a handful of large companies, often through alliances, with a growing concentration that accelerated after the 2008 Financial Crisis (see Table 1.7), which led to important turmoil in the shipping industry until the bankruptcy of Hanjin Shipping in 2017. Other companies adopted the strategy of slow steaming, defined by decreasing travel speed and increased vessel size, to increase economies of scale, save fuel and money, and, at the same time, send to scrap their older or smaller vessels. Certain shipping lines, to avoid financial losses, have even reincorporated favourable sailing winds in their route operations to save even more fuel and money. In the meantime, China deploys new strategies to build a round-the-world Maritime Silk Road (Wang et al., 2018), trying to bypass Panama Canal costs through the project of a new Nicaragua Canal, and to bypass the Suez Canal through a railway land bridge through Israel, as well as a through Myanmar to avoid the Malacca Straits. The pattern of global shipping is thus still changing very

Table 1.7 The four major alliances on the east–west trades in 2015.

Alliance	Carriers	Country	Market share (%) on the Asia-North Europe trade
G6	American President Line	Singapore	24
	Hapag Lloyd	Germany	
	Hyundai Merchant Marine	South Korea	
	Mtsui OSK Line	Japan	
	NYK Line	Japan	
	Orient Overseas Container Line	Hong Kong	
CKYHE	COSCO	China	24
	K Line	Japan	
	Yang Ming	Taiwan	
	Hanjin Shipping	South Korea	
	Evergreen	Taiwan	
2M	Maersk	Denmark	31
	MSC	Switzerland	
Ocean Three	CMA-CGM	France	21
	China shipping (CSCL)	China	
	UASC	Qatar	

From Frémont, A., 2015. A geo-history of maritime networks since 1945. The case of the Compagnie Générale Transatlantique's transformation into CMA-CGM. In: Ducruet C. (Ed.), Maritime Networks: Spatial Structures and Time Dynamics. In: Routledge Studies in Transport Analysis, Routledge, London and New York, pp. 37–49.

fast and a lot of research remains to be done to foresee what will be routes of the future. For the rest, the Arctic passage, regardless of Canadian or Russian, is still not a reality and is a very minor priority for container shipping given all the geopolitical, technical, and financial issues.

A glimpse of the current (or recent) world pattern of maritime container flows is proposed in Fig. 1.6, in which the major routes and trunk lines' distribution across the globe are shown. Regional integration in East Asia and the major link between Europe and Asia are responsible for the concentration in these parts of the world, notwithstanding an important share of Trans-Pacific trade, but it is more imbalanced due to the large proportion of boxes returning empty from NA to Asia (see also Table 1.2).

The aforementioned changes and dynamics are examined in this section in two ways. First, we recall the main operational aspects of shipping lines to give a better understanding on how they design their various liner shipping services. Second, we provide some empirical evidence on how these economic and operational aspects have affected the distribution of global shipping flows in the last decade.

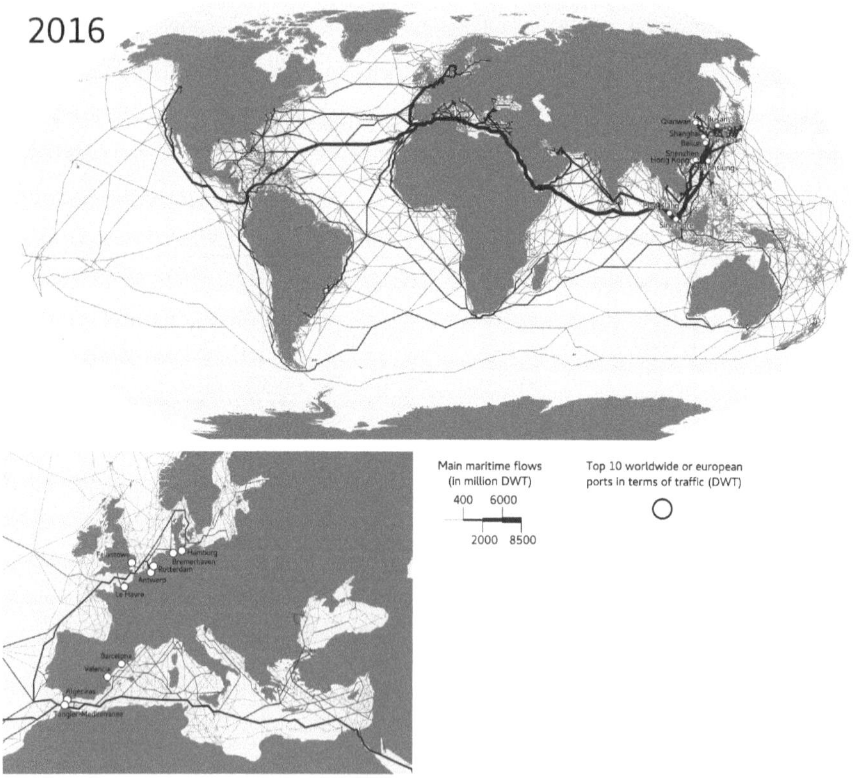

Fig. 1.6 Global maritime container flows in 2016.
From Ducruet C., Berli J., Bunel M., 2018. Geography vs. topology in the evolution of the global container shipping network (1977–2016). In: Wilmsmeier G., Monios J. (Eds.), Geographies of Maritime Transport. Edward Elgar Publishing (forthcoming).

Operational aspects of liner shipping services

Apparently very complex, the way shipping lines organise their services is based on numerous factors but a few key principles are shown in Fig. 1.7. All is a matter of business profitability, but with the exception, compared with other businesses of geographic aspects that shipping lines cannot avoid. They thus analyse the route along which ships should be deployed (fleet), and in the meantime, zoom on the route to select the best ports of call according to multiple criteria that are so much researched in the literature. Why this port and not another is not a straightforward question, which has to do its efficiency, performance, technical quality and capacity, handling costs, presence of specific arrangement with terminal operators, and finally its proximity to the end markets (Tiwari et al., 2003; Tongzon, 2009). Shipping lines are often said to be 'footloose' in terms of port selection because they keep a certain margin in the case of disruption in transport chains (Achurra-Gonzalez et al., 2017), due to many causes (dockers' strike, natural disasters, etc.). Then come the choice of service shipping lines offer to their customers (mainly, shippers) to satisfy their needs. Volatility, freight rate fluctuation, and seasonality also come into play.

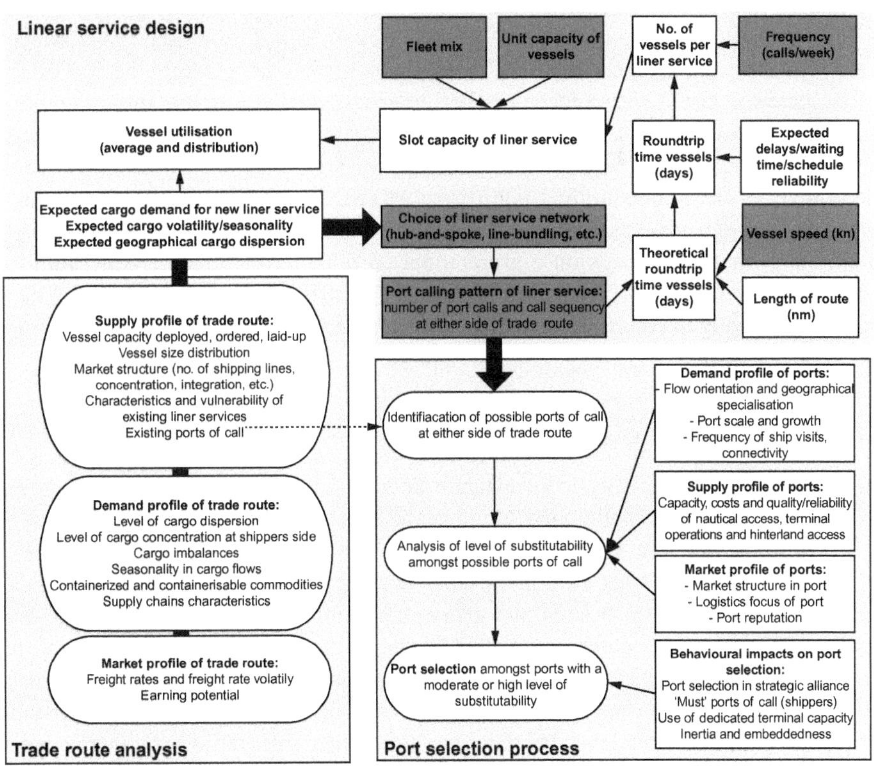

Fig. 1.7 The process of liner service design.
From Ducruet C., Notteboom T.E., 2012. Developing liner service networks in container shipping. In: Song, D.W., Panayides, P. (Eds.), Maritime Logistics: A Complete Guide to Effective Shipping and Port Management, Kogan, pp. 77–100.

The main configuration is based on 'bundling' which is a key driver of container service design. This can take place at two levels: individual liner service and bundling by combining/linking two or more liner services. The first aims to "collect container cargo by calling at various ports along the route instead of focusing on an end-to-end service" (Ducruet and Notteboom, 2012). This service is a set of several round trips of several vessels having in common calling patterns (i.e. order of port calls) and time intervals (i.e. frequency) between consecutive port calls. The overlap of all these roundtrips provides an optimal calling frequency. It is important to note that bundling can be symmetric or asymmetric; in the latter case, different ports of calls are used on the way back. In general, carriers select about five ports of call per loop, keeping in mind that increases in vessel size may even decrease the number of ports of call. Two extreme forms of line bundling are round-the-world services and pendulum services.

Another option is to bundle container lines by combining two or more liner services as follows: hub-and-spoke network (hub/feeder), interlining, and relay. On their side, governments and port authorities invested heavily in the development of specific redistribution nodes called 'intermediate hubs' (Rodrigue and Notteboom, 2010). Such hubs provide good nautical accessibility, proximity to main shipping lanes, and ownership, in whole or in part, by carriers or multinational terminal operators, most of these being located along the East–West circumterrestrial trunk line, 'in-between' main producer, or consumer markets. Their pivotal role complements the one of so-called 'gateway ports' within those markets to access final consumers through hinterland services. Container carriers use both gateways and hubs to design their services in the most efficient way possible, but it was demonstrated that the same node is often 'dedicated' to one main shipping line (Frémont and Soppé, 2007), just like terminal operators, through concession agreements (Notteboom et al., 2012). Another strategy of shipping lines is to focus on a preferential corridor development by investing in gateways and inland ports (Franc and Van der Horst, 2010), often through vertical integration, especially in countries where the transport sector is more liberalised (Ducruet and Van der Horst, 2009).

The evolution of flow patterns

One simple and classic way to investigate how have container port traffic patterns evolved under the aforementioned circumstances is to look at two famous concentration indices, namely Gini coefficient and Herfindahl-Hirschman Index (HHI) (Fig. 1.8). The latter index exhibits a clear tendency toward a de-concentration as many more ports are constructed and adopts the 'container revolution' through successive diffusion waves across the world (Guerrero and Rodrigue, 2014). Thus, the bulk of global container port traffic is less and less concentrated in the top of the world hierarchy overtime. In opposition, but without being contradictory to the previous observation, there is a tendency (cf. Gini coefficient) for this traffic to be increasingly concentrated across space, despite a less clear-cut trend and several fluctuations. For example, although the concentration (Gini coefficient) of port traffic in Asia had been stable between 1985 and 2005, the gaps within Southeast, South and West Asia, and

Fig. 1.8 Global container port traffic concentration, 1977–2016.
Based on Lloyd's List Intelligence data. NB: calculations realized using Wessa software, https://www.wessa.net.

containerised backward regions, have been expanding (Itoh, 2012). We cannot observe a paramount concentration but this was the trend until the early 2000s as containerisation has been highly selective and concentrated around large hubs and gateways, until a process of de-concentration occurred, as already exemplified by the pioneering work of Hayuth (1981) on the matter. The global shift of manufacturing from the Western to the Asian world is also responsible for such a mixed evidence, including the 'China effect', especially since its integration in the WTO in 2001.

Another but complementary way to understand the evolution of the global pattern of container flows is to apply one of the simplest graph-theoretical algorithms to the unweighted interport matrix of vessel flows (Fig. 1.9), namely the Gamma index, often coined 'density' in the network-analytical literature, i.e. the proportion of observed links (or 'edges') in the maximum possible number of links in the network (see Ducruet and Lugo, 2013 for a review of transportation network measures). The clear decline proves that at least until the late 1990s and early 2000s, global container flows went through a process of rationalisation and simplification, i.e. a lesser number of linkages compared with the number of ports, resulting in a star-like configuration, or hub-and-spokes, as a consequence of the aforementioned strategies of shipping lines when designing their networks and selecting large hubs. Despite a revival and re-densification of the network afterwards especially just after the global financial crisis, the latter shifted again toward centralisation so that the last value of 2016 is the lowest of the time series. This means a lot for global trade and connectivity in terms of vulnerability vs robustness, as the routes of the past have been replaced by ever-more efficient and optimal routes centred around large hubs and gateways, but at the expense of smaller, medium-size ports that cannot access the rest of the network without passing through this recently installed redistribution platforms.

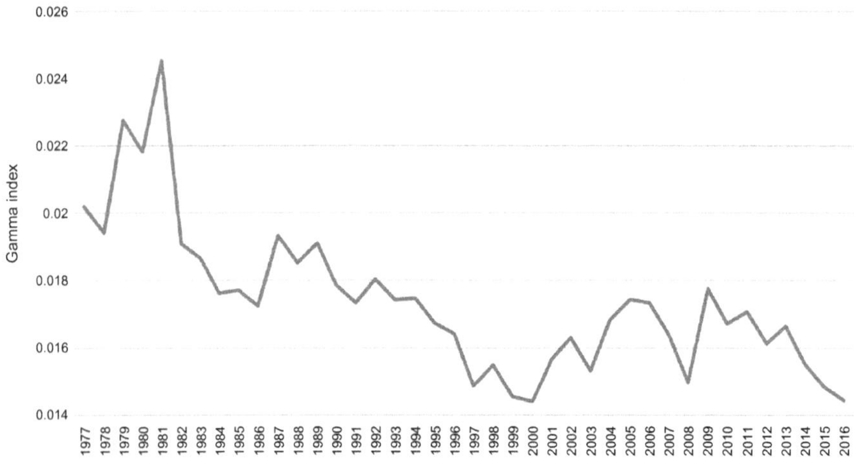

Fig. 1.9 Global container network concentration, 1977–2016.
Based on Lloyd's List Intelligence data.

Conclusion

This chapter recalled and demonstrated deep changes in the way maritime transport had been reorganised with the ongoing advent of containerisation in the past decades up to the present time. This multifaceted approach to containerisation is not so common as often, specific aspects are well covered and analysed by scholars and professionals but without offering an all-encompassing view. A review of the complex and changing relationships between containerisation (technological change) and economic development, port and shipping line operations, and related impacts on former ways of doing things is necessary before widening the approach to other segments of the global value and supply chain, such as hinterlands and shipping networks, as described in the following chapters. We wish this book to become a useful if not a key reference to scholars, students, and also to experts and practitioners for a better understanding of past, current, and future transformations of transport and logistics systems serving our world as a whole.

Appendix

Table 1.A1 Average growth rates by countries/regions.

	(A) Container handling volumes							
	1975–80 (%)	**1980–85** (%)	**1985–90** (%)	**1990–95** (%)	**1995–2000** (%)	**2000–05** (%)	**2005–10** (%)	**2010–15** (%)
United States	10.5	6.3	5.7	4.6	7.5	7.2	4.6	1.0
Canada	12.1	7.5	7.2	3.1	16.3	7.4	3.2	4.0
Germany	15.3	8.8	7.8	6.4	11.8	12.1	−0.2	9.3
Spain	22.8	16.7	5.2	10.7	12.9	9.7	6.8	3.0
Netherlands	12.6	6.2	6.5	5.4	5.7	8.3	4.2	2.1
Belgium	13.7	10.3	5.3	8.8	12.1	9.4	7.5	0.4
Italy	31.3	4.8	3.5	10.9	18.6	7.4	−0.1	1.0
United Kingdom	10.5	5.1	7.0	3.2	6.7	5.6	1.8	5.8
France	22.4	7.2	1.3	2.4	11.6	6.6	3.1	4.2
Sweden	9.4	8.9	0.2	9.7	3.9	7.9	−0.5	3.4
Ireland	8.2	1.9	7.6	8.8	9.4	6.2	−3.7	2.3
Denmark	39.2	6.1	−1.4	4.4	4.5	14.4	−37.2	−1.0
China	65.3	53.1	26.8	93.4	19.5	11.1	16.6	6.9
Hong Kong	13.3	9.4	17.5	19.9	7.7	4.6	1.2	−3.0
Japan	13.6	10.3	7.7	6.0	4.6	5.5	1.6	2.3
Korea	32.8	13.3	13.9	14.1	15.0	11.1	4.5	6.6
Taiwan	28.8	13.8	12.7	7.6	−20.9	4.1	0.6	2.5
Singapore	33.2	13.3	25.3	17.8	7.6	6.6	5.2	1.8
Malaysia	21.2	18.1	18.2	18.6	17.8	22.0	8.5	5.7
Indonesia	49.2	23.6	33.5	17.4	18.4	7.9	11.0	8.4
Thailand	94.7	16.3	22.0	12.8	10.4	10.0	5.4	5.1
Vietnam						16.8	18.7	8.7
Philippines	36.3	8.8	17.2	8.3	10.4	3.7	7.5	7.3
India	224.8	22.9	12.0	15.1	12.6	15.3	13.1	5.6

Continued

Table 1.A1 Continued

(A) Container handling volumes

	1975–80 (%)	1980–85 (%)	1985–90 (%)	1990–95 (%)	1995–2000 (%)	2000–05 (%)	2005–10 (%)	2010–15 (%)
Brazil	74.7	31.6	3.1	16.0	12.8	19.9	5.1	6.4
Australia	10.4	3.7	4.1	6.9	9.3	8.0	4.9	3.4
Turkey		117.7	15.2	16.3	42.1	16.3	14.1	7.4
United Arab Emirates	58.7	16.6	17.6	17.9	27.2	14.6	9.7	7.1
Saudi Arabia	80.9	3.8	−3.4	7.7	6.7	75.1	7.5	8.2

(B) Gross domestic products (GDP)

	1970–75 (%)	1975–80 (%)	1980–85	1985–90 (%)	1990–95 (%)	1995–2000 (%)	2000–05 (%)	2005–10 (%)	2010–15 (%)
United States	2.7	3.7	3.4	3.4	2.6	4.3	2.5	0.8	2.2
Canada	4.4	3.7	2.7	2.6	1.7	4.0	2.6	1.2	2.2
Germany	2.4	3.4	1.4	3.3	2.1	1.9	0.6	1.3	1.7
Spain	5.3	2.0	1.4	4.5	1.5	4.1	3.4	1.1	−0.2
Netherlands	3.3	2.6	1.1	3.4	2.3	4.3	1.3	1.3	0.8
Belgium	3.7	3.2	0.9	3.1	1.6	2.9	1.8	1.4	1.0
Italy	3.2	4.5	1.7	3.1	1.3	2.0	0.9	−0.3	−0.6
United Kingdom	2.1	2.3	2.4	3.5	1.6	3.3	2.8	0.4	2.1
France	3.9	3.4	1.6	3.4	1.3	2.9	1.7	0.8	1.0
Sweden	2.6	1.4	2.0	2.4	0.7	3.6	2.6	1.7	2.1
Ireland	4.9	4.6	2.6	4.7	4.7	9.4	5.6	0.8	7.7
Denmark	1.7	2.7	2.8	1.5	2.3	3.0	1.3	0.3	1.1
China	5.9	6.6	10.7	8.0	12.3	8.6	9.8	11.3	7.9
Hong Kong	6.6	11.6	5.8	7.8	5.3	2.7	4.3	4.0	3.0

Japan	4.6	4.4	4.3	5.0	1.6	1.1	1.2	0.2	1.0
Korea	10.0	8.6	9.4	10.5	8.4	5.7	4.7	4.1	3.0
Taiwan									
Singapore	9.6	8.6	6.9	8.7	8.7	5.7	4.9	6.9	4.1
Malaysia	7.2	8.6	5.2	6.9	9.5	5.0	4.8	4.6	5.3
Indonesia	7.0	7.9	4.8	6.3	7.1	1.0	4.7	5.7	5.5
Thailand	5.8	8.0	5.4	10.3	8.2	0.9	5.5	3.8	2.9
Vietnam			3.8	4.8	8.2	7.0	6.9	6.3	5.9
Philippines	5.8	6.1	−1.1	4.7	2.2	3.6	4.6	5.0	5.9
India	2.9	3.2	5.2	6.0	5.1	6.1	6.7	8.3	6.8
Brazil	10.3	6.7	1.2	2.3	3.1	2.1	2.9	4.5	1.1
Australia	3.2	2.8	2.9	4.0	2.4	4.2	3.2	2.8	2.7
Turkey	5.8	2.5	4.9	5.7	3.3	4.1	4.9	3.4	7.1
United Arab Emirates		16.2	−1.3	3.3	3.8	5.6	5.4	2.5	4.9
Saudi Arabia	15.0	7.5	−9.9	7.6	3.7	1.7	4.1	2.8	5.2

(A) West Germany and East Germany are integrated as Germany before 1990. And, Hong Kong is excluded in China for constant discussion.
(B) The GDP data of Taiwan is not available on the World Bank Open Data.

Table 1.A2 The relative shares for global total by countries/regions.

(A) Container handling volumes									
	1975 (%)	1980 (%)	1985 (%)	1990 (%)	1995 (%)	2000 (%)	2005 (%)	2010 (%)	2015 (%)
United States	30.3	23.1	20.6	17.8	13.9	11.8	9.8	8.3	7.0
Canada	2.5	2.0	1.9	1.8	1.3	1.3	1.1	0.9	0.8
Germany	4.2	4.0	4.0	3.8	3.2	3.3	3.5	2.3	2.8
Spain	1.5	1.9	2.7	2.3	2.3	2.5	2.3	2.3	2.1
Netherlands	6.5	5.5	5.0	4.4	3.6	2.8	2.4	2.1	1.8
Belgium	2.8	2.5	2.6	2.2	2.1	2.2	2.0	2.0	1.6
Italy	1.8	3.3	2.7	2.1	2.2	3.0	2.5	1.8	1.5
United Kingdom	8.0	6.1	5.2	4.7	3.4	2.8	2.1	1.5	1.6
France	2.3	2.9	2.7	1.8	1.2	1.3	1.0	0.8	0.8
Sweden	1.2	0.8	0.8	0.6	0.5	0.4	0.3	0.2	0.2
Ireland	1.2	0.6	0.5	0.4	0.4	0.3	0.2	0.1	0.1
Denmark	1.2	0.9	0.8	0.4	0.3	0.2	0.2	0.1	0.1
China	0.0	0.1	0.8	1.4	12.6	17.7	17.2	25.4	28.3
Hong Kong	4.6	3.9	4.1	6.0	9.1	7.8	5.8	4.3	2.9
Japan	10.7	9.2	9.9	9.3	7.7	5.7	4.4	3.3	2.9
Korea	1.1	1.8	2.2	2.7	3.3	3.9	3.9	3.4	3.7
Taiwan	2.7	4.4	5.5	6.4	5.7	4.5	3.3	2.4	2.1
Singapore	1.3	2.5	3.0	6.1	8.6	7.4	5.9	5.3	4.6
Malaysia	0.4	0.5	0.7	1.0	1.5	2.0	3.1	3.3	3.5
Indonesia	0.0	0.2	0.4	1.1	1.5	1.6	1.4	1.5	1.7
Thailand	0.1	0.5	0.7	1.3	1.4	1.4	1.3	1.2	1.2
Vietnam	0.0	0.0	0.0	0.0	0.0	0.5	0.6	1.1	1.3
Philippines	0.5	1.2	1.1	1.6	1.4	1.3	0.9	0.9	1.0
India	0.0	0.4	0.7	0.8	1.0	1.1	1.3	1.7	1.7
Brazil	0.3	0.4	1.1	0.8	1.0	1.0	1.4	1.3	1.4
Australia	4.3	3.2	2.5	1.9	1.7	1.5	1.3	1.2	1.1

Turkey	0.0	0.0	0.3	0.4	0.5	0.7	0.8	1.1	1.2
United Arab Emirates	0.0	0.9	1.3	1.8	2.6	2.2	2.5	2.8	3.1
Saudi Arabia	0.0	2.2	1.7	0.9	0.8	0.6	1.0	1.0	1.1
Others	10.5	14.9	14.4	14.1	5.1	7.2	16.4	16.6	16.5

(B) Gross domestic products (GDP)

	1975	1980	1985	1990	1995	2000	2005	2010	2015
United States	23.7	23.5	24.3	23.9	24.4	25.4	24.8	22.7	22.0
Canada	2.8	2.8	2.8	2.7	2.6	2.7	2.6	2.4	2.4
Germany	7.5	7.3	6.9	6.8	6.7	6.2	5.5	5.2	4.9
Spain	2.6	2.3	2.2	2.3	2.2	2.3	2.3	2.2	1.9
Netherlands	1.6	1.5	1.4	1.4	1.4	1.5	1.4	1.3	1.2
Belgium	1.0	1.0	0.9	0.9	0.8	0.8	0.8	0.7	0.7
Italy	4.8	5.0	4.7	4.6	4.4	4.1	3.7	3.2	2.7
United Kingdom	4.8	4.4	4.4	4.3	4.2	4.2	4.1	3.7	3.6
France	5.5	5.4	5.1	5.0	4.8	4.7	4.4	4.0	3.7
Sweden	1.1	0.9	0.9	0.8	0.8	0.8	0.8	0.7	0.7
Ireland	0.2	0.2	0.2	0.2	0.2	0.3	0.4	0.3	0.4
Denmark	0.7	0.7	0.7	0.6	0.6	0.6	0.5	0.5	0.5
China	1.1	1.2	1.8	2.2	3.5	4.5	6.1	9.2	11.8
Hong Kong	0.1	0.2	0.2	0.3	0.3	0.3	0.3	0.3	0.3
Japan	10.4	10.7	11.6	12.4	12.0	10.7	9.8	8.6	7.9
Korea	0.4	0.5	0.7	1.0	1.3	1.4	1.5	1.7	1.7
Taiwan	0.0	0.0	0.0	0.0	0.0	0.0	0.0	0.0	0.0
Singapore	0.1	0.1	0.1	0.2	0.2	0.3	0.3	0.4	0.4
Malaysia	0.1	0.2	0.2	0.2	0.3	0.3	0.4	0.4	0.4
Indonesia	0.5	0.7	0.7	0.8	1.0	0.9	1.0	1.1	1.3
Thailand	0.2	0.2	0.3	0.4	0.5	0.4	0.5	0.5	0.5
Vietnam	0.0	0.0	0.1	0.1	0.1	0.1	0.1	0.2	0.2

Continued

Table 1.A2 Continued

	(B) Gross domestic products (GDP)								
	1975	**1980**	**1985**	**1990**	**1995**	**2000**	**2005**	**2010**	**2015**
Philippines	0.3	0.3	0.2	0.2	0.2	0.3	0.3	0.3	0.4
India	1.0	1.0	1.1	1.2	1.4	1.6	1.9	2.5	3.0
Brazil	3.2	3.6	3.4	3.1	3.3	3.1	3.1	3.3	3.1
Australia	1.7	1.6	1.6	1.6	1.6	1.7	1.7	1.7	1.7
Turkey	0.8	0.8	0.9	1.0	1.0	1.0	1.1	1.2	1.4
United Arab Emirates	0.2	0.4	0.3	0.3	0.4	0.4	0.4	0.4	0.5
Saudi Arabia	1.1	1.3	0.7	0.8	0.8	0.8	0.8	0.8	0.9
Others	22.3	22.3	21.6	20.6	18.6	18.5	19.3	20.2	19.8

(A) West Germany and East Germany are integrated as Germany before 1990. And, Hong Kong is excluded in China for constant discussion.
(B) The GDP data of Taiwan is not available on the World Bank Open Data.

References

Achurra-Gonzalez, P., Angeloudis, P., Zavitsas, K., Niknejad, A., Graham, D.J., 2017. Attacker-defender modelling of vulnerability in maritime logistics corridors. In: Ducruet (Ed.), Advances in Shipping Data Analysis and Modeling. Tracking and Mapping Maritime Flows in the Age of Big Data. Routledge Studies in Transport Analysis, Routledge, London & New York, pp. 297–315.

Bernhofen, D.M., El-Sahli, Z., Kneller, R., 2013. Estimating the effects of the container revolution on world trade. In: Lund University Working Paper 2013:4. Department of Economics, School of Economics and Management.

Cullinane, K.P.B., Khanna, M., 2000. Economies of scale in large containerships: optimal size and geographical implications. J. Transp. Geogr. 8 (3), 181–195.

Doi, M. (Ed.), 2003. Economics of Ports and Regions. Taga-Shuppan, Tokyo (in Japanese).

Ducruet, C., Berli, J., 2018. Mapping the globe: the patterns of mega-ships. Port Technol. Int. 77, 94–96.

Ducruet, C., Lugo, I., 2013. In: Rodrigue, J.P., Notteboom, T.E., Shaw, J. (Eds.), Structure and dynamics of transportation networks: models, concepts, and applications. SAGE Publications, The SAGE Handbook of Transport Studies, pp. 347–364.

Ducruet, C., Notteboom, T.E., 2012. Developing liner service networks in container shipping. In: Song, D.W., Panayides, P. (Eds.), Maritime Logistics: A Complete Guide to Effective Shipping and Port Management. Kogan, pp. 77–100.

Ducruet, C., Van der Horst, M.R., 2009. Transport integration at European ports: measuring the role and position of intermediaries. Eur. J. Transp. Infrastruct. Res. 9 (2), 121–142.

Franc, P., Van der Horst, M.R., 2010. Analyzing hinterland service integration by shipping lines and terminal operators in the Hamburg-Le Havre range. J. Transp. Geogr. 18 (4), 557–566.

Frémont, A., 2007. Global maritime networks: the case of Maersk. J. Transp. Geogr. 15 (6), 431–442.

Frémont, A., 2015. A geo-history of maritime networks since 1945. The case of the Compagnie Générale Transatlantique's transformation into CMA-CGM. In: Ducruet, C. (Ed.), Maritime Networks: Spatial Structures and Time Dynamics. Routledge Studies in Transport Analysis, Routledge, London and New York, pp. 37–49.

Frémont, A., Soppé, M., 2007. Northern European range: shipping line concentration and port hierarchy. In: Wang, J.J., Olivier, D., Notteboom, T.E., Slack, B. (Eds.), Ports, Cities, and Global Supply Chains. Ashgate, Aldershot, pp. 105–120.

Guerrero, D., Rodrigue, J.P., 2014. The waves of containerization: shifts in global maritime transportation. J. Transp. Geogr. 35, 151–164.

Hayuth, Y., 1981. Containerization and the load center concept. Econ. Geogr. 57 (2), 160–176.

Hoshino, H., 1995. The impacts of containerization to international logistics. In: The Containerization, p. 272 (in Japanese).

International Transport Handbook, 2017. Ocean Commerce Ltd. (in Japanese). see https://www.gov-book.or.jp/book/detail.php?product_id=350401.

Itoh, H., 2002. Efficiency changes at major container ports in Japan: a window application of Data Envelopment Analysis. Rev. Urb. Reg. Dev. Stud. 14 (2), 133–152.

Itoh, H., 2012. Structural changes in port cargo flow distribution in Asian container port systems. In: The Proceeding at the International Association of Maritime Economist (IAME) Annual Conference, 6–8 September, 2012, Taipei, Taiwan.

Kurihara, Y., 2014. The Global Trend of Port Service Industry. Mitsui & Co. Global Strategic Studies Institute (in Japanese).

Levinson, M., 2006. The Box: How the Shipping Container Made the World Smaller and the World Economy Bigger. Princeton University Press.
Morel, J.C., Ducruet, C., 2015. Interview—the man who brought containerisation to Europe. Portus Online, 29. http://portusonline.org/fr/interview-the-man-who-brought-containerisation-to-europe/.
Notteboom, T.E., 2016. The adaptive capacity of container ports in an era of mega vessels: the case of upstream seaports Antwerp and Hamburg. J. Transp. Geogr. 54, 295–309.
Notteboom, T.E., Pallis, A.A., Farrell, S., 2012. Guest editorial: terminal concessions in seaports revisited. Marit. Policy Manag. 39 (1), 1–5.
Ohmae, K., 1985. Triad Power: The Coming Shape of Global Competition. Flammarion, Paris (translation in French).
Rodrigue, J.P., Notteboom, T.E., 2010. Foreland-based regionalization: integrating intermediate hubs with port hinterlands. Res. Transp. Econ. 27 (1), 19–29.
Slack, B., Frémont, A., 2009. Fifty years of organisational change in container shipping: regional shift and the role of family firms. GeoJournal 74 (1), 23–34.
Tiwari, P., Itoh, H., Doi, M., 2003. Shippers' port and carrier selection behaviour in China: a discrete choice analysis. Marit. Policy Manag. 5 (1), 23–39.
Tongzon, J., 2001. Efficiency measurement of selected Australian and other international ports using data envelopment analysis. Transp. Res. A 35, 107–122.
Tongzon, J., 2009. Port choice and freight forwarders. Transp. Res. E 45, 186–195.
Wang, J.J., Chen, M.C., 2010. From a hub port city to a global supply chain management center: a case study of Hong Kong. J. Transp. Geogr. 18 (1), 104–115.
Wang, L., Zhu, Y., Ducruet, C., Bunel, M., Lau, Y.Y., 2018. From hierarchy to networking: the evolution of the 21st century Maritime Silk Road' container shipping system. Transp. Rev. 38 (4), 416–435.
Xu, H., Itoh, H., 2018. Density economies and transport geography: evidence from the container shipping industry. J. Urban Econ. 105, 121–132.

A global analysis of hinterlands from a European perspective

2

David Guerrero
AME-SPLOTT, Univ Gustave Eiffel, IFSTTAR

Introduction

This chapter examines the main determinants of hinterland evolution through a literature review. It explores the possibilities and challenges for interregional comparisons, and suggests a tentative common framework. The chapter mainly deals with European issues, though complementing them by integrating studies done on other regions. The literature review covers mainly the last two decades, during which containerisation have reached maturity in most world regions.

The chapter is organised as follows. The first section introduces some historical elements to understand the specificity of European hinterlands, mostly shaped during the 16th–18th centuries. The second section presents the current situation of European ports compared with the rest of the world. The third section reviews current determinants of hinterland expansion and shrinkage in various regional contexts. Finally, the conclusion discusses the need for pushing further the elaboration of a common framework notwithstanding challenges for interregional comparison.

Historical origin of European hinterlands

The current spatial configuration of European hinterlands has been largely prefigured at the early days of capitalism, during the 16th–18th centuries. Major ports such Antwerp and Rotterdam benefitted from their position within and old-established megalopolis stretching from the South of England to the North of Italy. In the early 16th century, with the development of Atlantic trade, Antwerp was preferred over Lisbon by the most important European merchant families, mostly German (Braudel, 1982). In the early 18th century, the Dutch were considered "*the Carryers of the World, the middle Persons in Trade, the Factors and Brokers of Europe*" (Defoe et al., 1728, quoted by Braudel, 1982, p. 239). At that period, "*a large proportion of colonial produce was bought by the Dutch at the auctions held by the East India Company; they also bought a great deal of tobacco, sugar, sometimes grain,* [...] *woollen cloths* – [...] *were stored in warehouses in Rotterdam and Amsterdam before being re-exported, chiefly to Germany*" (Defoe et al., 1728, quoted by Braudel, 1982, p. 261).

The later development of inland connections along the Rhine—waterways, during the 16th–18th centuries and railways, during the 19th century—further consolidated the dominance of Northern port cities, to the detriment of Northern Italian urban

Global Logistics Network Modelling and Policy. https://doi.org/10.1016/B978-0-12-814060-4.00002-2

centres. Hamburg also greatly benefitted from Atlantic trade, rising in the 17th century as an important European financial and trading centre, connecting England, Germany, the Baltics, Central and Eastern Europe, and the Mediterranean. Its emergence as a global trade centre was also closely connected to the development of an urban hierarchy of commercial cities within Germany (Lindberg, 2008).

In Rotterdam, Antwerp, and Hamburg, historical networks undoubtedly favoured early competition for common inland markets, prefiguring the contemporary competitive context of containerisation. Regions such as the East of France (Demangeon, 1918), the South of Germany, and many countries in the Central and Eastern Europe (Morgan, 1948) historically relied on several ports for their maritime trade. The port orientation of these regions often changed depending not just on market conditions, but also on the economic policies favouring or limiting the development of certain ports. At the end of WW1, Demangeon warned against protectionist policies aiming to promote the development of national ports by introducing special taxes to trade, "*artificially constraining the movement of foreign cargo trying to reach the sea through the shortest path*" (Demangeon, 1918, p. 314, translated by the author). Because French manufacturers could not always find the materials they needed within France, they paid surcharges to import them through the port of Antwerp. This put French manufacturers in a position of inferiority with regard to their German competitors (Demangeon, 1918, p. 337).

The importance of political barriers was later confirmed by Morgan (1948), who considered them as one of the main factors shaping hinterlands, together with market conditions. Based on the cases of Hamburg and Bremen, Morgan stressed on the need to analyse separately the regions next to the ports, often captive (*primary or 'captive' hinterland*) and those largely disputed between several ports (*secondary or 'contested' hinterland*). Eastern France, Southern Germany, and many Central and Eastern European countries are good examples of secondary hinterlands.

Before closing this historical parenthesis, there are two key lessons which can be drawn:

- The spatial organisation of hinterlands in mainland Europe has been strongly shaped by commercial and transport networks developed in the 16th–18th centuries. Although there has been some geographical disconnection between ports and trade centres during the 20th century (Ducruet, 2006), hinterlands remain path dependent.
- Despite an early development of liberalism in many sectors, European states have often promoted protectionist port policies. With the end of WW2 and the later development of a common market, internal customs barriers were progressively eliminated. However, certain habits remained, as evidenced by the way in which the French trade passing through Antwerp and Rotterdam is still called 'diverted' in many official documents (Court of Auditors/Cour des Comptes, 2006). France is probably not the only country employing this concept.

Analysing global hinterlands in a contemporary context

In the last three decades, with an increasing participation of emerging economies to global trade and the spread of containerisation, the position of European ports had rapidly weakened (Guerrero and Rodrigue, 2014). In 1985, three European ports were

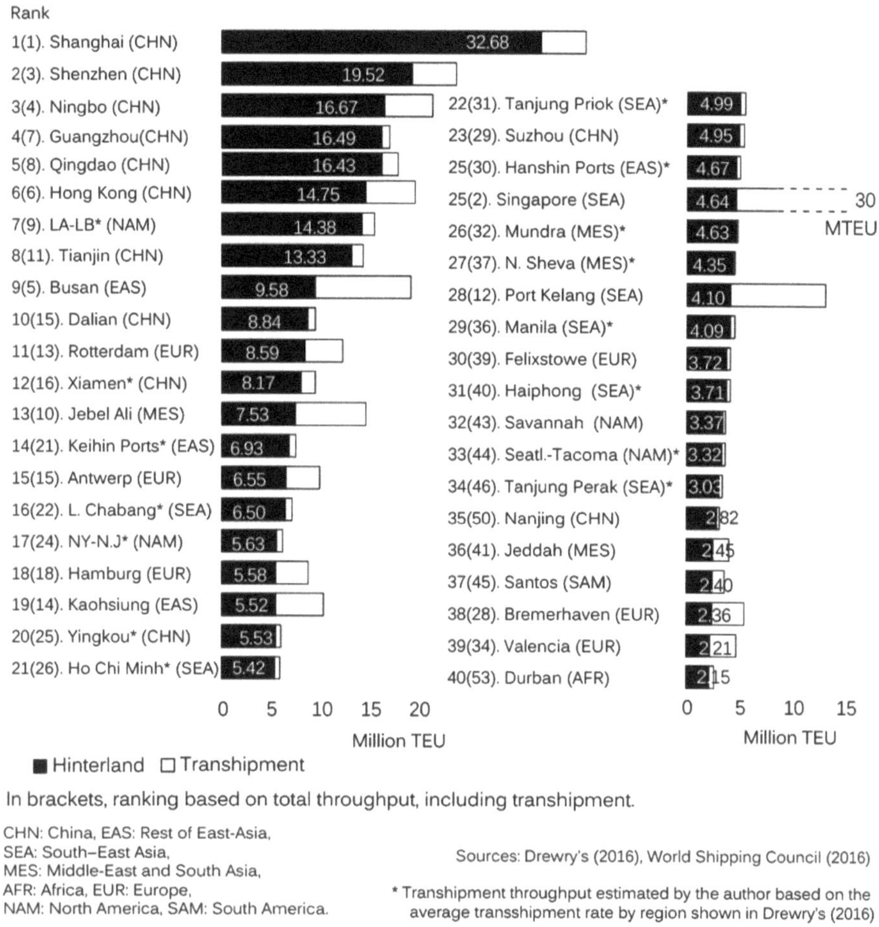

Fig. 2.1 Top global port gateways in 2016.
Data from Drewry Maritime Research, 2017. Container Market Annual Review and Forecast 2017/18.

in a global top-10 ports dominated by Rotterdam. In 2016, the top-10 ranking was dominated by six Chinese ports (Fig. 2.1) and the first European port ranked 11th.

This ranking (Fig. 2.1), based on the hinterland throughput of ports, brings to light a hierarchy of container ports slightly different than the usual one including transshipment traffic (rank in brackets). On the one hand, it relativizes the importance of transshipment hubs in Asia (e.g. Singapore), the Middle-East (e.g. Jebel Ali), and the Mediterranean (e.g. Valencia), with limited inland markets. So-called 'pure hubs' such as Tanjung Pelepas (19), Colombo (26), and Algeciras (32) do not even show up, although the latter also serves as a gateway port for the capital region Madrid. On the other hand, it highlights cases where a relatively weak transshipment function is partially offset by large inland markets. Chinese ports, already prominent when including transshipment, are even stronger as gateways. The same happens to top United States and Japanese ports, albeit on a lesser scale. Within the top-40 ports, the best represented regions are China (12), Rest of Asia (12), Europe (6), Middle East and South Asia (4), and North America (4). The minor contributions of South America (1) and Africa (1), which reflect the marginal contribution of these regions to the global container market.

Focusing on hinterland throughput does not imply to deny the advantages of sea-sea transshipment for ports, which usually facilitate the achievement of scale economies in cargo handling, intraport competition, and so on, eventually contributing to expand their inland markets. However, transshipment throughput is very different from the one generated by imports and exports. Moreover, transshipment throughput implies doubly counting, which artificially increases its importance. In view of these issues, there is a strong need to separate both types of throughput, in order to enable interport comparisons.

Determinants of hinterland expansion and shrinkage

A number of factors affect the extent and degree of competition of port hinterlands. Understanding these factors is an issue that has been widely studied in the academic literature.

One of the key factors that influence the success of ports is their geographic proximity to inland markets. This was demonstrated in several regional contexts such as Europe (Charlier, 1990; Moura et al., 2017; Guerrero, 2014), Japan (Itoh, 2013; Xu and Itoh, 2018), United States (Pitts, 1994; Malchow and Kanafani, 2004), South America (Tiller and Thill, 2017), and China (Wang et al., 2017). However, the impact of distance on inland flows varies considerably depending on the shape and the geographic character of countries. Some studies using gravity models showed that the sensitivity of flows to inland distance is lower in large countries with broad inland regions, such as China (Wang et al., 2017) and United States (Pitts, 1994), compared with much smaller countries such as France (Guerrero, 2014, 2018) or Italy (Ferrari et al., 2011). In these countries the lack of alternatives to road haulage has resulted in relatively high values of inland friction. The theoretical explanation of this parameter is however difficult since an increase or decrease in friction can result from very different factors, which usually go far beyond the sphere of action of port authorities (Anderson et al., 2009).

The aim of this study is not so much to provide theoretical explanation of the effects of distance on hinterlands (an interesting review on this topic has been recently done by Tiller and Thill, 2017) but to draw up a panorama of the factors influencing hinterland expansion or shrinkage in different regions of the world.

Variations in port choice behaviour

One of the most studied aspects affecting hinterland development is the different port choice behaviour depending on the characteristics of decision makers, either shippers or freight forwarders. Perceptions are usually analysed through (a) surveys on declared preferences (Tiwari et al., 2003; Tongzon, 2009; Nir et al., 2003) or (b) data on revealed preferences. The latter can be individual shipment data obtained from bills of lading (Malchow and Kanafani, 2004; Steven and Corsi, 2012; Martinez et al., 2016; Kashiha et al., 2016), or commodity flow surveys (Gouvernal et al., 2010). Whilst the declared preferences approach is better adapted to regions where data is scarce, studies on revealed preferences analyse the real choices of shippers and forwarders. Of course, the two approaches can lead to different results, as discussed by Tongzon (1995). These differences may arise from the respondent's profile, for example, top managers who are not familiar with what happens precisely on the ground. On the other side, quantitative surveys on revealed preferences hardly provide accurate information on key variables like the inland origin or destination of cargo within the hinterland, or the places where the containers are stuffed and unstuffed (Itoh and Tiwari, 2002). Yet, these two variables are essential to define the extent of hinterlands and to understand port choice.

Large versus small shippers

The size of the shipper turns to be an important factor affecting hinterlands. Whilst small-size shippers appear to be more concerned by cost factors, larger shippers seem more focused on the speed of delivery. This has been confirmed within the United States (Steven and Corsi, 2012) and Europe (Kashiha et al., 2016), using data from bills of lading of US deep-sea shipments. The main explanation for these differences in port choice behaviour rely on the different perceptions of the actors involved in international transport chains. Large shippers would have a broader vision of the costs of end-to-end transport chains than small shippers. The latter better focus on cost reduction only in the inland segment, overlooking factors like port efficiency (Kashiha et al., 2016) or the diversity and frequency of liner services (Tiwari et al., 2003). A significant presence of large shippers in certain regions would then favour larger hinterlands. The gains obtained by choosing most efficient ports and routes would eventually compensate additional costs of longer inland haulage. A high proportion of maritime trade in the United States (Steven and Corsi, 2012), Japan (Itoh et al., 2002), and Europe (Kashiha et al., 2016) is generated by large manufacturing and retail firms. This seems to be less the case in developing regions such as Africa (Pedersen, 2003).

Shippers versus freight forwarders

Another source of differences in hinterland structures is the degree of involvement of freight forwarders in the organisation of maritime shipments. The results of research surveys in Asia and Europe show that freight forwarders are more focused on cost reduction on the land segment than large shippers (Tiwari et al., 2003; De Langen, 2007; Tongzon, 2009). The latter are less focused on a single segment than on the whole transport chain, being more concerned by indirect costs such as unreliability, damage, and adverse reputation effects (Tongzon, 1995).

However, there may be differences between large and small freight forwarders. Fig. 2.2 shows a highly concentrated ocean freight forwarding market, with eight top

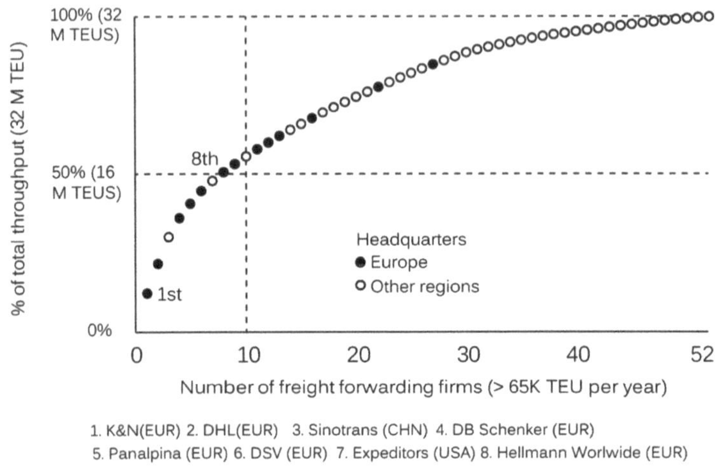

Fig. 2.2 Ocean freight forwarders dealing with more than 65K TEU per year. Data from A&A's 2018 Top 25 Global Freight Forwarders List. Ranked by 2017 Logistics Gross Revenue/Turnoverand Freight Forwarding Volumes* (https://www.3plogistics.com/3pl-market-info-resources/3pl-market-information/aas-top-25-global-freight-forwarders-list/).

firms concentrating more than 50% of the TEU throughput. Obviously, these large firms, mostly European, are in good position to negotiate preferential rates with several shipping companies, eventually being less dependent on specific companies and ports than small freight forwarders. These large freight forwarders can, depending on the needs of their clients, select more efficient door-to-door solutions, even if it does imply not choosing the shortest path. It was also demonstrated that in Europe, the concentration of small- and medium-sized freight forwarders in port city-regions was a symptom of supply chain inefficiency in a context of less-liberalised countries such as France and Eastern Europe (Ducruet and Van der Horst, 2009).

Freight forwarders make most of their business in Europe, Asia, and North America (Fig. 2.3). Their secondary markets are South America, and, to a much lesser extent, Africa. Their importance can also considerably differ within the same region. For example, South Africa is an exception within Africa, with many international freight forwarders. Despite the importance of the volumes handled by freight forwarders in Asia, their negotiation power against large shippers appears to be lower than that in Europe or the United States (Itoh and Tiwari, 2002). This results from the strong presence of large shippers, usually negotiating freight rates directly with maritime companies, and then reducing the profit opportunities for freight forwarders.

Intermodal connections

In Europe, hinterland intermodal connections are relatively well developed, particularly from Northern Range ports. The setting of strategies such as rail shuttles (Gouvernal and Daydou, 2005) and barge services (Frémont and Soppé, 2017; Franc

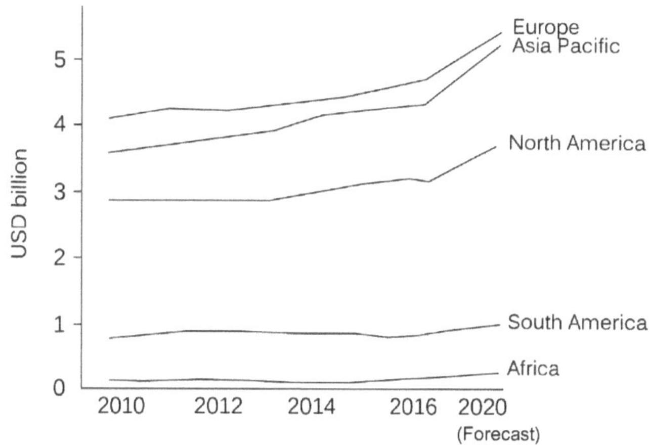

Fig. 2.3 Global freight forwarding market, by world region.
Data from Transport Intelligence, 2017. Global Freight Forwarding 2017 Report. https://www.ti-insight.com/product/global-freight-forwarding/.

and Van der Horst, 2010) contributed to expanding the hinterlands of the most efficient container ports. This was demonstrated by De Langen (2007), showing how the port choice of Austrian shippers and freight forwarders substantially changed in favour of Northern Range ports, following an upgrade on connections with the Rhine waterway system. In France as well, it has been shown, through a spatial interaction model of foreign trade flows, that when waterway and rail connections are available the impedance of inland distance is lower (Guerrero, 2018). However, not all types of traffic are equally sensitive to the existence of efficient alternatives to the road. The presence of high-capacity waterway connections between a port and an inland region has a positive impact on cargo flows, particularly for high value-added goods. This finding is counterintuitive, since waterway connections are often perceived as more adapted to low-value added cargo. Inland waterways are also used to convey high value-added products to the congested metropolitan areas. It is also a mean to delay deliveries to retailers, to save warehousing costs, and have the possibility to accelerate the flow by shifting from waterway to road if needed (Guerrero, 2018).

North America is probably the region where intermodal connections of ports are most developed, both in terms of distance and modal share of alternative modes. Efficient railway connections are used to connect the West and East coast, creating large areas of competition between ports. Though US railways are old, the development of intermodal services is relatively recent, mainly during the 1980s (Monios and Lambert, 2013). As compared to other regions the world, the conditions of demand (geographically concentrated, long interurban distances) and supply (long double-stack trains, competition between rail companies owning their own networks) make rail particularly competitive against road (Douet and Gouvernal, 2004). In this context, containers are often carried over long distances by rail to reach their final destinations. However, the international trade of North America is increasingly unbalanced, which makes it very difficult to find haulage cargo for containers. In order to avoid extra-container rental charges, and to improve the efficiency of loading, sometimes containers are unstaffed in warehouses at the vicinity of ports and stuffed in 52′ containers. Three 40′ containers can be staffed on two 52′ containers. According to Rodrigue and Notteboom (2009), about 25% of the rail cargo moved by rail is transferred to domestic containers.

In China, the development of intermodalism varies considerably across the country. Waterway transportation of containers is highly developed along the Yangtze River (more than 2000 km), with many river container terminals reaching altogether about 9 million TEU/year capacity (Veenstra and Notteboom, 2011). But the capacities and traffic remain highly concentrated on a relatively short segment, namely between Shanghai and Nanjing, the latter located 350 km from the sea. Besides the Yangtze, the port of Shanghai also handles a substantial part of the 1200 km distant Beijing [between 16% and 24% during 1994–2012 (Yang et al., 2016)]. But the depth of Shanghai's hinterland is exceptional within the context of China. The hinterlands of ports remain, in general, limited, and very concentrated in coastal regions (Lee et al., 2008; Wang et al., 2017).

In other developing regions, the integration between shipping lines and land transport remains very limited. In the context of South America, competitive hinterlands remain scarce, partly because of the lack of modal alternatives (Tiller and Thill, 2017; Ng et al., 2013). In Africa, the degree of development of intermodalism is slightly higher, with some East–West (and in a lesser extent North–South) railway lines connecting ports with landlocked countries. The development of intermodalism in Africa has been encouraged by overseas exporters aiming to protect their high-value manufactured goods all over the journey. However, the degree of integration of shipping lines with inland transport remains low, with a majority of containers being unstaffed and staffed in warehouses at the vicinity of ports. As reported by Pedersen (2003), in Ghana only 5% of the inbound containers continue inland service. In East Africa, the situation is slightly better, with a higher proportion of containers going inland (Hoyle and Charlier, 1995; Charlier, 1996). About 28% of Mombasa port's throughput value (Kenya) is generated by landlocked countries (TTCANC, 2017). In South Africa, the level of integration between rail and maritime transport is probably the highest in the region, with several ports serving the wealthy Gauteng province (Fraser and Notteboom, 2014). It is however difficult to obtain accurate data on the number of containers going inland, as compared to those which are stuffed and unstuffed in warehouses surrounding the ports.

Borders and natural barriers

The importance of borders and natural barriers has been recognised by early scholars. Sargent, for example, considered that natural characteristics played an important role in differentiating between hinterlands in different regions of the world (Sargent, 1938). The case of South America is used to illustrate the strong impedance of elevation to hinterland expansion. The exception are the plains in Argentina where the hinterlands for cereal exports look like the traditional (semi)circular areas organised around the ports (Sargent, 1938). Rivers also matter, since hinterlands are often organised along watersheds, which usually provide good conditions for the development of transport corridors. Even when the rivers are not navigable the valleys provide good conditions for land transportation (Comtois, 2012). More marginally, natural disasters such earthquakes or floods can temporarily obstruct inland connections or block certain ports. They can, in certain cases, accelerate the long-term decline of a port, as showed by Xu and Itoh (2018) for Kobe after the 1996 earthquake. In Europe, it was shown that the commodity diversity of ports' traffic was higher at island and peripheral locations such as behind the mountains and at peninsulas, as such ports are 'naturally protected' from competition across plains (Ducruet et al., 2010).

Another important factor influencing port hinterlands is the development of infrastructure overcoming physical obstacles. The recent upgrade of the Panama Canal, allowing the transit of larger container vessels (up to 14,000 TEUs), is expected to favour the expansion of US East Coast ports to the detriment of West coast ones, as showed by Martinez et al. (2016). The expected shifts, estimated for shippers located in Pittsburgh (600 km inland from the East Coast), are however different depending on the overseas origin of cargo: East Asian shipments would be more affected than Southeast Asian ones.

Borders also play an important role in the configuration of hinterlands, even within Europe, several decades after the introduction of single market. Borders have uneven impacts depending on the geographical position of countries: coastal or landlocked (Kashiha et al., 2016). The former are more constrained by inland distance than the latter. Whilst the countries with coastal borders are tied to their national logistics networks, the countries with landlocked borders have historically expanded their logistics networks to other countries in order to get access to the sea (Kashiha et al., 2016). The impact of borders on hinterland flows would also vary depending on other factors such the ports considered, or the direction of flows, import or export. For example, the effect of the alpine border would be stronger for Italian ports than for those of the Northern Range, both competing for the North Italian market (Ferrari et al., 2011). The latter differences are not just limited to language proficiency but also to commercial skills and the ability to cope with different cultures.

Traffic imbalances and empty container repositioning issues

Commercial asymmetries are an important obstacle to hinterland expansion. They affect all the regions of the world: mainly on East–West routes, but in North–South ones as well, albeit to a lesser extent. As shown in Fig. 2.4, the largest East–West imbalances take place between the Far East and its partners: North America, Europe, and the Middle-East. However, in relative terms, the most imbalanced routes are those where oil economies of the Mid-East are involved: inbound volumes are on average 2–3 times larger than outbound ones. Comparatively, North–South trades are less imbalanced (Fig. 2.5), and are globally in favour of Northern regions. The largest imbalances (in absolute terms) take place also on the routes involving the Far East: Australasia, Africa, and to a lesser extent Latin America. Important imbalances take place as well on Europe-related trades. In relative terms, the most unbalanced trades

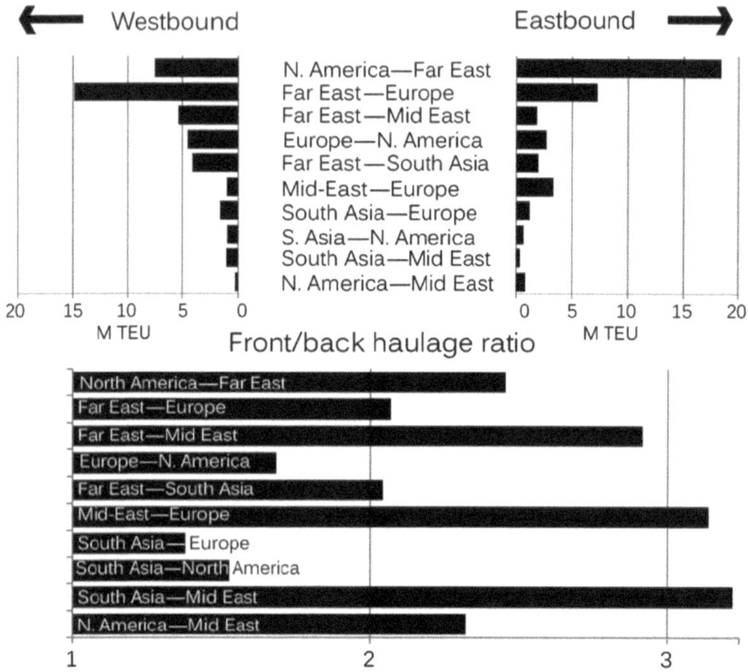

Fig. 2.4 Main East–West containerised flows in 2016.
Data from Drewry Maritime Research, 2017. Container Market Annual Review and Forecast 2017/18.

are also those involving Australasia, and secondarily Africa, both regions having a limited potential for generating containerised exports.

These imbalances result in large volumes of empty containers and transport inefficiencies impacting haul rates. In regions with limited potential for exporting containers, shippers and freight forwarders often prefer unstuffing containers near ports to avoid extra charges for container rental. Moreover, in these areas, shipping companies often prefer to get their empty containers back as soon as possible to maximise their utilisation (Rodrigue and Notteboom, 2009).

Port-centric logistics facilities are viewed by some as a solution to palliate part of the problems resulting from traffic imbalances. They imply container unstuffing at ports and redistribution of unpacked goods from port-based distribution centres (DCs). The elimination of running empty inland containers results in a reduction in inland transport costs (Mangan et al., 2008; Ng and Liu, 2014). However, when the inputs of DCs are sourced not exclusively from overseas, the location at ports, generally at the margins of countries, may not be optimal (Monios and Wilmsmeier, 2012; Acciaro and McKinnon, 2015). In developed regions such Europe, another problem related to port-centric logistics is that the gains obtained with the reduction in empty hauls would be partially offset by the increase in volume of cargo resulting from the use of pallets and individualised packages (Acciaro and McKinnon, 2015).

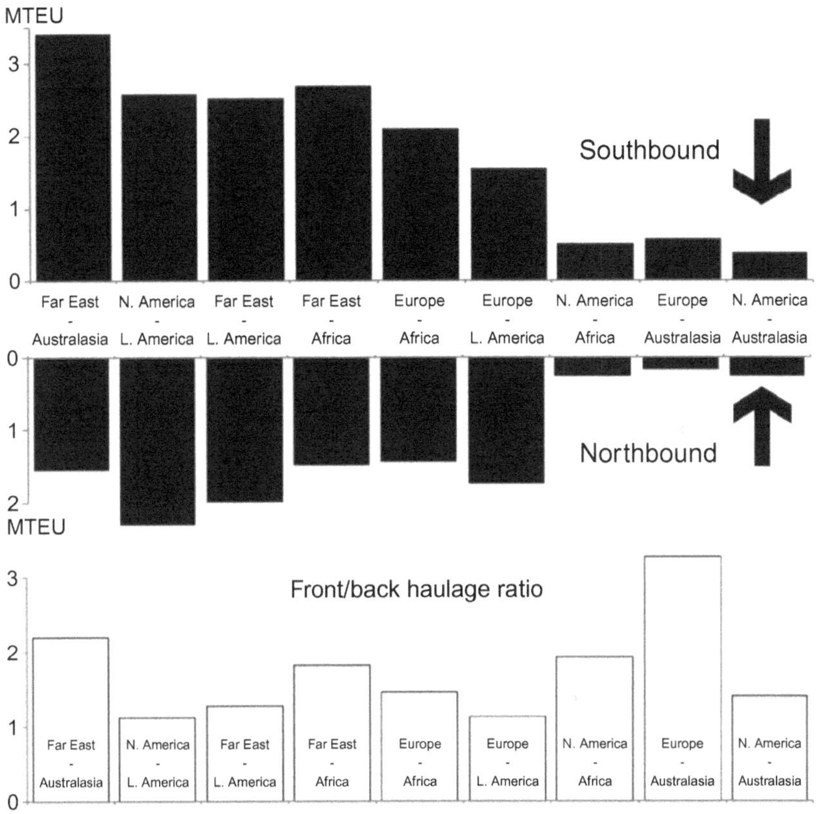

Fig. 2.5 Main North–South containerised flows in 2016.
Data from Drewry Maritime Research, 2017. Container Market Annual Review and Forecast 2017/18.

This list of determinants is not exhaustive. Some other 'forces' impacting hinterland shrinkage or expansion are presented in Fig. 2.6. This scheme shows that the forces driving hinterland shrinkage such as elevation or border crossing (b1–b3) generally affect more developing regions than developed ones. However, there are factors affecting both types of regions, as, for example, traffic imbalances or the likely increase in fuel cost or hinterland congestion issues (b4, b5, b6). Forces contributing to hinterland expansion mainly concern developed regions like North America, Europe, and some parts of China, where economies of scale achieved on maritime transport and port handling can be achieved on the land side as well by efficient intermodal integration (a2–a4, a6). The intermodal issue is probably less important in South America or Africa. The factors inducing more time-based competition (a5, a8) primarily concern developed countries and high value-added cargo. Traffic imbalances affect many regions, both developed and developing ones. America and Europe share the common issue of single market development (a1), which is less relevant in other parts of the world. In some cases, sharing a common language or culture could eventually

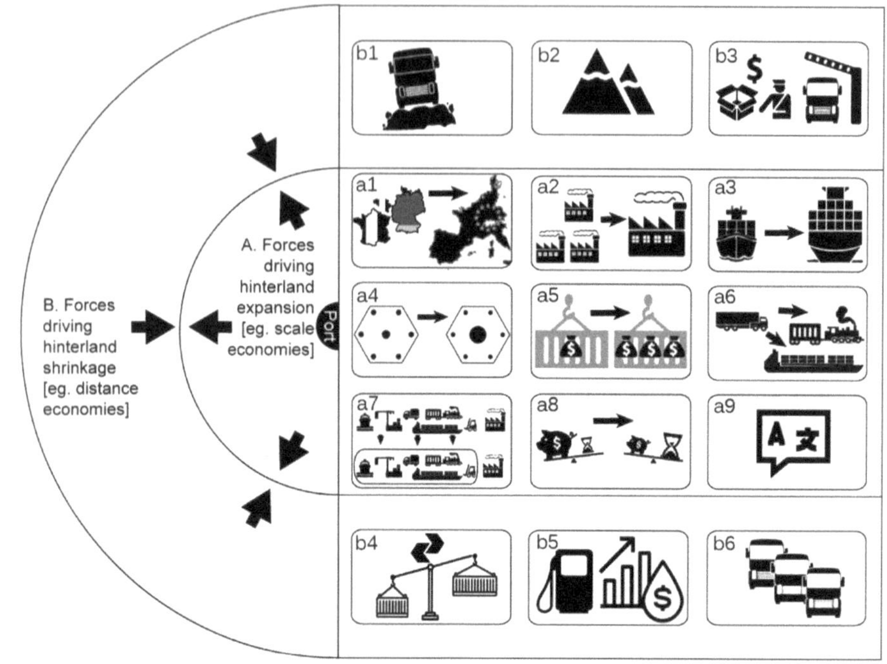

A: Forces driving hinterland expansion
a1: single market, a2: concentration of shippers, a3: increasing size of maritime vessels, a4: geographical concentration of demand, a5: increase of value density of cargo, a6: existence of efficient alternatives to road transport, a7: integration of carriers, a8: development of time-based competition, a9: existence of common business culture/ language.

B. Forces driving hinterland shrinkage
b1: poor inland transport conditions b2: physical barriers, b3: customs duties and compliance issues, b4: trade asymmetries and empty container repositioning issues, b5: increase of energy cost, b6: congestion.

Fig. 2.6 An overview of forces driving hinterland shrinkage and expansion.
Source: Own elaboration.

facilitate hinterland expansion, although counterexamples of this can be found as well (De Langen, 2007; Acciaro et al., 2017).

Conclusion

This chapter provided a literature review of hinterland developments in different regions of the world. It identified some factors of hinterland development enabling interregional comparisons. The review is not exhaustive and tends to overlook factors such as governance or regulation of competition, which are difficult to measure or to characterise at the global scale. Table 2.1 provides a synthesis of the main factors which have been reviewed. It shows that in most of the regions analysed, hinterlands change slowly. The main exception is China, where the rapid development of inland urban areas can favour further development of intermodalism.

Table 2.1 Characteristics of hinterlands in several world regions.

	North America	Western Europe	Japan	China	South America	Africa
Inland Competition	+++	++	+	+/++	−−−	−−
Intermodalism	+++	++	−	+/++	−	+
Large Shippers	+++	++	+++	++	−	−−−
Large Freight Forwarders	++	+++	+	+/++	+	−−/−
Low Inland Barriers	+++	+	−	−	−−−	−−
Throughput Balance	−−	+	++	++	+++	−−−
Related Works	Slack (1990), Fleming (1989), Hayuth (1988), Pitts (1994), Malchow and Kanafani (2004), Douet and Gouvernal (2004), Levine et al. (2009), Steven and Corsi (2012), and Martinez et al. (2016)	De Langen (2007), Garcia-Alonso and Sanchez-Soriano (2009), Ferrari et al. (2011), Monios and Wilmsmeier (2012), Guerrero (2014, 2018), and Moura et al. (2017)	Itoh et al. (2002), Itoh (2013), and Xu and Itoh (2018)	Tiwari et al. (2003), Lee et al. (2008), Monios and Wang (2013), Yang et al. (2016), and Wang et al. (2017)	Ng et al. (2013), Wilmsmeier and Monios (2016), and Tiller and Thill (2017)	Hoyle and Charlier (1995), Charlier (1996), Pedersen (2003), and Fraser and Notteboom (2014)

+, Well developed; − underdeveloped.
Source: Own elaboration.

Traffic imbalances and the places used for stuffing and stuffing containers appear to be important factors in hinterland development, which have been largely overlooked by the literature. Further research on these topics would contribute to a better understanding of the barriers to intermodal integration and hinterland expansion. Another promising avenue for further research is the in-depth analysis of activities generating maritime trade flows, to understand the ways in which shippers and freight forwarders interact with maritime companies in the choice of routes and ports.

Other regions such as South Asia, the Middle East, Australia, and Central America have not been studied here. Even if the level of integration of land and maritime transport is considered to be low in most of these areas, more research would be needed to understand shippers and forwarders and the ways in which they deal with unbalances.

The lack of reliable data remains a big challenge for the study of port hinterlands. This problem becomes more acute within the context of developing countries where exhaustive and comprehensive surveys on freight flows are extremely rare. Carrying out qualitative surveys from the actors involved in freight entails difficulties, but seems more adapted to this specific context.

References

Acciaro, M., McKinnon, A., 2015. Efficient hinterland transport infrastructure and services for large container ports. International Transport Forum. In, 75–101. https://doi.org/10.1787/9789282107850-en.

Acciaro, M., Bardi, A., Cusano, M.I., Ferrari, C., Tei, A., 2017. Contested port hinterlands: an empirical survey on Adriatic seaports. Case Stud. Transp. Policy 5 (2), 342–350.

Anderson, C.M., Opaluch, J.J., Grigalunas, T.A., 2009. The demand for import services at US container ports. Marit. Econ. Logist. 11 (2), 156–185.

Braudel, F., 1982. Civilization and Capitalism, 15th–18th Century, Vol. III: The Perspective of the World. vol. 3 University of California Press.

Charlier, J., 1990. L'arrière-pays national du port du Havre: une approche macro-géographique. In: L'espace Géographique, pp. 325–334.

Charlier, J., 1996. Multimodalisme et désenclavement en Afrique Sub-Saharienne. Cah. Géogr. Trop. 4, 117–128.

Comtois, C., 2012. Définition et périmètre des grands corridors de transport fluvio-maritime. In: Alix, Y. (Ed.), Les corridors de Transport. EMS Editions.

Court of Auditors/Cour des Comptes, 2006. Les ports français face aux mutations du transport maritime: L'urgence de l'action. La documentation Française. 187 p.

De Langen, P.W., 2007. Port competition and selection in contestable hinterlands; the case of Austria. Eur. J. Transp. Infrastruct. Res. 7 (1), 1–14.

Defoe, D., Cutler, N., Halley, E., 1728. Atlas Maritimus and Commercialis; or, A General View of the World.

Demangeon, A., 1918. Anvers. Ann. Georgr. vol. 27 (148/149), 307–339.

Douet, M., Gouvernal, E., 2004. Quels enseignements peut-on tirer pour l'Europe de l'analyse de la réglementation et de l'intermodalité aux Etats-Unis? Transports, 426.

Ducruet, C., 2006. Port-city relationships in Europe and Asia. J. Int. Logist. Trade 4 (2), 13–35.

Ducruet, C., Van der Horst, M.R., 2009. Transport integration at European ports: measuring the role and position of intermediaries. Eur. J. Transp. Infrastruct. Res. 9 (2), 121–142.

Ducruet, C., Koster, H.R.A., Van der Beek, D.J., 2010. Commodity variety and seaport performance. Reg. Stud. 44 (9), 1221–1240.
Ferrari, C., Parola, F., Gattorna, E., 2011. Measuring the quality of port hinterland accessibility: the Ligurian case. Transp. Policy 18 (2), 382–391.
Fleming, D.K., 1989. On the beaten track: a view of US West-Coast container port competition. Marit. Policy Manag. 16 (2), 93–107.
Franc, P., Van der Horst, M., 2010. Understanding hinterland service integration by shipping lines and terminal operators: a theoretical and empirical analysis. J. Transp. Geogr. 18 (4), 557–566.
Fraser, D., Notteboom, T., 2014. A strategic appraisal of the attractiveness of seaport-based transport corridors: the Southern African case. J. Transp. Geogr. 36, 53–68.
Frémont, A., Soppé, M., 2017. Northern European range: shipping line concentration and port hierarchy. Ports, Cities, and Global Supply Chains. Routledge, pp. 121–136.
Garcia-Alonso, L., Sanchez-Soriano, J., 2009. Port selection from a hinterland perspective. Marit. Econ. Logist. 11 (3), 260–269.
Gouvernal, E., Daydou, J., 2005. Container railfreight services in north-west Europe: diversity of organizational forms in a liberalizing environment. Transp. Rev. 25 (5), 557–571.
Gouvernal, E., Slack, B., Franc, P., 2010. Short sea and deep sea shipping markets in France. J. Transp. Geogr. 18 (1), 97–103.
Guerrero, D., 2014. Deep-sea hinterlands: some empirical evidence of the spatial impact of containerization. J. Transp. Geogr. 35, 84–94.
Guerrero, D., 2018. Impacts of transport connections on port hinterlands. Reg. Stud. https://doi.org/10.1080/00343404.2018.1474192.
Guerrero, D., Rodrigue, J.P., 2014. The waves of containerization: shifts in global maritime transportation. J. Transp. Geogr. 34, 151–164.
Hayuth, Y., 1988. Rationalization and deconcentration of the US container port system. Prof. Geogr. 40 (3), 279–288.
Hoyle, B., Charlier, J., 1995. Inter-port competition in developing countries: an East African case study. J. Transp. Geogr. 3 (2), 87–103.
Itoh, H., 2013. Market area analysis of port in Japan: an application of fuzzy clustering. In: International Conference of the International Association of Maritime Economists, Marseille, France.
Itoh, H., Tiwari, P., 2002. An analysis of cargo transportation behaviour in Kita Kanto (Japan). Int. J. Transp. Econ. 29 (3), 1000–1017.
Itoh, H., Tiwari, P., Doi, M., 2002. An analysis of cargo transportation behaviour in Kita Kanto (Japan). Int. J. Transp. Econ. 319–335.
Kashiha, M., Thill, J.C., Depken II, C.A., 2016. Shipping route choice across geographies: coastal vs. landlocked countries. Transport. Res. E Log. 91, 1–14.
Lee, S.W., Song, D.W., Ducruet, C., 2008. A tale of Asia's world ports: the spatial evolution in global hub port cities. Geoforum 39 (1), 372–385.
Levine, B., Nozick, L., Jones, D., 2009. Estimating an origin–destination table for US imports of waterborne containerized freight. Transp. Res. E Log. Transp. Rev. 45 (4), 611–626.
Lindberg, E., 2008. The rise of Hamburg as a global marketplace in the seventeenth century: a comparative political economy perspective. Comp. Stud. Soc. Hist. 50 (3), 641–662.
Malchow, M.B., Kanafani, A., 2004. A disaggregate analysis of port selection. Transport. Res. E Log. 40 (4), 317–337.
Mangan, J., Lalwani, C., Fynes, B., 2008. Port-centric logistics. Int. J. Logist. Manag. 19 (1), 29–41.

Martinez, C., Steven, A.B., Dresner, M., 2016. East Coast vs. West Coast: the impact of the Panama Canal's expansion on the routing of Asian imports into the United States. Transport. Res. E Log. 91, 274–289.

Monios, J., Lambert, B., 2013. Intermodal freight corridor development in the United States. In: Dry Ports—A Global Perspective, Challenges and Developments in Serving Hinterlands, pp. 197–218.

Monios, J., Wang, Y., 2013. Spatial and institutional characteristics of inland port development in China. GeoJournal 78 (5), 897–913.

Monios, J., Wilmsmeier, G., 2012. Port-centric logistics, dry ports and offshore logistics hubs: strategies to overcome double peripherality? Marit. Policy Manag. 39 (2), 207–226.

Morgan, F.W., 1948. The pre-war hinterlands of the German North Sea ports. Trans. Pap. 14, 45–55.

Moura, T.G.Z., Garcia-Alonso, L., Salas-Olmedo, M.H., 2017. Delimiting the scope of the hinterland of ports: proposal and case study. J. Transp. Geogr. 65, 35–43.

Ng, A., Liu, J., 2014. Port-Focal Logistics and Global Supply Chains. Palgrave-McMillan.

Ng, A.K., Padilha, F., Pallis, A.A., 2013. Institutions, bureaucratic and logistical roles of dry ports: the Brazilian experiences. J. Transp. Geogr. 27, 46–55.

Nir, A.S., Lin, K., Liang, G.S., 2003. Port choice behaviour-from the perspective of the shipper. Marit. Policy Manag. 30 (2), 165–173.

Pedersen, P., 2003. Development of freight transport and logistics in sub-Saharan Africa: Taaffe, Morrill and Gould revisited. Transp. Rev. 23 (3), 275–297.

Pitts, T.C., 1994. Inter-Port Competition and Cargo Tributary Areas for International Containerized Exports from the United States. doctoral thesis, State University of New York at Buffalo.

Rodrigue, J.P., Notteboom, T., 2009. The terminalization of supply chains: reassessing the role of terminals in port/hinterland logistical relationships. Marit. Policy Manag. 36 (2), 165–183.

Sargent, A.J., 1938. Seaports & Hinterlands. A. and C. Black.

Slack, B., 1990. Intermodal transportation in North America and the development of inland load centers. Prof. Geogr. 42 (1), 72–83.

Steven, A.B., Corsi, T.M., 2012. Choosing a port: an analysis of containerized imports into the US. Transport. Res. E Log. 48 (4), 881–895.

Tiller, K.C., Thill, J.C., 2017. Spatial patterns of landside trade impedance in containerized South American exports. J. Transp. Geogr. 58, 272–285.

Tiwari, P., Itoh, H., Doi, M., 2003. Shippers' port and carrier selection behaviour in China: a discrete choice analysis. Marit. Econ. Logist. 5 (1), 23–39.

Tongzon, J.L., 1995. Determinants of port performance and efficiency. Transp. Res. A Policy Pract. 29 (3), 245–252.

Tongzon, J.L., 2009. Port choice and freight forwarders. Transport. Res. E Log. 45 (1), 186–195.

TTCANC, 2017. Northern Corridor Transport Observatory Report. 11th Issue, November. http://top.ttcanc.org/download_doc.php?docid=1520500390883l9398.

Veenstra, A., Notteboom, T., 2011. The development of the Yangtze River container port system. J. Transp. Geogr. 19 (4), 772–781.

Wang, L., Goodchild, A., Wang, Y., 2017. The effect of distance on cargo flows: a case study of Chinese imports and their hinterland destinations. Marit. Econ. Logist., 1–20.

Wilmsmeier, G., Monios, J., 2016. Institutional structure and agency in the governance of spatial diversification of port system evolution in Latin America. J. Transp. Geogr. 51, 294–307.

Xu, H., Itoh, H., 2018. Density economies and transport geography: evidence from the container shipping industry. J. Urban Econ. 105, 121–132.

Yang, J., Luo, M., Ji, A., 2016. Analyzing the spatial–temporal evolution of a gateway's hinterland: a case study of Shanghai, China. Transport. Res. E Log. 95, 355–367.

Cross-border logistics practices, policies, and its impact

3

Masahiko Furuichi
International Association of Ports and Harbors (IAPH), The University of Tokyo

Introduction

The global logistics network continues to expand between oceanic and inland spaces, by connecting with maritime shipping, hinterland transport, and inventory management in warehouses. In particular, the maritime container transport network significantly evolved to such an extent that it has increased its importance in the global supply chain.

Container transport between shippers and consignees, however, has been studied mainly from the ocean shipping lines perspective, focusing on the delivery of containers between ports. Container terminal operators provide loading/unloading services to/from ocean shipping lines and inland transporters. Inland transporters deliver containers between ports and warehouses in the case of Less than Container Loaded (LCL), and between ports and final destinations (i.e. consignees), in the case of Full Container Loaded (FCL). Multimodal container transport services between shippers and consignees remain fragmented amongst specialised service providers (Ducruet and Van der Horst, 2009). Furthermore, inland transporters often provide inventory management services through their own warehouses in the global supply chain.

The modelling of the global logistics network needs to integrate both maritime shipping and inland transport into a combined perspective, taking into account the whole management of multimodal container transport between shippers and consignees. This chapter focuses on the practical manner of multimodal container transport chain, especially from a viewpoint of cross-border logistics in the hinterland.

Logistics performance and liner shipping connectivity

Globalised production processes together with growing division of labour fostered international, intraindustry trade of both raw materials and intermediate products, thereby requiring a higher logistics service quality, in terms of just-in-time (JIT) deliveries' reliability, service frequency, and lower transport costs. Developing countries that want to participate in global production processes need higher quality of logistics services in the region and better access to global markets. Furthermore, national competitiveness deeply relies on the quality of domestic and international logistics services.

Both the Logistics Performance Index (LPI) developed by World Bank and the Liner Shipping Connectivity Index (LSCI) developed by UNCTAD aim at providing

Global Logistics Network Modelling and Policy. https://doi.org/10.1016/B978-0-12-814060-4.00003-4

information on countries' trade competitiveness in logistics and maritime container transport in different ways. The LPI in particular aims to reflect the quality of logistics services/environments of each country. Liner shipping connectivity is a crucial determinant of bilateral trade between trade partners, because most cargoes except dry bulk, liquid bulk (i.e. manufactured goods), intermediate products, and even raw materials, are increasingly containerised due to the globalisation of maritime container transport services. The LSCI has been proposed to measure an integration level of countries into the existing liner shipping network as a proxy of the accessibility to the global markets.

Logistics performance index[a]

Logistics is understood as a network of services that supports the physical movement of goods, trade across borders, and commerce within countries (Rodrigue et al., 2013). A wide variety of activities beyond transportation are observed, such as warehousing, brokerage, express delivery, terminal operations, and related data and information management. Consignees are required to avoid a high uncertainty as to when and how deliveries are to be completed in a global supply chain. Reliability is more important than speedy delivery, for which many shippers are willing to pay a premium. Supply chain predictability is a primary quality of shipment compared with delivery time and cost.

The LPI analyses countries through six components, namely:

(1) Customs: efficiency of customs and border management operation
(2) Infrastructure: quality of trade and transport infrastructure
(3) Ease: ease of competitively priced international shipments
(4) Quality: quality of logistics services
(5) Timeliness: frequency with which shipments are delivered to consignees within the scheduled or expected delivery time
(6) Tracking and tracing: ability to track and trace shipment.

These components were selected following the theoretical and empirical analyses and the practical experiences in international freight forwarding. The six LPI components are classified into two main categories:

(i) Policy regulation, which includes main inputs to the supply chain (i.e. customs, infrastructure, and quality).
(ii) Supply chain performance (i.e. timeliness, ease, and tracking and tracing).

The standard statistical techniques are applied to LPI by aggregating the data into a composite indicator, which enables comparison amongst countries, regions, and income groups.

The top 10 countries ranked by LPI in 2018 are high-income countries, most of which are in Europe (see Table 3.1), because these countries are traditionally dominant in the global supply chain services. On the other hand, the bottom 10 countries ranked by LPI in 2018 are mostly low-income and lower-middle-income countries in Africa or isolated regions (see Table 3.2). These are either fragile economies affected by armed conflict, natural disasters, and lack of political stability, or landlocked countries facing diseconomies of scale in connecting global supply chains due

[a] Arvis et al. (2018).

Table 3.1 Top 10 LPI economies in 2018.

Economy	2018		2016		2014		2012	
	Rank	Score	Rank	Score	Rank	Score	Rank	Score
Germany	1	4.20	1	4.23	1	4.12	4	4.03
Sweden	2	4.05	3	4.20	6	3.96	13	3.85
Belgium	3	4.04	6	4.11	3	4.04	7	3.98
Austria	4	4.03	7	4.10	22	3.65	11	3.89
Japan	5	4.03	12	3.97	10	3.91	8	3.93
Netherlands	6	4.03	4	4.19	2	4.05	5	4.02
Singapore	7	4.00	5	4.14	5	4.00	1	4.13
Denmark	8	3.99	17	3.82	17	3.78	6	4.02
United Kingdom	9	3.99	8	4.07	4	4.01	10	3.90
Finland	10	3.97	15	3.92	24	3.62	3	4.05

Reproduced with permission from Arvis, J.F., Ojala, L., Wiederer, C., Shepherd, B., Raj, A., Dairabayeva, K., 2018. Connecting to Compete 2018: Trade Logistics in the Global Economy, World Bank.

Table 3.2 Bottom 10 LPI economies in 2018.

Economy	2018		2016		2014		2012	
	Rank	**Score**	**Rank**	**Score**	**Rank**	**Score**	**Rank**	**Score**
Afghanistan	160	1.95	150	2.14	158	2.07	135	2.30
Angola	159	2.05	139	2.24	112	2.54	138	2.28
Burundi	158	2.06	107	2.51	107	2.57	155	1.61
Niger	157	2.07	100	2.56	130	2.39	87	2.69
Sierra Leone	156	2.08	155	2.03	N.A.	N.A.	150	2.08
Eritrea	155	2.09	144	2.17	156	2.08	147	2.11
Libya	154	2.11	137	2.26	118	2.50	137	2.28
Haiti	153	2.11	159	1.72	144	2.27	153	2.03
Zimbabwe	152	2.12	151	2.08	137	2.34	103	2.55
Central African Rep.	151	2.15	N.A.	N.A.	134	2.36	98	2.57

Reproduced with permission from Arvis, J.F., Ojala, L., Wiederer, C., Shepherd, B., Raj, A., Dairabayeva, K., 2018. Connecting to Compete 2018: Trade Logistics in the Global Economy, World Bank.

to natural isolation from the maritime transport services. A recent application is the Mediterranean region including both North African and Southern European countries (Arvis et al., 2018). The LPI is a useful tool when understanding a level of comprehensive logistics performance of a country as at-a-glance picture.

When looking at the cumulative distribution of the LPI scores, five LPI quintile groups are revealed. Each group contains the same number of countries within a quintile range of LPI score (see Fig. 3.1). The top LPI quintile is the so-called 'logistics-friendly' which includes top-performing countries; most of the countries are in the high-income group. The second LPI quintile is the so-called 'consistent performers' of which scores are rated based on logistics performance in their income group. The third and fourth LPI quintile is recognised as 'partial performers' of which logistics constraints are relatively significant, often seen in low- and middle-income countries. Finally, the bottom LPI quintile is the so-called 'logistics-unfriendly' of which logistics constraints are severe, such as the least developed countries (Arvis et al., 2018).

Liner shipping connectivity index[b]

The LSCI is an indicator of the position of a country amongst those countries that are connected to the global liner shipping networks. The LSCI is calculated based on the following five components:

(1) the number of containerships that are deployed on services to/from a country's ports;
(2) the total annual container-carrying capacity of those ships;

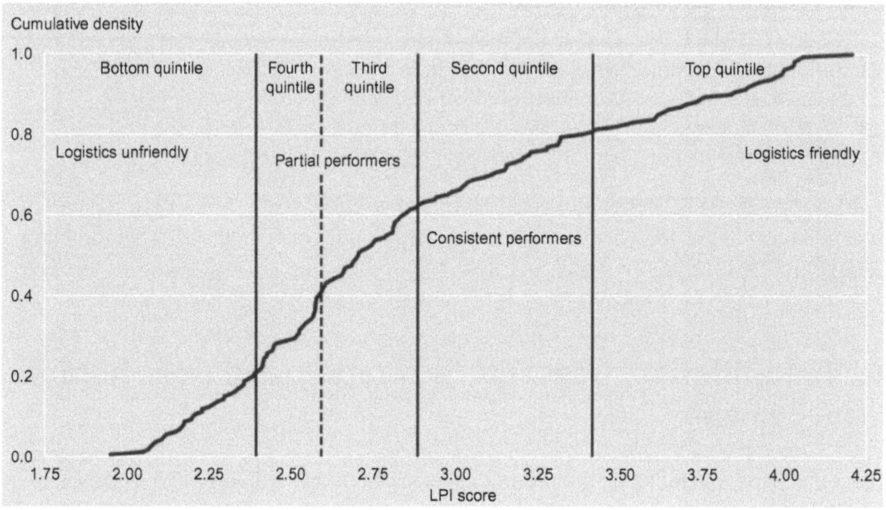

Fig. 3.1 Cumulative distribution of LPI scores in 2018.
Reproduced with permission from Arvis, J.F., Ojala, L., Wiederer, C., Shepherd, B., Raj, A., Dairabayeva, K., 2018. Connecting to Compete 2018: Trade Logistics in the Global Economy, World Bank.

[b] UNCTAD (2018).

(3) the maximum vessel size;
(4) the number of services; and
(5) the number of companies that deploy those containerships.

The data are derived from Containerisation International Online (until 2015) and MDS Transmodal (from 2016 onwards). The base year of LSCI is 2004, and its base value is calculated from countries' maximum score for 2004. For each of the five components for the target year:

(1) a country's value for the target year is divided by the maximum value of that component in 2004;
(2) the averages of the five components are calculated for the target year;
(3) each average for the target year is then divided by each maximum average for 2004 and multiplied by 100; and
(4) LSCI generates the value 100 for the country with the highest average index of the five components in 2004.

Top economy of the LSCI in 2018 was China, which was best connected to the global liner shipping network, followed by Singapore, the Republic of Korea, Hong Kong SAR, and Malaysia (see Table 3.3). The LSCI is closely related to trade costs and trade competitiveness, and reflects changes in both demand and services provided by shipping carriers, which depend on their vessel deployment strategy and their responses to port investments and reforms in the container ports of countries (UNCTAD, 2018) (Table 3.4).

Liner Shipping Bilateral Connectivity Index ($LSBCI_{jk}$) between country j and country k is defined for a country pair jk by calculating the following five components (Fugazza and Hoffmann, 2017):

(1) the theoretical minimum number of transshipments required from country j to country k;
(2) the number of common direct connections in each country pair jk;
(3) the geometric mean of the number of direct connections in each country pair jk;
(4) the level of competition on services that connect country pair jk; and
(5) the size of the largest ships on the weakest (thinnest) route in each country pair jk.

Most top-20 LSBCI economies are observed within Europe and within Eastern and South-Eastern Asia. However, China is relatively well connected also with the United States (see Table 3.5).

Trade facilitation, transport facilitation, and cross-border management

The efficient movement of goods and services between shippers and consignees is essentially supported by (1) trade facilitation, (2) transport facilitation, (3) efficient cross-border management, and (4) adequate logistics infrastructure. These logistics environments result in reduced trade costs, transit time, and bottlenecks for all stakeholders within the global supply chain. Since traders require seamless supply chains to compete in the global economy, participating economies are required to provide efficient logistics environments.

Table 3.3 Top 10 LSCI economies in 2018.

Economy	2018		2016		2014		2012	
	Rank	Score	Rank	Score	Rank	Score	Rank	Score
China	1	187.8	1	169.2	1	165.1	1	156.2
Singapore	2	133.9	2	118.5	3	113.2	3	113.2
Korea Rep.	3	118.8	3	111.4	4	108.1	4	101.7
Hong Kong (China)	4	113.5	5	106.9	2	116.0	2	117.2
Malaysia	5	109.9	4	108.9	5	104.0	5	99.7
Netherlands	6	98.0	9	89.9	7	94.2	8	88.9
Germany	7	97.1	6	94.1	8	94.0	7	90.6
United States	8	96.7	8	92.0	6	95.1	6	91.7
United Kingdom	9	95.6	7	92.3	9	88.0	9	84.0
Belgium	10	91.1	10	84.6	10	80.8	10	78.9

Reproduced with permission from UNCTAD Website https://unctad.org/en/PublicationsLibrary/UNCTAD_Liner_Shipping_Connectivity_Index_2004-2018.xlsx (Accessed 15/07/2019).

Table 3.4 Bottom 10 LSCI economies in 2018.

Economy	2018		2016		2014		2012	
	Rank	Score	Rank	Score	Rank	Score	Rank	Score
Norfolk Island	1	0.6	1	0.5	–	–	–	–
Christmas Island	2	0.9	–	–	–	–	–	–
Cayman Island	3	1.2	4	1.2	1	1.1	23	4.1
Bermuda	4	1.5	5	1.5	4	1.5	3	1.6
Tuvalu	5	1.6	12	2.4	–	–	–	–
Wallis and Futuna Islands	6	1.6	10	2.3	–	–	–	–
Nauru	7	1.9	7	1.9	–	–	–	–
Cook Islands	8	2.0	8	2.2	–	–	–	–
Greenland	9	2.3	13	2.4	6	2.3	7	2.3
Timor-Leste	10	2.5	14	2.5	–	–	–	–

Reproduced with permission from UNCTAD Website https://unctad.org/en/PublicationsLibrary/UNCTAD_Liner_Shipping_Connectivity_Index_2004-2018.xlsx (Accessed 15/07/2019).

Table 3.5 Top 20 LSBCI country pars in 2014.

Rank	Exporter	Importer	Score	Rank	Exporter	Importer	Score
1	Netherlands	United Kingdom	0.86	11	Singapore	China	0.81
2	Netherlands	Germany	0.86	12	Korea	Hong Kong	0.80
3	United Kingdom	Belgium	0.95	13	United Kingdom	France	0.80
4	Hong Kong	China	0.85	14	Netherlands	France	0.79
5	Korea	China	0.85	15	Italy	Spain	0.79
6	Netherlands	Belgium	0.84	16	France	Belgium	0.79
7	Germany	Belgium	0.83	17	France	Spain	0.78
8	Malaysia	China	0.83	18	France	Germany	0.78
9	United Kingdom	Germany	0.82	19	Malaysia	Hong Kong	0.77
10	Singapore	Malaysia	0.82	20	United States	China	0.76

Reproduced with permission from UNCTAD, 2016. Bilateral Liner Shipping Connectivity Since 2006. Policy Issues in International Trade and Commodities Research Study Series No. 72.

Trade facilitation[c]

Trade facilitation may reduce trade costs and transit time needed to export and import goods across the borders, by simplifying paperwork, modernising procedures, and harmonising customs requirements. Reductions in trade costs and transit time can vary between countries seamlessly linked to an integrated global production chain or being unlinked to the chain.

The WTO Trade Facilitation Agreement (TFA) became effective on 22 February 2017. With the trade facilitation agreement, WTO members aim at

(1) expediting 'the movement, release, and clearance of goods, including goods in transit';
(2) facilitating the 'effective cooperation amongst members on trade facilitation and customs clearance issues'; and
(3) enhancing 'assistance and support for capacity building' for developing and least developed country members.

The full implementation of the TFA is estimated to reduce global trade costs by an average of 14.3%. African countries and least-developed countries (LDCs) are forecasted to gain the biggest average reductions in trade costs, 16.5% and 16.7%, respectively (Moïséi and Sorescu, 2013). Full implementation has also been found to potentially reduce the average import time by a day and a half (by 47%). Reduction in export time may be even more significant, which is estimated at 2 days (by an average of 91%) (Hillberry and Zhang, 2015).

By reducing trade costs and transit time, the TFA is expected to encourage the existing exporters whilst enabling new companies to export for the first time. Furthermore, the TFA is forecasted to increase the annual growth of world export by up to 2.7% and the annual world GDP growth by more than 0.5% for the period of 2015–30. Developing countries are expected to enjoy larger gains than the global average. Overall, two-thirds of all benefits are predicted to be delivered to the developing and least-developed world (WTO, 2015).

The fundamental principles of trade facilitation consist of four pillars: (1) transparency, (2) simplification, (3) harmonisation, and (4) standardisation (see Fig. 3.2). Transparency implemented within governmental and administrative actions promotes openness and accountability. Simplification is a series of actions, i.e. eliminating all unnecessary processes and procedures as well as duplications in trade formalities. Harmonisation is the alignment of national procedures, operations, and documents with international conventions, standards, and practices. Standardisation is a series of processes of developing formats, documents, and information internationally agreed by various parties. Full cooperation is essential to achieve these principles, between government authorities and the business community.

From the supply chain perspective, the Buy-Ship-Pay (BSP) Model developed by UN/CEFACT (United Nations Centre for Trade Facilitation and Electronic Business) is an example of the theoretical patterns. It demonstrates the supply chain as a sequence of business processes that can be grouped into the domains of (1) Buy, (2) Ship, and (3) Pay (see Fig. 3.3).

[c] UNECE Website http://tfig.unece.org/details.html (Accessed 15/07/2019).

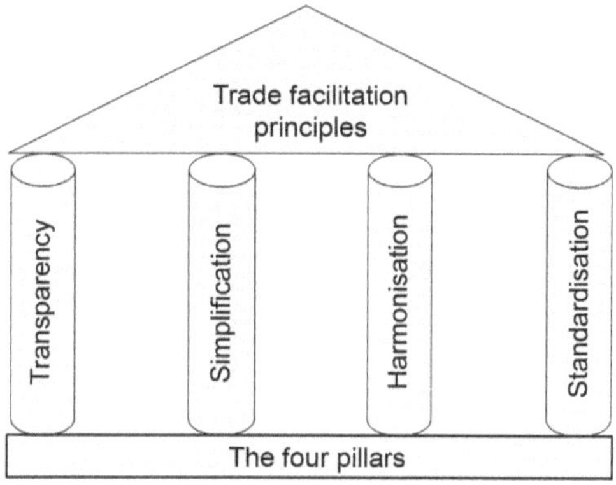

Fig. 3.2 Four pillars of trade facilitation.
Reproduced with permission from UNECE Website http://tfig.unece.org/details.html (Accessed 15/07/2019).

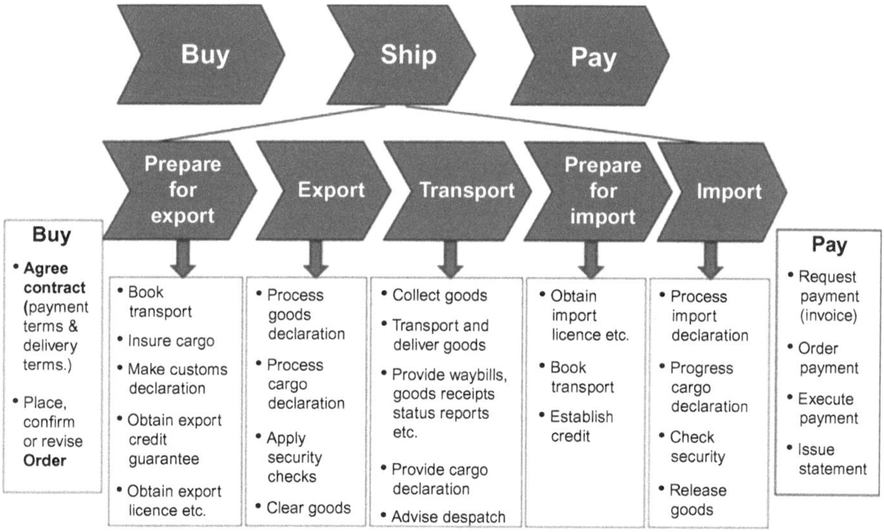

Fig. 3.3 UN/CEFACT Buy Ship Pay Reference Model.
Reproduced with permission from The UN/CEFACT Buy-Ship-Pay Reference Models http://tfig.unece.org/contents/buy-ship-pay-model.htm (Accessed 15/07/2019).

Transport facilitation

TIR convention[d]

Work on the TIR transit system (i.e. Transports Internationaux Routiers or International Road Transports) commenced immediately after the World War II under the support of UNECE (United Nations Economic Commission for Europe). The first TIR Agreement was concluded in 1949 amongst a small number of European countries.

[d] UNECE (2018).

The negotiation of a TIR Convention was adopted in 1959 by the Inland Transport Committee and entered into force in 1960, following the success of the first TIR agreement. The first TIR Convention was revised during a Review Conference in November 1975, to give effect to technical advances and changing customs and transport requirements. The revised TIR Convention of 1975 became effective in 1978.

The TIR Convention has 73 Contracting Parties, including the European Union. It covers a large part of the Eurasian continent, reaches out to North Africa, and has Contracting Parties in North and South America. China, India, Pakistan, the State of Palestine, and Qatar are the latest countries having recently acceded to the Convention.

There is no doubt that harmonised border crossing procedures, such as the TIR procedure, are essential to leverage the benefits of infrastructure projects. Furthermore, efficient and secured international transit not only contributes to improving the global supply chain but also avoids wasting precious resources at border crossings. The implementation of the TIR Convention can also help countries to meet the numerous transit-related objectives of the WTO Trade Facilitation Agreement, which became effective on 22 February 2017. Concrete and recent examples include China's intention to develop, under the umbrella of the One Belt, One Road strategy (OBOR), cross-border freight corridors through Myanmar and Israel to avoid the Malacca Strait and the Suez Canal, respectively, as well as the Panama Canal via a possible new canal through Nicaragua. Previous examples were successful such as Vancouver's Asia-Pacific corridor initiative in Canada. In regions marked by strong border blockade like the MENA (Middle East and North Africa), the deep-sea traffic of developing countries such as Algeria is routed through external maritime hubs such as Marsaxlokk (Malta) and Tangier-Med (Morocco). The same occurs in Southeast Asia for countries whose external trade, dominantly maritime, is routed through the Singapore hub, until the launch of port expansion projects such as Tanjung Priok (Jakarta) and Port Klang, Tanjung Pelepas (Malaysia). Even closed economies like North Korea opted for a strategy to ensure transit trade through its ports such as Rajin for Russian, Japanese, and South Korea trade (Jo and Ducruet, 2007).

In order to ensure a minimum interference 'en route' and offer maximum safeguards to customs administrations for border-crossing goods, the TIR system provides the following five basic requirements (see Fig. 3.4):

(1) Goods should move in customs secured vehicles or containers.
(2) Throughout the border-crossing delivery, duties and taxes at risk should be covered by an internationally valid guarantee.
(3) Goods should be accompanied by an internationally accepted customs document (TIR Carnet), serving as a customs control document in the countries of departure, transit, and destination.
(4) Customs control measures taken in the country of departure should be accepted by all countries of transit and destination.

UNECE definition of transport facilitation[e]

Transport facilitation is a part of trade facilitation and defined by UNECE as the simplification and harmonisation of international transport procedures and the information

[e] UNECE Website http://www.unece.org/trans/theme_facilitation.html (Accessed 15/07/2019).

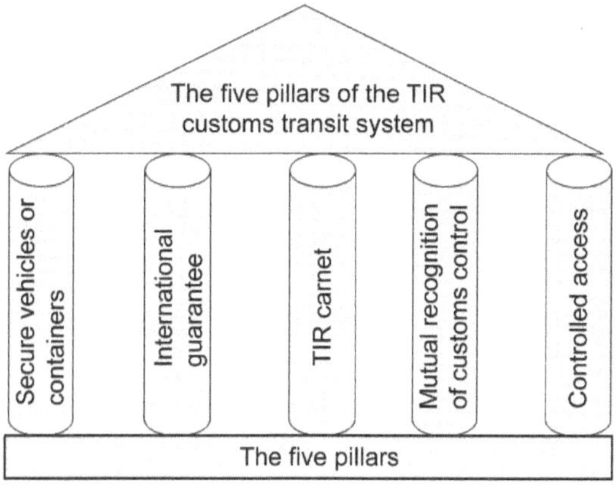

Fig. 3.4 Five pillars of TIR system.
Reproduced with permission from UNECE, 2018. TIR Handbook.

flows associated with them. It aims at increasing efficiency by performing complex operations as rationally as possible, whilst keeping a delicate balance between the requirements of the transport industry and the national economy. On the other hand, it needs to conform with indispensable governmental regulations related to national health and security, customs duties and taxes, etc.

The significance of transport is recognised in international trade procedures. In fact, goods are not able to move by themselves without transport services and, in practice, transport operators are confronted with trade barriers, such as extensive border controls and lengthy border crossing procedures. Nevertheless, such procedures may be seen as beneficial in terms of security and environment (e.g. dangerous goods) by the destination market as a matter of risk prevention. Yet such cumbersome procedures may result in 'unfair competition' as transport players of ports and freight forwarders divert the traffic toward ports in proximity. For instance, Antwerp in Belgium is the main port serving the French market (Guerrero, 2014), partly due to the absence of a harmonised European port policy.

One-stop border post[f]

The One-Stop Border Post (OSBP) concept, which was applied as a trade facilitation tool at borders, promotes a coordinated and integrated movement of people and goods across borders, and improving security. Both travellers and goods can avoid stopping twice to undertake border crossing formalities. This is the concept of the OSBP which relies on joint controls of both countries beyond the border to minimise routine activities and duplications (NEPAD, 2016).

The OSBP concept started its operation in Africa in the 2000s. The East African Community (EAC) together with the Northern Corridor Transit and Transport Coordination Authority facilitated the East African Transport and Trade Facilitation

[f] NEPAD (2016).

Project in 2004. The Chirundu OSBP—serving Zambia and Zimbabwe—is the first fully operated OSBP in Africa. Following the Chirundu OSBP, the OSBPs have been extended rapidly and widely across Africa to encourage trade growth in Africa. More than 80 OSBPs and Joint Border Posts (JBPs) are now at the planning or implementation stage in African. The OSBPs at the planning or implementation stage in African are summarised in the tables in Appendix.

Logistics infrastructure investment needs to 2030/2040

Logistics infrastructure is critically important for economic and social development of the world. All cargoes that are imported or exported by air or maritime transport across the borders need to be delivered to the inland destinations through roads and railways. By making full use of such transport infrastructure, both producers and consumers are able to reach the markets they need to trade their goods and services. Therefore, logistics infrastructure is essential to boost people's quality of life and nations' economic development.

Future logistics infrastructure investment needs over the several decades will depend on the current level of infrastructure, expected demand growth over the period, and the additional capacity required in the future. OECD (2011), PIDA (2011), and Global Infrastructure Hub (2017) have challenged significantly difficult tasks of estimating those infrastructure needs in the future.

Logistics infrastructure investment needs to 2030[g]

The assessments of OECD (2011) revealed that global infrastructure investment is estimated at approximately USD 8.82 trillion for airports, ports, rail, and oil and gas (transport and distribution) for the period of 2015–30 and USD 11.28 trillion for the period 2009–30 (see Table 3.6). When focusing on airport, port, and railway sectors, the global infrastructure investment needs summed up to USD 8.03 trillion for the period 2009–30. Estimated global infrastructure investment needs mainly rely on fast-growing developing economies, reflecting the extensive new infrastructure they

Table 3.6 Global logistics infrastructure investment needs to 2030 (unit: USD Trillion).

Infrastructure	2009–30	2015–30
1. Airports	2.2	1.8
2. Ports	0.83	0.63
3. Railways	5.0	4.06
4. Oil and gas	3.25	2.33
Total	11.28	8.82

Reproduced with permission from OECD, 2011. Strategic Transport Infrastructure Needs to 2030: Main Findings International Futures Programme, pp. 4–19.

[g] OECD (2011).

will require and the increased maintenance needs which will accrue approximately 10 years after the initial investment.

International gateways and trade corridors promote national and regional competitiveness, productivity, employment, quality of life, and a sustainable environment (Hall et al., 2011). The future growth in freight demand will emerge along the major interregional trade lanes, which will be carried by extra-large aircraft and container vessels at lowest cost. And the major international gateway airports and ports will need to prepare both the high-volume capacity and the longer runway and deeper fairway and basins to properly accommodate these extra-large aircraft and container vessels, respectively.

More and more freight movement is expected to proceed to the cities and industrial areas in their hinterlands via inland connections from major international gateways. For these reasons, each country's key international gateways and inland trade corridor infrastructure have become even more important to their national economies than ever before.

Logistics infrastructure investment needs to 2040[h,i]

According to Global Infrastructure Hub (2017), global infrastructure investment needs are estimated at approximately USD 94 trillion under the investment need scenario, whilst USD 79 trillion under the current trend scenario, for the period 2016–40. Largest sectors in global infrastructure investment needs of electricity and road account for 67% of the total needs. Airport, port, and railway sectors in total account for approximately 17% of total needs resulting in USD 15.7 trillion under the investment need scenario for the period 2016–40 (see Table 3.7).

When looking at regional distribution of infrastructure investment needs under the investment need scenario, Asia, Americas, and Africa accounts for 54%, 22%, and 6.4%, respectively, of the total needs. Whilst African infrastructure market remains

Table 3.7 Global infrastructure investment needs to 2040 (unit: USD Trillion).

Sector	2016–40	Region	2016–40
1. Road	33.5	Asia	50.8
2. Electricity	29.5	Americas	20.7
3. Rail	10.4	Europe	14.9
4. Telecoms	8.6	Africa	6.0
5. Water	6.7	Oceania	1.6
6. Airport	2.7	–	–
7. Port	2.6	–	–
Total	94		94

Reproduced with permission from Global Infrastructure Hub, 2017. Global Infrastructure Outlook: Infrastructure Investment Needs 50 Countries, 7 Sectors to 2040, Global Infrastructure Hub.

[h] Global Infrastructure Hub (2017).
[i] PIDA (2011).

small, Africa shows a considerable potential growth of the infrastructure investment needs in the future.

Infrastructure encourages economic growth by directly boosting economic activities and by underpinning productivity. Infrastructure supports economic activities by enhancing productivity throughout an economy. Road and railway networks enable easier and cheaper transportation of goods in the hinterland. Airports and ports may provide better delivery services to connect producers and consumers in foreland across the borders.

As indicated by Global Infrastructure Hub (2017), the world's fastest growing economies are located in Africa, quickly positioning themselves as the world's leading 'resource frontier'. Continued growth and prosperity of Africa will boost the infrastructure demand, which may lead to the largest constraint to its sustainable development. Africa's continental infrastructure investment needs were estimated at USD 360 billion up to 2040 (PIDA, 2011). This may be an ambitious but achievable target, if concerted and coordinated efforts are made by all concerned stakeholders.

Expansion of 11 cross-border railways needs to be achieved to meet demand by 2040. Similarly, regional railway lines need to be linked to new and expanded port development. Furthermore, building new modern railway lines to link landlocked countries to the sea can be justified, assuming that the development of the additional port capacities is concentrated in a several efficient locations.

Priority transport infrastructure investment projects and the OSBP projects are presented in Tables 3.A1–3.A6 in Appendix.

Conclusions

This chapter demonstrated how logistics infrastructure alone could not achieve efficient global movement of goods from shippers to consignees without globally efficient logistics environment. Logistics performance generally represents logistics environment of the hinterland, whilst liner shipping connectivity represents connectivity/accessibility to the global liner shipping network. Both indicators jointly demonstrate accessibility to the global market of each economy. Furthermore, trade facilitation, transport facilitation, and cross-border management are of more importance than logistics infrastructure alone, when focusing on the totally efficient logistics environment from the global supply chain perspective. There is no doubt that global logistics network modelling needs to integrate both maritime shipping and the inland transport into a combined model, which takes trade facilitation, transport facilitation, and cross-border management into account.

Appendix

Table 3.A1 List of OSBP projects in Eastern Africa (as of January 2016).

Border crossing	Corridor	Country A	Country B	Traffic (trucks/ day)	Funding source
1. Namanga/ Namanga	North-Central Interlink	Kenya	Tanzania	–	JICA and AfDB
2. Rusumo/Rusumo	Central	Tanzania	Rwanda	–	JICA
3. Malaba/Malaba	Northern	Kenya	Uganda	–	JICA, WB, and TMEA
4. Taveta/Holili	North-Central Interlink	Kenya	Tanzania	15	JICA, WB, and TMEA
5. Lunga Lunga/ Horo Horo	–	Kenya	Tanzania	–	JICA and WB
6. Isibania/Sirari	–	Kenya	Tanzania	–	JICA and WB
7. Busia/Busia	Northern	Kenya	Uganda	–	JICA and TMEA
8. Mutukula/ Mutukula	Central	Uganda	Tanzania	–	JICA and TMEA
9. Namba/Gasenti I	Northern	Rwanda	Burundi	–	JICA and AfDB
10. Gatuna/Katuna	Northern	Uganda	Rwanda	–	JICA and WB
11. Kobero/Kabanga	Central	Burundi	Tanzania	–	TMEA
12. Akanyaru/ Kanyaru	–	Rwanda	Burundi	–	AfDB
13. Kagitumba/ Mirama Hills	–	Rwanda	Uganda	–	WB, AfDB, and TMEA
14. Ruhwa/Ruhwa	–	Rwanda	Burundi	–	AfDB
15. Bibia/ Elegu-Nimule	Northern	Uganda	South Sudan	–	TMEA
16. Gisenyi/Goma	–	Rwanda	DRC	–	TMEA
17. Mpondwe	Northern	Uganda	DRC	–	TMEA
18. Rusizi/Bakavu	Central	Burundi	DRC	–	–
19. Nadapal	Northern	South Sudan	Kenya	–	–
20. Moyale	Northern	Ethiopia	Kenya	–	AfDB
21. South Sudan/ Sudan	Northern	South Sudan	Sudan	–	–
22. Rubavu/Goma	Northern	DRC	Rwanda	–	–
23. Galafi	Djibouti	Djibouti	Ethiopia	–	–
24. Gallabat/Metema	Djibouti	Ethiopia	Sudan	–	–
25. Nimule	South Sudan	Uganda	South Sudan	–	–

From NEPAD, 2016. One-Stop Border Post Sourcebook, second ed. and JICA Website https://www.jica.go.jp/english/news/field/2017/180124_01.html (Accessed 22/09/2019).

Table 3.A2 List of OSBP projects in Southern Africa (as of January 2016).

Border crossing	Corridor	Country A	Country B	Traffic (trucks/day)	Funding source
1. Chirundu	North–South	Zambia	Zimbabwe	300–400	JICA, DFID and IOM
2. Kazungula	North–South	Zambia	Botswana	115	JICA, AfDB and IOM
3. Pandamatenga	–	Zambia	Botswana	–	–
4. Mamouno/Trans Kalahari	Mamouno/Trans Kalahari	Namibia	Botswana	100	JICA
5. Tunduma /Nakonde	Dar es Salaam/North–South	Zambia	Tanzania	870	TMEA, TMSA, AfDB and IOM
6. Mwami/Mchinji	Nacala	Zambia	Malawi	100	JICA, AfDB and IOM
7. Mandimba/Chiponde	Nacala	Mozambique	Malawi	20	JICA
8. Wenela/Katima Mulilo	Trans Caprivi	Namibia	Zambia	50	IOM
9. Oshikango/Santa Clara	Trans Cunene	Namibia	Angola	100	–
10. Lebombo/Ressano Garcia	Maputo	South Africa	Mozambique	250–600	IOM
11. Machipanda/Forbes	Beira/Nacala	Mozambique	Zimbabwe	70	–
12. Nyampanda/Cuchimano	Beira/Nacala	Zimbabwe	Mozambique	–	–
13. Zobue/Mwanza	Beira/Nacala	Mozambique	Malawi	100–150	–
14. Colomue/Dedza	North–South	Mozambique	Malawi	100–200	–
15. Martin's Drift	North–South	South Africa	Botswana	–	–
16. Beitbridge/Messina	North–South	South Africa	Zimbabwe	–	–
17. Kasumbalesa	North–South	Zambia	DRC	1–2.5	IOM
18. Kasumulu/songwe	Dar es Salaam/North–South	Malawi	Tanzania	–	JICA, TMEA and WB
19. Mtambaswala/Namoto	Mtwara	Tanzania	Mozambique	–	IOM and EU
20. Plumtree/Ramokgwebane	–	Zimbabwe	Botswana	2000	–
21. Pioneer's Gate/Skilpadeshek	Mamuno/Trans Kalahari	South Africa	Botswana	–	–
22. Oshoek/Ngwenya	–	South Africa	Swaziland	–	–
23. Lavumisa	–	South Africa	Swaziland	–	–
24. Victoria Falls	–	Zimbabwe	Zambia	–	–

From NEPAD, 2016. One-Stop Border Post Sourcebook, second ed. and JICA Website https://www.jica.go.jp/english/news/field/2017/180124_01.html (Accessed 22/09/2019).

Table 3.A3 List of OSBP projects in Western Africa (as of January 2016).

Border crossing	Corridor	Country A	Country B	Traffic (trucks/day)	Funding source
1. Cinkanse	Lome-Ouagadougou-Bamako	Burkina Faso	Togo	–	JICA and UEMOA
2. Kantchari/Makalondi	Lome-Ouagadougou-Niamey	Burkina Faso	Niger	–	–
3. Moussala	Dakar-Bamako-Niamey	Senegal	Mali	–	–
4. Paga/Dakola	Tema-Ouagadougou-Bamako	Ghana	Burkina Faso	–	–
5. Laleraba	Abidjan-Ouagadougou	Cote d'Ivoire	Burkina Faso	–	–
6. Kidira/Diboli	Dakar-Bamako-Niamey	Senegal	Mali	–	–
7. Kololo/Heremakono	Dakar-Bamako-Niamey	Burkina Faso	Mali	–	–
8. NOE/Elubo	Abidjan-Lagos	Ghana	Cote d'Ivoire	–	–
9. Pogo/Zegou	Abidjan-Ouagadougou-Bamako	Cote d'Ivoire	Mali	–	–
10. Danane	–	Cote d'Ivoire	Guinia	–	–
11. Tabou	–	Cote d'Ivoire	Liberia	–	–
12. Nigoni	–	Cote d'Ivoire	Mali	–	–
13. Akanu/Noepe	Abidjan-Lagos	Ghana	Togo	–	–
14. Krake/Seme	Abidjan-Lagos	Bennin	Nigeria	–	–
15. Hillacondji/Sanveekondji	Abidjan-Lagos	Togo	Bennin	–	–
16. Gaya/Malanville	Cotonou-Niamey	Niger	Bennin	–	WB
17. Trans Gambia	–	Gambia	Senegal	–	AfDB
18. Kouremale	–	Guinea	Mali	–	–
19. Mpack	–	Senegal	Guinea Bissau	–	UEMOA
20. Labezanga	–	Mali	Niger	–	–
21. Kantchari/Makalondi	–	Burkina Faso	Niger	–	–
22. Petel Kole	–	Niger	Mali	–	–
23. Boundou/Fourdou	–	Senegal	Guinea	–	UEMOA and AfDB
24. Mfum	Enugu-Bamenda	Nigeria	Cameroon	–	AfDB

From NEPAD, 2016. One-Stop Border Post Sourcebook, second ed. and JICA Website https://www.jica.go.jp/english/news/field/2017/180124_01.html (Accessed 22/09/2019).

Table 3.A4 List of OSBP projects in Central Africa (as of January 2016).

Border crossing	Corridor	Country A	Country B	Traffic (trucks/day)	Funding source
1. Brazzaville/ Kinshasa	Pointe Noire-Brazzaville-Kinshasa-Bangui-N'Djamena	DRC	Republic of Congo	–	–
2. Koussere	Douala-Bangui-Douala-N'Djamena	Cameroon	Chad	–	–
3. Koutère	Douala-Bangui-Douala-N'Djamena	Cameroon	Chad	–	–
4. Garoua Boulai	Pointe Noire-Brazzaville-Kinshasa-Bangui-N'Djamena	Central Africa	Cameroon	–	–
5. Campo	Douala-Bangui-Douala-N'Djamena	Equatorial Guinea	Cameroon	–	–
6. Doussala	Douala-Nyanga-Kibangou-Dolisie-Libreville-Brazzaville	Republic of Congo	Gabon	–	AfDB

From NEPAD, 2016. One-Stop Border Post Sourcebook, second ed. and JICA Website https://www.jica.go.jp/english/news/field/2017/180124_01.html (Accessed 22/09/2019)

Table 3.A5 List of OSBP projects in Northern Africa (as of January 2016).

Border crossing	Corridor	Country A	Country B	Traffic (trucks/day)	Funding source
1. Dakla/Nouadhibou	Trans African Highway I	Mauritania	Morocco	–	AfDB
2. Oujda Tlemeen	Trans African Highway I	Morocco	Algeria	–	AfDB
3. Ghardimaou	Trans African Highway I	Tunisia	Algeria	–	AfDB
4. Ras Adjir	Trans African Highway I	Tunisia	Libya	–	AfDB
5. Musaid-Soloum	Trans African Highway I	Libya	Egypt	–	AfDB

From NEPAD, 2016. One-Stop Border Post Sourcebook, second ed. and JICA Website https://www.jica.go.jp/english/news/field/2017/180124_01.html (Accessed 22/09/2019)

Table 3.A6 List of priority transport infrastructure investment in Africa.

Region	Corridor	Description	Countries	Stage	Total cost (USD million)
Eastern Africa	Northern	Modernise the highest priority multimodal corridor in Eastern Africa. Facilitate goods across the borders between Kenya, Uganda, Rwanda, Burundi, and DRC with a spur to South Sudan.	Kenya, Uganda, Rwanda, Burundi	S3/S4	1000
Eastern Africa	North–South	Modernise the highest priority multimodal corridor in Southern Africa, and facilitate goods across the borders between South Africa, Botswana, Zimbabwe, Zambia, Malawi, and DRC.	DRC, Zambia, Zimbabwe, South Africa, Mozambique	S3/S4	2325
Eastern Africa	Djibouti-Addis	Revive the rail system in a high-priority multimodal corridor in Eastern Africa, and increase the flow of goods across the border between Djibouti and Ethiopia.	Djibouti, Ethiopia	S3/S4	1000
Eastern Africa	Central	Modernise the third priority corridor in Eastern Africa, and facilitate goods across the borders between Tanzania, Uganda, Rwanda, Burundi, and DRC.	Tanzania, Uganda, Rwanda, Burundi, DRC	S3/S4	840
Eastern Africa	Beira-Nacala	Modernise and upgrade the rail and port system serving a major coal export through the Beira Corridor and the Nacala Corridor.	Mozambique, Malawi	S3/S4	450
Eastern Africa	Lamu Gateway Port	Develop sufficient port capacity to accommodate future demand from both domestic sources and landlocked countries, with priority given to Lamu Gateway Port.	Kenya, Uganda, Rwanda, Burundi	S3/S4	5900
Southern Africa	Southern Africa Hub-Port and Rail	Develop sufficient port capacity to accommodate future demand from both domestic sources and landlocked countries.	SADAC Members	S1	2270
Western Africa	Abidjan-Lagos	Modernise heavily used corridor in Western Africa. Trade facilitation, OSBPs, capacity enhancement, and implementation of PPP for five countries.	Nigeria, Bennin, Togo, Ghana, Cote d'Ivoire	S3/S4	290

Continued

Table 3.A6 Continued

Region	Corridor	Description	Countries	Stage	Total cost (USD million)
Western Africa	Dakar-Niamey	Modernise heavily used corridor in Western Africa. Trade facilitation, OSBPs, and capacity enhancement.	Senegal, Mali, Burkina Faso, Niger	S3/S4	590
Western Africa	Praia-Dakar-Abidjan	Improve maritime transport and the connection between island and mainland countries by a new maritime service between ports and a modern information system by linking the maritime service with ports and roads. Trade facilitation, OSBPs, capacity enhancement.	Carvo Verde, Senegal, Gambia, Guinea Bissau, Guinea, Sierra Leone, Liberia, Cote d'Ivoire	S2/S4	150
Western Africa	Abidjan-Ouagadougou/ Bamako	Modernise and rehabilitate multimodal corridor damaged by civil war in Côte d'Ivoire.	Cote d'Ivoire, Burkina Faso, Mali	S3/S4	540
Western Africa	Western Africa Hub Port and Rail	Address future capacity problems in Western Africa's ports with two components: (a) a regional hub port and rail linkage master plan and (b) port expansion.	15 countries (PMAWCA)	S1	2140
Central Africa	Pointe Noire, Brazzaville/ Kinshasa, Bangui, N'Djamena	Revive river transport in the Congo-Ubangi River Basin, and modernise road transport along the corridor.	Republic of Congo, DRC, CAR	S3/S4	300
Central Africa	Central Africa Hub Port and Rail	Responds to emerging capacity problems in Central Africa's ports through two components: (a) a regional hub port and rail linkage master plan; (b) port expansion.	Cameroon, Chad, CAR, Rep. of Congo, DRC, Gabon, Burundi	S1	1400
Northern Africa	Trans Maghreb Highway	Improve goods across the Maghreb, where trade is limited by artificial barriers. Design and implement a smart corridor system along the highway and install one-stop border posts.	Morocco, Egypt, Algeria, Tunisia, Libya	S3/S4	75

Remarks: *S1*, concept proposal; *S2*, feasibility/needs assessment; *S3*, programme/project structuring; *S4*, implementation and operation.
Data: PIDA (2011).

References

Arvis, J.F., Ojala, L., Wiederer, C., Shepherd, B., Raj, A., Dairabayeva, K., 2018. Connecting to Compete 2018: Trade Logistics in the Global Economy. World Bank.

Ducruet, C., Van der Horst, M.R., 2009. Transport integration at European ports: measuring the role and position of intermediaries. Eur. J. Transp. Infrastruct. Res. 9 (2), 121–142.

Fugazza, M., Hoffmann, J., 2017. Liner shipping connectivity as determinant of trade. J. Shipping Trade 2 (1). https://link.springer.com/article/10.1186/s41072-017-0019-5.

Global Infrastructure Hub, 2017. Global Infrastructure Outlook: Infrastructure Investment Needs 50 Countries, 7 Sectors to 2040. Global Infrastructure Hub.

Guerrero, D., 2014. Deep-sea hinterlands: some empirical evidence of the spatial impact of containerisation. J. Transp. Geogr. 35, 84–94.

Hall, P.V., McCalla, R.J., Comtois, C., Slack, B., 2011. Integrating Seaports and Trade Corridors. Routledge, London and New York.

Hillberry, R., Zhang, X., 2015. Policy and performance in customs: evaluating the trade facilitation agreement. In: Policy Research Working Paper. World Bank, p. 7211.

Jo, J.C., Ducruet, C., 2007. Rajin-Seonbong, new gateway of Northeast Asia. Ann. Reg. Sci. 41 (4), 927–950.

Moïséi, E., Sorescu, S., 2013. Trade facilitation indicators: the potential impact of trade facilitation on developing countries' trade. In: OECD Policy Paper No.144. OECD.

NEPAD, 2016. One-Stop Border Post Sourcebook, second ed.

OECD, 2011. Strategic Transport Infrastructure Needs to 2030: Main Findings International Futures Programme, pp. 4–19.

PIDA, 2011. Infrastructure Outlook 2040. The Programme for Infrastructure Development in Africa.

Rodrigue, J.P., Comtois, C., Slack, B., 2013. The Geography of Transport Systems. Routledge, London and New York.

UNCTAD, 2018. Review of Maritime Transport.

UNECE, 2018. TIR Handbook.

WTO, 2015. World Trade Report 2015.

Basics of container demand forecast

4

Ryuichi Shibasaki
The University of Tokyo

Introduction

The classical methodology of transport demand forecast is a four-step demand forecast approach. This is a heuristic method to compute the traffic demand in a stepwise manner as shown in Fig. 4.1. One of the outputs of the four-step model framework is a traffic volume or flow passing through links in the transport network during a given period. For example, in the case of a container cargo demand forecast model, an output is typically the annual cargo volume (on a TEU or tonnage basis) transported from one port to another. Meanwhile, inputs into the model framework are typically socio-economic information such as gross domestic product (GDP) and population.

The first step is called 'traffic generation/attraction', where socio-economic indices such as GDP and population are input. The amount of traffic demand generated from each zone (origin) and the amount of traffic demand attracted into each zone (destination) are output in the first step. The second step is called 'traffic distribution', where the traffic volumes generated from origins and those attracted to destinations are input; then, the amount of traffic demand for each combination of origins and destinations is output. A set of outputs from this step is often called an origin–destination (OD) matrix. The third step is called transport 'modal choice', where the traffic volume of an OD pair is input; then, the traffic volume by transport mode is output. A percentage of traffic volume transported by a specific transport mode out of the total OD traffic volume is called the transport modal share. Finally, the fourth step is called 'traffic assignment' or 'network assignment', where the traffic volume transported by a specific transport mode in an OD pair is input, and the traffic volume passing through each link in the transport network of the transport mode is output.

One of the unique characteristics of the four-step transport demand forecast system is that the input of the second, third, or fourth step is controlled or constrained by the output of the first, second, or third step, respectively. On the other hand, all four steps may not be necessarily considered fully in the traffic demand forecast process. Rather, in some cases, only some of them are considered, based on the assumption that the traffic demand in the other steps is given. A mathematical model can be formulated to forecast the outputs in each step if necessary, by various methods based on different theories.

One of the typical methods for modelling is econometric modelling; by this method, mathematical equations are formulated for explaining the traffic volume with explanatory variables and their unknown parameters are estimated with observed data

Global Logistics Network Modelling and Policy. https://doi.org/10.1016/B978-0-12-814060-4.00004-6

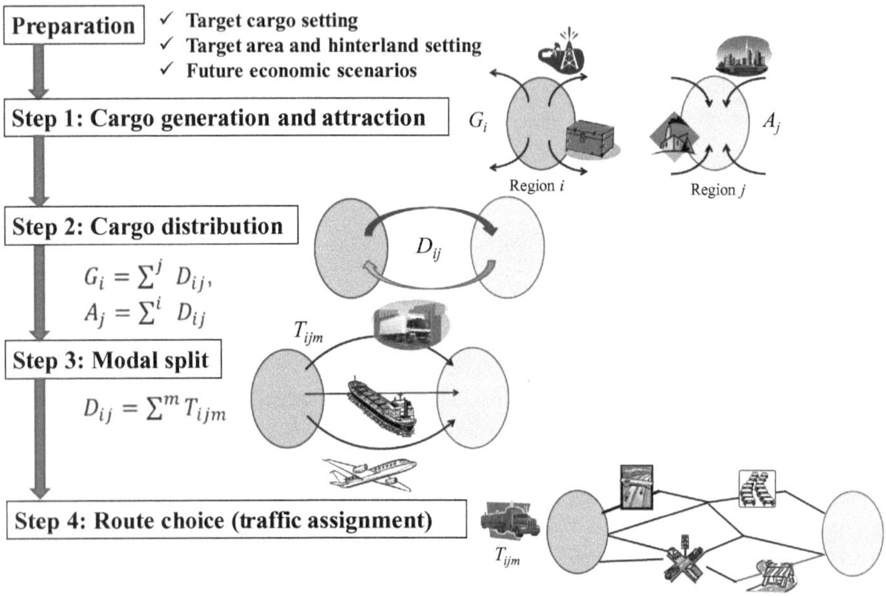

Fig. 4.1 Chart of the four-step estimation methodology.
Source: Own compilation.

by using statistical methods such as an ordinary least squared method and a maximisation of the likelihood function. Another typical method is to solve an optimisation problem that describes the behaviours of players, such as forwarders, shippers, and customers, in the model systems to assumingly optimise an objective function related to social welfare and/or social cost. It is typically assumed that models are given when the traffic demand is forecasted. This means that traffic demand structure and/or the preferences of stakeholders in the model system are assumed to be fixed across different scenarios.

Preparation

Target cargo setting

The following discussions focus on container transport, because it is the main topic of this book. Container cargo contributed only 13% of the international maritime cargo shipping in 2016 on a tonnage basis according to the World Trade Service (WTS) data (provided by IHS, Inc., 2017). In other words, natural resources, such as oil, gas, and ores, and other bulky cargo, including iron and steel, grain, and finished cars, are transported by dedicated ships such as dry bulk carriers, tankers, and pure car carriers. Table 4.1 lists the containerised rate of each type of major bulk cargo; this rate varies greatly among commodities. Therefore, the methodology for the demand forecast of international cargo depends on the target commodities.

Table 4.1 Containerised rate and total volume of the world for each major bulk cargo (2016, million tonne).

Commodity	Rate	Volume	Commodity	Rate	Volume
Crude Petroleum	0.0%	1888	Basic Iron and Steel	12.6%	67
Ores, Iron, and Manganese	0.1%	1584	Hay, Fodder, Bran, Oilcake, Etc.	14.3%	85
Coal	0.2%	1062	Sugar, Beet or Cane	16.2%	55
Petroleum Refineries	0.3%	836	Wood of Coniferous Species	17.6%	67
Briquettes and Coke	0.3%	98	Flat-Rolled Products of Iron and Steel	19.3%	141
Natural Gas	0.4%	413	Cement and Lime	19.5%	109
Pitch Coke, Petroleum Coke, Bitumen, etc.	0.9%	74	Scrap Metal	19.8%	56
Ores, Non-Ferrous, excl. Manganese	1.7%	228	Inorganic Chemical Compounds	20.2%	90
Corn (Maize)	2.3%	110	Iron and Steel, n.e.s.	24.8%	105
Wheat	2.5%	134	Organic Chemicals	27.5%	125
Soybeans	5.1%	134	Wood of Non-Coniferous Species	36.9%	55
Sands, Pebbles, Gravel, and Crushed Stone	5.7%	64	Chemical Products, nec.	41.6%	78
Stone, Clay, and Other Crude Minerals	7.5%	193	Plastics in Primary Forms and Synthetic Rubber	79.6%	113
Fertilisers and Pesticides	9.7%	122			

Source: Own compilation based on the WTS data.

Generally, container cargo and other types of maritime cargo are separately forecasted because of several reasons. One reason is that container terminals and major bulk terminals are generally different, and hence, their respective development is also separately planned even within the same port. Another reason is the difference in the structure of their demand. For instance, containers, which generally carry consumer goods and intermediates, are exported by manufacturers and demanded by consumers and various types of manufacturers; on the other hand, major bulk cargo are transported from specific (i.e. limited number of) exporters, such as mines, oil and gas extractors, and heavy manufacturers, to specific large importers, such as power plants, petrochemical complexes, and iron mills. If focusing on containers, the general economic situation in each country of the world should be more considered, whereas for major bulk cargo, the world trends of specific commodity markets are of greater concern. If forecasting the demand of some cargo that can be carried by both containers and other means, containerised rates and the possibility of their changes in future should be considered in the third step (modal split) of the four-step estimation methodology.

In addition, the modal split between air, land, and maritime transport is often given as a pre-condition. If the competition between different modes is considered in the forecast, this consideration is made in the third step of the four-step estimation methodology. On the other hand, some transport means are exclusive; for example, pipeline transport, which is one of the means of land transport, is exclusively used for oil and gas transport. Further, air transport is often considered only for some specific valuable cargo because its transport cost per tonne-kilometre is very high and because the space for cargo transport is limited. Table 4.2 lists the major cargo transported by air and their rates of air transport. Note that the table is summarised on a value basis. In some cases, the unit price of cargo transported by air is significantly larger than that transported by other means even if it is categorised in the same commodity. Therefore, air transport is often excluded from the initial stage itself for the forecast of international cargo shipping demand for other modes.

Target area and hinterland setting

The need to forecast cargo shipping demand frequently arises in specific projects such as infrastructure investment and launch of new shipping services. Forecasting the demand of specific infrastructure (e.g. a port) or service necessitates the target area to be preliminary decided. Although any cargo of the world can theoretically use the infrastructure or service, its major target countries can be sometimes geographically limited.

However, in this book, the authors recommend considering maritime container shipping globally because all the liner shipping networks of the world provided by various shipping companies influence each other. Therefore, particularly for a global-scale project such as hub port development and land bridge construction across a continent—some of which are mentioned in Part 3 of this book—the maritime container shipping network should be considered on a global scale.

Another issue is on the setting of the hinterland (or foreland) of the target cargo. The hinterland of most bulky cargo is normally limited around the port of export or

Table 4.2 Air transport rate and total value of the world for each major airborne cargo (2016, million USD).

Commodity	Rate	Value	Commodity	Rate	Value
Computers	61.4%	60	Aircraft and Parts	29.2%	34
Precious Metals	47.6%	127	Other Communications Equipment	29.0%	95
Telephones, Microphones, etc.	45.6%	13	Electricity Distribution and Control Apparatus	19.9%	18
Optical and Measuring Equipment, Meters and Counters	39.2%	64	Electric Engines, Generators and Transformers	18.4%	17
Computer Equipment and Parts	38.0%	55	Electrical Equipment, n.e.s.	15.7%	11
Pharmaceutical Goods Excl. Antibiotics	36.7%	51	Insulated Wire and Cable; Accumulators and Batteries	15.2%	11
Electronic Valves, Tubes, Semiconductors and Other Electronic Components	35.9%	248	Special Industrial Machinery, n.e.s.	15.0%	18
Medical Instruments, Appliances and Diagnostic Apparatus	34.8%	17	Goods not classified by kind	12.4%	231
Other Manufacturing, nes.	32.9%	76	Organic Chemicals	9.4%	14

Source: Own compilation based on the WTS data.

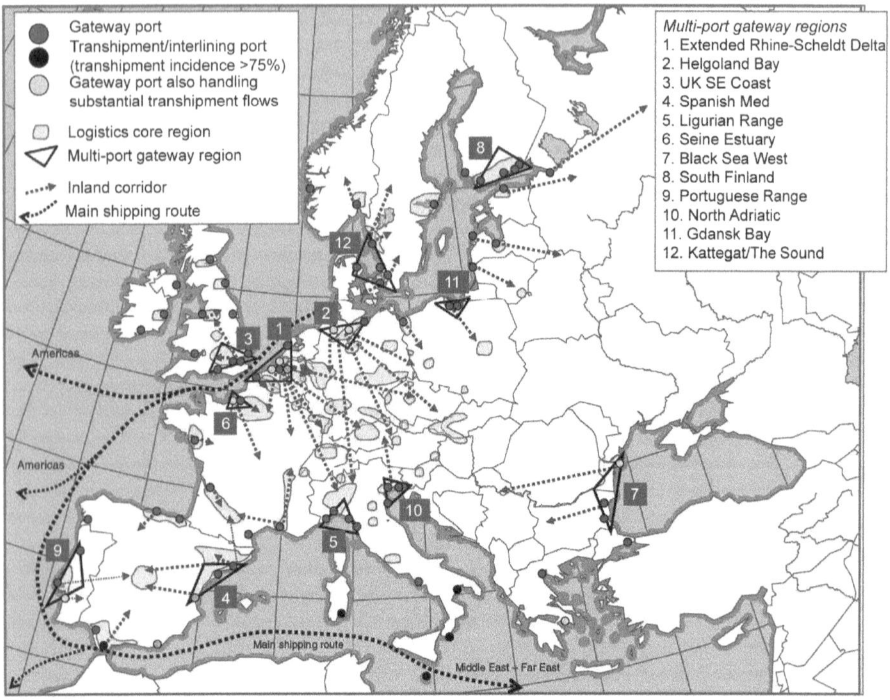

Fig. 4.2 Example of hinterland competition: Hamburg–Le Havre range.
Source: Notteboom, T.E., 2010. Concentration and the formation of multi-port gateway regions in the European container port system: an update. J. Transp. Geogr. 18, 567–583.

import because the number of players who export and import bulky cargo is limited and most of them are located around the port. In such a situation, the hinterland of each port is not duplicated, and no competition can be observed between ports; then, it is not necessary to consider any issue of port choice for export and import by cargo owners.

On the other hand, the hinterland of container cargo is generally larger than that of bulky cargo because of the higher intermodalism of the former; regarding container cargo, the hinterland sometimes extends beyond national borders and is frequently duplicated with neighbouring ports (see Chapter 2). Therefore, gateway seaports for containers often face heavy competition with neighbouring ports. The most famous competition between container ports is observed in Northern Europe, in the so-called 'Hamburg–Le Havre range' (see Fig. 4.2). This competition involves not only several major container seaports in Europe, such as Le Havre, Antwerp, Rotterdam, Bremerhaven, and Hamburg, but also most hinterland transport modes (including trucks, railways, and inland waterways) for providing better connection with multiple hinterland countries such as France, Belgium, Netherland, Germany, Switzerland, Austria and other eastern European countries. Another example is the access to seaports to/from Central Asian countries; as described in Chapter 13, the possible gateway is broadly spread across the Eurasian continent, including Far East Russia, China,

South Asia, Middle East, the Black Sea, and the Baltic Sea region (see Fig. 13.2 in Chapter 13). If considering the gateway competition with neighbouring ports, it is necessary to properly define the range of the whole hinterland. Additionally, in a hinterland analysis, a zoning system is generally introduced, i.e. each country should be generally divided into detailed zones. The more detailed the zoning, the better it is for more detailed analysis though this generally indicates a trade-off with the geographical resolution of the dataset.

Future economic scenarios

When forecasting the future demand of cargo shipping, the future trends of socio-economic indices, such as population and GDP (or gross regional domestic products, GRDP), in the target area should be acquired or estimated in advance as inputs of the demand forecasting methodology. The United Nations provides future population of each country of the world, whereas several international organisation and consulting companies provide estimates of the future GDP of world's major countries.

In addition, in the demand forecast of cargo shipping, future changes in the economic and industrial structure of countries and future trends in international economic and trade policies, including consequences of negotiation in the World Trade Organisation (WTO) and conclusions of free trade agreement and economic partnership agreements between multiple countries, may be considered. However, developing an economic model that can endogenise such future changes is difficult because of huge uncertainties; therefore, such changes are usually given exogenously in a model as input variables. If they are considered as given inputs, the explainable scenarios that reasonably describe their future change should be prepared.

Step 1: Cargo attraction and generation

Relationship with GDP

It is well known that the cargo shipping demand generated from and attracted into a region is generally related to the scale and contents of economic activities in the region in question. For example, the amount of export and import containers (i.e. excluding transshipment containers) handled in a country (or port) is often strongly related to the GDP of that country (or GRDP of the region) because any container cargo generated from and attracted into the country (or the regional hinterland of the port) should be handled in the ports of the country (or in the port in question) in the absence of international (or domestic) duplication of the hinterland. Further, the shipping demand of a specific bulky cargo is strongly related to the scale of the specific industry to use them in the country (or region).

Fig. 4.3 shows the relationship between the container cargo throughput (which is available in the World Bank (WB) (n.d.) website) and the amount of GDP on a purchasing-power-parity (PPP) US dollar constant price basis (which is estimated from the World Economic Outlook Database provided by the International Monetary

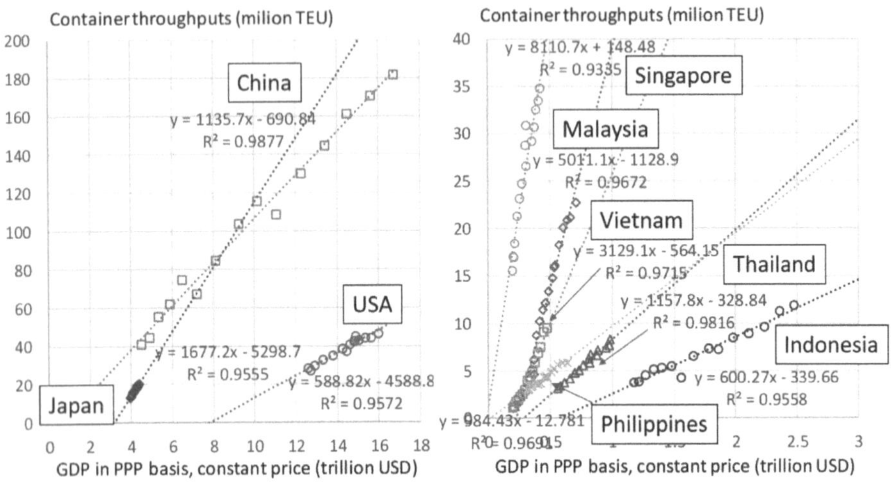

US, China and Japan ASEAN 6

Fig. 4.3 Relationship between container throughput and GDP of each country from 2000 to 2014. *Source:* Own compilation based on the statistics provided by WB and IMF.

Funds (IMF) as of October 2016) of three economic giants of the world (left figure) and six countries in Southeast Asia (right figure) from 2000 to 2014. In all the countries represented in these figures, the container cargo throughput in a year is strongly positively correlated (multiple correlation coefficient for every country is more than 0.9) to the amount of GDP in that year though the slope of the approximation for each country is different. The left figure in Fig. 4.3 shows that the relative significance of container throughput against the amount of GDP in China is greater than that for the US and Japan because the Chinese economy has a greater dependence on international trade—especially exporting containers—compared to the other two countries.

The right figure in Fig. 4.3 that shows data for six Southeast Asian countries is more suggestive. The slope of the approximate line for Singapore is the largest, followed by Malaysia because the statistics of container throughput provided by the WB includes transshipped containers, which contribute significant portions in the total throughputs of these two countries. For Philippines, the container throughput includes much domestic container shipping because Philippines consists of many islands served by coastal shipping. On the other hand, the slope for Indonesia is the least because the Indonesian economy heavily depends on the production of natural resources such as crude oil and natural gas not manufacturing; therefore, container throughput in Indonesia is less important and relatively smaller though Indonesia's GDP is larger than the GDPs of other Southeast Asian countries. The other two countries—Vietnam and Thailand—lie in the middle and their multiple correlation coefficients are the highest because container throughputs observed in these countries are the purest if focusing on the relationship between international export/import containers and the amount of economic activities (i.e. GDP) of the country.

Elasticity of GDP amount to cargo throughput

One of the reasons for the difference between the slopes of Vietnam and Thailand (the right figure in Fig. 4.3) is the difference in the stages of economic development (another reason may be the difference in the length of coastal lines between two countries). It is generally known that the elasticity of GDP amount to cargo throughput (i.e. the increasing rate of cargo throughput to the rate of increase of GDP) increases when the economy of the country grows rapidly (i.e. in the early-middle stage of the economic development). Table 4.3 lists the elasticity of GDP (on a constant price of local currency basis) to cargo throughput from 2008 to 2014 for the countries shown in Fig. 4.3 and three other countries in Southeast Asia (i.e. Brunei Darussalam, Cambodia, and Myanmar). As shown in the table, the elasticity for Vietnam is the highest among these countries, followed by Brunei Darussalam. However, the elasticity is lower not only in developed countries such as the US, Japan, and Singapore and in the countries in the latter-middle stage of economic development, such as China and Thailand, but also in countries in very early stages of economic development, such as Myanmar and Cambodia.

Note that the above discussions in this section are all based on the total (i.e. sum of export and import) throughput of the country. The explanation variables of the cargo throughput may be different between export and import cargo. The amount of export containers is often more purely proportional to the amount of GDP, whereas that of import containers tends to be correlated also to the population; therefore, a multiple regression analysis is sometimes applied.

Step 2: Cargo distribution

Gravity model

The second step of the four-step demand forecasting methodology is to estimate the amount of distribution (i.e. the breakdown by partner countries/regions) of the cargo generated from or attracted into the country or region, which is estimated in the first step.

Table 4.3 Value of elasticity of GDP amount to cargo throughput from 2008 to 2014.

Country	Elasticity	Country	Elasticity	Country	Elasticity
United States	1.017	Brunei Darussalam	1.405	Philippines	0.959
China	0.949	Cambodia	0.797	Singapore	0.827
Japan	1.079	Indonesia	1.156	Thailand	1.031
		Malaysia	1.081	Vietnam	1.548
		Myanmar	0.920		

Source: Own compilation from the statistics provided by WB and IMF.

In this step, a gravity model, which is an analogical application of Newton's law of gravity, is often applied.

$$d_{ij} = \exp(\alpha) \cdot (G_i)^{\beta} \cdot (A_j)^{\gamma} \cdot \exp(-\delta \cdot t_{ij}), \tag{4.1}$$

where d_{ij} is the amount of cargo distributed from a generating region (i.e. the origin) i to the attracting region (the destination) j; G_i is the amount of cargo generated from an origin i; A_j is the amount of cargo attracted into a destination j; t_{ij} is 'transport resistance' from an origin i to a destination j, and α, β, γ, and δ are unknown parameters to be estimated so as to best fit the actual record. The amount of cargo distributed, d_{ij}, is usually called the 'OD cargo volume', constituting each element of the OD matrix. Eq. (4.1) represents the OD cargo volume increases as the amount of cargo generated and/or attracted increase and it decreases as the transport resistance between regions increases.

The unknown parameters included in Eq. (4.1) are generally estimated by a multiple regression analysis after taking the logarithm of both sides in Eq. (4.1).

$$\log d_{ij} = \alpha + \beta \cdot \log G_i + \gamma \cdot \log A_j - \delta \cdot t_{ij}. \tag{4.2}$$

An example of the application of the gravity model is shown as follows: It is applied to the world container movement between 65 countries/regions, corresponding to the model analysis presented in the following chapters of this book. The observed number of containers generated, attracted, and distributed (G_i, A_j, and d_{ij}) is acquired from the WTS database on a TEU basis as of 2013, whereas the transport resistance, t_{ij}, is represented by the maritime shipping distance (in nautical miles) between the representative ports for each country/region. The maritime shipping distance is acquired from Toriumi (2010), which is also used in the model analysis in the following chapters of this book.

Table 4.4 lists the estimated results of unknown parameters. Not only the sign condition of every estimated parameter is satisfied, but also the t-value of each estimated parameter is sufficiently large because the model describes a huge sample with only very few parameters. The multiple correlation coefficient is also sufficiently large considering the size of the sample.

Fig. 4.4 compares the estimated OD cargo volume with the observed value. Compared with the figure (employing a logarithmic scale) shown in the left in Fig. 4.4, the estimated OD cargo volumes (displayed in a linear scale) in the right in Fig. 4.4 seem greatly different. The major differences between them are observed in terms of the number of containers distributed to and from China. Namely, the amounts of con-

Table 4.4 Estimated unknown parameters in an exemplified gravity model.

Parameter	α	β	γ	δ	**Sample number**	R^2
coefficient	−23.60	1.250	1.176	$2.593*10^{-4}$	3502	0.6945
t-value	−59.01	71.55	52.69	29.88		

Source: Own compilation.

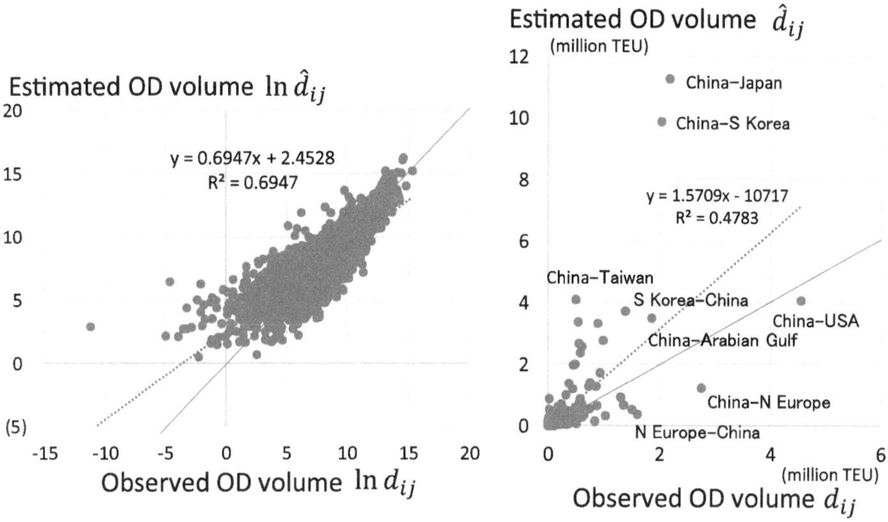

Fig. 4.4 Comparison of the estimated and observed OD cargo volume in 2013 on a logarithmic and linear scale.
Source: Own compilation.

tainers distributed to and from China are overestimated in short-distance trade, such as the trade with Japan, South Korea, and Taiwan, and underestimated in the long-distance trade, such as the trade with the US and Europe.

Two possible implications can be observed from this result. One is that the shipping distance should be substituted by other indices such as shipping cost or ocean freight charges as a transport resistance parameter. This reflects the reality that the shipping cost in trunk routes, which connect East Asia, Europe, and North America, is relatively lower than that in other routes regardless of the shipping distance because much larger containerships are used in such trunk routes based on the economies of scale. Furthermore, the freight charge should be carefully applied, if it is used, because it is generally decided under the balance of supply and demand rather than the shipping cost.

Another implication from the results shown in Fig. 4.4 is that globalisation of the world economy with less shipping cost makes the gravity model less significant. In other words, the gravity model may be less applicable to the current container shipping market of the world than to the shipping market of the past (see also Guerrero et al., 2015). At any rate, the application of the model and interpretation of the estimated results should be always carefully performed though the gravity model is still a useful tool for some purposes.

If considering to break the OD cargo volume to/from a country down into detailed zones such as state (or province) level or port level for the detailed analysis, the difference in trade partners for each zone should be considered. Fig. 4.5 describes the trade partner countries of each region in China in export. The figure shows that even at the municipal level, the geographic distance affects trade patterns. For instance, the northwest region is well connected with Russia and Central Asian countries, the shares

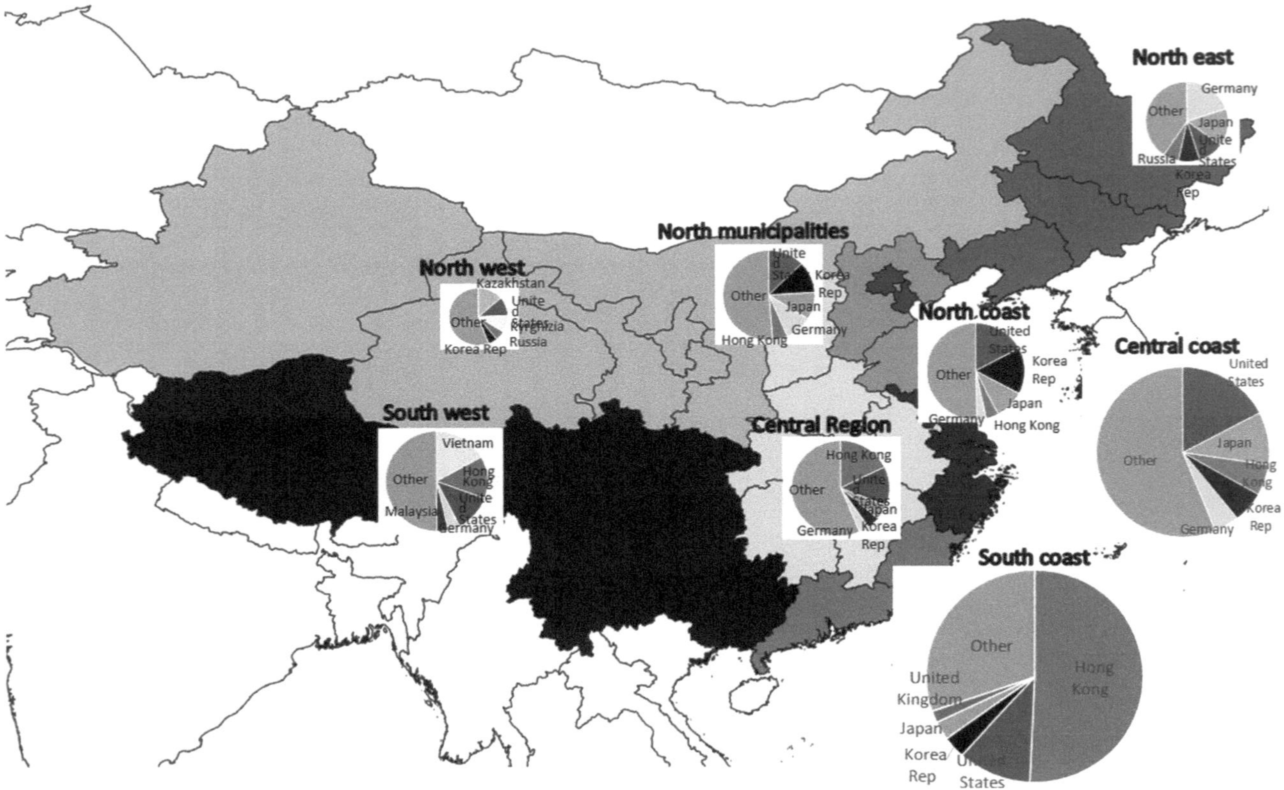

Fig. 4.5 Example of trip distribution: Chinese trade partners by region (export trade on a value basis, 2015).
Source: Own compilation based on China Customs Statistics Information Center, 2015. China Customs Statistics 2015.

of South Korea in the northeast and north coast region are larger than those in other regions and the shares of Hong Kong and Southeast Asian countries in the south coast and southwest region are larger than those in other regions. In other words, a gravity model may be also applicable for estimating the OD cargo volume on a municipal basis.

However, the problem is that countries where such detailed observed data of trade on a municipal basis are available are very limited; in most cases, there are no other means than proportionally dividing the OD cargo volume to/from a country based on a regional economic index such as GRP or cargo throughput for each port, but it does not provide any information on the partner country or region. Details of the regional (municipal) zonal division for each country are described in the chapters of Part 3.

Coordination of estimation errors: Present pattern method

The calculation results of the gravity model, i.e. the sum of the amount of cargo distributed by import and export countries ($\sum_i d_{ij}$ and $\sum_j d_{ij}$), do not always correspond to the amount of cargo attracted (A_j) and generated (G_i), which are estimated in the first step of the four-step estimation methodology and input into the gravity model. This is because any preserved laws on cargo flows are not considered in the gravity model. Therefore, the present pattern methods to coordinate the estimation errors should be applied to the estimation results of the gravity model. Eq. (4.3) represents a coordination achieved by the Frater method, which is one of the typical current pattern methods.

$$D_{ij} = k \cdot d_{ij} \cdot \frac{G_i}{g_i} \cdot \frac{A_j}{a_j} \cdot \frac{1}{2} \cdot \left(\frac{g_i}{\sum_j d_{ij} \cdot \frac{A_j}{a_j}} + \frac{a_j}{\sum_i d_{ij} \cdot \frac{G_i}{g_i}} \right), \tag{4.3}$$

$$s.t.\ a_j = \sum_i d_{ij} \tag{4.4}$$

and

$$g_i = \sum_j d_{ij}, \tag{4.5}$$

where D_{ij} is the revised amount of distribution cargo and k is the coordination coefficient, which is calibrated so that the estimation error (sum of squared errors in each distribution amount) can be minimised. By repeatedly applying the above equations and replacing $d_{ij}^{(n)}$ in the n-th calculation by $D_{ij}^{(n-1)}$ calculated in the $(n-1)$th iteration, the converged distribution amount $\widehat{D_{ij}}$ is finally estimated.

Use of economic model as a substitute

An alternative method often applied to forecast cargo shipping demand is to directly estimate the distribution amount of trade by means of an economic model(s) with several economic variables. In addition to the econometric models such as time-series analysis, an international economic model, which can spatially consider the balance (i.e. equilibrium) of supply and demand in the market of each good and their international exchanges (i.e. trade), is often developed, applied, or used, based on the microeconomic theory. These models are termed applied (or spatial computerised) general equilibrium (AGE or SCGE) models.

A typical SCGE model for trade analysis is a Global Trade Analysis Project (GTAP) model, which was originally developed by Hertel and colleagues in Purdue University and Monash University in the 1990s. The standard GTAP model is a multi-region, multi-sector CGE model with perfect competition and constant returns to scale (Hertel, 1997, 2013). As shown in Fig. 4.6, each region consists of producers set by commodities output and regional household, which is a composite of private households, a government, and a global bank. Each regional household provides primary factors to producers, including labour, capital, and land, and then purchases final consumption goods from the producers. Each producer outputs a goods by inputting primary factors provided in the region and intermediate goods (including all types of possible commodities) output from producers by commodities. Intermediate and final consumption goods can be imported from other regions through a global shipping industry. The choice of goods and primary factors to be input by producers is formulated to minimise the producing cost under the same production level to the output, whereas the choice of goods for consumption by private households is formulated to minimise

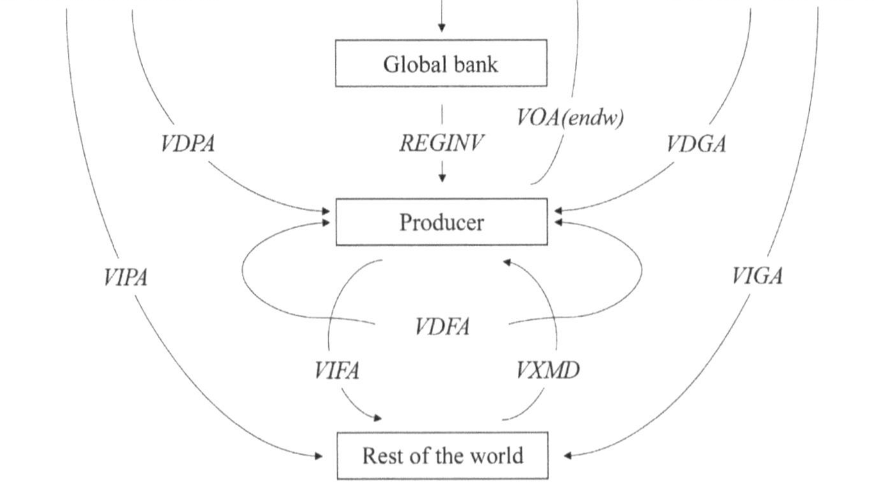

Fig. 4.6 Structure of the GTAP Model.
Source: Own compilation based on Hertel (1997).

expenditure under the same utility level. One feature of the GTAP model is that the choice of the trade partner (import country) is also endogenised in the model based on the Armington assumption, not only the choice whether each good is imported or domestically produced (Hertel, 1997); in other words, bilateral trade for each good can be explicitly considered.

Another feature of the GTAP model is operability. The team not only provides their programme in their unique language (called 'GEMPACK') and interface (called 'RunGTAP'), for easily applying the model for practitioners and governmental officers to measure the impact of related polices such as import tax reduction, but also provides the database containing complete information on bilateral trade, transportation and protection linkages. The latest database is the GTAP 10 Data Base, released in July 2019, comprising four reference years (2004, 2007, 2011, and 2014), 141 regions and 65 production sectors.

The GTAP model is mainly used for estimating the impact of increase or decrease in tax in each country, especially on the import tariff. Many efforts to constantly or bilaterally reduce the import tariff are challenged under various kinds of international cooperative frameworks, including not only the WTO, but also economic and customs unions such as the EU and free trade and economic partnership agreement such as the North America Free Trade Agreement and Trans-Pacific Strategic Economic Partnership. Many international organisations such as the Asia-Pacific Economic Cooperation (APEC) and individual governments estimated the changes in trade pattern for each commodity and the impact on a country's economy such as the GDP, from the calculation of the change in market equilibrium by the GTAP model if the tariff is exogenously reduced.

Further, the GTAP model can estimate the impact of policies to reduce the international shipping costs, such as investments in logistics infrastructure and facilitation in national borders, by incorporating a coefficient on the technological advancement of international shipping. This coefficient is defined for each combination of export and import countries by commodities (*atmfsd*). For example, Shibasaki et al. (2018) applied it into an impact analysis of the Northern Sea Route in liquefied natural gas trade. Such simulations can be regarded as the reflection of the changes in the transport pattern, which are estimated in the third step (modal split) and/or fourth step (route choice) of the four-step estimation methodology, to the changes in economy. In other words, the model can estimate the induced shipping demand by implementing logistics policies.

Another application of the GTAP model is forecasting the future amount of trade by inputting the increasing rate of economic indices, including primary factors, total factor productivity and other coefficients on technological advancement. Shibasaki and Watanabe (2012) forecasted the future amount of trade in each member economy of the APEC until 2025 by four future scenarios based on economic indices in each member economy and international economic policies. The details of future forecast using the GTAP model are described in Chapter 9.

Converting from value to weight basis

Note that the outputs of economic model including the GTAP model are generally described on a value basis, not a tonnage (or TEU) basis. Therefore, the outputs should be converted into a tonnage (or TEU) basis for further traffic analyses which are usually based on the volume. There are two approaches for the conversion. One is to estimate converting coefficients (USD/ton or USD/TEU) from a value basis to a tonnage (or TEU) basis by commodity, by using the existing trade statistics. Note that a volume unit is often different by commodity (e.g. ton, kilogram, and number) in some trade statistics. Furthermore, multiplying multiple absolute values (i.e. estimated trade values and converting coefficients) from different data sources often produces significant errors due to differences in the definition of regions and commodities.

Another approach to estimate the outputs on a volume basis is to use the increasing rate of the outputs that are estimated from the economic model, instead of the absolute value. Then, it is multiplied by the current shipping demand on a volume basis, which is estimated from the gravity model or other means, for each combination of the origin and destination and for each commodity. This approach often gives more consistent estimation results than the first approach.

Step 3: Modal split

As stated in 'Target Cargo Setting' in 'Preparation' section, the share by the transport mode or shape (i.e. container or bulk cargo) in international shipping is often given as a prerequisite condition in the demand forecast. In other words, these shares are implicitly assumed to be unaffected by any policies considered in the demand forecast model. For example, in Japan, 97% of international cargo on a tonnage basis are transported by maritime shipping and usually fixed in the demand forecast; only 3% are transported by airplane, and no international land transport are observed because Japan is a country surrounded by sea on all sides.

Another note that the third step (modal split) and fourth step (route assignment) of the four-step estimation methodology are occasionally integrated for the estimation as in the case of the model that the authors develop in this book (as stated after Part 2). In other words, the authors' model focuses on the choice of transport mode and route if the cargo shipping demand from origin to destination is given; this shipping demand is estimated in the second step of the aforementioned four-step estimation methodology. Therefore, the following part of this chapter includes only an introduction of each representative model (the logit model and network assignment model), which also underlies the authors' model, and does not contain any example of application of the model (which will be described in Part 2).

Logit model

One of major models that can describe the choice of transport mode or shape for cargo is a logit model introduced by Macfadden for discrete choice analysis. In the logit model, each cargo owner (or forwarder) n who needs to ship a cargo from origin i to destination j (for which the shipping demand is estimated in the second step of the four-step estimation methodology) is assumed to choose a mode (or shape) m where a utility, U_{ijnm}, of mode m from origin i to destination j for cargo owner n is the greatest relative to the utilities, U_{ijnm}', of any other mode m'. Namely,

$$U_{ijnm} > U_{ijnm'}, \quad \forall m \in M_{ij}, \forall m' \in M_{ij}, m \neq m', \forall ij \in \Omega, \tag{4.6}$$

where M_{ij} is the choice set of shipping modes and Ω is the set of origin and destination. The utility in each route assumingly consists of a deterministic term V_{ijm}, and an unobservable (or error) term ε_{ijnm}, namely,

$$U_{ijnm} = V_{ijm} + \varepsilon_{ijnm}. \tag{4.7}$$

In the transportation problem, a deterministic term V_{ijm} is generally defined as the shipping cost C_{ijm} and/or shipping time T_{ijm}. In addition, note that V_{ijm}, C_{ijm}, and T_{ijm} are the same for the shipping by mode m from origin i to destination j, irrespective of the cargo owners. Namely,

$$V_{ijm} = -C_{ijm}, \tag{4.8a}$$

$$V_{ijm} = -T_{ijm}, \text{ or} \tag{4.8b}$$

$$V_{ijm} = -G_{ijm} = -\left(C_{ijm} + vt \cdot T_{ijm}\right), \tag{4.8c}$$

where G_{ijm} is a generalised cost, including both the shipping cost and time, and vt is the value of time, which is a coefficient for converting shipping time into a monetary term. Note that utility U and the deterministic term V are more favourable as they increases, whereas generalised cost G including shipping cost and time is less favourable as it increases.

The distributions of the error term ε in Eq. (4.7) are assumed to be independent of each other in the following expansions; in other words, the random utility theory is adopted. The error term's distribution $f(\eta)$ of variable η generally follows the normal distribution as

$$f_N(\eta) = \frac{1}{\sqrt{2\pi\sigma}} \exp\left[-\frac{(\eta-\mu)^2}{2\sigma^2}\right], \tag{4.9}$$

where μ is the average and σ is the variance of variable η. The model where the distribution of the error term follows the normal distribution is called a probit model; however, it is difficult to solve analytically the probability that the utility of route h is the greatest. Therefore, in the following expansion, the distribution of the error term approximately follows a Gumbel distribution, as shown in Eq. (4.10), instead of a normal distribution:

$$f_G(\eta) = \theta \cdot \exp\left(-\theta \cdot (\eta - \beta)\right) \cdot \exp\left[-\exp\left(-\theta \cdot (\eta - \beta)\right)\right], \tag{4.10}$$

where θ is a distribution parameter, and β is mode of variable η. The variance of variable η is expressed as $\pi^2/6\theta^2$.

For simplicity, consider a case in which there exist two choices (i.e. binary choice). The probability P_{ij1} that mode 1 is chosen is

$$\begin{aligned} P_{ij1} &= \Pr\left[U_{ijn1} > U_{ijn2}\right] \\ &= \Pr\left[V_{ij1} + \varepsilon_{ijn1} > V_{ij2} + \varepsilon_{ijn2}\right] \\ &= \Pr\left[\varepsilon_{ijn1} = \eta\right] \cdot \Pr\left[\varepsilon_{ijn2} < \eta + V_{ij1} - V_{ij2}\right], -\infty < \eta < \infty, \end{aligned} \tag{4.11}$$

because of the assumption of the random utility theory. If the error term's distribution follows the Gumbel distribution, Eq. (4.11) can be rewritten as follows:

$$\begin{aligned} P_{ij1} &= \int_{-\infty}^{\infty} f_G(\eta) \cdot \Phi_G\left(\eta + V_{ij1} - V_{ij2}\right) \cdot d\eta \\ &= \int_{-\infty}^{\infty} \theta \cdot \exp(-\theta \cdot \eta) \cdot \Phi_G(\eta) \cdot \Phi_G\left(\eta + V_{ij1} - V_{ij2}\right) \cdot d\eta \\ &= \int_0^1 \frac{\theta \cdot \exp(-\theta \cdot \eta) \cdot y}{\theta \cdot \exp(-\theta \cdot \eta) \cdot y \cdot \left(1 + \exp\left(V_{ij1} - V_{ij2}\right)\right)} \cdot dy \\ &= \left[\frac{y}{1 + \exp\left\{\theta \cdot \left(V_{ij1} - V_{ij2}\right)\right\}}\right]_0^1 \\ &= \frac{\exp\left(\theta \cdot V_{ij1}\right)}{\exp\left(\theta \cdot V_{ij1}\right) + \exp\left(\theta \cdot V_{ij2}\right)}. \end{aligned} \tag{4.12}$$

Generally, if the error term's distribution follows the Gumbel distribution with distribution parameter θ, the probability P_{ijm} that the utility of mode m, or U_{ijm}, is the greatest among those of any other modes $U_{ijm'}$ is described as follows:

$$P_{ijm} = \exp\left(\theta \cdot V_{ijm}\right) \Big/ \sum_{m \in M_{ij}} \exp\left(\theta \cdot V_{ijm}\right). \tag{4.13}$$

Fig. 4.7 shows the shape of the probabilistic function described as Eq. (4.12) in the binary choice. If the deterministic term of mode 1, V_{ij1}, is sufficiently larger than that of mode 2, V_{ij2}, a probability that mode 1 is chosen is nearly 100%, whereas if V_{ij1} is significantly smaller than V_{ij2}, a probability for mode 1 is nearly 0%. If V_{ij1} and V_{ij2} are the same value (i.e. no difference between them), the probability of each mode being chosen is 50%. These are favourable results that the probabilistic function should meet. In addition, the figure shows the relationship with the significance of distribution parameter θ. The larger the value of θ, the sharper is the probabilistic function; i.e. more sensitive is the function to the differences in deterministic terms. On the other hand, if θ is smaller, the probabilistic function becomes flatter; i.e. it is less sensitive to the difference in deterministic terms.

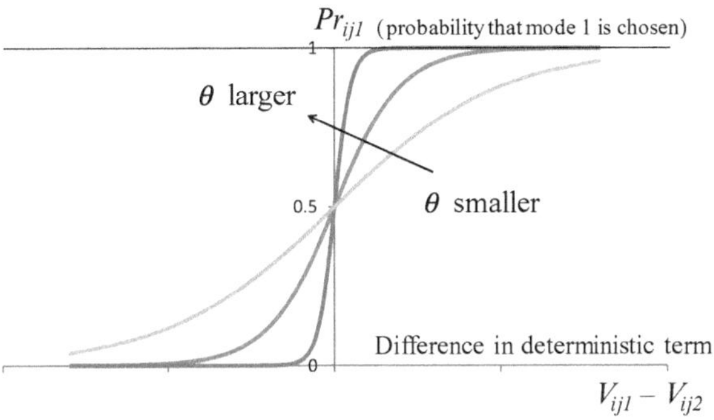

Fig. 4.7 Shape of the probabilistic function in the logit model and relationship to the significance of distribution parameter θ.
Source: Own compilation.

The unknown parameters (e.g. θ and vt) included in the model can be generally estimated by the maximum likelihood methodology, which estimates them so as to maximise the likelihood L defined in the following equation.

$$\max L, \tag{4.14}$$

$$s.t. \quad L = \prod_{ij \in \Omega} \prod_{n \in N_{ij}} \prod_{m \in M_{ij}} \left(\widehat{P}_{ijm} \right)^{\delta_{ijnm}}. \tag{4.15}$$

Here, $\widehat{P}_{ijm}$ is the estimated probability that a mode m is chosen for shipping from origin i to destination j; N_{ij} is a set of cargo owners for shipping from origin i to destination j. Further, δ_{ijnm} is the Kronecker delta; $\delta_{ijnm}=1$ if a cargo owner n actually chooses a mode m and $\delta_{ijnm}=0$ otherwise.

Step 4: Route choice

The last step of the four-step demand forecast methodology is the route choice if the cargo shipping demand from the origin to destination by transport mode is given. The route choice problem in the (maritime) logistics field can be roughly classified based on two viewpoints—shipper's (cargo owner's) choice and carrier's (shipping company's) choice.

In this step, an assignment model in a physical transport network is frequently applied. The 'network' generally consists of two components, namely, nodes and links (or arcs). Nodes are a set of points, and the links are a set of line segments connecting these points. The physical network is not considered in some models; for example, if a logit model is applied to the route choice problem, each route is regarded as an individual alternative in the discrete choice set.

Shipper's route choice model

If considering the route choice from the shipper's viewpoint, shipping networks are usually given. The shipper (or forwarder as its agent) will choose the shipping route, including the port to export or import and shipping company, under the condition that the level of service on the network as indicated by parameters such as service frequency, capacity, speed, or shipping time, and freight charge are given by the shipping company. In reality, shippers do not care about the actual shipping route but only about the total cost (i.e. freight charge) and time.

Shipping companies decide the actual shipping route so as to minimise the shipping cost under the condition that they can provide attractive freight charges and shipping time to the shippers for differentiating themselves from their competitors. In case that the transport market is sufficiently competitive with numerous operators such as in the conventional truck market, the transport route can be approximately decided by shippers because transport companies cannot pursuit the non-zero profit and have no power of influence on the market as a price taker. Because the international maritime container shipping market has recently become more oligopolistic with 10 large companies integrated into three groups for operation or alliances (see Chapter 1 for details), it is more complicated to describe the relationship between the shippers' behaviour in choosing the port and shipping company and the shipping companies' behaviour in choosing the actual shipping route. Therefore, the authors developed a general route choice model of shippers on the maritime container shipping network (see Chapter 6) under the condition that the freight charge is given (therefore it is not included in the maritime shipping assignment model) and that the fixed shipping time planned by the shipping company can be sometimes changed because of additional waiting time due to overloading in a port.

Two major aspects should be considered if developing the route choice model for a shipping network (i.e. network assignment model): capacity constraint and handling of minor elements that are not explicitly considered in the model (unlike shipping cost and time). As described in Table 4.5, a simple all-or-nothing assignment, in which all cargoes are assigned the shortest path, can be applied, if neither capacity constraints in any links, nor any minor elements are considered. In the case that the effect of minor elements is considered (often as an error term with randomness), some stochastic approach, such as the logit model, is necessary. Dial's assignment (1971), which is a

Table 4.5 Typical network assignment model.

	Deterministic	**Stochastic**
No capacity constraint	All-or-nothing assignment based on a shortest path search	Stochastic network assignment
With capacity constraint	Network equilibrium assignment (user equilibrium assignment)	Stochastic network equilibrium assignment (stochastic user equilibrium assignment)

Source: Own compilation.

typical stochastic network assignment methodology that provides results similar to those of the logit model, is applied in the master model of this book (see Chapter 7). In the case that only capacity constraints in some links of the given network are considered, a network equilibrium assignment methodology should be applied; such a method is described later in this chapter. Finally, a stochastic network equilibrium assignment methodology should be applied when both the capacity constraint and stochastic elements are considered; this approach combines the network equilibrium assignment and stochastic assignment, though the conventional solution algorithm is different from that of the deterministic network equilibrium assignment.

The following text describes the basic idea of the network equilibrium assignment model, which is also applied in this book for the maritime shipping model (in Chapter 6) and regional transport submodel (in Chapter 7). For detailed description of the network equilibrium assignment model, refer Sheffi (1985) or JSCE (1998) for a description in Japanese.

User equilibrium assignment based on Wardrop's first principle

In the following discussion, only shipping time is considered as the cost of each link. An approach of the network equilibrium assignment is based on the principles proposed by Wardrop (1952). The user equilibrium (UE) assignment is a mathematical expression of Wardrop's first principle: "*The journey times on all the routes actually used are equal, and less than those which would be experienced by a single vehicle on any unused route*". The equilibrium is reached under the assumption that each user non-cooperatively seeks to minimise their transportation cost and no user may lower their transportation cost through unilateral action under equilibrium. Note that the application of Wardrop's second principle that "*the average journey time is a minimum*" is called as a system optimum assignment implying that each user behaves cooperatively in choosing their own route to ensure the most efficient use of the whole system.

The mathematical formulation of the Wardrop's first principle is as follows. Consider a simple network shown in Fig. 4.8. Note that a link a is defined to connect two nodes,

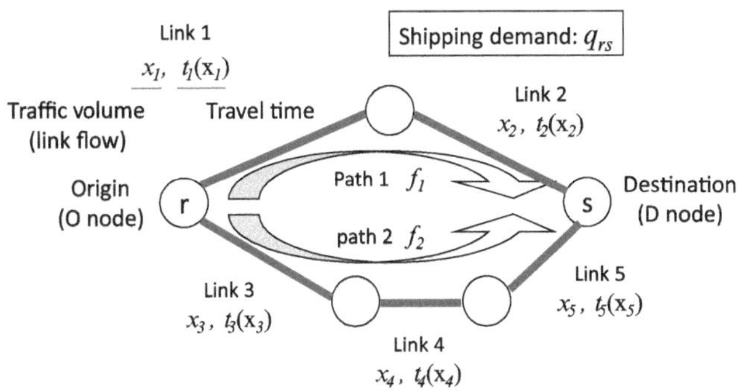

Fig. 4.8 Sample network, including a description of relationship between the link and path.
Source: Own compilation.

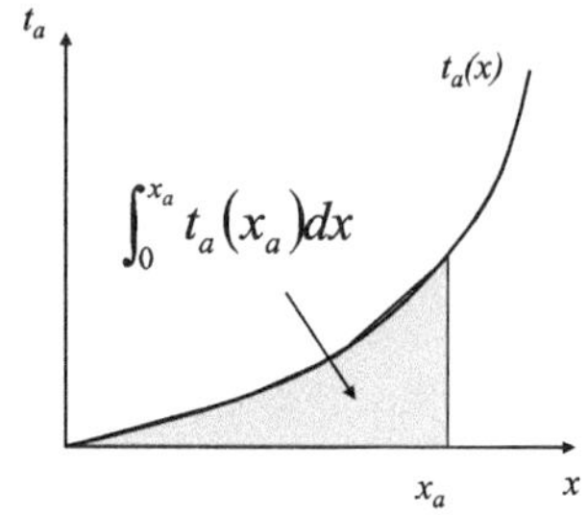

Fig. 4.9 Cost (shipping time) function t_a, link flow x_a, and integration of cost function. *Source:* Own compilation.

whereas a path k is defined as a possible route from origin r to destination s. Further, x_a is a cargo flow of link a; t_a is a shipping time of cargo in link a; f_k is a cargo flow of path k and q_{rs} is cargo shipping demand from origin r to destination s. Note that (some of) the shipping time function, t_a, is dependent on link flow x_a as shown in Fig. 4.9. The congestion, which is expressed by the fact that as the shipping time increases with the link flow because of capacity constraint, is essential for the UE model.

First, from the relationship between link a and path k, the following preservation law of link flow should be identically formulated by using the Kronecker delta, δ_{ak}^{rs} (= 1 if link a is included in a path k and 0 otherwise).

$$x_a = \sum_{(r,s)\in R\times S}\sum_{k\in K^{rs}} \delta_{a,k}^{rs} \cdot f_k^{rs}, \quad \forall a, \tag{4.16}$$

where

$$\delta_{a,k}^{rs} = \begin{cases} 1 \text{ if } a \in k \\ 0 \text{ if } a \notin k \end{cases}; \tag{4.17}$$

R is the set of the export port; S is the set of the import port; and K^{rs} is the set of paths for OD pair rs.

Another identical equation is a preservation law of cargo shipping demand:

$$\sum_{k\in K_{rs}} f_k^{rs} - q_m^{rs} = 0, \quad \forall r,s. \tag{4.18}$$

Further, each path flow should be non-negative.

$$f_k^{rs} \geq 0, \quad \forall k,r,s. \tag{4.19}$$

In the network shown in Fig. 4.8, Wardrop's first principle "*the journey times in all routes actually used are equal and not more than those which would be experienced by a single vehicle on any unused route*" is mathematically formulated as follows also by using the Kronecker delta.

$$\sum_a \delta_{a,k}^{rs} \cdot t_a(x_a) = u_{rs} \left(\text{if } f_k^{rs} > 0\right) \tag{4.20a}$$

$$\sum_a \delta_{a,k}^{rs} \cdot t_a(x_a) \geq u_{rs} \left(if \ f_k^{rs} = 0 \right), \tag{4.20b}$$

where u_{rs} is the minimum path travel time from origin r to destination s, which is referred in Wardrop's first principle. Eq. (4.20a) represents the first part of Wardrop's first principle that the journey times for all routes actually used, u_{rs}, are equal. Eq. (4.20b) represents the latter part that u_{rs} are not more than those which would be experienced by a single vehicle on any unused route.

Eqs. (4.20a), (4.20b) are proved to be mathematically equal to Eq. (4.21) such that Eqs. (4.16), (4.18), and (4.19) are satisfied, although detailed proof is not provided herein.

$$\min_x z(x) = \sum_{a \in A} \int_0^{x_a} t(x_a) dx, \tag{4.21}$$

$$s.t. (4.16), (4.18) \text{and} (4.19),$$

where A is the set of links, and $z(.)$ is the objective function. Because $\int_0^{xa} t_a(x_a)dx$ is an area shown in Fig. 4.9, the first-order differential by each link flow x_a of the objective function z is the shipping time function itself. That is,

$$\frac{\partial z}{\partial x_a} = t_a(x_a). \tag{4.22}$$

Then, the second-order differential by each link flow is expressed as follows:

$$\frac{\partial^2 z}{\partial x_a^2} = \frac{dt_a(x_a)}{dx_a} \tag{4.23}$$

$$\frac{\partial^2 z}{\partial x_a \partial x_b} = 0 \qquad (a \neq b).$$

Therefore, if a link cost function $t_a(x_a)$ is an increasing function (or a constant) to a flow of the link x_a, the objective function z is convex. Because all the constraint conditions formulated in Eqs. (4.16), (4.18), and (4.19) are also convex, the problem defined in Eq. (4.21) has a unique solution. If a unique solution is guaranteed, any algorithm to solve the problem can be applied, though the calculation time of each algorithm is significantly different. In the UE assignment in this book, the Frank–Wolfe algorithm, which is one of the most popular algorithms to solve the UE problem, is applied; please refer Sheffi (1985) or JSCE (1998) for detail.

Carrier's route choice model

The main focus of the model developed from the viewpoint of carriers is to clarify how the shipping network is structured; this model is not described in this book. Many papers have been published, mainly in the field of operational research, as summarised in several review papers such as Ronen (1983, 1993), Christiansen et al. (2004, 2007, 2013), Meng et al. (2014), and Lee and Song (2017). More papers have been published based on this viewpoint than for the shipper's viewpoint.

A shipping company has to decide many decision variables in its operation. Under the general condition that the shipping demand from an export port to an import port is given (though sometimes it is changeable), each shipping company makes various decisions at the strategic, tactical, and operational levels. At the strategic level, a company decides whether it joins the shipping market or not and whether it cooperates with or charters a slot from other companies. At the tactics level, it has to decide not only the size of a vessel to be operated and the frequency (or the number of vessels operated) of the service, but also the ports to call with their order in case of container shipping and the ports to which the containers are transshipped. Even at the operational level, it decides the navigation route, vessel speed (which significantly depends on fuel price), and how cargo is loaded in a vessel.

Because a shipping company deals with a large number of decision variables, the models describing the behaviour of the company generally focus on a single or a few variable(s) under the condition that the other variables are exogenously given. Most existing research focus on the decisions at the tactical level, such as the vessel size and ports of call, or at the operational level such as vessel speed, by applying many types of operational research methodologies such as integer programming and heuristic methods (including evolutional algorithms). Several researches at the strategic level such as network design have also recently been published. Moreover, some of them are extended to include an intermodal transport network; for example, Shibasaki et al. (2010), Meng and Wang (2011), and Wang and Meng (2017) focusing on East Asia, Shibasaki and Watanabe (2012) focusing on the APEC region, and Tran et al. (2017) applying to the Trans-Atlantic region. However, these models are difficult to be applied to the entire global liner service network while ensuring agreement with the current situation. Liu et al. (2014) solved the network design problem by combining simple cost minimisation models for structuring the network and a logit model of port selection. Wang et al. (2016) developed an intermodal network assignment model considering the distribution of cargo owners on the hinterland. Such models based on new concepts are also in progress for the empirical, global-scale analysis.

Conclusion

This chapter introduced the basics of the container demand forecast, according to the flow of four-step traffic estimation methodology. The network assignment model, which will be introduced in the parts of this book afterward, is classified as an integrated model of the third step (transport mode choice) and fourth step (route choice), if the cargo shipping demand from origin to destination estimated in the second step is given. Moreover, if the future shipping demand is input into the network assignment model in the future simulation, another model to forecast the future shipping demand is necessary, which will be also introduced in Chapter 9 of Part 2 and some chapters of Part 3.

References

Christiansen, M., Fagerholt, K., Ronen, D., 2004. Ship routing and scheduling: status and perspectives. Transp. Sci. 38 (1), 1–18.

Christiansen, M., Fagerholt, K., Nygreen, B., Ronen, D., 2007. Maritime transportation. In: Barnhart, C., Laporte, G. (Eds.), Handbook in Operations Research and Management, Volume 14: Transportation. North Holland, Amsterdam, pp. 189–284.

Christiansen, M., Fagerholt, K., Nygreen, B., Ronen, D., 2013. Ship routing and scheduling: in the new millennium. Eur. J. Oper. Res. 228, 467–483.

Dial, R.B., 1971. A probabilistic multipath traffic assignment algorithm which obviates path enumeration. Transp. Res. 5 (2), 83–111.

Guerrero, D., Grasland, C., Ducruet, C., 2015. Explaining international trade flows with shipping-based distances. In: Ducruet, C. (Ed.), Maritime Networks: Spatial Structures and Time Dynamics, Routledge Studies in Transport Analysis. Routledge, London and New York, pp. 303–321.

Hertel, T.W. (Ed.), 1997. Global Trade Analysis: Modeling and Applications. Cambridge University Press, New York.

Hertel, T., 2013. Global applied general equilibrium analysis using the global trade analysis project framework. In: Dixon, P.B., Jorgenson, D. (Eds.), Handbook of Computable General Equilibrium Modeling. vol. 1. Elsevier, Amsterdam, pp. 815–876.

IHS, Inc, 2017. World Trade Service (WTS) database. [online] (Accessed 14 December 2017).

International Monetary Funds (IMF), 2016. World Economic Outlook Database. http://www.imf.org/external/pubs/ft/weo/2016/02/weodata/index.aspx.

Japan Society of Civil Engineers, the (JSCE), 1998. Equilibrium Analysis of Transport Network. (in Japanese).

Lee, C.Y., Song, D.P., 2017. Ocean container transport in global supply chains: overview and research opportunities. Transp. Res. 95B, 442–474. https://doi.org/10.1016/j.trb.2016.05.001.

Liu, Z., Meng, Q., Wang, S., Sun, Z., 2014. Global intermodal liner shipping network design. Transp. Res. 61E, 28–39. https://doi.org/10.1016/j.tre.2013.10.006.

Meng, Q., Wang, X., 2011. Intermodal hub-and-spoke network design: incorporating multiple stakeholders and multi-type containers. Transp. Res. 45B, 724–742.

Meng, Q., Wang, S., Andersson, H., Thun, K., 2014. Containership routing and scheduling in liner shipping: overview and future research directions. Transp. Sci. 48 (2), 265–280.

Ronen, D., 1983. Cargo ships routing and scheduling: survey of models and problems. Eur. J. Oper. Res. 12 (2), 119–126.

Ronen, D., 1993. Ship scheduling: the last decade. Eur. J. Oper. Res. 71 (3), 325–333.

Sheffi, Y., 1985. Urban Transportation Networks: Equilibrium Analysis with Mathematical Programming Methods. Prentice-Hall, Inc., New Jersey.

Shibasaki, R., Watanabe, T., 2012. Future forecast of trade amount and international cargo flow in the APEC region: an application of trade-logistics forecasting model. Asian Transp. Stud. 2 (2), 75–89.

Shibasaki, R., Watanabe, T., Araki, D., 2010. How is model accuracy improved by usage of statistics? An example of international freight simulation model in East Asia. Asian Transp. Stud. 1 (1), 33–46.

Shibasaki, R., Usami, T., Furuichi, M., Teranishi, H., Kato, H., 2018. How do the new shipping routes affect Asian LNG markets and economy? Case of the Northern Sea route and Panama Canal expansion. Marit. Policy Manag. 45 (4), 543–566. https://doi.org/10.1080/03088839.2018.1445309.

Toriumi, S., 2010. Pattern analysis of containerships using maritime shipping network. J. Oper. Res. Soc. Jpn. 55 (6), 359–367 (in Japanese).

Tran, N.K., Haasis, H.D., Buer, T., 2017. Container shipping route design incorporating the costs of shipping, inland/feeder transport, inventory and CO_2 emission. Marit. Econ. Logist. 19, 667–694.

Wang, X., Meng, Q., 2017. Discrete intermodal freight transportation network design with route choice behavior of intermodal operators. Transp. Res. 95B, 76–104. https://doi.org/10.1016/j.trb.2016.11.001.

Wang, X., Meng, Q., Miao, L., 2016. Delimiting port hinterlands based on intermodal network flows: model and algorithm. Transp. Res. 88E, 32–51. https://doi.org/10.1016/j.tre.2016.02.004.

Wardrop, J.G., 1952. Some theoretical aspects of road traffic research. Proc. Inst. Civil Eng. 2 (1), 325–362.

World Bank (WB) n.d. website: https://data.worldbank.org/indicator/IS.SHP.GOOD.TU.

Part Two

Model & data

Basic concept

5

Ryuichi Shibasaki
The University of Tokyo

Model concept

The purpose of this book is to provide a tool for simulating the global container cargo flow and provide examples to estimate the impact of different logistics policies, including infrastructure investment and facilitation of land national borders across the world. Before introducing example applications to the real field in Part 3, the structure of the simulation model and dataset intended to be used are explained in the chapters of Part 2 including this chapter.

The simulation model used in this book focuses on how container cargo is moved on a given global, intermodal transport network and how it will change with a change in policies. In consideration of its practical application, the simulation model is developed, based on two principal preconditions. First, the cargo shipping demand from each origin to destination is provided as an input to the simulation model. In other words, the shipping demand induced by the improvement in transport networks is not considered in this model. If it needs to be considered, then the demand should be preliminarily estimated by other economic models such as the GTAP model by inputting the saved transport cost and/or time calculated from the simulation model of an earlier time. Similarly, if a future policy simulation based on the future network is implemented, the future shipping demand should be preliminarily estimated by other models. The method of estimating the cargo shipping demand (container OD matrix), including future demand, is described in Chapter 9.

The second precondition of the model is the transport network itself, including both maritime and land networks, which is elaborated upon in Chapter 8. As described in Chapter 4, many studies have been conducted on route choice models of maritime container shipping companies as well as network design problems of intermodal cargo transport. In contrast, the model in this book is developed from the cargo owners' perspective under a given networks. As indicated by Haralambides (2008), the vessel operation and cargo allocation problems from perspectives of the carriers and shippers, respectively, have different objectives and often conflict with each other. There are relatively few studies focusing on the allocation of maritime container cargo from the shippers' perspective, with a few exceptions, such as Malchow and Kanafani (2001) applying a multinomial logit model to explain the selection of a port for each shipment exported from the US, Bell et al. (2011) applying a frequency-based traffic assignment model to the maritime container assignment problem on a given liner shipping network with frequency and other strategic variables, Tavasszy et al. (2011) assigning global container cargo into an intermodal network, including land and maritime shipping by a path size logit model but not considering any actual liner shipping service,

Global Logistics Network Modelling and Policy. https://doi.org/10.1016/B978-0-12-814060-4.00005-8

ITF-OECD (2015) developing a shortest path search model on a global intermodal network, even including air transport, Jones et al. (2011) and Wang et al. (2018) developing an intermodal network model with capacity constraint of shippers and carriers in the US based on the freight network equilibrium model (FNEM), first proposed by Harker and Friesz (1985) and Friesz et al. (1986), but not considering maritime shipping, Fan et al. (2009, 2012) and Meng and Wang (2011) developing a model to minimise the total shipping cost, including both the land and maritime networks and applied it to the North American and East Asian intermodal network, respectively, and Lin and Huang (2017) applying an user equilibrium network assignment model to the trans-Pacific route. Additionally, the ocean freight charge is included in the shipping cost for the shipper, which is different from the actual maritime shipping cost of the carrier. This is not considered in any of the reference literature introduced above.

One of the features of our model is to include both maritime container shipping (MCS) and land transport (LT) networks, considering the characteristics of each mode of transport. For example, the service frequency of rail, ferry, and maritime shipping is considered by defining the expected waiting time as a half of the inverse of service frequency, whereas it is not necessary to consider the expected waiting time for truck transport. Another example that reflects the actual characteristics of transport modes is the difference in the definition of freight charge; only the ocean freight charge is estimated on a path basis from an export port to an import port as mentioned above, whereas the freight charge of other modes can be defined on a link basis by approximating it to be proportional to the transport distance. By such simultaneously considering both the MCS and LT networks, the impact of both policies related to the MCS and LT can be compared.

The cargo is allocated on the transport network, based on the freight charge and shipping time in principle. The shipping time is converted into a monetary term by multiplying the coefficient, which is called the value of time. In that sense, the model can examine the impact of all policy options, which can alter into shipping costs and/or times, such as discounting the port and rail freight charge, enlargement of containership, increasing the frequency of liner and rail service, reducing port and rail handling time, reducing the land border-crossing time and cost, and speeding up the land transport service by constructing new expressways and railways. Additionally, because the congestion (i.e. capacity constraint) in each mode is considered in the model as a delayed transport time, policies to reduce congestion, such as rail, road, and vessel capacity expansions, can also be examined in the model simulation. The current settings of all variables included in the model are listed and described in Chapter 8.

In contrast, the model does not include any variables that cannot be expressed as either the cost or time of transport. Moreover, it does not explicitly include any variables that are difficult to numerically determine. However, the author believes that almost all variables affecting the mode and route choice of shippers can be expressed in terms of cost or time of transport, because all decisions related to logistics activities should be economical. Furthermore, to consider the variables that are difficult to numerically determine, the model adopts the random utility theory that can incorporate an unobservable term from the model developer, as mentioned later in Chapter 7.

The following sections of this chapter describe the entire structure of the simulation model, whereas Chapters 6 and 7 describe the details of each part of the model.

Entire structure of model

The simulation model in this book is to assign container flows under the given global, intermodal container transport network and fixed cargo shipping demand from the shippers' perspective. Fig. 5.1 shows the entire structure of the model, which is a two-layered network assignment model: a super-network for intermodal shipping in the upper layer and two real networks representing each MCS and LT in the lower layer. Note that the LT network is called the hinterland transport (HT) network in some chapters

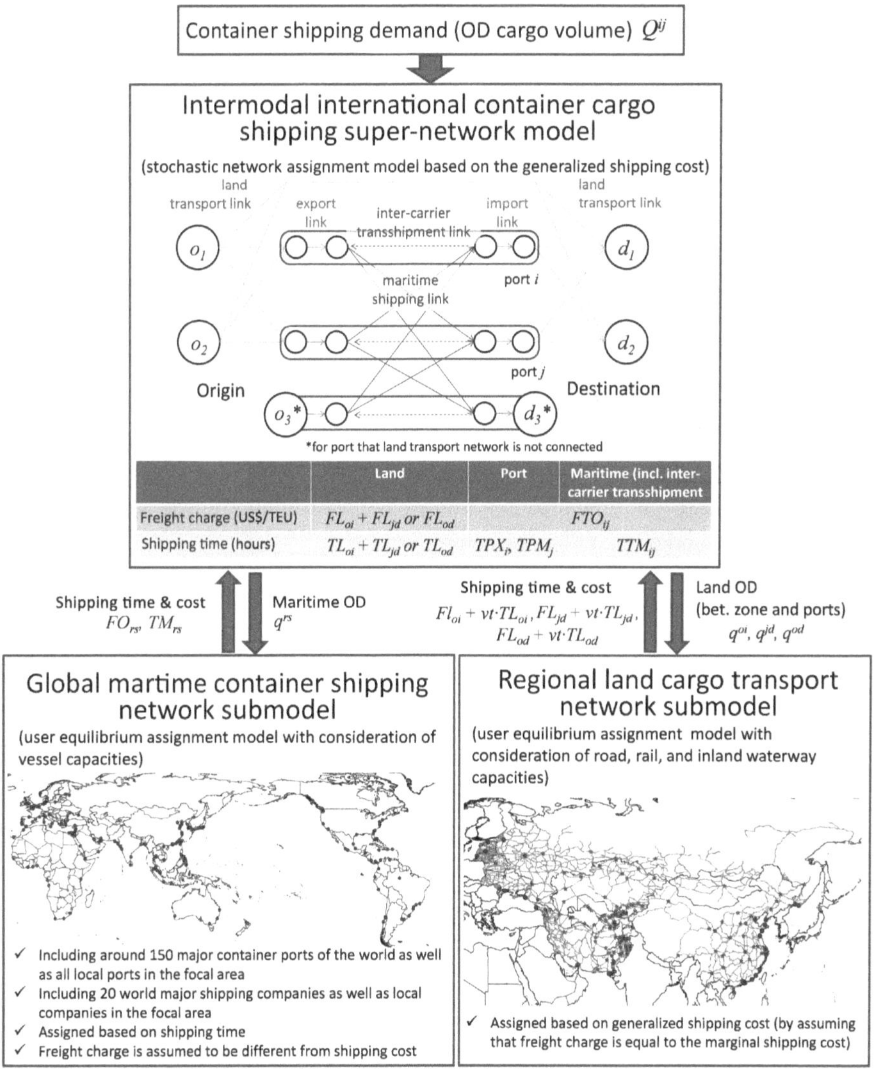

Fig. 5.1 Entire structure of the model used in this book.
Source: Own compilation.

in Part 3, which does not consider the international land cargo shipping demand. Full containers are only considered as the fixed regional cargo shipping demand.

The super-network model in the upper layer includes outputs of real network submodels in the lower layer, namely, the freight charge and shipping time for MCS and LT (or HT). The maritime and land cargo shipping demand, which are inputs of the two submodels in the lower layer, are cargo flows of the MCS and LT link in the super-network model. There are two major reasons why the model is divided into two layers. One is in consideration of ocean freight charge in the MCS, which is different from the shipping cost and estimated with reference to the path, rather than link-based shipping cost, to better reflect reality. The other reason is to incorporate a stochastic approach with capacity constraints. Applying the stochastic user equilibrium assignment to an integrated network is an alternative approach; however, its results can be affected by the density of real network due to independent axioms, if a Gumbel distribution is assumed for errors of the utility function; it will not be applicable to a huge network, if normal distribution is assumed, instead.

In principle, the cargo is allocated, based on generalised cost of each link in the network. The model includes two features for improving the fitness: the stochastic approach and the capacity constraint. The network assignment model is developed based on generalised cost; however, there may be other factors for shippers to choose the shipping route, mode, and company such as lot size, safety and security issues, historical background (sank cost), and environmental awareness. Therefore, a stochastic approach based on random utility theory should be applied. The capacity constraint of each transport mode is considered in the real networks in the lower layer. The capacities of roads, railways, and vessels (for both inland waterways and maritime shipping) are considered in each submodel, which is introduced in Chapters 6 and 7. However, the capacity of port (i.e. container terminal) is not considered in the present model, which would be worked upon for the future.

The submodels in the lower layer are divided into MCS and LT, being that they are physically separated and have a difference in their definition of freight charge. Generally, the freight charge of trucks and railways is approximated to be proportional to the transport distance, but this is not always true in the MCS. The ocean freight charge in the MCS depends on the balance between cargo shipping demand and vessel supply; put differently, the ocean freight charge may be decided on a path basis from port of export to port of import.

Because each submodel in the lower layer contains a capacity constraint of the link, a network equilibrium assignment model is applied. Additionally, any link costs included in each submodel are either an increasing or constant function of the cargo flow, as explained in the following chapters; therefore, each submodel in the lower layer has a unique solution. It should be noted that cargo flow of the LT submodel is allocated, based on generalised cost, whereas that of the MCS submodel is based on shipping time; because the ocean freight charge (which is an expense for the shippers) is decided on a path basis, as mentioned above. The MCS submodel includes all the liner shipping services of the world that are operated by major container shipping companies. Of all the models mentioned in Part 3, the LT submodel is developed in regions, which are focused on each simulation.

It is difficult to simultaneously solve the entire model, including both the upper and lower layers; therefore, the model calculation is separated for the two layers. However, a unique solution is not guaranteed for such iterative calculations; meaning the model solution depends on the algorithm. The calculation procedure of the model is described in Chapter 7.

Other model features and future works

1. The current model considers only laden (full) containers and container-equivalent cargo, and not empty containers, to simplify the model. Optimising the relocation of empty containers is an important topic from both business and research perspectives; therefore, incorporating the movement of empty containers and cargo is one of the future works for model expansion.
2. The current model finds the converged solution on a global level; therefore, the model is more suitable to examine policies that affect the global international transport networks, such as investment in seaports and facilitation in national land borders, rather than focusing on more detailed transport networks in a particular region or city. Improving the applicability of this model at the detailed network level is one of the future challenges. Applying a detailed stepwise assignment than an equilibrium approach is worth considering.
3. The current model is developed, based on annual cargo flow. That is to say, it cannot consider any shorter (daily, monthly, and seasonal) fluctuations of demand and supply within a year. For example, consideration of a seasonal difference in supply is essential if the potential of the Northern Sea Route across the Arctic Sea is examined. Such seasonal (or much shorter) fluctuations should be considered in future models.
4. Furthermore, the model focuses on maritime shipping and land transport of containers and container-equivalent cargo. Thus, other types of maritime cargo such as dry bulk and liquid bulk as well as air and pipeline transport are not included. These modes and cargo types should be also included in the future expanded model. In addition, the difference in contents of containers should be also considered.
5. The programming of the model is done using FORTRAN, because the model has been developed over a period of 15 years. The model calculations require a dual CPU computer with 32.0GB RAM and the calculation time is approximately a few hours with it.

Structure of Part 2

Chapter 6 focuses on the MCS submodel, which can be used independently for the MCS simulation if necessary, as shown in some simulations in Part 3. A procedure of a model calculation, including a convergence check, is also shown. Along with the model that is developed, based on equilibrium of shipping time, the methodology to estimate ocean freight charge is also explained.

Chapter 7 describes the super-network model in the upper layer as well as the LT submodel in the lower layer. The interaction of upper and lower layers, including the calculation procedure, is also explained and convergence of the iterative calculations is checked.

Other two chapters are related to the data input into the model. Chapter 8 focuses on the network data for MCS and LT. Not only is the physical condition of each network derived from the external data source, but coefficients for the cost and congestion functions are also described. Chapter 9 describes the estimation methodology of container shipping demand on both port and regional basis, which are input into the model, both for the present and future.

References

Bell, M.G.H., Liu, X., Angeloudis, P., Fonzone, A., Hosseinloo, S.H., 2011. A frequency-based maritime container assignment model. Transp. Res. 45B, 1152–1161.

Fan, L., Wilson, W.W., Tolliver, D., 2009. Logistical rivalries and port competition for container flows to US markets: impacts of changes in Canada's logistics system and expansion of the Panama Canal. Marit. Econ. Logist. 11 (4), 327–357.

Fan, L., Wilson, W.W., Dahl, B., 2012. Congestion, port expansion and spatial competition for US container imports. Transp. Res. 48E, 1121–1136.

Friesz, T.L., Gottfried, J.A., Morlok, E.K., 1986. A sequential shipper-carrier network model for predicting freight flows. Transp. Sci. 20 (2), 80–91.

Haralambides, H.E., 2008. Structure and operations in the liner shipping industry. In: Hensher, D.A., Button, K.J. (Eds.), Handbook of Transport Modelling, second ed. Pergamon, Oxford, pp. 761–775.

Harker, P.T., Friesz, T.L., 1985. The use of equilibrium network models in logistics management: with application to the US coal industry. Transp. Res. 19B (5), 457–470.

International Transport Forum (ITF)-OECD, 2015. ITF Transport Outlook 2015 (online) http://www.keepeek.com/Digital-Asset-Management/oecd/transport/itf-transport-outlook-2015_9789282107782-en#page1. (Accessed 29 April 2018).

Jones, D.A., Farkas, J.L., Bernstein, O., Davis, C.E., Turk, A., Turnquist, M.A., Nozick, L.K., Levine, B., Rawls, C.G., Ostrowski, S.D., Sawaya, W., 2011. US import/export container flow modeling and disruption analysis. Res. Transp. Econ. 32 (1), 3–14. https://doi.org/10.1016/j.retrec.2011.06.003.

Lin, D.Y., Huang, K.L., 2017. An equilibrium-based network model for international container flows. Marit. Policy Manage. 44 (8), 1034–1055. https://doi.org/10.1080/03088839.2017.1371855.

Malcow, M., Kanafani, A., 2001. A disaggregate analysis of factors influencing port selection. Marit. Policy Manage. 28 (3), 265–277.

Meng, Q., Wang, X., 2011. Intermodal hub-and-spoke network design: incorporating multiple stakeholders and multi-type containers. Transp. Res. 45B, 724–742. https://doi.org/10.1016/j.trb.2010.11.002.

Tavasszy, L., Minderhoud, M., Perrin, J.F., Notteboom, T., 2011. A strategic network choice model for global container flows: specification, estimation and application. J. Transp. Geogr. 19, 1163–1172.

Wang, H., Nozick, L., Xu, N., Gearhart, J., 2018. Modeling ocean, rail, and truck transportation flows to support policy analysis. Marit. Econ. Logist. 20, 327–357.

Global maritime container shipping model

6

Ryuichi Shibasaki
The University of Tokyo

Model framework

The maritime container shipping (MCS) model is not only functioned as a submodel in the lower layer of the entire simulation model, which is introduced in this book as mentioned in the previous chapter, but also used as an independent model for some specific simulations on the maritime-related policies. The MCS (sub-)model allocates container cargo to the global liner shipping network, which is structured, based on containership movement data. In this model, the author assumes that each container of each maritime origin–destination (OD) pair chooses a route to minimise the total transit time; in other words, the shipper chooses a carrier considering only the transit time and not the freight charge. It is based on the understanding of the MCS market that the international MCS market is oligopolistic, but that the ocean freight charge for an OD pair can approximate the same for all carriers if service is provided and used, because of severe price competition.

As vessels for each liner service (LS) have their own capacities, there is congestion (diseconomy of scale) if containers concentrate on a specific LS. Therefore, to consider traffic congestion in the link, a user equilibrium (UE) assignment is applied, which is formulated as

$$\min_{x} z(x) = \sum_{a \in A} \int_0^{x_a} t(x_a)\, dx, \tag{6.1}$$

$$s.t.\ x_a = \sum_{(r,s) \in R \times S} \sum_{k \in K^{rs}} \delta_{a,k}^{rs} \cdot f_k^{rs}, \forall a, \tag{6.2}$$

$$\sum_{k \in K^{rs}} f_k^{rs} - q_m^{rs} = 0, \forall r, s, \text{and} \tag{6.3}$$

$$f_k^{rs} \geq 0, \quad \forall k, r, s, \tag{6.4}$$

where

a = link
A = set of links
x_a = flow of the link a
$t(.)$ = cost function of each link
$z(.)$ = objective function
r = maritime origin (export port)

Global Logistics Network Modelling and Policy. https://doi.org/10.1016/B978-0-12-814060-4.00006-X

s = maritime destination (import port)
R, S = set of the export and import ports, respectively
k = path
K^{rs} = sets of paths for OD pair rs
f_k^{rs} = flows on path k
$\delta_{a,k}^{rs}$= Kronecker deltas
q_m^{rs} = cargo demand of the MCS (TEU/year) from export port r to import port s

Kronecker delta δ_{ak}^{rs} is defined as

$$\delta_{a,k}^{rs} = \begin{cases} 1 \text{ if } a \in k \\ 0 \text{ if } a \notin k \end{cases}. \tag{6.5}$$

Regarding the networks, only the navigating link has a flow-dependent cost function as described later, whereas the cost functions of other links have flow-independent cost functions. According to the UE assignment definition, MCS time, TM_{rs}, from export port r to import port s, is defined as

$$TM_{rs} = \min_k \left\{ \sum_{a \in k} t(x_a) \right\}. \tag{6.6}$$

The UE problem shown in Eq. (6.1) is solved using the traditional Frank–Wolfe algorithm presented in Sheffi (1985). The network structure of the MCS model, including each LS, is shown in Fig. 6.1. Each container of the shipper chooses a link from the maritime origin node of an export port to the maritime destination node of an import port.

Shipping time function

Navigation link

A navigation link connects each port on the sea, set by each LS. The link cost includes the shipping time and congestion due to the vessel capacity constraint:

$$t_n(x_a) = \left(\frac{\text{lm}_a}{v_a} + \gamma_a^s \cdot TS + \gamma_a^p \cdot TP \right) + TW_{a'} \cdot b1 \left(\frac{x_a}{cap_a \cdot freq_a} \right)^{b2}, \tag{6.7}$$

where

t_n = shipping time (hour) of the navigation link a
lm_a = sea distance (nautical mile)
v_a = vessel speed (knot)
γ_a^s, γ_a^p = dummy variable for the Suez and Panama Canal transit (1 if link a passes through the canal and 0 otherwise), respectively
TS, TP = additional time for the Suez and Panama Canal transit, respectively
a' = loading link in the departure port of the navigation link a

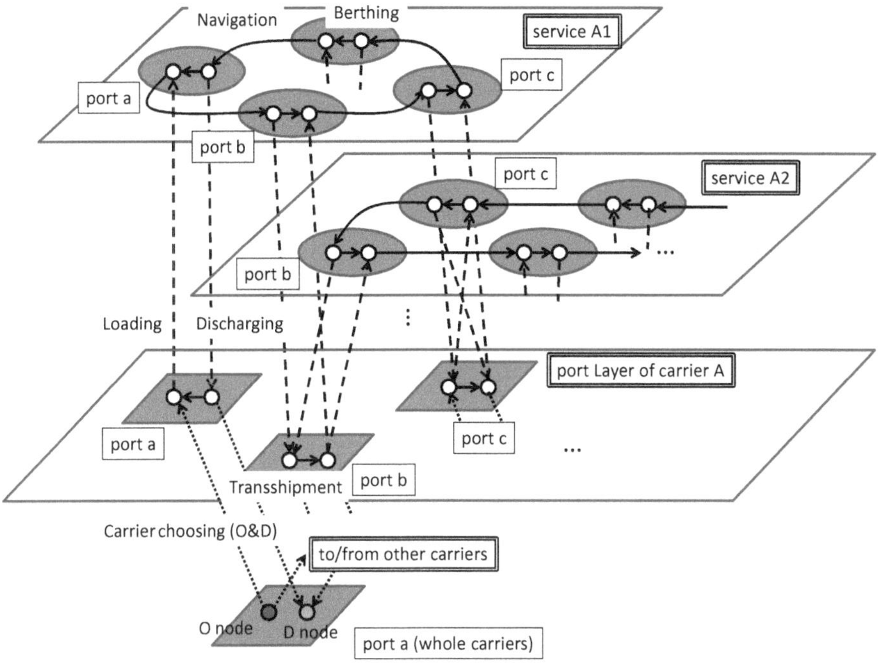

Fig. 6.1 MCS model network structure.
Source: Shibasaki, R., Iijima, T., Kawakami, T., Kadono, T., Shishido, T., 2017. Network assignment model of integrating maritime and hinterland container shipping: application to Central America. Marit. Econ. Logist. 19 (2), 234–273. DOI:10.1057/s41278-016-0055-3.

TW_a' = expected waiting time (hour) for loading in the loading link a'
cap_a = average vessel capacity (TEU/vessel) of the LS for each shipping company
$freq_a$ = service frequency (vessels/year)
$b1$, $b2$ = parameters related to congestion

The first term of the equation is the shipping time without any congestion, including the transit time of the Suez and Panama Canals. The second term represents the delay time due to congestion. The delay time is obtained by multiplying the waiting time for loading as shown in Eq. (6.8) by the congestion function, which is a function of the load factor. The average load factor is defined as the rate of the annual link flow for the annual vessel capacity of the link. The expected waiting time is defined as

$$TW_{a'} = \frac{1}{2} \cdot \frac{YH}{freq_a}, \tag{6.8}$$

where
YH = constant for conversion from one year to hours, in this case: 52 (weeks/year) × 7 (days/week) × 24 (hours/day) = 8736 (hours/year)

The term $YH/freq_a$ represents the duration for each LS expressed in hours. The expected waiting time is assumed to be half of that value.

Loading link

The loading link goes from a port layer to each LS in each port. The link cost is defined as the expected waiting time for departure related to the frequency of each LS. Note that the loading time is not considered in this model because the export lead time (CY-cut time) is generally between half a day and a few days and much longer than the loading time. This export lead time is necessary when a container comes to a terminal from the hinterland before loading onto a containership, and it is considered in the super-network model (see Chapter 7):

$$t_l\left(x_a\right) = TW_a, \tag{6.9}$$

where

t_l = loading time (hour) of the loading link

Discharging, berthing, and transshipment link

A discharging link connects each LS to a port layer in each port inversely with the loading link. A berthing link is used for a container on board a vessel in the port without neither discharged nor loaded, set by each LS. A transshipment link is passed if a container is transshipped from one LS to another. It is assumed in this model that transshipment is allowed only within the same liner shipping company. Transshipment between different companies is considered in the super-network model in Chapter 7.

The cost of these links is defined as

$$t_d\left(x_a\right) = SSN, \tag{6.10}$$

$$t_b\left(x_a\right) = TBE_a, \quad \text{and} \tag{6.11}$$

$$t_r\left(x_a\right) = TPR_a, \tag{6.12}$$

where

t_d = discharging time (hour) of the discharging link
t_b = berthing time (hour) of the berthing link
t_r = transshipment time (hour) of the transshipment link
SSN = sufficient small number (actually, set at 0.01 h)
TBE_a = berthing time (hour) of the port a
TPR_a = transshipment time (hour) in the same liner shipping company in port a

Note that the discharging time is not considered in this model either, because the import lead time is considered instead in the super-network model in Chapter 7.

Carrier choosing link

This model does not allow for MCS using multiple liner shipping companies. Therefore, the carrier choosing link is set to avoid transshipment of the container between carriers. The link cost is defined as

$$t_c(x_a) = SSN, \tag{6.13}$$

where

t_c = carrier choosing time (hour) of the carrier choosing link

Shipping cost function

The shipping cost of each link included in the MCS model is defined per TEU as follows. Note that the following descriptions and formulations (including an estimation of the ocean freight charge) are not directly related to the global MCS model described above as Eqs. (6.1)–(6.4); nevertheless, they are necessary for the calculation of the intermodal transport network model, which is introduced in the next chapter.

Navigating link

The cost of navigation is defined as the sum of the fuel cost, capital cost, operation cost, and canal toll as

$$c_n(x_a) = \left\{(FC_a + CC_a + OC_a)\cdot\frac{lm_a / v_a}{24} + \gamma_a^s \cdot CS + \gamma_a^p \cdot CP\right\}\cdot\frac{cap_a}{Vcap_a}\bigg/\frac{x_a}{freq_a}, \tag{6.14}$$

where

c_n = cost function (USD/TEU) of the navigating link
FC_a = fuel cost (USD/vessel/day) of the containership
CC_a = capital cost (USD/vessel/day) of the containership
OC_a = other operation costs (USD/vessel/day) of the containership
CS = toll (USD/vessel) for the Suez Canal transit
CP = toll (USD/TEU) for the Panama Canal transit
$Vcap_a$ = average vessel capacity (TEU/vessel) of the LS

The term $x_a/freq_a$ represents the average number of containers transported in one vessel. Each item of the navigation cost (fuel cost, capital cost, operation cost, and canal toll) is calculated as follows, derived from Shibasaki et al. (2016).

The fuel cost is defined as

$$FC_a = FP \cdot FR_a, \tag{6.15}$$

where

FP = fuel price (USD/t)
FR_a = fuel consumption rate (tonne/day) of the containership

It is common knowledge in marine engineering that the wave resistance a vessel encounters can approximate to be proportionate to the product of the square of vessel speed and its cross-section under water. Additionally, the fuel consumption rate is proportionate to the product of wave resistance and ship speed. Therefore, the fuel consumption rate is defined as

$$FR_a = c_1 \cdot DWT_a^{\frac{2}{3}} \cdot v_a^3, \tag{6.16}$$

where

c_1 = coefficient defined by the vessel type
DWT_a = dead weight tonnage of the vessel

Coefficient c_1 is estimated to be $6.49*10^{-6}$ from the regression of actual consumption data covering all vessels built in Japan (see Fig. 6.2).

The capital cost of the containership is defined as

$$CC_a = VP_a \cdot \frac{ir}{\left\{1-(1+ir)^{-PP}\right\}} \cdot \frac{1}{365 \cdot ODR}, \tag{6.17}$$

where

VP_a = vessel price (USD/vessel) of the containership
ir = interest rate
PP = project period
ODR = operation day rate

The term $ir/\{1-(1+ir)^{-PP}\}$ represents the annual payment rate by compound interest calculation. The price of the containership can be approximately related with the vessel size as

$$VP_a = c_2 \cdot DWT_a + c_3, \tag{6.18}$$

where
c_2, c_3 = coefficients defined by the vessel type

Coefficients c_2 and c_3 are estimated through regression analysis at 8.37×10^2 and 4.46×10^6, respectively, as shown in Fig. 6.3.

The containership operation cost, including manning, insurance, stores, spares, lubricating oil, R&D, and administration, is also approximately related with vessel size as

$$OC_a = c_4 \cdot DWT_a + c_5, \tag{6.19}$$

where

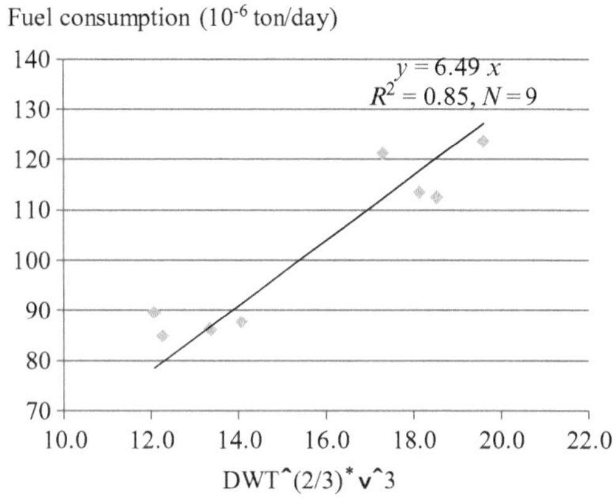

Fig. 6.2 Relationship between vessel size, speed, and fuel consumption rate.
Source: Own compilation based on The Japan Shipping Exchange, Inc., 2012. Register of ships 2012 (in Japanese).

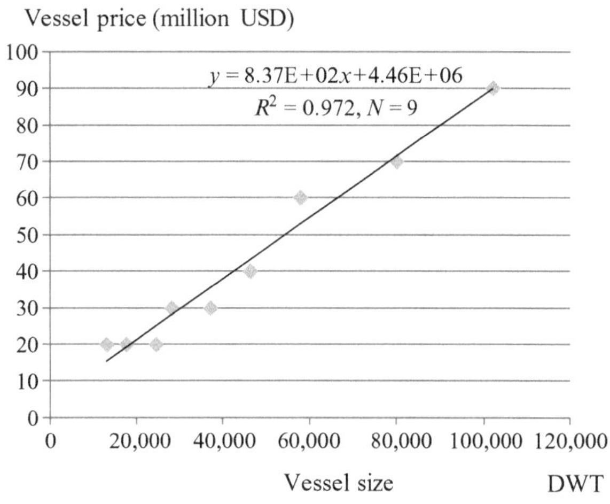

Fig. 6.3 Relationship between vessel size and price.
Source: Own compilation based on Drewry Maritime Research, 2011. Ship Operating Costs Annual Review and Forecast 2011/12. Drewry Maritime Research: London, UK.

c_4, c_5 = coefficients defined by the vessel type

Coefficients c_4 and c_5 are also estimated through regression analysis at 6.66×10^{-2} and 3.98×10^3, respectively, as shown in Fig. 6.4.

The canal toll is defined as

$$CT_a = SDRrate \cdot \left(c_6\left(scnrt_a\right) \cdot scnrt_a + c_7\left(scnrt_a\right)\right) \text{for the Suez Canal and} \quad (6.20)$$

$$CT_a = c_8 \cdot Vcap_a \text{ for the Panama Canal,} \quad (6.21)$$

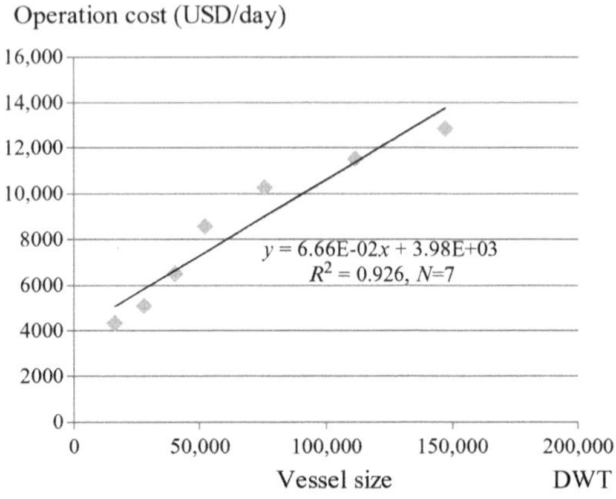

Fig. 6.4 Relationship between vessel size and operation cost.
Source: Own compilation based on Drewry Maritime Research, 2011. Ship Operating Costs Annual Review and Forecast 2011/12. Drewry Maritime Research: London, UK.

where

SDRrate = conversion rate from SDR (special drawing rights defined by the International Monetary Fund) to USD
$scnrt_a$ = Suez Canal's containership net registered tonnage
c_6, c_7 = coefficient functions established by the Suez Canal Authority
c_8 = coefficient established by the Panama Canal Authority

Coefficient functions c_6 and c_7 are set by the Suez Canal's net registered tonnage, as shown in Table 6.1. The toll per TEU for the Suez Canal decreases as the size of the vessel increases. On the other hand, the toll for the Panama Canal is proportionally set to vessel capacity on a TEU basis for containership; therefore, coefficient c_8 is constant at 72.0 (USD/TEU capacity) as of 2013, irrespective of the vessel size. The conversion rate from SDR to USD, *SDRrate*, is set to 1.5 from the 2013 average.

Table 6.1 Coefficient set by the Suez Canal's containership net registered tonnage in Eq. (6.20).

$scnrt_a$			
From	**To**	c_6	c_7
0	5000	7.65	0
5000	10,000	5.00	5000
10,000	20,000	4.00	10,000
20,000	40,000	2.80	20,000
40,000	70,000	2.60	40,000
70,000	120,000	2.05	70,000
120,000		1.95	120,000

Source: Suez Canal Authority's website (as of 2013).

The Suez Canal's net registered tonnage of containership can also be approximated to the deadweight tonnage as

$$scnrt_a = c_9 \cdot DWT_a + c_{10}, \tag{6.22}$$

where

c_9, c_{10}=coefficients defined by the vessel type

Coefficients c_9 and c_{10} are estimated by regression analysis at 1.01×10^0 and 9.99×10^3, respectively, as shown in Fig. 6.5.

Loading, discharging, berthing, transshipment, and carrier choosing link

The port charge and terminal handling charge are considered in these links, as

$$c_l(x_a) = SSN, \tag{6.23}$$

$$c_d(x_a) = SSN, \tag{6.24}$$

$$c_b(x_a) = SSN, \tag{6.25}$$

$$c_r(x_a) = CR_a, \tag{6.26}$$

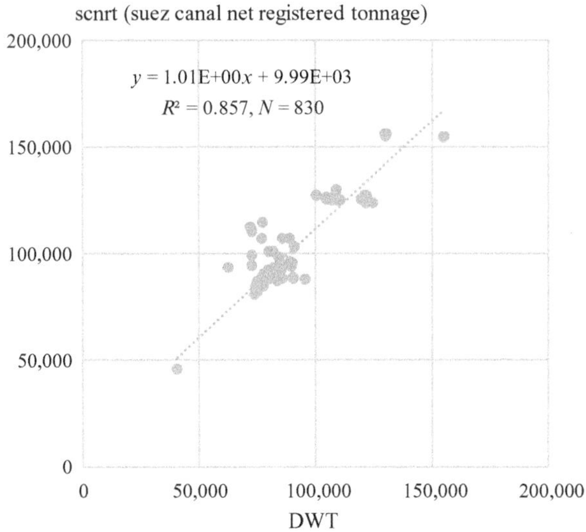

Fig. 6.5 Relationship between deadweight tonnage and Suez Canal's containership net registered tonnage.
Source: Own compilation based on the Suez Canal Authority's data.

$$c_{cx}\left(x_a\right) = CPX_a \text{ and} \tag{6.27}$$

$$c_{cm}\left(x_a\right) = CPM_a, \tag{6.28}$$

where

c_l, c_d, c_b, c_r, c_{cx}, c_{cm} = cost function (USD/TEU) of the loading, discharging, berthing, transshipment, carrier choosing export, and carrier choosing import link, respectively
SSN = sufficient small number (in this model, it is assumed that SSN = 0.01 USD)
CR_a = container handling charges (USD/TEU) if the container cargo is transshipped by the same liner shipping company at port a
CPX_a = container handling charges (USD/TEU) if the container cargo is loaded at port a
CPM_a = container handling charges (USD/TEU) if the container cargo is discharged at port a

It is an empirical fact that the handling charge for transshipment is less than double that for loading or discharging. Therefore, to avoid giving a negative link cost to the transshipment link, handling charges are imposed on the carrier choosing links, but not on the loading and discharging link. Reflecting the transshipment charge is usually set between the average and the sum of load and discharge charges, this model assumes it as

$$CR_r = 0.75 \cdot \left(CPX_r + CPM_r\right). \tag{6.29}$$

Estimation of ocean freight charge

The ocean freight charge from an origin (export port) to a destination (import port) provided by the carrier is generally different from the summation of the marginal monetary cost along the shipping route for the carrier, reflecting the balance of demand and supply in the market. Furthermore, the shipping cost may differ for each shipping company, mainly because the vessel size and shipping routes are different among companies. Hereinafter, the MCS market is assumed to be individually established for each combination of origin and destination ports (i.e. for each OD pair), although markets are related to each other. Under this assumption, individual MCS markets connecting specific export and import ports may be relatively easy to enter and leave for the liner shipping companies that already operate containerships in the region. In this case, each liner shipping company may decide whether to participate in each shipping market based on the average shipping cost including sank cost (e.g. vessel cost), not marginal shipping cost, so as to avoid negative profit.

Therefore, the ocean freight charge, FO_{rs}, for each liner shipping company on the market from export port r to import port s, which is assumed to be the same among companies present in the market as mentioned earlier in this chapter, is set so that the profit of the company with the highest average shipping cost is zero, a minimum level for the company to join the market. Namely,

$$FO_{rs} = \max_{g \in G^{rs}} AC_g^{rs}, \tag{6.30}$$

where

$AC^{rs}{}_g$ = average cost of liner shipping company g from export port r to import port s
G^{rs} = set of the liner shipping company that participates in shipping from export port r to import port s

The G^{rs} set is defined as

$$g \in G^{rs} \text{ if } TM_g^{rs} = TM_{rs} \text{ and } g \notin G^{rs} \text{ if } TM_g^{rs} > TM_{rs}, \tag{6.31}$$

where

$TM^{rs}{}_g$ = minimum shipping time from export port r to import port s of liner shipping company g

The minimum shipping time for each company is defined as

$$TM_g^{rs} = \min_k \left\{ \sum_{a \in k} t(x_a) \right\}, \forall k \in K_g^{rs}, \tag{6.32}$$

where

$K^{rs}{}_g$ = path set from export port r to import port s of liner shipping company g

The average shipping cost is defined as

$$AC_g^{rs} = \sum_{a \in k_g} c(x_a), \tag{6.33}$$

where

k_g = path to minimise the generalised shipping cost from export port r to import port s of shipping company g

k_g is defined as

$$k_g = \arg\left[\min_k \left\{ \sum_{a \in k} \left[c(x_a) + vt \cdot t(x_a) \right] \right\} \right], \forall k \in K_g^{rs}. \tag{6.34}$$

Model performance

Unknown parameter estimation

The global MCS model described in this chapter as Eqs. (6.1)–(6.4) includes two unknown parameters, *b1* and *b2*, which are related to the congestion included in Eq. (6.7). These unknown parameters are estimated to best fit the estimation results to the actual data as of 2010 (Shibasaki et al., 2013). Because this chapter's model is developed to describe container movement under a given global LS network and MCS cargo demand (OD matrix) between ports, the amount of laden containers transshipped at each port is used as a criterion to check model fitness. Namely, these two unknown parameters are estimated to best fit the observed transshipment rate, which is derived by dividing the number of transshipment containers by the total throughput for each major hub port defined in Chapter 8.

The author adopts the steepest descent method to estimate unknown parameters by inputting all necessary data, including the MCS network (see Chapter 8 for necessary items) and MCS demand (see Chapter 9 as well) into the model, to minimise the sum of squares of difference between the observed and estimated transshipment rates in each major hub port. Because the analytic calculation of the first-order differentiation of the objective value is difficult, the steepest direction is judged from the changes of the objective value if each parameter is minimally changed.

Based on approximately 50 repeated calculations, the parameters are estimated as $b1 = 2.308$ and $b2 = 1.017$. The estimated values imply that the congestion function is mostly linear to the load factor of the LS and that if the load factor is 100% (i.e. $x_a/(cap_a \cdot freq_a) = 1$), the equivalent additional time resulting from congestion is slightly longer than the duration time of the LS ($YH/freq_a$). This is a reasonable setting because congestion is normally observed in the port as a left-behind, as previously stated.

Model calculation and convergence

The following example is based on the calculation of the global MCS model applied to Central Asia described in Chapter 13. The number of links in the network is 82,721 for 187 ports. The calculation time for one iteration of the Frank–Wolfe algorithm is 3–5 min, using a Windows laptop computer with an Intel® Core™ i7 vPro-5600U™ Processor and 8.00 GB of RAM.

The convergence rates of each iterative calculation (the sum of squares of differences in the link flow calculated in the iteration from the one in the previous iteration) are shown in Fig. 6.6A. Note that in the calculation shown in this figure, the global MCS model is incorporated as a submodel of the intermodal simulation model described in Chapter 7. The model is thus calculated several times.

Fig. 6.6B shows the comparison between the calculated link flow and the link flow in the previous iteration when the convergence rate first becomes less than 10^{-3}. Considering these results and calculation time, 10^{-3} is sufficient as a criterion for the judgement of convergence.

Fitness to the observed transshipment throughput

Fig. 6.7 shows an example of the comparison between the observed and model-estimated laden containers transshipped in major hub ports where the annual transshipped container throughput is more than one million TEU (for the model described in Chapter 13). The two charts in the figure represent the transshipment rate (left, A) and volume (right, B). As shown in the figure, transshipped laden containers are well estimated by the model as a whole, in terms of both the transshipment rate and volume. The largest difference in terms of the transshipment rate is observed in Lianyunggang port in China, where the estimated transshipment rate is zero. The reason for the significant

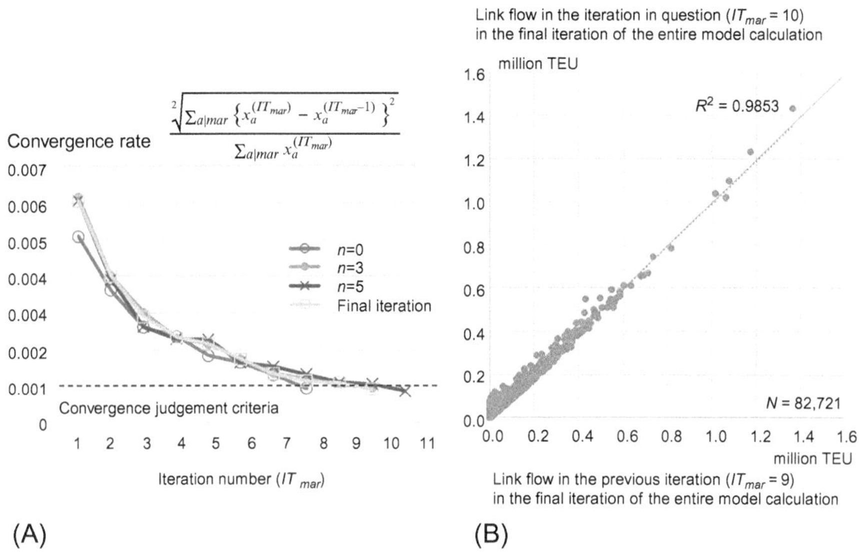

Fig. 6.6 Convergence of the MCS model (example of the Central Asian model). (A) Convergence rate of each iteration. (B) Link flow around convergence judgement criteria. Source: Own compilation.

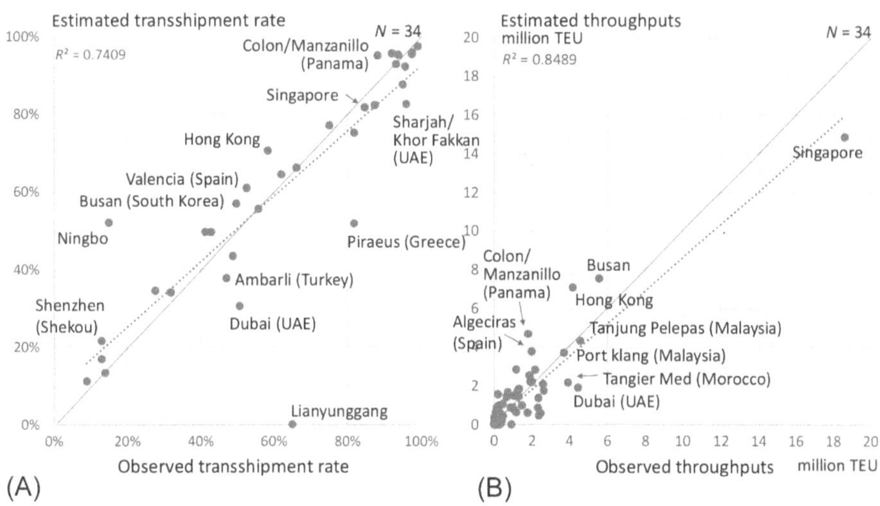

Fig. 6.7 Example of comparisons between observed and model-estimated laden containers transshipped in major hub ports. (A) Transshipment rate. (B) Transshipment throughput. Source: Shibasaki, R. Tanabe, S., Kato, H., Lee, T.-W., 2019. Could Gwadar port in Pakistan be a new gateway? A network simulation approach in the context of the belt and road initiative. Sustainability 11 (20), 5757.

underestimation is that most domestic feeder services to/from Lianyunggang port are supplied by other small carriers that are not considered in the model.[a]

On the other hand, the overestimation of the transshipment rate in Ningbo port is due to the decrease of the denominator (i.e. the underestimation of export and import container throughput due to shifting to Shanghai port, which will be discussed in Chapter 13). Furthermore, the transshipment rate is underestimated in some ports in the Middle East, such as Dubai and Sharjah/Khor Fakkan in the UAE, and in the east Mediterranean, such as Ambarili in Turkey and Piraeus in Greece. Container movement in these areas should be further examined by involving additional data collection and close investigations into the regional LS network and container shipping demand.

Conclusion

This chapter described the global MCS network assignment model. As mentioned in Chapter 5, this model constitutes a part of the entire intermodal network assignment model that is introduced in the next chapter. However, if necessary, the model from this chapter can be applied independently if the analysis focuses on the MCS network (see Chapter 14 as an example).

As shown in Fig. 6.7, as well as in several verifications of model performance introduced in some chapters of Part 3, the developed MCS model can successfully reproduce actual global laden container movement as a whole, including transshipment rate and volume at major hub ports. However, to simplify the problem, the model involved many assumptions; for example, it considered only the capacity constraint of the containership, but not of the container terminal.

Another controversial assumption was that containers were assigned, based on the shipping time only, not considering the ocean freight charge, mainly because there was less agreement with the observed data if the link-based shipping cost was considered in the assignment model according to the author's exploratory calculations. Therefore, the ocean freight charge was estimated separately from the MCS model, as described in the latter part of this chapter. Moreover, it is necessary to further clarify the characteristics of the model that selects the MCS route, including the liner shipping company and the transshipment port. In the real MCS market, the liner shipping company decides the transshipment port, whereas shippers (cargo owners) choose the liner shipping company; therefore, these selections should be separately modelled from different viewpoints (i.e. shippers and carriers).

From the carrier's viewpoint, endogenising the behaviour to structure the LS network of a liner shipping company is also important, particularly in future simulations that aim to consider the change of the LS network. Although many models are proposed from this viewpoint, as described in Chapter 4, the author considers that a

[a] Actually, Lianyunggang port has not been included in the list of major hub ports in the *Container Forecaster Annual Review* (published by Drewry Maritime Research) since 2015. The author considers there was a problem related to the observed amount in Lianyunggang port.

lot more work remains to be done to incorporate into the author's model, which should include many decision variables for carriers to be simultaneously optimised.

From the shippers' viewpoint, it is also important to incorporate the difference in contents of cargo. In the model developed in this chapter, the author implicitly assumed that the contents of all containers are the same. In reality, some containers (e.g. auto parts and perishable goods) prioritise a faster shipping time, whereas others prefer cheaper freight. An assignment model that takes the commodity type into consideration should be further applied to the problem.

References

Sheffi, Y., 1985. Urban Transportation Networks: Equilibrium Analysis with Mathematical Programming Methods. Prentice-Hall, Inc., New Jersey.

Shibasaki, R., Azuma, T., Watanabe, T., Toriumi, S., 2013. A container cargo assignment model on a real international maritime shipping network and application to the Suez Canal transit analysis. In: Proceedings of the International Association of Maritime Economists Annual Conference (IAME 2013), 03–05 July 2013, Marseille, France.

Shibasaki, R., Azuma, T., Yoshida, T., 2016. Route choice of containership on a global scale and model development: focusing on the Suez Canal. Int. J. Transp. Econ. 43 (3), 263–288.

Intermodal transport super-network model

7

Ryuichi Shibasaki
The University of Tokyo

This chapter describes models other than the maritime container shipping (MCS) network assignment (sub-)model, which was introduced in the previous chapter, including intermodal transport (IMT) super-network assignment model and land transport (LT) network assignment submodel. The network structures of these models are slightly different in each model introduced in each chapter of Part 3, as will be described in this chapter. In the most extreme case, the models in this chapter are not considered for the region where LT can be negligible, such as the Pacific Islands (see Chapter 14).

Model framework

In the upper layer, each shipper is assumed to choose the transport route, including the ports to be used for export and import if necessary, on the IMT network described in Fig. 7.1. The freight charges and time for MCS and LT are given. Note that the LT link to a direct connection between origin and destination is not considered in some models introduced in Part 3 (i.e. Chapters 10–13 and 15); in other words, it is considered only in the model applied in Chapter 16. Furthermore, inter-carrier transshipment links are added later to better describe the actual MCS market. Thus, they are not considered in the early models (Chapters 10–12).

For the transport of cargo from origin, o, to destination, d, a path, h, is chosen for a cargo, l, to maximise utility.

$$U_{hl}^{od} > U_{h'l}^{od}, \forall h \in H^{od}, \forall h' \in H^{od}, h \neq h', (o,d) \in O \times D, \tag{7.1}$$

$$s.t.\ U_{hl}^{od} = V_h^{od} + \varepsilon_{hl}^{od}, \tag{7.2}$$

where

U_{hl}^{od} = utility of cargo l on path h from origin o to destination d
H^{od} = path choice set of transport from origin o to destination d
O = set of origins
D = set of destinations
V_h^{od} = deterministic term of path h from origin o to destination d
ε_{hl}^{od} = error term of cargo l of path h from origin o to destination d

Global Logistics Network Modelling and Policy. https://doi.org/10.1016/B978-0-12-814060-4.00007-1

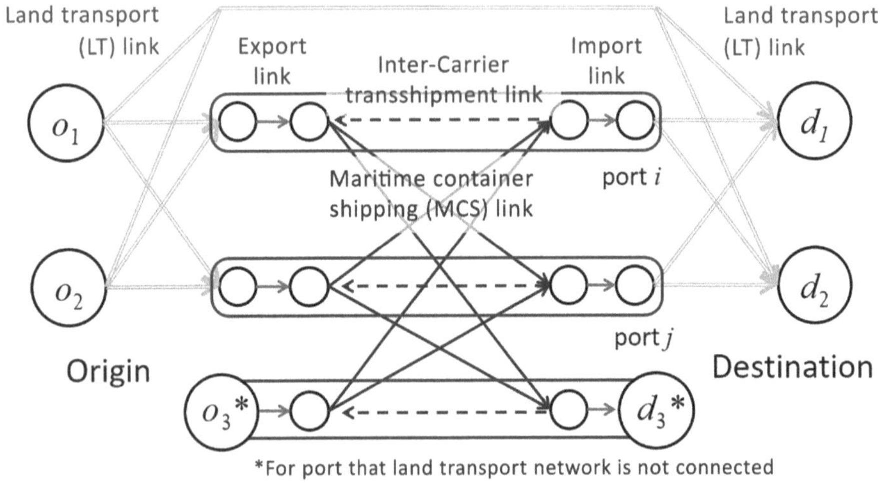

Fig. 7.1 IMT network of the model.
Source: Own compilation.

If the error term follows a Gumbel distribution, the choice of the shipper is formulated as

$$F_h^{od} = Q^{od} \cdot \frac{\exp\left(\theta \cdot V_h^{od}\right)}{\exp\left(\theta \cdot V_h^{od}\right) + \sum_{h' \in H^{od}} \exp\left(\theta \cdot V_{h'}^{od}\right)}, \tag{7.3}$$

where

F_h^{od} = cargo volume on path h from origin o to destination d
Q^{od} = cargo shipping demand (TEU) from origin o to destination d
θ = distribution parameter

as shown in Chapter 4. The deterministic term for each path is expressed as the summation of freight charge and cost related to transport time.

$$V_h^{od} = -\left(FL_{oi} + FTO_{ij} + FL_{jd}\right) - vt \cdot \left(TL_{oi} + TPX_i + TTM_{ij} + TPM_j + TL_{jd}\right), \quad \forall i, j \in h, \text{if } h \text{ includes MCS link, or} \tag{7.4a}$$

$$V_h^{od} = -FL_{od} - vt \cdot TL_{od}, \text{if } h \text{ does not include MCS link,} \tag{7.4b}$$

where

vt = value of time (USD/TEU/h) for shipper
FL_{oi}, FL_{jd}, FL_{od} = freight charges (USD/TEU) of LT from origin o to port i, port j to destination d and origin o to destination d, respectively
FTO_{ij} = total ocean freight charge (USD/TEU) from port i to port j, including port charges
TL_{oi}, TL_{jd}, TL_{od} = LT time (hour) from origin o to port i, port j to destination d and origin o to destination d, respectively

TPX_i = lead time (hour) if exporting from port i
TTM_{ij} = total MCS time (hour) from port i to port j
TPM_j = lead time (hour) if importing in port j

Note that all port charges are included in the total ocean freight charge so that the total ocean freight charge can include export, import, and transshipment ports.

In addition, inter-carrier transshipment is incorporated in some models in Part 3, which is illustrated with a dotted line in Fig. 7.1 (note that intra-carrier transshipment is considered in the MCS submodel in Chapter 6). This is introduced for a more realistic consideration of the actual MCS market, including small local liner shipping companies in each region. In sum, the total ocean freight charge and total MCS time are given as

$$FTO_{ij} = \sum_{(r,s)\in h} FO_{rs} + \sum_{r\in h, r\neq i, r\neq j} \left(\tau \cdot CR_r - CPX_r - CPM_r\right), \forall i, j \in h, \text{and} \tag{7.5}$$

$$TTM_{ij} = \sum_{(r,s)\in h} TM_{rs} + \sum_{r\in h, r\neq i, r\neq j} \tau \cdot TPR_r, \forall i, j \in h, \tag{7.6}$$

where

FO_{rs} = ocean freight charge (USD/TEU) in a liner shipping company from port r to port s in path h, including port charges
CR_r = container handling charges (USD/TEU) if the container cargo is transshipped in the same company in port r (defined in Eq. (6.29) in Chapter 6)
τ = multiplier of inter-carrier transshipment ($\tau > 1$)
CPX_r, CPM_r = container handling charges (USD/TEU) if the container cargo is loaded and discharged, respectively. Note that they should be subtracted in each inter-carrier transshipment port because they are included in each ocean freight charge as mentioned above, although they are not actually imposed.
TM_{rs} = total MCS time (hour) from port r to port s in path h

The ocean freight charge and MCS time are obtained from the computation results of the MCS submodel in the lower layer, whereas the land freight charges and LT time are computed from the LT submodel. The container handling charges and lead time for export and import are set in the next chapter (see Table 8.A2).

A cargo flow of each link in the IMT model represents inputs (i.e. cargo shipping demand) of the submodels in the lower layer, which is given as

$$q_m^{rs} = x_{rs}, \tag{7.7}$$

$$q_l^{oi} = x_{oi}, q_l^{jd} = x_{jd}, \text{ and } q_l^{od} = x_{od}, \tag{7.8}$$

where

q_m^{rs} = cargo demand of the MCS (TEU/year) from export port r to import port s
x_{rs} = cargo flow (TEU/year) of the MCS link
q_l^{oi}, q_l^{jd}, q_l^{od} = cargo demand of the LT (TEU/year) from origin o to export port i, import port j to destination d and origin o to destination d, respectively
x_{oi}, x_{jd}, x_{od} = cargo flows (TEU/year) of the LT link

Note that if the direct LT link from origin o to destination d is not considered, q_l^{od} and x_{od} should be zero.

Regional land transport submodel

The LT submodel in the lower layer is defined as the problem to allocate container cargo to the LT network, including road, rail, and inland waterways with the capacity constraint of each mode. It is assumed that the cargo shipping demands, q_l^{oi}, q_l^{jd}, and q_l^{od}, between inland origin o (or import port j) and destination d (or export port i) are given. The model is formulated as a UE problem, as with the MCS (sub-)model. Note that the market is expected to be sufficiently competitive as many enterprises, such as trailer operating companies, should participate in the LT market. The shipper can then choose a transport mode and route to minimise the total generalised cost, including transport time and freight charges, given as

$$\min_{x} z'(x) = \sum_{a \in A} \int_0^{x_a} u(x_a)\, dx, \tag{7.9}$$

$$s.t.\ x_a = \sum_{(o,i) \in O \times I} \sum_{k \in K^{oi}} \delta_{a,k}^{oi} \cdot f_k^{oi} + \sum_{(j,d) \in J \times D} \sum_{k \in K^{jd}} \delta_{a,k}^{jd} \cdot f_k^{jd} + \sum_{(o,d) \in O \times D} \sum_{k \in K^{od}} \delta_{a,k}^{od} \cdot f_k^{od}, \forall a, \tag{7.10}$$

$$\sum_{k \in K^{oi}} f_k^{oi} - q_l^{oi} = 0, \quad \forall o, i, \tag{7.11a}$$

$$\sum_{k \in K^{jd}} f_k^{jd} - q_l^{jd} = 0, \quad \forall d, j, \tag{7.11b}$$

$$\sum_{k \in K^{od}} f_k^{od} - q_l^{od} = 0 \quad \forall o, d, \tag{7.11c}$$

$$f_k^{oi} \geq 0, f_k^{jd} \geq 0, \quad \text{and} \quad f_k^{od} \geq 0 \forall o, d, i, j, k, \tag{7.12}$$

where

a = Link
A = set of links
x_a = flow of the link a
$u(.)$ = cost function of each link
$z'(.)$ = objective function
I, J = set of the export and import ports, respectively
k = path
K^{oi}, K^{jd}, K^{od} = sets of paths for OD pair oi, jd, and od, respectively
f_k^{oi}, f_k^{jd}, f_k^{od} = flows on path k for OD pair oi, jd, and od, respectively
$\delta_{a,k}^{oi}$, $\delta_{a,k}^{jd}$, $\delta_{a,k}^{od}$ = Kronecker deltas

Note that the cost function of each LT link is defined as a generalised cost rather than transport time. The Kronecker deltas are written as

$$\delta_{a,k}^{oi} = \begin{matrix} 1 \text{ if link } a \text{ is on path } k \text{ between OD pair } oi \\ 0 \quad \text{otherwise} \end{matrix}, \tag{7.13a}$$

$$\delta_{a,k}^{jd} = \begin{matrix} 1 \text{ if link } a \text{ is on path } k \text{ between OD pair } jd \\ 0 \quad \text{otherwise} \end{matrix}, \text{ and} \tag{7.13b}$$

$$\delta_{a,k}^{od} = \begin{matrix} 1 \text{ if link } a \text{ is on path } k \text{ between OD pair } od \\ 0 \quad \text{otherwise} \end{matrix}. \tag{7.13c}$$

Regarding the networks, each road, rail, and ferry links have different flow-dependent cost functions, respectively, as shown in the following section. According to the UE assignment definition, the generalised cost of LT is defined as

$$GL_{oi} = \min_k \left\{ \sum_{a \in k} u(x_a) \right\}, k \in K^{oi}, \tag{7.14a}$$

$$GL_{jd} = \min_k \left\{ \sum_{a \in k} u(x_a) \right\}, k \in K^{jd}, \text{ and} \tag{7.14b}$$

$$GL_{od} = \min_k \left\{ \sum_{a \in k} u(x_a) \right\}, k \in k^{od} \tag{7.14c}$$

These generalised costs are related to the freight charge and time of LT included in Eqs. (7.4a), (7.4b) as

$$GL_{oi} = FL_{oi} + vt \cdot TL_{oi}, \tag{7.15a}$$

$$GL_{jd} = FL_{jd} + vt \cdot TL_{jd}, \text{ and} \tag{7.15b}$$

$$GL_{od} = FL_{od} + vt \cdot TL_{od}. \tag{7.15c}$$

Similar to the MCS network submodel in the previous chapter, the UE problem shown in Eq. (7.9) is solved using traditional Frank–Wolfe algorithm. Note that if the direct LT link from origin o to destination d is not considered, Eq. (7.10) is replaced by

$$x_a = \sum_{(oi) \in O \times I} \sum_{k \in K^{oi}} \delta_{a,k}^{oi} \cdot f_k^{oi} + \sum_{(jd) \in J \times D} \sum_{k \in K^{jd}} \delta_{a,k}^{jd} \cdot f_k^{jd}, \tag{7.10'}$$

and Eqs. (7.4b), (7.11c), (7.13c), (7.14c), and (7.15c) are excluded.

Furthermore, in earlier models, any capacities in the LT network are not considered because the networks in these models are slightly simplified (for details, see Chapters 10 and 11). The shortest path search is applied to these models instead, based on the generalised cost.

Generalised cost function

Fig. 7.2 shows the network structure of the LT submodel. Road and rail network are connected with a rail connection link, whereas a ferry link is directly connected with a road or rail link. Note that the cargo origin and destination are connected with only road networks by an O link and D link rather than rail because the last one mile of container transport should be served by a trailer. Furthermore, for considering the border barrier effect, additional transport and time costs are added to each cost function if a road, rail, or ferry link crosses the national border. Note that only road transport is considered in the earlier simplified models in Chapters 10 and 11.[a]

O/D link

As the generalised cost, u_O (or u_D), of an O (or D) link that connects inland origin or import port (or inland destination or export port) with the road network, the fixed cost of trailers is only considered as

$$u_O\left(x_a\right) = u_D\left(x_a\right) = CFRo/2, \tag{7.16}$$

where

$CFRo$ = fixed charge (USD/TEU) of container transport by trailer

The fixed term should be considered only once whenever a trailer is used. Therefore, the fixed cost is equally divided into O and D links.

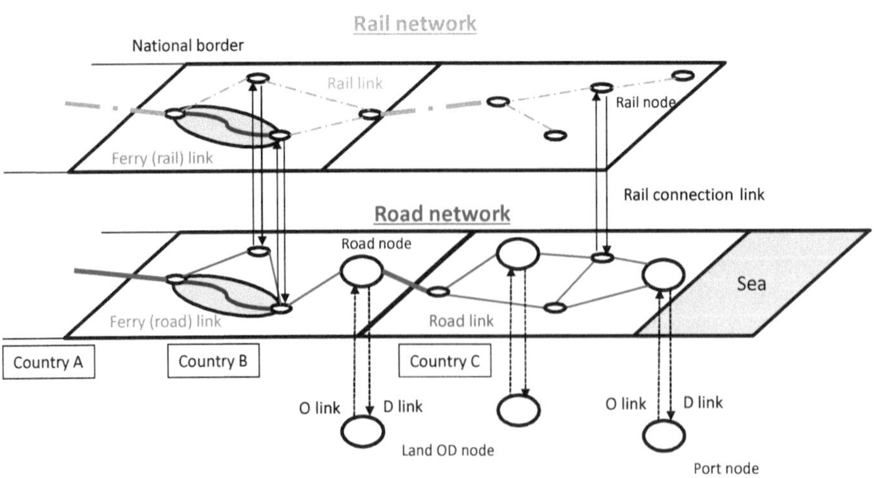

Fig. 7.2 LT submodel network structure.
Source: Own compilation.

[a] The river shipping in Chapter 11 is considered in the MCS submodel (see Chapter 11 for details).

Road link

The generalised cost of trailer transport on road, u_{ro}, consists of the trailer freight charge and time cost, including congestion, as

$$u_{ro}\left(x_a\right)=CVRo\cdot\left(2\cdot ll_a\right)+vt\cdot\frac{ll_a}{vRo_a}\cdot\left\{1+b3\cdot\left(\frac{x_a}{capRo_a}\right)^{b4}\right\}, \tag{7.17}$$

where

$CVRo$ = coefficient of operational charge (USD/km/TEU) of container trailer in proportion to the distance
ll_a = land distance (km) of link a
vRo_a = trailer speed (km/h)
$capRo_a$ = annual road capacity (TEU/year) for container trailer
$b3$, $b4$ = parameters related to road congestion

The first term of the equation is the monetary transport cost whereas the second term represents time cost, including delay due to the congestion. Note that the transport distance in the operational cost calculation is doubled (i.e. $2\cdot ll_a$), because a trailer usually returns to the departure point without any cargo and a trailer operating company charges the repositioning cost to shippers. The delay time is expressed as a power function of the average congestion rate, which is defined as the rate of the annual link flow for the annual road capacity of the link.

Rail connecting link

The generalised cost of rail and road connections, u_{rc}, also consists of operation and time cost, given as

$$u_{rc}\left(x_a\right)=\frac{CFRo}{2}+CVRo\cdot\left(2\cdot ll_a\right)+\frac{CFRa}{2}+vt\cdot\left\{\frac{ll_a}{vRo_a}+THRa_a+\frac{TWRa_a}{2}\right\}, \tag{7.18}$$

where

$CFRa$ = fixed charge (USD/TEU) of container transport by rail
$THRa_a$ = handling time (hour) for loading or discharging cargo onto/from the rail
$TWRa_a$ = expected waiting time (hour) for the rail loading

Note that not only the rail cost, such as fixed rail transport cost, the handling time at the rail station, and expected waiting time, but the access transport cost and time by trailers are also included in the rail connecting link. Additionally, the fixed cost of trailers is considered, because an additional trailer should be arranged after a container is discharged from a train. The fixed costs of trailer and rail transport and expected waiting time are equally divided in the links when a container is loaded onto and discharged from a train.

The expected waiting time for the rail loading is given as

$$TWRa_a=\frac{1}{2}\cdot\frac{YH}{freqRa_a}, \tag{7.19}$$

where

$freqRa_a$ = annual number of trains (train/year) for container transport

The term $YH/freqRa_a$ represents the duration in hours for each rail service. The expected waiting time is assumed to be half of that value.

Rail link

The generalised cost of rail transport, u_{ra}, also consists of the operation and time costs, including congestion, as

$$u_{ra}(x_a) = CVRa \cdot ll_a + vt \cdot \left\{ \frac{ll_a}{vRa_a} + TWRa_{a'} \cdot b5 \cdot \left(\frac{x_a}{capRa_a \cdot freqRa_a} \right)^{b6} \right\}, \tag{7.20}$$

where

$CVRa$ = coefficient of operational charge (USD/km/TEU) of container transport by rail in proportion to the distance
vRa_a = railway speed (km/h)
$TWRa_{a'}$ = expected waiting time (hour) for the rail loading onto rail connection link a' (shown in Eq. (7.19))
$capRa_a$ = rail capacity (TEU/train) for container transport per train
$b5$, $b6$ = parameters related to rail congestion

In contrast to transport by trailers, the transport distance in the operational cost calculation is not doubled. Additionally, because rail capacity is defined per train, it is multiplied by the annual frequency of rail service to calculate the annual rail capacity. The congestion function is multiplied by the expected waiting time, which is similar with that in the navigation link (see Eq. (6.7)) in the MCS (sub-)model, not the running time like that in the road link expressed in Eq. (7.17).

Ferry (inland waterway transport) link

The generalised cost of ferry (or inland waterway transport), u_{fe}, also consists of the operation cost and time cost, including congestion, as

$$u_{fe}(x_a) = (CFFe + CVFe \cdot ll_a) + vt \cdot \left[\frac{ll_a}{vFe_a} + 2 \cdot THFe_a + TWFe_a \cdot \left\{ 1 + b7 \cdot \left(\frac{x_a}{capFe_a \cdot freqFe_a} \right)^{b8} \right\} \right], \tag{7.21}$$

where

$CFFe$ = fixed charge (USD/TEU) of container shipping by ferry
$CVFe$ = coefficient of operational charge (USD/km/TEU) of container shipping by ferry in proportion to the distance
vFe_a = speed (km/h) of vessel
$THFe_a$ = handling time (hour) for loading or discharging cargo onto/from vessel
$TWFe_a$ = expected waiting time (hour) for the ferry service
$capFe_a$ = capacity (TEU/vessel) for container shipping per vessel

$freqFe_a$ = annual frequency (vessel/year) of the service
$b7, b8$ = parameters related to vessel congestion

Note that the generalised cost of a ferry includes both the fixed charge and operational charge in proportion to the distance, as well as the handling time and expected waiting time at the port, because a ferry service corresponds to one link, unlike railways where a service usually consists of multiple links. The expected waiting time for the vessel loading is given as

$$TWFe_a = \frac{1}{2} \cdot \frac{YH}{freqFe_a}. \quad (7.22)$$

Additional cost at the national border

If a road, rail, or ferry link crosses the national border, additional transport and time costs are added to each cost function as

$$u(x_a) = u(x_a) + \lambda_a \cdot (CBO_a + vt \cdot TBO_a), \quad (7.23)$$

where

λ_a = dummy variable, which is set at more than 0 if link a crosses the national border
CBO_a = additional monetary cost (USD/TEU) in border-crossing
TBO_a = additional time (hour/TEU) in border-crossing

Model calculation and convergence

Procedure of model calculation

It is difficult to simultaneously solve the stochastic network assignment model of the IMT super-network in the upper layer (formulated in Eq. (7.3)) with two UE assignment submodels (formulated in Eq. (6.1) in Chapter 6 and Eq. (7.9)) of the real MCS and LT network in the lower layer. Additionally, a unique solution is not guaranteed whereas each submodel in the lower layer has a unique solution if the cargo shipping demand of each submodel, q^{rs}, q^{oi}, q^{jd}, or q^{od}, is fixed.

Therefore, the model calculation is separated for the upper and lower layers. The stochastic network assignment model for the upper layer is calculated if the freight charge and time, which are the outputs of the submodels in the lower layer, are fixed, whereas each UE submodel is calculated if the cargo shipping demand, which is the outputs (i.e. link flows) of the super-network model, are fixed. These calculations are repeated until the outputs reach convergence. A more specific procedure is given below. Note that n is the iteration number of the calculation.

Step 0: Initial iteration calculation ($n=0$)

The model calculation starts with the calculation of two submodels in the lower layer. An initial calculation of the MCS submodel is performed by inputting the initial demand of MCS, $q^{rs(0)}$, which is obtained as the port-basis OD matrix in the container shipping demand estimation described in Chapter 9. On the other hand, an initial calculation of the LT submodel is performed on a zero-flow basis by the shortest path search, because the initial demand is not available. Subsequently, an initial calculation of the IMT super-network model in the upper level is performed based on an algorithm proposed by Dial (1971), by inputting the freight charge and time in the MCS and LT, $FO_{rs}^{(0)}$, $TM_{rs}^{(0)}$, $GL_{oi}^{(0)}$, $GL_{jd}^{(0)}$, and $GL_{od}^{(0)}$, which are outputs of both submodels, as well as the regional-basis OD matrix, Q^{od}. Let $n = n + 1$ and proceed to Step 1.

Step 1: Recalculation of the LT submodel based on initial land cargo shipping demand (n = 1)

Because the outputs of the MCS submodel in Step 0 are based on the initial demand of MCS, $q^{rs(0)}$, whereas those of the LT submodel are based on a zero-flow basis, the LT submodel is once performed by inputting the land cargo shipping demand, $q^{oi(1)}$, $q^{jd(1)}$, and $q^{od(1)}$, which are the outputs of the initial calculation of the IMT super-network model in Step 0. Subsequently, by inputting the calculated freight charge and time (generalised cost) in the LT, $GL_{oi}^{(1)}$, $GL_{jd}^{(1)}$, and $GL_{od}^{(1)}$, and initial MCS time, $TM_{rs}^{(0)}$, the calculation of the IMT super-network model is performed again. Note that ocean freight charge, $FO_{rs}^{(1)}$, is estimated by a procedure shown later. Let $n = n + 1$ and proceed to Step 2.

Step 2: Repetitive calculation in nth iteration (if n is an even number)

Because the calculation of the MCS submodel takes longer time, use the MCS time in the previous iteration, $TM_{rs}^{(n-1)}$, rather than the calculation from the MCS submodel in the nth iterative calculation if n is an even number. The LT submodel is performed by inputting the land cargo shipping demand, $q^{oi(n)}$, $q^{jd(n)}$, and $q^{od(n)}$, which are outputs of the IMT super-network model in the previous iteration. Subsequently, by inputting the calculated generalised cost in the LT, $GL_{oi}^{(n)}$, $GL_{jd}^{(n)}$, and $GL_{od}^{(n)}$, and MCS time in the previous iteration, $TM_{rs}^{(n-1)}$, the calculation of the IMT super-network model is performed. The ocean freight charge, $FO_{rs}^{(n)}$, is estimated by a procedure shown later. If the outputs do not converge, let $n = n + 1$, and proceed to Step 3.

Step 3: Repetitive calculation in nth iteration (if n is an odd number)

The MCS submodel, as well as the LT submodel, are performed in the nth iterative calculation if n is an odd number, by inputting the cargo shipping demand, $q^{rs(n)}$, $q^{oi(n)}$, $q^{jd(n)}$, and $q^{od(n)}$, which are outputs of the IMT super-network model in the previous iteration. Subsequently, the calculation of the IMT super-network model is performed by inputting the calculated MCS time, $TM_{rs}^{(n)}$, and generalised cost in the LT, $GL_{oi}^{(n)}$, $GL_{jd}^{(n)}$, and $GL_{od}^{(n)}$. Meanwhile, ocean freight charge, $FO_{rs}^{(n)}$, is estimated by a procedure shown later. Let $n = n + 1$ and proceed to Step 2.

Step 4: Convergence check

After the calculation in Step 2, if the outputs converge in comparison to those in the previous iterations or the number of iterative calculations reaches the upper limit, the calculation is finished. The import and export amount of container cargo handled in each port, $\sum_{o \in O} q^{oi(n)}$ and $\sum_{d \in D} q^{jd(n)}$, are used as the 'outputs' for convergence check for saving calculation time. At the end of the model calculation, both submodels are calculated again to obtain the cargo flow in the real network under the final solution.

Ocean freight charge calculation after Step 1

The ocean freight charge defined in Eq. (6.30) in Chapter 6 would fluctuate even by a trivial change of link flow because it includes inverse functions of container flow (see Eq. (6.14)). Thus, the prompt calculation of convergence is hampered. Therefore, the ocean freight charge in the calculation after Step 1 is calculated by the elasticity of shipping demand of export and import containers to freight charge, γ_e and γ_i, and previous freight charge, $FO_{rs}^{(n-1)}$, as follows:

$$FO_{rs}^{(n)} = \left\{ \frac{q^{rs(n)}}{q^{rs(n-1)}} \right\}^{\gamma_e} \cdot FO_{rs}^{(n-1)} \tag{7.24a}$$

if exporting cargo from the region which considers the LT network,

$$FO_{rs}^{(n)} = \left\{ \frac{q^{rs(n)}}{q^{rs(n-1)}} \right\}^{\gamma_i} \cdot FO_{rs}^{(n-1)} \tag{7.24b}$$

if importing cargo to the region which considers the LT network and

$$FO_{rs}^{(n)} = FO_{rs}^{(n-1)} \quad \text{for other cargo.} \tag{7.24c}$$

The elasticity of shipping demand of export and import containers to ocean freight charge, γ_e and γ_i, is set to be 0.00207 and 0.0394, respectively. They are estimated from an average change of ocean freight charge, where each change is calculated by Eq. (6.30), if the demand per unit is changed by each partner port for the cargo exported from and imported into Acajutla port, El Salvador, using the model incorporated in Chapter 10 (see Shibasaki et al., 2017).

Unknown parameter estimation

The IMT super-network model described as Eq. (7.3) includes two unknown parameters, the distribution parameter for stochastic assignment, θ, included in Eq. (7.3), and time value for the shipper, vt, included in Eqs. (7.4a) and (7.4b). Furthermore, if the capacities of the LT networks are considered, the LT network submodel described as Eqs. (7.9)–(7.12) includes six unknown parameters; *b3* and *b4* for the road transport congestion function in Eq. (7.17), *b5* and *b6* for rail transport congestion function in Eq. (7.20), and *b7* and *b8* for the ferry shipping congestion function in Eq. (7.21).

In principle, these unknown parameters should be estimated to best fit the estimation results to the observed data. However, it is difficult to simultaneously estimate using steepest descent method (as applied in unknown parameter estimation in Chapter 6) and other estimation methods because the model in this chapter contains too many (i.e. eight) parameters. Therefore, the authors calibrated θ at 0.05, *vt* at 0.5 (US$/TEU/h), *b3* and *b4* at 1.0 and 3.0, respectively, *b5* and *b6* at 2.0 each and *b7* and *b8* at 2.0 each, by trial-and-error calculations using the model in South Asia with 2013 data (see Chapter 12 and Shibasaki and Kawasaki, 2020). Note that different values were set for θ and *vt* in the earlier models (which are incorporated in Chapters 10 and 11) based on 2010 data, and that *b3* to *b8* were not considered in these models because the capacities of the LT networks were not considered.

Model calculation and observed convergence

The number of links included in each network of the models described in each chapter of Part 3 is very large. For example, the model focused in Central Asia (Chapter 13) containers 82,721, 34,180, and 42,554 links for the MCS, LT, and IMT network, respectively, and the average time for iterative calculation is 2 to 2.5 h, using a Windows computer with an Intel® Core™ i7-4790U™ Processor and 16.0 GB of RAM.

Fig. 7.3A indicates an example of the convergence rates of each iterative calculation (the sum of squares of the differences between the link flow calculated in the iteration and that in the previous iteration) in the LT submodel (in case of the model used in Chapter 13 for Central Asia) and Fig. 7.3B compares the calculated link flow with that in the previous iteration when the convergence rate first becomes less than 10^{-3}. These figures reveal that the LT submodel promptly converges after the second iterative calculations. The difference in convergence from the MCS submodel, which is shown in Fig. 6.6 in Chapter 6, is mainly because of the sufficient capacities of the links in the LT network.

Fig. 7.4A indicates the convergence rates of each iterative calculation (the sum of squares of the differences between the laden container throughput for export and import in each port calculated in the iteration, $\sum_{o \in O} q^{oi(n)}$ and $\sum_{d \in D} q^{jd(n)}$, and those in the previous iteration, $\sum_{o \in O} q^{oi(n-1)}$ and $\sum_{d \in D} q^{jd(n-1)}$) in the IMT super-network model and Fig. 7.4B compares the calculated throughput in each port with that in the previous iteration when the convergence rate first becomes less than $5.0 * 10^{-3}$. Considering these results and the calculation time, the above criteria are appropriate in judging that the model calculation is sufficiently converged.

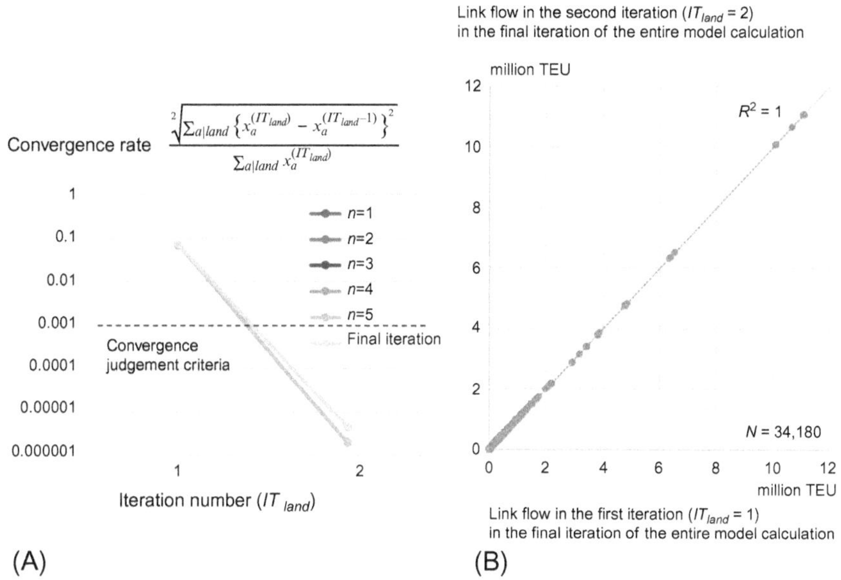

Fig. 7.3 Convergence of the LT submodel (example of the Central Asian model). (A) Convergence rate of each iteration. (B) Link flow around convergence judgement criteria. Source: Own compilation.

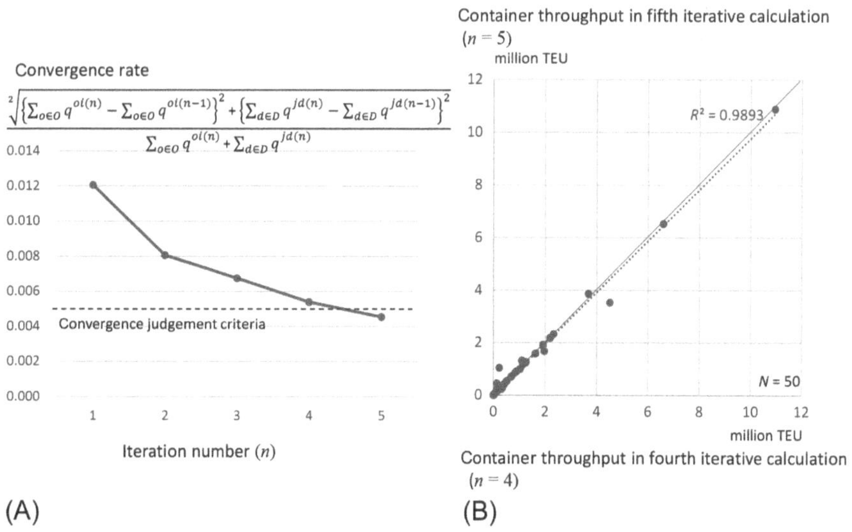

Fig. 7.4 Convergence of the IMT super-network model (example of the Central Asian model). (A) Convergence rate of each iteration. (B) Change in laden container throughput for export/import in each port where the LT network is considered around convergence judgement criteria. Source: Own compilation.

Conclusion

This chapter described the ITM super-network assignment model as well as LT network assignment submodel, including road, rail, and inland waterway transport. Together with the MCS network assignment submodel, which is introduced in Chapter 6, the two-layered container cargo assignment model was all described. Because the theoretical convergence of the calculation in the entire system is not guaranteed, it was ex-post confirmed that the calculation converged, at least in terms of the number of laden containers exported from and imported into each port, as shown in Fig. 7.4.

In the following chapters, the data to be input into this model, including the MCS and LT networks and the cargo shipping demand among regions, are introduced. Subsequently, the model is applied into several regions of the world to confirm its agreement with the observed data, and then used for policy simulations in each chapter of Part 3. In these chapters, the model will sufficiently describe the actual maritime container (and container-equivalent land) cargo flow.

In summary, the key points to develop a model, which can sufficiently describe the actual container movement, are to incorporate stochastic approach and consider the capacity constraints in each transport mode. In other words, the related parameters (i.e. the distribution parameter, θ, capacity of each link and parameters on congestion function by transport mode, $b1$ to $b8$) should be more carefully determined throughout the sensitivity analysis.

References

Dial, R., 1971. A probabilistic multipath traffic assignment algorithm which obviates path enumeration. Transp. Res. 5, 83–111.

Shibasaki, R., Kawasaki, T., 2020. International intermodal container shipping network in South Asia: modelling and policy simulations. Int. J. Shipp. Transp. Logist. (accepted).

Shibasaki, R., Iijima, T., Kawakami, T., Kadono, T., Shishido, T., 2017. Network assignment model of integrating maritime and hinterland container shipping: application to Central America. Marit. Econ. Logist. 19 (2), 234–273. https://doi.org/10.1057/s41278-016-0055-3.

Data [1] maritime container shipping and land transport network

8

Ryuichi Shibasaki
The University of Tokyo

As described in Chapter 5, two data types should mainly be prepared for inputting into this model: the networks for maritime container shipping (MCS) and land transport (LT), and the cargo shipping demand among regions. In this chapter, how the data on the transport network are generally prepared is explained. It should be noted that the data introduced in this chapter are principal as of 2013, although the years of data used in the model simulations introduced in Part 3 differ among simulations. Another point to be noted is that the model focuses on the shipping of containers or container-equivalent cargo in certain simulations.

Ports: Intersection between MCS and LT networks

Seaports play an important role as intersections of intermodal transport connecting MCS and LT networks. Basically, the major container seaports of the world and the global liner service (LS) network operated by global liner shipping companies (LSCs) are considered in any simulations, as these are currently connected on a global scale and difficult to separate into specific regions. Moreover, smaller ports and local LS networks (operated by local LSCs) are included for regions on which the simulations specifically focus and the LT network is included. Additional local ports and LS network for each simulation are described in each chapter of Part 3.

Ports considered in simulation model

Specifically, all container seaports for which the international throughput was over 500,000 TEU/year (including empty containers, but excluding domestic containers) are generally considered. A list of container seaports as of 2013[a] and their locations are presented in Table 8.A1 and Fig. 8.1. The total number is 173. Creating such a list

[a] Currently our project team have already prepared and used the data till 2016 or more recent years, although the detailed data is not included in this book. To keep updating the database is another tough and money-consuming task.

Global Logistics Network Modelling and Policy. https://doi.org/10.1016/B978-0-12-814060-4.00008-3

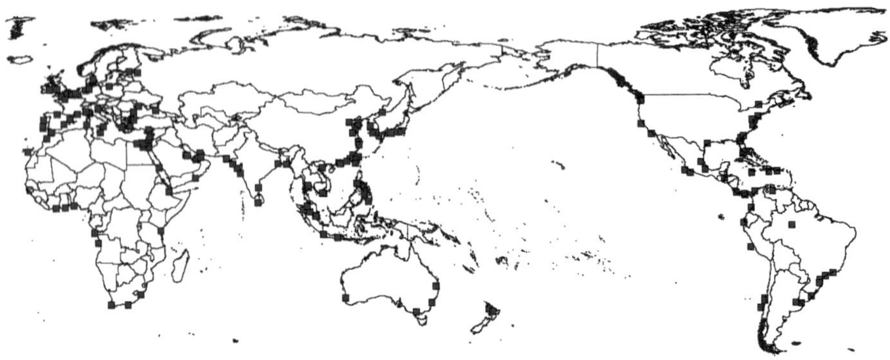

Fig. 8.1 Major container seaports of the world for which international throughput was more than 500,000 TEU as of 2013.
Source: Own compilation.

has become more difficult, as certain data (such as CI-online) are no longer available. Therefore, the list is created by integrating multiple data sources, including:

(a) *Container Forecaster Annual Review* published by Drewry Maritime Research;
(b) *Containerisation International Top 100 Ports* online published by Lloyd's List for the top 100 container ports;
(c) *China Port Yearbook* for ports in Mainland China;
(d) *Global Container Terminal Operators*, annually published by Drewry Maritime Research, which is available for each container terminal in which the world's major terminal operators are in operation;
(e) websites for each port or terminal; and
(f) substitution with previous records in cases where data are not available from any of the sources listed above;

where the suffixes for each source (a to f) corresponding to these are indicated in Table 8.A1.

Moreover, certain special treatments are considered for several ports, as follows:

- Differentiating between international and domestic throughputs is generally challenging, owing to data availability. Domestic container throughputs can be ignored in most countries but cannot be neglected for certain countries. Among these, domestic container throughputs in certain Chinese ports are significant and thus excluded from the total throughputs, which are available from the data source (c) in the above list. As a result, 12 Chinese ports, including Yingkou, Rizhao, and Quanzhou, for which the total throughput was more than 500,000 TEU/year but the international throughput was less than 500,000 TEU/year, are excluded. For ports in other countries, all of the containers are assumed to be transported internationally, owing to data availability.
- Furthermore, certain ports (e.g. Chinese ports such as Suzhou, Nanjing, and Zhongshan, Makassar in Indonesia, Honolulu in the United States, and Duisburg in Germany) are eliminated because no or very few containership movement data on vessels calling at these ports are available from a LS database (that is, the MDS Containership Databank), which is used for developing LS networks in the following section.

- The port of Shenzhen in China is divided into two parts: (i) Yantian terminal and (ii) Shekou and other terminals, because these two terminals are located on opposite sides of the port of Hong Kong, and neither are negligible. By separating these into two ports, it becomes easier to develop an MCS network. Meanwhile, certain ports that are closely located are integrated (such as Singapore and Jurong; Puerto Manzanillo and Cristobal in Panama; Alexandria and El Dekheila in Egypt; and Odessa and Illichivsk in Ukraine) to reduce the number of ports included in the model, although these are generally separated in port statistics.

Estimation of transshipment and empty container amount

The container throughput summarised above includes transshipment containers. Local (i.e. export and import) and transshipped containers are separated in this book based on requirements. Table 8.A1 also indicates the transshipment amount of each port, which are estimated according to the following steps:

- Regarding the major transshipment hubs which are listed in the *Container Forecaster Annual Review*, as 33 container ports of the world annually handle over one million transshipped containers, the described transshipment rate is used. The transshipment rate is assumed to be same for all the terminals in a port even if they are not included in the list above.
- Regarding other container ports, the average transshipment rate by world region which are also listed in the *Container Forecaster Annual Review* is used. Note that first the total container throughput and transshipment amount in the region are calculated from the tables provided in the *Container Forecaster Annual Review*, then both total amounts (i.e. total and transshipment containers) of major hub ports in the region are deducted from above amounts; subsequently, the 'average transshipment rate in other ports' in the region is calculated by dividing the total transshipment amount in other ports by total container throughput in other ports in the region.

The amount of empty containers is also necessary to deduct from the total throughput because the model focuses on only the laden containers. Although it is varied throughout the world, the rate of empty containers is assumed to be uniformly 24.0%, which is also acquired from the *Container Forecaster Annual Review*, owing to data availability.

Parameter setting

Table 8.1 summarises the parameters relating to the seaports (i.e. policy variables on seaports) set in this model. Specifically, the lead times at the terminal when exporting and importing, TPX_r and TPM_s, as well as the handling charges at the container terminal for exporting and importing, CPX_r and CPM_s, are set according to country, as indicated in Table 8.A2 from the Doing Business website (trading across borders) provided by the World Bank, although these values are currently not available. The transshipment time, TPR_a, is estimated by the authors for each port, as indicated in Table 8.A1, judging from the comprehensive level of service in the port [note that the handling charges for transshipment are defined by CPX_r and CPM_s, as indicated in

Table 8.1 Parameters relating to seaports set in model.

Parameter	Description	Equations included	Coefficient and unit	Source
TPX_r, TPM_s	Export and import lead times	(7.4)	Set in hours by country. See Table 8.A2	World Bank (Doing Business website)
CPX_r, CPM_s	Export and import handling charges	(6.27), (6.28), (6.29), (7.5)	Set in USD per TEU by country. See Table 8.A2	
TPR_a	Transshipment time	(6.12)	Set in hours by port. See Table 8.A1	Authors
TBE_a	Berthing time hours	(6.11)	12h irrespective of ports	
τ	Multiplier of inter-carrier transshipment	(7.5), (7.6)	2.0	Authors by model calibration

Source: Own compilation.

Eq. (7.7)]. However, the berthing time, TBE_a, is constantly assumed to be 12 (h) for each port of every LS, owing to data availability. Moreover, the multiplier of the inter-carrier transshipment, τ, is set to 2.0 by model calibration.

Note that the data on terminal capacity is not included in the present model as described in Chapter 6. To include it is a future challenge because there is a need to simulate the port congestion and policy to solve it.

Global MCS network

LS data

The global LS network in the model is developed based on the MDS Containership Databank data (hereinafter, 'MDS data'), which provide information for each containership on not only the service name, (joint) operator and slot-chartered LSC(s), list of ports of call and their order, but also the vessel speed, v_a, average vessel capacity, $Vcap_a$, and frequency, $freq_a$ (see Table 8.3). By aggregating these vessel-basis data into service-basis data; that is, a carrier vessel group that navigates the same route and calls at the same ports as one regular service, the LS network is structured.

All LSs are separated according to the LSC in the model simulation. For an exclusive LS supplied by one LSC, the entire vessel capacity will be reserved for the LSC in question. For an LS jointly operated by multiple companies (which can be determined from the information on partners in the MDS data), the vessel capacity $Vcap_a$ is divided by that of each LSC (cap_a), according to the numbers of joint operator and slot-chartered LSCs. For the sake of simplicity of the model, the authors assume no ex-post accommodation of spaces between LSCs; that is, the vessel capacity is strictly divided into each LSC. Specifically, the vessel capacity is equally divided by the number of partner (joint operator) LSCs, including the operator LSC. For an LS with slot-charted LSC(s), the vessel capacity is assigned to these as half of each operator capacity. For example, for an LS with one operator, three partner and two slot-chartered LSCs, the assigned capacity is 20% of the vessel capacity for the operator and each partner LSC, and 10% of the capacity for each slot-chartered LSC.

Meanwhile, the MDS data does not include any information on the actual schedule (the exact day and time for arriving and departing for each port). Therefore, the schedule for the connection in the transshipment ports between the mother and feeder vessel is not considered in the model. In the model, the expected waiting time for departure following transshipment is assumed to be half of the duration time of the service that will be loaded from the transshipment port, as described in Eq. (6.8) in Chapter 6 if the local export cargo are loaded.

Furthermore, the MDS data does not provide any information on the containers transported, including the load factor (the rate of the number of containers on board to the capacity) during navigation and the number of containers loaded and discharged at each port. To the best of the authors' knowledge, no data sources that can comprehensively grasp such information types are available in the world.[b]

Fig. 8.2 presents examples of an LS network for major and local LSCs around South Asia as of June 2013, estimated by the authors from the MDS data using the above methodology. Note that the figures are depicted on the basis of vessel capacity, including the slot-chartered LSs. It is found that an LSC provides numerous LSs, and the LS network differs significantly among LSCs. Moreover, most LSs call at more than three ports constituting a loop (see also Chapter 1), whereas most regional LSs call at several ports located along a specific sea area (e.g. the Arabian Sea and Bay of Bengal).

LSCs considered

As the model focuses on the flow on the global MCS network and the transshipment of containers in hub ports, several LSs provided by smaller, local LSCs are eliminated for the sake of calculation simplicity. In particular, the model includes the 20 largest LSCs in the world, as well as several (10–20) local LSCs that have an LS network in the region on which the simulation focuses. The list of major LSCs as of 2013 is presented in Table 8.2 (note that certain LSCs in the list no longer exists owing to mergers or bankruptcy). Among the 2603 LSs of the world as of June 2013 that are included in the MDS data, 1036 LSs were operated or slot-chartered by at least one major LSC, covering 64.0% of the global annual vessel capacity.

[b] This proves the necessity of the model to describe container movements developed by the authors. However, in other words, validation of the developed model is thus not possible by means of comparison with the observed container flows; therefore, validation from other viewpoints is necessary.

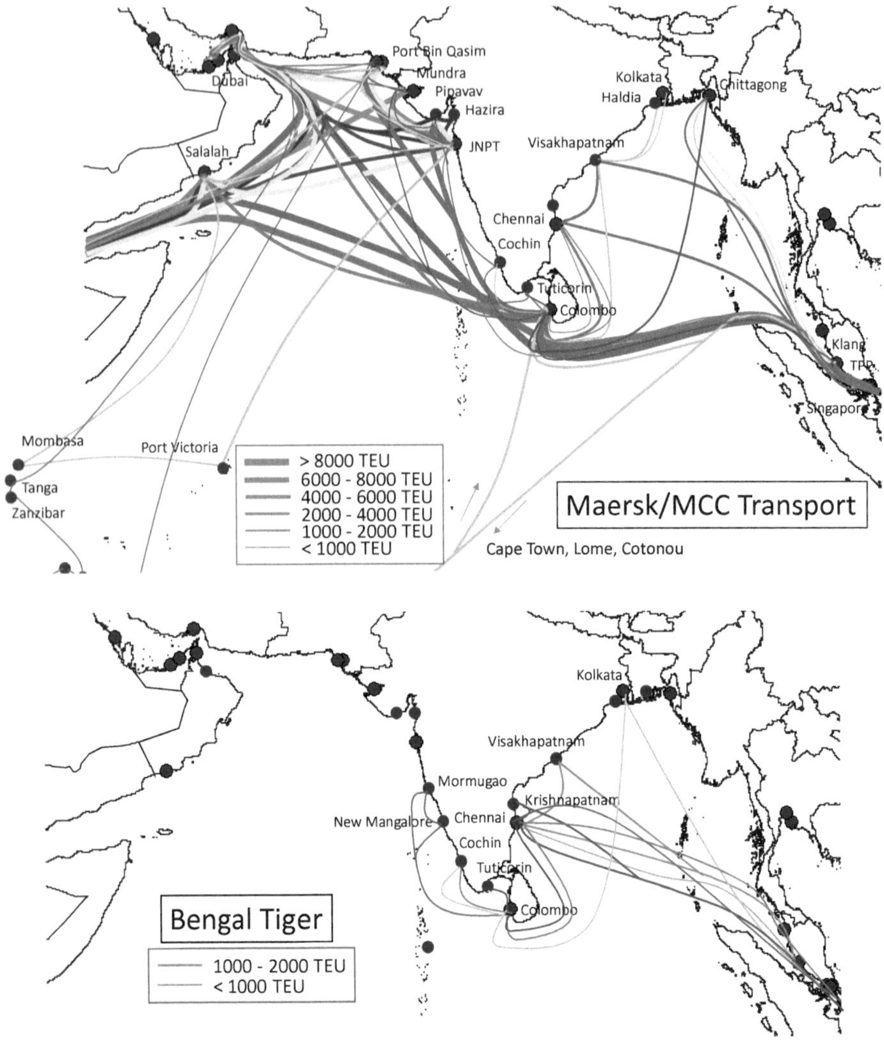

Fig. 8.2 Examples of LS network: weekly capacity in South Asia for two LSCs (as of June 2013).
Source: Own compilation based on the MDS Containership Databank.

Parameter setting

Table 8.3 presents the parameters that should be prepared for inputting into the MCS submodel as variables relating to the MCS network. The parameters that should be prepared from sources other than the MDS data are explained in the following.

The sea distance between ports, lm_a, is generally obtained from the work of Toriumi (2010), which is calculated based on the assumption that every containership passes through the shortest route on the sea out of the preset navigation routes. The dummy variables for the Suez and Panama Canal transits, γ_a^s and γ_a^p, are also acquired

Table 8.2 World major (top 20) LSCs and their capacities (as of 2013).

No.	Group	Group name	Included carriers	Annual capacity ('000 TEU)	Company share
1	Group A	Maersk	Maersk Line, Norfolkline Ferries, Safmarine Container Lines, MCC Transport, Mercosul Line	17,208	9.7
2	Group B	MSC	Mediterranean Shipping Co (MSC)	15,994	9.0
3	Group C	CMA-CGM	CMA-CGM, ANL Container Line, China Navigatrion Co. (CNC Line), Campagrie Marocaine de Navigation (Comanav), Delmas, MacAndrews, FAS, Gemartrans, OT Africa Line, US Lines	12,954	7.3
4	Group D	Evergreen	Evergreen Marine, Italia Marittima (LT), Jatsu Marine	7167	4.0
5	Group E	Hapag-Lloyd	Hapag-Lloyd, CP Ships	4808	2.7
6	Group F	APL	APL	4640	2.6
7	Group G	CSAV	CSAV (Compania Sud Americana de Vapores), CSAV Norasis Liner Services	2378	1.3
8	Group H	Cosco	Cosco Container Lines, Shanghai Panasia	5854	3.3
9	Group I	Hanjin	Hanjin Shipping, Senator Lines	4411	2.5
10	Group J	CSCL	China Shipping Container Lines (CSCL), Shanghai Puhai	5191	2.9
11	Group K	MOL	Mitsui-OSK Lines, Meimon Taiyo Ferry, Shosen Mitsui Ferry	4480	2.5
12	Group L	NYK	Nippon Yusen Kaisha (NYK), Tokyo Senpaku Kaisha (TSK), NYK-Hinode Line, NYKLauritzenCool, Kinkai Yusen	3706	2.1
13	Group M	OOCL	Orient Overseas Container Line (OOCL)	4599	2.6
14	Group N	Hamburg-Sud	Hamburg-Sud, Alianca Transportes Maritimos, Crowley Liner Services, Ybarra y Cia Sudamerica	3208	1.8

(Continued)

Table 8.2 World major (top 20) LSCs and their capacities (as of 2013)—cont'd

No.	Group	Group name	Included carriers	Annual capacity ('000 TEU)	Company share
15	Group O	K-Line	Kawasaki Kisen Kaisha, Kawasaki Kinkai Kisen Kaisha	3033	1.7
16	Group P	Yang Ming	Yang Ming Marine Transport Corp, Kuang Ming Shipping	3717	2.1
17	Group Q	ZIM	Zim Integrated Shipping Services, Gold Star Line, Laurel Navigation	2825	1.6
18	Group R	HMM	Hyundai Merchant Marine	3176	1.8
19	Group S	PIL	Pacific International Lines (PIL), Advance Container Line, Pacific Direct Line Ltd	2998	1.7
20	Group T	UASC	United Arab Shipping Co (UASC)	2025	1.1
Others				63,926	36.0
Total				178,299	100.0%

Source: Own compilation based on the MDS Containership Databank.

therefrom. Moreover, it is partly interpolated by online databases that can determine the shipping distance on the sea, such as Sea-Distance.org (n.d.) and SeaRates.com (n.d.). It should be noted that the shipping distance, even between the same combination of ports, often differs from the database, because the setting of the points on the sea that should pass in the calculation differs.

The parameters for the calculation of the MCS cost (e.g. the fuel price and coefficients on the cost functions for each shipping cost element), as well as the additional time for canal transit, are derived from Shibasaki et al. (2016), which calculates the shipping cost from the viewpoint of operation of the Suez Canal, as described in Chapter 6. The parameters for the canal tolls are derived from the websites of each authority.

The parameters for the congestion function of MCS, *b1* and *b2*, are calibrated as the best fit to the observed transshipment rates in the major hub ports, provided in the *Container Forecaster Annual Review*. Specifically, these parameters are set to 2.308 and 1.017, respectively, as mentioned in Chapter 6. Furthermore, the elasticity of shipping demand of export and import containers to ocean freight charge, γ_e and γ_i, is estimated to be at 0.00207 and 0.0394, respectively, which are estimated from an average change of ocean freight charge [calculated by Eq. (6.30)] when the demand per unit is changed by each partner port for the cargo to/from a specific port (i.e. the port of Acajutla, El Salvador using the model introduced in Chapter 10), as described in Chapter 7.

Table 8.3 Parameters related to LS set in model.

Parameter	Description	Equations included	Coefficient and unit	Source
a	MCS link	Any equations	Set by LS from list of ports to call	MDS
v_a	Vessel speed	(6.7), (6.14), (6.16)	Set in knots by LS	Containership Databank
$Vcap_a$	Average vessel capacity	(6.14), (6.21)	Set in TEU by LS	
cap_a	Average capacity by LSC	(6.7), (6.14)	Set in TEU by LS by LSC	
$freq_a$	Service frequency	(6.7), (6.8), (6.14)	Set in number per year by LS	
DWT_a	Vessel size (Dead Weight Tonnage)	(6.16), (6.18), (6.19), (6.22)	Set in DWT by LS	
lm_a	MCS distance	(6.7), (6.14)	Set in nautical miles by combination of ports	Toriumi (2010) and other online databases
γ_a^s	Dummy variable for Suez Canal transit	(6.7), (6.14)	Set to 1 if transit Suez Canal and 0 otherwise by combination of ports	
γ_a^p	Dummy variable for Panama Canal transit	(6.7), (6.14)	Set to 1 if transit Panama Canal and 0 otherwise by combination of ports	
TS	Additional time for Suez Canal transit	(6.7)	24 h	Shibasaki et al. (2016)
TP	Additional time for Panama Canal transit	(6.7)	24 h	
FP	Fuel price	(6.15) For fuel cost calculation	600 USD per ton (2013)	
c_1	Coefficient for fuel assumption	(6.16)	6.49×10^{-6}; see Chapter 6	
ir	Interest rate	(6.17) For vessel cost calculation	2% per year	
PP	Project period	(6.17)	15 years	
ODR	Operation day rate	(6.17)	90%	
c_2, c_3	Coefficients for vessel price	(6.18)	8.37×10^2 and 4.46×10^6; see Chapter 6	
c_4, c_5	Coefficients for vessel operation cost	(6.19) For vessel operation cost calculation	6.66×10^{-2} and 3.98×10^3; see Chapter 6	

Continued

Table 8.3 Parameters related to LS set in model—cont'd

Parameter	Description	Equations included		Coefficient and unit	Source
c_9, c_{10}	Coefficients for Suez Canal net registered tonnage	(6.22)	For Suez Canal toll calculation	1.01×10^0 and 9.99×10^3; see Chapter 6	
SDRrate	Conversion rate from SDR to USD	(6.20)		1.5 SDR per USD (2013)	
c_6, c_7	Coefficients for Suez Canal toll	(6.20)		See Table 6.1 in Chapter 6	Suez Canal Authority
c_8	Coefficient for Panama Canal toll	(6.21)	For Panama Canal toll calculation	72.0 USD per TEU capacity (2013)	Panama Canal Authority
$b1$, $b2$	Coefficient for congestion function of MCS	(6.7)		2.308 and 1.017; see Chapter 6	Estimated to best fit to observed results
γ_e, γ_i	Elasticity of shipping demand of export and import containers to ocean freight charge	(7.24a), (7.24b)		0.00207 and 0.0394; see Chapter 7	Average change of ocean freight charge when the demand per unit is changed

Source: Own compilation.

Regional LT network

LT network

A regional (hinter-)land transport network is prepared for each simulation to cover the area on which the simulation focuses. The network generally includes national borders on land, considering international competitions among gateway seaports and/or between MCS and LT. Therefore, the area covered is often widespread; for example, an LT network across the Eurasian continent including more than 20 countries is considered in the simulations in Chapter 13 (gateway seaport competitions of Central Asian cargo) and Chapter 16 (competition between LT and MCS network under China's Belt & Road Initiative), as illustrated in Fig. 8.3.

As stated in Chapter 7, road networks, rail networks, and inland waterways (or sometimes coastal shipping) are generally included in the LT network. Road and rail networks anywhere in the world are generally derived from GIS software, namely, the ADC WorldMap, which provides not only the land distance, ll_a, but also the road classification. A ferry network for serving inland waterways or coastal shipping is accordingly added by the authors, with information collected from field interview surveys, if necessary.

Parameter setting

Table 8.4 displays the parameters that should be prepared for inputting into the (hiter-) land transport submodel as variables relating to the LT network. The parameters that should be prepared from sources other than the ADC WorldMap are explained in the following. As a matter of course, these parameters can be set differently by the regions on which the simulation focuses if necessary, although the conventional settings estimated from the authors' field and literature surveys and calibration results are described as follows as well as in Table 8.4.

The trailer speed, vRo_a, and annual road capacity, $capRo_a$, for each road link are set to 60, 50, and 40 (km/h) and 10,000,000, 2,000,000, and 500,000 (TEU/year), respectively, according to the road classification ('motorway', 'primary route', and 'important route') in the ADC WorldMap. Meanwhile, both the railway and ferry speed, vRa_a and vFe_a, are normally set to 20 km/h, considering that these services in developing countries, on which the simulations dealt with in this book typically focus, are sometimes obsolete and less competitive. Moreover, the capacity per train and vessel, $capRa_a$ and $capFe_a$, and frequency of trains for container shipping and ferry, $freqRa_a$ and $freqFe_a$, are defined differently by country (see each chapter in Part 3 for details).

The freight charge of each mode and handling time at the rail station and ferry port are also determined, based on the interview survey and model calibration. Specifically, the fixed and operational charges of the trailer, *CFRo* and *CVRo*, are normally set to 60.0 (USD/TEU) and 1.0 (USD/km/TEU), respectively. However, the fixed charges of the rail and ferry are not normally considered; that is, *CFRa* and *CFFe* are set to 0 (USD/TEU), while both operational charges, *CVRa* and *CVFe*, are set to 0.5

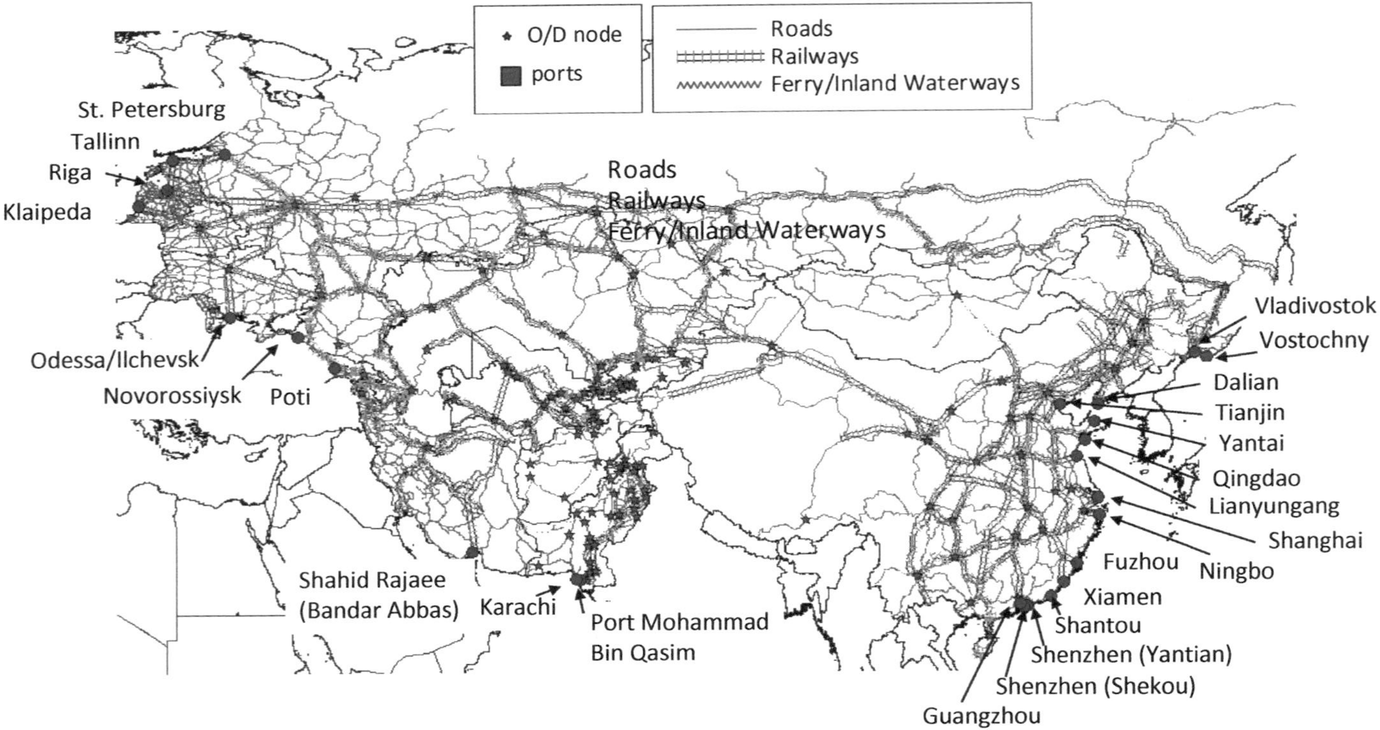

Fig. 8.3 Example of LT network included in the model: across the Eurasian continent.
Source: Shibasaki, R. Tanabe, S., Kato, H., Lee, T.-W., 2019. Could Gwadar port in pakistan be a new gateway? A network simulation approach in the context of the Belt and Road Initiative. Sustainability 11 (20), 5757.

Table 8.4 Parameters relating to LT network set in model.

Parameter	Description	Equations included	Coefficient and unit	Source
a	LT link	Any equations	Set by mode from list of ports to call	ADC World Map and estimated by authors from other sources
ll_a	LT distance	(7.17), (7.18), (7.20), (7.21)	Set in km for each link	
vRo_a	Truck speed of container (or container-equivalent) trailers	(7.17), (7.18)	60 (for motorway), 50 (for primary route), and 40 (for important route) km/h by road classification	
$capRo_a$	Annual road capacity for container (or container-equivalent) trailers	(7.17)	10.0 (for motorway), 2.0 (for primary route) and 0.5 (for important route) million TEU/year by road classification	
vRa_a	Rail speed of trains for container shipping	(7.20)	20 km/h	Authors
vFe_a	Ferry (or barge) speed for container shipping	(7.21)	20 km/h (in some models, 10 km/h)	
$capRa_a$	Maximum container capacity per train	(7.20)	Set in TEU per train by country	Estimated from field survey results and other sources (see each chapter in Part 3)
$freqRa_a$	Frequency of trains for container shipping	(7.19), (7.20)	Set in trains per year by link or country	
$capFe_a$	Maximum container capacity per vessel	(7.21)	Set in TEU per vessel by link	
$freqFe_a$	Frequency of ferry service	(7.21), (7.22)	Set in vessels per year by link	
$CFRo$	Fixed charge of container trailer	(7.16), (7.18)	60.0 (USD/TEU)	Estimated by authors from interview survey, other sources and calibration
$CVRo$	Coefficient for operational charge of container trailer	(7.17), (7.18)	1.0 (USD/km/TEU)	
$CFRa$	Fixed charge of container train	(7.18)	0 (USD/TEU)	
$CFFe$	Fixed charge of ferry	(7.21)	0 (USD/TEU)	

Continued

Table 8.4 Parameters relating to LT network set in model—cont'd

Parameter	Description	Equations included	Coefficient and unit	Source
$CVRa$	Coefficient for operational charge of container train	(7.20)	0.5 (USD/km/TEU)	
$CVFe$	Coefficient for operational charge of ferry	(7.21)	0.5 (USD/km/TEU)	
$THRa_a$	Handling times for rail	(7.18)	24.0 (h)	
$THFe_a$	Handling times for ferry	(7.21)	24.0 (h)	
CBO_a	Additional cost in border crossing	(7.23)	Set in hours by country by import/export. See Table 8.A2	World Bank (Doing Business website)
TBO_a	Additional time in border crossing	(7.23)	Set in USD per TEU by country by import/export. See Table 8.A2	
λ_a	Dummy variable for crossing national border	(7.23)	Set by country pair, based on bilateral relationship. See each chapter in Part 3	Estimated as best fit to observed results
$b3, b4$	Coefficients for congestion function of road transport	(7.17)	1.0 and 3.0	
$b5, b6$	Coefficients for congestion function of rail transport	(7.20)	2.0 and 2.0	
$b7, b8$	Coefficients for congestion function of ferry transport	(7.21)	2.0 and 2.0	

Source: Own compilation based on Shibasaki, R., Kawasaki, T., 2020. International intermodal container shipping network in South Asia: modelling and policy simulations. Int. J. Ship. Transp. Logist. (accepted).

(USD/km/TEU). Both handling times for loading or discharging cargo into or from the rail and vessel, $THRa_a$ and $THFe_a$, are uniquely set to 24.0 (h), irrespective of the link.

The additional cost and time in the border crossing, CBO_a and TBO_a, are obtained from the summation of the cost and time for 'documents preparation' and 'customs clearance and technical control' on the Doing Business website (trading across borders) provided by the World Bank, as indicated in Table 8.A2. It should be noted that these variables are defined for exporting and importing, respectively.

Furthermore, the dummy variable for crossing the national border, λ_a, is set by country pair, based on the model calibration, with consideration of bilateral relationships such as customs union and incidents. Note that this value should be normally between 0 and 1, because the simulations in this book generally consider bonded transport if cargo crosses a land national border. Its setting is detailed in each chapter of Part 3.

Finally, the parameters for each congestion function are calibrated as the best fit to the observed throughputs of the export and import containers in each gateway seaport (see Chapter 7): namely, *b3* and *b4* for the road transport congestion function at 1.0 and 3.0, respectively; *b5* and *b6* for rail transport at 2.0 for both; *b7* and *b8* for the international ferry at 2.0 for both.

Conclusion

All the descriptions in this chapter focus on setting networks and parameters for describing the status quo (i.e. as of 2013) using the proposed model. Many variables are included in both MCS and LT network and set based on the available information or to best fit the observed data.

If conducting the policy simulations in each chapter of Part 3, some of the settings will be changed, as necessary. Because the changes vary with the simulations, each change will be described in each chapter of Part 3. For example, if simulating a new container port construction, the author not only add a new port in the port list, but several LSs to call at the new port are exogenously prepared, considering the geographical condition and current LS network for each LSC (see Chapters 10, 12, and 13). Similarly, if simulating the increase in the level of LS (e.g. speed) in rail transport or inland waterway and reduction of border barriers, in principle, their capacity should also increase by increasing the service frequency for meeting the expected traffic increase (see Chapters 12, 13, and 16). This is because the outputs will fluctuate more as the effects of the capacity constraint become severe.

Determining how each parameter should be set in future simulations is a challenging task. Most other parameters—than those under the author's examination in the simulation—cannot be easily changed because there are no appropriate grounds for such a change. However, the capacity of the service in both MCS and LT (especially rail and inland waterway) should be increased according to the increase in the amount of shipping demand, as mentioned in the previous paragraph. Such 'balanced' changes among parameters are required for acquiring reasonable results from the simulations.

Appendices

See Tables 8.A1 and 8.A2.

Table 8.A1 Container ports generally included in model simulations and variables set by port (example of 2013).

No	Port name	Country	Country/region in the WTS	Annual throughput ('000 TEU) $XXM_p + XPR_p$	Annual throughput ('000 TEU) source	Transshipment rate $\frac{XPR_p}{XXM_p + XPR_p}$	Transshipment time[a] TPR_a (hours)
1	Tokyo	Japan	Japan	4861	b	9.7%[b]	24
2	Yokohama	Japan	Japan	2888	a	9.7%[b]	24
3	Shimizu	Japan	Japan	499	e	9.7%[b]	24
4	Nagoya	Japan	Japan	2709	a	9.7%[b]	24
5	Osaka	Japan	Japan	2485	a	9.7%[b]	24
6	Kobe	Japan	Japan	2553	b	9.7%[b]	24
7	Hakata	Japan	Japan	868	e	9.7%[b]	24
8	Vladivostok	Russia	South Korea/FE Russia	817	a	9.7%[b]	48
9	**Busan**	**South Korea**	**South Korea/FE Russia**	**17,686**	**a**	**49.5%**	**12**
10	Yeosu/Gwangyang	South Korea	South Korea/FE Russia	2285	b	9.7%[b]	12
11	Pyongtaek	South Korea	South Korea/FE Russia	518	a	9.7%[b]	24
12	Incheon	South Korea	South Korea/FE Russia	2160	a	9.7%[b]	24
13	Dalian	China	China	5909[c]	a	8.3%[b]	48
14	Tianjin/Xingang	China	China	7417[c]	a	8.3%[b]	48
15	Yantai	China	China	541[c]	a	8.3%[b]	48
16	Qingdao	China	China	11,182[c]	a	8.3%[b]	24
17	**Lianyungang**	**China**	**China**	**3265**[c]	**a**	**65.0%**	**24**
18	**Shanghai**	**China**	**China**	**28,911**[c]	**a**	**14.0%**	**24**
19	**Ningbo**	**China**	**China**	**15,967**[c]	**a**	**15.0%**	**24**
20	Fuzhou	China	China	1206[c]	a	8.3%[b]	48
21	Xiamen	China	China	5125[c]	a	8.3%[b]	24
22	Shantou	China	China	553[c]	a	8.3%[b]	48

23	**Shenzhen (Yantian)**	**China**	**China**	**10,796[c]**	**c**	**13.0%**	**24**
24	**Shenzhen (Shekou, Chiwan, Dachan Bay)**	**China**	**China**	**10,644[c]**	**c**		**24**
25	**Guangzhou (Nansha, Huangpu)**	**China**	**China**	**6096[c]**	**a**	**8.3%**	**24**
26	**Hong Kong**	**Hong Kong**	**China**	**22,352**	**a**	**58.6%**	**12**
27	Keelung	Taiwan	Taiwan	1613	a	9.7%[b]	24
28	Taipei New Port	Taiwan	Taiwan	1029	b	9.7%[b]	24
29	Taichung	Taiwan	Taiwan	1468	a	9.7%[b]	24
30	**Kaohsiung**	**Taiwan**	**Taiwan**	**9938**	**a**	**46.6%**	**24**
31	Manila	Philippines	Philippines	3770	b	9.5%[b]	48
32	Cebu	Philippines	Philippines	555	f1(2012)	9.5%[b]	48
33	Davao	Philippines	Philippines	569	e	9.5%[b]	48
34	Haiphong	Vietnam	Vietnam	1040	e	9.5%[b]	48
35	Ho Chi Minh	Vietnam	Vietnam	5542	b	9.5%[b]	48
36	Cai Mep/Vung Tau	Vietnam	Vietnam	1268	d	9.5%[b]	24
37	Laem Chabang	Thailand	Thailand	6041	a	9.5%[b]	24
38	Bangkok	Thailand	Thailand	1505	a	9.5%[b]	24
39	Pasir Gudang	Malaysia	Malaysia	801	f1(2012)	9.5%[b]	24
40	**Tanjung Pelepas**	**Malaysia**	**Malaysia**	**7628**	**b**	**91.3%**	**12**
41	**Port Klang**	**Malaysia**	**Malaysia**	**10,350**	**a**	**63.5%**	**24**
42	Penang	Malaysia	Malaysia	1238	b	9.5%[b]	24
43	**Singapore/Jurong**	**Singapore**	**Singapore**	**32,579**	**a**	**84.8%**	**12**
44	Tanjung Perak (Surabaya)	Indonesia	Indonesia	3001	b	9.5%[b]	48
45	Tanjung Priok (Jakarta)	Indonesia	Indonesia	6590	b	9.5%[b]	48

Continued

Table 8.A1 Container ports generally included in model simulations and variables set by port (example of 2013)—cont'd

No	Port name	Country	Country/region in the WTS	Annual throughput ('000 TEU) XXM_p+XPR_p	source	Transshipment rate $\frac{XPR_p}{XXM_p+XPR_p}$	Transshipment time[a] TPR_a (hours)
46	Belawan	Indonesia	Indonesia	899	e	9.5%[b]	48
47	Chittagong	Bangladesh	Bangladesh	1540	b	3.6%[b]	72
48	Kolkata	India	India	575	a	3.6%[b]	72
49	Chennai/Madras	India	India	1485	g	3.6%[b]	72
50	Jawaharlal Nehru (JNPT)	India	India	4120	g	3.6%[b]	72
51	Pipavav	India	India	661	e	3.6%[b]	72
52	Mundra	India	India	2156	e	3.6%[b]	72
53	**Colombo**	**Sri Lanka**	**Indian Subcontinent Islands**	**4306**	**b**	**74.8%**	**48**
54	Port Mohammad Bin Qasim	Pakistan	Pakistan	768	a	3.6%[b]	72
55	Karachi	Pakistan	Pakistan	1586	a	3.6%[b]	72
56	St Petersburg	Russia	Baltics	2514	a	9.7%[b]	48
57	Prince Rupert	Canada	Canada Pacific Coast	539	a	8.3%[b]	24
58	Vancouver BC	Canada	Canada Pacific Coast	2825	a	8.3%[b]	24
59	Seattle	USA	USA_North Pacific	1575	a	8.3%[b]	24
60	Tacoma	USA	USA_North Pacific	1892	a	8.3%[b]	24
61	Oakland	USA	USA_South Pacific	2346	a	8.3%[b]	24
62	Los Angeles	USA	USA_South Pacific	7869	a	8.3%[b]	24
63	Long Beach	USA	USA_South Pacific	6731	a	8.3%[b]	24
64	Manzanillo (Mexico)	Mexico	Mexico & Central America Pacific	2136	a	8.3%[b]	24

65	Lazaro Cardenas	Mexico	Mexico & Central America Pacific	1051	a	8.3%[b]	24
66	**Balboa**	**Panama**	**Costa Rica and Panama**	**3064**	**a**	**91.3%**	**24**
67	**Manzanillo (Panama)/ Cristobal/Colon**	**Panama**	**Costa Rica and Panama**	**3356**	**a**	**84.6%**	**24**
68	Puerto Limon	Costa Rica	Costa Rica and Panama	1037	a	25.4%[b]	48
69	Puerto Cortes	Honduras	Mexico & Central America Pacific	571	d	25.4%[b]	48
70	Veracruz	Mexico	Mexico Gulf Coast	867	a	4.1%[b]	24
71	Altamira	Mexico	Mexico Gulf Coast	598	a	4.1%[b]	24
72	San Juan	USA (Puerto Rico)	USA_Gulf Coast, South Atlantic & Carib	1270	b	25.4%[b]	48
73	Caucedo	Dominican Rep	USA_Gulf Coast, South Atlantic & Carib	1083	d	25.4%[b]	48
74	**Kingston**	**Jamaica**	**USA_Gulf Coast, South Atlantic & Carib**	**1672**	**a**	**82.5%**	**48**
75	**Freeport**	**Bahamas**	**USA_Gulf Coast, South Atlantic & Carib**	**1400**	**b**	**99.0%**	**48**
76	Houston	USA	USA_Gulf Coast, South Atlantic & Carib	1951	a	4.1%[b]	24
77	Miami	USA	USA_Gulf Coast, South Atlantic & Carib	901	a	7.5%[b]	24
78	Port Everglades	USA	USA_Gulf Coast, South Atlantic & Carib	928	a	7.5%[b]	24
79	Jacksonville	USA	USA_Gulf Coast, South Atlantic & Carib	925	a	7.5%[b]	24

Continued

Table 8.A1 Container ports generally included in model simulations and variables set by port (example of 2013)—cont'd

No	Port name	Country	Country/region in the WTS	Annual throughput ('000 TEU) XXM_p+XPR_p	Annual throughput ('000 TEU) source	Transshipment rate $\frac{XPR_p}{XXM_p+XPR_p}$	Transshipment time[a] TPR_a (hours)
80	Savannah	USA	USA_Gulf Coast, South Atlantic & Carib	3034	a	7.5%[b]	24
81	Charleston	USA	USA_Gulf Coast, South Atlantic & Carib	1601	a	7.5%[b]	24
82	Virginia (Hampton Roads)	USA	USA_Gulf Coast, South Atlantic & Carib	2224	a	7.5%[b]	24
83	Baltimore	USA	USA_North Atlantic	705	a	7.5%[b]	24
84	New York/New Jersey	USA	USA_North Atlantic	5467	a	7.5%[b]	24
85	Montreal	Canada	Canada Atlantic Coast	1357	a	7.5%[b]	24
86	Buenaventura	Colombia	Colombia Pacific Coast	533	e	9.0%[b]	48
87	Guayaquil	Ecuador	Ecuador	1518	b	9.0%[b]	48
88	Callao	Peru	Peru	1856	a	9.0%[b]	48
89	Valparaiso	Chile	Chile	910	a	9.0%[b]	48
90	San Antonio	Chile	Chile	1197	a	9.0%[b]	48
91	San Vicente(Concepcion)	Chile	Chile	453	d	9.0%[b]	48
92	**Cartagena**	**Colombia**	**Colombia Atlantic Coast**	**1865**	**a**	**56.0%**	**48**
93	Puerto Cabello	Venezuela	Venezuela	750	e	25.4%[b]	48
94	Manaus	Brazil	Brazil	545	a	10.3%[b]	48
95	Rio De Janeiro	Brazil	Brazil	506	f1(2012)	10.3%[b]	48
96	Santos	Brazil	Brazil	3446	b	10.3%[b]	48
97	Paranagua	Brazil	Brazil	739	a	10.3%[b]	48

98	Navegantes	Brazil	Brazil	706	d	10.3%[b]	48
99	Itajai	Brazil	Brazil	1105	a	10.3%[b]	48
100	Rio Grande	Brazil	Brazil	622	a	10.3%[b]	48
101	Montevideo	Uruguay	Other Southeast Coast of South America	804	a	10.3%[b]	48
102	Buenos Aires	Argentina	Argentina	1651	b	10.3%[b]	48
103	Shahid Rajaee (Bandar Abbas)	Iran	Arabian Gulf	1763	b	4.1%[b]	48
104	Dammam	Saudi Arabia	Arabian Gulf	1674	a	4.1%[b]	48
105	Khalifa Bin Salman	Bahrain	Arabian Gulf	430	d	4.1%[b]	48
106	Mina Zayed (Abu Dhabi)	UAE	Arabian Gulf	787	f2(2012)	4.1%[b]	24
107	**Dubai/Jebel Ali**	**UAE**	**Arabian Gulf**	**13,600**	**a**	**50.5%**	**24**
108	**Khor Fakkan/ Sharjah Combined**	**UAE**	**Arabian Gulf**	**3800**	**b**	**96.0%**	**24**
109	**Salalah**	**Oman**	**Arabian Gulf**	**3343**	**a**	**97.5%**	**24**
110	**Jeddah**	**Saudi Arabia**	**Arabian Gulf**	**4561**	**a**	**41.0%**	**48**
111	Aqaba	Jordan	Other Mediterranean	883	d	4.1%[b]	48
112	El Sokhna	Egypt	Egypt	511	d	14.7%[b]	48
113	**Port Said**	**Egypt**	**Egypt**	**4100**	**b**	**86.2%**	**24**
114	Damietta	Egypt	Egypt	747	d	14.7%[b]	48
115	Alexandria/El Dekheila	Egypt	Egypt	1508	b	14.7%[b]	48
116	**Tangier/Tangier Med**	**Morocco**	**West Mediterranean**	**2558**	**b**	**96.7%**	**24**
117	Casablanca	Morocco	West Mediterranean	825	a	10.0%[b]	48
118	Las Palmas De Gran Canaria	Spain (Canary Is)	West Mediterranean	1017	a	18.9%[b]	24

Continued

Table 8.A1 Container ports generally included in model simulations and variables set by port (example of 2013)—cont'd

No	Port name	Country	Country/region in the WTS	Annual throughput ('000 TEU)		Transshipment rate	Transshipment time[a]
				$XXM_p + XPR_p$	source	$\frac{XPR_p}{XXM_p + XPR_p}$	TPR_a (hours)
119	Ashdod	Israel	Israel	1182	a	14.7%[b]	24
120	Haifa	Israel	Israel	1357	a	14.7%[b]	24
121	Beirut	Lebanon	Other Mediterranean	1117	a	14.7%[b]	48
122	Mersin	Turkey	Turkey	1367	d	14.7%[b]	48
123	Izmir	Turkey	Turkey	720	f3(2010)	14.7%[b]	48
124	**Ambarli/Istanbul/ Marport/ Kumport/ Haydarpasa**	**Turkey**	**Turkey**	**3378**	**b**	**45.6%**	**48**
125	Constantza	Romania	Romania	634	d	14.7%[b]	48
126	Odessa/Illichivsk	Ukraine	Ukraine	535	d	14.7%[b]	48
127	Novorossiysk	Russia	Russia Black Sea	732	a	14.7%[b]	48
128	**Piraeus**	**Greece**	**Greece**	**3164**	**b**	**82.0%**	**24**
129	Koper	Slovenia	Slovenia	600	a	14.7%[b]	48
130	**Marsaxlokk**	**Malta**	**Malta**	**2750**	**b**	**95.7%**	**24**
131	Cagliari	Italy	Italy	656	d	19.8%[b]	24
132	**Gioia Tauro**	**Italy**	**Italy**	**3087**	**b**	**94.5%**	**24**
133	Leghorn (Livorno)	Italy	Italy	559	e	19.8%[b]	24
134	La Spezia	Italy	Italy	1300	a	19.8%[b]	24
135	Genoa	Italy	Italy	1988	a	19.8%[b]	24
136	Marseilles/Fos	France	France Mediterranean	1098	a	19.8%[b]	24
137	Barcelona	Spain	West Mediterranean	1720	a	19.8%[b]	24

138	**Valencia**	**Spain**	**West Mediterranean**	**4328**	**a**	**49.9%**	**24**
139	**Algeciras**	**Spain**	**West Mediterranean**	**4345**	**a**	**91.0%**	**24**
140	Felixstowe	UK	United Kingdom	3740	b	10.2%[b]	24
141	London (Tilbury)/ Thamesport	UK	United Kingdom	950	e	10.2%[b]	24
142	Southampton	UK	United Kingdom	1491	b	10.2%[b]	24
143	Liverpool	UK	United Kingdom	650	d	10.2%[b]	24
144	Dublin	Eire	Ireland	517	a	10.2%[b]	24
145	Sines	Portugal	West Mediterranean	931	d	10.2%[b]	24
146	Lisbon	Portugal	West Mediterranean	549	a	10.2%[b]	24
147	Leixoes	Portugal	West Mediterranean	626	a	10.2%[b]	24
148	Bilbao	Spain	France/Spain North Atlantic	607	a	10.2%[b]	24
149	Le Havre	France	France/Spain North Atlantic	2600	a	10.2%[b]	24
150	Zeebrugge	Belgium	North Sea	2026	a	10.2%[b]	24
151	**Antwerp**	**Belgium**	**North Sea**	**8578**	**a**	**28.5%**	**24**
152	**Rotterdam**	**Netherlands**	**North Sea**	**11,621**	**a**	**31.0%**	**24**
153	**Bremen/ Bremerhaven**	**Germany**	**North Sea**	**5831**	**a**	**61.0%**	**24**
154	**Hamburg**	**Germany**	**North Sea**	**9257**	**a**	**41.9%**	**24**
155	Gdansk	Poland	Poland	1178	a	9.7%[b]	24
156	Kotka	Finland	Finland	627	a	9.7%[b]	24
157	Gothenburg	Sweden	North Sea	859	a	9.7%[b]	24
158	Abidjan	Cote dIvoire	West Africa	676	d	18.9%[b]	48
159	Tema	Ghana	West Africa	670	d	18.9%[b]	48
160	Lagos/Apapa/Tin Can Island	Nigeria	West Africa	1106	d	18.9%[b]	48

Continued

Table 8.A1 Container ports generally included in model simulations and variables set by port (example of 2013)—cont'd

No	Port name	Country	Country/region in the WTS	Annual throughput ('000 TEU)		Transshipment rate	Transshipment time[a]
				$XXM_p + XPR_p$	source	$\frac{XPR_p}{XXM_p + XPR_p}$	TPR_a (hours)
161	Point Noire	Congo, R.	Central Africa	585	d	18.9%[b]	48
162	Luanda	Angola	Angola	650	d	18.9%[b]	48
163	Cape Town	South Africa	Southern Africa	921	a	21.6%[b]	24
164	Port Elizabeth/Coega	South Africa	Southern Africa	775	a	21.6%[b]	24
165	Durban	South Africa	Southern Africa	2633	a	21.6%[b]	24
166	Mombasa	Kenya	East Africa—Center	875	d	21.5%[b]	48
167	Djibouti	Djibouti	East Africa—North	780	d	21.5%[b]	48
168	Brisbane	Australia	Australia	1085	a	4.8%[b]	24
169	Sydney	Australia	Australia	2153	a	4.8%[b]	24
170	Melbourne	Australia	Australia	2492	a	4.8%[b]	24
171	Fremantle	Australia	Australia	703	e	4.8%[b]	24
172	Auckland	New Zealand	New Zealand	819	e	4.8%[b]	24
173	Tauranga	New Zealand	New Zealand	800	a	4.8%[b]	24

Bold with grey-coloured ports: major transshipment ports indicated in Drewry Maritime Research (2014).

[a] Authors' estimation.

[b] Estimated by authors based on average transshipment rate by region indicated in Drewry Maritime Research (2014).

[c] International containers only.

Source: Own compilation based on a. Drewry Maritime Research, 2014. Container Forecaster 2014 Annual Review. London, UK; b. Lloyd's List. 2016 Containerisation International Top 100 Container Ports 2014. (online). (Accessed 31 March 2016); c. China Port Yearbook Publishers, 2014. Chinese Port Yearbook 2014. Beijing, China; d. Drewry Maritime Research, 2014. Global Container Terminal Operators, Annual Report 2014. London, UK; e. website of each port or terminal; f. substitution with past records: f1. Drewry Maritime Research, 2013. Container Forecaster 2013 Annual Review. London, UK; f2. Lloyd's List Intelligence. 2018 Seasearcher – Places. (online) (Accessed 29 April 2018); f3. Informa Group, 2012. Containerisasion International Yearbook 2010. London, UK.

Table 8.A2 Variables set by countries.

No.	Country name	Port handling				Border crossing							
		Days		Cost(US$/TEU)		Days TBO_a				Cost (US$/TEU) CBO_a			
						Export		Import		Export		Import	
		Export TPX_r	Import TPM_s	Export CPX_a,	Import CPM_a	D.C.[a]	C.C.[a]	D.C.	C.C.	D.C.	C.C.	D.C.	C.C.
1	Japan	2	2	250	250	5	2	5	2	120	75	140	135
2	South Korea	2	2	100	100	3	1	2	1	55	15	65	30
4	Mongolia	–	–	–	–	28	2	28	4	145	160	110	150
5	China	3	3	140	140	14	2	15	4	305	80	260	80
6	Hong Kong	2	1	265	265	2	1	2	1	105	0	100	0
8	Taiwan	2	2	180	180	5	1	5	1	175	100	240	100
9	Philippines	3	3	225	200	8	2	8	2	105	85	90	185
10	Vietnam	3	4	150	175	12	4	12	4	160	100	130	95
13	Thailand	3	2	160	160	8	1	8	2	175	50	135	255
14	Malaysia	2	2	120	120	5	1	3	1	85	60	120	60
15	Singapore	1	1	150	150	2	1	1	1	120	50	100	50
17	Indonesia	2	4	165	165	11	1	13	4	165	125	210	125
20	Bangladesh	7.5	10.5	450	650	14	3	22	3	225	150	370	150
23	India	3	6	175	200	8	2	8	4	420	144	468	144
24	Sri Lanka	3	2	185	185	12	2	11	2	135	160	190	285
241	Maldives	6	8	500	550	9	4	9	4	375	200	460	200
25	Pakistan	4	3	115	150	11	3	11	2	96	200	130	220
26	Russia	3	2	480	490	13	1	12	2	200	550	285	650
27	Kazakhstan	–	–	–	–	21	9	21	9	330	425	310	425
28	Uzbekistan	–	–	–	–	31	5	46	8	285	200	335	200
29	Turkmenistan	–	–	–	–	30	30	30	30	240	240	240	240

Continued

Table 8.A2 Variables set by countries—cont'd

No.	Country name	Port handling				Border crossing							
		Days		Cost(US$/TEU)		Days TBO_a				Cost (US$/TEU) CBO_a			
						Export		Import		Export		Import	
		Export TPX_r	Import TPM_s	Export CPX_a,	Import CPM_a	D.C.[a]	C.C.[a]	D.C.	C.C.	D.C.	C.C.	D.C.	C.C.
30	Tajikistan	–	–	–	–	20	3	20	4	700	550	580	420
31	Kyrgyz	–	–	–	–	23	3	25	11	210	300	280	420
32	Afghanistan	–	–	–	–	44	8	49	7	570	300	680	300
33	United States	2	1	400	420	2	1	2	1	230	60	205	90
34	Canada	1	2	600	650	4	1	3	1	295	35	205	75
35	Mexico	2	3	200	300	5	2	4	2	200	150	290	200
36	Costa Rica	3	3	220	250	6	2	7	2	240	105	215	155
39	Panama	1	1	65	265	5	1	6	1	160	50	150	200
41	Honduras	1	2	50	215	8	2	8	4	260	135	255	130
43	Puerto Rico	3	3	450	500	6	4	8	2	250	275	250	275
44	Dominican Rep.	1	2	325	410	3	2	5	2	215	200	235	200
45	Bahamas	4	2	200	950	10	3	7	3	375	130	300	220
46	Jamaica	3	2	495	740	10	4	10	3	450	235	490	550
48	Peru	3	5	330	395	5	2	7	3	150	130	150	185
49	Chile	3	3	210	210	7	2	5	2	220	100	170	100
51	Ecuador	2	4	360	320	10	4	15	4	375	200	350	250
52	Colombia	3	2	170	150	5	2	6	2	300	350	250	170
53	Venezuela	12	15	800	800	34	7	54	10	690	500	695	700
54	Argentina	2	3	550	800	6	2	22	3	450	150	610	400
55	Brazil	3	3	500	500	6	3	8	4	325	400	275	450
56	Uruguay	3	3	350	450	8	2	7	3	325	250	440	250
60	Iran	4	5	225	250	12	2	24	2	270	175	330	220
61	Bahrain	2	3	110	110	6	2	8	3	380	70	380	110

65	UAE	1	1	190	190	4	1	4	1	230	30	190	30
66	Saudi Arabia	4	3	75	174	6	1	6	6	145	115	135	200
67	Oman	3	2	135	105	5	1	5	1	285	65	250	65
69	Jordan	3	2	110	130	6	2	8	3	135	80	385	65
70	Israel	3	3	200	200	4	1	4	1	110	110	120	70
71	Lebanon	4	6	125	400	11	3	16	6	370	285	315	400
74	Turkey	3	3	270	355	6	2	8	2	220	200	280	200
75	Armenia	–	–	–	–	9	1	9	2	260	75	190	135
76	Azerbaijan	–	–	–	–	15	4	14	3	515	425	535	425
77	Georgia	2	2	800	800	4	1	5	1	255	200	255	240
78	Greece	2	3	300	380	11	1	8	2	160	230	140	265
79	Italy	3	3	345	345	11	2	10	2	180	145	130	145
80	Portugal	5	5	260	260	7	1	6	1	195	125	200	265
81	Spain	2	2	250	250	4	1	4	2	30	18	30	18
86	Romania	3	2	300	300	7	1	8	1	410	75	420	75
87	Slovenia	3	3	200	200	10	1	9	1	135	60	195	85
88	Morocco	2	2	250	350	6	1	10	2	125	100	300	150
94	Egypt	2	3	170	250	7	1	8	2	125	100	210	100
96	Malta	2	2	275	410	6	1	4	2	280	50	260	50
98	Belgium	2	2	300	300	3	1	4	1	190	100	270	100
100	Finland	2	2	160	160	4	1	3	1	170	85	180	85
101	France	3	3	315	315	4	1	5	1	310	80	300	150
102	Germany	2	1	250	250	4	1	3	1	175	30	185	55
103	United Kingdom	2	1	205	205	3	1	2	1	175	75	180	75
104	Ireland	1	2	220	253	5	1	5	1	205	185	165	70
106	Netherlands	1	1	260	250	4	1	3	1	160	90	220	90
107	Sweden	2	2	200	200	3	1	2	1	120	55	130	55

Continued

Table 8.A2 Variables set by countries—cont'd

No.	Country name	Port handling				Border crossing							
		Days		Cost(US$/TEU)		Days TBO_a				Cost (US$/TEU) CBO_a			
		Export TPX_r	Import TPM_s	Export CPX_a,	Import CPM_a	Export		Import		Export		Import	
						D.C.[a]	C.C.[a]	D.C.	C.C.	D.C.	C.C.	D.C.	C.C.
111	Poland	3	2	140	140	10	1	9	1	145	65	120	65
113	Estonia	1	1	280	330	3	1	2	1	220	25	200	25
114	Latvia	3	3	150	220	5	1	4	2	200	100	315	100
115	Lithuania	2	2	190	200	4	1	3	1	100	60	130	70
116	Ukraine	3	3	430	600	22	1	20	2	250	300	555	350
117	Belarus	–	–	–	–	8	1	22	1	160	160	565	200
118	Moldova	–	–	–	–	11	3	13	4	314	200	338	234
130	Ghana	3	8	100	100	10	4	26	5	125	150	310	450
131	Cote d'Ivoire	3	6	800	1000	15	5	19	7	290	200	410	300
133	Nigeria	4	5	450	605	12	3	14	12	280	350	330	360
136	Cameroon	4	6	330	480	3	8	7	11	306	983	849	1407
139	Rep. of Congo.	4	6	365	900	32	8	34	10	790	400	690	400
141	Gabon	4	7	400	500	3	4	5	4	200	1633	170	1320
142	Angola	6	8	400	500	25	5	25	7	560	400	825	400
143	Democratic Rep. of Congo	–	–	–	–	29	21	9	25	2500	2223	875	3039
147	Djibouti	3	3	270	270	13	2	11	2	295	170	320	170

150	Kennya	6	8	375	390	12	4	11	3	305	375	250	510
151	Madagascar	2	2	225	550	15	2	14	3	200	270	190	315
152	Malawi	–	–	–	–	3	4	3	3	342	243	162	143
153	Mauritius	2	2	175	175	5	1	5	1	285	75	295	100
154	Mozambique	4	5	320	400	12	2	16	2	230	250	490	340
157	Tanzania	4	7	320	540	8	4	13	5	270	250	575	250
158	Zambia	–	–	–	–	5	6	6	7	200	370	175	380
159	Zimbabwe	–	–	–	–	4	3	3	10	170	285	150	562
160	Botswana	–	–	–	–	1	0.3	0.1	0.2	179	317	67	98
161	South Africa	4	9	285	450	8	2	7	2	355	65	405	125
162	Namibia	4	5	330	470	4	5	0.1	0.3	348	745	63	145
215	Lesotho	–	–	–	–	0.1	0.2	0.1	0.2	90	150	90	150
216	Le Union	7	8	215	220	5	3	4	4	260	130	225	130
218	Swaziland	–	–	–	–	0.2	0.1	0.2	0.2	76	134	76	134
219	Seychelles	7	8	215	220	5	3	4	4	260	130	225	130
163	Australia	1	2	400	400	5	1	3	1	285	65	200	170
164	New Zealand	2	1	300	300	5	1	5	1	220	50	175	50

[a] *D.C.*, document preparation; *C.C.*, customs clearance; Grey-coloured countries: landlocked countries.
Source: Own compilation based on World Bank Group, 2015. Doing Business Website 'Trading Across Borders' 2015 Rank. (online). http://www.doingbusiness.org/data/exploretopics/trading-across-borders (Accessed 15 September 2015) and Shibasaki, R., Kawasaki, T., 2020. International intermodal container shipping network in South Asia: modelling and policy simulations. Int. J. Ship. Transp. Logist. (accepted).

References

Drewry Maritime Research, 2014. Container Forecaster 2014 Annual Review. UK, London.

Sea-Distances.org. http://www.sea-distances.org/ (online) (Accessed 31 December 2016).

SeaRates.com. https://www.searates.com/ (online) (Accessed 31 December 2016).

Shibasaki, R., Azuma, T., Yoshida, T., 2016. Route choice of containership on a global scale and model development: focusing on the Suez Canal. Int. J. Transp. Econ. 43 (3), 263–288.

Toriumi, S., 2010. Pattern analysis of containerships using maritime shipping network. J. Oper. Res. Soc. Jpn. 55 (6), 359–367. (in Japanese).

Data [2] container shipping demand for the present and future

9

Ryuichi Shibasaki
The University of Tokyo

The shipping demands for container cargo from origins to destinations (OD matrix) are preliminarily given in the models proposed in this book, as one of its essential inputs (see Chapter 5). In other words, the OD matrix should be preliminarily estimated from other data sources. This chapter first introduces how the present container OD matrix is estimated from existing sources. Then, the latter part of this chapter describes how the future container OD matrix is forecasted—a prerequisite for future simulations.

Present demand

Fig. 9.1 describes the estimation process for the current container OD matrix. As emphasised in each chapter of this book, the model targets global movement of international container cargo, including the global maritime container shipping (MCS) and regional land transport (LT) networks. Therefore, the 'global-basis' container OD matrix is taken to be the input. The World Trade Service (WTS) database by IHS Markit Ltd. (n.d.) reports the cargo shipping demand for 116 countries/regions (except for 'others', where the data of certain countries/regions are not clear) at the time of this manuscript's writing. The main feature of the WTS data—compared with other databases such as the United Nations Commodity Trade Statistics Database (United Nations, n.d.), hereafter referred to as 'UN Comtrade Database'—is that the cargo shipping demand is provided by cargo type, such as maritime container, maritime dry bulk, land/pipeline cargo, and air cargo. These data are not only provided based on value but also based on the tonnage and even TEU for container cargo.

The estimation process for container OD matrix differs by country, whether or not (hinter-)land transport network is considered (see Fig. 9.1). Because only the MCS network is considered for the countries where the LT network is not considered, maritime container OD cargo volume between the selected ports (from ports r to s), q_{rs}, should be estimated for these countries. On the other hand, for the countries where the LT network is considered, maritime container (and land container-equivalent if necessary) OD cargo volume, Q_{ij}, should be estimated on a zonal basis from region i to j. This estimation is done *only* after each country is properly divided into several zones (traffic analysis zone: TAZ) according to the boundaries of the local municipality (e.g. state, province, and prefecture). For a country pair between one where LT network is not considered and another where it is, the shipping demand in the former country is divided on a port basis, whereas that in the latter country is divided on a TAZ basis, by following each division step.

In the following part of this section, each estimation process is detailed.

Global Logistics Network Modelling and Policy. https://doi.org/10.1016/B978-0-12-814060-4.00009-5

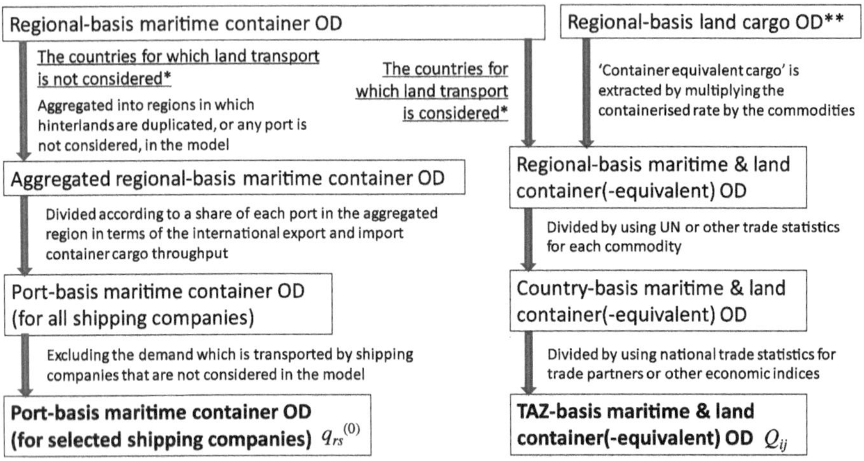

Fig. 9.1 Flowchart of the estimation process of container OD matrix. Note: *For the OD pair between regions where the LT network is considered in one side and not considered in another side, the figures on each side should be treated by following each rule, respectively. ** Optional.
Source: Own compilation.

Estimation of port-basis maritime container OD matrix

Regional aggregation

First, TEU-basis regional maritime container OD matrix, which is acquired from the WTS data where the world is divided into 116 regions as mentioned above, is aggregated into several regions. The aggregation is necessary because some regions are landlocked without any coastal lines nor seaports, or no seaports which the model would specifically consider (i.e. seaports where the international container throughput is more than 500,000 TEU/year; see Chapter 8). Regional aggregation is also required because the hinterlands of container seaports are duplicated in some regions across the national borders. In these regions, the total import/export container throughput of a country does not often coincide with the total container shipping demand of the country—even the transshipment and empty containers are subtracted from the throughput. For example, the north European ports located in the Hamburg–Le Havre range internationally compete to attract hinterland containers (see Chapters 2 and 4). In this case, the throughput of the Rotterdam port becomes larger than Dutch shipping demand.

The latest regional aggregation based on the 2013 data is summarised in Table 9.A1. There are 62 aggregated regions: Some regions such as the Arabian Gulf, North Sea, US Gulf and Carib, West Africa, and West Mediterranean are broadly aggregated because the difference between the throughput and shipping demand in each region occurs owing to hinterland duplication. Further, the authors have not focused on any simulations introduced in this book—In other words, such rough regional aggregations may be resolved in future after intensive application of the model is advanced for any region. On the other hand, as described in Table 9.A1, the container shipping demand is separated from the start into multiple coastal zones in some countries, namely, Canada, Colombia, France, Mexico, Russia, Spain, and the United States. These countries have multiple coastal zones facing different sea areas (e.g. the Atlantic and Pacific Sea). Their trade partner countries are significantly different for each coastal zone. Besides separating the United States into six regions by default in the WTS data, only TEU-basis container shipping demand, without any commodity classifications, is separately provided for other countries listed above by the WTS data.

Table 9.A1 also compares the amount of container shipping demand and throughput in each aggregated region. Whereas the container shipping demand is estimated from the WTS data, the import and export container throughput is estimated by summarising the throughput in ports (see Table 8.A1) after subtracting the amount of transshipped and empty containers. As shown in Table 9.A1, significant differences between the amount of container shipping demand and throughput are still observed in some aggregated regions. For example, the import/export container throughputs are significantly larger than the shipping demands in Hong Kong, Malta, and the Philippines as described in italic letters in the table, but they are significantly smaller than the shipping demand in Central Africa, Ukraine, and West Africa as described in bold letters.

Division into port-basis OD matrix

The aggregated regional OD matrix is then divided into port-basis OD matrix according to the port's share of the local container cargo throughput for the aggregated region. Note that the share of each port is calculated based on the total amount of export and import containers, that is, the transshipment containers are excluded because they are not related to the regional shipping demand. Another meaning is that there actually exists a difference in the shares between the export and import containers, but it is not considered due to data availability. If the difference in shares of each port between export and import containers can be considered further, the estimated OD matrix can more accurately approximate reality thus the model precision will be improved.

Furthermore, the domestic shipping demand is not considered in the models of this book. In other words, the OD cargo volume between ports in the same country should be zero. Therefore, for the OD pair which includes the port pair of export and import in the same country, the port share should be calculated after excluding these pairs. Note that the shipping demand between European countries (including Russia, Central Asian countries, and South Caucasus countries) cannot be considered in the models of this book because they are not included in the WTS data. Therefore, the OD cargo volume between European ports should be zero in the above calculation.

Excluding the shipping demand transported by companies not included in the model

The containers transported by liner shipping companies (LSCs) not included in the model are finally excluded. This is necessary for the balanced calculation of the model between the vessel capacity and number of containers shipped in each liner service. This is obtained by first subtracting the amount of shipping demand transported by LSCs, which is not considered in the model, from the total amount, based on the share of the vessel capacity of LSCs arriving at and departing from each port. Subsequently, the Frater method (see Chapter 4) is applied to adjust errors, by inputting the total amount of shipping demand for each port for the target LSCs as given and the OD matrix estimated in the first step as initial inputs. Hitherto, initial MCS demand, $q_{rs}^{(0)}$, from ports r to s is obtained.

Estimation of TAZ-basis container(-equivalent) OD matrix

Regional-basis OD matrix

For the countries to consider the LT network, the regional-basis maritime container OD matrix provided in the WTS data is again considered. Furthermore, if necessary, the land container-equivalent cargo OD volume is also added for the countries where the LT network is considered.[a] Note that the amount of 'container-equivalent land cargo' is approximated by multiplying the amount of land cargo by a maritime-based containerised rate for each commodity because the land cargo OD matrix provided in the WTS data is not categorised by cargo type unlike maritime cargo (i.e. container, dry bulk, liquid bulk, and neo bulk/general cargo). This amount is estimated for each commodity as the tonnage share of maritime containers amongst all types of maritime cargo.

Division into country-basis OD matrix

For the countries which are originally integrated into aggregated regions in the WTS database, these OD matrices should be divided into each country by using the exported/imported shares in terms of trade amount amongst the countries included. The exported/imported trade amount for each country is acquired from the UN Comtrade Database although it is provided on a value basis. Note that, in 2019, the WTS data was revised to provide each country's data separately; therefore, this manipulation is currently not necessary.

Division into TAZ-basis OD matrix

Finally, the country-basis OD matrices estimated above are divided into a subcountry level (TAZ) such as province, federal district, and oblast, based on the available statistics which represent the economy of each TAZ. Note that the more detailed zoning enables the more detailed analysis; however, this generally indicates a trade-off with the geographical resolution of the dataset (see Chapter 4).

Because the available statistics representing a subcountry level economy are very different amongst countries, the data used in each county are listed in each chapter of Part 3. In the worst case, some countries have only demographic statistics at the subcountry level. Generally, the demographic distribution is significantly different from the distribution of economic activity in the country. Furthermore, the gross regional domestic products (GRDP) is occasionally not proportional to the amount of containers demanded, especially for the countries/regions whose economy heavily depends on the production of natural resources, as discussed in Chapter 4. However, if no other indices are available, even population can be used as an index for TAZ-basis division, rather than no division at all (Shibasaki et al., 2010).

[a] In Part 3, it is considered only in the simulation in Chapter 16.

Future demand

Some policy simulations which will be introduced in Part 3 focus on the future situation based on the future cargo shipping demand, but not the current demand. This is because some policies, such as infrastructure investment, take longer to realise; thus, they are planned to meet the future demand, which is expected to significantly increase from the current amount, particularly in developing countries. Therefore, the future cargo shipping demand should be also taken as an input of the model.

There are several principles to incorporate future shipping demand. The simplest way is to use the demand estimated from others; for example, the WTS data also provides the future cargo shipping demand between countries by transport mode, cargo type, and commodity until 2030. They are partly used in the simulation in the Pacific Islands (see Chapter 14). The other example is to use the output of the Geographical Simulation Model (GSM) developed by the Institution of Developing Economies, the Japan External Trade Organisation (IDE-JETRO), and the Economic Research Institute for ASEAN and East Asia (ERIA). The GSM output is available for six trade goods, namely, agriculture, automobile, electricity, textiles, food processing, and other manufacturing, based on value, but not tonnage. Further details are available in Kumagai et al. (2013) and Isono et al. (2016). The GSM provides not only the current values but also the future values of regional trade. They are also used in the simulation for South Asia (see Chapter 12).

Another way to incorporate future demand is self-estimation. The simplest way is to use the elasticity of the GDP amount to the cargo throughput in the country (see Chapter 4). The future demand can be estimated multiplying the current demand by the elasticity and the future GDP. This is applied in the future simulation for Central America (Chapter 10). One alternative is to apply the gravity model, which can estimate the growth rate differently for each OD pair, as described in Chapter 4; it is also partly applied in Chapter 14.

A more complicated, but sophisticated, method is to use the output of an economic model, such as the GTAP model (see Chapter 4). The GTAP model can be applied to forecast the future amount of bilateral trade by using a recursive dynamic approach that incrementally inputs the increasing rate to the next term of each economic index, including the primary factors, total factor productivity, other coefficients on technological advancement, and the accumulated capital in the previous term. These stepwise calculations can be done by repeatedly inputting them into the static GTAP model (RunGTAP) by using the command 'use updated database from the last simulation', or using the RunDyn software, which packages the recursive calculation of the GTAP model during the designated years. Although no example of the application of the GTAP model is included in this book, the authors used it to forecast the future trade amount. For example, Shibasaki and Watanabe (2012) forecasted the future amount of trade in each member economy of the APEC until 2025 by using four future scenarios based on economic indices in each APEC member economy and international economic policies. Fig. 9.2 graphically illustrates the example of the output on the future trade amount by Shibasaki and Watanabe (2012).

Note that the outputs of the economic model, including the GSM and the GTAP model, are generally described on a value basis, but not tonnage (or TEU) basis. Therefore, the

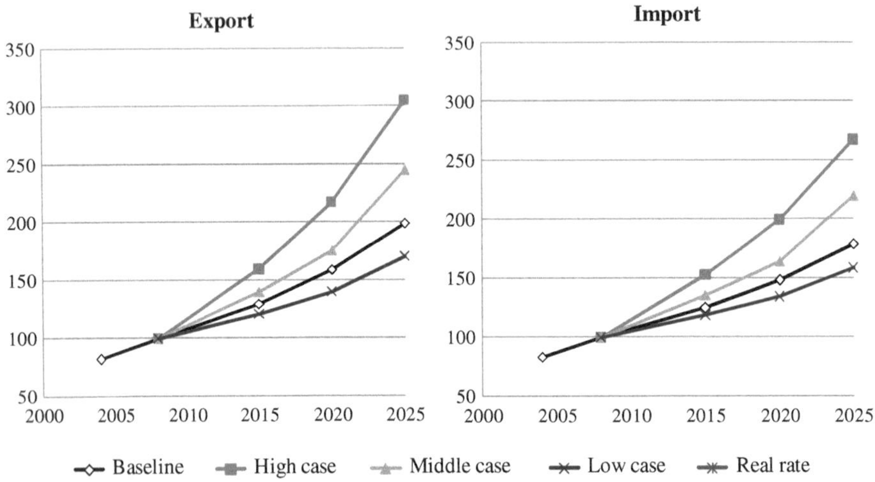

Fig. 9.2 Example of the future trade amount in the APEC region estimated using the GTAP model by economic scenario (2008 estimated results = 100).
Source: Shibasaki, R., Watanabe, T., 2012. Future forecast of trade amount and international cargo flow in the APEC region: an application of trade-logistics forecasting model. Asian Transp. Stud. 2(2), 75–89.

outputs should be converted into tonnage (or TEU) (see Chapter 4). Moreover, when increasing the shipping demand to be input into the model, the capacity of MCS and LT network should be increased to meet the increased demand (see Chapter 8).

Conclusion

The cargo shipping demand is a vital input of the simulation models introduced in this book. Based on the authors' experience, the model output's precision improves as the OD matrix is more precise. Particularly, the division of the OD matrix into the TAZ level from the country level considering a characteristic of each zone is important. The regional distribution of the container and container-equivalent cargo shipping demand should be paid very careful attention because it is generally different from other types of cargo. It is also important to estimate the future shipping demand. If considering the regional difference in the increasing rate of demand, extreme differentiation is better to be avoided because it could severely deteriorate the balance of the current shipping network.

When considering further expansions of the model, a more detailed or expanded cargo shipping demand should be estimated accordingly. For example, the container shipping demand by commodity should be established when creating a model which can consider the difference in the contents of the cargo with different values of the time. Further, the cargo shipping demand transported by cargo types other than containers (i.e. air transport, pipeline, bulk carrier, or tanker) should also be established if the model is expanded to include such types of cargo. The abovementioned demands can be estimated from the WTS data within the range of provided commodity and cargo-type classification.

Appendix

Table 9.A1 Regional aggregation of container ports generally included in the model simulations and comparison in terms of annual throughput (example of 2013).

Country or region after aggregation	Annual export and import laden container throughputs (2013, thousand TEU)			Equivalent countries and regions in the WTS data	Number of ports included
	A. Amount from WTS data	B. Amount by summarising throughputs described in Table 8.A1 (excluding transshipment and empty containers)	B/A		
Angola	425	400	0.94	Angola	1
Arabian Gulf	10,336	10,734	1.04	Bahrain; Central Asia; Kuwait; Other Western Asia; Qatar; Saudi Arabia; Southern Arabian Peninsula; United Arab Emirates	8
Argentina	1203	1126	0.94	Argentina	1
Australia	4023	4656	1.16	Australia; Pacific Islands	4
Baltics	2421	1724	0.71	Baltics; Belarus; Russia Baltics	1
Bangladesh	1190	1128	0.95	Bangladesh	1
Brazil	5137	5228	1.02	Brazil; Other Northeast Coast of South America	7
Canada Atlantic Coast	1447	954	0.66	Canada Atlantic Coast; Great Lakes (United States)	1

Continued

Table 9.A1 Continued

Country or region after aggregation	Annual export and import laden container throughputs (2013, thousand TEU)			Equivalent countries and regions in the WTS data	Number of ports included
	A. Amount from WTS data	B. Amount by summarising throughputs described in Table 8.A1 (excluding transshipment and empty containers)	B/A		
Canada Pacific Coast	1966	2344	1.19	Canada Pacific Coast	2
Central Africa	610	360	**0.59**	Central Africa—North; Central Africa—South	1
Chile	2127	1770	0.83	Bolivia; Chile	3
China	48,652	70,694	1.45	China	13
Colombia Atlantic Coast	889	623	0.70	Colombia Atlantic Coast;	1
Colombia Pacific Coast	243	369	1.52	Colombia Pacific Coast	1
Costa Rica and Panama	1092	1184	1.08	Costa Rica and Panama	3
East Africa—Centre	773	522	0.68	East Africa—Centre; Kenya	1
East Africa—North	631	465	0.74	East Africa—North	1
Ecuador	772	1050	1.36	Ecuador	1
Egypt	2434	2223	0.91	Egypt; Libya	4
Finland	569	430	0.76	Finland	1
France Mediterranean	655	669	1.02	France Mediterranean	1
France/Spain North Atlantic	2361	2189	0.93	France Atlantic/North Sea; Spain North Atlantic	2
Greece	618	433	0.70	Albania; Cyprus; Greece; Other Europe	1
Hong Kong	3352	7038	*2.10*	Hong Kong	1

India	5053	6732	1.33	India	5
Indian Subcontinent Islands	518	825	1.59	Indian Subcontinent Islands	1
Indonesia	6715	7215	1.07	Indonesia	3
Ireland	279	353	1.27	Ireland	1
Israel	1395	1646	1.18	Israel	2
Italy	3204	2872	0.90	Italy	5
Japan	11,561	11,576	1.00	Japan	7
Malaysia	4225	4778	1.13	Malaysia	4
Malta	56	90	*1.61*	Malta	1
Mexico and Central America Pacific	2585	2544	0.98	Mexico Pacific; Belize and Guatemala; El Salvador, Honduras, and Nicaragua	3
Mexico Gulf Coast	1226	1068	0.87	Mexico Gulf Coast	2
New Zealand	1337	1172	0.88	New Zealand	2
North Sea	15,223	18,542	1.22	Austria; Belgium; Czech Republic; Denmark; Germany; Netherlands; Norway; Slovak Republic; Sweden; Switzerland	6
Other Mediterranean	1021	1368	1.34	Other Mediterranean	2
Other Southeast Coast of South America	444	548	1.23	Other Southeast Coast of South America	1
Pakistan	1430	1724	1.21	Afghanistan, Bhutan, and Nepal; Pakistan	2
Peru	1253	1284	1.02	Peru	1
Philippines	2016	3366	*1.67*	Philippines	3
Poland	826	808	0.98	Poland	1

Continued

Table 9.A1 Continued

Country or region after aggregation	Annual export and import laden container throughputs (2013, thousand TEU)			Equivalent countries and regions in the WTS data	Number of ports included
	A. Amount from WTS data	B. Amount by summarising throughputs described in Table 8.A1 (excluding transshipment and empty containers)	B/A		
Romania	486	411	0.85	Bulgaria; Moldova; Romania	1
Russia Black Sea	680	475	0.70	Russia Black Sea; South Caucasus	1
Singapore	3900	3763	0.96	Singapore; Other Southeast Asia	1
Slovenia	435	389	0.90	Croatia; Hungary; Slovenia	1
South Korea	10,801	10,761	1.00	South Korea; Other Northeast Asia; Russia Pacific	5
Southern Africa	2740	2580	0.94	East Africa—South; Southern Africa; Southern African Islands	3
Taiwan	5652	6852	1.21	Taiwan	4
Thailand	6105	5190	0.85	Thailand	2
Turkey	3135	2749	0.88	Turkey	3
Ukraine	633	347	**0.55**	Ukraine	1
United Kingdom	3717	4662	1.25	United Kingdom	4

US Gulf and Carib	11,051	9745	0.88	Greater Antilles, Bahamas, and Bermuda; Gulf (United States); Lesser Antilles; South Atlantic (United States)	11
United States_North Atlantic	6993	4337	0.62	North Atlantic (United States)	2
United States_North Pacific	2474	2415	0.98	North Pacific (United States)	2
United States_South Pacific	11,052	11,806	1.07	South Pacific (United States)	3
Venezuela	541	425	0.79	Venezuela	1
Vietnam	4755	5399	1.14	Vietnam	3
West Africa	3309	1511	**0.46**	Benin and Togo; Burkina Faso, Mali, and Niger; Cote dIvoire; Ghana; Nigeria; Other Western Africa; Senegal	3
West Mediterranean	5112	5684	1.11	Algeria; Morocco; Portugal; Spain Med/ South Atlantic; Tunisia	9
Total	237,861	266,325	1.12	116 zones (except for 'others', in which certain countries/regions are not clear)	173

Source: Own compilation.

References

IHS, Inc., World Trade Service (WTS) Database (Accessed 21-05-2015).

Isono, I., Kumagai, S., Hayakawa, K., Keola, S., Tsubota, K., Gokan, T., 2016. Comparing the economic impacts of Asian integration by computational simulation analysis. In: IDE Discussion Paper No.567.

Kumagai, S., Hayakawa, K., Isono, I., Keola, S., Tsubota, K., 2013. Geographical simulation analysis for logistics enhancement in Asia. Econ. Model. 34, 145–153.

Shibasaki, R., Watanabe, T., 2012. Future forecast of trade amount and international cargo flow in the APEC region: an application of trade-logistics forecasting model. Asian Transp. Stud. 2 (2), 75–89.

Shibasaki, R., Watanabe, T., Araki, D., 2010. How is model accuracy improved by usage of statistics? An example of international freight simulation model in East Asia. Asian Transp. Stud. 1 (1), 33–46.

United Nations, UN Comtrade Database. https://comtrade.un.org/ (Accessed May 21, 2015).

Part Three

Application to the developing world

Central America: Small countries with active border-crossing transport on land

10

Ryuichi Shibasaki[a], *Takashi Kadono*[b], *and Taiji Kawakami*[c]
[a]The University of Tokyo, [b]NEWJEC Inc., [c]Ministry of Land, Infrastructure, Transport and Tourism, Japan

Introduction

Central America (CAM) is the southern isthmian portion of the North American continent that connects it to the South American continent. The region is bordered by Mexico to the north, Colombia to the southeast, the Caribbean Sea to the east, and the Pacific Ocean to the west. It consists of seven countries: Belize, Costa Rica, El Salvador, Guatemala, Honduras, Nicaragua, and Panama. One of the unique geographic characteristics of the region is that there are two coastlines, one touching the Atlantic (Caribbean) and the other touching the Pacific, with many ports serving as points of connection with various trading partners.

The total volume of international container cargo handled in this region is smaller than that of other regions, but the unique geographic circumstances of CAM has led to an imbalance between the two coastlines in terms of container traffic at the ports. The ports on the Pacific side are located on a trunk line connecting the Pacific coast of North America with South America and the Atlantic coast of North America with East Asia, because of the access provided by the Panama Canal at the southernmost part of the region (see Fig. 10.1). However, the volume of containers handled in these ports is relatively small. By contrast, the ports on the Atlantic side are located within the inner part of the Gulf of Mexico, far from the trunk line, but the volume of containers handled in the ports is significantly larger than that of the Pacific ports because the major trade partners—the US East Coast ports—are located across the Mexican Gulf. This imbalance in the maritime container shipping (MCS) network requires the land transport of cargo across national borders.

This chapter reviews the conditions of local ports, MCS flows, and liner services (LSs) in the region. It also presents a case study in which a scenario is applied to cargo flow analysis using a simulation model customised for the region. In particular, the scenario focuses on La Union port, newly constructed in El Salvador. Because La Union port has a long access channel that requires constant maintenance dredging, it is important to set the channel depth properly, considering the characteristics of the LSs provided by the different sizes of containership.

Global Logistics Network Modelling and Policy. https://doi.org/10.1016/B978-0-12-814060-4.00010-1

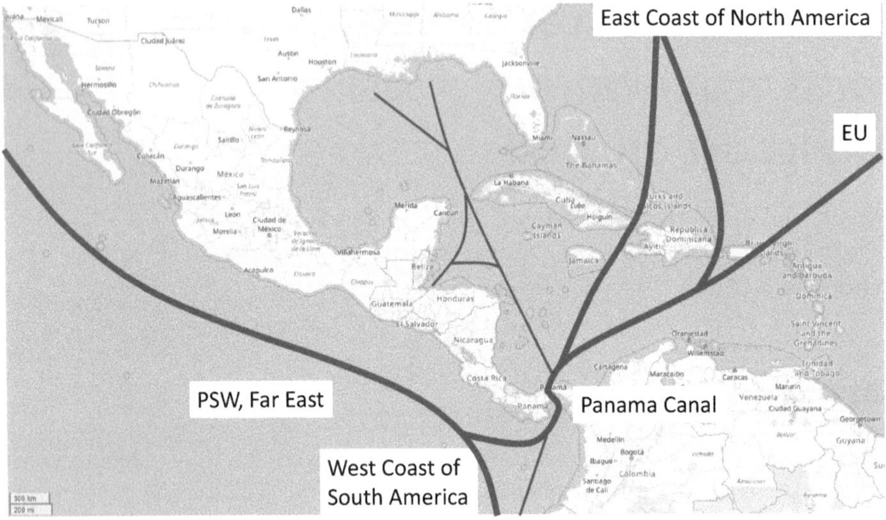

Fig. 10.1 Location of CAM and surrounding MCS routes.
Source: Own compilation based on OpenStreetMap.

Ports and maritime container cargo movements in Central America

Fig. 10.2 depicts the locations of major container ports in the CAM region with their throughputs in 2016. The container throughputs of Colon–Cristobal and Balboa ports, which are at either end of the Panama Canal and function as international hubs, show distinctive features (more than 3 million TEU/year; see the left of Fig. 10.3). Manzanillo and Lazaro Cardenas ports, which are located off the Pacific Coast of Mexico, also function as regional hubs along the trunk routes connecting the Panama Canal/South America and North America/East Asia.

Between Panama and Mexico, there are five CAM countries (CAM5): Guatemala, Honduras, El Salvador, Nicaragua, and Costa Rica. As shown in Fig. 10.2, Puerto Limón, on the Caribbean coast of Costa Rica, ranked first amongst the CAM5 container ports in 2016, with a throughput of 1,177,386 TEUs. It was followed by Puerto Cortés (Honduras, hereinafter, 'Cortés'), Santo Tomás de Castilla (Guatemala, hereinafter, 'Sto. Tomás'), and Puerto Barrios (Guatemala), which are all located on the Caribbean coastline. Puerto Quetzal (Guatemala, hereinafter 'Quetzal'), the largest container port amongst the Pacific coastal ports in the CAM5 countries, is the fifth largest overall in the CAM5 countries. The left of Fig. 10.4 shows the changes in container throughput of the CAM5 ports from 1991 to 2010. Container movements in the region have been increasing, and container throughput at the region's ports is reflective of this trend. The right of Fig. 10.4 also shows the change in container throughput of the CAM5 ports by the coast. Approximately four-fifths of the container throughput of the CAM5 ports is handled by ports on the Caribbean coast, with the remaining one-fifth handled by ports on the Pacific coast.

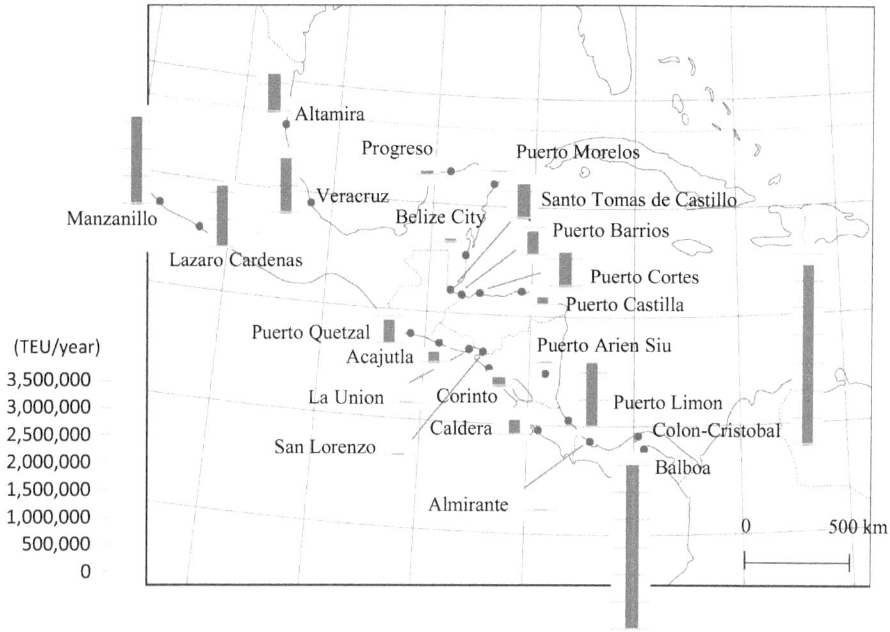

Fig. 10.2 Major container ports in CAM and their throughput (2016).
Source: Own compilation based on COCATRAM and other websites.

Fig. 10.3 Examples of CAM container ports (2013). Note: Balboa (Panama, *left*) is a fully equipped modern transshipment hub, whilst in Acajutla (El Salvador, *right*) containers are loaded and discharged by ship gears because it has neither dedicated terminals nor quay cranes.
Source: Authors.

Amongst the ports on the Pacific coast, Quetzal is the largest, followed by Cardela (Costa Rica), Acajutla (El Salvador), and Corinto (Nicaragua) (see the left of Fig. 10.4). San Lorenzo (Honduras) has only handled containers since 2011, and La Union (El Salvador) also started an LS in 2011 but it was suspended in 2013. Interviews with local liner shipping companies (LSCs) along the CAM5 Pacific coast produced several findings. First, almost all LSCs held the view that the ports on the CAM5 Pacific

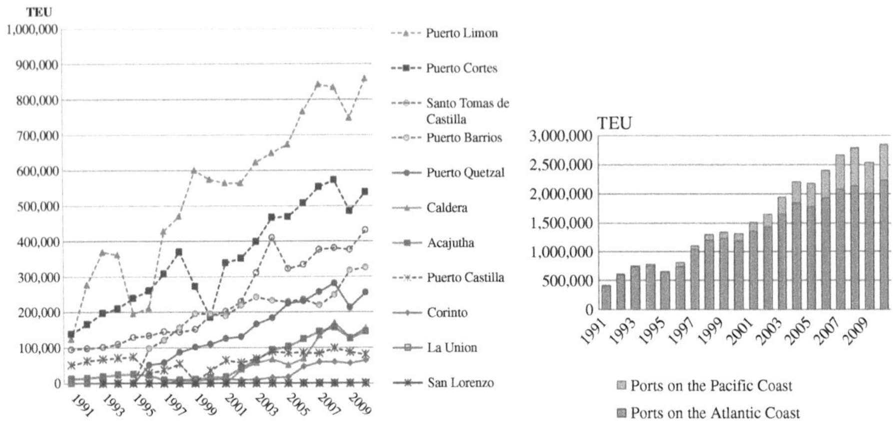

Fig. 10.4 Historical container throughput at each port *(left)* and by coast *(right)* in CAM5. *Source:* Japan International Cooperation Agency (JICA), 2014. Final report for special technical assistance for maintenance dredging of the port of La Union in the Republic of El Salvador.

coast are positioned as feeder ports in Mexico- or Panama-based LS networks (see the right of Fig. 10.3 which shows Acajutla port as an example of a feeder port). Second, many LSCs provided weekly services to Quetzal, Acajutla, Corinto, and Caldera, although some LSCs selected one or a few ports of them as calling ports according to expected cargo volume. Third, many LSs were carried out through joint operations by several LSCs because the cargo volume that one LSC could collect was not enough for it to operate by itself. Fourth, various sizes of containerships were deployed in the LS of this region. As of 2013, the largest vessel capacity was 2800 TEU (LOA: 200 m, draught: 11 m) and the smallest was 670 TEU (draught 8.5 m). Several LSCs had a view that the size of the vessels deployed in this region would not be affected by the expansion of the Panama Canal which was completed in 2016. Fifth, all LSCs pointed out that the CAM market is small and would not expand rapidly in the near future. Most cargoes exported to or imported from the EU and the east coast of North America use ports on the Caribbean coast and such cargoes are not often transported through the Panama Canal. Finally, LSs provided along the CAM5 Pacific coast were classified into three patterns: (i) a feeder service under a Mexican or North American west coast port-based LS network, under which vessels call at ports in every country; (ii) a feeder service under a Panamanian port-based LS network, under which vessels call at ports in every country; and (iii) a 'way-port' service under the route between Asia and North/South American west coast ports and a feeder service network, under which vessels of the way-port service call at selected port(s).

Fig. 10.5 shows the relationship between the seaports used for export/import and the origin/destination countries of containers in this region by the trading partner. This is estimated from the statistics provided by the CAM Commission of Maritime Transport (COCATRAM) (n.d.) and additional knowledge acquired from the interviews with the LSCs.

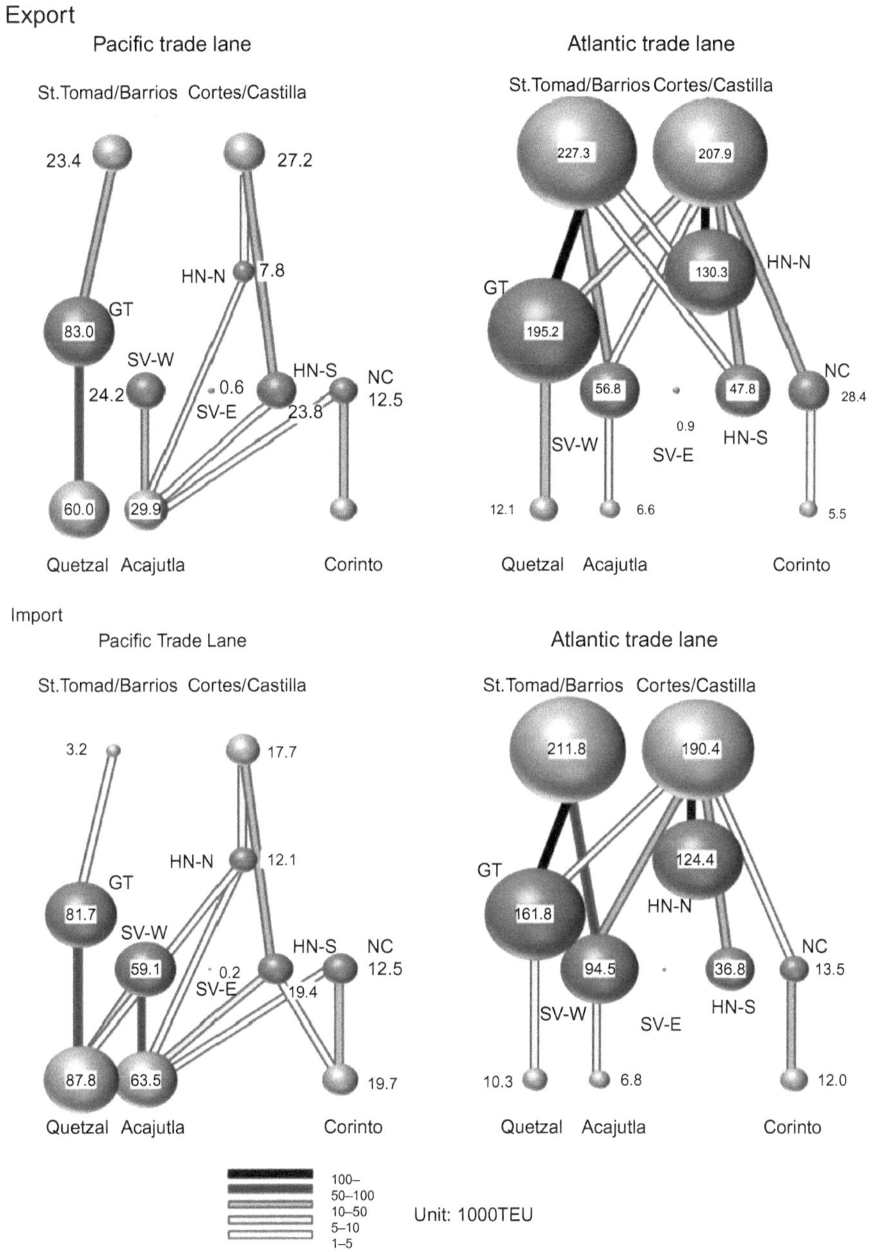

Fig. 10.5 Gateway seaports and estimated flow in 2010 for containers to/from CAM4 countries (Guatemala, El Salvador, Honduras, and Nicaragua) by trade partner.
Source: Lacayo, M., Romero, J., Callejas, P., Canales, M., Castillo, A., Kadono, T., Shibasaki, R., 2014. Analysis on changes of container market share Among Central American ports caused by Salvadoran port development. In: 33rd PIANC Congress, 1–5 June 2014, San Francisco, CA.

Fig. 10.6 Trailers across land national borders in CAM (2013). Note: Transiting a narrow road in a mountainous area near a border between El Salvador and Honduras *(left)* and a queue waiting to enter Honduras from El Salvador on the Pacific coast *(right)*.
Source: Authors.

Regarding the role of the ports on each coast, cargoes to/from Asia and the west coast of North and South America are mainly exported or imported through the ports on the Pacific coast. Meanwhile, cargoes to/from Europe and the east coast of North and South America are generally exported or imported through the ports on the Caribbean coast. In other words, in principle, cargoes to/from CAM5 countries are not transported via the Panama Canal.

In general, cargoes produced and consumed in a country are exported or imported mainly through the ports of that country. However, most Salvadoran cargoes to/from Europe and the east coast of North and South America are transported across land national borders and exported or imported through the ports of Honduras (Cortés) and Guatemala (Sto. Tomás) (see Fig. 10.6) because El Salvador has no coast along the Caribbean Sea. In addition, certain Nicaraguan cargoes to/from Europe and the east coast of North and South America are exported or imported through the ports of Honduras (Cortés) and Costa Rica (Puerto Limón) because, whilst Nicaragua has a long coastline along the Caribbean Sea, it has no major coastal ports along that coast. Similarly, some Honduran cargoes to/from Asia and the west coast of North and South America are exported or imported through the ports along the Pacific coast of neighbouring countries, such as Acajutla in El Salvador, because there are only small ports in Honduras along the Pacific coast.

Data

Ports

The model in this chapter is developed as of 2010. In addition to the world's major container ports with throughputs of more than 500,000 TEU/year, including transshipment and empty containers (as discussed in Chapter 8), several other ports are considered for a detailed regional analysis: Quetzal, Acajutla, La Union, Corinto, Cardela, and Sto. Tomás (see Fig. 10.7), as well as one port from the United States

Fig. 10.7 Ports and HT network in CAM adopted in the model.
Source: Own compilation based on Japan International Cooperation Agency (JICA), 2014 Final report for special technical assistance for maintenance dredging of the port of La Union in the Republic of El Salvador.

(New Orleans). Note that La Union port is also included for future scenario simulation although its container cargo throughput was zero as of 2010. The container handling charge for export and import, CPX_r and CPM_s, in the ports of four CAM countries (CAM4: Guatemala, El Salvador, Honduras, and Nicaragua), where the hinterland transport (HT) is considered, are set as shown in Table 10.1 by scrutinising the tariff of each port, regardless of the setting shown in Table 8.A2 of Chapter 8. The transshipment time, TPR_a, for each CAM4 port, which is estimated by the authors, is also shown in Table 10.1.

MCS network

The MCS network is made for the 20 largest LSCs of the world, as shown in Chapter 8, plus an additional eight small and medium-sized LSCs (CCNI, CFS, Dole, Great White Fleet, Horn Line/Network Shipping, Gulf Africa Line/Nordana, Seaboard, and Streamline) that have an LS network in CAM. As a result, of the 2857 LSs from the MDS Containership Databank (as of May 2010), 809 LSs are included in the model. Although the number of LSs included in the model is less than 30% of the total, 58% of the annual vessel capacity of the world is covered by the model, because larger LSCs provide more significant LSs across the world.

Table 10.2 shows the LSs that called regularly at Acajutla port, the focus of the policy simulation section, as of May 2010. Of these, four LSs operated by single LSC provided a feeder service from transshipment port(s) located in Mexico and/or Panama

Table 10.1 Settings of container handling charge and transshipment time in CAM4 ports.

No.	Port name	Country	Container handling charge for export CPX_a (US$/TEU)	Container handling charge for import CPM_a (US$/TEU)	Transshipment time TPR_a (h)
65-1	Quetzal	Guatemala	117.65	117.65	24
65-2	Acajutla	El Salvador	73.48	73.48	48
65-3	La Union	El Salvador	65.79	65.79	24
65-4	Corinto	Nicaragua	58.82	58.82	48
69	Cortés	Honduras	64.70	64.70	24
69-1	Sto. Tomás	Guatemala	64.70	64.70	48

Source: Own compilation.

Table 10.2 Existing LSs calling at the port of Acajutla as of May 2010.

No	Service name	Operator (s)	Average TEU capacity	Frequency	Calling ports
1	WCCA	Maersk	1695	Twice a week (westbound and eastbound)	Los Angeles—Manzanillo (Mexico)—Lazaro Cardenas—Quetzal—Acajutla—Corinto—Balboa—Caldera—Corinto—Acajutla—Quetzal—Lazaro Cardenas—Manzanillo (Mexico) —Los Angeles
2	MAYA	MSC	1232	Biweekly	Quetzal—Balboa—Caldera—Acajutla—Quetzal
3	MALEX	NYK	1610	Weekly	Manzanillo—Quetzal—Acajutla—Corinto—Caldea—Quetzal—Manzanillo
4		APL	1118	Weekly	Lazaro Cardenas—Acajutla—Quetzal—Lazaro Cardenas
5	ACSA	CMA-CGM/CSCL/CCNI	2516	Weekly	Japanese ports—Hong Kong—Chinese ports—Busan—Manzanillo (Mexico)—Quetzal—Acajutla—Corinto—Buenaventura (Colombia)—Peru ports—Chile ports—Quetzal—Manzanillo (Mexico)—Japanese ports
6	ANDEX2	CSAV	2599	4 times every 9 weeks (in other 5 times calling at Caldera instead of Acajutla)	Manzanillo (Mexico)—Lazaro Cardenas—Acajutla—Buenaventura (Colombia)—Ecuador ports—Lazaro Cardenas—Manzanillo (Mexico)

Source: Own compilation based on Japan International Cooperation Agency (JICA), 2014. Final report for special technical assistance for maintenance dredging of the port of La Union in the Republic of El Salvador.

using smaller vessels. The other two LSs (No. 5 and 6) provided a way-port service using relatively larger vessels coming from East Asian countries, such as China and Japan, or travelling from the west coast of North America to the west coast of South America, including Colombia, Ecuador, Peru, and Chile.

HT network

In the following model, the HT network is only considered for CAM4 countries. Note that Costa Rica is excluded because the cargo to/from Costa Rica is considered to be shipped through the country's own ports, located on both coasts. The HT network is shown in Fig. 10.7. Given their geographical characteristics, El Salvador and Honduras are each divided into two regions. All representative nodes (O and D nodes) of the region and all CAM4 ports are connected by road links with each other, whilst any land connections between CAM4 and neighbouring countries (e.g. Mexico and Costa Rica) are excluded. The driving time and cost of each link are set based on interviews with forwarders in CAM, as shown in Table 10.3. The additional time and costs are considered when crossing land national borders (see Table 8.A2 of Chapter 8) and the coefficient of the land border barrier, λ_a, which is multiplied by the border-crossing time and cost [see Eq. (7.23)], is set at 0.3 in the simulations in this chapter. In some links, more than two land borders must be crossed. Note that the capacity and congestion in the HT network are not considered in this model.

Container cargo shipping demand (OD matrix)

The demand for container cargo shipping, Q_{ij}, from region i to j as of 2010 is estimated as follows. First, the demand for container cargo shipping (OD matrix) between countries or regions on a TEU basis is obtained from the World Trade Service (WTS) database, as with many cases introduced in this book.

Second, the country-basis OD matrix (except for the cargo to/from CAM) acquired above is converted into a port-basis (i.e. MCS-basis) OD matrix according to the ports' respective share of the local container cargo throughput of the country/region. Meanwhile, the OD matrix to/from the CAM region is first divided by coast (Caribbean or Pacific) by using the data shown in Fig. 10.5, because the characteristics of each coast are different, as described in the previous section, and then allocated to each port. Note that this division is used for estimating the initial port-basis OD matrix, $q_{rs}^{(0)}$, which is one of the model inputs as explained in Chapter 7.

Finally, the port-basis OD matrix to/from the ports in CAM4 countries is integrated and then allocated again to each country to include the HT network. The integrated OD matrix is allocated according to trade value by partner region using the UN Comtrade Database and data provided by the Secretariat for CAM Economic Integration (SIECA) (n.d.). In addition, Salvadoran and Honduras cargoes are each separated into two zones by a constant ratio (El Salvador West: 94%, El Salvador East: 6%; Honduras North: 70%, Honduras South: 30%).

A future container OD matrix for the year 2030 is estimated by multiplying the current OD matrix by multipliers set for each CAM4 country by imports/exports, as

Table 10.3 Driving time and cost between zone representatives and ports.

Ports	Guatemala		El Salvador		Honduras	Nicaragua
Zone representative	**Quetzal**	**Sto. Tomás**	**Acajutla**	**La Union**	**Cortés**	**Corinto**
Driving time (h)						
Guatemala	1.7	5.0	3.2	7.0	6.3	11.0
El Salvador West	4.5	6.4	1.4	3.1	6.8	7.1
El Salvador East	6.7	8.6	3.7	0.8	6.7	4.9
Honduras North	8.6	2.5	7.2	5.5	1.0	8.6
Honduras South	12.0	6.0	6.8	3.1	5.0	4.6
Nicaragua	14.1	12.3	11.1	7.3	11.3	2.3
Driving cost (US$/TEU)						
Guatemala	151.5	450.0	286.5	630.0	568.5	990.0
El Salvador West	403.5	571.5	129.0	277.5	612.0	637.5
El Salvador East	603.0	774.0	328.5	73.5	600.0	439.5
Honduras North	774.0	220.5	649.5	495.0	88.2	771.0
Honduras South	1080.0	541.5	615.0	274.5	451.5	417.0
Nicaragua	1270.5	1102.5	996.0	655.5	1012.5	207.0

Source: Shibasaki, R., Iijima, T., Kawakami, T., Kadono, T., Shishido, T., 2017. Network assignment model of integrating maritime and hinterland container shipping: application to Central America. Marit. Econ. Logist. 19(2), 234–273.

Table 10.4 Multipliers set for the 2030 container OD matrix from the 2010 matrix.

	Export	Import
El Salvador	7.76	3.95
Guatemala	2.77	2.78
Honduras	2.48	2.52
Nicaragua	3.49	3.43

Source: CEPA.

shown in Table 10.4. These multipliers were estimated by the Executive Autonomous Port Commission (CEPA), by multiplying the forecasted GDP growth rate (provided by the IMF and other sources) by the current elasticity of MCS demand against the GDP growth rate acquired. Note that the OD matrix between any countries of the world other than CAM4 countries does not change from the amount in 2010 in the future simulations of this chapter. This can be justified because the MCS market in CAM4 countries is sufficiently small compared to the entire MCS market of the world.

Calculation results

Container cargo throughput

The authors calibrated the distribution parameter for stochastic assignment, θ, to be 0.01, and time value for the shipper, vt, to be 8.0 (US$/TEU/h) to best fit the estimation results to the observed data. Fig. 10.8 shows a comparison of the observed and model-estimated annual laden container throughput for export, import, and transshipment in five ports of CAM4 countries. The figure shows that the developed model well describes the cargo throughput in each port in terms of the export, import, and transshipped laden containers; namely, both Atlantic ports have a larger throughput than do the three Pacific ports, whilst of the Pacific ports, Quetzal is the largest, followed by Acajutla and Corinto.

Fig. 10.9 also compares the observed and model-estimated shares of export and import laden containers by the location of partner countries or regions (i.e. Atlantic countries/regions, including the east coast of North and South America and Europe; Pacific countries/regions, including the west coast of North and South America and East Asia). The observed share is derived from the results shown in Fig. 10.5. Fig. 10.9 indicates that in most cases, each port is more specialised in a certain direction in the model estimation (i.e. the Pacific ports specialise in trade with Pacific countries/regions, whilst the Atlantic ports specialise in trade with Atlantic countries/regions). This indicates that the laden container is more sensitive to shipping costs and time in the model. The Atlantic export cargo from Quetzal is exceptionally noteworthy in this regard; the Atlantic export cargo partly includes cargo exported to European countries to which the distance from Pacific and Atlantic ports in CAM is not significantly different.

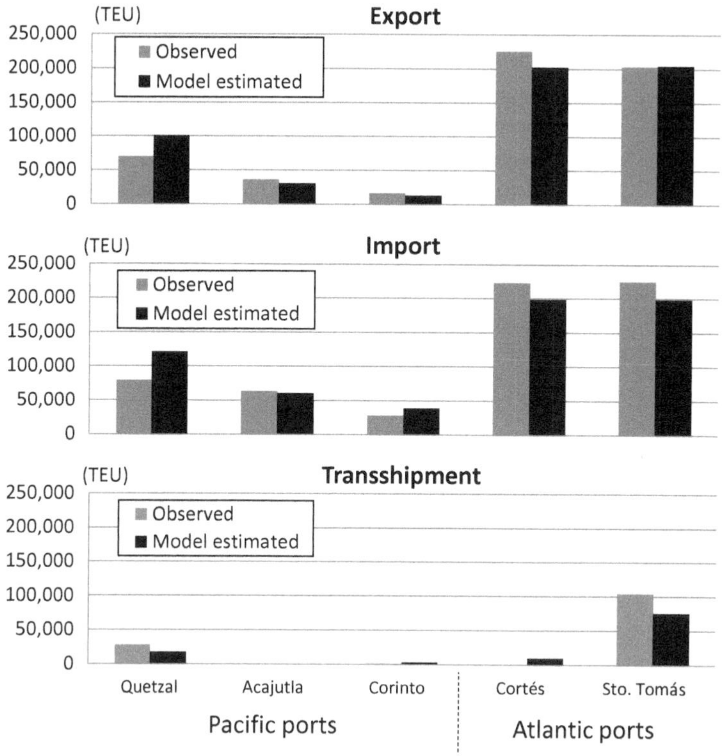

Fig. 10.8 Comparison of the observed and model-estimated annual laden container throughput for ports in CAM4 countries.
Source: Own compilation based on Japan International Cooperation Agency (JICA), 2014. Final report for special technical assistance for maintenance dredging of the port of La Union in the Republic of El Salvador.

Land container flow

Fig. 10.10 shows a comparison of the observed and model-estimated annual land laden container flow between ports and ODs in CAM4 countries. The observed flow by each combination is also derived from Fig. 10.5. The model can effectively predict the observed flows, judging from the correlation coefficient shown for both exports and imports although some differences are observed between the observed data and the model-estimated results.

Container flow by LSCs

The model can calculate the amount of MCS either by LSC or by service. Fig. 10.11 shows the observed share (which was provided by the CEPA) and model-estimated share in laden container throughputs by LSCs in Acajutla port. The figure also shows the share in vessel capacity by LSCs to call at Acajutla port for reference. The figure indicates

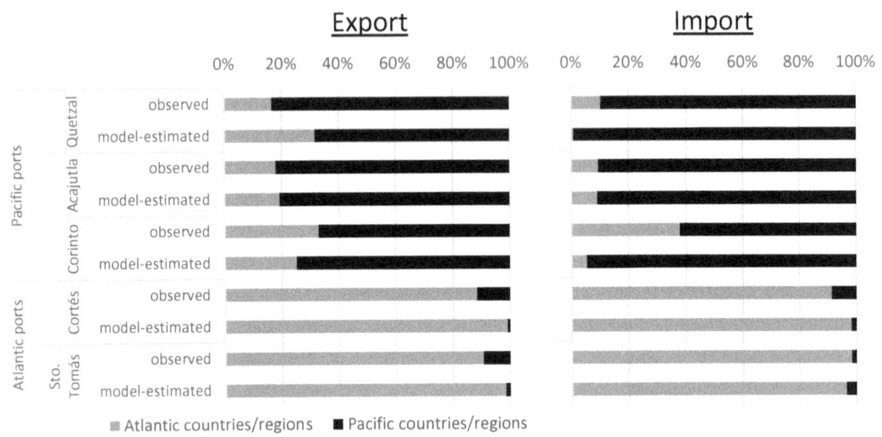

Fig. 10.9 Comparison of the observed and model-estimated share of laden containers by partner country/region for ports in CAM4 countries.
Source: Own compilation.

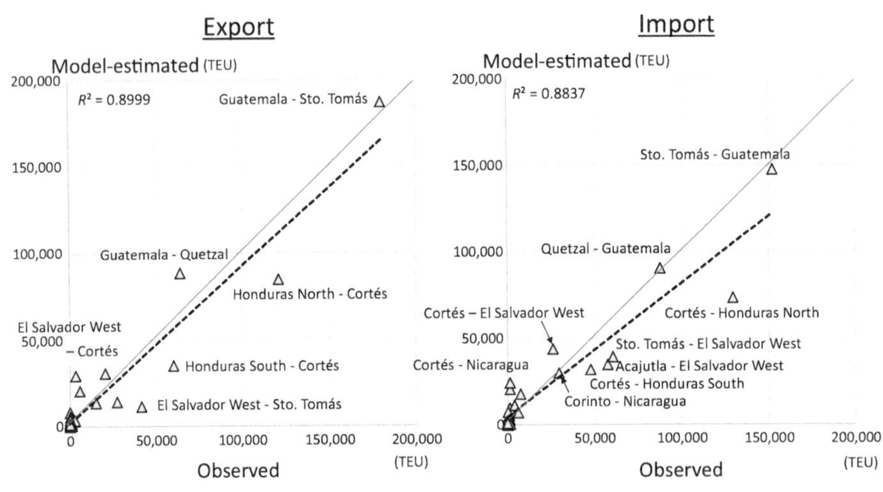

Fig. 10.10 Comparison of the observed and model-estimated annual laden container flow between ports and origin/destination in CAM4 countries.
Source: Own compilation based on Shibasaki, R., Iijima, T., Kawakami, T., Kadono, T., Shishido, T., 2017. Network assignment model of integrating maritime and hinterland container shipping: application to Central America. Marit. Econ. Logist. 19(2), 234–273.

that the model-estimated throughput share is closer to the observed one than the share in vessel capacity for most LSCs except for NYK. This may imply that the model is useful from the viewpoint of analysing the efficiency of LSCs. Some LSCs, such as APL, can be said to be operating efficiently because their throughput share exceeds the vessel capacity share, whilst other LSCs, such as MSC, CSAV, and China Shipping, can be said to be operating inefficiently because they are handling far less than their capacity.

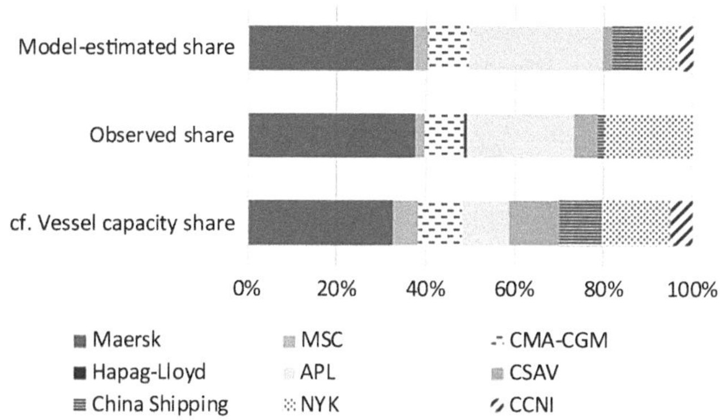

Fig. 10.11 Comparison of the observed and model-estimated shares in annual laden container throughput of LSCs in Acajutla port (2010).
Source: Own compilation based on Shibasaki, R., Iijima, T., Kawakami, T., Kadono, T., Shishido, T., 2017. Network assignment model of integrating maritime and hinterland container shipping: application to Central America. Marit. Econ. Logist. 19(2), 234–273.

Policy simulations

Opening La Union port

In 2012, there was one LS, jointly operated by APL and Hamburg-Sud, calling weekly at La Union port, with no LS previously available. It started from Lazaro Cardenas (Mexico) and called in sequence at La Union, Caldera, Balboa, Paita (Peru), Guayaquil (Ecuador), Balboa, and Quetzal, before returning to Lazaro Cardenas. The vessel capacity was 1324 TEU, with a maximum draught of 9.5 m. In the following simulation, this service is added to the 2010 MCS network and container cargo shipping demand as of 2010 is input. Because the depth of the access channel of La Union port is insufficient, additional waiting time for a vessel intending to enter the port is considered to be 3.3 (hours), according to a calculation based on the tidal wave simulation.

Fig. 10.12 indicates the increased or decreased amount of annual laden containers in this scenario (S1) in CAM4 ports from the base scenario which result is shown in the previous section for reproducing the actual situation. The estimated annual amount of laden containers in La Union port is 3563 and 10,265 TEU for exports and imports, respectively, which are zero in the base scenario. The observed figures for the containers in 2012 were 8081 and 10,317 TEU, according to the statistics, but these included empty containers. Because many empty containers may be included in the export containers for repositioning imported containers and offsetting the imbalance of imports and exports, the authors consider that the model can predict the number of laden containers in La Union port as a whole.

Another finding from the figure is that laden containers using La Union port are mainly shifted from foreign Pacific ports (i.e. Corinto and Quetzal), whilst the throughput in Acajutla port also slightly increases for both exports and imports (note that the throughputs in two Atlantic ports, Cortés and Sto. Tomás, affect each other because they are closely located).

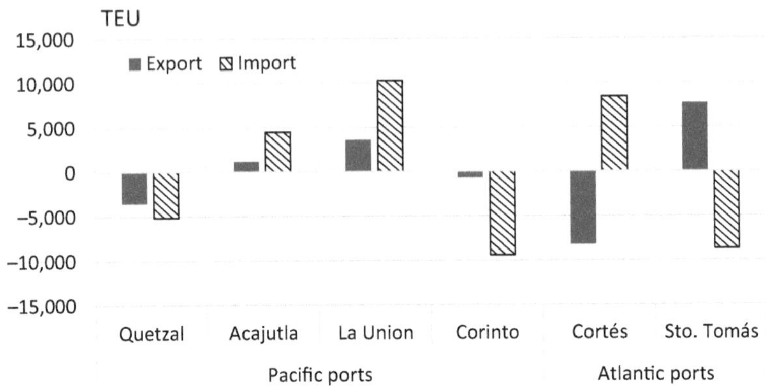

Fig. 10.12 Difference in annual laden container throughputs of S1 from the base scenario in CAM4 ports.
Source: Own compilation.

Future forecast based on several liner shipping service network scenarios

In the following future simulations, the focus is also on La Union port. It is important for La Union port to estimate the optimal channel depth based on the throughput forecasted by channel depth, in other words, by the maximum size of containerships to be allowed to call at the port. Therefore, the objective of the following simulations is to forecast the container throughput in La Union port by channel depth.

Port scenario

Table 10.5 shows the future level of service assumed in CAM4 ports including La Union as of 2030. Note that the maximum draught of La Union port is differently

Table 10.5 Future settings of level of service in each CAM4 port (as of 2030).

Port name	Maximum draught (m)	Lead time: Export TPX_r (h)	Lead time: Import TPM_r (h)	Transshipment time TPR_r (h)
Quetzal	**−14**	**48**	24	**24**
Acajutla	−11	60	48	48
La Union	Set by scenario	**48**	**24**	**24**
Corinto	**−12**	**48**	**24**	48
Caldera	**−12**	48	24	48
Cortés	−12	48	24	**24**
Sto. Tomás	**−12**	**48**	24	48

Bold and underlined columns are improved from the current level.
Source: Own compilation based on Japan International Cooperation Agency (JICA), 2014. Final report for special technical assistance for maintenance dredging of the port of La Union in the Republic of El Salvador.

set by scenario. Some of the improvements are a reflection of our interview survey results, such as a project for a new container terminal development in Quetzal, whilst others are based on the authors' assumptions; for example, all CAM4 ports except for Acajutla can accommodate a vessel with a – 12 m draught at maximum. Furthermore, in Acajutla port, the limitation in the handling capacity for containers (i.e. 200,000 TEU/year) is taken into account. The container handling charge, CPX_r and CPM_r, is unchanged during these periods for all ports, except for La Union port, for which a policy simulation on raising its tariff will be shown later.

LS scenario

According to our interview survey, an LSC has the following options for responding to an increase in demand:

(a) Increase frequency, for example, from biweekly to weekly or from weekly to twice a week by calling in both directions of navigation.
(b) Increase the number of ports to call at in the region, such as calling at both Acajutla and La Union in El Salvador.
(c) Enlarge the vessel size. In cases where a port does not have enough water depth, a vessel may shift to a neighbouring port. On the other hand, if the area has a large amount of cargo demand but there are no ports with sufficient water depth, an LSC may introduce a smaller vessel.

Based on the above outcomes, possible future options to alter the level of each LS that currently calls at either of the Salvadoran ports are considered (including not only the six LSs to call at Acajutla port shown in Table 10.2, but also the one LS to call at La Union port assumed in the previous scenario), as summarised in Table 10.6.

Furthermore, to examine La Union port's potential as a transshipment hub, the authors assume additional scenarios in which one trunk service, which is jointly operated by two LSCs (company A and B) to connect East Asia and the east coast of North America with larger containership, also call at La Union port (see Table 10.7 for the detailed setting). Because the schedule of the service should be maintained, a long waiting time (more than 0.2 h) is not allowed if a vessel enters into La Union port. Thus, for example, the original vessel with a capacity of 4230 TEU and a 13 m draught can access La Union port only if the depth of the navigation channel is – 14 m or more.

At the same time, a new feeder service connecting with the above trunk route in La Union port by transshipment is considered, as shown in Table 10.7. Based on the exploratory simulations, the new feeder service is assumed to operate between La Union and Balboa ports by way of Corinto and Caldera ports. Shipping company A is chosen as an operator of both the trunk route and feeder service because it does not have any LSs in the Pacific coast of CAM as of 2010. Consequently, the effect of its entry into the MCS market on the Pacific coast of CAM can be measured clearly.

Simulation results

By combining the options of each LS listed in Tables 10.6 and 10.7, 10–15 viable scenarios are constructed for each depth of channel in La Union port, then a model

Table 10.6 Possible options in the level of LSs calling at Salvadoran ports to be changed from the current level.

No.	Operator	TEU capacity and draught	Frequency of calling at Acajutla and La Union for each
1	Maersk	✓ 1240 (9 m) ✓ 1695 (10 m) ✓ 2480 (11 m) ✓ 3350 (12 m)	✓ Weekly for both westbound and eastbound; N.A. ✓ Weekly for westbound; weekly for eastbound ✓ Weekly for eastbound; weekly for westbound ✓ N.A.; weekly for both westbound and eastbound
2	MSC	✓ 1232 (9 m) ✓ 1790 (10 m) ✓ 2480 (11 m) ✓ 3350 (12 m)	✓ Biweekly; N.A. ✓ Weekly; N.A. ✓ Weekly; weekly (additional call) ✓ N.A.; weekly
3	NYK	✓ 1240 (9 m) ✓ 1610 (10 m) ✓ 2480 (11 m) ✓ 3350 (12 m)	✓ Weekly; N.A. ✓ Weekly for westbound; weekly for eastbound (additional call) ✓ Weekly for eastbound; weekly for westbound (additional call) ✓ N.A.; weekly
4	APL(1)	✓ 1118 (9 m) ✓ 1790 (10 m) ✓ 2480 (11 m) ✓ 3350 (12 m)	✓ Weekly; N.A. ✓ N.A.; weekly
5	CMA-CGM/ CSCL/CCNI	✓ 1790 (10 m) ✓ 2516 (11 m) ✓ 3350 (12 m)	✓ Weekly; N.A. ✓ N.A.; weekly
6	CSAV	✓ 1790 (10 m) ✓ 2599 (11 m) ✓ 3350 (12 m)	✓ 4 times every 9 weeks; N.A. ✓ N.A.; 4 times every 9 weeks
7	APL(2)/ Hamburg-Sud	✓ 1324 (9 m) ✓ 1790 (10 m) ✓ 2480 (11 m) ✓ 3350 (12 m)	✓ N.A.; N.A. ✓ N.A.; weekly

Underlined columns are the current level; *N.A.*, not available.
Source: Own compilation based on Japan International Cooperation Agency (JICA), 2014. Final report for special technical assistance for maintenance dredging of the port of La Union in the Republic of El Salvador.

calculation is conducted for each scenario. The calculated result is checked from the following viewpoints:

(1) whether the model calculation converges;
(2) whether each LSC can collect enough cargo; and
(3) whether the load factor of each LS is not too small or too large.

As a result, 22 scenarios out of 121 viable scenarios are selected as possible scenarios. Note that the possibility of each scenario occurring is very different; however,

Table 10.7 Possible options of a transshipment service to call at La Union port.

Operator	Service	TEU capacity and draught	Frequency	Ports to call
A/B	Trunk route	✓ 3350 (12 m) ✓ 4230 (13 m)	Weekly for both westbound and eastbound	East Asian ports—Lazaro Cardenas—La Union—Miami—New York—Miami—La Union—Lazaro Cardenas—East Asian ports
A	Feeder	✓ 3350 (12 m)	Weekly	La Union—Corinto—Caldera—Balboa—Caldera—Corinto—La Union

Underlined columns are the current level.
Source: Own compilation based on Japan International Cooperation Agency (JICA), 2014. Final report for special technical assistance for maintenance dredging of the port of La Union in the Republic of El Salvador.

it is very difficult to predict which scenario is most likely to occur. Another point is that each scenario is set under a given depth of navigation channel with the maximum capacity of the vessel that can navigate the access channel. In other words, if a scenario is judged 'possible' at a particular depth of the channel, it is also automatically possible for any case with a deeper channel.

Fig. 10.13 indicates the estimated annual throughput of laden containers in La Union port, as well as Acajutla port for reference purposes, for each possible scenario by the depth of the navigation channel. Dredging is not necessary if the channel depth

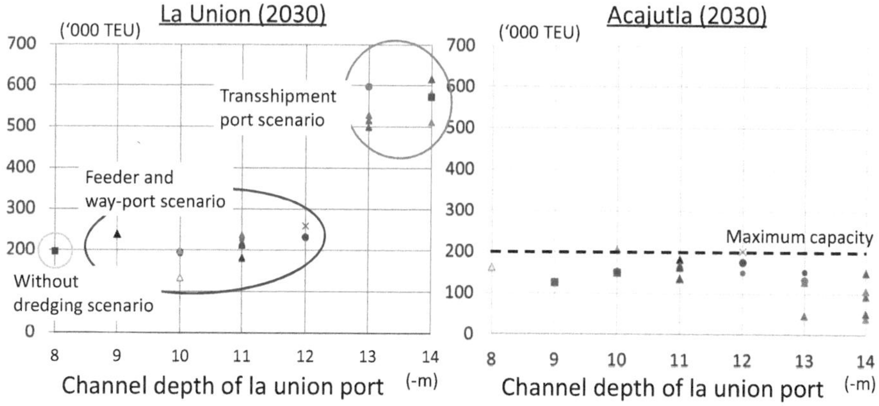

Fig. 10.13 Estimated laden container throughputs in Salvadoran ports for each possible scenario in 2030.
Source: Own compilation based on Japan International Cooperation Agency (JICA), 2014. Final report for special technical assistance for maintenance dredging of the port of La Union in the Republic of El Salvador.

is – 8 m, whilst the transshipment service can be provided only if the channel depth is maintained at – 13 m or deeper.

The figure indicates that around 500,000–600,000 TEU laden containers are annually expected as the throughput of La Union port if the trunk service successfully calls, whilst 150,000–250,000 TEU/year are expected if only the feeder and way-port services call. The throughputs in Acajutla port often reach capacity (i.e. 200,000 TEU/year) if La Union port remains the feeder and way-port service port, whilst they are less than capacity if La Union port becomes the regional transshipment hub.

Tariff increase in La Union port

After concluding the concession contract to operate La Union port with a private terminal operator, the tariff for container handling in the port is expected to increase for securing the budget of constant maintenance dredging. Specifically, the authors assume the new tariff of container handling in La Union port to be 185.76 US$/TEU, which is almost three times the current tariff (65.79 US$/TEU) shown in Table 10.1.

Fig. 10.14 indicates the estimated annual throughputs of laden containers in La Union port in 2030 in the case of tariff increases there. Compared to the results shown in Fig. 10.13 without tariff increases, the maximum estimated throughput in La Union port in each channel depth is smaller by around 100,000 TEU/year at most, but the difference is not large if compared to the expected amount of increase in income.

To examine sensitivity to the tariff increase, a simulation of the barrier reduction at the land national border is calculated. Fig. 10.15 depicts an example of the impact of a tariff increase on the annual export and import laden container throughput in CAM4

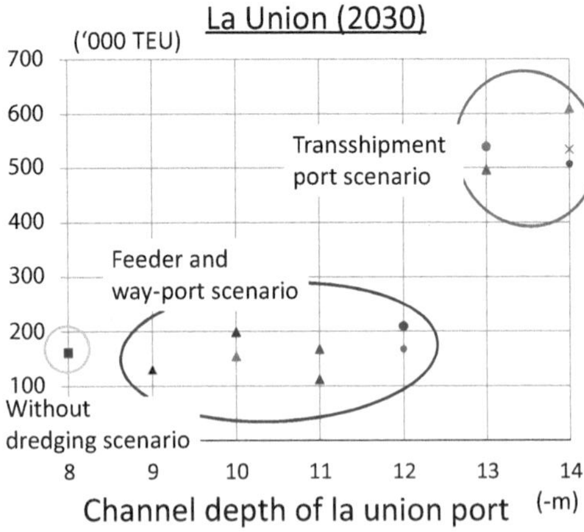

Fig. 10.14 Estimated laden container throughputs in La Union port with tariff increases for each possible scenario in 2030.
Source: Own compilation based on Japan International Cooperation Agency (JICA), 2014. Final report for special technical assistance for maintenance dredging of the port of La Union in the Republic of El Salvador.

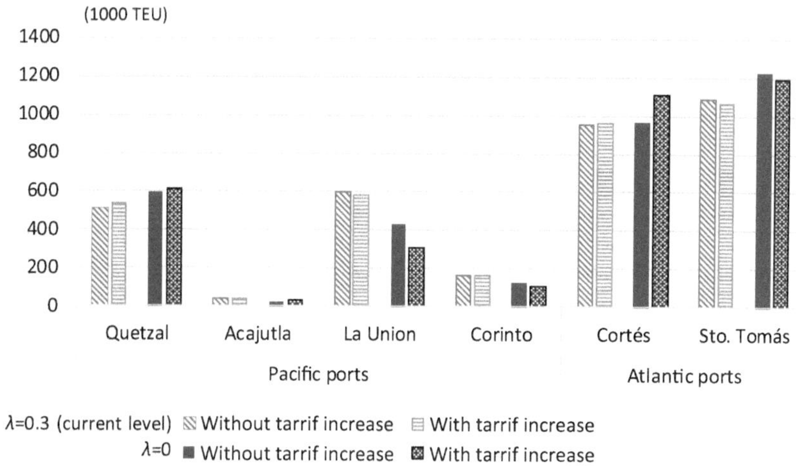

Fig. 10.15 Estimated export/import laden container throughputs in La Union port after tariff increases for each possible scenario in 2030.
Source: Own compilation based on Japan International Cooperation Agency (JICA), 2014. Final report for special technical assistance for maintenance dredging of the port of La Union in the Republic of El Salvador.

ports, with a representative scenario featuring −14 m channel depth in La Union port in 2030. As shown in the figure, in a case where the current land national border level is assumed (i.e. $\lambda_a = 0.3$), the change in container throughput in La Union port from an increased tariff is very small. However, the impact of the tariff increase is more significant if any barriers at the land national borders between CA4 countries are removed (i.e. $\lambda_a = 0$).

Conclusion

This chapter applied the intermodal container cargo flow simulation model, which was developed in the previous chapters, to CAM. Because it consists of several small countries but has two coastlines (i.e. the Atlantic and Pacific Oceans) with different characteristics in the MCS network, many container flows across land national borders can be observed.

After confirming that the outputs estimated from the model agreed well with the observed data, three kinds of simulations were examined that focused on La Union port, a new container terminal in El Salvador. The first simulation showed that the model can reproduce well the observed annual container throughput in La Union port if new LS were to call at the port. Then, the future annual laden container throughput for La Union port was forecasted for each possible scenario of the LS network calling at the port, under the condition of the annual throughput in Acajutla port being capped. The annual laden container throughputs were estimated at more than 500,000 TEU in 2030 if La Union port functioned as a regional hub, achieved by maintaining an access channel sufficiently deep (−13 m or deeper), whilst throughputs would be 250,000 TEU at most if the port remained a feeder and way-port service port.

Furthermore, the simulation result indicates that if the cargo-handling tariff in La Union port was increased to almost three times the current level after concluding concession contracts, the annual estimated laden container throughputs would decrease by 100,000 TEU at most, in comparison with throughputs where on increase in the tariff has not taken place. These were not negligible differences, but the expected amount of increased income would be rather larger because the tariff was assumed to be increased three times its current rate. However, if any barriers to crossing the land national borders were eliminated, a tariff increase would more severely affect the container throughput in La Union port.

The authors consider that these simulation results are reasonable (i.e. the model well describes the real choice behaviour of container cargo owners or their agencies) and that the model can reveal the potential of La Union port if the access channel maintains a reasonable depth. However, the model does not include implicitly the choice behaviour of LCSs, including decisions on vessel size and ports to call; the LS network is simply given as a preliminary input into each scenario calculation. Endogenising the behaviour of LSCs is one of the further significant challenges, as similarly mentioned in many chapters of this book. Another challenge is to consider the capacity constraint not only in the MCS network but also in the HT network, as well as in port operations. A model with capacity constraints in the HT network is applied to many simulations in Chapter 12 and other chapters below (except for Chapter 14).

References

Central American Commission of Maritime Transport (COCATRAM), n.d. Statistics for Maritime Shipping and Ports, http://www.cocatram.org.ni/redmarport.html.

Secretariat for Central American Economic Integration (SIECA), n.d. Trade Statistics in Central America.

Greater Mekong Subregion: Is the Mekong River shipping competitive with other modes?

11

Ryuichi Shibasaki[a], *Takashi Shimada*[b], *and Masaru Suzuki*[c]
[a]The University of Tokyo, [b]The Overseas Coastal Area Development Institute of Japan (OCDI), [c]Nikken Kogaku Co., Ltd.

Introduction

The Greater Mekong Subregion (GMS) is an international region of the Mekong River Basin in Southeast Asia. The Mekong River is a transboundary river in Southeast Asia, which starts from the Tibetan Plateau, running through China's Yunnan Province, Myanmar, Lao PDR, Thailand, Cambodia, and Vietnam. For many years, the Mekong River has been an important waterway for people and goods between villages/towns along the river, as well as functioned as international trade routes, connecting the six Mekong countries. Recently, it has been extended to the rest of the world.

Meanwhile, in the GMS, Cambodia, China, Lao PDR, Myanmar, Thailand, and Vietnam entered into a programme of regional economic cooperation, designed to enhance economic relations among the countries, under the initiative of Asian Development Bank (ADB). The GMS programme has adopted a three-pronged strategy: strengthening connectivity through physical infrastructure and the development of economic corridors; improving competitiveness through market integration and the facilitation of cross-border trade and travel; and building a sense of community by addressing shared social and environmental concerns (ADB, 2017). Since 1998, the GMS programme has been introducing economic corridors to promote economic growth and development. These economic corridors are investment areas, usually running along major highways, which connect centres of economic activities in the GMS; three main economic corridors are being developed—the North–South Economic Corridor, the Southern Economic Corridor, and the East–West Economic Corridor. The GMS programme also includes the development of ports in Cambodia, Myanmar, Thailand, and Vietnam.

This chapter reviews the current status of international transport in the GMS, particularly traffic routes to/from Cambodia, and analyses the potential impact of infrastructure improvement and other related policies on international logistics in Cambodia. Particularly, we focus on the inland waterway transport (IWT) connecting Phnom Penh (PP) and Ho Chi Minh (HCM) via the Lower Mekong Delta and identify obstacles and issues which need to be addressed for enhancing barge navigation along the Mekong River, including infrastructure, administration, and operation of the IWT.

Global Logistics Network Modelling and Policy. https://doi.org/10.1016/B978-0-12-814060-4.00011-3

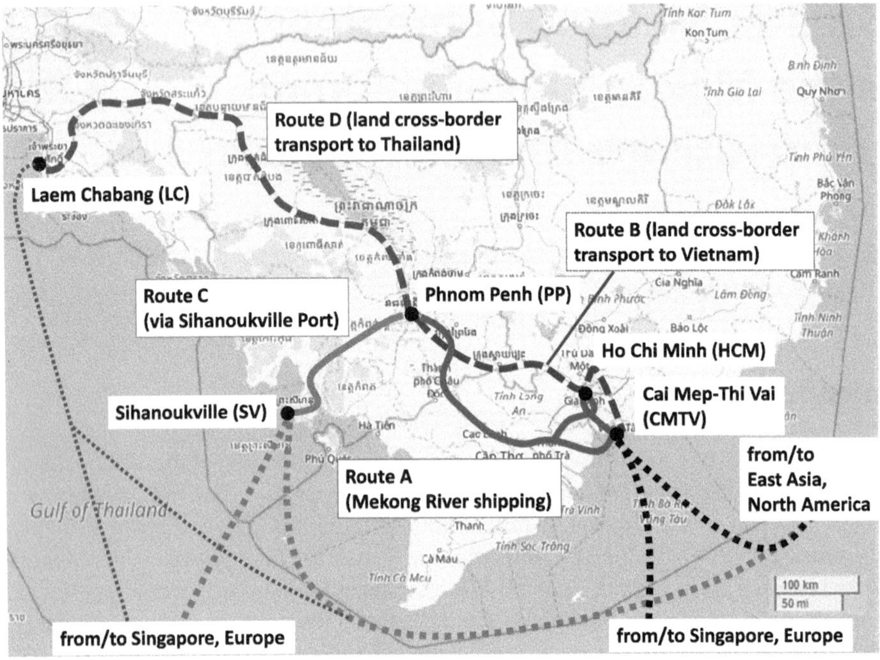

Fig. 11.1 Major transport routes of international cargo from/to Phnom Penh.
Source: Own compilation based on open street map.

Thus, simulations of the proposed measures to facilitate IWT are examined by using the intermodal cargo flow assignment model.

The international cargo to/from PP has multiple transport alternatives (Fig. 11.1). In other words, the international transport of Cambodian cargo is one of the most complicated in Asian countries, in terms of the gateway port choice for export and import, including both its own seaports and neighbouring countries', as well as both road and river shipping as hinterland transport (HT) means.

International container transport in Cambodia

There are four major routes for international maritime container shipping (MCS) to/from PP, as shown in Fig. 11.1. Route A: Mekong River shipping route via PP river port, connecting to Vietnamese ports including HCM or Cai Mep/Thi Vai (CMTV); Route B: international road transport route directly connecting to Vietnamese ports; Route C: domestic road transport route to use Sihanoukville (SV) port; and Route D: international road transport route connecting to Laem Chabang (LC) port in Thailand. Major conditions for each transport route are summarised in Fig. 11.2.

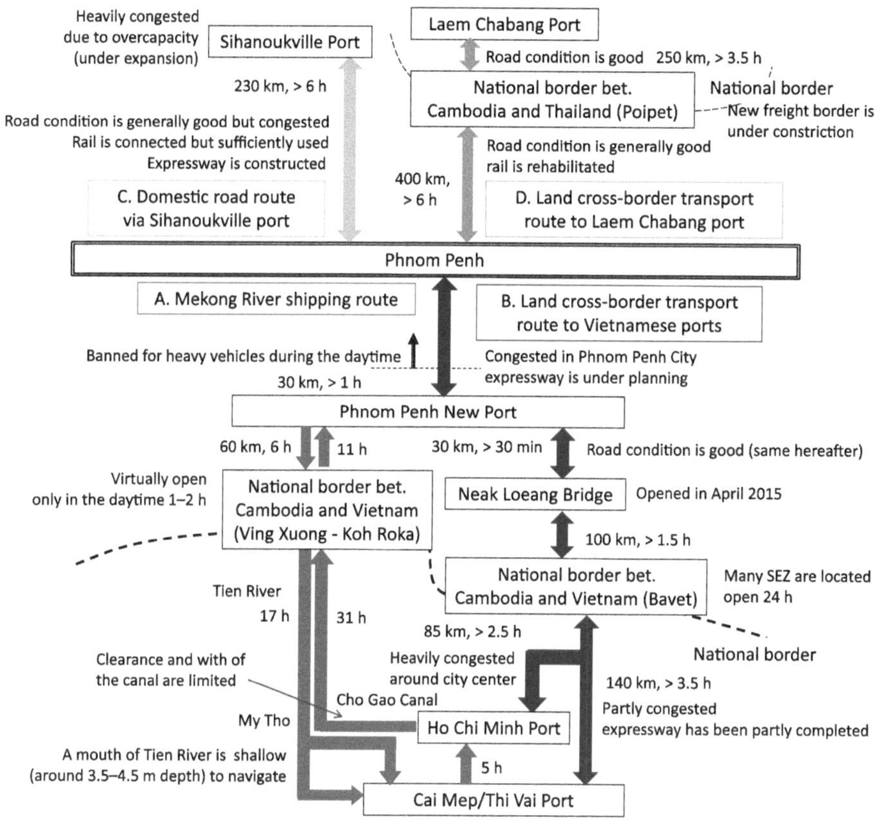

Fig. 11.2 Major conditions for each international transport route from/to PP (as of 2019). *Source:* Revised Shibasaki et al. (2017) as of 2019.

Mekong River route (via Phnom Penh river port)

Most barges departing PP river port navigate the Mekong River, the Tien River (local name of the Mekong River in Vietnam), and the Cho Gao Canal to avoid navigating the ocean, heading to the Vietnamese ports, including CMTV and HCM (Fig. 11.3). Using IWT via the Mekong River has several advantages compared to road transport. For example, one barge can transport more than 100 TEU containers per voyage, which makes IWT more economical than road transport. The cost of transporting cargo by IWT from PP port to Cat Lai port in HCM city is half that by road transport. Even during the dry season, the water level of the lower and middle sections of the Mekong River is sufficient for IWT. Therefore, shipping companies offer regular and reliable transport services all year round.

PP new port was opened in 2010 (Fig. 11.4) and the second phase development was completed in 2018. The terminal is located about 24 km away from the city centre of PP, along National Highway No. 1 (NH1) on the way to Neak Loeung and the national

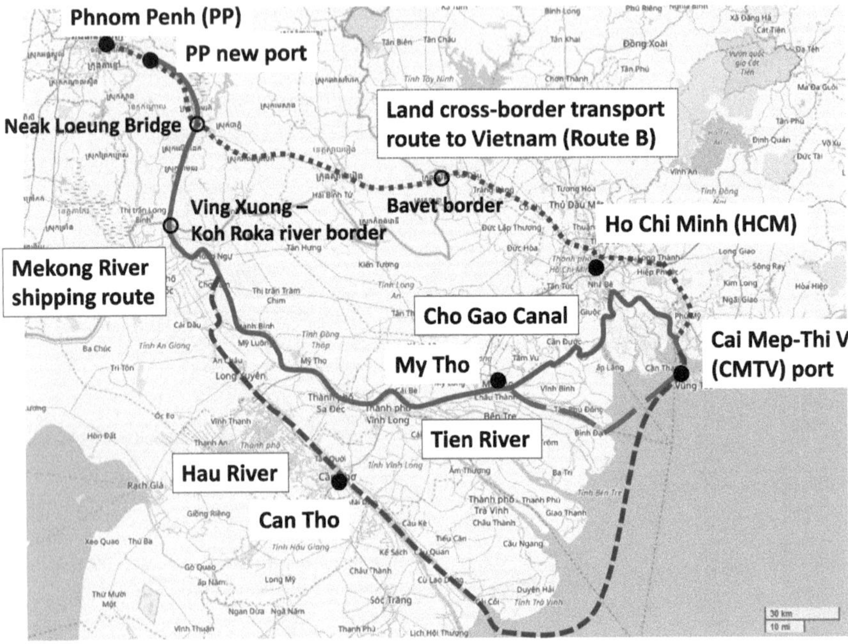

Fig. 11.3 Navigation routes of Mekong River shipping.
Source: Own compilation based on open street map.

Fig. 11.4 PP new port (left) and barge leaving the port (right) (2015).
Source: Authors.

border of Vietnam. The volume of containers handled in 2018 was 213,571 TEU (export: 111,212 TEU, import: 102,359 TEU), which has been increasing in recent years. The port authority (Phnom Penh Autonomous Port) consider the annual throughput will soon reach the terminal capacity (300,000 TEU per year); therefore, the third phase development is under planned.

The land access to PP new port has been improved, including the construction of the ring road in PP city, whereas the government instituted a ban on heavy vehicles entering the city centre in the daytime. Main obstacles in IWT are clearance of bridges and electric cables in the Cho Gao Canal (Fig. 11.5) and water depth of river mouth. For transiting the Cho Gao Canal, some barges have to wait for low tide and then charge ballast water to deepen vessel draught. The alternative route to avoid the Cho

Fig. 11.5 Electric cables and bridge above Cho Gao canal (above) and transit by barge (bottom) (2015).
Source: Authors.

Gao Canal is via the river mouth of the Mekong (see Fig. 11.3); however, a barge has to wait 11 h for high tide because the river mouth is very shallow. Another problem is administration procedures at a national border. Although almost all the necessary procedures for customs clearance and international trade are completed before leaving PP port, the barges need to stop at the border and have their documents checked. Moreover, the border gates of both countries are virtually open during the daytime although they should be open throughout the day juridically.

CMTV port has been developed in the deepwater area near the ocean, about 80 km away from HCM city, as an outer port to HCM port. The first terminal was opened in 2009. Most of the liner services (LSs) which call at CMTV port are connecting to the United States or Europe with large containerships because CMTV port has the deep berths to accommodate them. Meanwhile, HCM port still retains many LSs in intra-Asian routes connecting to Southeast and Northeast Asia, including ports of Japan, China, Taiwan, Hong Kong, Singapore, and Malaysia, despite its shallow berths.

International road route to Vietnam (via Ho Chi Minh or Cai Mep port)

The distance between PP and HCM port is around 240 km along the Asian Highway No. 1 (AH1), whereas that between PP and CMTV port is around 300 km. A Neak Loeung Bridge was opened in 2015 along the route across the Mekong River (Fig. 11.6 left). Before this, there was a ferry service to cross the river (Fig. 11.6 right); although it was frequently served, trailers with containers had to wait for a few hours on average for boarding because the ferry could accommodate only one large vehicle per trip.

Around the national border (i.e. Bavet), along the AH1 between Cambodia and Vietnam, there are several special economic zones (SEZ) and dry ports (Fig. 11.7). Some of the international trailers connecting to Vietnam originate from and are bound for these SEZs and dry ports. The annual container trailers transiting to Cambodia from Vietnam across the Bavet border were 78,672 in 2018, which is fairly larger than that to Vietnam from Cambodia (24,649).

Own seaport route (via Sihanoukville port)

The distance between PP and SV ports is around 230 km, which is similar to that between PP and HCM. The condition of NH4 is generally good; however, only a single lane for each direction of traffic is available in most sections, thus the travel speed was

Fig. 11.6 Neak Loeung Bridge (left, 2015) and a ferry before it was opened (right, 2014).
Source: Authors.

Fig. 11.7 Trailers waiting to enter Cambodia at Bavet border (left) and a dry port in Bavet (right) (2014).
Source: Authors.

significantly lower because of the recent rapid growth in truck and trailer traffic (see Fig. 11.8 left). The railways connecting PP and SV have also been rehabilitated with the assistance of the ADB. Since operations restarted in 2013, the railways transported 30,000–40,000 TEU containers annually; however, its potential has not been fully exploited, mainly because it takes too much time in both handling at dry port and operations (see Fig. 11.8 right).

SV port is the most important seaport in Cambodia (Fig. 11.9). The number of containers handled in SV port in 2018 was 541,228 TEU (export: 264,475 TEU, import: 276,753 TEU). The recent container throughputs in SV port have increased by 15%–20% per year, and already exceeded the terminal capacity (500,000 TEU). Fig. 11.10 shows the LSs that call at SV port as of August 2016. Three out of a total of 13 LSs are feeder services in the Thailand Bay, connecting to Singapore with smaller vessels (with an average capacity of 1000 TEU), whereas the remaining LSs connect to ports in Northeast Asia (such as Japan and China), Southeast Asia, and Bangladesh including larger vessels (with an average capacity of 1500 TEU).

For alleviating the congestion at roads and ports, new expressway and container terminal are planned to be constructed in the coming years, which is financially and technically supported by China and Japan, respectively. The first phase of the new container terminal development project with a 14.5 m deep berth, which can accommodate around 5000 TEU class containerships, will be completed by 2023.

Fig. 11.8 Road traffic between PP and SV (left, 2019) and inland container depot in PP (right, 2014).
Source: Authors.

Fig. 11.9 Sihanoukville port (left) and open-top containers imported (right) (2019).
Source: Authors.

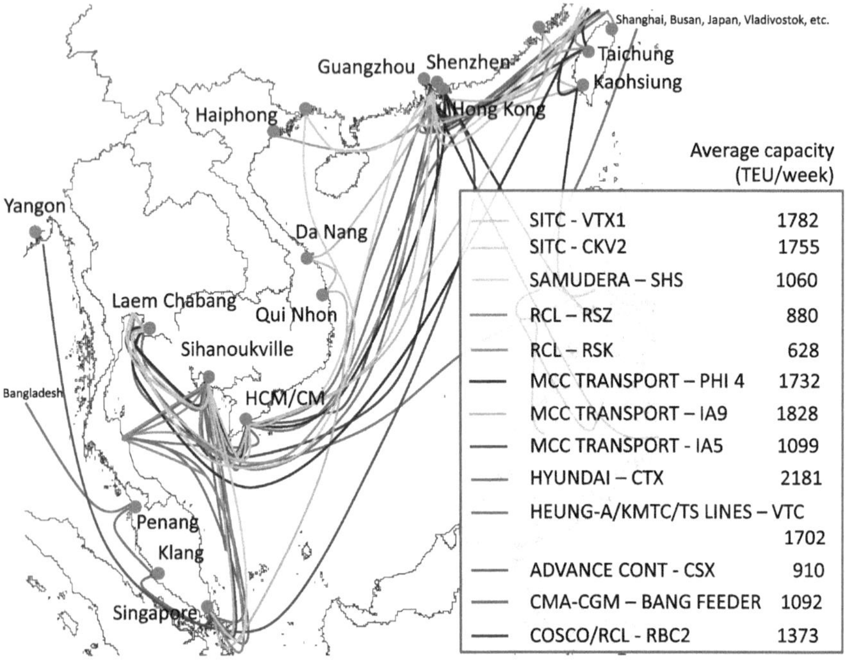

Fig. 11.10 LSs in SV port as of August 2016.
Source: Own compilation based on the MDS Containership Databank.

International road route through Thailand (via Laem Chabang port)

The distance between PP and LC port in Thailand is around 650km along the Cambodian NH5, through the national border of Poipet/Aranyaprathet (Fig. 11.11 left). The road condition is generally acceptable although only a single lane for each direction of traffic is available in most sections. Although trailers have been able to cross the border of the two countries without any transshipment since June 2012, very few of them have actually taken advantage of this arrangement and direct connect with PP and Thai cities such as Bangkok. The reasons are that the quota is very small and each country needs transport over long distances.

There is no PP cargo using LC port for export to or import from other countries through this route. However, there are some instances in which cargo that originated from the border area in Cambodia (in the Poipet SEZ; Fig. 11.11 right) used LC port as an export port to Europe. Additionally, some companies use the route between PP and Bangkok as part of their international division of labour strategy; for example, the materials for auto parts are imported from Thailand to an assembly factory near PP and the finished parts are returned to Thailand through this route.

Currently, railway connection between PP and Bangkok is planned and has been already opened in most sections other than that across the border (Fig. 11.12 left); however, it is not sufficiently used for freight transport because it is not fully connected.

Fig. 11.11 National border (Poipet) with Thailand (left, 2019) and factory located in SEZ in Poipet (right, 2015).
Source: Authors.

Fig. 11.12 Rehabilitated railroad (left) and planned new border (right) (2019).
Source: Authors.

Furthermore, a new border dedicated in the freight transport is planned near Poipet, for reducing the congestion of the existing border (Fig. 11.12 right); it is expected to open within a few years.

Container transport share of each route

Fig. 11.13 illustrates the amount of Cambodian international container cargo transported in each route (or gateway port) from 2007 to 2018. Note that the number of containers using Route D is not shown in the figure due to the lack of available data.

Fig. 11.13 indicates that the number of export containers in all routes is rapidly increasing after 2013. After 2013, the containers exported from SV port account for 55%–60%, followed by the IWT via PP port (29%–32%) and the direct road transport connecting to Vietnamese ports (12%–13%). On the other hand, before 2012, the share of export containers from SV port drastically declined from 88% in 2007 to 55% in 2012, whereas that of PP port and Vietnamese ports rapidly increased from 7.5% to 33%, and from 4.6% to 12%, respectively. Meanwhile, Fig. 11.13 also indicates that the number of import containers in all routes is gradually increasing although the increasing rate varies among the routes. As a result, the share of SV port and PP port in import containers is gradually decreasing from 79% in 2007 to 54% in 2018 and

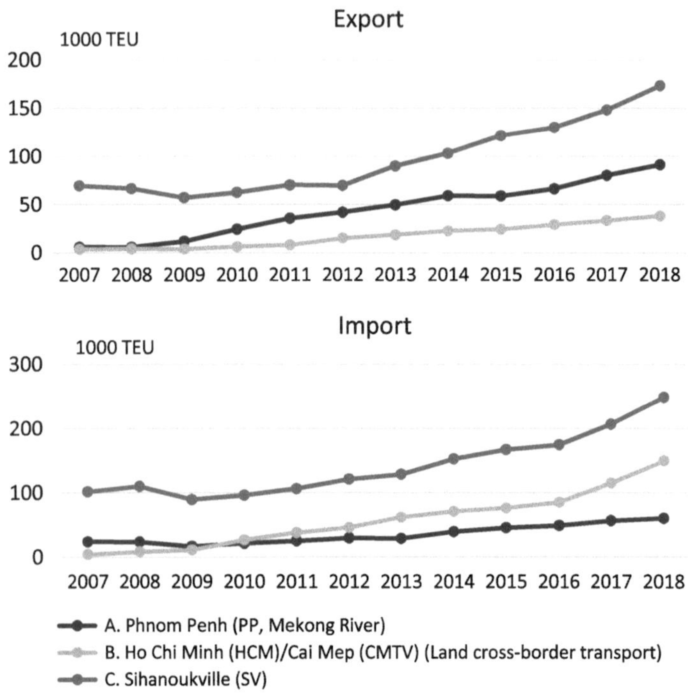

Fig. 11.13 Number of Cambodian international laden containers transported by route (gateway port).
Source: Shibasaki et al. (2017) before 2012 and statistics from each port and custom after 2013.

from 18% to 13% respectively, whereas that of Vietnamese ports is increasing from 3.1% to 33%.

This probably reflects the result of several policies introduced to improve the conditions of the road and river transports between PP and HCM, including not only infrastructure investment, such as the construction of new container terminal and road improvement, but also cross-border facilitation, such as reducing protocol time and cost and increasing the quota of the number of trailers that can directly pass through without any transshipment at the border.

Fig. 11.13 also reveals that over half of Cambodian international containers use SV port for both export and import on one hand and that the share of IWT is larger than that of international road transport in terms of export, whereas the share of IWT is smaller than that of international road transport in terms of import on the other hand. One possible explanation for this difference between export and import is that import cargo comes from various Northeast Asian ports through various liner shipping companies (LSCs) in the intra-Asian routes, mainly via HCM port. Meanwhile, export cargo mainly goes to the United States or Europe via CMTV port; in other words, the destination ports and LSCs providing the LSs are relatively limited, as will be discussed in detail later.

Data

Ports and MCS network

The model in this chapter has been developed as of 2010. In addition to the major global container ports where throughput was over 500,000 TEU per year including transshipment and empty containers (introduced in Chapter 8), two Cambodian ports (SV and PP) and two neighbouring countries' ports (Songkhla in Thailand and Kuantan in Malaysia) along the Thailand Bay are added. The container handling charge for export and import, CPX_a and CPM_a, the lead time for export and import, TPX_a and TPM_r, and the transshipment time, TPR_a, in the additional ports are set as shown in Table 11.1.

The MCS network has been created for the 20 largest LSCs of the world, plus ten additional LSCs of middle and small classes (i.e. Heung-A, Hub Line, KMTC, Nam Sung, RCL, SinoKor, SITC, STX Pan Ocean, TS Lines, and Wan Hai), which have LS network in Southeast Asia. The LSs or capacities that are not included in any of these 30 LSCs as operators, partners, or slot charters are excluded. In addition, ports that are not included in the model are eliminated. As a result, out of the 2857 LSs from the MDS Containership Databank (as of May 2010), 892 are included in the model. Although the number of LSs included in the model is about one-third of the total, 62% of the annual vessel capacity of the world is covered by the model, because larger LSCs provide more significant LSs across the world.

In addition, any LSCs calling at PP port are not included in the MDS Containership Databank; therefore, one hypothetically integrated service calling at PP, CMTV, HCM, and returning to PP is added. The frequency and average vessel capacity of the hypothetical service are assumed to be 15 per week and 82 TEU, respectively.

HT network

The HT network is only considered for Cambodia and neighbouring countries, as shown in Fig. 11.14. This network is structured based on APEC (2010) and Shibasaki and Watanabe (2012). Note that any railway links and future road links are not included in the calculation network of the model at this time due to lack of data although they are included in the figure.

The driving time and cost are calculated from the shortest path search based on the HT network shown in Fig. 11.14. Note that this model focuses on the border-crossing time and cost only in the transit country (i.e. Vietnam or Thailand) by land and river shipping, and not Cambodia. The coefficient of the border barrier, λ_a, which is multiplied by the border-crossing time and cost, is set at 0.4 for road transport and 0.1 for IWT in the simulations in this chapter.

Container cargo shipping demand (OD matrix)

The demand for container cargo shipping, Q_{ij}, from region i to j is estimated as follows: First, the demand for container cargo shipping (OD matrix) between countries or regions on a TEU-basis is obtained from the World Trade Service (WTS) database,

Table 11.1 Settings of level of service in added ports.

Ports			Lead time (export) TPX_a	Lead time (import) TPM_a	Transshipment time TPR_a	Container handling charge (export) CPX_a	Container handling charge (import) CPM_a
No.	**Port name**	**Country**	**(hours)**	**(hours)**	**(hours)**	**(US$/TEU)**	**(US$/TEU)**
31-1	Phnom Penh	Cambodia	72	96	48	100	225
31-2	Sihanoukville	Cambodia	48	48	48	100	225
33-1	Songkhla	Thailand	72	48	24	160	160
33-2	Kuantan	Malaysia	48	48	48	120	120

Source: Own compilation based on World Bank and the authors' interview survey.

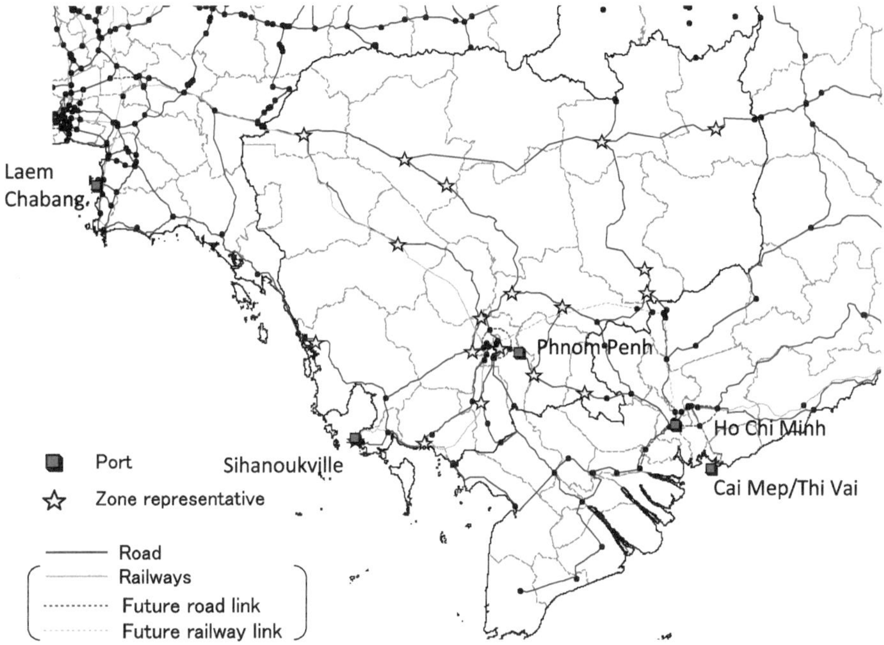

Fig. 11.14 HT network included in the model.
Source: Own compilation.

as with all cases introduced in this book. However, the container OD matrix to/from 'Other Asia' in the WTS database included not only the cargo to/from Cambodia but also other Asian countries such as Brunei Darussalam, Lao DPR, Mongolia, Myanmar, North Korea, and Papua New Guinea when the model was developed; therefore, it needed to be divided as per each country. For greater precision, it was divided according to the trade value data by commodity (on a WTS commodity classification basis); the share of each country by partner country, by commodity on a value basis was calculated from the UN Comtrade Database.

Second, the country-basis OD matrix acquired above is further divided into a port-basis OD matrix according to the port's share of the local container cargo throughput in the country/region. Finally, because Cambodian cargo is considered to include HT as discussed, whereas other cargoes of the world are only considered for MCS, the container OD matrix to/from Cambodia, which is estimated in the first step, is once again divided into 24 provinces according to the index that represents the regional economy. Due to limited available data, the amount of sales in each province is assumed to be an index for the regional division.

Fig. 11.15 show the share of partner regions for Cambodian international container cargo on a TEU basis, which is estimated in the first step. This shows that the partner regions for export from Cambodia are significantly different from those imported; the export cargo from Cambodia is mainly bound for North America and Europe, whereas the import cargo into Cambodia is mainly from Northeast and Southeast Asia.

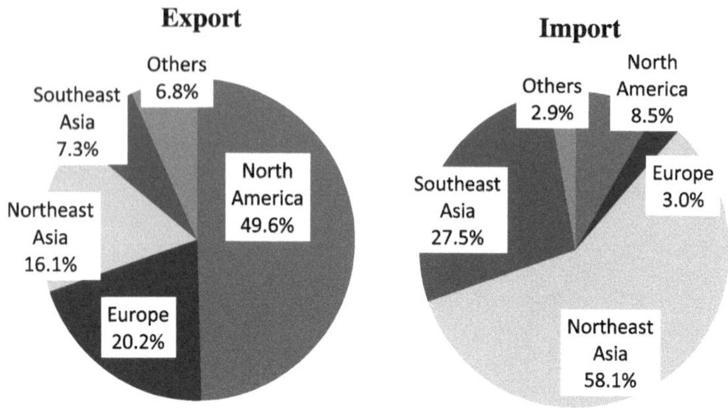

Total: 82,400 TEU Total: 112,000 TEU

Fig. 11.15 Shares of Cambodian international container cargo in 2010 by partner region. *Source:* Own compilation.

The main commodity for export is 'wearing apparel' (which holds a 54.4% share of Cambodian exports in terms of value) and 'postcards, calendars, and other printed materials' (31.8%), whereas that for import is 'textiles' (which holds a 38.2% share of Cambodian imports in terms of value).

Calculation results

The authors calibrated the distribution parameter for stochastic assignment, θ, to be 0.003, and time value for the shipper, vt, to be 5.0 (US$/TEU/h) to best fit the estimation results to the observed data in terms of the share of transport routes for Cambodian international laden containers. Fig. 11.16 shows the observed vs model-estimated shares by transport route (or gateway port) in 2010, for Cambodian export and import laden containers. They represent that the model can estimate the shares of two Vietnamese ports (i.e. HCM and CMTV) separately, and include the share of LC port, which is not actually available.

They unveil that the shares of SV port are underestimated in terms of both export and import by about 10 percentage points, whereas the shares of PP port (i.e. Mekong River shipping) are both overestimated. The share of Vietnamese ports (i.e. the sum of HCM and CMTV ports) is overestimated in export, whereas that of import is underestimated. As shown in Fig. 11.13, the shares of SV port are decreasing as time goes on; in other words, the model seems to predict the share in the near future in the case that cargo owners are more sensitive to choosing the more cost-effective transport route.

The difference between the observed and estimated shares of Vietnamese ports implies that the model does not sufficiently describe the differences in feature of export and import, compared with the actual. However, the model describes the differences in

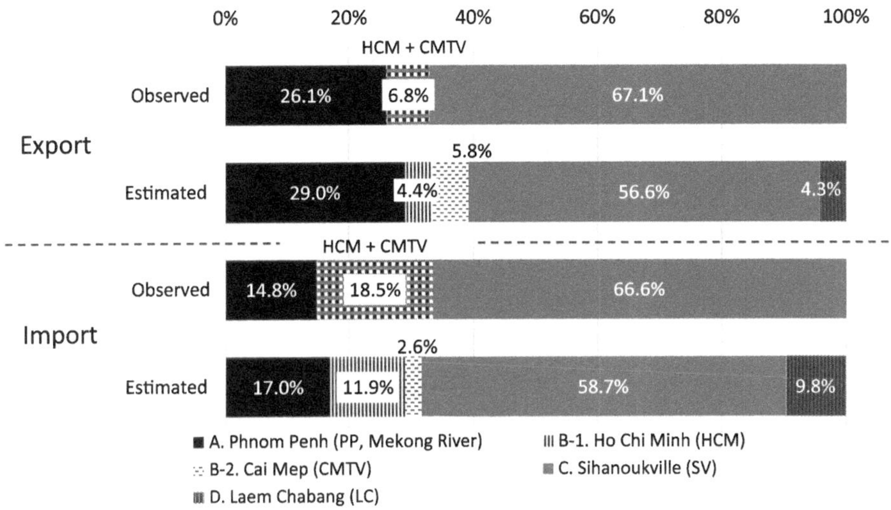

Fig. 11.16 Observed and estimated share of transport routes (gateway ports) for Cambodian international laden containers in 2010.
Source: Own compilation based on Shibasaki et al. (2017) and Shimada et al. (2017).

the features of HCM and CMTV port, as discussed earlier: CMTV port is mainly used for export, whereas HCM port, for import. One reason for which the model cannot sufficiently describe the difference in export and import may be that the model does not consider the difference in lot sizes of export and import cargo; export cargo tends to be shipped by barge because it mainly comprises of final goods with larger lot sizes, whereas import cargo tends to be transported by trailers because it mainly comprises of materials with smaller lot sizes.

Fig. 11.17 illustrates the estimated trailer flow of Cambodian international laden containers in 2010. The greatest traffic volume is estimated between PP and SV on NH4, followed by the section between PP and HCM on NH1. Fig. 11.18 shows the estimated shares by transport route (or gateway port) in 2010, for international laden containers in five regions of Cambodia. The figure shows that the shares of each transport route vary across regions, especially in terms of imports; for example, the shares of PP port in the central region are larger than those of other regions for both export and import owing to its proximity to PP port, whereas the shares of SV port in the coastal region are larger than those of other regions for both export and import owing to the similar reason.

Fig. 11.19 shows the estimated shares by transport route (or gateway port) in 2010 for international laden containers by partner regions, for the total amount of export and import. It represents that the shares of each transport route significantly vary across partner regions. For example, the share of SV port is quite large (about 90% of the total amount) for container OD cargo volume between Cambodia and Southeast Asia or Europe. This implies that SV port has a significant advantage in terms of southbound or westbound shipment from Cambodia, with strong connections to Singapore

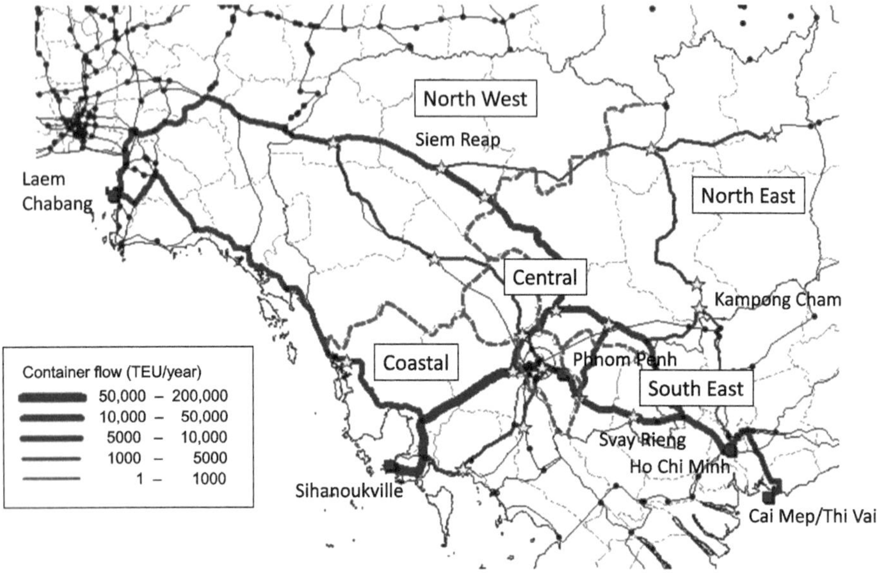

Fig. 11.17 Estimated trailer flow of Cambodian international laden container in 2010.
Source: Own compilation.

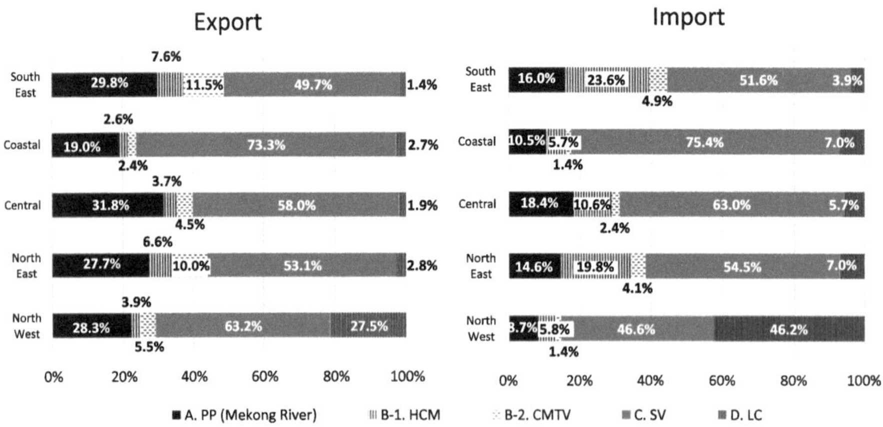

Fig. 11.18 Estimated share of gateway ports for the international laden container by region in Cambodia in 2010. Note: Each region is defined in Fig. 11.17.
Source: Own compilation.

and other hub ports in Southeast Asia through feeder services. On the other hand, for the container OD cargo volume between Cambodia and North America, the sum of CMTV and PP ports is about half the total amount. Because most cargo to North America using PP port is transshipped from CMTV port into mother vessels that directly sail to North America on the trunk route, CMTV port functions as a gateway

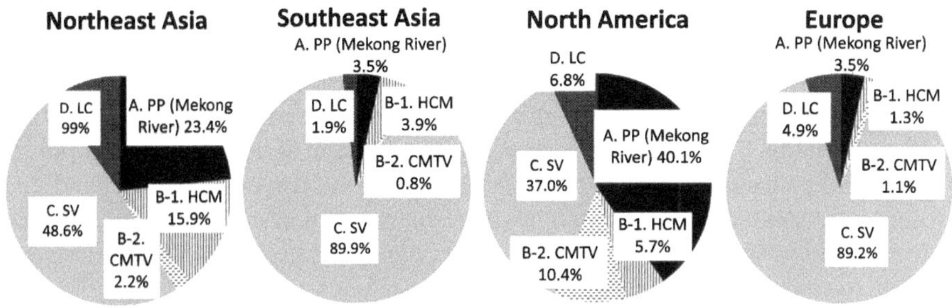

Fig. 11.19 Estimated share of gateway ports for international laden container by partner region in 2010 (sum of export and import).
Source: Own compilation.

for Cambodian container cargo to North America. As well, HCM port functions as a gateway for Cambodian container cargo from Northeast Asia, based on the strong connection with Northeast Asian ports through many LSs (note that PP port is also functioned as a feeder port of HCM port).

Fig. 11.19 also reveals potential reasons that the share of PP port (i.e. Mekong River shipping) in terms of export containers is larger than that of import containers, whereas the share of Vietnamese ports (i.e. international land transport) in terms of export is smaller than that of import. Because the export containers are mainly bound for North America, as shown in Fig. 11.15, CMTV port is the main gateway of Cambodian export containers, as discussed above. Meanwhile, because the import containers are mainly from Northeast Asia, HCM port is the main gateway of Cambodian import containers. Note that owing to the structure of the cost function, the share of IWT generally becomes larger than that of land transport as the shipping distance is longer. Additionally, as described earlier, the feeder services in the Mekong River directly head for CMTV port after leaving PP port, without calling at HCM port on the way, which becomes more advantageous as compared to the trailer transport to CMTV port. Thus, the share of IWT connecting to CMTV port is significantly larger than that connecting to HCM port.

Policy simulations

Base scenario

For implementing simulations on the IWT, changes to the road network in the region from 2010 to 2016 are introduced into the model as a base scenario, including

- opening Neak Loeung Bridge across Mekong River,
- completion of improvement of Cambodian NH1 road from PP City to PP new port,
- truck ban during daytime in PP city,
- partial opening of PP ring road No. 2, and
- partial opening of the expressway from HCM city to CMTV port.

Figs 11.20 and 11.21 show the share of the transport route estimated in each scenario for Cambodian export and import laden containers. No significant differences are found between the shares estimated in the previous section and the base scenario. The reasons may be explained through two kinds of 'trade-offs'; first, between the positive impact on road transport, such as the construction of new links and improvement of traffic speed vs a negative impact including the truck ban; and second, the impacts on each route, i.e. a change of road network conditions may affect not only the route to southern Vietnam by international road transport but also all the international shipping routes to/from PP city and neighbouring regions.

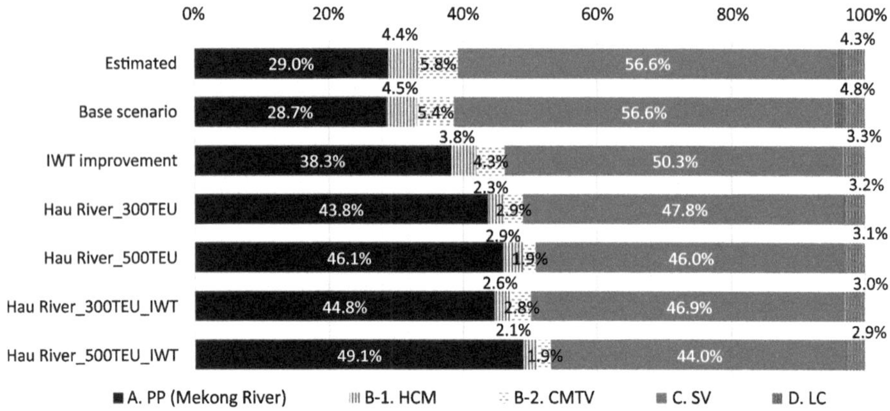

Fig. 11.20 Estimated share of transport routes for Cambodian export laden containers in each scenario.
Source: Own compilation.

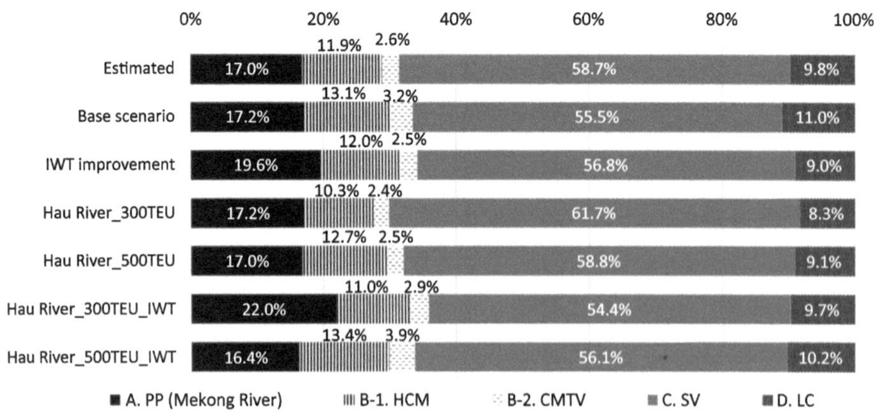

Fig. 11.21 Estimated share of transport routes for Cambodian import laden containers in each scenario.
Source: Own compilation.

Improvement of existing IWT route

This scenario assumes that the travel time for a barge on a round trip between PP new port and south Vietnamese ports (including HCM and CMTV) is reduced to 5 days, owing to a real 24-h-a-day operation at the national border. Note that a barge should spare 7 days for the same round trip in the base scenario due to the time loss at the national border during the nighttime. This means that two-sevenths of the capital cost of shipping, CC_a, included in the shipping cost function can be saved in this scenario. In addition, the barge size is assumed to be doubled (i.e. 170 TEU), the maximum size with which a barge can transit without any infrastructure improvement of the river and canals.

Figs 11.20 and 11.21 summarise the estimated shares by transport route for Cambodian export and import laden containers in this scenario (see 'IWT improvement'). Fig. 11.20 reveals that the share of PP port for the export containers is significantly larger than that in the base scenario, which is mainly a shift from SV port. On the other hand, Fig. 11.21 reveals that the share of PP port for import containers is also larger than that in the base scenario but their rates of increase are relatively lower than those of export cargo. In other words, if the advantage of the IWT is enhanced, then the gap in the shares between export and import may be further widened.

Use of new IWT route

These scenarios assume the use of the Hau River (Basac River, see Fig. 11.3), instead of the Tien River (Mekong River), which can accommodate a larger barge with faster speed. In these scenarios, the barge size is enlarged to 300 TEU or 500 TEU; therefore, if sufficient amount of cargo uses this service, the shipping cost is expected to decrease owing to the scale of economy. On the other hand, the shipping time is assumed to be unchanged as the previous scenarios, because the speed increase by vessel enlargement will be cancelled out by the increase in shipping distance. Another advantage of the Hau River route is that the cargo of Can Tho, the largest city in the Mekong Delta and located along the Hau River, can be attracted; however, it is not considered in these simulations because Vietnamese cargo is not considered in this model.

Figs 11.20 and 11.21 also summarise the estimated shares by transport route for Cambodian export and import laden containers in two scenarios ('Hau River_300TEU' and 'Hau River_500TEU'). Additionally, combined two scenarios with Hau River usage and a real 24-h-a-day operation at the national border are also calculated (see the results on 'Hau River_300TEU_IWT' and 'Hau River_500TEU_IWT').

Fig. 11.20 reveals that the shares of PP port by IWT for the export containers are also significantly larger than that in the base scenario and that in the case with a 500 TEU barge is much larger than that in the case with a 300 TEU barge. Additionally, much larger shares of PP port are expected in both combined scenarios. On the other hand, as with the previous scenario on the existing IWT improvement, the shares of SV port in these scenarios are smaller than that in the base scenario.

Meanwhile, the difference in shares for import containers shown in Fig. 11.21 is a little complicated. Basically, the new IWT route has no significant impact on the share of PP port. Only in the combined 'Hau River_300TEU_IWT' scenario, the share of PP port is larger than those in other scenarios, whereas that in the combined 'Hau River_500TEU_IWT' scenario is slightly smaller than those in other scenarios. There are two implications from these estimation results. The first one is that the enlargement of a barge is effective for import containers only if the shortening the shipping cost (caused by a real 24-h-a-day operation at the national border) is accompanied. Another one is that too much enlargement of a barge (i.e. 500 TEU) may have an adverse effect because the sufficient amount of import containers cannot be collected.

Conclusion

This chapter focused on the multiple transport routes of Cambodian international containers and carried out policy simulations to improve the condition of each transport route, such as the improvement of IWT and change of road network conditions, by using an assignment model of Cambodian international container cargo on the large-scale intermodal network. The model could describe well the actual shares of the route of HT (or gateway port) for Cambodian international laden containers. Moreover, the expected impacts in the simulation were reasonable as a whole; the improvement of IWT would increase its share, whereas the change of road transport conditions would bring both positive and negative impacts on its share.

This chapter proposed ways to increase the competitiveness of IWT based on the model simulation. Vietnam and Cambodia signed an agreement on the facilitation of IWT in 2008, including the use of ports along the Mekong River in both countries and the facilitation of border-crossing procedures although this has not been realised. The simulations described in this chapter only focused on the reduction in shipping cost through the real 24-h-a-day operation at the national border and the vessel enlargement; however, more additional facilitations between two countries, including the improvement of river channels and canals, could reduce the shipping cost and time. Additionally, increasing the number of ports covered by IWT services in both Cambodia and Vietnam, not only PP port, will raise the viability of the service although the cabotage issue should be solved.

Furthermore, the developed model should also be improved for the land transport network, for example, by including railway transport as well as planned new expressways. In addition, the authors will expand the model that include all land transport networks in the ASEAN region, which will enable us to compare quantitatively several kinds of international logistics policies in multiple countries of the ASEAN, to support the discussion for economic integration and cooperation in the ASEAN countries.

References

Asian Development Bank (ADB), 2017. Trade Facilitation and Better Connectivity for an Inclusive Asia and Pacific, Manila, the Philippines. Asian Development Bank.

Asia-Pacific Economic Cooperation (APEC), 2010. Impacts of trade and transport policy on international cargo shipping and economic activities. In: Presented at the 33rd Transportation Working Group Meeting (TPTWG), Tokyo. http://www.apec-tptwg.org.cn/new/Archives/tpt-wg34/Maritime/Final/draft_final_report_revised%20(2).pdf. (Accessed 1 January 2016).

Shibasaki, R., Watanabe, T., 2012. Future forecast of trade amount and international cargo flow in the APEC region: an application of trade-logistics forecasting model. Asian Transp. Stud. 2, 75–89.

Shibasaki, R., Shimada, T., Suzuki, M., 2017. Intermodal route choice of international containers in southern Mekong region and policy simulation using the large-scale intermodal network assignment model. In: Kobayashi, K., Rashid, K.A., Furuichi, M., Anderson, W.P. (Eds.), Economic Integration and Regional Development: The ASEAN Economic Community. Routledge Studies in the Modern World Economy, Routledge, pp. 162–176 (Chapter 12).

Shimada, T., Shibasaki, R., Kume, H., Suzuki, M., 2017. Implications for better container transit in the lower Mekong River through the field survey and policy simulation. Int. J. Oper. Res. 14 (3), 89–106.

South Asia: Impact simulations of logistics projects in India, Bangladesh, and Sri Lanka

12

Ryuichi Shibasaki[a] *and Tomoya Kawasaki*[b]
[a]The University of Tokyo, [b]Tokyo Institute of Technology

Introduction

South Asia (SAS) represents the southern region of the Asian continent, which is topographically dominated by the Indian Plate, bounded on the south by the Indian Ocean and on land by West Asia, Central Asia, East Asia, and Southeast Asia. The current territories of Afghanistan, Bangladesh, Bhutan, the Maldives, Nepal, India, Pakistan, and Sri Lanka form SAS.

SAS is expected to remain the fastest-growing region in the world. Growth in SAS is forecasted to increase to 6.9% in 2018, mainly reflecting fading disruptions to economic activity in India (World Bank, 2018). However, SAS countries are falling behind the global trend in the maritime shipping market. For instance, according to the World Bank's rankings of the business environment in terms of international trade, the SAS countries were placed 69th (Sri Lanka) at best, followed by 126th (India) and 140th (Bangladesh) (World Bank, 2015). The total throughput of international maritime containers exported from or imported into SAS is also lower than in other regions. For example, total container throughput in India was 10.7 million TEU in 2013, whereas throughput in China was 174.1 million TEU (World Bank, 2015).

Moreover, the current transshipment rate in the region is poor. Fig. 12.1 shows the rate of transshipped containers for regions located along the trunk line connecting Europe and East Asia. The transshipment rate in SAS is lower than that in other areas, especially after 2010. The significant increase in SAS in the 1980s was caused by the development of Colombo port (Sri Lanka), which is still a unique, major hub port in the region. Although the transshipment rates in other regions have increased in recent years, the rate in SAS has seemingly reached a plateau since the 1990s.

This chapter first reviews the current maritime container shipping (MCS) market in SAS. Next, the input data for estimating the model that represent the regional container cargo flow and computation results are presented. Then, six policy simulations are shown using the developed model. The empirical analysis and simulations focus mainly on Bangladesh, India, and Sri Lanka in SAS.

Global Logistics Network Modelling and Policy. https://doi.org/10.1016/B978-0-12-814060-4.00012-5

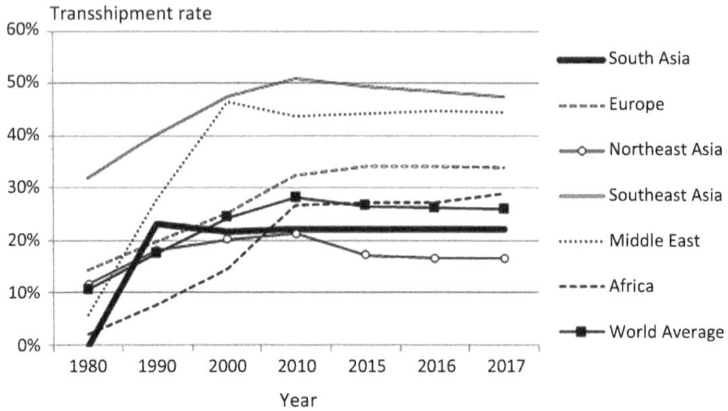

Fig. 12.1 Transshipment rate by region along trunk line between Europe and East Asia. *Note*: 'World Average' includes other regions (e.g. North America and Latin America) that are not shown in the figure.
Source: Own compilation based on each year's 'Container Forecaster & Annual Review' reported by Drewry Maritime Research.

Ports and container cargo flow in South Asia

Fig. 12.2 represents container ports and main liner service (LS) routes in the SAS region. SAS ports are covered by multiple trade lanes for areas such as East Asia, the Middle East, the Indian Ocean Islands (IOI), Africa, the Mediterranean, and Europe; access to the east coast of North America (ECNA) is also relatively easy by stretching further from Europe. Thus, they are located in the middle of the busiest East–West maritime trade lane.

Fig. 12.3 depicts the container throughputs in major SAS ports. Currently, the throughputs in most of major SAS ports including Colombo, Mundra (India), Karachi (Pakistan), and Chittagong (Bangladesh) rapidly increase reflecting the current economic and trade growth of SAS, whereas those in Jawaharlal Nehru Port (JNP) and Chennai in India stagnate in recent years, mainly due to capacity constraint which will be mentioned later.

Fig. 12.4 illustrates LS network of selected major liner shipping companies (LSCs) as of June 2013. Fig. 12.5 also schematically illustrates the LS network in and around SAS, including neighbouring major hub ports in Southeast Asia and Middle East. As a hub port in the region, Colombo port attracts the largest vessel capacities of 8.3 million TEU/year for mainlines and 1.9 million TEU/year for feeders.

Fig. 12.4 also clarifies that not only the East–West maritime trade lane is busiest and largest, but also the shape of the network is different between the Arabian Sea (i.e. westside of SAS) and the Bay of Bengal (eastside of SAS). In the Arabian Sea, many LSCs provide a larger amount of containership capacity to call at west coast ports of India (e.g. JNP, Mundra, and Pipavav) and Pakistani ports (i.e. Karachi and Bin Qasim). For example, European LSCs such as Maersk provides LSs to directly connect these ports with Europe and ECNA,

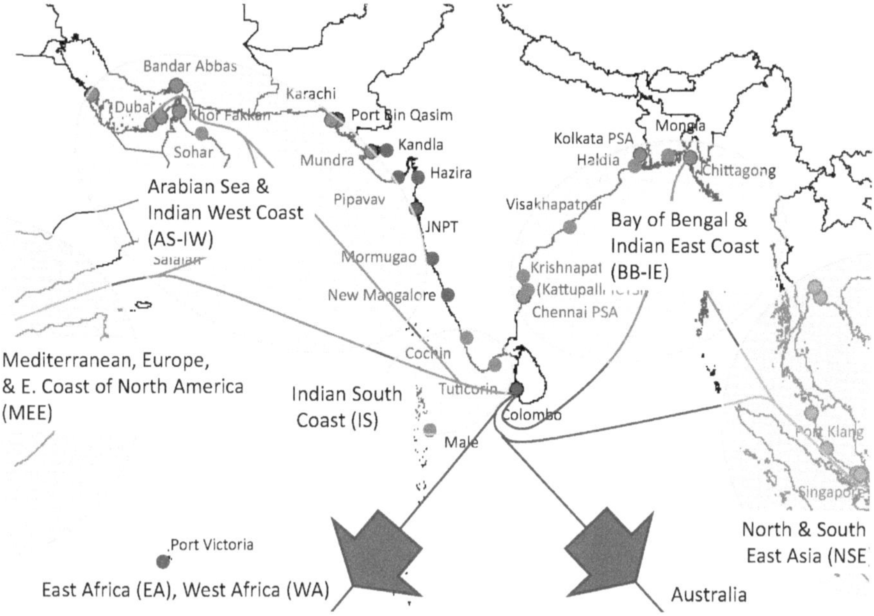

Fig. 12.2 Container ports and main international MCS routes in SAS.
Source: Shibasaki, R., Kawasaki, T., 2019. A transhipment hub in South Asia and its competition: application of network equilibrium assignment model for global maritime container shipping. Asian Transp. Stud. 5 (4), 546–569.

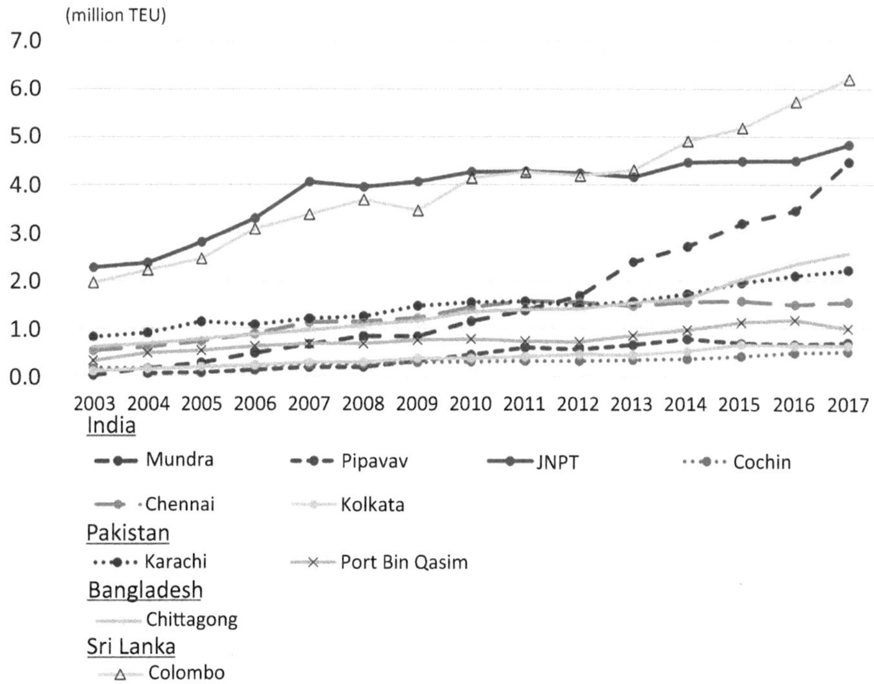

Fig. 12.3 Container throughput in major SAS ports.
Source: Own compilation based on various statistics.

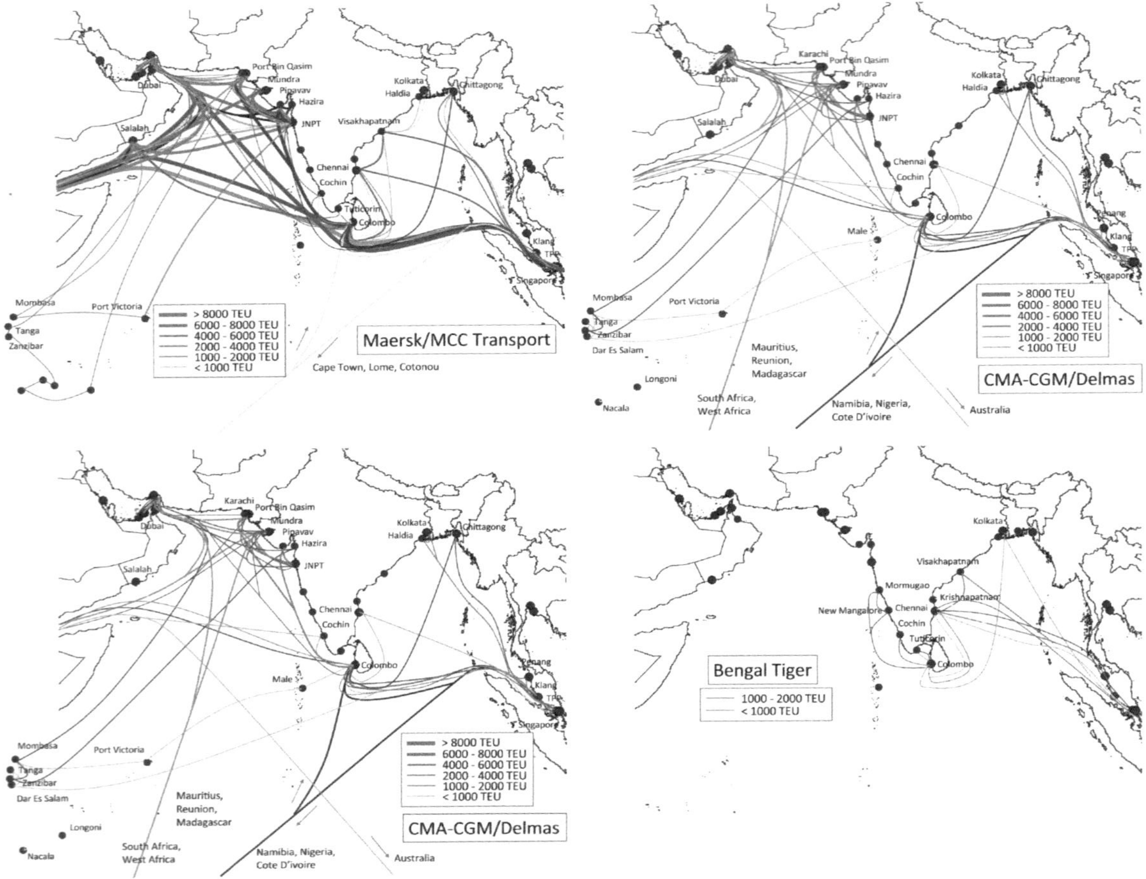

Fig. 12.4 LS network related to SAS (as of June 2013) for selected LSCs (on a weekly TEU capacity basis).
Source: Own compilation based on the MDS Containership Databank.

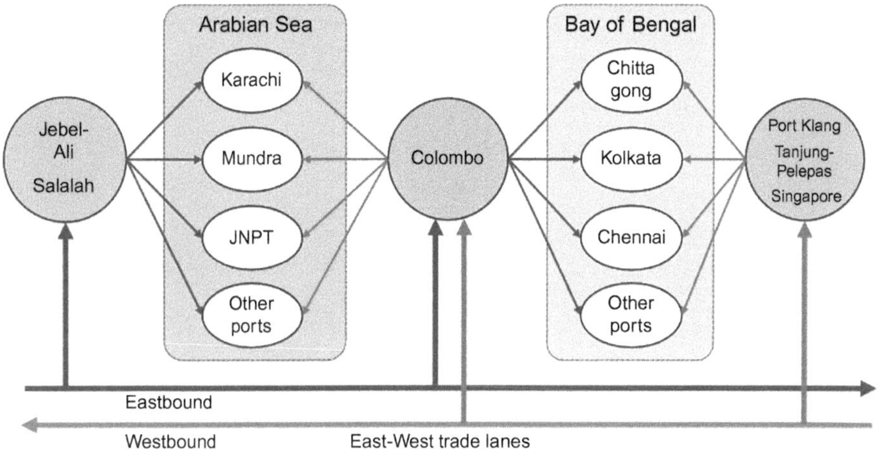

Fig. 12.5 Schematic LS network in and around SAS.
Source: Japan International Cooperation Agency (JICA), 2016. Data collection survey on cross-border maritime traffic in South Asia, final report. Available at http://open_jicareport.jica.go.jp/pdf/12251989.pdf (Accessed 24 August 2018).

whereas Asian LSCs such as NYK provides those to directly connect with East Asia. Furthermore, CMA-CGM, a French LSC laying stress on the African network, provides the most complicated LSs across the Arabian Sea and Indian Ocean to connect with East Africa. On the other hand, in the ports along the Bay of Bengal (e.g. Chittagong, Kolkata, Vishakhapatnam, and Chennai), feeder services to connect with hub ports in Southeast Asia (e.g. Singapore, Tanjung Pelepas, and Klang in Malaysia) or Colombo port are mainly provided by local LSCs such as Bengal Tiger, rather than direct LSs with other regions.

The MCS market and the transshipment container volume in SAS are expected to grow further. In reality, several projects to establish new transshipment hubs by renovating existing ports (such as Tuticorin and Cochin) or constructing greenfield ports (such as Hambantota, Vizhinjam, and Colachel) have been proposed and partly implemented in Sri Lanka and south India, based on an expected increase in future demand for transshipped containers to/from SAS. At the same time, the existing transshipment hub, Colombo port, has also attempted to improve its level of service by developing new terminals (i.e. South Harbour, see Fig. 12.6). After completion of these projects, the MCS market in SAS will change and the competition among the transshipment hubs in the region would become more severe.

Numerous projects to improve the level of logistics infrastructure, including not only port infrastructure but also hinterland transport (HT) infrastructures, such as road, rail, and inland waterway transport (IWT), are planned in SAS. For example, JNP, which is located in the suburbs of Mumbai, is the largest container port in India and notorious as a heavily congested port, especially on the landside (see Fig. 12.7). It has been expanding its third container terminal, operated by DP World (United Arab Emirates),

Fig. 12.6 South terminal of Colombo South Harbour (2015).
Source: Authors.

Fig. 12.7 Road traffic congestion at the gate of JNP (*upper-left*, 2015), present and future-planned berths in JNP (*upper-right*, 2015), and terminals in JNP (*below*, 2019). Note: Because the fourth container terminal *(near side of the upper-right picture)* is planned to align with the existing berths, congestion on the landside, shown in the upper-left picture, is not expected to be drastically resolved by the new terminal construction.
Source: Authors.

completed in 2016, and contracted to develop the fourth container terminal with PSA (Singapore) with a concession in 2014 because the port capacity for container handling has already reached its limit, as indicated in Fig. 12.3. Moreover, to decrease shipping time and increase the reliability of hinterland rail transport, a western dedicated freight corridor (WDFC), which connects JNP and Delhi, the capital of India, through a newly constructed rail dedicated to freight transport, has been planned and constructed with the support from JICA (Fig. 12.8).

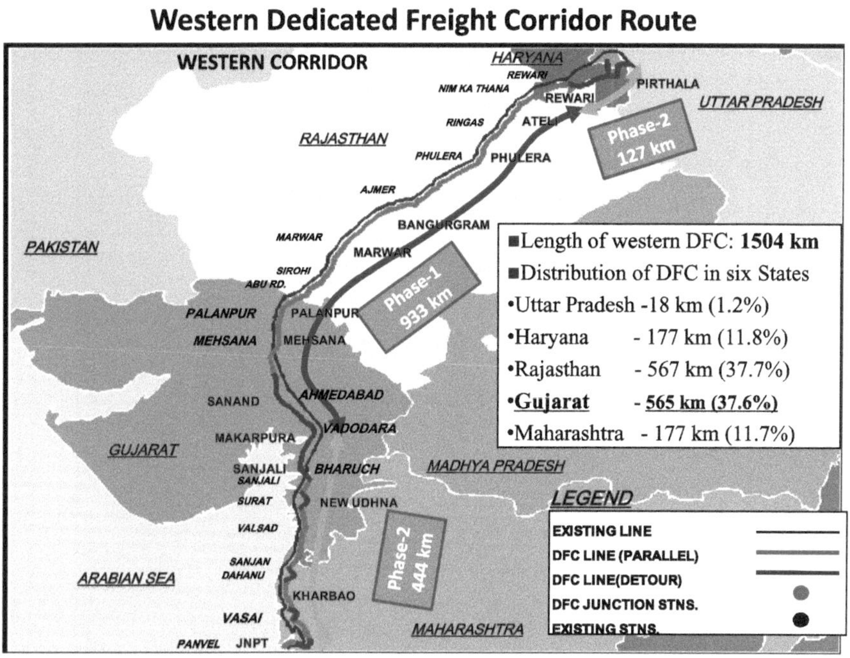

Fig. 12.8 WDFC of India: Connecting Delhi (Dadri) and Mumbai (JNP).
Source: Dedicated Freight Corridor Corporation of India Ltd. (DFCCIL).

Data

Ports and MCS network

The model in this chapter is developed based on 2013 data. In principle, all local container ports in Bangladesh, India, Maldives, Pakistan, and Sri Lanka, where one or more international LS(s) calls, are introduced into the model in addition to the world's major container ports introduced in Chapter 8. Similarly, some local ports around SAS, such as Myanmar, Oman, Mozambique, Tanzania, and the IOI countries are also incorporated in the model. Table 12.1 summarises a list of ports added in this analysis (also see Fig. 12.2 for location).

A MCS network is constructed of the 20 largest LSCs in the world plus 14 small and medium-sized LSCs that have LS networks in SAS. A list of the local LSCs is shown in Table 12.2. Consequently, 932 LSs are included in the model, covering 66% of the global annual vessel capacity. Several ports are not included in the MDS Containership Databank or are not sufficiently covered (less than 30% of the total capacity) by the above LSCs considered in the model; therefore, they are eliminated from the port list described in Table 12.1.

Table 12.1 List of ports added in the model in this chapter.

No.	Port name	Country	Country/region in the WTS	Annual throughput ('000 TEU)		Transshipment time[a] TPR_a (hours)
					source	
43-1	Yangon (Rangoon)	Myanmar	Other Southeast Asia	233	g	48
47-1	Mongla	Bangladesh	Bangladesh	45	i	72
48-1	Haldia	India	India	113	g	72
48-2	Visakhapatnam	India	India	262	g	72
48-3	Krishnapatnam	India	India	59	e	72
49-1	Tuticorin	India	India	469	g	72
49-2	Cochin	India	India	324	g	72
49-3	New Manglore	India	India	46	g	72
49-4	Mormugao	India	India	20	g	72
50-1	Hazira	India	India	50	i	72
51-1	Kandla	India	India	130	g	72
53-1	Male	Maldives	Indian Subcontinent Islands	80	h	72
108-1	Sohar/Mina Qabos (Mascut)	Oman	Arabian Gulf	331	i (2014)	24
165-1	Maputo	Mozambique	Southern Africa	113	i	48
165-2	Nacala	Mozambique	Southern Africa	83	i	48
165-3	Dar es Salam/Zanzibar	Tanzania	East Africa	526	h	48
167-1	Port Victoria	Seychelles	Southern African Islands	41	i	48
167-2	Mutsamudu/Moroni	Comoros	Southern African Islands	46	i (2010)	48
167-3	Toamasina	Madagascar	Southern African Islands	173	h	48
167-4	Pointe des Galets	Reunion (France)	Southern African Islands	213	i	48
167-5	Port Louis	Mauritius	Southern African Islands	622	h	48

[a] Authors' estimation.

Source: Own compilation based on (e) websites for each port or terminal; (g) Indian Ports Association. 2015 Traffic at Major Ports (online). http://www.ipa.nic.in/oper.htm (Accessed 15 September 2015); (h) World Bank, 2015. Container port traffic (TEU: 20 foot equivalent units) (online), http://data.worldbank.org/indicator/IS.SHP.GOOD.TU/countries (Accessed 15 September 2015), for the port in the country where only one container port is considered in operation; and (i) Logistics Capacity Assessment 2015 Wiki (online), http://dlca.logcluster.org/display/public/DLCA/LCA+Homepage (Accessed 15 September 2015), as well as the results from authors' interview survey with port trusts and other organisations.

Table 12.2 List of local LSCs added in the model in this chapter.

No.	Group	Group name	Included carriers	Annual TEU capacity (Authors' estimation from the MDS Containership Databank, '000 TEU)	Share of the world
1–20	Group A to T		Top 20 companies	114,538	64.2%
21	Group U	X-Press	X-Press Feeders	2193	1.2%
22	Group V	Bengal Tiger	Bengal Tiger Line	426	0.2%
23	Group W	OEL	Orient Express Lines	450	0.3%
24	Group X	Emirates	Emirates Shipping Line	477	0.3%
25	Group Y	Wan Hai	Wan Hai Lines	2267	1.3%
26	Group Z	SCI	Shipping Corp of India	186	0.1%
27	Group AA	DAL	DAL Deutsche Afrika-Linien	165	0.1%
28	Group AB	Hub	Hubline	366	0.2%
29	Group AC	RCL	Regional Container Lines	1061	0.6%
30	Group AD	Samudera	Samudera Indonesia	780	0.5%
31	Group AE	Shreyas	Shreyas Shipping	219	0.1%
32	Group AF	Simatech	Simatech Shipping	766	0.4%
33	Group AG	STX	STX Pan Ocean Shipping	611	0.4%
34	Group AH	Far Shipping	Far Shipping	57	0.0%
Others				53,738	30.1%
Total (including top 20 companies)				173,192	100.0%

Note: The table contains all the LSs described in the MDS Containership Databank, including those which do not call at any ports considered in the model.
Source: Own compilation based on the MDS Containership Databank.

HT network

The HT network is considered only in three SAS countries (i.e. Bangladesh, India, and Sri Lanka), as well as each representative link connecting Bhutan and Nepal with their Indian borders, as shown in Fig. 12.9. After all road networks and selected rail links are extracted from the ADC WorldMap, following the description in Chapter 8, an IWT link connecting Dhaka and Chittagong in Bangladesh is added. Furthermore, two international ferry links (i.e. for both road and rail) connecting India (Tuticorin) with Sri Lanka (Colombo) are included for policy simulation as introduced later.

The trailer speed, vRo_a, and annual road capacity, $capRo_a$, for each road link are also typically set according to the description in Chapter 8. Note that the capacities for several primary routes near congested ports (i.e. JNP and Chennai) are limited to the same level as those for the important route because heavy congestion is observed on the roads around these ports (Motono et al., 2014).

Meanwhile, the railway speed, vRa_a, and ferry and IWT speed, vFe_a, are set at 20 and 10 km/h, respectively. Further, capacity per train, $capRa_a$, and vessel, $capFe_a$, are defined as 90 (in India) or 62 (in Bangladesh and Sri Lanka) for rail (TEU/train) and

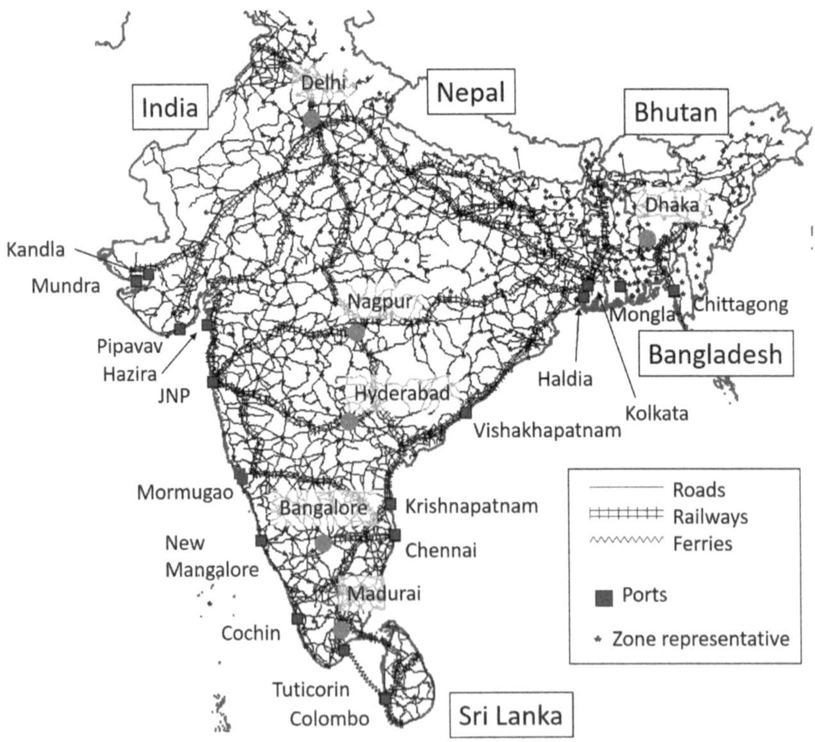

Fig. 12.9 HT network in the model.
Source: Own compilation.

128 for the ferry and IWT (TEU/vessel), respectively, based on the authors' interview survey (e.g. a standard Indian railway has 45-wagon trains, in which each wagon can carry a 40 ft container).

The freight charge for each mode and the handling time at the rail stations and ferry/IWT ports are typically assumed as shown in Table 8.4 of Chapter 8. The annual numbers of trains and ferries for container transport, $freqRa_a$ and $freqFe_a$, are set by link, according to the interview survey, the website of the operating company (such as Container Corporation of India (CONCOR, 2015), a rail company operating container trains in India), and model calibration.

The coefficient of the border barrier, λ_a, which is multiplied by the border-crossing time and cost, is set at 1.0 in the simulations in this chapter because almost no cross-border trade can be observed in international MCS in this region.

Container cargo shipping demand (OD matrix)

The demand of container cargo shipping, Q_{ij}, from region i to j is estimated as follows. First, the demand of container cargo shipping (OD matrix) between countries or regions on a TEU-basis is obtained from the World Trade Service (WTS) database, as with all cases in this book. Second, the country-basis OD matrix acquired above is divided into a port-basis OD matrix according to the port's share of the local container cargo throughput of the country/region. The third step is to estimate the regional-basis OD matrix, Q_{ij}, for five SAS countries. The port-basis OD matrices acquired in the second step are again aggregated into a country basis. Next, they are divided into a regional basis, namely, India is divided into 575 administrative districts; Bangladesh into 64 districts; Sri Lanka into nine provinces; whereas Bhutan and Nepal are not divided.

The country-basis OD matrix is divided into a regional basis with the share of each region for each trade partner, as acquired from the Indian Customs Data and the output of the Geographical Simulation Model (GSM, see Chapter 9). Because the export/import value in India by trade partner is available for each state from the Indian Customs Data, the container shipping demand to/from India is first divided according to the shares by state for each trade partner. Figs 12.10 and 12.11 show breakdown by customs including seaports, airports, and ICDs, of Indian export and import cargo by states as of March 2013. Subsequently, container cargo shipping demand to/from Indian states, Bangladesh, and Sri Lanka are divided into the detailed regions by share according to the GSM output.

The share of each region for each trade partner is estimated from the output of the GSM as follows. First, the containerised rate of the cargo to/from the three countries is computed on a tonnage basis by trade partner country for each goods item in the GSM, using the WTS data, where the commodity specifications are more detailed than these in the GSM and where breakdowns of containerised and non-containerised cargo are available. Then, the containerised rates estimated by trade partner country and by goods item are multiplied by the values in the OD matrix estimated from the GSM. Finally, the share of containerised cargo by trade partner country for each region is computed and estimated for all goods items after the aggregation of the container OD matrix above.

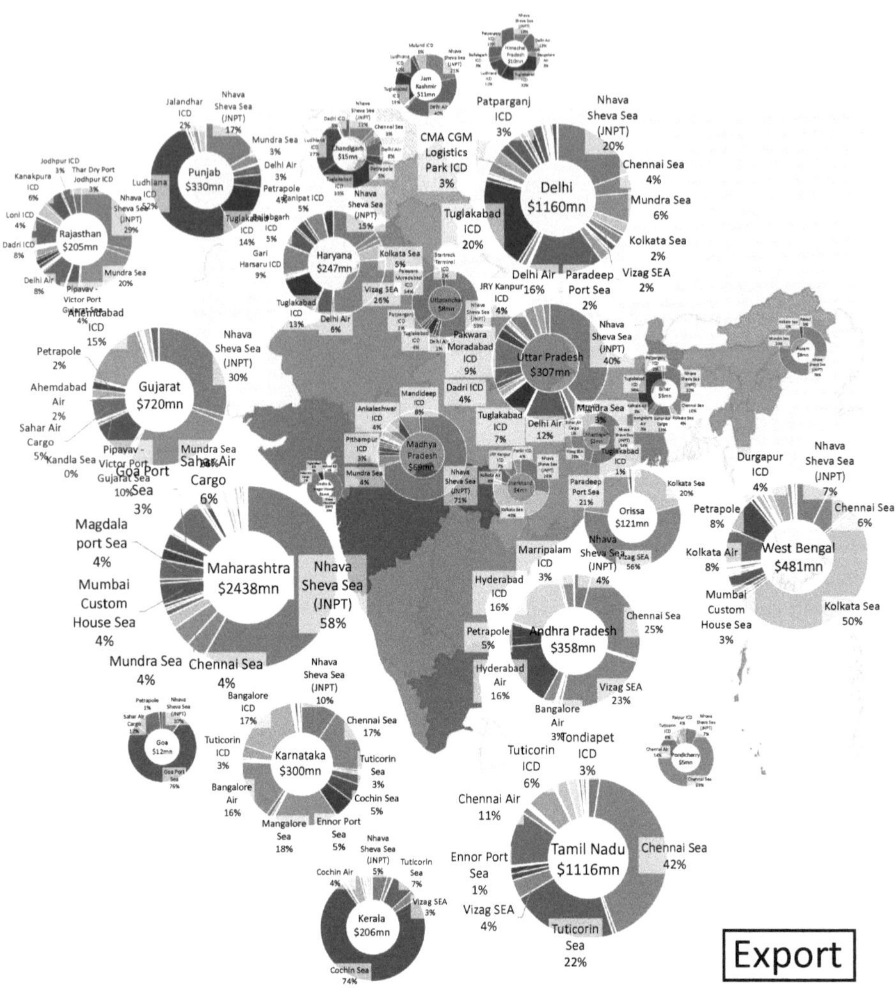

Fig. 12.10 Breakdown of Indian export cargo by states (March 2013) by customs including seaports, airports, and ICDs.
Source: Own compilation based on Indian Customs Data provided by Export Genius Inc.

In some of the following simulations, the future shipping demand (OD matrix) as of 2030 is input. In such simulations, the output by the GSM is used as the growth rate of the OD matrix. The GSM OD matrix is provided for the year of 2010 and 2030, and these two OD matrices are identical in terms of zones and goods items (six types). Here, the growth rate is computed for each of the OD pairs due to the fact that growth rates vary across OD pairs. Note that the average growth rate to 2030 from 2010 in the value-based amount of shipping demand in all OD pairs of the world is estimated at 2.43. Similarly, in OD pairs to/from Bangladesh, India, and Sri Lanka is estimated at 3.51, 4.06, and 3.89, respectively.

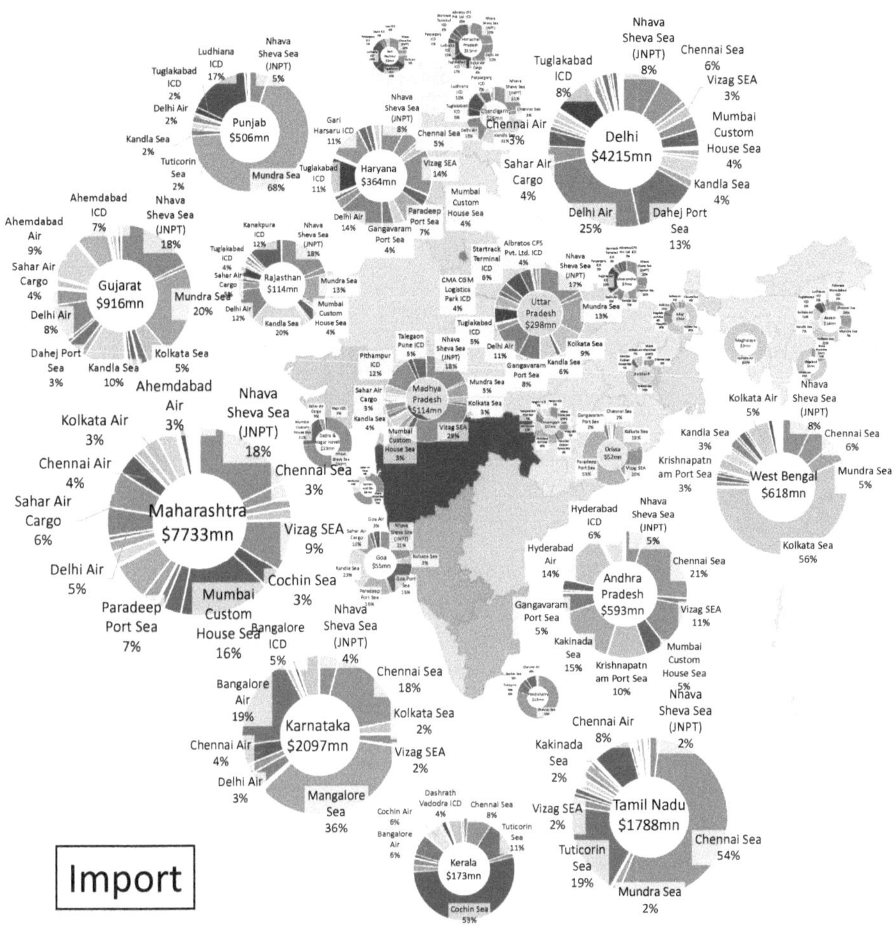

Fig. 12.11 Breakdown of Indian import cargo by states (March 2013) by customs including seaports, airports, and ICDs.
Source: Own compilation based on Indian Customs Data provided by Export Genius Inc.

Calculation results

Transshipment containers at Colombo port

Table 12.3 summarises the laden container flows transshipped at Colombo port, which are estimated in the model as of 2013. Geographic locations for each region categorised in Table 12.3 are shown in Fig. 12.2. The results indicate that containers transshipped from neighbouring regions (i.e. the Bay of Bengal and the Arabic Sea, including the Indian Coast) through long-distance LSs, including trunk lines both westbound (Mediterranean, Europe, and the ECNA, or MEE) and eastbound (Northeast and

Table 12.3 Estimated flow of laden containers transshipped at Colombo port (1000 TEU, 2013).

From	To									
	BB-IE	IS	AS-IW	NSA	MEE	EA	WA	Australia	Others	Total
Bay of Bengal and Indian East Coast (BB-IE)	4.5	0.6	63.3	7.7	178.8	11.1	8.5	6.8		281.3
Indian South Coast (IS)	2.1	0	15.8	56.1	39.2	3.1	6.8			123.1
Arabian Sea and Indian West Coast (AS-IW)	39.5	4.6	7.6	20.7	121.0	13.0	27.4		0	233.9
Northeast and Southeast Asia (NSA)	6.7	60.6	7.9	0	15.0	7.0	0		2.1	99.2
Mediterranean, Europe, and East Coast of North America (MEE)	109.0	21.3	11.6	13.9	65.2	5.0	0.4	21.1		247.7
East Africa (EA)	4.0	1.0	0.6	0.001	12.1	0	0.007		0.03	17.8
West Africa (WA)	11.7	0.7	3.8	0.05	9.0	0.02				25.2
Australia	1.3		6.0	0.3	40.1	0.9				48.6
Others				0		0.1				0.1
Total	178.9	88.7	116.6	98.8	480.5	40.2	43.2	27.9	2.2	1077.0

Source: Shibasaki, R., Kawasaki, T., 2019. A transhipment hub in South Asia and its competition: application of network equilibrium assignment model for global maritime container shipping. Asian Transp. Stud. 5 (4), 546–569.

Southeast Asia), as well as African LSs comprise 493,524 TEU (the grey-coloured cells with underlined figures in Table 12.3), which represents almost half of the total transshipped containers. Meanwhile, the table also reveals that containers with the reverse routes carry in total 238,941 TEU (the grey-coloured cells without underlined figures), which is less than half of the other direction. Intra-regional transshipment is much lower (137,992 TEU, the thick-framed cells). These estimated results are consistent with the findings from the authors' interview surveys with LSCs in SAS. Note that transshipment between MEE and other long-distance LSs is not negligible (181,958 TEU, the thick-dotted-framed cells), partly because this may include containers to/from the Arabian Gulf, where MEE services call at these ports.

Container throughput in SAS ports and hinterland cargo flow

Fig. 12.12 compares the observed versus estimated laden container throughput at SAS ports as of 2013 (excluding transshipment containers). Further, Fig. 12.13 compares the shares of container throughput by region. From these figures, laden container throughput handled by each port, which is estimated by the model, aligns with the overall observed data, although the balance in the same region (e.g. JNP, Pipavav, and Mundra in west coast ports in India) should be further improved.

Fig. 12.14 illustrates the annual laden container flows of the HT network, including road, rail, international ferry, and IWT in SAS. As shown in the figure, several significant cities are located away from the coast and long-distance transport is used, such as between Delhi and JNP or Mundra port, Nagpur/Hyderabad and JNP, and Bangalore and Chennai port. Another observation is that the national border affects transport significantly. For example, the amount shipped by international ferry between India and Sri Lanka is very small. Another example can be observed in the 'Seven Sisters' in India (Indian north-eastern states), where almost all cargo to/from is shipped for import from/export to Kolkata or Haldia port via the Siliguri Corridor

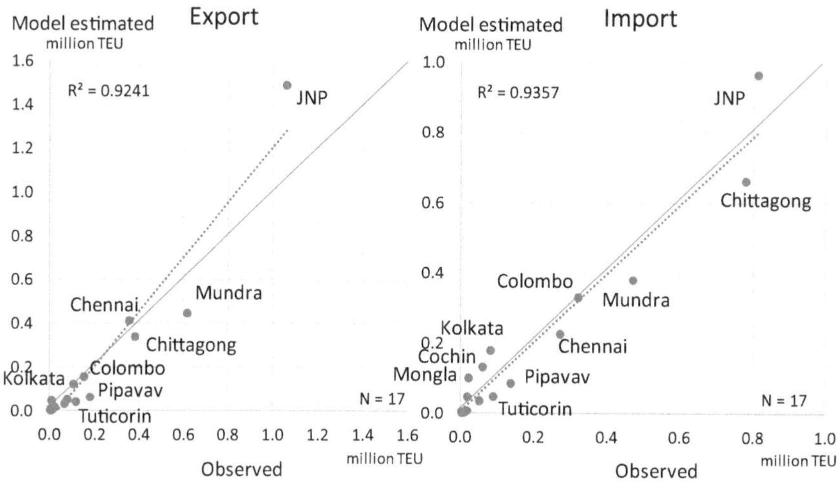

Fig. 12.12 Comparison between observed and model estimated laden container throughput at SAS ports (as of 2013, excluding transshipment containers).
Source: Shibasaki, R., Kawasaki, T., 2020. International intermodal container shipping network in South Asia: Modelling and policy simulations. Int. J. Ship. Transp. Logist. (accepted).

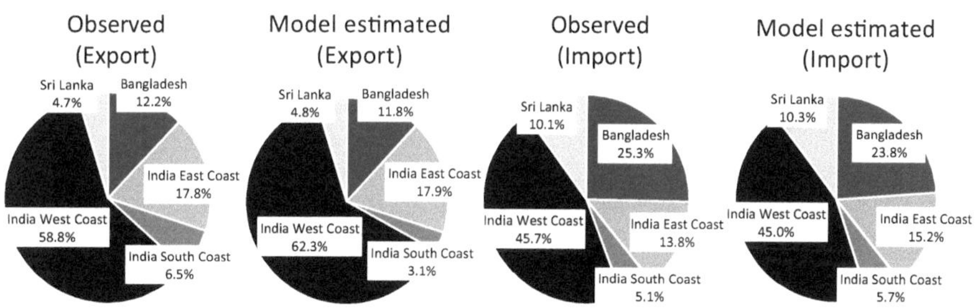

Fig. 12.13 Comparison between observed and model estimated shares of laden container throughput by region at SAS ports. *India East Coast: Kolkata, Haldia, Vishakhapatnam, Krishnapatnam, Chennai; India South Coast: Tuticorin, Cochin, New Mangalore; India West Coast: Mormugao, JNP, Hazira, Pipavav, Kandla, Mundra.
Source: Own compilation based on Shibasaki, R., Kawasaki, T., 2020. International intermodal container shipping network in South Asia: Modelling and policy simulations. Int. J. Ship. Transp. Logist. (accepted).

(which is known as the 'Chicken Neck of India'), and not Chittagong port, which is the closest to these states and larger in terms of container throughput.

Fig. 12.15 compares the observed and model estimated laden container shares by transport mode between Dhaka and Chittagong in Bangladesh. It indicates that the model describes the actual modal split well.

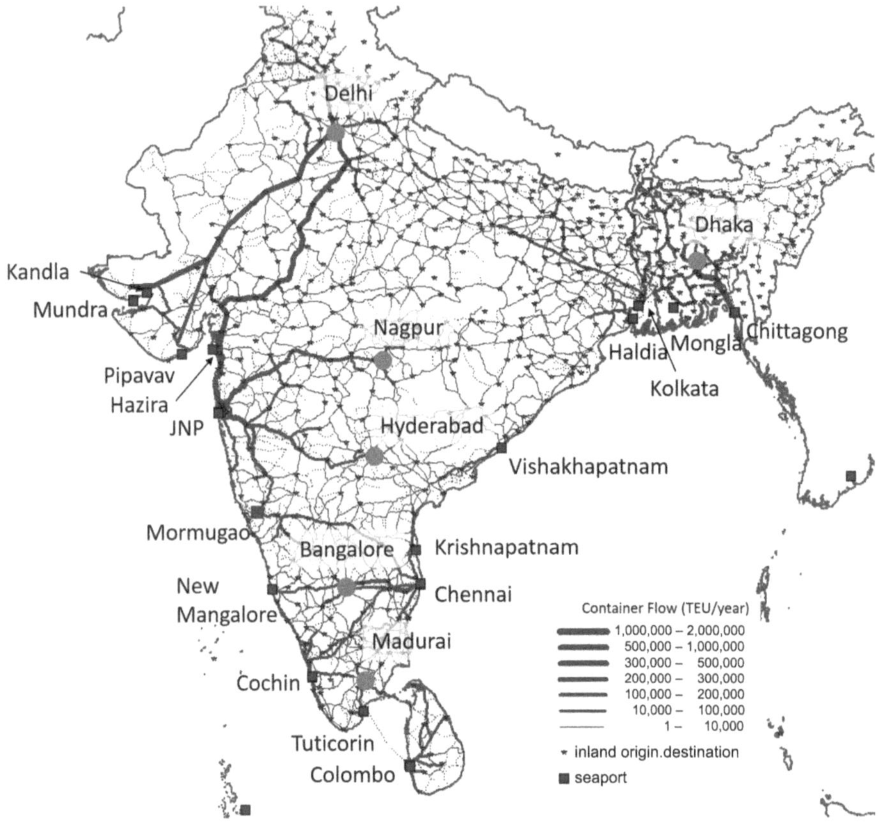

Fig. 12.14 Estimated laden container flows of the HT network (as of 2013).
Source: Shibasaki, R., Kawasaki, T., 2020. International intermodal container shipping network in South Asia: Modelling and policy simulations. Int. J. Ship. Transp. Logist. (accepted).

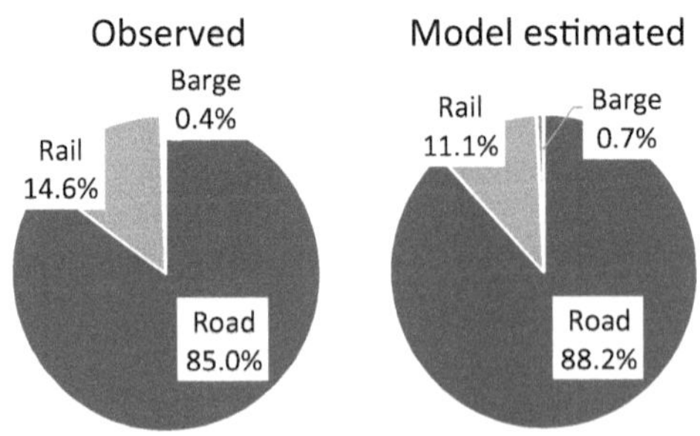

Fig. 12.15 Comparison between the observed and model estimated laden container shares in transport modes between Dhaka and Chittagong.
Source: Own compilation based on Shibasaki, R., Kawasaki, T., 2020. International intermodal container shipping network in South Asia: Modelling and policy simulations. Int. J. Ship. Transp. Logist. (accepted).

Policy simulations

Six policy scenarios are simulated as follows. The current shipping demand (as of 2013) is input for the first four scenarios, whereas the future shipping demand (as of 2030) is input for the last two scenarios.

Improvement of transshipment service at Colombo port

Although the cargo handling at Colombo port is more efficient than that at other ports in SAS, its quality of port facilities and service (e.g. berth depth, yard area, warehouses, and crane efficiency) is still poorer than those at the world's major hub ports in neighbouring regions such as Singapore and Tanjung Pelepas (Malaysia). Therefore, a new port area (South Harbour) with deeper berths (– 18 to 20m) is developed next to the existing container terminals and the first terminal opens in 2013, as shown in Fig. 12.6. This simulation assumes that the quality of transshipment service at Colombo port is improved to the level of neighbouring major hub ports.

Fig. 12.16 summarises two simulation results related to the improvement of transshipment cargo handling at Colombo port; S1-1 assumes that the transshipment time at Colombo port is equal to that at other regional major ports such as Port Klang (Malaysia) and Shanghai (i.e. $TPR_{colombo}=24$); and S1-2 assumes the transshipment time at Colombo port is equal to that at other major transshipment hubs such as Singapore and Hong Kong (i.e. $TPR_{colombo}=12$). Note that the transshipment time at Colombo port in the base scenario is the current level ($TPR_{colombo}=48$). The results indicate that the laden containers annually transshipped at Colombo port increase by

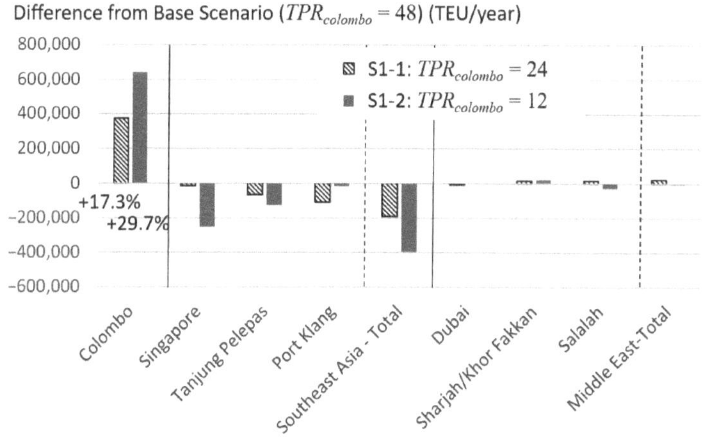

Fig. 12.16 Estimation results in the improvement scenarios for Colombo port: Changed number of laden containers transshipped at Colombo and neighbouring major hub ports. *Source*: Own compilation based on Shibasaki, R., Kawasaki, T., 2019. A transhipment hub in South Asia and its competition: application of network equilibrium assignment model for global maritime container shipping. Asian Transp. Stud. 5 (4), 546–569.

17.3% in S1-1 and by 29.7% in S1-2. Although the impact on each neighbouring major hub port is not consistent, the total number of containers transshipped in Southeast Asia hub ports decreases as the level of service in Colombo port improves; in contrast, the number in the Middle East hub ports is not affected significantly. In other words, the current major competitors of Colombo port are the Southeast Asia ports, not the Middle East ports.

Another observation is that the decrease in the number of laden containers transshipped in Southeast Asia ports is around half of the increase in the number of containers transshipped in Colombo port. In other words, half of the increased transshipped laden containers in Colombo port changes from direct shipping, that is, without any transshipment. This implies that the improvement level of service for transshipment handling at a port would increase the total number of transshipped containers in the region and not just shift them from competitors.

Construction of new transshipment hub in south India

A new hub port in south India, which is located between two existing container ports (Tuticorin and Cochin), is assumed to be constructed in this scenario (S2-1). It also assumes that the LSs operated or jointly operated by a LSC group, whose eight LSs call at Colombo port, totally shift from Colombo to the new hub port. Second, it assumes there are no container cargo shipping demands exported from or imported into the new port because it is located in a remote area in south India. Furthermore, the transshipment times in both Colombo and the new port are assumed to be 12 (hours) as a result of severe competition between them.

The simulation results are presented in Table 12.4. As the laden containers annually transshipped by the shifted LSC group account for over 30% at Colombo port, the number of laden containers transshipped at Colombo port declines significantly. Meanwhile, the number of laden containers transshipped at the new hub port is estimated to reach almost one-third of that at Colombo port. Note that the total number of laden containers transshipped at either Colombo or the new port is smaller than the results under S1-2, for the shifted group as well as the total for other groups; this is because several LSs that are partially slot-chartered by the group shifted remain to call at Colombo port, whereas a LS jointly operated by another group calls at the new port under S2-1.

Note that these computation results are based on simple assumptions regarding the changes in the LS network; thus, the mechanism of the change in LS networks by carriers should be endogenised in the model as a next step if more precise feasibility of the new container terminal requests is studied.

Hinterland improvement in JNP

In this simulation, capacities in the HT network around JNP, known as a congested port, especially on the landside, are improved through infrastructure investments. The first scenario (S3-1) assumes that its capacities are improved from the 'important route' level to the 'primary route' level in each direction. The second scenario (S3-2) assumes that rail capacity is also improved, reflecting the WDFC project cooper-

Table 12.4 Estimation results in the new transshipment hub scenario: the number of laden containers (1000 TEU, 2013) transshipped at Colombo and a new hub port in south India.

	S1-2: monopoly by Colombo port (where $TPR_{colombo}=12$)			**S2-1: construction of new transshipment hub (where $TPR_{colombo}=TPR_{newport}=12$)**					
	Number of transshipped containers			**Number of transshipped containers**			**Difference from S1-2**		
Ports	**Total in all LSCs**	**A LSC group**	**Other groups total**	**Total in all LSCs**	**A LSC group**	**Other groups total**	**Total in all LSCs**	**A LSC group**	**Other groups total**
Colombo	2793	788	2006	1963	0	1963			
A new hub port in south India	–	–	–	658	658	0			
Two ports total	2793	788	2006	2621	658	1963	−172	−130	−42

Source: Shibasaki, R., Kawasaki, T., 2019. A transhipment hub in South Asia and its competition: application of network equilibrium assignment model for global maritime container shipping. Asian Transp. Stud. 5 (4), 546–569.

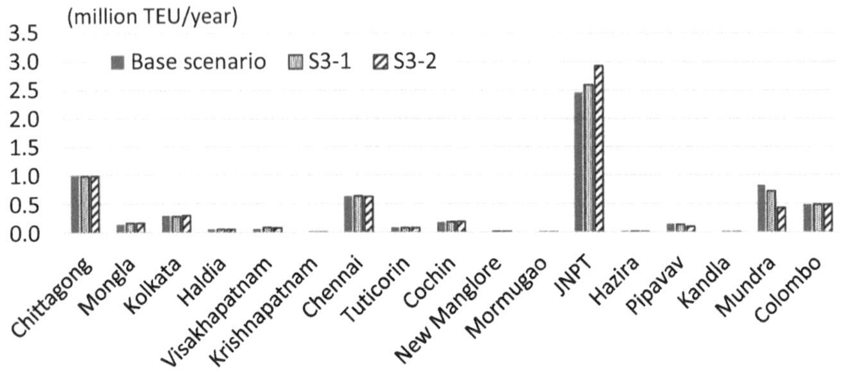

Fig. 12.17 Exported and imported number of laden containers at each port for hinterland improvement scenarios in JNP.
Source: own compilation.

ated by JICA, where the capacity per train, $capRa_a$, between Delhi and Indian West Coast ports (i.e. not only JNP but other ports, such as Hazira, Pipavav, and Mundra, are included) is assumed to increase 4 times (i.e. 360 TEU/train), by introducing a double-stack train whose length will be double, according to the interview survey. Moreover, railway speed, vRa_a, is also increased to 80 km/h, which is 4 times as fast as in the base scenario; operational charge, $CORa$, is reduced to 0.25 (USD/km/TEU), which is half of that in the base scenario; and service frequency in each link, $freqRa_a$, is increased 4 times that of the base scenario.

Fig. 12.17 illustrates the simulation results. It reveals that the number of laden containers annually handled in JNP in each scenario should increase by 130,000 and 470,000 TEU, respectively for each scenario from the base scenario. On the other hand, the numbers of containers in other west coast ports, such as Pipavav and Mundra, decline. Among them, the decreased amount in Mundra is significant. Note that even in S3-2, where the rail capacities to other west coast ports also increase, the cargo is estimated to shift from these ports to JNP. This implies that the current capacity constraint at JNP is significant, and some cargo currently handled in other ports stays away from JNP to avoid heavy congestion, which aligns with the results of the interview survey.

Ferry and rail improvements in south India and Sri Lanka

The initiation of international ferry service connecting south India (Tuticorin port) with Sri Lanka (Colombo port), which is currently suspended, is considered in this simulation. Further, rail connections in south India (between Chennai and Tuticorin) are also assumed to improve. The frequency of the ferry service, $freqFe_a$, is set at 28 (for trucks) and 7 (for rail) vessels/week and vessel speed, vFe_a, is 30 (km/h), whereas other levels of service are assumed to be the same as the existing IWT service in other areas. Moreover, the frequency of rail services in south India, $freqRa_a$, is assumed to increase significantly by 14 trains/week from 0.5 trains/week in the base scenario and the operational charge, $CORa$, to equate to half of the base scenario, whereas other levels of services are assumed to be the same as the existing ones, unlike in the WDFC scenario.

Furthermore, the additional cost and time for border crossing, CB_a and TB_a, between India and Sri Lanka are also assumed to decrease. Because a reduction in border-crossing barriers is considered to significantly affect the amount of international ferry shipping, three scenarios are prepared, namely, CB_a and TB_a are assumed to be half (S4-1), one-fourth (S4-2), and zero (S4-3) of the existing figures, respectively.

Fig. 12.18 summarises the estimated numbers of exported and imported laden containers annually handled in south Indian ports (i.e. Chennai, Tuticorin, and Cochin) and Colombo port for each of the three scenarios and the base scenario. Moreover, Fig. 12.19 illustrates the differences in annual laden container flows between S4-2 and the base scenario. The estimated load factors of the ferry link for three scenarios (S4-1 to S4-3) are 2.4%, 16.3%, and 60.6% (excluding empty containers), respectively, which reveal that the level of barrier reduction for border crossing causes significant differences in the use of the international ferry, as expected.

Another observation is that at both end ports of the ferry service (i.e. Tuticorin and Colombo), the total number of exported and imported laden containers shipped by containership and ferry increases as the border-crossing barrier is mitigated. The number of exported and imported laden containers at Colombo port seems to significantly increase in S4-3. However, the total number of laden containers handled at Colombo port, including transshipped containers, is nearly the same across the scenarios. In other words, some containers exported from/imported into south India with transshipment at Colombo port in the base scenario are shifted towards international ferry use and are handled at Colombo port as exported or imported containers in S4-3.

Multimodal use in HT in Bangladesh

Main transport mode between Dhaka and Chittagong in Bangladesh is currently road transport as shown in Fig. 12.15. In this route, the modal shift from truck to rail

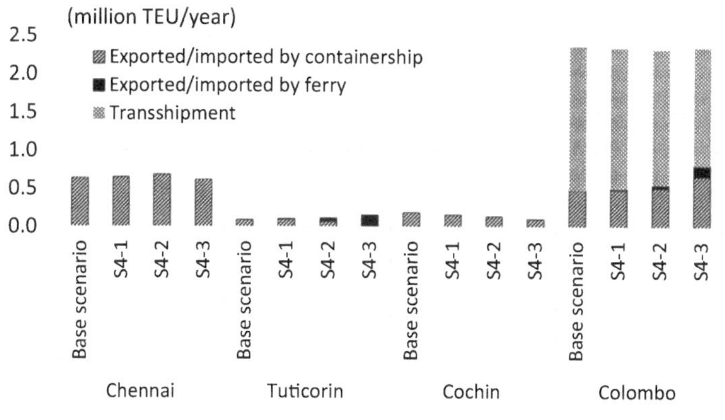

Fig. 12.18 Estimated number of laden containers at each port for international ferry and rail improvement scenarios in south India and Sri Lanka.
Source: Own compilation.

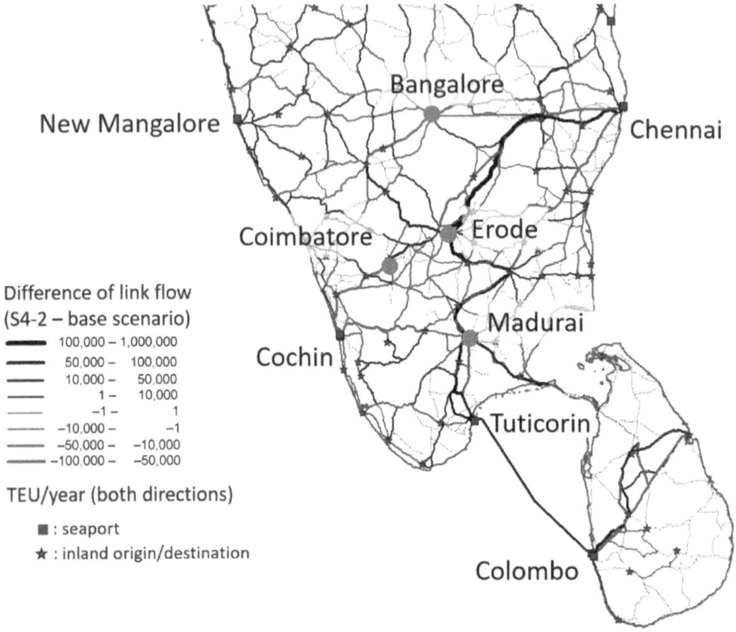

Fig. 12.19 Estimated difference in laden container flow of the HT network between S4-2 and the base scenario.
Source: Shibasaki, R., Kawasaki, T., 2020. International intermodal container shipping network in South Asia: Modelling and policy simulations. Int. J. Ship. Transp. Logist. (accepted).

or IWT is necessary due to severe congestion on road. Therefore, this simulation assumes the improvement of multimodal HT network including road, railways, and IWT in Bangladesh. Note that future shipping demand (container OD matrix) as of 2030 is input in this simulation. Scenarios for each transport mode are developed as follows.

First, roads between Dhaka and Chittagong are of only one lane for one-direction and pavement condition is relatively poor for several sections. It is assumed that road development for a four-lane road and rehabilitation of existing roads are to be carried out. Consequently, in the 2030 base case (S5-1), driving speed is improved to 60 km/h. In contrast, the above improvement is not realised in the 'without case' (S5-2).

Second, there are multiple single truck sections between Dhaka and Chittagong for the railway connection. Moreover, the rail departs only after the train becomes full (its capacity is 62 TEU). Thus, rail transport has the problem on reliability or punctuality. In S5-1, all the railway sections are double-tracked with higher frequency (4 trains/day) and speed (40 km/h). Further, new rail construction and significant improvement of existing railways connecting with the northeast states of India are assumed (1 train/day). In contrast, the above improvement is not achieved in S5-2.

Third, container haulage by barge between Pangaon ICT and Chittagong port is currently in operation, although traffic volumes are low. Assuming improvement of competitiveness between Pangaon ICT and Chittagong port, frequency and vessel speed will be increased up to 7 services/week and 20 km/h, respectively, in S5-1.

Besides, road improvement and rail construction to access Mongla port are incorporated in S5-1. Furthermore, a reduction of the border-crossing barrier is also incorporated to be half of the current level in S5-1. In this case, container flows from Indian north-eastern states will possibly enter Bangladesh and use Chittagong port. Fig. 12.20 illustrates improved or newly constructed links in the Bangladesh HT network assumed in S5-1.

Regarding any other roads and rails in SAS, capacity is assumed to increase by the average growth rate of shipping demand in each country (i.e. 3.51, 4.06, and 3.89 times of the current capacity for Bangladesh, India, and Sri Lanka, respectively) in both S5-1 and S5-2, for meeting with the increase in demand generally, as described in Chapter 8. Regarding the rail capacity, service frequency is assumed to increase. Similarly, the capacity of each LS is assumed to increase by the world average growth rate of shipping demand (2.43 times), by enlarging the vessel size in principle (if the increased vessel size reach the maximum (i.e. 24,000 TEU), the service frequency is proportionally increased).

The change in laden container flow of the Bangladesh HT network from S5-1 to S5-2 is shown in Fig. 12.21. As seen in the figure, barge usage is accelerated (+217,000

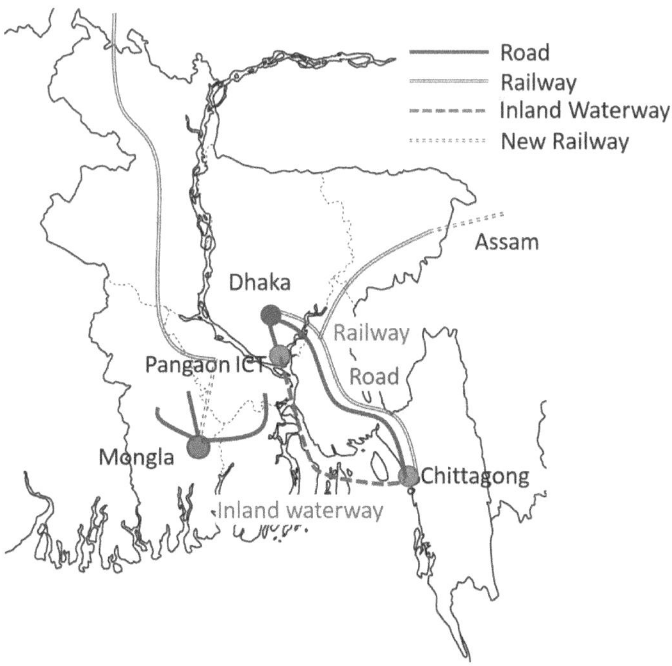

Fig. 12.20 HT network in Bangladesh to be improved.
Source: Japan International Cooperation Agency (JICA), 2016. Data collection survey on cross-border maritime traffic in South Asia, final report. Available at http://open_jicareport.jica.go.jp/pdf/12251989.pdf (Accessed 24 August 2018).

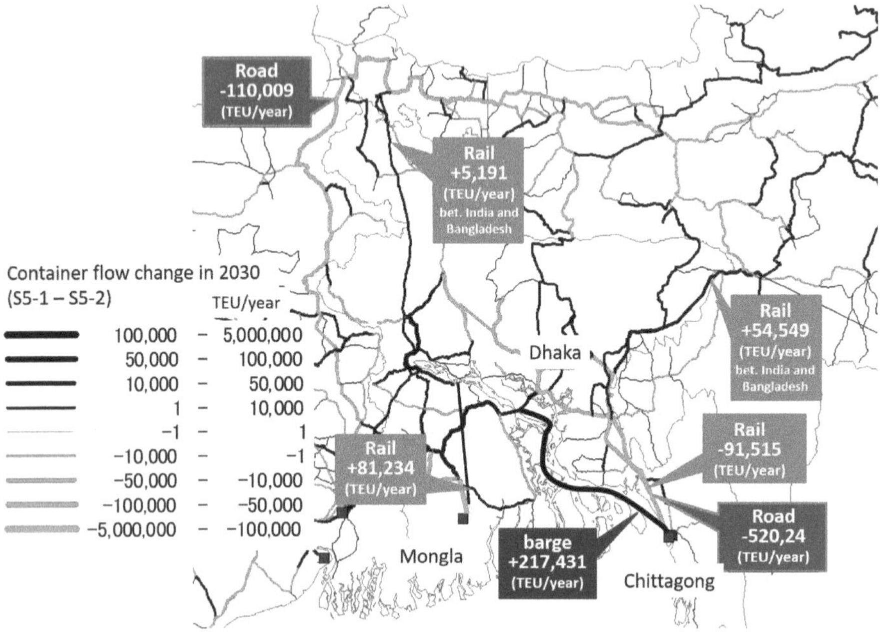

Fig. 12.21 Estimated change in laden container flow for Bangladesh HT investment between S5-1 and S5-2.
Source: Japan International Cooperation Agency (JICA), 2016. Data collection survey on cross-border maritime traffic in South Asia, final report. Available at http://open_jicareport.jica.go.jp/pdf/12251989.pdf (Accessed 24 August 2018).

TEU/year) whereas road (−520,000 TEU/year) and rail (−91,000 TEU/year) usage is decreased from S5-2 to S5-1.

Modal share between Dhaka and Chittagong in both scenarios are shown in Fig. 12.22. In S5-1, the share of road transport is expected to significantly decrease from the current level (see Fig. 12.15), whereas that of rail and IWT increase. If comparing to the shares in S5-2, the road share decreases by 12.3 percent points, whereas the share of IWT increases. These results imply that the progress of HT environment will encourage the shift from road to IWT because the improvement of the level of service in IWT is assumed to be the most significant in this scenario.

New port construction in east coast India

Andhra Pradesh (AP) state, situated along east coast India, was divided into new AP and Telangana state in 2014. Hyderabad, one of the major cities of India, was a capital of the old AP state (before separation) and is currently a capital of Telangana state, whereas a capital city of new AP state is planned to be newly constructed. Concurrently, a new container port is planned to be constructed in Machilipatnam, located near the new capital. This simulation assumes an opening of the new container port in Machilipatnam, as well as other two new ports in different cities (i.e. Kakinada and Bhavanapadu) in the new AP state (see Fig. 12.23). The future cargo shipping

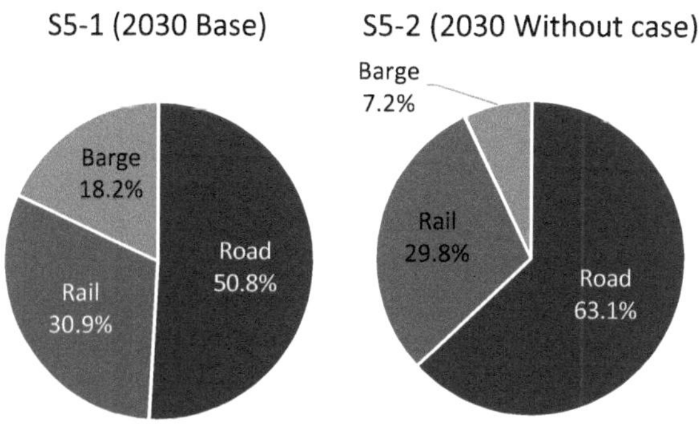

Fig. 12.22 Modal share of laden containers transported between Dhaka and Chittagong in Bangladesh HT investment scenarios.
Source: Japan International Cooperation Agency (JICA), 2016. Data collection survey on cross-border maritime traffic in South Asia, final report. Available at http://open_jicareport.jica.go.jp/pdf/12251989.pdf (Accessed 24 August 2018).

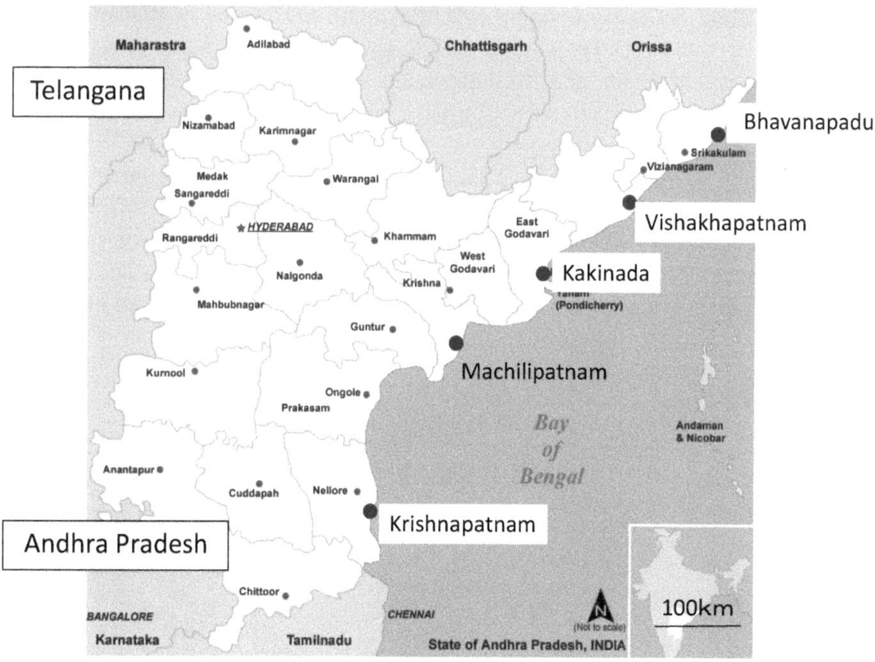

Fig. 12.23 Andhra Pradesh state and existing/future container ports.
Source: Own compilation based on a map contained in Wikipedia.

demand (as of 2030) is input and future capacities in vessels, roads, and railways are also assumed to correspondingly increase as with the previous simulation.

The LSs to call at new constructed ports are assumed as the following steps. First, the authors select all regional LSs to navigate off the coast of the port in question as candidates, which are defined as the LSs to call at both south neighbouring ports (e.g.

Chennai and Krishnapatnam) and north neighbouring ports (e.g. Vishakhapatnam, Kolkata, and Haldia). Then, the LSs whose average vessel size (as of 2030) exceed the maximum size that the port can accommodate are eliminated from the candidates. The maximum vessel sizes of Machilipatnam, Kakinada, and Bhavanapadu are assumed 7000, 4000, and 3000 TEU, respectively, based on a development plan of each port. As a result, 4 LSs are assumed to call at Machilipatnam port, whereas 3 LSs call at Kakinada and Bhavanapadu port, respectively.

Fig. 12.24 indicates the estimated number of laden containers annually handled in each AP port as well as its neighbouring ports (Kolkata/Haldia and Chennai) as of 2030. In the first scenario (S6-1), any new ports are not opened in the AP state; only two existing ports (Visakhapatnam and Krishnapatnam) are available. In the second scenario (S6-2), Machilipatnam port is assumed to open, whereas in the third scenario (S6-3), the other two ports (Bhavanapadu and Kakinada) start to handle containers. In S6-2, the total number of laden containers in all AP ports is expected to increase from that in S6-1, whereas each number of laden containers handled in all existing ports (Kolkata/Haldia, Visakhapatnam, Krishnapatnam, and Chennai) is expected to decrease.

Fig. 12.25 describes the difference in laden container flow of the HT network between S6-1 and S6-2. By opening Machilipatnam port, some containers to/from inland cities of India (i.e. Hyderabad and Nagpur) are estimated to shift to use it from JNP, Vishakhapatnam, and Krishnapatnam. Fig. 12.26 shows the hinterland of Machilipatnam port, which covers not only the central AP region but also Hyderabad and Nagpur.

Another finding from Fig. 12.24 is that the total number of laden containers in all AP ports in S6-3 becomes smaller than that in S6-2. This implies a possibility that distributive investment (i.e. opening three container terminals in the same state) may negatively affect by dispersing the power of the ports in the state to attract container cargo. This is mainly caused by increased MCS time due to the increased number of

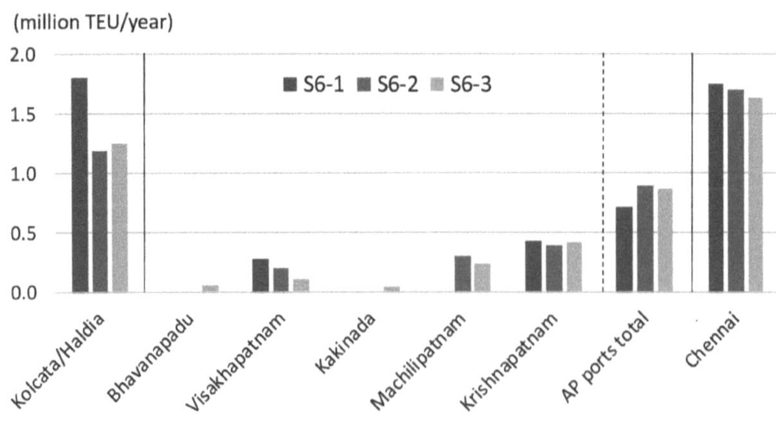

Fig. 12.24 Estimated number of laden containers at AP and its neighbouring ports for new ports construction scenario (as of 2030).
Source: Own compilation.

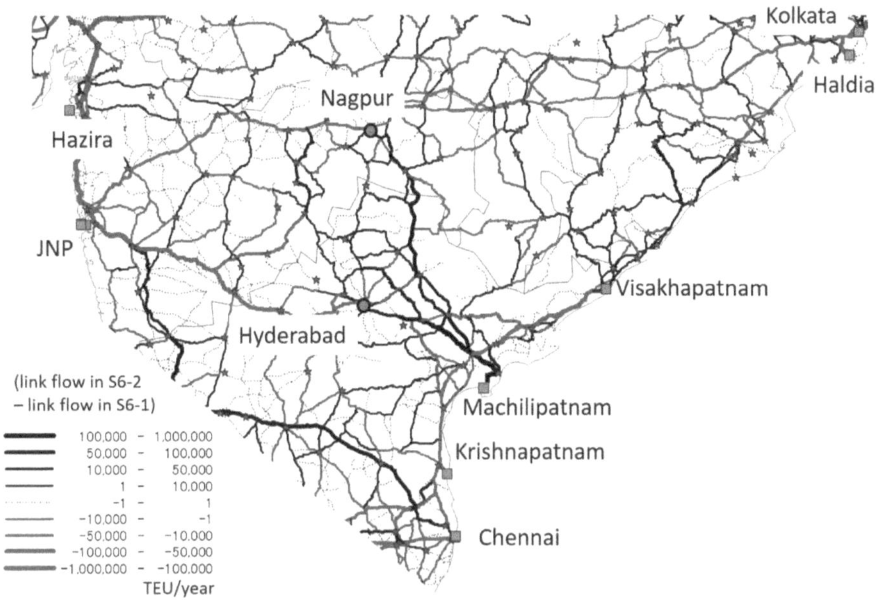

Fig. 12.25 Difference in laden container flow of the HT network between S6-1 and S6-2. *Source*: Own compilation.

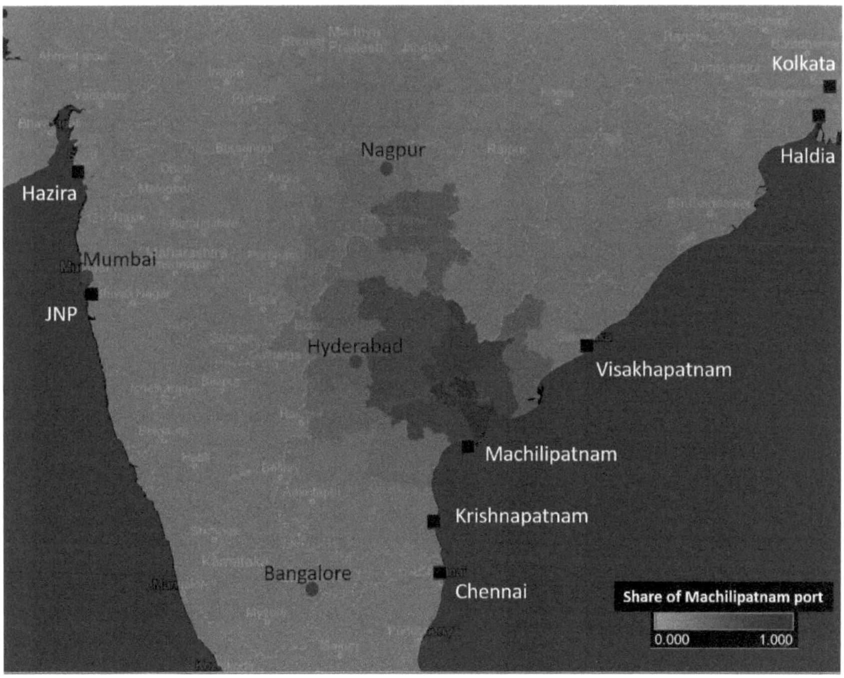

Fig. 12.26 Estimated hinterland of Machilipatnam port in S6-2. *Source*: Own compilation.

ports to call for each LS. In other words, if planning a new container terminal, collective investment is an important key to maintain the optimal container throughput.

Conclusion

This chapter focused on the international intermodal container shipping network in SAS, particularly India, Bangladesh, and Sri Lanka. The first two simulations focused on Colombo port as the regional hub in SAS. First, the number of containers transshipped at Colombo port would increase whereas it decrease in neighbouring hub ports, as the transshipment time at Colombo port decreases. Moreover, the construction of a new container hub in south India would cause a significant decrease in the number of containers transshipped at Colombo port, if the new port is able to successfully attract all LSs operated by the largest LSC. Because such a drastic shift from an old hub to a new one has occasionally been observed in the real international MCS market (e.g. the shift to Tanjung Pelepas from Singapore by Maersk and Evergreen in the 1990s), the shift to the new port from Colombo port is quite possible, although the simulation results shown here were based on the simplified model that the LS network was provided and its change was not endogenised.

Another four simulations related to the logistics policies on the HT in SAS were then examined; namely, the improvement of the level of HT infrastructure around JNP, the international connection by ferry between India and Sri Lanka, multimodal use in the HT between Dhaka and Chittagong in Bangladesh, and new port construction in the AP state of India. The simulation results showed that Indian West Coast ports would be very competitive in attracting cargo originating from/destined to Delhi and, among them, a capacity constraint of the HT around JNP would be critical. Another finding showed the reduction of cross-border barriers would be essential for promoting international ferry services between south India and Sri Lanka as well as the improvement in connecting rail services. Encouraging the use of IWT and rail between Dhaka and Chittagong could raise the share of these modes. The new container port in the AP state would encourage shifting cargo of inland cities such as Hyderabad and Nagpur to it, although the development of multiple ports in the same state might weaken the attractiveness as a whole of the state.

Throughout the scenario analyses, the authors found that some simulation results indicate a drastic change in cargo flow from the current status or normally expected future situation. In particular, a considerable improvement of HT network to promote multimodal transport, including rail, international ferry, and IWT, would significantly affect, not only the HT pattern but also the choice of export/import port, which reflects the fact that these modes are currently facing with many challenges as described in the former part of this chapter.

Meanwhile, there still remains some challenges on the model development. Particularly, in all simulations introduced in this chapter, the LS network is fixed or its change is exogenously assumed. In reality, if the regional shipping environment changes, each LSC should optimally change its LS network, including vessel size, frequency, and port of call. Such behaviour of LSCs should be incorporated in the model, although it would not be easy if the model should optimise the entire global shipping network.

References

Container Corporation of India Ltd. (CONCOR). (2015). International Service List (online). http://www.concorindia.com/ShowService.asp?typeID=International&typeRJ=Train (Accessed 15 December 2015).

Motono, I., Furuichi, M., Kimoto, H., Suzuki, S., 2014. A new concept of off-dock container traffic control for heavily conested ports. In: Presented at the Fifth International Conference on Transportation and Logistics (T-LOG 2014), Bangkok, Thailand, 28–30 July 2014.

World Bank, 2015. Container port traffic (TEU: 20 foot equivalent units) (online). http://data.worldbank.org/indicator/IS.SHP.GOOD.TU/countries. (Accessed September 15, 2015).

World Bank, 2018. Global Economic Prospects, June 2018: The Turning of the Tide? World Bank, Washington, DC. https://doi.org/10.1596/978-1-4648-1257-6.

World Bank Group, 2015. Doing Business Website "Trading Across Borders" 2015 Rank (online). http://www.doingbusiness.org/data/exploretopics/trading-across-borders. (Accessed September 15, 2015).

Central Asia: Typical landlocked region across Eurasian continent

13

Ryuichi Shibasaki, Satoshi Tanabe, and Hironori Kato
The University of Tokyo

Introduction

Central Asia (CAS) is the largest landlocked area of the world which has suffered from poor access to global markets. It consists of Afghanistan, Kazakhstan, Kyrgyz, Tajikistan, Turkmenistan, and Uzbekistan and is surrounded by Russia, China, Pakistan, Iran, and South Caucasian countries as shown in Fig. 13.1. All of these surrounding countries aim to become a gateway of maritime cargo to/from the CAS countries, as will be described in more detail later. As summarised in Table 13.1, major trade partners of the CAS countries are currently these adjacent countries. However, as Gallup et al. (1999) pointed out, the accessibility to the maritime shipping market is also important for their steady economic growth.

The importance of transport infrastructure in international trade of the CAS region has been pointed out by many studies such as Mirzohid and Galiba (2005), Kurmanalieva and Parpiev (2008), Vinokurov et al. (2009), Pomfret (2010), Gavrilis (2012), Kulipanova (2012), Maeno (2012), Mogilevskii (2012), Rastogi and Arvis (2014), Wang et al. (2015), and Sárvári and Szeidovitz (2016). China's Belt and Road Initiative (BRI), which promotes regional integration among Eurasian countries, can also be positioned as one of such movements (Yeh and Wharton, 2016). For example, regular container train services connecting Chinese inland cities with European cities (including Russian cities) have been rapidly increasing, not only in number but also in areas covered on both the Chinese and European sides. In the concept of 'Silk Road Economic Belt', transport routes to connect Iran, South Caucasian countries, Turkey, and other European countries via the CAS countries without transiting Russia are also included.

Meanwhile, the Central Asia Regional Economic Cooperation (CAREC) Program, led by the Asian Development Bank (ADB), has been promoting regional cooperation and development in CAS since the late 20th century. Currently, 11 partner countries are included in the CAREC program, not only the CAS countries but also China (the Xinjiang Uygur Autonomous Region or the XUAR), Mongolia, Pakistan, and South Caucasian countries (Azerbaijan and Georgia). Regarding China, the environment of international logistics in the XUAR is quite similar to that in the CAS countries, as it is located very far from Chinese seaports (more than 3000 km away). Therefore, acquiring access to another, closer gateway seaport (namely, in Pakistan) and overcoming the geographic handicap is key for the further economic development of the XUAR. Regarding the Pakistani seaports, the CAS region, as well as the XUAR, is a very prospective market to include as their hinterland for the future, because the linear distance

Global Logistics Network Modelling and Policy. https://doi.org/10.1016/B978-0-12-814060-4.00013-7

Fig. 13.1 Map of CAS.
Source: Own compilation based on https://n.freemap.jp/.

from most cities in the CAS region is the shortest among the competitive ports such as those in Russia, China, Iran, South Caucasus (Georgia), and the Baltic countries. However, because the level of infrastructure in Afghanistan is currently very insufficient, the current hinterland of the Pakistani seaports extends only to Afghanistan rather than Pakistan itself.

Many research papers and project reports have been published recently, discussing the impact of infrastructure investment and initiatives to promote connectivity and trade, and the resulting changes in freight traffic patterns in the CAS region. However, most of them only focus on a qualitative discussion for a specific area of the CAS region without any development nor use of analysing tools enabling the quantitative discussion on the policy impacts. In addition, only a few papers target the whole CAS region to comprehensively grasp the regional situation. These problems may be mainly due to data availability. Tanabe et al. (2016) developed a traffic flow simulation model of international land cargo to/from the CAS countries, but the maritime shipping was not explicitly included. Given this situation, as introduced in other chapters of this book, the authors developed a traffic flow simulation model to analyse the choice of ports and hinterland transport (HT) route of international containers on the global-scale intermodal transport network and applied it to developing countries where the availability of the data is not sufficient. Therefore, in this chapter, the authors apply the model to focus on the CAS countries and their surrounding region to quantitatively estimate the impact of related policies on the international container transport in these regions.

Table 13.1 Major trade partners of each CAS country (2015, million US $).

	Kazakhstan		Kyrgyz		Tajikistan		Uzbekistan		Turkmenistan		Afghanistan	
Import	Russia	10,529	Russia	1272	China	1796	China	2236	Turkey	1858	Iran	1807
	China	5088	China	1029	Russia	759.1	Russia	2221	Russia	843.9	Pakistan	1346
	Germany	1986	Kazakhstan	677.4	Kazakhstan	418.4	South Korea	1284	China	817.1	China	1044
	USA	1484	Turkey	164.3	Turkey	162.8	Kazakhstan	942.3	Germany	339.1	Turkmenistan	632.4
	Italy	1174	USA	122.1	Germany	49.6	Turkey	488.7	UAE	252.7	Kazakhstan	426.8
	Total	30,567	Total	4068	Total	N.A.	Total	N.A.	Total	N.A.	Total	7723
Export	Italy	8136	Switzerland	562.1	Turkey	203.8	Switzerland	1884	China	7828	Pakistan	226.6
	China	5480	Kazakhstan	227.7	Kazakhstan	164.9	China	1267	Afghanistan	632.4	India	188.9
	Netherlands	4981	Russia	157.3	Switzerland	140.9	Kazakhstan	725.7	Turkey	557.4	Iran	29.2
	Russia	4548	UAE	99	Afghanistan	91.5	Turkey	711.6	Italy	184.6	UAE	24.4
	France	2681	Uzbekistan	95	Algeria	54.6	Russia	575.8	Georgia	111.8	Russia	18.5
	Total	45,954	Total	1441	Total	N.A.	Total	N.A.	Total	N.A.	Total	571.4

Source: Own compilation based on the UN Comtrade Database 2015.

This chapter first presents the current situation regarding gateway competition among Eurasian ports for international maritime containers to/from CAS. Next, it describes the input data for developing a model and then, simulates policies on infrastructure investment and cross-border facilitation in the CAS region and its neighbours.

Gateway seaports and access routes of Central Asia

The major gateway seaports and access routes to them for international maritime cargo to and from the CAS region are described in Fig. 13.2. Currently, five regions are considered as gateways across the Eurasian continent: Far East Russia, China, the Arabian Sea (Iran and Pakistan), the Black Sea (Georgia and Russia), and the Baltic Sea (Russia and Baltic countries). Although they are varied across the continent, transport time and cost to access them are highly considerable in any routes. The status of each route is summarised respectively as follows, based on the authors' field and literature surveys. A few studies are additionally referred to collect a regional traffic data, such as Corridor Performance Measurement and Monitoring (CAREC, 2011, 2012), collecting transport time and cost at each road and railway section and border-crossing point from local freight forwarders; ADB (2015); Organisation for cooperation between Railways (OSJD, 2014); International Road Transport Union (IRU, 2011); Japan International Cooperation Agency (JICA, 2010a,b); and Japan External Trade Organisation (JETRO, 2010, 2013).

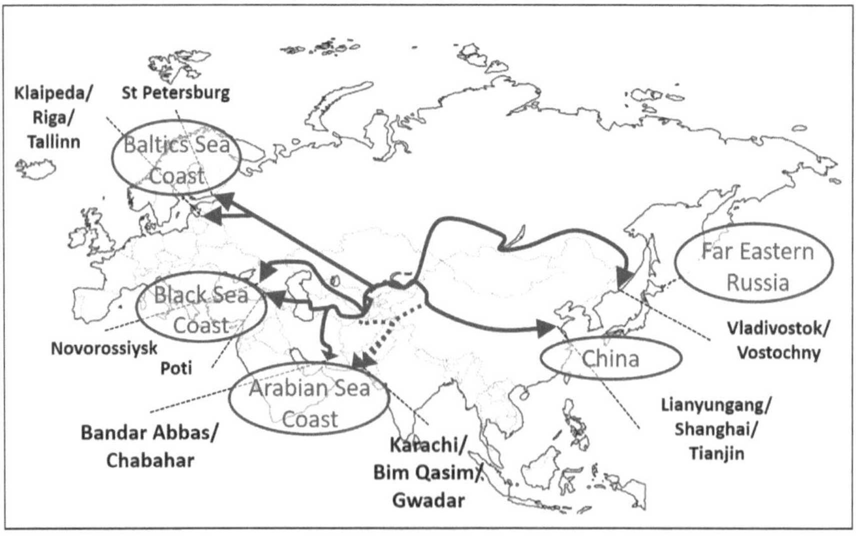

Fig. 13.2 Main access routes to seaports to and from CAS.
Source: Own compilation based on Shibasaki, R. Tanabe, S., Kato, H., Lee, T.-W., 2019. Could Gwadar port in Pakistan be a new gateway? A network simulation approach in the context of the belt and road initiative. Sustainability 11 (20), 5757.

Far Eastern Russia and China

There are two major routes to connect the CAS region with the Pacific coast: Trans-Siberian land bridge (TSLB) and China land bridge (CLB). These are the major transport routes to Far Eastern countries such as Japan and South Korea (which is virtually separated from the continent because any land connections via North Korea are currently not available), especially for the eastern area of the CAS region including two largest cities in CAS, namely, Almaty (Kazakhstan) and Tashkent (Uzbekistan).

Among them, TSLB is a traditional access route to the Pacific since the Soviet era, because the CAS countries were republics of the Union of Soviet Socialist Republics (USSR). TSLB still offers several advantages such as the same rail gauge, established custom and other trade systems. Furthermore, the Eurasian Economic Union (EEU, member countries: Russia, Belarus, Kazakhstan, Kyrgyz, and Armenia), which came into force in 2015, enables Kazakh and Kyrgyz cargo to transport inbond with Russian seaports. The disadvantages compared with CLB are a longer distance of land transport and the capacity limitation of Russian railways. The Trans-Siberian railway is an important trunk line connecting eastern and western Russia and focused more on the transport of fundamental goods for Russian economy such as coal and crude oil; therefore, the capacity for container trains is very limited (Shibasaki et al., 2013). Although the oil pipeline (Eastern Siberia–Pacific Ocean oil pipeline) connecting the production area with an export port in the Pacific has been recently completed, most of the rail capacity which was allocated to crude oil before has shifted to coal transport because of the recent boom in coal export.

The main gateway port of CLB is Lianyungang port (Fig. 13.3). The shorter distance to the seaport is a significant advantage compared with TSLB. Japanese cargo

Fig. 13.3 Container terminal of Lianyungang port with a sign to be the 'Starting Point of New Asia–Europe Land Bridge' (2006).
Source: Authors.

was transported to the CAS countries via this route in 1990s after the dissolution of the USSR. Recently this route has been also attracting their interests for transporting containers to/from Europe.

One of the major issues of this route is the difference in rail gauges between China and Kazakhstan. China uses the standard rail track gauge (1453 mm), whereas countries in the former USSR use wider gauge (1520 mm); therefore, containers need to be transshipped at the border. The facilities for unloading–reloading containers at the point used to often fail, which caused backlogs of containers at both borders. Severe delays in transshipping sometimes require a transit route for Japanese containers to be shifted from CLB to TSLB (JETRO, 2010). The Alashankou border used to be the only main border crossing point between China and Kazakhstan, but the Khorgos border opened in 2012 with railway constructions in both countries and facilities development at the border, to solve the issues caused by their poor hard infrastructures. Thus, the rail capacity for regular container trains in China is currently not considered a significant issue, judging from the significant increase in the number of regular container trains on CLB. It is partly because of the massive construction of high-speed railways throughout China resulting in higher capacity in existing railways for freight transport.

Baltic Sea and Black Sea coast (Russia, Baltic countries, and Georgia)

Russia is a historical important trade partner for each CAS country as indicated in Table 13.1. The railway infrastructure extending from Moscow to the CAS region has been developed since the Soviet era and used for trade between Russia and the CAS region. As an extension of this main route, St. Petersburg port can be a potent gateway seaport of CAS maritime cargo, especially for the trade with Northern and Western European countries. There are several access routes connecting the major CAS cities with Moscow, by rail as well as by trucks, via northern or western Kazakhstan. Furthermore, some ports in the Baltic countries aim to become a gateway to the CAS region. For example, a Japanese trade firm acquired an 80% equity stake in the container terminal of Riga port (Latvia) in 2013, to be 'a gateway to Russia, Kazakhstan, and other burgeoning CIS (Commonwealth of Independent States) countries' (Mitsui & Co, 2013). Another competitor of this route is a direct land transport between Eastern/Western European countries and the CAS region by rail or by trucks, in addition to other maritime shipping routes described in this section.

The traditional transport route between the CAS region and the Black Sea was to use inland waterway connecting the Black Sea and the Caspian Sea along two rivers connected by the Volga–Don Canal, which is currently used for the cargo that is difficult to carry by land transport such as some project cargo. Meanwhile, current major gateway seaports in the Black Sea are Novorrosiysk (Russia) and Poti (Georgia), especially for the cargo to/from the western CAS region such as Turkmenistan and western Kazakhstan. Novorrosiysk port is not only the gateway to southern Russia, but also to the CAS region such as Kazakhstan and Uzbekistan, which is connected by rail and roads. The route for which Poti port is the gateway crosses the Caspian Sea between Baku (Azerbaijan) and Turkmenbashi (Turkmenistan) or Aktau (Kazakhstan). According to the authors' field survey, there are currently no significant issues on land connections between the Black Sea and the Caspian Sea; however, the international ferry in the Caspian Sea, has a significant problem on frequency and reliability, because it is not regularly operated and only departs each port when a vessel is fully loaded.

Arabian Sea coast (Iran and Pakistan)

The Arabian Sea is the nearest in the direct distance to the coast for almost CAS countries including the XUAR. However, there are many national borders and other barriers on this route, including conflict and unsafe areas such as Afghanistan, and geographical difficulty due to mountainous terrains such as the border between China and Pakistan (Pamir Mountains). To make matters worse, challenges may increase; for example, the border between Uzbekistan and Tajikistan temporally closed due to worsened bilateral relations.

Therefore, the sole existing gateway port in the Arabian Sea is Bandar Abbas in Iran (except for Afghanistan cargo going to Pakistani ports). Bandar Abbas port is the largest seaport in Iran, which is located at the mouth of the Persian Gulf (Arabian Gulf) opposite to Dubai port, the largest hub in the Middle East. According to the authors' field survey, maritime containers, as well as vehicles shipped from the Arabian Sea, are imported via Bandar Abbas port with shuttle ships after the transshipment in Dubai port. From Tashkent to Bandar Abbas port, around 1 week is necessary to transport by truck, which is almost half of the transport time to access the Chinese coastal ports (JETRO, 2010). However, Japanese companies cannot use Bandar Abbas port as the gateway, due to Japan's economic sanction on Iran. An alternative gateway in Iranian ports is Chabahar port, which is located near the border with Pakistan.

Pakistani ports including Karachi, Muhammad Bin Qasim, and Gwadar have also recently been expected to function as a new gateway to CAS countries, and hinterland infrastructure is planned to be developed, although the hinterlands of these ports are currently within Pakistan and some areas in Afghanistan, due to a lack of infrastructure as well as a safety problem if crossing Afghanistan. In Afghanistan, the circular railway network is proposed to connect the CAS region (Pleis et al., 2014, see Fig. 13.4). In Pakistan, around 60% of total international cargo in Pakistani ports are destined into and originated from the Punjab province, which is far from the Arabian Sea coast (more than 1200 km away). However, railway services are not sufficient for the current transport demand. The Ministry of Railways of Pakistan Government (2016) shows that the share of rail transportation mode is only 4% of total freight flow. It can be attributed to insufficient railroad freight cars, the single line section existing, and the low capacity on port access. The poor surface condition of national roads is also one of the major issues.

Meanwhile, Gwadar port in Balochistan Province is situated 120 km east from the border with Iran. The development of Gwadar port is distinctively associated to the China–Pakistan Economic Corridor (CPEC) initiatives. The CPEC is planned to connect Gwadar port with the Kashi city of the XUAR and to provide China

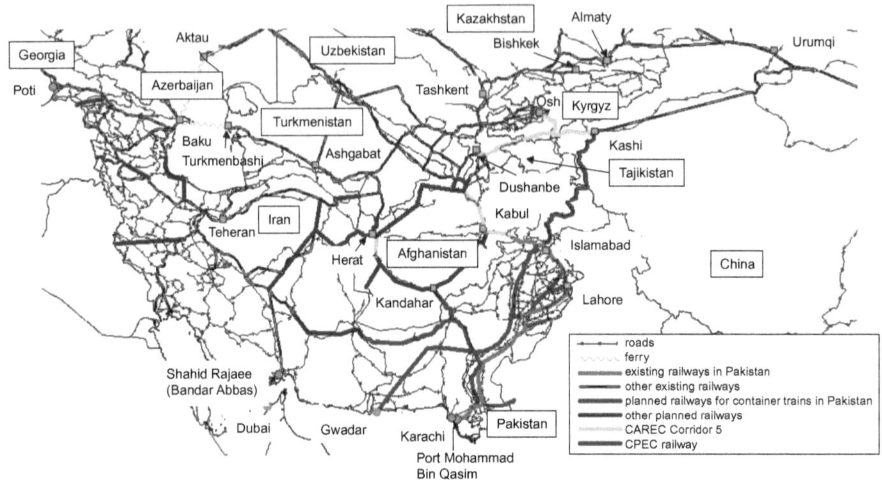

Fig. 13.4 Railway and other HT network in Arabian Sea coastal ports.
Source: Shibasaki, R. Tanabe, S., Kato, H., Lee, T.-W., 2019. Could Gwadar port in Pakistan be a new gateway? A network simulation approach in the context of the belt and road initiative. Sustainability, 11 (20), 5757.

direct access to the Arabian Sea by bypassing the Strait of Malacca and the South China Sea. Because it is expected to support the trade demand of the XUAR and CAS countries with global markets, the Pakistani government gives high priority to the development of an access road and railway to connect with the northern parts of the nation. CAREC also identifies it as one of their designated rail corridors as described in Fig. 13.4.

Data

Ports

The model in this chapter is developed based on 2013 data. In addition to major container ports of the world, five gateway ports which are mentioned in the previous section but not included above as major ports, i.e. Vostochny (Russia), Poti (Georgia), Klaipeda (Lithuania), Riga (Latvia), and Tallinn (Estonia), are included as shown in Table 13.2. Note that Gwadar port (Pakistan) is not included in the current maritime container shipping (MCS) network because there were no containers recorded to be loaded or discharged in 2013. Furthermore, several local container ports along the Arabian Sea and the Indian Ocean such as Sohar (Oman), Dar es Salaam (Tanzania), and Port Louis (Mauritius) are included as also shown in Table 13.2, because the authors will simulate the policies related to the CAS gateway ports along the Arabian Sea later. As a result, the total number of container ports considered in the model is 187.

Table 13.2 List of ports added in the model in this chapter.

No	Port name	Country	Country/region in the WTS	Annual throughput		Transshipment time[a] TPR_a (hours)
					source	
8-1	Vostochny	Russia	South Korea/FE Russia	475	a	48
49-1	Cochin	India	India	324	g	72
53-1	Male	Maldives	Indian Subcontinent Islands	80	h	72
108-1	Sohar/ Mina Qabos (Mascut)	Oman	Arabian Gulf	331	i (2014)	24
127-1	Poti	Georgia	South Caucasus	285	e (2012)	48
155-1	Klaipeda	Lithuania	Baltics	403	a	24
155-2	Riga	Latvia	Baltics	381	e	24
155-3	Tallinn	Estonia	Baltics	254	e	24
165-1	Nacala	Mozambique	Southern Africa	83	i	48
165-2	Dar es Salam/ Zanzibar	Tanzania	East Africa—Center	526	h	48
167-1	Port Victoria	Seychelles	Southern African Islands	41	i	48
167-2	Toamasina	Madagascar	Southern African Islands	173	h	48
167-3	Pointe des Galets	Reunion (France)	Southern African Islands	213	i	48
167-4	Port Louis	Mauritius	Southern African Islands	622	h	48

[a] Authors' estimation.

Source: Own compilation based on (a) Drewry Maritime Research, 2014. Container Forecaster 2014 Annual Review. London, UK; (e) websites for each port or terminal; (g) Indian Ports Association. 2015. Traffic at Major Ports (online). http://www.ipa.nic.in/oper.htm (Accessed 15 September 2015); (h) World Bank, 2015. Container port traffic (TEU: 20 foot equivalent units) (online), http://data.worldbank.org/indicator/IS.SHP.GOOD.TU/countries (Accessed 15 September 2015) for the port in the country where only one container port is considered in operation; and (i) Logistics Capacity Assessment 2015 Wiki (online), http://dlca.logcluster.org/display/public/DLCA/LCA+Homepage (Accessed 15 September 2015), as well as the results from authors' interview survey with port trusts and other organisations.

Global MCS network

As with other simulations in this book, the MCS network is developed based on the MDS Containership Databank data (as of 2013). Other than 20 worlds' largest liner shipping companies (see Table 8.2 in Chapter 8), 17 local companies providing a liner service (LS) network in CAS gateway ports (i.e. X-Press Feeders, TS Lines, Sinotrans, HDS Lines, SITC, United Feeder Services, Emirates Shipping, ARKAS Line, Simatech Shipping, Unifeeder, FESCO, DFDS, Orient Express Lines, Team Lines, Samskip, Containerships, and Finnlines) are included, resulting 1018 LSs to cover 67% of the global annual vessel capacity and 87% of the annual capacity of vessels which call at the CAS gateway seaports.

Regional HT network

The HT network is considered in 19 countries in the Eurasian continent (i.e. Afghanistan, Armenia, Azerbaijan, Belarus, China, Estonia, Georgia, Iran, Kazakhstan, Kyrgyz, Latvia, Lithuania, Mongolia, Pakistan, Russia, Tajikistan, Turkmenistan, Ukraine, and Uzbekistan), which cover all CAS countries as well as all potential CAS gateway seaports (see Fig. 8.3 of Chapter 8). In addition to extracting all road networks and selected rail links from the ADC WorldMap as with the other simulations of this book, international ferry links in the Caspian Sea are added for both road and rail to connect Baku with Turkmenbashi or Aktau.

Capacity per train is differently defined by country as summarised in Table 13.3, whereas capacity per vessel is defined as 1000 (TEU/vessel). Table 13.3 also shows settings on the frequency of trains for container transport for each country, whereas the frequency of international ferry is set at 3 (services/week) between Baku and Turkmenbashi and 1 (service/week) between Baku and Aktau, based on the authors' interview survey, although they are actually irregular services.

The coefficient of the land border barrier, λ_a, which is multiplied by the border-crossing time and cost, is set by country pair as shown in Table 13.4, considering the bilateral relationship such as customs union and incidents as of 2013.

Container shipping demand between regions (OD matrix)

The regional OD matrix and initial MCS demand are estimated as follows. First, the container OD matrix between countries or regions on a TEU-basis is obtained from the WTS data. In addition, the breakdown of 'Central Asia' into Kazakhstan, Kyrgyz, Tajikistan, Turkmenistan, and Uzbekistan and 'South Caucasus' into Armenia, Azerbaijan, and Georgia in 2011 is preprovided, which is used for dividing the demand of Central Asia and South Caucasus as of 2013 into each country with the share by partner country/region.

A traffic analysis zone (TAZ) basis OD matrix for 19 countries in the Eurasian continent is estimated by dividing the aggregated regional OD matrix into a sub-country level, based on the available statistics which represent an economy of each TAZ (for detail, see Shibasaki et al., 2019). Note that because Afghanistan is integrated with Bhutan and Nepal in the original WTS data, they are divided into each country by the

Table 13.3 Settings on rail capacity and service frequency in the model.

Country	Region	Source	Capacity (TEU/train)	Service frequency (trains/week)
China	Coastal area	a	90	67
	Inland area	a	90	30
Russia	Trans-Siberian railway	b	112	28 or 56
	CLB	b	112	10
	Black Sea	c	112	0.5
Baltic states		d	112	2.5
Belarus and Ukraine		b	62	2.5
Central Asia	CLB and sections between Chinese border and Tashkent	b	50	20
	Others	e	50	10
Iran		f	90	50
Pakistan		g	90	5
South Caucasus and others		h	62	12

Source: Own compilation based on each source: (a) Fukuyama, H., 2014. Development of Chinese railway container transit and Eurasian land bridge. Maritime Transp. Res. Yamagata Maritime Inst. 63, 73–82 (in Japanese); (b) authors' assumption; (c) Yankee Russia, 2015. Novorossiysk: The Largest Sea Port in Russia; (d) FESCO, n.d. Container Lines Schedule website; (e) authors' interview survey in Uzbekistan; (f) Container Shipping Co. (CONSHIP), n.d. website; (g) authors' interview survey in Karachi and Bin Qasim in Pakistan; and (h) ADB, 2014. Georgia Transport Sector Assessment, Strategy, and Road Map.

Table 13.4 Settings on the multiplier in land border crossing.

Country pair	λ_a	Note
Between Russia, Kazakhstan, and Belarus (member countries of Eurasian Customs Union as of 2013)	0.25	Eurasian Customs Union was upgraded to EEU in 2015 with newly joining Armenia and Kyrgyz
Between Russia and Georgia	10.0	Current worsening between Russia and Ukraine is not considered in the model, because it started in 2014
Others	0.5	

Source: Own compilation.

exported and imported shares in terms of trade amount among three countries which are acquired from the UN Comtrade Database.

Another notation is that the OD matrix between European countries (including Russia, South Caucasus countries, and CAS countries except Afghanistan) are not considered in the model, due to the availability of data. In other words, we assume any cargo between these countries is transported on land, not using MCS.

Calculation results

Exported and imported container throughput in each port

Figs 13.5 and 13.6 compare the observed and model estimated export and import annual laden container throughputs for CAS gateway seaports. The left charts (figures A) of both figures show the throughputs in all seaports included in the HT submodel. Note that some Chinese ports [namely, Qingdao and Yantai; Shanghai and Ningbo; Fuzhou, Xiamen, and Shantou; Shenzhen (two terminals) and Guangzhou] are integrated into the figures. In other words, more detailed zoning (prefecture or county level, not province level) of TAZ is necessary if the model precision on Chinese ports should be improved. The right charts (figures B) of both figures show the throughputs in 12 seaports other than Chinese ports for more focusing on them.

From these figures, the observed throughputs of export and import laden containers are well estimated as a whole. The difference between them is observed in several Chinese ports such as Shenzhen–Guangzhou and Qingdao–Yantai in export and Tianjin and Shanghai–Ningbo in import. Further, especially in export, some trade-offs are observed among the ports which are closely located, such as Karachi and Bin Qasim in Pakistan and Vladivostok and Vostochny in Far Eastern Russia. The precise differentiations between these ports are difficult in the current zoning system because both ports are located in the same TAZ.

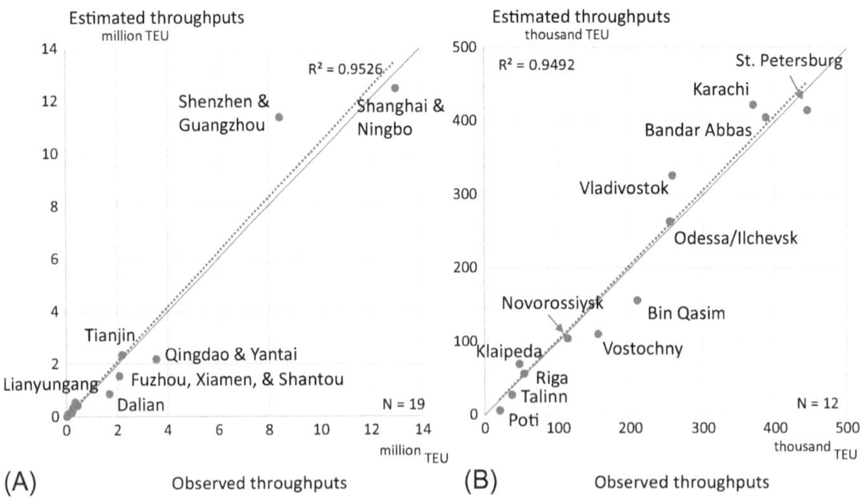

Fig. 13.5 Comparison between observed and model estimated *export* laden container throughputs in gateway seaports (as of 2013). (A) All seaports included in the HT submodel. (B) 12 seaports except for Chinese ports.
Source: Shibasaki, R. Tanabe, S., Kato, H., Lee, T.-W., 2019. Could Gwadar port in Pakistan be a new gateway? A network simulation approach in the context of the belt and road initiative. Sustainability 11 (20), 5757.

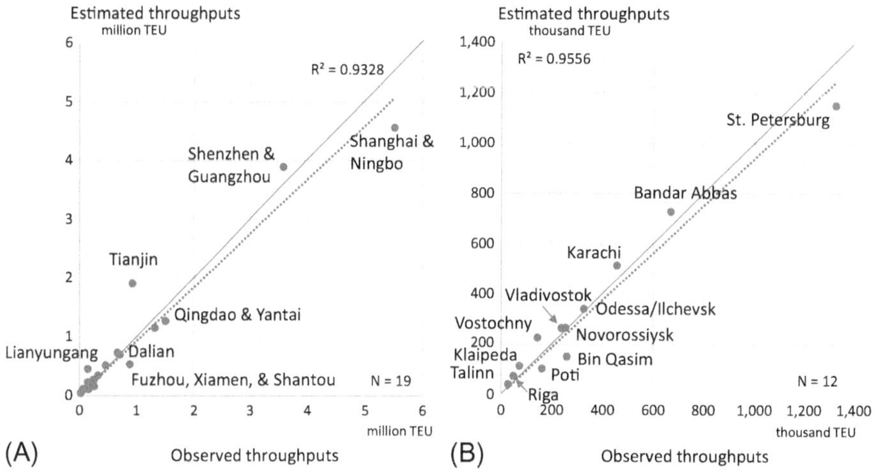

Fig. 13.6 Comparison between observed and model estimated *import* laden container throughputs in gateway seaports (as of 2013). (A) All seaports included in the HT submodel. (B) 12 seaports except for Chinese ports.
Source: Shibasaki, R. Tanabe, S., Kato, H., Lee, T.-W., 2019. Could Gwadar port in Pakistan be a new gateway? A network simulation approach in the context of the belt and road initiative. Sustainability 11 (20), 5757.

Container flow in the HT network

Fig. 13.7 describes the annual laden container flow in the HT network estimated by the model. Although it cannot be compared with the observed flow because it is not available, the major transport routes are estimated to be used. Furthermore, railways are used in most of the long-distance transport. The annual number of laden containers that use international ferries across the Caspian Sea are estimated at 3100 TEU between Kazakhstan and Azerbaijan but zero between Turkmenistan and Azerbaijan, including both directions. Although it cannot be compared with the observed number of laden containers shipped by these ferries because it is also not available, the reasons of the zero flow estimated between Turkmenistan and Azerbaijan are considered that the major cargoes for these ferries are bulky cargo, not containers, according to the authors' field survey, and that the containers to/from European countries are not included in the model due to data availability as mentioned.

The results shown in Figs 13.5 and 13.6 indicate that the model can well estimate the observed laden container throughput in each port. This may be logical particularly if only one port in a country is considered in the model, because its hinterland covers almost the whole country and the number of containers to/from landlocked countries is generally small (e.g. the total number of containers to/from the six CAS countries in 2013 is 32 and 147 thousand TEU for export and import, respectively). Therefore, we examine the breakdown of trade partner countries/regions and export/import seaports estimated by the model for each CAS country as well as the Siberian Federal District (SFD) in Russia, as summarised in Fig. 13.8. Note again that the container cargo

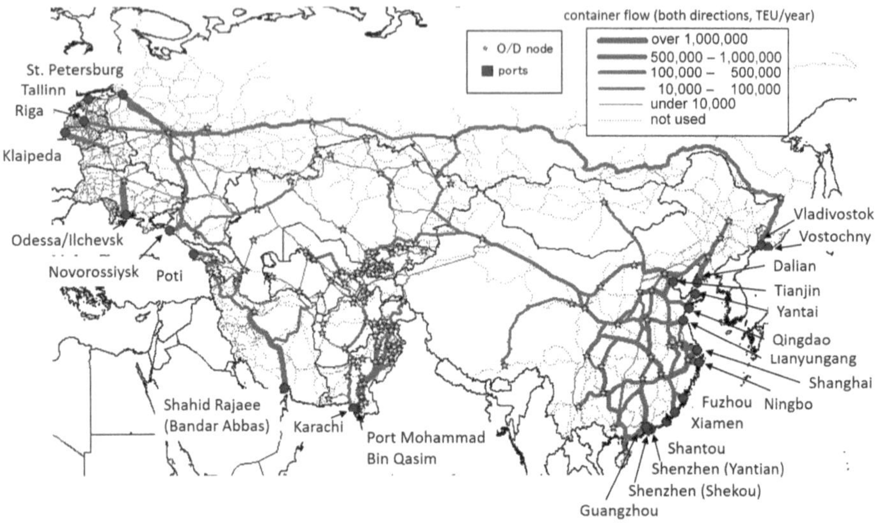

Fig. 13.7 Estimated annual laden container flow in the HT network (as of 2013).
Source: Own compilation.

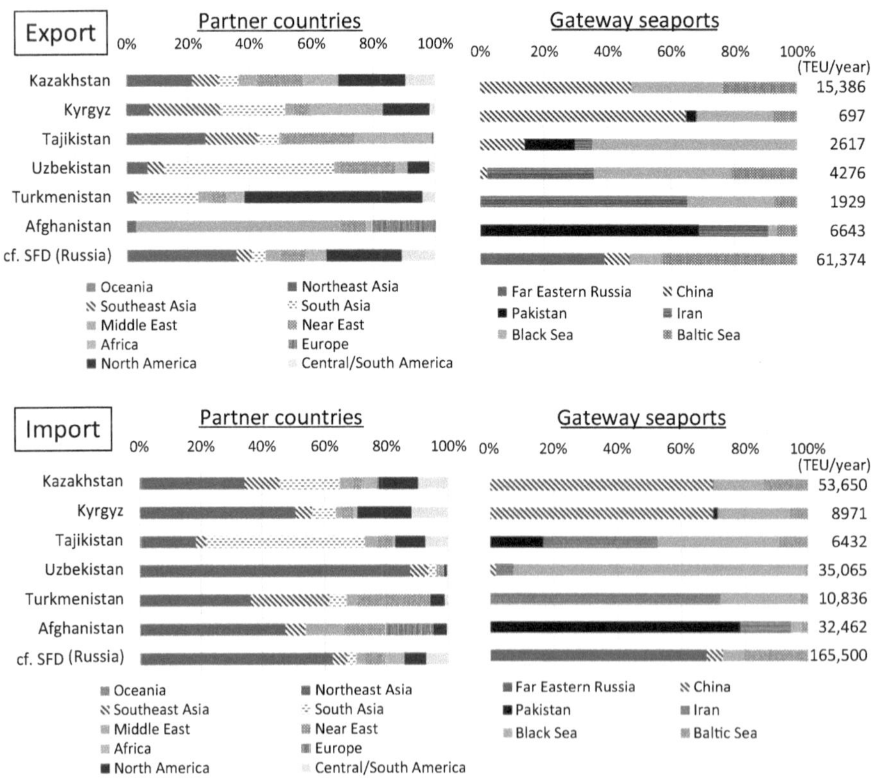

Fig. 13.8 Breakdown of trade partner countries/regions and gateway seaports for each CAS country and the SFD (Russia) cargo (TEU-basis, as of 2013).
Source: Own compilation from the WTS data (for partner countries) and the model calculation results (for gateway seaports).

shipping demand between CAS countries (including Russia) except for Afghanistan and European countries is not included in the model.

It is found that the export/import seaports (gateway seaports) are significantly different by each CAS country and by partner countries. Namely, the export/import laden containers to/from Kazakhstan and Kyrgyz with (North- and South-)East Asia, South Asia, and the Middle East (hereinafter, it is named 'the eastern regions') are considered to use Chinese ports, whereas those with the Near East, Africa, and the American continent ('the western regions') mainly use the Black Sea ports and the Baltic Sea ports. In these countries, the choice of gateway seaports includes the overland crossing of the Eurasian continent in the east–west direction. Similarly, the export/import laden containers to/from the SFD in Russia with the eastern regions are considered to use mainly Far Eastern Russian ports but scarcely Chinese ports, whereas those with the western regions use mainly the Baltic Sea ports and partly the Black Sea ports (supposedly for those with the Near East).

The choice of gateway seaports into the other four CAS countries seems different. The main gateways seaports for these cargoes are Pakistani ports, an Iranian port, and the Black Sea ports. In particular, the laden containers imported into Uzbekistan and those exported from/imported into Turkmenistan and Afghanistan tend to use ports in a specific region. Namely, the containers imported into Uzbekistan are mostly transported via the Black Sea ports, whereas most of them are coming from the Far East. Meanwhile, the export/import laden containers to/from Turkmenistan mainly use an Iranian port, not only for those with the eastern regions but also with North America (especially for the export containers); only those with the Near East and neighbouring countries are considered to use the Black Sea ports. Further, the export/import laden containers to/from Afghanistan use mainly Pakistani ports but sometimes an Iranian port; from the observation that the containers to/from the western Afghan regions mainly use an Iranian port, the choice may be made by region within Afghanistan rather than the partner countries. The usage pattern in gateway seaports for laden containers exported from/imported into Tajikistan and those exported from Uzbekistan is in the middle of the overland crossing of the Eurasian continent and by the regional specification.

These observed differences in gateway seaports and trade partner countries are considered reasonable as well as consistent with the authors' quantitative observations. Exceptionally, the share of Chinese ports in the laden containers imported into Uzbekistan is small, although a significant portion of it is containers imported from the Far East. The other difference found by the authors' interview survey is that some of Tajik laden containers are estimated to use Pakistani ports, although only Afghanistan is currently the hinterland of Pakistani ports; this may be caused by the insufficient consideration of unsafe situations for logistics within Afghanistan in the model.

Policy simulations

Decline of barriers in CAREC national border

There are three scenarios prepared on the decrease of the barrier at the land national borders. The first scenario (S1-0) reflects the actual change in the border barriers since

2013, the year considered in the model; namely, accessions of Kyrgyz and Armenia to the EEU and worsening relations between Russia and Ukraine. Between the member countries of EEU, a multiplier in border crossing is set at 0.25. A multiplier between Russia and Ukraine is set at a similar level of that between Russia and Georgia (namely, 10.0). The second and third scenarios consider the decline of barriers in all national land borders between the CAREC member countries, in addition to the change considered in S1-0, to be half of the current level (S1-1) and zero (S1-2), reflecting various kinds of facilitations and applications implemented.

Fig. 13.9 summarises the changes by regions in annual export and import laden container throughputs estimated in three scenarios, from those estimated in the base scenario (S0), which is shown in the previous section. The figure reveals that the first scenario (S1-0) has no significant impact on any container throughputs because the number of borders with reduced barriers is very limited. Meanwhile, in the second and third scenarios, the laden container throughputs in the ports in the CAREC countries (i.e. China, Pakistan, and Georgia in the Black Sea) are expected to significantly increase, whereas those in other countries' ports decrease. Note that the laden container throughputs in Poti port (Georgia) are expected to increase in both scenarios (namely, the number of export and import laden containers is expected to increase to 398 and 30,197 TEU/year, respectively, in S1-1 and to 446 and 38,850 TEU/year in S1-2), although the total throughputs in the Black Sea ports are mostly cancelled out by the decrease in those in Novorossiysk port.

Another finding is that the changes observed in S1-2 are larger than those in S1-1 in most cases; in other words, the impact to the laden container throughput will be larger as the barriers in national borders decline, although the changed amounts are nonlinear and varied between export/import and between regions. The exception is observed in imports into the Black Sea ports and the Baltic Sea ports because some Russian cargo is 'swinging' between Novorossiysk and St. Petersburg ports in S1-2. It may imply that the impact input in S1-2 is too significant against the model settings (e.g. capacity

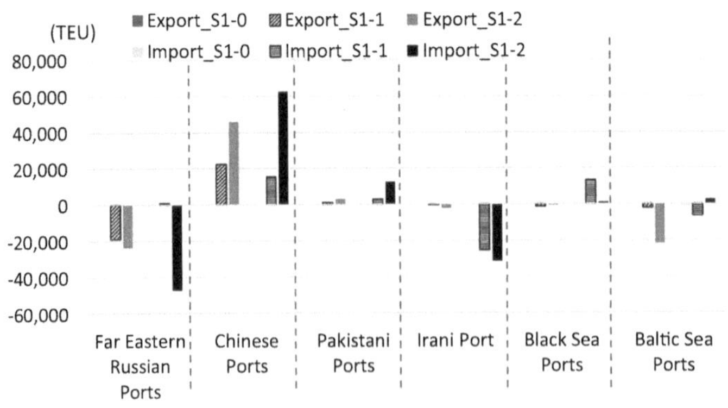

Fig. 13.9 Estimated change in annual export/import laden container throughputs for each region in the scenarios of national border barrier decrease.
Source: Own compilation.

constraint), especially in European Russia. Therefore, more careful settings will be necessary on the European side including the Baltic Sea and the Black Sea, if the model uscr would like to further focus on these regions.

Fig. 13.10 indicates a breakdown of export/import gateway seaports for each CAS country estimated in S1-2. Note that the figure includes not only the SFD in Russia but also the XUAR in China. Compared with the breakdown in S0 which is shown in Fig. 13.8, several differences are observed as follows. Regarding the Kazakh and Kyrgyz cargo, the shares of Pakistani ports as well as an Iranian port are expected to increase for both export and import. This means that the alternatives in gateway seaports are increased (especially to the south, not only in the east–west direction) thanks to the decline of border barriers. Regarding the Tajik and Uzbek cargo, the shares of Chinese and Pakistani ports are expected to increase, shifting from an Iranian port and the Black Sea ports. In particular, drastic changes can be observed in the Uzbek cargo; export containers are shifted to Pakistani ports and import containers are to Chinese ports. It may reflect that Uzbekistan is one of only two 'double-landlocked countries' in the world, which is a country only surrounded by landlocked countries. Therefore, the Uzbek cargo is very sensitive to the change of border barriers and its gateway seaport can easily swing. The difference in main gateway seaports between export and import cargo is owing to the difference in their partner countries as indicated in Fig. 13.8.

Regarding the Turkmen and Afghan cargo, the share of an Iranian port is expected to decrease in S1-2, mainly shifting to the Black Sea ports as well as to Chinese and the Baltic Sea ports in some cases. In particular, the export cargo from Turkmenistan significantly shifts to the Black Sea and the Baltic Sea ports, because most of them are exported to North America, as indicated in Fig. 13.8. Reflecting this, the annual number of laden containers shipped by international ferries in the Caspian Sea is also expected to increase in S1-2 to 1120 TEU between Turkmenistan and Azerbaijan and

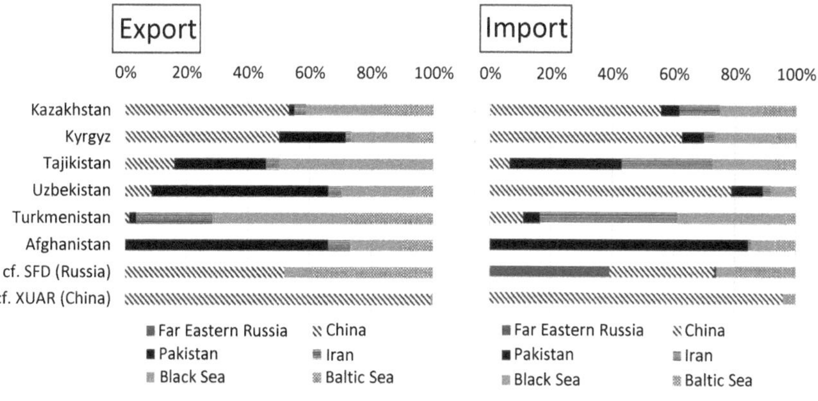

Fig. 13.10 Breakdown of gateway seaports for each CAS country, the SFD (Russia), and the XUAR (China) cargo in S1-2 (TEU-basis).
Source: Own compilation.

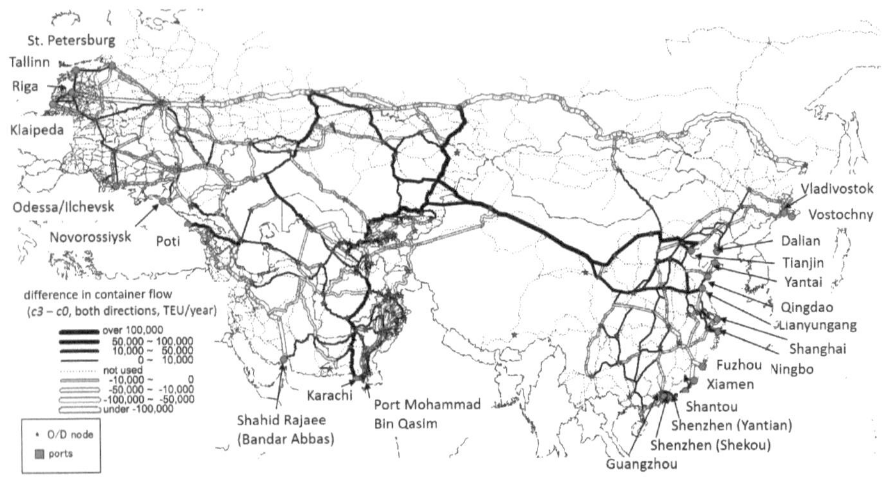

Fig. 13.11 Estimated difference in annual laden container flow between the base scenario (S0) and removal scenario in CAREC border (S1-2).
Source: Own compilation.

573 TEU between Kazakhstan and Azerbaijan (also see Fig. 13.11, which describes the change in each link flow in the HT network from S0 to S1-2).

Moreover, a significant portion of export/import laden containers to/from the SFD is expected to shift from Far Eastern Russian ports to Chinese ports. In particular, the number of import containers shifted in S1-2 is significant, as indicated in Fig. 13.9. Similarly, in Fig. 13.11, the significant shift can be observed from TSLB to CLB. Furthermore, some cargos to/from the XUAR in China are expected to use the Baltic Sea and the Black Sea ports in S1-2 across several borders (note that in S0, all containers to/from the XUAR are expected to use Chinese ports). Although the changes in share are small, the absolute numbers are not negligible (1490 TEU for export and 5937 TEU for import are considered to annually shift to other countries' ports).

New Eurasian rail connections from northeast China

As mentioned earlier, the number of regular container train services connecting Chinese cities and Europe or Russia has been recently increasing. Among them, trains from the northeast area (including Dalian, Shenyang, and Tianjin) via Trans-Siberian railway are rapidly increasing. Reflecting this, the scenarios (S2-0 to S2-2) are prepared on the launch of new container train services from northeast China. For each scenario, the authors assume that two new rail connections between northeast China and the SFD in Russia via Manzhouli (which is a Chinese border city in Inner Mongolia Autonomous Region with Russia) and Mongolia are provided with the similar level of service (in both frequency and capacity) as the other existing services in the same country as shown in Table 13.3. The differences of three scenarios are on the barriers of national border in the CAREC countries: the current level of barriers in all national borders (S2-0), the barriers declined to half in all CAREC borders (S2-1), and no border barriers in all CAREC borders (S2-2), which correspond to each scenario (i.e. S1-0, S1-1, and S1-2) dealt with in the previous subsection.

Fig. 13.12 summarises the annual number of laden containers (in both directions) estimated to transit either TSLB or CLB route by the above scenarios. The numbering of each route in the figure also corresponds to the number shown in Fig. 13.13. The results indicate that the total transit number of laden containers estimated in S2-0, S2-1, and S2-2 is larger than that in S1-0, S1-1, and S1-2, respectively; in other words, the total number of laden containers to transit TSLB or CLB will increase by implementing new rail connections in any scenarios on the CAREC border barrier. Furthermore, the share of CLB is expected to increase as the barriers in the CAREC borders decline, although its share becomes smaller by new rail connections in any scenarios on the border barrier. By implementing new rail connections, the most used route is the second one, TSLB via Manzhouli, especially in cases where the border barrier is relatively higher. Meanwhile, the third route, TSLB via Mongolia, is expected to be partly used only if the border barrier becomes smaller. These estimation results are generally consistent with the observed situation on the regular container train service.

Fig. 13.13 describes the difference estimated in annual laden container flow between S2-1 and S1-1. By implementing the new Eurasian rail connection in northeast China, the number of laden containers to transit the second and third route is expected to increase, whereas those in the first and fourth route decrease. Note that most Mongolian cargos (which mostly use Chinese ports as the gateway seaports) are also expected to shift from road to rail after the implementation of new container rail services in Mongolia.

New port operation and international rail connections in Pakistan

The scenarios in this subsection examine Pakistani policies to become a gateway for the maritime containers to/from the CAS countries (including the XUAR in China)

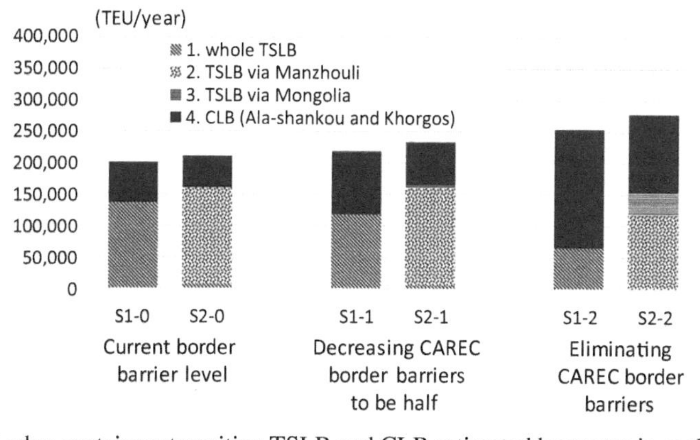

Fig. 13.12 Laden containers transiting TSLB and CLB estimated by routes in each scenario. Source: Own compilation.

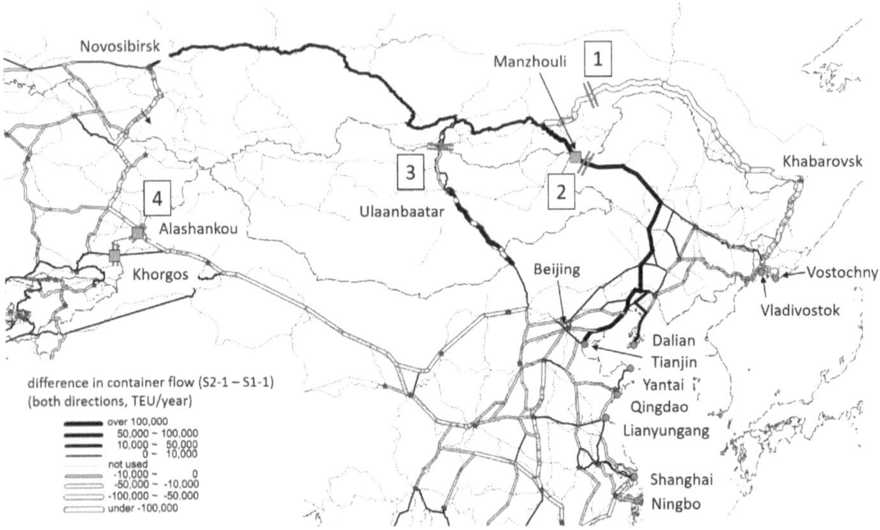

Fig. 13.13 Estimated difference in annual laden container flow between the new Eurasian rail connection scenario (S2-1) and the corresponding scenario before connections (S1-1).
Source: Own compilation.

after its accession to CAREC in 2010, such as the development of Gwadar port and infrastructure investment along the CAREC corridor as well as the CPEC. In all the following scenarios, all regional (i.e. within the Arabic Sea including off the coast of East Africa) LSs on the way (i.e. which are considered to navigate off its coast) are assumed to call at Gwadar port. Specifically, 12 LSs which call at any of the other Pakistani ports and any of the Persian Gulf ports are assumed to additionally call at Gwadar port among more than 1000 LSs of the world.

Moreover, several scenarios on the rail network development in and around Pakistan are assumed. In the first scenario (S3-0), no additional developments of Pakistani railway are assumed, based on the current level of barriers in all land national borders (S1-0); namely, only the call of LSs at Gwadar port is considered. The second scenario (S3-1) assumes the capacity improvement of the existing rail network to increase the service frequency from 5 to 50 trains/week as well as the new rail construction with the frequency of 10 trains/week in the whole of Pakistan (except for the CPEC). Furthermore, the third scenario (S3-2) assumes the opening of the railway in the CAREC corridor 5 outside Pakistan (except for the CPEC) with the frequency of 10 trains/week. Similarly, the opening of other planned railways in Afghanistan, Iran, and South Caucasus countries are assumed with the frequency at the level of the existing rail service in each country (see Fig. 13.4). The fourth scenario (S3-3) assumes the additional opening of the CPEC railway with the frequency of 10 trains/week. Two further scenarios additionally assume the barrier reductions of the CAREC border to be half of the current level (S3-4) and zero (S3-5).

The changed amounts in these scenarios in the estimated annual export and import laden container throughputs from those estimated in the reference scenario (S1-0) are summarised by regions in Fig. 13.14. Furthermore, Fig. 13.15 indicates the annual throughputs of export and import laden containers estimated in Gwadar port and breakdowns by origin and destination country in the hinterland. Note that the results shown in the last scenario (S3-5) are the reference because the model calculation is not converged within the given number of iterations (10 times at maximum).

Based on Fig. 13.15, the laden container throughputs in Gwadar port are estimated to be more than 10,000 TEU/year for both export and import, respectively, if the railway is connected (S3-1 or more). Fig. 13.15 also indicates that most laden containers using Gwadar port in S3-1 and S3-2 originate from/are destined into Pakistan, whereas some import laden containers are transported to Afghanistan and Tajikistan. The results shown in Fig. 13.14 also support that they are mainly shifted from other

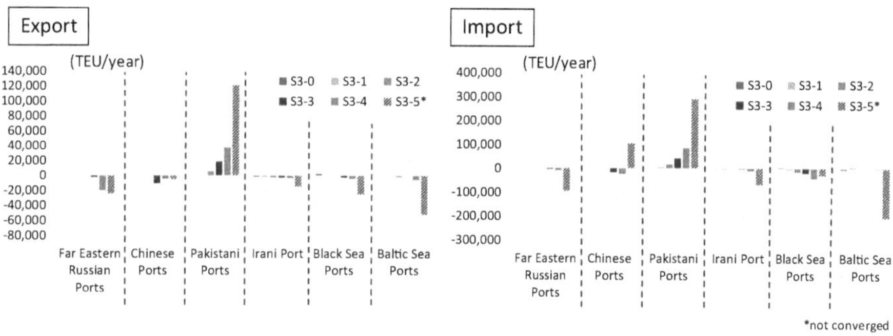

Fig. 13.14 Estimated change in laden container throughputs for each region in the scenarios related to the Pakistani port and HT connections from the reference scenario (S1-0).
Source: Own compilation based on Shibasaki, R. Tanabe, S., Kato, H., Lee, T.-W., 2019. Could Gwadar port in Pakistan be a new gateway? A network simulation approach in the context of the belt and road initiative. Sustainability 11 (20), 5757.

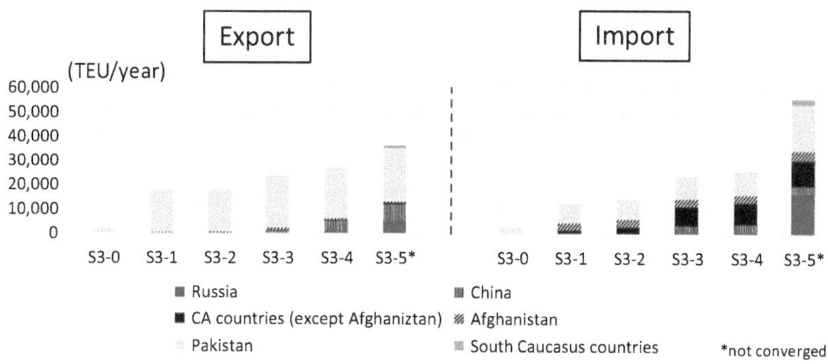

Fig. 13.15 Estimated laden container throughput in Gwadar port by origin/destination countries in the hinterland for each scenario.
Source: Own compilation based on Shibasaki, R. Tanabe, S., Kato, H., Lee, T.-W., 2019. Could Gwadar port in Pakistan be a new gateway? A network simulation approach in the context of the belt and road initiative. Sustainability 11 (20), 5757.

Pakistani ports because the regional throughputs have not changed from those in S1-0. Meanwhile, in S3-3 which assumes the opening of the CPEC railways, in addition that the annual laden container throughputs in Gwadar port are expected to further increase (at more than 20,000 TEU/year for each export and import), the total throughputs in Pakistani ports are expected to increase.

Moreover, the total annual laden container throughputs in Pakistan, as well as the throughput in Gwadar port, are expected to further increase by the reduction of border barriers in S3-4 and S3-5. The increased amount of total throughput in Pakistani ports is significantly larger than those estimated in S1-1 and S1-2 (see Fig. 13.9), particularly remarkable in the case that the border barriers are thoroughly eliminated (in S3-5), although the model calculation is not converged. In such a scenario, even some laden containers to/from Russia and South Caucasus countries are expected to use Pakistani ports.

Conclusion

The current gateway seaports of each CAS country are mainly located in five regions across the Eurasian continent: Far East Russia, China, the Arabian Sea, the Black Sea, and the Baltic Sea. Although the observed share of each region as gateway seaports is not available, the share estimated by the model was likely to correspond to the observed situation described by many stakeholders in the CAS region, as well as what can be estimated from qualitative and quantitative information fragmentally available. For example, the containers to/from the northern countries (namely, Kazakhstan, Kyrgyz, and the SFD in Russia) are split to use the ports in the east (namely, China or Far Eastern Russia) for the trade with the eastern regions, and to use the ports in the west (namely, in the Baltic Sea or the Black Sea) for the trade with the western regions. On the other hand, the containers to/from the southern countries (namely, Turkmenistan and Afghanistan) use a specific port(s), such as an Iranian port for the Turkmen cargo and Pakistani ports for the Afghan cargo, with some exceptions, such as the Black Sea ports for the trade between Turkey and Turkmenistan, and an Iranian port for the cargo in the western region of Afghanistan. The containers to/from the central countries (namely, Tajikistan and Uzbekistan) are a hybrid of them, although the model estimated them to use the Black Sea ports and Pakistani ports more than generally expected.

This chapter applied the developed model to three different kinds of related policies: the decline of national border barriers, launching of new container train service, and launching of new container port. The decline of border barriers in the CAREC countries would significantly impact the choice of gateway seaports for each CAS country as well as inland regions of Russia and China, resulting in an increase in throughputs in Chinese ports and decrease in those in Far Eastern Russian ports, an Iranian port, and the Baltic Sea ports. The new container train services connecting northeast China and Russia would be competitive among the existing routes (i.e. TSLB and CLB); in particular, the transport route via Manzhouli would have a potential to carry more than 150,000 TEU/year for laden containers in both directions, which corresponds to the actually observed situation that many regular container train services start operation via these routes. The route via Mongolia would be also promising if the border barriers in these countries are sufficiently decreased. Gwadar port could handle more than 30,000 TEU/year total for export and import laden containers if the rail is connected, which enables several LSs to call at the port sustainably. If the rail is even connected with not only the CAS countries via CAREC corridor 5, but also China via the CPEC, much of the containers would use Gwadar port, which can be partly shifted from the ports in other countries, not only the existing Pakistani ports. The port throughputs would synergistically increase if the border barriers are declined at the same time. Further simulations related to this region are described in Chapter 16, which focuses on the impact of China's BRI policy on the land transport across the Eurasian continent.

References

Asian Development Bank (ADB), 2015. Final Report TA-8586 (REG): Facilitation of Regional Transit Trade in CAREC.

Central Asia Regional Economic Cooperation, the (CAREC), 2011. CAREC CPMM Corridor Performance Measurement and Monitoring Annual Report 2011.

Central Asia Regional Economic Cooperation, the (CAREC), 2012. CAREC CPMM Corridor Performance Measurement and Monitoring Annual Report 2012.

Gallup, J.L., Sachs, J.D., Mellinger, A.D., 1999. Geography and economic development. Int. Region. Sci. Rev. 22 (2), 179–232.

Gavrilis, G., 2012. Central Asia's border Woes and the Impacts of International Assistance. Central Eurasia Projects. Occasional Paper Series, 6.

International Road Transport Union (IRU), 2011. Analysis of goods flow in Central Asia and border regions of China: Executive summary.

Japan External Trade Organization (JETRO), 2010. Rail and Air Transport to the Moscow and Siberia Region via Russian Far Eastern Customs. (in Japanese).

Japan External Trade Organization (JETRO), 2013. Logistics Condition in Uzbekistan. (in Japanese).

Japan International Cooperation Agency (JICA), 2010a. JICA Survey on the Logistics in Tajikistan (online). http://open_jicareport.jica.go.jp/700/700/700_156_11989407.html. (Accessed 29 April 2018 (in Japanese).

Japan International Cooperation Agency (JICA), 2010b. JICA Survey on the Logistics in Turkmenistan (online). http://open_jicareport.jica.go.jp/pdf/11991213_01.pdf. (Accessed 29 April 2018 (in Japanese).

Kulipanova, E., 2012. International Transport in Central Asia: Understanding the Patterns of (non-) Cooperation. University of Central Asia. Institute of Public Policy and Administration Working Paper, 2.

Kurmanalieva, E., Parpiev, Z., 2008. Geography and Trade in Central Asia.

Maeno, T., 2012. Pattern of trade in Central Asia and analysis of the determinants of export. In: Kiyo of Research Institute of Economic Science, p. 41. College of Economics, Nihon University, 169–190 (in Japanese).

Ministry of Railways, Government of Pakistan, 2016. Pakistan Railways Year Book 2014–15.

Mirzohid, R., Galiba, U., 2005. Central Asian Nations and Border Issues. Conflict Studies Research Centre, Defense Academy of the United Kingdom.
Mitsui & Co, 2013. Portek to acquire Latvian port terminal company. Press release on 30-04-2013.
Mogilevskii, R., 2012. Trends and patterns in foreign trade of Central Asian countries. University of Central Asia. Institute of Public Policy and Administration Working Paper, 1.
Organization for Co-operation between Railways (OSJD), 2014. OSJD Bulletin of Statistical Data on Railway Transport for 2014.
Pleis, L., Lliteras, R., David, W., Bain, M., Hendrickson, S., 2014. The Afghanistan National Railway: A Plan of Opportunity.
Pomfret, R., 2010. Trade and transport in Central Asia. Glob. J. Emerg. Mark. Econ. 2, 237–256.
Rastogi, C., Arvis, J., 2014. The Eurasian connection supply-chain efficiency along the Modern Silk Route through Central Asia. International Bank for Reconstruction and Development, World Bank, Washington, D.C.
Sárvári, B., Szeidovitz, A., 2016. The political economics of the new silk road. Baltic J. Eur. Stud. 6 (1), 3–27.
Shibasaki, R., Naruse, S., Arai, H., Kawada, T., 2013. Industries, intermodal logistics and ports in far eastern Russia: Current status and future projects. In: Proceedings of the Eastern Asia Society for Transportation Studies, vol. 9. (presented on 9–12 September 2013, Taipei),.
Shibasaki, R., Tanabe, S., Kato, H., Lee, T.-W., 2019. Could Gwadar port in Pakistan be a new gateway? A network simulation approach in the context of the Belt and Road Initiative. Sustainability 11 (20), 5757.
Tanabe, S., Shibasaki, R., Kato, H., 2016. Impact assessment model of international transportation infrastructure development: focusing on trade and freight traffic in Central Asia. Asian Transp. Stud. 4 (1), 159–177.
Vinokurov, E., Dzhadraliyev, M., Shcherbanin, Y., 2009. The EurAsEC Transport Corridors. Eurasian Development Bank.
Wang, Z., Dong, S., Li, Z., Li, Y., Li, J., Cheng, H., 2015. Traffic patterns in the silk road economic belt and construction modes for a traffic economic belt across continental plates. J. Resour. Ecol. 6 (2), 79–86.
Yeh, E.T., Wharton, E., 2016. Going west and going out: discourses, migrants, and models in Chinese development. Eurasian Geogr. Econ. 57 (3), 286–315.

Pacific Islands: Small and dispersed 'sea-locked' islands

14

Takashi Riku, Ryuichi Shibasaki, and Hironori Kato
The University of Tokyo

Introduction

The Pacific region consists of numerous islands dispersed in the southwest of the Pacific Ocean (Fig. 14.1). The region has suffered from the high costs of participating in international trade due to its remoteness from the world's major markets. The dispersed nature of the region also leads to expensive transport, especially if connecting the smaller remote islands. Such geographic disadvantages and the consequent high trade costs have hampered the economic development of the Pacific Islands (PIs) (ADB-ADBI, 2015). This situation resembles that of landlocked countries as we discussed in the previous chapter; thus, the PI region could be referred to as a 'sea-locked' area. The development of the international transport infrastructure is essential for the economic growth of sea-locked developing countries, as it is typically mentioned in the context of landlocked regions (Collier, 2008; World Bank, 2009).

International transport in the Pacific region relies heavily on maritime shipping (MS). On a tonnage basis, over 80% of all international cargo (both container and non-container) in this region is transported by ship, according to the World Trade Service (WTS) database (IHS Global Insight, 2013). Another important trend in the Pacific region is containerisation, which is seen in other parts of the world as well. Fig. 14.2 shows the amount of international maritime cargo to/from the PIs on a tonnage basis by cargo type from the year 2000 to 2030. The data up to 2013 were observed data and from 2014 onwards, they are estimated by the WTS. Fig. 14.2 shows that although the majority of international cargo is still dry bulk, the volume of container cargo is increasing. In 2000, the amount of container cargo was 1.7 million tons, accounting for 12.1% of the total seaborne cargo. In 2013, the number had risen to 2.8 million tons, accounting for 17.3%. This trend is expected to continue in the future, and the amount of container cargo for 2030 is estimated at 4.2 million tons, accounting for 19.0%.

Given the inevitable dependence of the PIs on MS and the increasing importance of container shipping, this study focuses on maritime container shipping (MCS) in the Pacific Region. This chapter begins with a brief description of local ports, maritime container flows, and shipping networks in the region. Next, with a simulation model customised for the Pacific Region as an application of the global MCS model described in Chapter 6, this chapter presents case studies assuming different scenarios related to MCS. Note that the simulation model in this chapter only focuses on the MCS, neither land transport network nor intermodal transport super-network model described in Chapter 7 is not included.

Global Logistics Network Modelling and Policy. https://doi.org/10.1016/B978-0-12-814060-4.00014-9

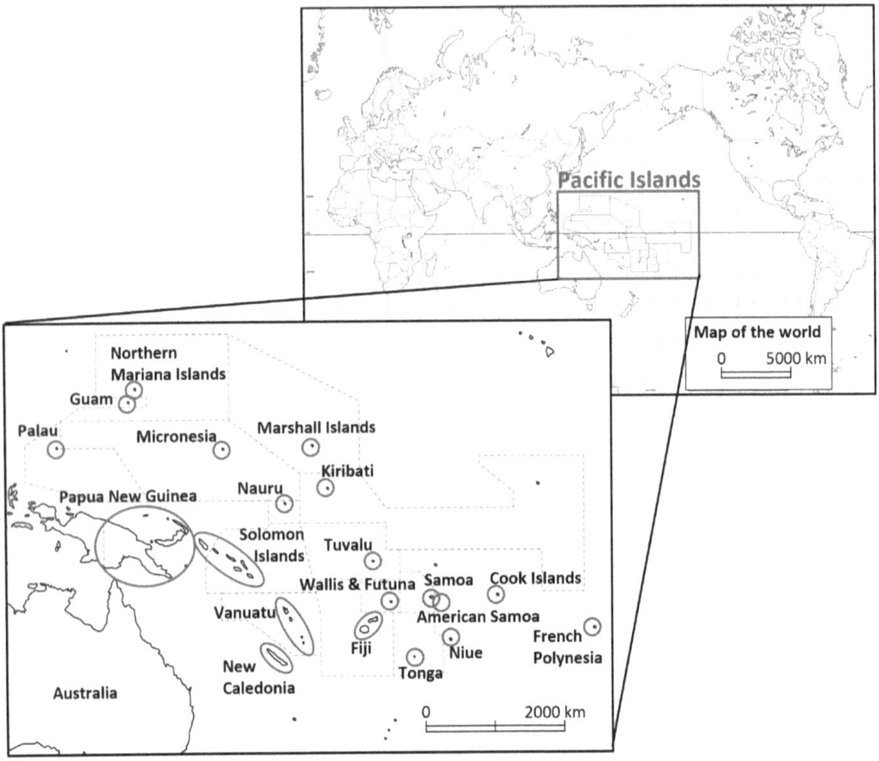

Fig. 14.1 PIs of study.
Source: Own compilation.

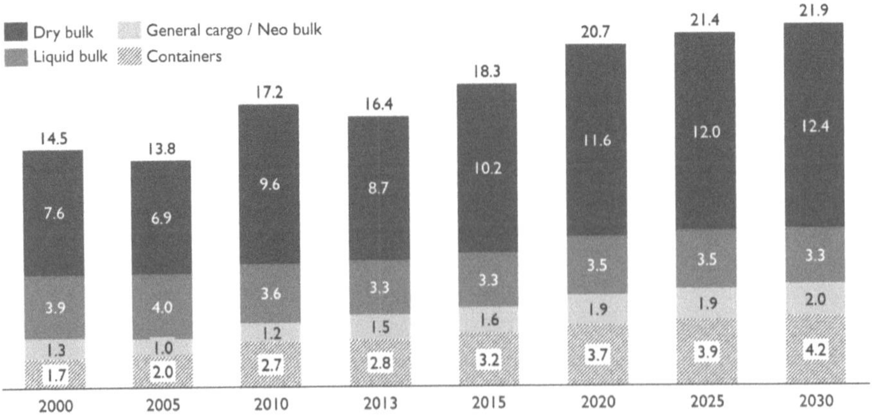

Fig. 14.2 International maritime cargo to/from the PIs by cargo type (million tons).
Source: Own compilation based on the WTS database.

Ports and maritime container cargo movement in the Pacific region

Fig. 14.3 illustrates the local ports in this study as well as the liner service (LS) network in the Pacific Region as of 2013.

Each PI typically has one or two international ports. The infrastructure at the PI ports ranges from simple wharves to better-equipped facilities with relatively high capabilities for cargo handling. Even at ports where capacity is considered sufficient, serious problems are often observed regarding the operational performance of the port infrastructure (ADB, 2007). The Pacific Regional Transport Study (AusAID, 2004), which was prepared for the PI Forum in 2004, pointed out that the majority of PI ports were built in the 1950s or 1960s, which was before the containerisation of international cargo. Consequently, the infrastructure at PI ports fails to meet today's international transport demands. The current situation is that containerships typically need to load and discharge containers at PI ports using their ship gear due to a lack of modern cargo-handling equipment in the ports. On the quayside, obsolete facilities and deteriorated surfaces often prevent smooth stevedoring. Wharves that cannot withstand the weight of a forklift and heavy containers occasionally require double handling of containers. Examples of local ports in the Pacific Region are shown in Figs. 14.4–14.6. In recent years, second-tier PI ports such as Honiara and Port Vila are quickly developing their infrastructure due to heavy congestion and an increase in cruise ships.

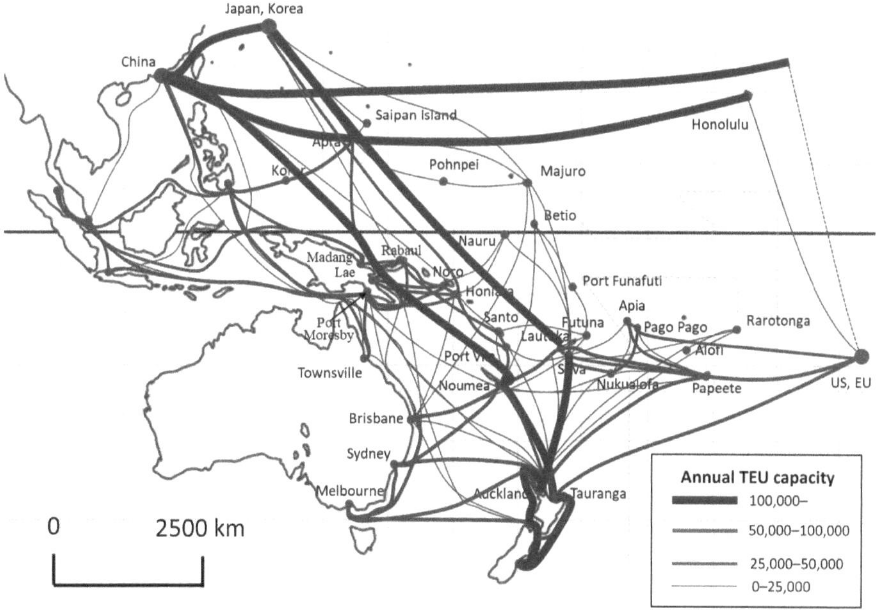

Fig. 14.3 Local ports and LS network in the Pacific Region.
Source: Own compilation based on the MDS Containership Databank (as of June 2013).

Fig. 14.4 Suva port in Fiji (2015). Note: This is the regional hub port and one of the most equipped ports in the region: unloading of cargo with ship gear due to lack of port equipment (left image), container trailers (middle image), and the one and only mobile crane (right image).
Source: Authors.

Fig. 14.5 Honiara port in Solomon Islands (2015). Note: Berthing containership (left image), unloading of cargo with ship gear (middle image), and the on-going construction of a new international wharf (right image).
Source: Authors.

Fig. 14.6 Port Vila in Vanuatu (2015). Note: Berthing containership and shallow coral reef close to the port (left image), forklift and container trailer in operation (middle image), and an abandoned ship sunk by Cyclone Pam and a cruiser at the back (right image).
Source: Authors.

The major MCS routes available in the region are connected with Asia, North America, Europe, and Oceania and within the PI. Two main streams exist in the region: (1) connecting North America/Europe with Australia/New Zealand via several Polynesian ports including Papeete (French Polynesia), Suva (Fiji), and Noumea (New Caledonia); and (2) connecting East/Southeast Asia with Australia/New Zealand via several Melanesian ports including Lae (PNG), Honiara (Solomon Islands), Port Vila (Vanuatu), and Suva. Regional hub ports are Australian ports, such as Brisbane, New Zealand ports, such as Auckland and Tauranga, and several PI ports such as Suva, Lae, Noumea, and Papeete.

The PIs have few domestic industries aside from agriculture and fishery. As a result, the people in the PIs rely heavily on imported goods for almost all aspects of life: daily goods, commodities, food, clothing, medicine, and machinery, for example. In fact, the PIs rank high on the list of countries with large trade deficits. Reflecting on this situation, the containers arriving at the PIs carry various goods from Asia or America/Europe. The authors have even spotted cars arriving inside containers at the PIs. In contrast, containers leaving the PIs are often empty. The small number of full containers carry fish, such as tuna, or other niche products for export such as Fiji Water, coffee, and dried tropical fruits. The issue of empty containers is not only problematic in terms of transport efficiency but also seriously undermines the profitability of liner shipping companies (LSCs) operating in this region. It is often the case that LSCs receive subsidies from PI governments to maintain operations in the region. Without these LSs, the local residents would lose their livelihoods.

Data

Ports and MCS network

The model in this chapter only focuses on MCS based on 2013 data, not including hinterland transport. Ports in the world with an annual container throughput of over 500,000 TEU are included as with other models in this book. To simulate container flows in more detail in the Pacific region, the model also includes all international ports in sovereign PIs with at least one appearance in the MDS Containership Databank and relatively large ports in the United States (US) and French territories. As a result, 200 ports in the world are included in the model: 27 PI ports and 173 ports in other regions.

In the model, the transshipment time of a port reflects its service level. The transshipment time TPR_a of PI ports is newly determined to be 144 (hours). In addition, in this study, the time for transshipment between different LSCs is assumed to be four times longer (i.e. $\tau = 4.0$) than the time for transshipment within the same LSC.

The MCS network in this model is constructed based on the MDS Containership Databank. The model includes 32 LSCs: the top 20 of the largest LSCs in the world and 12 local LSCs that provide LSs to the PIs. The model includes 904 LSs, and the developed network consists of 274,088 links. Fig. 14.3 shown in the previous section provides a visual of the constructed MCS network as of 2013. Table 14.1 provides a list of local LSCs added in the model.

Table 14.1 LSCs added in the model.

No.	Group	Group name	Included carriers	Annual TEU capacity (author's estimation from the MDS Containership Databank, '000 TEU)	Share of the world (%)
21	Group U	Hub	Hub Line	276	0.2
22	Group V	Kyowa	Kyowa, FSM Line, Western Pacific Shipping	14	0.0
23	Group W	Matson	Matson	426	0.3
24	Group X	Swire	Swire Shipping	141	0.1
25	Group Y	PDL	Pacific Direct Line	30	0.0
26	Group Z	Sofrana	Sofrana	50	0.0
27	Group AA	Horizon	Horizon Lines	442	0.4
28	Group AB	Mariana Express	Mariana Express	292	0.2
29	Group AC	Carpenters	Carpenters Shipping	65	0.1
30	Group AD	Pacific Forum	Pacific Forum	28	0.0
31	Group AE	NGPL	NGPL	15	0.0
32	Group AF	Marfret	Marfret	271	0.2
Others				14,586	11.7
Total (including Top 20 LSCs)				124,263	100.0

Source: Own compilation.

Container cargo shipping demand (OD matrix)

The estimation of the TEU-basis container cargo shipping demand between ports is calculated separately for the following three patterns: OD matrix between non-PI ports, OD matrix between a PI port and a non-PI port, and OD matrix among PI ports. After the estimations, the total OD matrix is adjusted using the Frater method with the initial total shipping demand for each port as given.

First, following Chapter 9, the container cargo shipping demand between non-PI ports is estimated based on the WTS database, which provides actual and anticipated container cargo shipping demand between countries and regions for each year from 2000 to 2030.

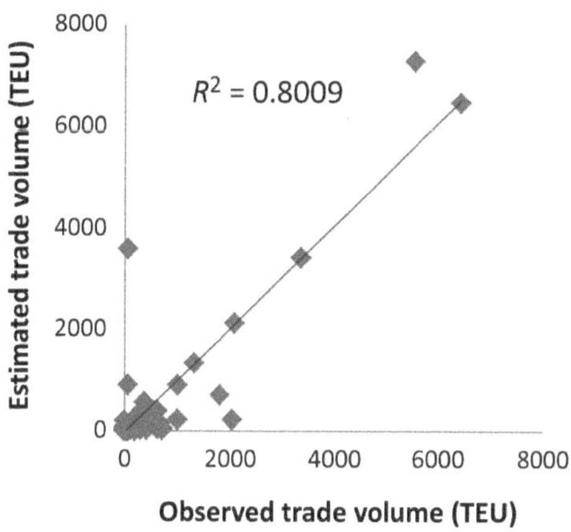

Fig. 14.7 Estimated versus observed trade volume among the PIs.
Source: Own compilation.

Second, the shipping demand between each PI port and non-PI port is estimated by breaking down the shipping demand between non-PI ports and the 'New Zealand' (for Tokelau and Niue), 'Other Southeast Asia' (for PNG), or 'Pacific Islands' (for other PI) WTS regions, estimated in the first step. Because annual throughput data are not available for most PI ports, the regional share for the PIs is computed from the UN Comtrade Database by trade partner considering the commodities.

Third, regarding the shipping demand among the PIs, the UN Comtrade Database are used where available because the WTS data do not contain intra-regional containers. Regarding the OD pairs where the UN Comtrade Database are not available, which includes the majority of the OD pairs in the PIs, the shipping demand is estimated with a gravity model developed based on the UN Comtrade Database. The model is basically a function of the GDPs of the origin and destination countries and the shipping time between the two countries. The model also takes into consideration Free Trade Agreements (FTAs) and historical colonial relationships.

Fig. 14.7 shows the estimated versus observed trade volumes among PIs in terms of TEU. The gravity model reproduces well the observed trade volumes. Note that the multiple correlation coefficient in Fig. 14.7 is calculated based on the TEU-basis trade volumes converted from the gravity model's raw output.

Calculation results

Table 14.2 shows the annual volumes of export and import laden containers input into the model and estimated volumes of transshipped laden containers and transshipment rates for each Pacific port, as of 2013. Ports in PNG are observed to have relatively

Table 14.2 Input annual volumes of export and import laden containers and estimated volumes of transshipped laden containers and transshipment rates for PI ports (as of 2013).

Ports	Export and import laden containers (TEU/year)			Estimated transshipment laden containers (TEU/year)	Estimated transshipment rate (%)
	Total	Export	Import		
Apra (Guam)	15,721	3687	12,034	5222	24.9
Saipan Island (Northern Mariana Islands)	4428	1608	2820	0	0.0
Koror (Palau)	4305	1586	2719	0	0.0
Pohnpei (Micronesia)	5870	2298	3572	0	0.0
Majuro (Marshall Islands)	14,943	8850	6093	400	2.6
Betio (Kiribati)	7750	4306	3443	0	0.0
Port Funafuti (Tuvalu)	1774	1087	687	0	0.0
Nauru (Nauru)	5197	111	5086	3	0.1
Futuna (Wallis and Futuna)	1324	24	1300	0	0.0
Lae (PNG)	79,768	33,483	46,284	7124	8.2
Madang (PNG)	33,002	13,853	19,149	0	0.0
Port Moresby (PNG)	100,987	42,391	58,596	3	0.0
Rabaul (PNG)	35,739	15,002	20,737	177	0.5
Honiara (Solomon Islands)	7766	1546	6220	12	0.1
Noro (Solomon Islands)	303	68	235	4	1.3
Port Vila (Vanuatu)	17,619	11,396	6223	136	0.8

Santo (Vanuatu)	170	40	129	30	15.2
Lautoka (Fiji)	14,266	8500	5766	737	4.9
Suva (Fiji)	60,825	33,829	26,995	10,719	15.0
Noumea (New Caledonia)	61,650	22,815	38,835	2701	4.2
Apia (Samoa)	9839	1112	8727	0	0.0
Pago Pago (American Samoa)	8788	1304	7485	0	0.0
Nukualofa (Tonga)	5313	389	4924	3	0.0
Alofi (Niue)	1287	1054	233	0	0.0
Rarotonga (Cook Islands)	9844	7758	2086	6	0.1
Papeete (French Polynesia)	30,932	1082	29,851	8920	22.4
Pacific Island ports Total	539,410	219,182	320,228	36,198	6.3
Brisbane (Australia)	555,252	208,790	346,462	185,999	25.1
Sydney (Australia)	1,164,196	437,854	726,342	129,775	10.0
Melbourne (Australia)	1,253,422	470,635	782,787	104,468	7.7
Auckland (New Zealand)	353,796	133,063	220,733	14,271	3.9
Tauranga (New Zealand)	596,593	354,264	242,329	178,889	23.1

Source: Own compilation.

large amounts in export and import laden containers, with Port Moresby at the top, handling 100,987 TEU/year. Relatively large amounts of export and import laden containers are observed in other PI ports such as Apra, Port Vila, Lautoka, Suva, Noumea, and Papeete. Majuro handles an unexpectedly large amount of export and import laden containers at 14,943 TEU/year; this is due to a large amount of trade in fish recorded in the UN Comtrade Database. Furthermore, the unexpectedly small amount of export and import laden containers at Honiara may be the result of non-containerised MS such as tankers being dominant in the country's international trade. The total amount of the export and import laden containers handled at PI ports is estimated at 539,410 TEU/year.

Another finding from Table 14.2 is that imports greatly exceed exports in many PI ports; this is consistent with existing literature and the results of the field survey by the authors. The same trend is observed in the Pacific region as a whole, which handles 324,729 TEU of import laden container volume whereas exports represent only 223,750 TEU. However, in some PI ports, such as Port Vila, Suva, and Rarotonga, exports largely exceed imports. The validity of the estimation results regarding such cases should be addressed in future studies, ideally by acquiring sufficient data.

Transshipped laden containers in the Pacific region are concentrated in many major PI ports. Suva, which is identified as a regional hub port, is estimated to handle the largest volume in transshipped laden containers at 10,719 TEU/year, which is widely used by PI containers. Papeete and Lae follow, at 8920 and 7124 TEU, respectively. Apra and Noumea handle some transshipped laden containers as well, and the other PIs handle little or no transshipped laden containers. The total transshipment rate for the PI region is 6.3%. According to the estimation results, laden containers to and from East Asia are transshipped at Apra, Lae, Suva, and Noumea; containers to and from Southeast Asia at Lae; and containers to and from Europe/North America at Papeete. These results are consistent with existing literature and information collected through a field survey conducted by the authors.

Table 14.3 shows the estimated versus observed volumes and rates of transshipment laden containers at selected Pacific ports, where the observed transshipment data for

Table 14.3 Comparison between estimated and observed volumes and rates of transshipped laden containers at selected PI ports.

Ports	Estimated		Observed	
	Volume (TEU/year)	Percentage	Volume (TEU/year)	Percentage
Auckland (New Zealand)	178,889	23.1	173,404[a]	17.9
Apra (Guam)	5222	24.9	42,000[b,c]	28.0
Suva (Fiji)	10,719	15.0	11,007[d]	10.0

[a] Own compilation based on data acquired from Ports of Auckland (POAL) (2015).
[b] Data acquired from Brinckerhoff (2013).
[c] Empty containers are included.
[d] Data acquired from Fiji Ports Corporation Ltd. (2015).

PIs are available. The estimated volume at Suva fits the observed volume well whereas the estimated volume at Apra is significantly lower than the observed one. One possible reason for this is that a significant number of empty containers are included in the observed data. The observed data for Suva do not include empty containers. The above suggests that the developed model describes the maritime container flows in the Pacific region reasonably well.

Policy simulations

Overview

This study assumes four scenarios for the year 2030; the 'baseline scenario', 'port underdevelopment scenario', 'Honolulu shipping service scenario', and 'vessel enlargement scenario'.

In the baseline scenario, the service frequency and vessel capacity of LSs are increased by the growth of the world's shipping demands. In addition, the transshipment time at all PI ports is assumed to improve. The port underdevelopment scenario assumes that the port development of an arbitrary single PI port is delayed relative to the baseline scenario; the transshipment time at this PI port remains unchanged from 2013. The Honolulu shipping service scenario assumes the commencement of a new LS connecting the north and south of the Pacific region in order to complement the existing MCS network in the Pacific region. The vessel enlargement scenario assumes the further enlargement of vessels calling on Australia and New Zealand.

The results of the latter three scenarios are compared with that of the baseline scenario in terms of MCS patterns, which are represented by transshipment volume at each port, as well as the performance of regional LSs, which is represented by shipping time.

Two indices for performance evaluation are calculated based on the average weighted shipping time for containers to and from each PI port, AT_r (hour), which is defined as

$$AT_r = \frac{\sum_s \left(q^{rs} t^{rs*} + q^{sr} t^{rs*}\right)}{\sum_s \left(q^{rs} + q^{sr}\right)}$$

where t^{rs*} is the converged shipping time from port r to port s. Note that the converged shipping time should be the same among the different paths for each OD pair.

The first performance index is the regional average weighted shipping time, AT_P (hour), which represents inefficiency with respect to shipping time. This is defined as

$$AT_P = \frac{\sum_{r \in R_P} \sum_s \left(q^{rs} t^{rs*} + q^{sr} t^{sr*}\right)}{\sum_{r \in R_P} \sum_s \left(q^{rs} + q^{sr}\right)}$$

where R_p is the set of PI ports. A reduction in AT_P would mean that the MCS network in the Pacific region as a whole is able to transport container cargo more efficiently.

The second performance index is the standard deviation of average shipping time σ_{AT} (hour), which represents inequality among PI ports regarding shipping time. It is defined as.

$$\sigma_{AT} = \sqrt{\frac{1}{N_{R_p}} \sum_{r \in R_p} \left(AT_r - \overline{AT} \right)^2}$$

where N_{Rp} is the number of ports in the set of PI ports R_p (i.e., $N_{Rp}=26$) and $\overline{AT}$ is the unweighted average of the average shipping time AT_r. A reduction in σ_{AT} would mean that the difference in shipping time between the advanced PI ports and the relatively lagging PI ports have become smaller. $\overline{AT}$ is calculated as

$$\overline{AT} = \frac{1}{N_{R_p}} \sum_{r \in R_p} AT_r$$

Baseline scenario

The baseline scenario assumes a future situation in 2030 where the international MCS market changes at a business-as-usual growth rate. The container cargo shipping demand (OD matrix) in this scenario is estimated based on the WTS projection data for 2030 where the world's total shipping demand is 2.22 times larger than in 2013. To retain the global balance between shipping demand and LS provision, the annual TEU capacity of each LS is increased at this growth rate. First, the annual service frequency is doubled under the condition that it does not exceed 52, which is a weekly service. Next, the vessel size is enlarged so that the annual TEU capacity of the service is 2.22 times larger.

Regarding the Pacific region (except PNG), the WTS projection data basically use 1.47 as the average growth rate of shipping demand, which is lower than the world average. Regarding PNG, however, 3.51 is used as the average growth rate. Because this study focuses on the Pacific region, 1.47 is used for LSs that call at PI ports but do not call at PNG ports and 3.51 is used for LSs that call at PNG ports. The service frequency and vessel capacity are enlarged in the same way as described above and, consequently, the service frequency tends to increase in the Pacific region whereas the vessel size tends to increase in the rest of the world.

In the baseline scenario, it is assumed as well that the transshipment time at every PI port improves from 144 to 72h, which is still the longest in the world. Moreover, regarding the shipping demand among PI ports, the FTA dummy variable in the gravity model, is set at 1 for every OD pair, which represents future regional integration in the form of the FTA among the PIs. It should be noted that although named the 'baseline scenario', it is intended as one possible future and is not intended as a fixed and definitive future situation.

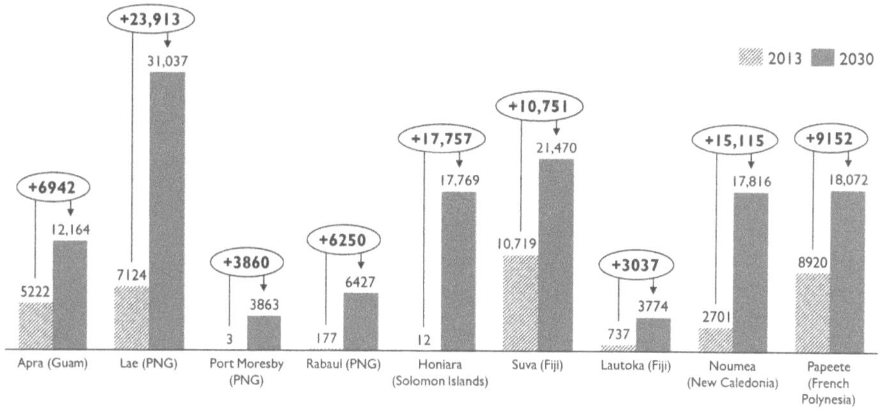

Fig. 14.8 Comparison between transshipment volumes of laden containers in 2013 and 2030 (baseline scenario) in PI ports with significant changes (TEU/year).
Source: Own compilation.

Fig. 14.8 shows the comparison between estimated transshipment volumes of laden containers in 2013 and those in 2030 (baseline scenario) at PI ports. Large increases in transshipments are observed in major PI ports, namely, Lae, Honiara, Suva, Noumea, and Papeete. The increases are particularly large in Lae, Honiara, and Noumea. The results imply that Southeast Asia and East Asia are expected to be important trade partners in the Pacific region.

Port underdevelopment scenario

This scenario assumes 23 cases where the transshipment time for a single PI port remains unchanged from 2013 whereas the transshipment times of all the other PI ports improve. In other words, this scenario assumes that one PI fails to develop its port. As in the baseline scenario, the transshipment time at the improved PI ports is assumed to be 72 h whereas that at the underdeveloped PI port is assumed to be 144 h. Note that the PI ports with zero transshipment in the baseline scenario are excluded from the following analysis.

As a result, a significant impact is observed if major PI ports are underdeveloped, namely Lae, Honiara, Suva, Lautoka, Noumea, and Papeete. Fig. 14.9 shows that the delay of port development at each major PI port would significantly decrease transshipped laden containers at that port. In most ports, transshipment would be reduced by over 50%. This indicates that major PI ports would suffer from a significant loss of transshipped laden containers if they fail to develop their ports on time.

In terms of transshipped laden containers, Lae would suffer the most from the underdevelopment of its port, losing over 20,000 TEU/year. The majority of Lae's transshipment would shift to Australian ports, particularly Brisbane, and some transshipment would shift to Madang, Rabaul, and Honiara (Fig. 14.10). This implies that a strong competitive relationship exists between Lae and Brisbane whereas Madang,

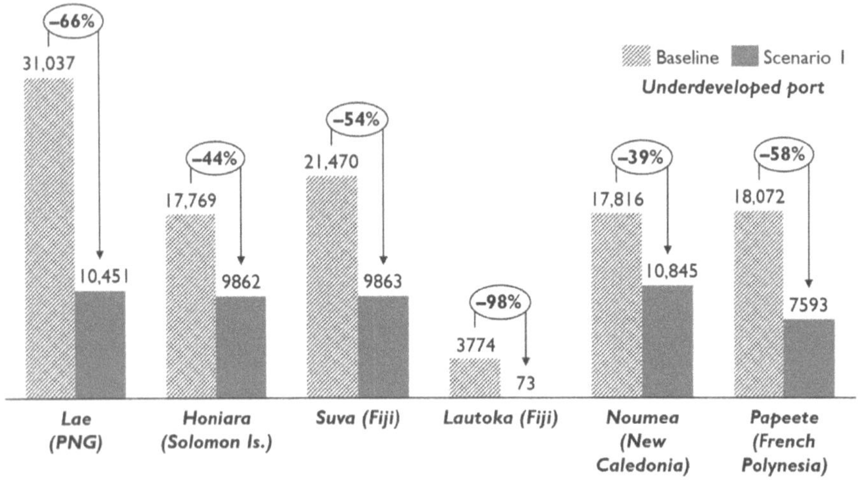

Fig. 14.9 Decrease in transshipped laden containers at major PI ports if that specific port is underdeveloped (TEU/year).
Source: Own compilation.

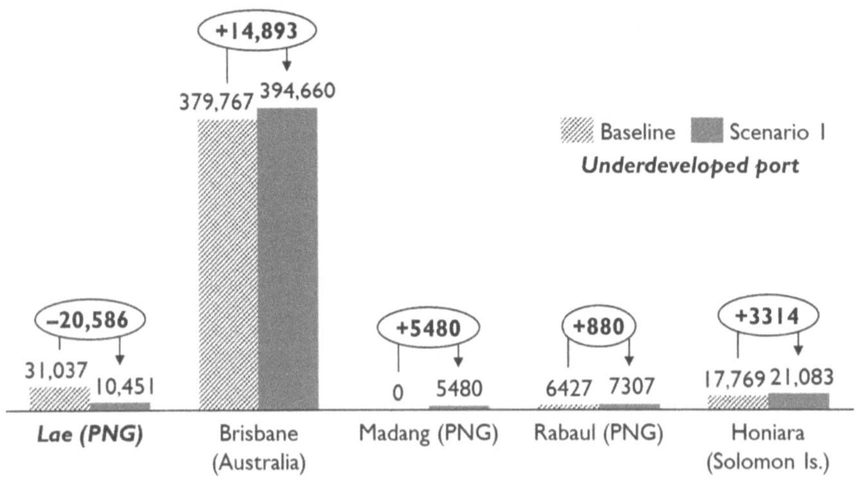

Fig. 14.10 Changes in transshipped laden container volume if Lae is underdeveloped (TEU/year).
Source: Own compilation.

Rabaul, and Honiara are minor competitors for Lae. In contrast, if Honiara is underdeveloped, the transshipment at Australian and New Zealand ports would decrease significantly, suggesting that Honiara is in a complementary relationship with Australian and New Zealand ports (Fig. 14.11).

Fig. 14.12 shows the changes in the regional performance indices from the baseline scenario if each PI port is underdeveloped. The inefficiency (AT_P) increases

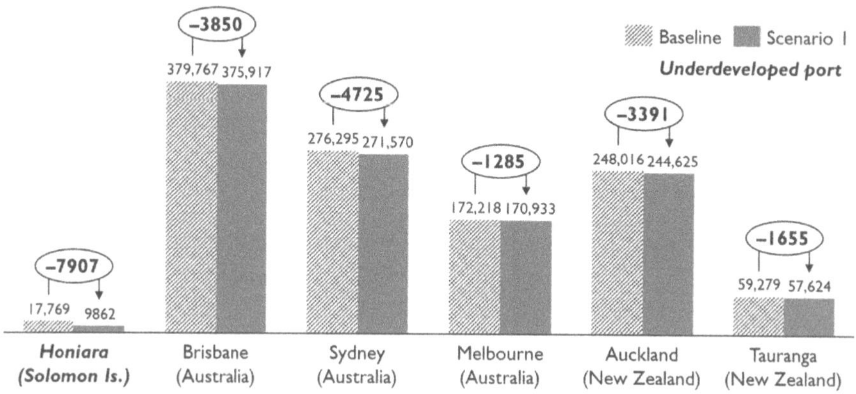

Fig. 14.11 Changes in transshipped laden container volume if Honiara is underdeveloped (TEU/year).
Source: Own compilation.

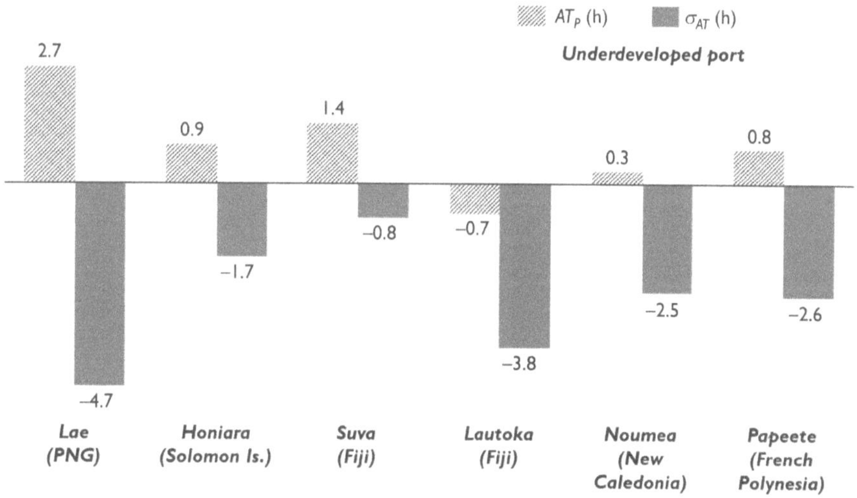

Fig. 14.12 Changes in the inefficiency (AT_P) and inequality (σ_{AT}) of regional MCS if major PI ports are underdeveloped.
Source: Own compilation.

the most if Lae is underdeveloped, followed by Suva. Delays in the development of regional hub ports with large container handling would negatively affect the whole region in terms of average shipping time per container. The container handling volume at Lae is particularly large due to the heavily used Asia–Pacific LSs often calling at Lae.

Next, most cases show possible changes in the inequality (σ_{AT}) if a PI port is underdeveloped. This is because the underdevelopment of a relatively convenient port results in increases in shipping time for all or most PIs, particularly the underdeveloped port itself. Thus, the variance of shipping time within the region becomes smaller. The largest change in inequality is seen if Lae is underdeveloped because the decrease in transshipment is significant. This is followed by Lautoka, as many LSs to remote PIs in the Polynesian and Micronesian area call at Lautoka.

Honolulu shipping service scenario

The Honolulu shipping service scenario assumes the commencement of a new LS directly connecting the north and south of the Pacific region. In the current MCS network, the LSs in the Micronesian area are scarce compared with those in Polynesia and Melanesia, and no LSs stretch in the northeast–southwest direction. Due to the lack of such an LS, the northern PI ports (e.g. Pohnpei, Majuro, Betio, Nauru, and Port Funafuti) tend to be isolated with very few LSs available for use. Not only is this situation inconvenient for the residents, but it is also a significant risk for their livelihoods. For example, if an LS is cancelled due to disasters, containers filled with daily necessities cannot reach the isolated PI ports for as long as several months.

This scenario introduces a new LS attempting to complement the current network. Given that Matson (an LSC) operates a large LS connecting Honolulu, Apra, and China, this scenario assumes that Matson begins a similar LS connecting Honolulu and Brisbane with PI ports in between. The ports of call for this new LS are assumed to be Honolulu, Majuro, Betio, Nauru, a candidate port, and Brisbane. For the candidate port, five ports are raised as possible options based on their geographic locations: Lae, Honiara, Port Vila, Noumea, and Suva. Fig. 14.13 shows the new LS in thick dotted lines.

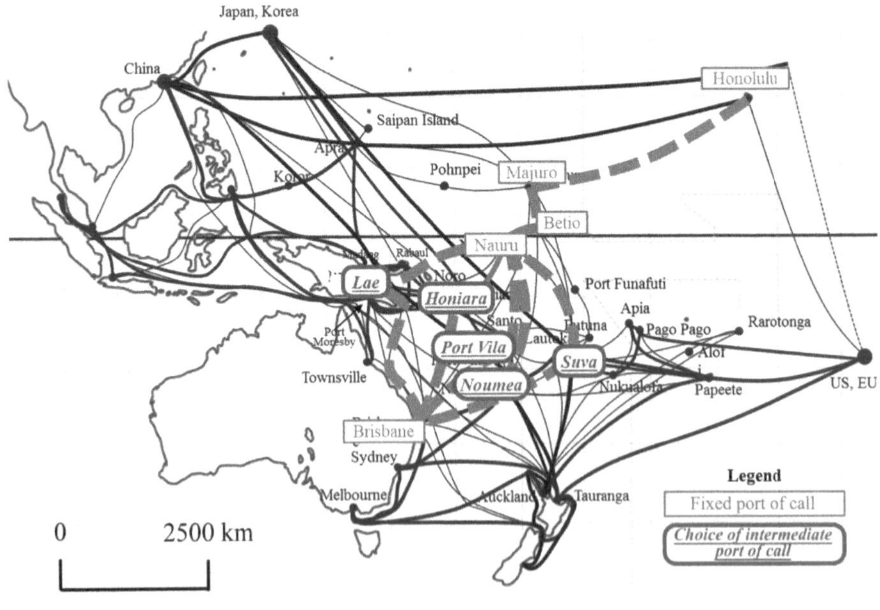

Fig. 14.13 New LS under the Honolulu shipping service scenario.
Source: Own compilation.

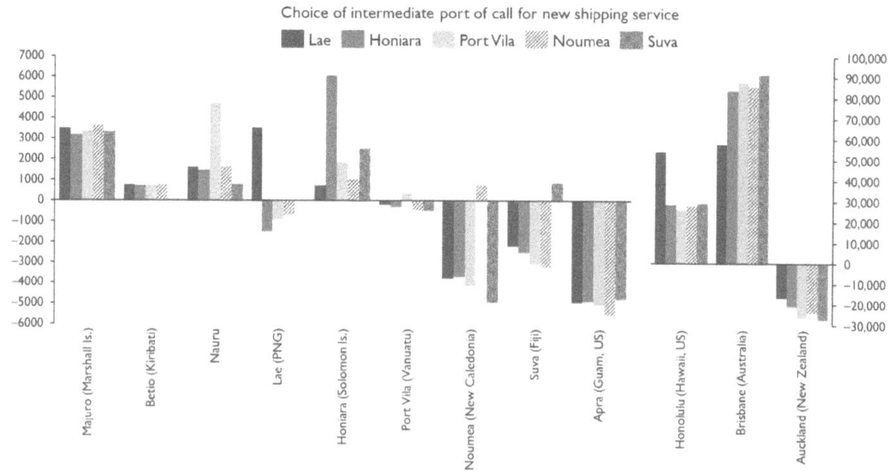

Fig. 14.14 Changes in transshipped laden container volume compared to the baseline scenario depending on the intermediate port of call for the new LS (TEU/year).
Source: Own compilation.

Fig. 14.14 shows the annual transshipped laden container volumes and their change from the baseline scenario at many Pacific ports in each case of this scenario. Only the ports of call of the new LS and ports with significant changes are shown in the figure. In all cases, the amount of transshipment may increase at every port of call, and significant increases may be observed at both ends of the new LS, namely Honolulu and Brisbane. The new LS would successfully call for the mass of laden containers transported between Europe/North America and Southeast Asia/Australia whereas simultaneously catering to the container cargo shipping demand of the PIs at a microlevel.

In the baseline scenario, some laden containers into Lae or neighbour PI ports are transshipped at Brisbane from Europe/North America. However, if the new LS calls on Lae, these containers, which had been transshipped at Brisbane, are newly transshipped at Honolulu, shortening the geologic distance and shipping time. As a result, if Lae is selected as the port of call of the new LS, the predicted increase in transshipped laden containers is relatively small in Brisbane and relatively large in Honolulu compared with the other cases.

Fig. 14.15 shows the changes in the performance indices from the baseline scenario in each case. The predicted changes under this scenario are more significant than those under the port underdevelopment scenario because the commencement of a new LS involves more containers than a scenario concerning only one PI port. For all indices, the results show that the new LS would contribute best to the region if Suva is selected as the port of call, although the differences between the cases are not significantly large. This result suggests that the new LS should call on Suva, the regional hub port, to contribute best to improving regional LS levels.

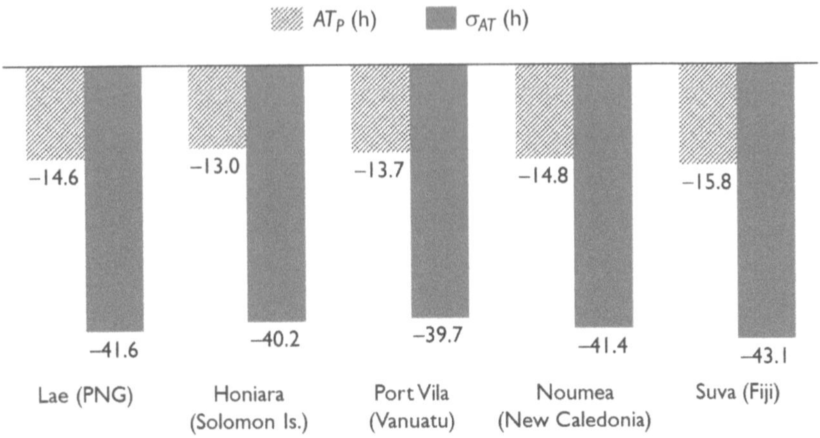

Fig. 14.15 Changes in the inefficiency (AT_P) and inequality (σ_{AT}) of regional MCS depending on the intermediate port of call for the new LS.
Source: Own compilation.

Vessel enlargement scenario

The vessel enlargement scenario assumes further enlargement of vessels where their capacity becomes 3-times larger in LSs between Australia/New Zealand and non-Pacific regions (hereafter referred to as AU/NZ services), reflecting the recent acceleration of containership enlargement in trunk routes of the world and its spillover to other local routes known as the 'cascade effect' (OECD, 2015). Note that the vessel size of each LS is already assumed to be enlarged in the baseline scenario as described earlier; thus, this scenario assumes additional enlargement in specific LSs.

This scenario consists of three cases. In Case 1, further vessel enlargement is assumed only for AU/NZ services not calling at PI ports. Those calling at PI ports remain unchanged from the baseline scenario. In Case 2, further vessel enlargement is assumed for all AU/NZ services including those calling at PI ports. This means that PI ports are able to accommodate larger vessels due to adequate investments in port infrastructure. Last, Case 3 assumes that all AU/NZ services are further enlarged and those with a vessel capacity of over 4000 TEU, the largest vessel size in the Pacific region in the baseline scenario, skip the PI ports. This means that PI ports are unable to accommodate larger vessels due to a lack of adequate investment.

Figs. 14.16–14.18 show the transshipped laden container volumes and changes from the baseline scenario for major Pacific ports in each case. In Case 1, Australian and New Zealand ports (AU/NZ ports) would experience significant increases in transshipment, particularly Brisbane; this is caused by East/Southeast Asian containers transported to and from AU/NZ ports. Although shifts in transshipment to AU/NZ ports are predicted in some PI ports, the numbers are small.

In Case 2, transshipment at Lae would significantly decrease whereas transshipment at the AU/NZ ports, particularly Brisbane and Tauranga, would significantly increase. In this case, Brisbane would partly substitute Lae as a transshipment port for

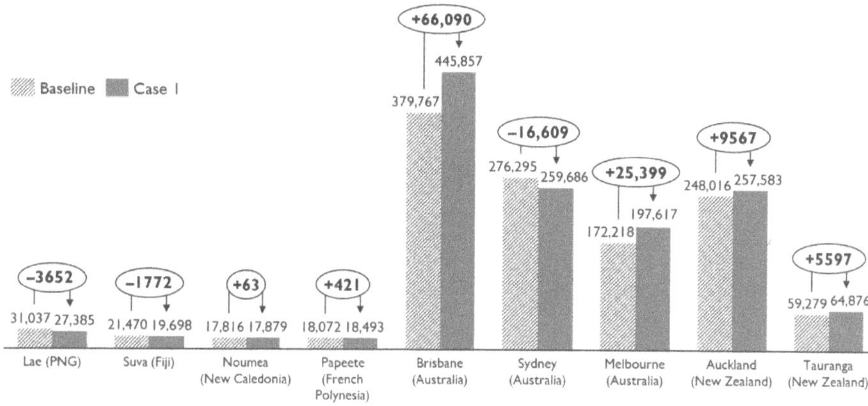

Fig. 14.16 Comparison of transshipped laden container volumes between the baseline scenario and Case 1 (TEU/year).
Source: Own compilation.

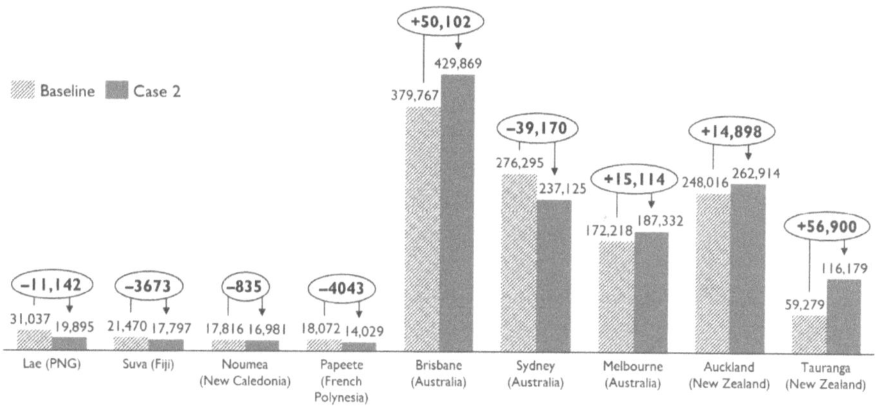

Fig. 14.17 Comparison of transshipped laden container volumes between the baseline scenario and Case 2 (TEU/year).
Source: Own compilation.

laden containers shipped between East/Southeast Asia and the PI ports. As a result, transshipment would shift from Lae to Brisbane. A similar situation is shown in the southern part of the Pacific region as well. Tauranga would be used frequently for laden containers shipped between Europe/North America and the PI ports. As a result, large shifts in transshipment would be seen from Suva and Papeete to Tauranga.

In Case 3, all PI ports listed in the figure would experience significant decreases in transshipment whereas the predicted increases in transshipment at the AU/NZ ports would be even more significant than in the other two cases. In this case, the majority of laden containers to and from East/Southeast Asia must get in or out of the Pacific region via the limited available LSs connecting the PI ports and Brisbane or Auckland, where they are transshipped. Similarly, a large number of laden containers to and

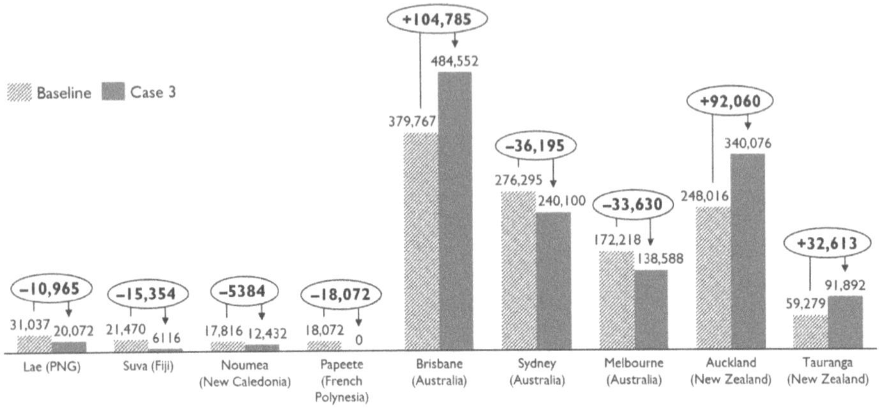

Fig. 14.18 Comparison of transshipped laden container volumes between the baseline scenario and Case 3 (TEU/year).
Source: Own compilation.

from Europe/North America, which used to be transported within the PI region to/from Tauranga, must get in or out of the PI region through Brisbane or Auckland before or after being transshipped at Tauranga. Particularly, Papeete may no longer function as a regional hub port and may lose all transshipped laden containers, which would have a devastating effect on laden containers transported between Europe/North America and the PIs.

Table 14.4 shows the changes from the baseline scenario in average weighted shipping time AT_r in each case in terms of hours and percentage. Reductions in average shipping time would be predicted in many PI ports in every case because the enlargement of vessels would result in less congestion, thus, shortening shipping times. Case 2 shows larger reductions than Case 1, which implies that most PI ports would benefit from the vessel enlargement if Pacific ports could accommodate larger vessels. Furthermore, significant reductions in Case 2 would be observed in relatively remote PIs, indicating that vessel enlargement inside the region would be an effective approach for reducing inequality in terms of shipping time. In Case 3, shipping time to/from southern PI ports, including Suva, Noumea, and Papeete, would increase. This is because Suva, Noumea, and Papeete, skipped by AU/NZ services, would be unable to accommodate the enlarged vessels. This implies that southern PI ports would suffer the most if LSs skip PI ports. Particularly, the damage would be exceptionally large at Papeete, due to its high dependence on AU/NZ services.

Conclusion

The PI region is a 'sea-locked' area with numerous small islands dispersed in a geographically large area. As is true for landlocked regions, the development of international transport infrastructure is essential for the economic growth of PI countries. However, a significant difference between the PI region and landlocked regions is that MS is the major mode of international transport in the PI region, not land transport.

Table 14.4 Changes in AT_r in the vessel enlargement scenario.

Ports	Case 1		Case 2		Case 3	
	Change (hour)	Percentage	Change (hour)	Percentage	Change (hour)	Percentage
Apra (Guam)	−2	−0.3	−10	−1.4	−12	−1.6
Saipan Island (*N. Mariana* Is.)	10	2.0	9	1.9	3	0.6
Koror (Palau)	12	2.2	13	2.4	3	0.5
Pohnpei (Micronesia)	−71	−4.8	−97	−6.6	−78	−5.3
Majuro (Marshall Is.)	−13	−1.1	−94	−8.6	−98	−9.1
Betio (Kiribati)	−16	−1.4	−107	−10.4	−105	−10.3
Port Funafuti (Tuvalu)	−13	−1.3	−138	−15.9	−141	−16.2
Nauru	−8	−0.7	−218	−21.4	−206	−20.2
Futuna (Wallis & Futuna)	−2	−0.1	−100	−8.4	−50	−4.2
Lae (PNG)	−2	−0.2	−20	−2.8	−19	−2.5
Madang (PNG)	−7	−0.8	−44	−5.3	−44	−5.3
Port Moresby (PNG)	−3	−0.5	−20	−2.7	−16	−2.2
Rabaul (PNG)	−2	−0.2	−26	−3.1	−22	−2.7
Honiara (Solomon Is.)	−2	−0.2	−26	−3.7	−18	−2.4
Noro (Solomon Is.)	−4	−0.4	−24	−2.7	−21	−2.3
Port Vila (Vanuatu)	−5	−0.7	−37	−5.3	−17	−2.5
Santo (Vanuatu)	−5	−0.9	−49	−8.3	−10	−1.6
Lautoka (Fiji)	−10	−1.2	−72	−9.5	5	0.6
Suva (Fiji)	−15	−2.0	−81	−11.1	62	7.2
Noumea (New Caledonia)	−9	−1.5	−52	−9.4	20	3.2
Apia (Samoa)	−12	−1.2	−86	−9.2	46	4.5
Pago Pago (American Samoa)	−11	−1.0	−88	−8.2	43	3.6
Nukualofa (Tonga)	−15	−1.4	−117	−12.0	9	0.8
Alofi (Niue)	−63	−4.7	−241	−20.9	−236	−20.3
Rarotonga (Cook Is.)	4	0.4	−131	−12.1	−133	−12.2
Papeete (French Polynesia)	−13	−1.7	−58	−8.5	481	37.7

Source: Own compilation.

Once built, roads are there to stay for several decades, if not more. On the other hand, there are no 'roads' for MS; shipping routes could change anytime. Thus, for sea-locked countries, MS policy is a top priority for governments, and continued monitoring of the changing environment and timely interventions are key for economic growth. Moreover, many of the key LSs in the PI region are operated either by globally major LSCs or governments of PI countries. This situation further strengthens the need for governments to effectively monitor and control the regional MCS network.

Despite MS policy being critical for regional economic growth, understanding the current status and the mechanisms of MS in the PI region is difficult due to the lack of even the most basic data. As a result, very few studies have taken a quantitative approach to analyse MS policy in the PI region. This study aimed to provide a breakthrough in this field by building a simulation model for container cargo flows within the PI region and by using the model to analyse MCS policies.

This chapter applied a global MCS model to the Pacific region with four scenarios in 2030 to examine the potential impacts of different policies and investments: the baseline scenario, the port underdevelopment scenario, the Honolulu shipping service scenario, and the vessel enlargement scenario.

The baseline scenario showed that Southeast and East Asia would become increasingly important trade partners for PIs in the future. The port underdevelopment scenario showed that major PIs would lose significant volumes of transshipped laden containers if they failed to develop their ports on time. This scenario also highlighted the competitive and complementary relationships in the Pacific region. A summary is provided in Fig. 14.19.

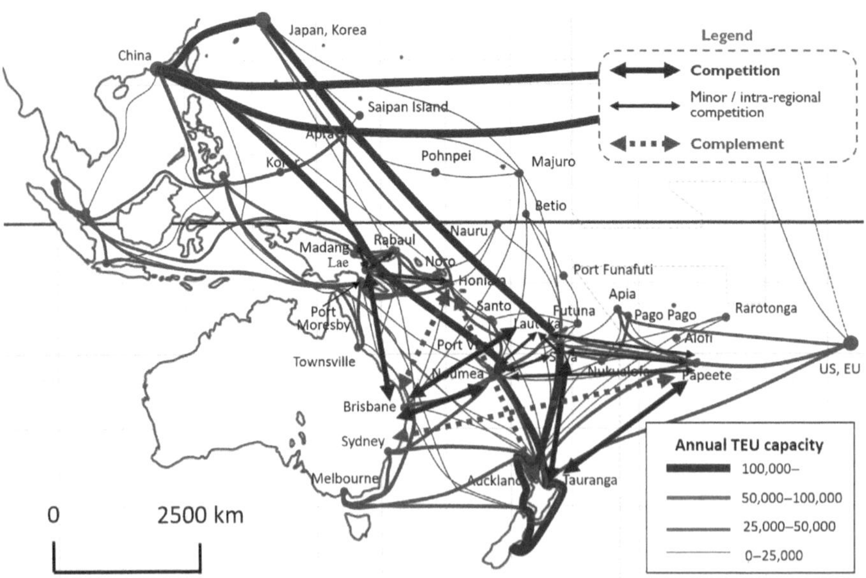

Fig. 14.19 Summary of competitive and complementary relationships between PI ports (including Australian and New Zealand ports).
Source: Own compilation.

From the viewpoint of regional efficiency, the scenario indicated that port development at Lae and Suva should be prioritised. However, the scenario also indicated that if these ports are improved, congestion at Lae may negatively affect regional equality.

The Honolulu shipping service scenario showed that the commencement of a new LS connecting the northeast and southwest of the Pacific region would successfully accommodate containers transported between Europe/North America and Southeast Asia/Australia whereas simultaneously catering to the needs of the PIs. The scenario also suggested that such an LS should call on Suva, the regional hub port, from the viewpoint of improving regional LS levels.

The vessel enlargement scenario showed that although transshipment handling may decrease, PIs would benefit from the global trend of vessel enlargement if their ports are able to accommodate larger vessels. However, if this should not be possible, the southern PI ports, particularly Papeete, would experience longer shipping times and lower accessibility.

Acknowledgments

This study was supported by many experts from port authorities, shipping companies, and regional organisations in Fiji, New Zealand, Solomon Islands, and Vanuatu as well as international organisations such as the Asian Development Bank (ADB) and the Secretariat of the Pacific Community (SPC). Our field survey was organised and supported by officials of Japan International Cooperation Agency (JICA). We thank them for their valuable comments and advice. This study is financially supported by Grants-in-Aid for Scientific Research (No. 25289159 and No. 15H01936).

References

Asian Development Bank, 2007. Oceanic Voyages: Shipping in the Pacific. Asian Development Bank, Manila.

Asian Development Bank, Asian Development Bank Institute, 2015. Pacific Opportunities: Leveraging Asia's Growth. Asian Development Bank, Manila.

Australian Agency for International Development (AusAID), 2004. Pacific Regional Transport Study. Country reports. Canberra.

Brinckerhoff, P., 2013. Jose D. Leon Guerrero Commercial Port of Guam Master Plan Update 2013 Report. Tamuning, Guam.

Collier, P., 2008. The Bottom Billion: Why the Poorest Countries Are Failing and What Can Be Done About It. Oxford University Press, New York.

Fiji Ports Corporation Ltd., 2015. Data Acquired From Interview Survey.

IHS Global Insight, 2013. World Trade Service. London.

OECD, 2015. The impact of mega-ships. In: Report Presented at the International Transport Forum, 1 May 2015.

Ports of Auckland (POAL), 2015. Data acquired from interview survey.

United Nations, 2016. The United Nations Commodity Trade Statistics Database. Available at http://comtrade.un.org/. (Last Accessed 3 February 2016).

World Bank, 2009. World Development Report 2009: Reshaping Economic Geography. Washington, DC.

Southern Africa: Overcoming corridor and border challenges for landlocked countries

15

Tomoya Kawasaki[a], *Masaya Kobayashi*[b], *and Ryuichi Shibasaki*[c]
[a]Tokyo Institute of Technology, [b]Nippon Express Co., Ltd., [c]The University of Tokyo

Introduction

The Southern Africa (SAF) region includes four large landlocked countries: Botswana, Malawi, Zambia, and Zimbabwe. These landlocked countries have a unique set of options when it comes to choosing a gateway seaport for trading with countries in other continents because they are surrounded by oceans on three sides: west, east, and south. Fig. 15.1 shows the locations of container seaports in the region. To the west are the seaports of Angola and Namibia, to the east are Tanzania and Mozambique, and to the south is South Africa. Because China and India are the region's main trade partners, the eastern seaports are generally preferred to those on the west coast. Minerals such as copper and cobalt mined in the Copper Belt—an area that stretches between northern Zambia and southern Democratic Republic of the Congo (DRC)—are mainly exported from these landlocked nations. Southern DRC is also virtually landlocked, because it is isolated from other areas of the DRC which touch the Atlantic Ocean for security reasons.

This chapter briefly reviews the current situation with regard to land transport and seaports in SAF before going on to explore two case studies related to transport in Zambia. The authors have used a model to simulate the development of transport infrastructure in these case studies and the implications of these developments for the region.

Regional seaports and land transport

Besides the four already mentioned large landlocked countries (Botswana, Malawi, Zambia, and Zimbabwe), the SAF region includes two smaller, landlocked countries—Lesotho and Swaziland (Eswatini)—and five coastal countries—Angola, Namibia, South Africa, Mozambique, and Tanzania. All of these countries are members of the Southern African Development Community (SADC), a regional organisation for political and socioeconomic cooperation. Because the multiple gateway seaports for three directions are available, port competition is not always observed between geographically close seaports.

Global Logistics Network Modelling and Policy. https://doi.org/10.1016/B978-0-12-814060-4.00015-0

Fig. 15.1 Seaports of SAF and surrounding countries.
Source: Own compilation.

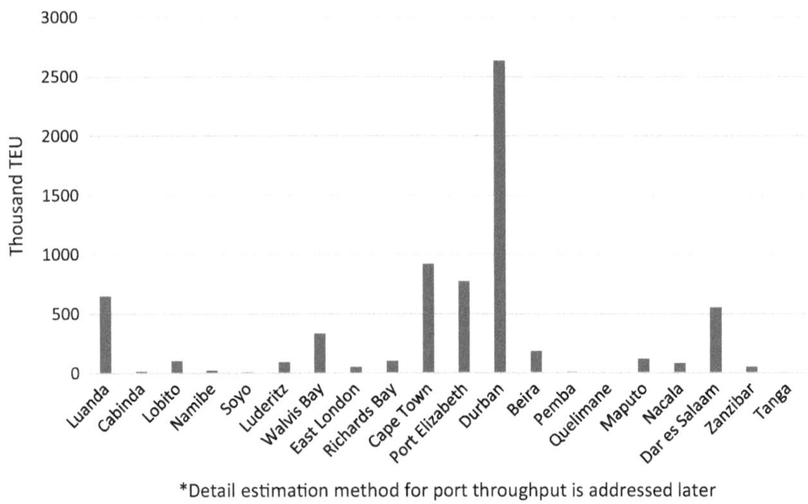

Fig. 15.2 Container throughput in seaports of SAF in 2013.
Source: Own compilation based on Drewry Maritime Research and the websites of port authority.

As shown in Fig. 15.2 (Drewry Maritime Research, 2014), taken together, container throughput of South African ports in 2013 (Durban, Port Elizabeth, and Cape Town) was higher than that of the rest of the region. Note that detail estimation method for port throughput is addressed in the 'Data' section. Throughput from Durban is particularly high because it is used not only for local South African cargo but also for cargo from several landlocked countries such as Zambia, Botswana, Zimbabwe, and Malawi. By contrast, Cape Town and Port Elizabeth are used mostly for local container cargo flow from South Africa. Angola's port of Luanda dealt with 650,000 TEU in 2013, which is the fourth largest volume in the region. Most of these containers were local, from Angola. Dar es Salaam (Tanzania) and Walvis Bay (Namibia) ports also deal with comparatively high container throughput and are also strong candidates for gateway ports of landlocked countries.

Gateway seaport choice of Zambian cargo is particularly complex because the number of port options is higher than that for other landlocked countries in the region. From Zambia, the options are Dar es Salaam and Beira (Mozambique) on the eastern coast, Durban on the southern coast, and Walvis Bay on the western coast. Because of the variety of choices, container cargo flow is observed at several Zambian borders. However, Zambian containers, in general, do not transit to Mozambique's ports (i.e. Beira). One of the reasons for this is that the draught of Beira port is only 8–12 m which cannot accommodate vessels with more than 40,000 DWT. Besides, language difference between the shipper's country and coastal country influences their seaport choice. Zambian containers are less used in the Angola route mainly because of security problems in the Lobito corridor which passes through southern DRC.

In this chapter, the authors analyse the route/seaport choice of landlocked countries focusing on Zambia where the route/seaport choice problem is more complex than in the other countries. The actual situation of gateway seaport choice by Zambia and economic corridor in SAF is reviewed as follows. The actual situation is revealed mainly by interview surveys with local freight forwarders and governmental organisations from September 5 to 11, 2017, including four organisations in South Africa, four organisations in Zambia, and three organisations in Mozambique.

Southern coast (South Africa)

Most Zambian maritime containers are likely to be imported or exported via Dar es Salaam in Tanzania or Durban in South Africa, which is located 2000 km away from the Zambian capital, Lusaka. Although hinterland transport (HT) distance from Lusaka to Durban is much greater than to other ports, the service level on the Lusaka/Durban route (including Durban port itself) is considered superior. South Africa's economy is much larger than other economies in this region; therefore, South Africa is the centre of regional trade (this is especially true for the city of Johannesburg). In general, the larger economic scale of the countries/region correlates to the higher service level of the logistics activities. Durban port is therefore the most popular gateway seaport for Zambia as well as for other landlocked countries in the region, according to interview surveys.

There are multiple routes available between Zambia and Durban port. The most popular route for maritime container shipping (MCS) is through Zimbabwe via the Chirundu (between Zambia and Zimbabwe) and Beitbridge (between Zimbabwe and South Africa) border. Fig. 15.3 shows the share of Zambian imported maritime container cargo for each border in 2014 according to data provided by the Zambian Revenue Authority (ZRA). Detailed data are available only for container imports, as the majority of Zambian trade is dominated by imported cargo. Fig. 15.3 shows that 37.9% of the total inflow of containers passed through the Chirundu border in 2014. The Livingstone border is also one of the points on the route to Durban port.

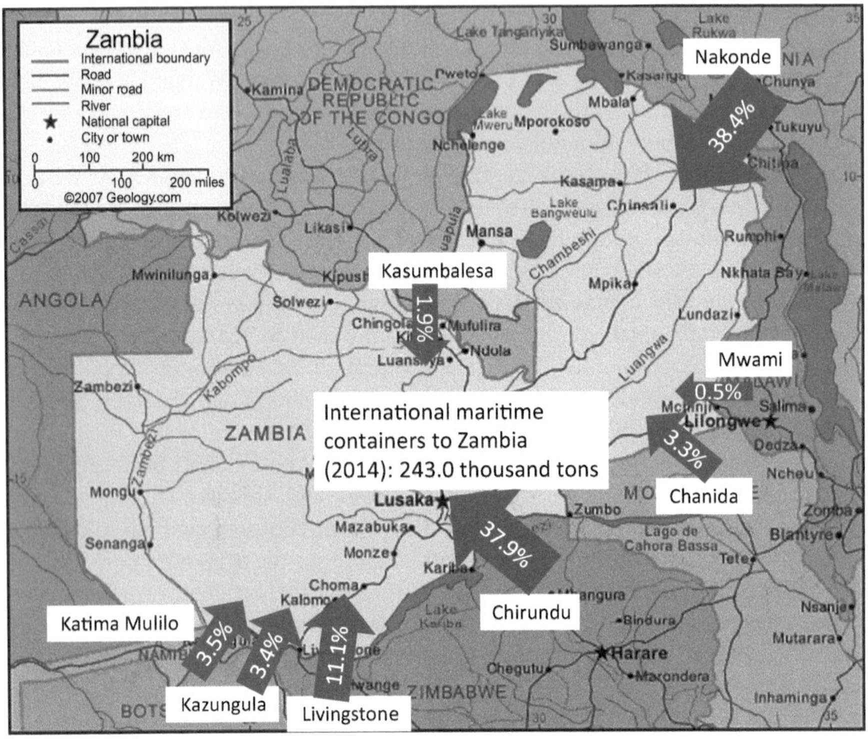

Fig. 15.3 Share of Zambian imported maritime container cargo for each border in 2014.
Note: Data only available for imports.
Source: Own compilation based on ZRA data.

As much as 49.0% (sum of the Chirundu and Livingstone borders) of total Zambian imported maritime container cargo could potentially flow in from the Zimbabwean route; however, according to interview surveys with Zambian freight forwarders, there are safety concerns and reports of bribery on Zimbabwean routes. For this reason, freight forwarders in the region prefer to pass through Botswana instead of Zimbabwe. Political corruption is less reported in Botswana than in other countries in this region. In Botswana, import/export time is shorter, and the cost of document preparation and customs clearance are lower (World Bank Group, 2015). Another reason that Botswana is often preferred is that its roads tend to be in good condition (i.e. fairly good pavement across flat terrain) even though several sections of the road, such as that between Nata and Francistown, need rehabilitation.

If using the Botswana route between Lusaka and Durban, trucks normally pass through either the Kazungula or Livingstone border. Via Kazungula, trucks must take a ferry to cross the Zambezi River. Transit time in this ferry is only 10 min, however, the ferry has a small capacity where only two trucks can be loaded at a time (as of 2017, there were only three ferries in operation). There are always long queues at the ferry landing. According to the Japan International Cooperation Agency (JICA) (2012), it takes approximately 30 h to cross this border (including queuing time).

To avoid this bottleneck, there is a 65-km route through Zimbabwean territory via the Livingstone border (Victoria Falls Bridge), finally reaching Botswana at the Kasane border. Nevertheless, the border crossing at Livingstone is still long, and there are weight limitations on the Livingstone Bridge.

A potential solution to many of the problems mentioned above is the construction of the Kazungula Bridge across the Zambezi River between Zambia and Botswana. This bridge is expected to be opened in December 2020. It incorporates road and railway functions as well as a One Stop Border Post (OSBP) to reduce border-crossing process to a single procedure. This development is expected to change container flows and gateway seaport choice in the region.

Eastern coast (Tanzania and Mozambique)

As shown in Fig. 15.3, the largest volume of Zambian import of maritime containers in 2014 flowed in via the Nakonde border from Dar es Salaam port, according to Customs data provided by the ZRA. One of the characteristics of this border is that the volume of container inflow is much higher than the outflow. One explanation is that Dar es Salaam port has a relatively shorter navigation distance to Zambia's main import partners, China and India, compared with other main seaports such as Durban and Walvis Bay. Approximately 75% of inflow cargo at the Nakonde border comprises Chinese and Indian maritime containers, according to ZRA data.

Rail is another option for freight forwarders along the Dar es Salaam–Zambia land haulage route via the Nakonde border. The Tanzania-Zambia Railway Authority (TAZARA) has been in operation since 1975. Until the 1990s, it was used mainly to transport metals, such as copper, manganese, and cobalt, mined in Zambia and southern DRC. However, this railway has deteriorated considerably, and slowdowns on the line and mechanical failures frequently occur. The average speed of the railway is approximately 35 km/h, although it is designed for speeds of 70 km/h. Therefore, as shown in Fig. 15.4, only 200,000 tons were transported in 2016, accounting for only

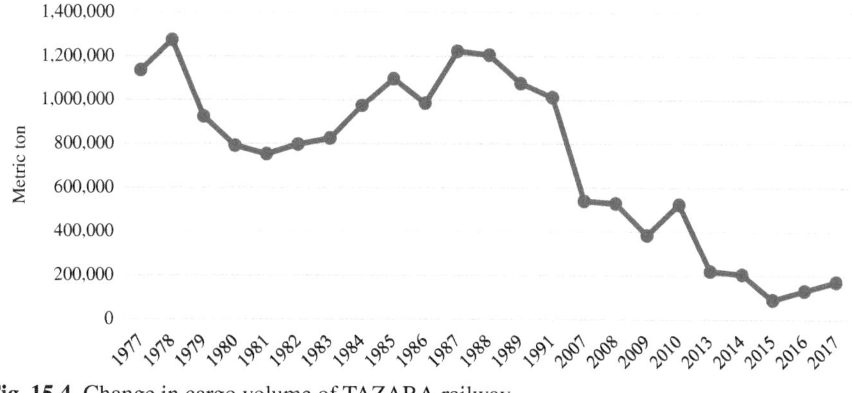

Fig. 15.4 Change in cargo volume of TAZARA railway.
Source: Own compilation from TAZARA railway website.

approximately 2% of total cargo volume through the Nakonde border. Consequently, the railway authorities of both Tanzania and Zambia are exploring the renovation of the TAZARA railway to encourage and free up trade via the Nakonde border. So far, 4 diesel locomotives and 18 wagons have been procured from China to expand the railway's capacity. Nevertheless, the railway needs to be completely overhauled in order to optimise the railway's potential. Once the TAZARA railway is renovated, the share of MCS volumes through the Nakonde border to Dar es Salaam port is expected to increase considerably.

Another seaport on the east coast, Beira in Mozambique, is the closest port to Lusaka. In general, the route via the Chirundu and Forbes borders (between Zimbabwe and Mozambique) is used for land haulage between Lusaka and Beira port. One of the main disadvantages of this route is that at least two border crossings are necessary, whereas on the Dar es Salaam route (via the Nakonde border), there is only one border crossing. Another drawback is that the sea depth of Beira port (maximum 12 m) is relatively shallow compared to other main seaports of SAF. The tide gap is also relatively high, forcing large vessels to wait for high tide offshore. Because the economy of central Mozambique is relatively small, approximately 90% of container cargo handled at Beira port is in transit from neighbouring countries. The Chanida border is another way of accessing Beira port. This route involves only one border crossing between Zambia and Mozambique, which gives it an advantage over the Chirundu border route. However, border facilities and road conditions near the Chanida border are currently too poor for shippers to choose this route. Only 3.3% of total Zambian imported maritime container cargo passes through the Chanida border (as shown in Fig. 15.3). According to interview surveys with several freight forwarders, language is a further barrier preventing the growth of the Mozambique route. Portuguese is used in Mozambique, whereas English is used in other SAF countries.

Western coast (Namibia and Angola)

Walvis Bay port in Namibia is the region's main seaport along the SAF western coast. Both Luanda and Lobito ports in Angola are closer in terms of distance to Lusaka; however, the Namibe corridor connecting Zambia with Angola has many unpaved sections and serious security problems. For these reasons, Angolan ports are seldom used for Zambian trade. There are serious security concerns along the route between Zambia and Angola via the DRC, especially in the DRC's Kasai region. Container cargos into and out of southern DRC (Lubumbashi, the DRC's second largest city, is located in this region) are forced to enter Zambia first before going to the gateway seaports in Namibia, South Africa, Mozambique, and Tanzania.

One of the advantages of using Walvis Bay port is that it requires only one border crossing via the Katima Mulilo border. Because China is an important trade partner for many SAF countries including Zambia, seaports on the eastern side are preferred as gateway seaports. However, Walvis Bay port is also valuable for trade with Europe and North America because it offers a shorter navigation distance and there is no need to pass through the Suez Canal.

Data

Ports and MCS network

The model used in this chapter is the same model used in other chapters. In addition to the 173 major world container ports where throughput was more than 500,000 TEU/year (including transshipment and empty containers as introduced in Chapter 8), seaports in SAF are added for this model. All local container seaports of SAF, as well as those of several surrounding countries, are included in the model as illustrated in Fig. 15.1. Madagascar, Comoro Islands, and Reunion are included as target ports because several liner services (LSs) that call at SAF countries also call in at these islands ports. As a result, 27 seaports are added (16 ports in SAF and 11 in surrounding countries) in order to replicate a real MCS network as closely as possible.

The data of container throughput at target seaports is collected from multiple sources, including port authority websites and the reports of international organisations. Regarding seaports where the observed data of container throughput is unavailable, their container throughputs are estimated in the following way: first, the entire throughput of a country is collected from the United Nations Conference on Trade and Development (UNCTAD) database (UNCTADstat). The missing port throughput is obtained by deducting the known throughputs of particular ports from the entire throughput of a country. For example, the port throughput data for Walvis Bay is known and that for Namibia as a whole is available from UNCTADstat, however, the information is unavailable for Luderitz port. In this case, port throughput of Luderitz is obtained by deducting Walvis Bay port throughput from that of the whole of Namibia. Throughput of each seaport in SAF is shown in Fig. 15.2.

As described in Chapter 8, the LS network is made up of the 20 largest liner shipping companies (LSCs) of the world as well as 13 local LSCs that operate LSs in SAF presented in Table 15.1, bringing the total to 33. As the capacity of the LSCs that

Table 15.1 Selected local shipping lines in SAF.

No.	Company name	Capacity (thousand TEU/year)	World share (%)
21	Ocean Africa Line	107	0.1
22	NileDutch	424	0.2
23	UAFL	88	0.1
24	Gulf Africa Line	31	0.0
25	MACS	81	0.0
26	Höegh Lines	71	0.0
27	Conti Lines	15	0.0
28	Messina	234	0.1
29	DAL	366	0.2
30	Hafiz Darya Shipping	494	0.3
31	Wan Hai Lines	182	0.1
32	Grimaldi Ferries	744	0.4
33	Lin Lines	79	0.05

Source: Own compilation.

have not been selected is ignored, 939 out of 2569 LSs (as of June 2013) on the MDS Containership Databank are included in the model. Although the number included in the model is about one-third of the world total, 67.8% of the annual world vessel capacity is covered by the model, because larger LSCs provide more extensive LSs across the world. As mentioned earlier, because local LSCs in the SAF region and global shipping lines in the world are included in the model, shipping flows between SAF and the world can be well described.

HT network

In the model, the HT network includes 11 countries and 1 province: 5 coastal countries (Angola, Namibia, South Africa, Mozambique, and Tanzania), 4 landlocked countries (Zambia, Malawi, Botswana, and Zimbabwe), 2 small countries (Lesotho and Swaziland), and Katanga province of the DRC. Katanga is included because maritime containers generated in or going there usually do not make use of DRC ports owing to security concerns in three central DRC provinces (Kasai, Central Kasai, and East Kasai) between Katanga province and Banana port of the DRC and Cabinda port of Angola. The HT network is illustrated in Fig. 15.5. The figure also shows

Fig. 15.5 HT network and ports in SAF used in the model.
Source: Own compilation.

all representative nodes (origin and destination nodes) in the region as well as in all target countries. There is a total of 4173 links comprising 4024 road, 75 railway, and 1 inland ferry.

The HT network has been adapted from the ADC WorldMap; however, ferry service and some railway links are not included in this map. Such links have therefore been added manually, based on the authors' investigations (including interview surveys and literature surveys[a]). Specifically, the TAZARA railway and the ferry service at Kazungula on the Zambezi River between Zambia and Botswana have been included in the model. Truck speed vRo_a (60, 50, and 40 km/h for motorway, primary route, and important route, respectively) and road capacity $capRo_a$ (1.0, 0.5, and 0.1 million TEU/year) are given for each road link and based on road classification type. Road capacity is set as value smaller than other areas in order to distribute the cargoes for other links. Note that level of service (LOS) has been manually set by the authors and based on actual road conditions, which are identified by interview and field surveys. The authors assume a higher LOS for South African trunk highways regardless of the above classifications because the LOS of South African roads is superior to that of other countries in the region. Specifically, truck speed for motorways and primary routes in South Africa are set as 70 and 60 km/h, respectively. Speeds for railways vRa_a and inland waterways vFe_a in the entire SAF region are set as 20 and 10 km/h, respectively. The railway capacity $capRa_a$ is 80 TEU/train in South Africa and 60 TEU/train in other countries based on actual conditions.

The cost of each transport mode (including fixed and variable costs) for shippers is based on information from interview surveys conducted in SAF. Trucking cost is set at 100 USD/TEU as fixed cost (*CFRo*) and 0.5 USD/TEU-km as variable cost (*CORo*). Note that a lower variable cost is provided for South Africa because road conditions are better than that in neighbouring countries (this issue will be discussed later). For railways and inland waterways, the fixed cost is not included from the perspective of shipper, whereas variable costs are 1.1 USD/TEU-km for railway (*CORa*) and 0.3 USD/TEU-km for inland waterway (*COFe*), respectively. Loading/discharging time of railway $THRa_a$ is set at 24 h, whereas that of ferry $THFe_a$ is set at zero since a loaded container truck rolls directly onto the ferry (note that only one ferry link exists between Botswana and Zambia across the Zambezi River).

Estimation of container shipping demand (OD matrix)

Because the model is for maritime container assignment on a global scale (focusing on SAF), it is necessary to estimate world container cargo shipping demand (i.e. OD matrix). Note that outside of SAF, OD nodes normally designate major world ports. This means that the HT network is applied only to SAF. The overall OD matrix development procedure in this chapter is similar to that in other chapters in this book and details are explained in Chapter 9.

The total container cargo volume generated from and attracted into SAF should be equal to the total throughputs of all SAF target seaports defined in Fig. 15.1.

[a] JICA (2010) and World Bank (2009).

This means that total throughputs from SAF ports can be allocated to each country using the container cargo trade values of each one. The UN Comtrade Database provides trade values on a country-by-country basis. However, this database includes not maritime containers but other types of goods that fall outside the scope of the model. To extract container cargo trade values from the UN Comtrade Database, containerised rates are estimated with container trade volume and total trade volume (both calculated on a per ton-basis) by commodity and country. Using the estimated containerised rates, container cargo trade values for all country pairs are estimated. Based on these values, container cargo trade share at country level can be estimated, and then an OD matrix can be obtained.

Subsequently, the OD matrix at country level is subdivided into those at administrative levels (traffic analysis zone: TAZ) in each SAF country using socioeconomic indicators that are associated with container cargo generation and attraction. Note that partial customs data has been obtained from the ZRA and used to validate the simulation model. Although gross regional domestic product (GRDP) is one of the possible indicators for each TAZ, GRDP is available only for three countries: South Africa, Tanzania, and Mozambique. Then, the socioeconomic indicators presented in Table 15.2 are used instead if GRDP is not available. Note that the number of firms (in Zambia) and employees (in Zimbabwe) seems to be relatively correlated with economic activities, and thus, might be acceptable for use in allocating container cargo attraction/generation. In both Angola and Botswana, population is used as an indicator, which is unlikely to properly reflect the observed share of container cargo flows in the country. An important avenue for future work is to replicate real container cargo flows according to the model. Note that Lesotho, Swaziland, and Katanga state in the DRC are each set as a TAZ. Following administrative divisions presented in Table 15.2, SAF is divided into 108 TAZs.

Table 15.2 Indicators used to allocate generated/attracted container cargo volume in each country.

Country	Num. of TAZ	Zonal level	Index of each region
Angola	18	Provinces	Population (2014)
Botswana	10	Districts	Population (2010)
Malawi	3	Regions	Number of non-poverty (2005)
Mozambique	10	Provinces	Gross regional product (GRP) (2009)
Namibia	14	States	Household income (2010)
South Africa	9	States	GRP (2011)
Tanzania	22	Regions	GRP (2012)
Zambia	9	States	Number of firms (2016)
Zimbabwe	10	8 states and 2 cities	Number of employees (2014)
Total	108		

Source: Own compilation.

Calculation results

Parameter estimation

SAF is unique in that border crossing by land is frequently observed and several border facilitation programmes have been initiated. Thus, optimal border resistance λ_a (related to border-crossing time) is used to minimise the gap between observed and estimated container cargo flow.

In this study, the container inflow to Zambia is used as validation data (Zambia's imports are based on ZRA data). Based on this data, annual inflow cargo volume for each border in 2014 is shown in Fig. 15.2. Note that this data does not distinguish container cargo or non-container cargo based on weight, whereas simulation output is obtained on a TEU basis. Thus, reproducibility is checked by referring to the share of inflow volume crossing the border as shown in Fig. 15.2.

Regarding the estimation of optimal λ_a, the gap between observed S_i^a and estimated S_i^e container inflow share at each border i is expressed as root mean square error (RMSE) as follows:

$$\mathrm{RMSE} = \sqrt{\frac{1}{n}\sum_{k=1}^{n}\left(S_i^a - S_i^e\right)^2}$$

RMSEs corresponding to several λ_a ranging between 0.1 and 1.0 are calculated as shown in Fig. 15.6. The lowest RMSE is observed in Fig. 15.6 if $\lambda_a = 0.3$. Therefore, λ_a is determined as 0.3 for the SAF model.

Similarly, HT cost in South Africa is assumed to be lower than that in surrounding countries because of the good condition of its roads and highways. Therefore, the value of variable trucking cost (*CORo*) of South Africa is estimated separately by minimising RMSE. As reported in Table 15.3, *CORo* for South Africa is estimated

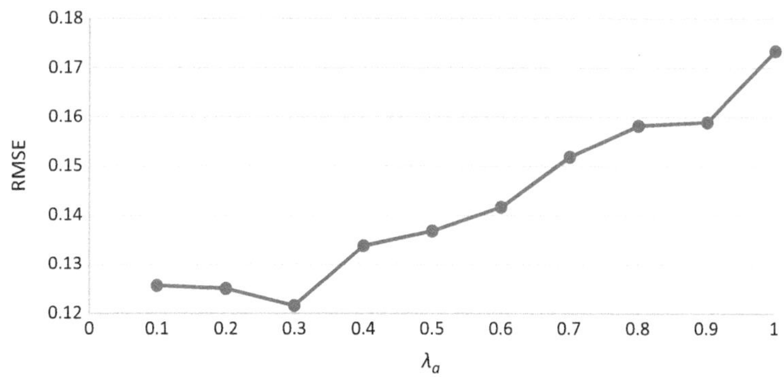

Fig. 15.6 Change in RMSE corresponding to λ_a.
Source: Own compilation.

Table 15.3 Value of RMSE by changing variable trucking cost (***CORo***) of South Africa.

***CORo* in South Africa (US$/km/TEU)**	**RMSE**
0.4	0.121
0.35	0.117
0.3	0.126

Note: CORo in other countries remains 0.5.
Source: Own compilation.

as 0.35, whereas *CORo* for the surrounding countries remains 0.5. Note that λ_a for South Africa is 0.3 as previously estimated.

This result implies that unit HT cost is lower in South Africa than in surrounding countries, and thus, shippers would be more likely to choose South Africa. There are also other important reasons why shippers might select South African seaports in and out of landlocked countries: they may find South African route easier to use because truck carriers in the region are often South African companies, and because the country plays an important role in internal regional trade. Besides, because South Africa is acting as an export country in the region, there are comparatively higher volumes of return cargoes to the landlocked countries. This generates an advantage for South Africa in terms of freight rate.

Container throughput at seaports

Fig. 15.7 shows the comparison between observed laden container volume at seaports in SAF countries and those estimated by the model. Judging from the correlation coefficient shown in the figure for both exports and imports, the model is able to replicate the actual situation well. However, some differences are found between

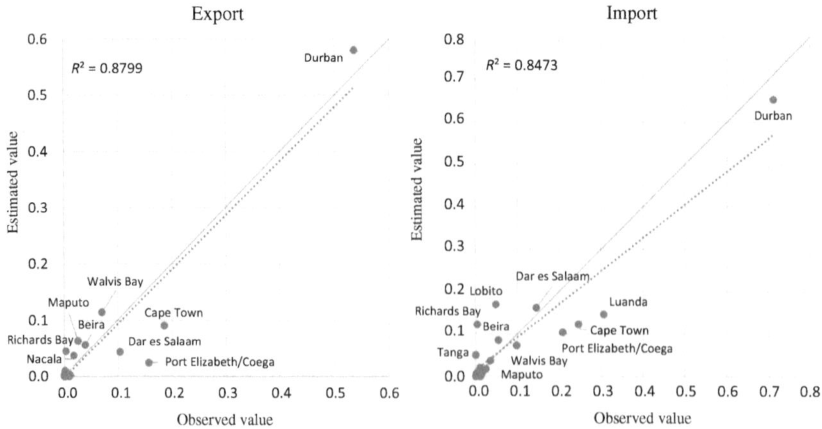

Fig. 15.7 Comparison of observed and model estimated laden container volume at seaports in SAF countries for export (left) and import (right) (million TEU/year).
Source: Own compilation.

the observed data and the estimated results. Some seaports, such as Dar es Salaam and Port Elizabeth, are not well replicated Exports at these ports receive a relatively high volume of laden containers and in particular are underestimated. By contrast, other smaller seaports, such as Beira and Richards Bay, are overestimated. In other words, there is a tendency toward overestimation at smaller ports and underestimation at larger ports. In the MCS network submodel, because congestion cost is relatively smaller than other costs, geographical distance in land is an influential factor for the port choice. This leads to more cargo at smaller but closer ports.

A similar trend can be observed in the case of exports, especially in Angola. Some of the laden containers actually dealt with in Luanda port appear to have been shifted to Lobito port. As mentioned earlier, the Lobito corridor, which connects Zambia to Lobito port via Katanga province in the DRC, has several problems such as substandard road quality and poor security. However, the model could not take these problems into full consideration, which potentially would have reduced the use of Lobito port.

Moreover, the development of an OD matrix in SAF might not incorporate a sufficient number of nodes and indicators of cargo allocation for each country. Notably, there are nine provinces in South Africa, thus the number of TAZs in South Africa is only nine. It would be better to increase the number of nodes so that detailed container flow could be replicated.

In the case of Dar es Salaam port, export volume is overestimated whereas import volume is underestimated. In the model, import and export throughputs of laden container are aggregated in the OD matrix development. It would be possible to increase model precision if export and import throughputs of laden container are collected separately and to develop a separate OD matrix for export and import cases.

The country-level observed and estimated container cargo flow through SAF seaports for exports is shown in Fig. 15.8 and for imports in Fig. 15.9, respectively. At country level, the model appears to work relatively well, although some differences are found between the observed data and the estimated results. Among them, exports from Mozambican ports are overestimated as shown in Fig. 15.8. These exports come

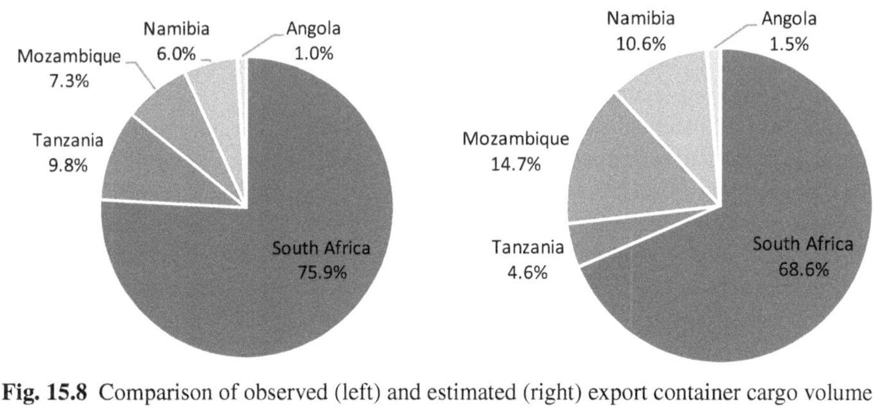

Fig. 15.8 Comparison of observed (left) and estimated (right) export container cargo volume per country.
Source: Own compilation.

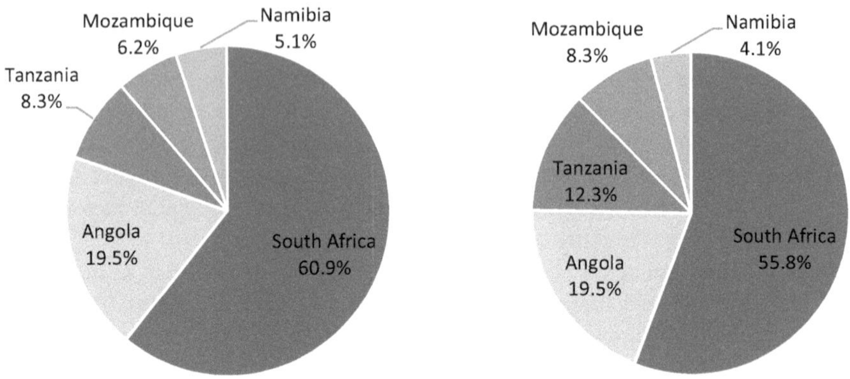

Fig. 15.9 Comparison of observed (left) and estimated (right) import container cargo volume per country.
Source: Own compilation.

mainly from Tanzanian and South African ports. Similarly, imports for Tanzania and Mozambique cargo are overestimated as shown in Fig. 15.9 and these come mainly from South Africa. Regarding South Africa, both import and export throughputs are estimated to be lower than the observed throughput. Although container assignment is generally a reasonable choice based on generalised cost, in reality, South African seaports are often selected on the basis of several unobservable variables such as better security, and the country's role in regional trade.

Land container flow

Fig. 15.10 illustrates annual hinterland container cargo flow estimated by developed model. The road links to Durban port receive high traffic volume especially from/to the Zimbabwe route. Container flow at the Kazungula and Livingstone border seems to be less compared to the Chirundu border, which might be a reasonable result. Thus, the Botswana route from/to Zambia is less used. On the other hand, container cargo flows at Trans-Caprivi Corridor through the Katima Mulilo border receive higher flows than the Botswana route. Container flow between the DRC and Angola is estimated to be few, which is well reflected in the real situation.

In the case of Zambia, Table 15.4 presents the share of the observed and the estimated laden container inflow volume at each border. Overall, the figure appears to show that the share of each border point is well replicated. However, a relatively large gap exists between the observed and the estimated share at the Nakonde (the border with Tanzania) and Chanida (the border with Mozambique) borders. Containers that are actually dealt with in Dar es Salaam port are shifted to Beira port in the model because there is a single border-crossing point between Zambia and Beira port through the Chanida border (although, in reality, facilities and road conditions are poor). Another reason is that the LOS at Beira port (which connects to Zambia via the Chanida border) is far inferior to that at Dar es Salaam port. This suggests that the route between the Chanida border and Zambia has the potential to become an important corridor for Zambian trade.

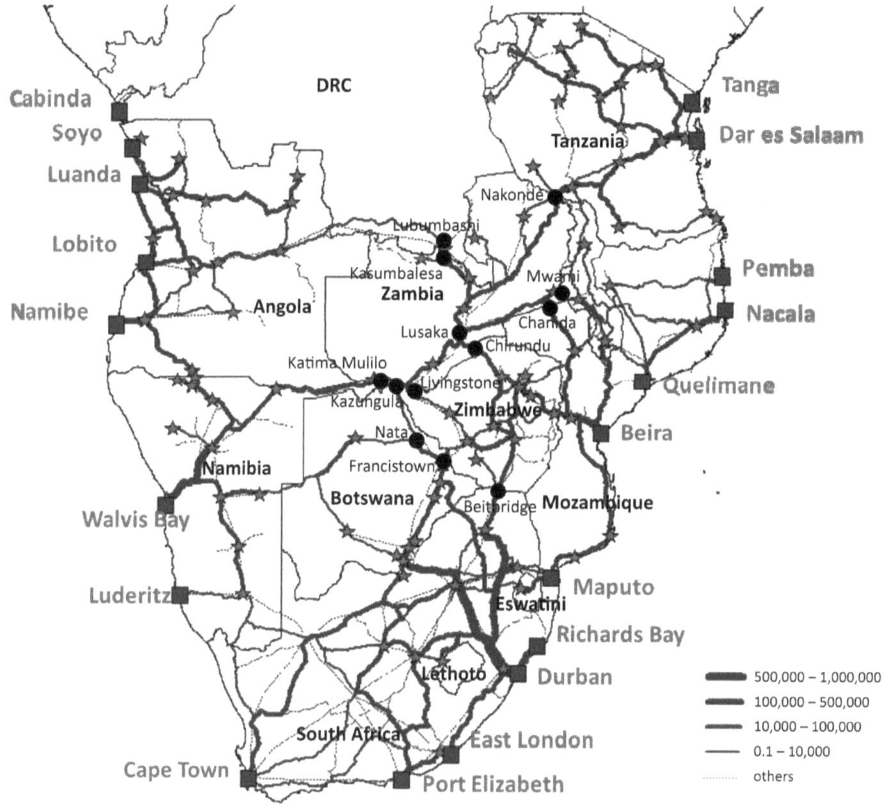

Fig. 15.10 Estimated annual container cargo flow in SAF (2013).
Source: Own compilation.

Table 15.4 Observed and estimated share of import laden container volume to Zambia by each border (as of 2014).

	Observed (A) (%)	Estimated (B) (%)	Difference (B – A) (pts)
Kasumbalesa and Ndola	1.9	3.8	1.9
Katima Mulilo	3.5	10.8	7.3
Kazungula	3.4	5.3	1.9
Chirundu and Livingstone	49.0	36.7	– 12.3
Chanida	3.3	24.3	21.0
Mwami	0.5	1.7	1.2
Nakonde	38.4	17.4	– 21.0

Source: Own compilation.

Policy simulations

Construction of Kazungula Bridge

The north–south corridor through Botswana is one of the most important corridors of SAF. The Zambezi River runs between the two countries (Zambia and Botswana) creating a bottleneck in this corridor. There is currently no bridge across the river and only three small ferry services are in operation. Each ferry has a capacity of only two trucks. According to JICA (2012), it takes an average of 30 h to cross this border with most of that time being spent queuing to cross the river (actual ferry transit time is 10 min). The African Development Bank and JICA are collaborating on a project to build a bridge for road and rail traffic which also includes a border facility. The bridge will be known as the Kazungula Bridge and is expected to be completed by the end of 2020. An OSBP will also be introduced so that the processing procedure is accelerated at the border. In the simulation model, construction of the Kazungula Bridge (including road and railway) and OSBP installation are assumed to be complete and ferry services discontinued.

In the model, the ferry link has been removed whereas road (vRo_a = 60 km/h) and railway [vRa_a = 20 km/h, $freqRa_a$ = 14 (train/week)] links have been added for the Kazungula Bridge. In addition, λ_a at the Kazungula border is set as 0.1 because the border-crossing cost would be reduced once the project is completed. Fig. 15.11 shows change in container cargo flow and port throughputs of laden containers after

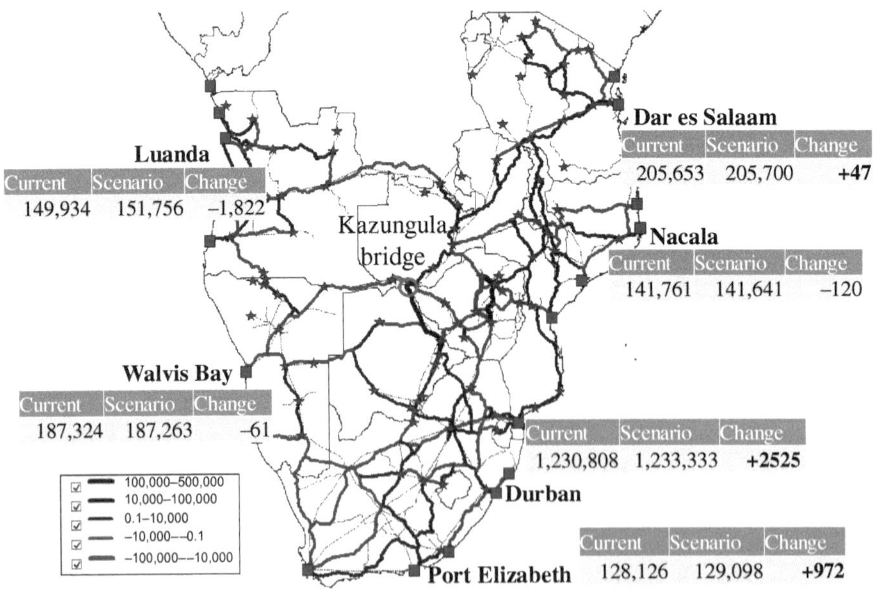

Fig. 15.11 Change in container cargo flow and port throughputs of laden containers after the opening of the Kazungula Bridge (TEU/year).
Source: Own compilation.

Table 15.5 Change in share of estimated cargo inflow volume at each border crossing point after the opening of the Kazungula Bridge.

	Current (%)	Scenario (%)	Change (pts)
Kasumbalesa & Ndola	3.8	3.8	0
Katima Mulilo	10.8	8.9	−1.9
Kazungula	5.3	10.2	+4.9
Chirundu & Livingstone	36.7	33.6	−3.1
Chanida	24.3	24.5	+0.13
Mwami	1.7	1.7	−0.02
Nakonde	17.4	17.3	−0.05

Source: Own compilation.

the opening of the Kazungula Bridge. Newly installed railway links are not used because the railway section is too short to be utilised for cross-border transport.

Table 15.5 presents changes in the share of cargo inflow volume at border-crossing points. Note that the share in Table 15.5 is estimated on the basis of only inflow laden containers to Zambia. Kazungula border share has increased by 4.9 points from its original share, whereas the share of the Katima Mulilo and Chirundu borders has decreased. This means Zambian import cargo has shifted from ports in Namibia and Mozambique to South African ports. However, as shown in Fig. 15.11, port throughput in South Africa shows a relatively large increase whereas those of Walvis Bay (Namibia) and Beira (Mozambique) have not significantly decreased. This is because export cargoes are not overly affected. Instead, throughput of Luanda port shows a slight decrease from the DRC's Katanga Province and Zambian containers shifting from Angolan ports to South African ports. Regarding the decrease in the Livingstone/Chirundu border's share, it has shifted to the Kazungula border. This means north–south container flow shifts from the Zimbabwe route to the Botswana route.

Renovation of TAZARA railway

The authors conducted a second simulation that assumes that the project to refurbish the TAZARA railway (jointly owned by the railway authorities of Tanzania and Zambia) has been completed. The railway crosses the border at Nakonde and branch lines connect to Zimbabwe and South Africa. As previously mentioned, both rolling stock and railway lines have deteriorated severely due to the poor maintenance since the railway was opened in 1975, and the average speed is currently approximately 20 km/h. The line is therefore not competitive. According to ZRA data, in 2014 the modal share of the railway was only 0.2% of the total mass passing through the Nakonde border. Thus, the simulation is an attempt to establish what the impact would be on freight throughput once the line is revitalised.

Table 15.6 Change in share of estimated cargo inflow volume at each border crossing point after the renovation of the TAZARA railway.

	Current (%)	Scenario (%)	Change (pts)
Kasumbalesa & Ndola	3.8	3.2	−0.6
Katima Mulilo	10.8	9.3	−1.5
Kazungula	5.3	5.2	−0.1
Chirundu & Livingstone	36.7	21.5	−15.2
Chanida	24.3	20.2	−4.1
Mwami	1.7	1.0	−0.7
Nakonde	17.4	39.6	22.2

Source: Own compilation.

The authors assumed the input parameters for the TAZARA railway as follows: railway link speed vRa_a is improved from 15 to 60 km/h, and frequency $freqRa_a$ from 7 to 14 times/week. In addition, because an OSBP is likely to be implemented as part of the project, border resistance λ_a at the Nakonde border is set as 0.1 (the same value set in the Kazungula border simulation where an OSBP is also planned). Table 15.6 presents a change in share of estimated cargo inflow volume at each border-crossing point of Zambia after the renovation of the TAZARA railway. The results of the simulation reveal that container cargo flow through the TAZARA railway increases the share of the Nakonde border by 22.2 points, whereas the share for all other borders decreased. The share of the Livingstone/Chirundu border shows the sharpest decrease of 15.2 points. This implies that container cargo flow is shifted from ports in South Africa or Mozambique (i.e. Beira port) to Tanzanian ports.

Fig. 15.12 shows the impact of the railway renovation on container cargo flows across HT links as well as on port throughputs. Container throughput to Dar es Salaam port increases, whereas that to Beira and Durban ports decreases. This implies that Dar es Salaam port is competing with Beira and Durban ports and can increase its competitiveness through the renovation of the TAZARA railway. From Fig. 15.12, the largest change in laden container throughputs is recorded as +26,621 TEU at Dar es Salaam port, whereas that from Fig. 15.11 is recorded as +2525 TEU at Durban port. These results suggest that railway transport is a more effective way to change cargo flow and gateway seaport choice than the development of the Kazungula Bridge. This is a reasonable result because the distance between Lusaka and Dar es Salaam is approximately 2000 km which is suitable for railway transport.

Conclusion

In this chapter, two scenario analyses were presented with a special focus on Zambia: the Kazungula Bridge project and the renovation of the TAZARA railway.

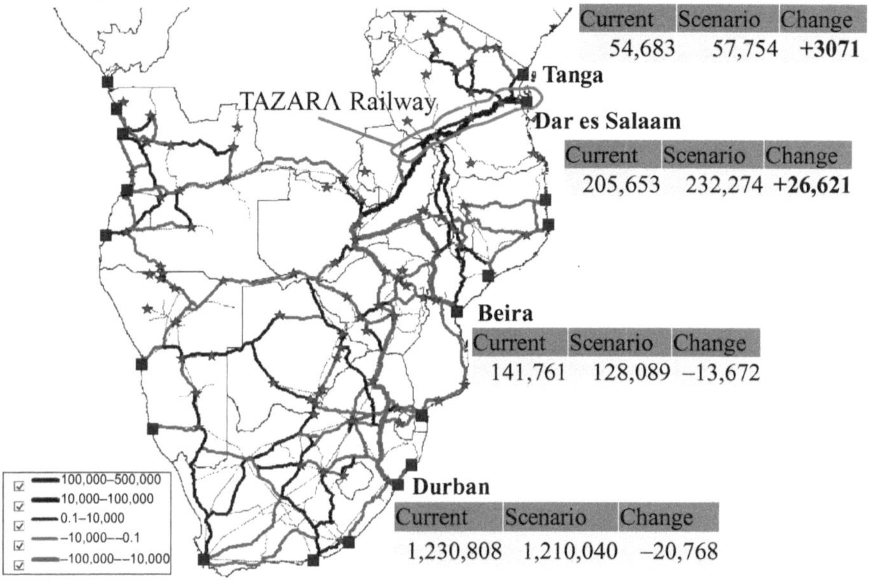

Fig. 15.12 Change in laden container throughput at each seaport after the renovation of the TAZARA railway (TEU/year).
Source: Own compilation.

A simulation model was used to forecast transport movement through the region's hinterland including worldwide container movements. The available seaports in this region are distributed along the Eastern, Western, and Southern African coasts. For this reason, LS networks are significant when it comes to choosing gateway seaports from landlocked countries.

In the developed model, output estimate was based on the observed laden container movement through SAF ports and the share of container cargo flows at the Zambian borders (although some overestimation and underestimation at these nodes was observed). Parameters for border resistance and unit transport cost in South Africa were estimated by minimising the gap between the observed and the estimated share of container cargo flow at the Zambian borders.

The simulation results showed that once the Kazungula Bridge is built, container cargo flow would shift from the Zimbabwean route and increase along the Botswana route. However, the choice of port would not change dramatically. Meanwhile, a more significant change was observed in port throughput in the simulation once the TAZARA railway renovation is completed. The results showed that laden container throughput at Dar es Salaam port would increase, whereas those at Beira and Durban ports would decrease. This implied that railway is effective for the SAF region to increase the share of cargo volume, because railway is far more competitive in terms of unit cost than road transport. Transport distance between landlocked countries and ports is suitable for railway transport in the SAF region.

References

Drewry Maritime Research, 2014. Container Forecaster, Annual Review. London, UK.

Japan International Cooperation Agency, 2010. Preparatory Survey for Southern Africa Integrated Regional Transport Program: Final Report, Tokyo, Japan.

Japan International Cooperation Agency, 2012. Newsrelease. https://www.jica.go.jp/press/2012/20121012_02.html.

World Bank, 2009. Transport Prices and Costs in Africa: A Review of the International Corridors. A Review of the International Corridors, Washington DC (Accessed September 2015).

World Bank Group, 2015. Doing Business Website "Trading Across Borders" Rank. Available online: http://www.doingbusiness.org/data/exploretopics/trading-across-borders (Accessed 15 September 2015).

Belt and Road Initiative: How does China's BRI encourage the use of international rail transport across the Eurasian continent?

16

Ryuichi Shibasaki, Kentaro Nishimura, Satoshi Tanabe, and Hironori Kato
The University of Tokyo

Introduction

A proliferation of container shipping resulting from the growth of the world economy has enhanced intermodality in global logistics. Maritime container shipping (MCS) and land transport (LT) are often complementarily connected in the global logistics network, whereas also occasionally competing against each other. This is the case even for long-distance transport across a continent, in which LT (sometimes called 'land bridge', LB) is sometimes available as an alternative transport route. The most well-known LB is a link crossing the North American continent (American LB); however, LBs across the Eurasian continent have also received more attention recently as China has heavily promoted the Belt and Road Initiative (BRI). Unlike the traditional Eurasian LB crossing the Siberia region of Russia, many routes have been proposed to connect China with the western countries of the Eurasian continent, including those in Europe, west Russia, and the South Caucasus, to support China's 'Silk Road Economic Belt (SREB)' policy, under which the amount of cargo actually transported by regular container railway services is rapidly increasing.

Recent studies have analysed the impacts of BRI, including effective allocation of dry ports (e.g. Zhao et al., 2018; Wei and Sheng, 2018; Wei et al., 2018) and MCS networks ('New Maritime Silk Road', such as Zeng et al., 2018; Ren et al., 2018; Qiu et al., 2018). Some of them have highlighted a specific LT route for crossing the Eurasian continent; for example Sheu and Kundu (2018) and Wang et al. (2018) studied a shortcut route via Pakistan for importing crude oil to China. Yang et al. (2018) and Shao et al. (2018) tried to deal more comprehensively with Eurasian LB(s), however, the former did not consider multiple routes of LBs and the latter did not consider the competition between MCS and LT routes. This chapter attempts to fill the research gap by using more detailed networks for both MCS and LT, and simulating policy impacts, considering the competition between multiple MCS and LT routes.

This chapter first reviews the intercontinental container railway networks and services to/from China that have been promoted under China's SREB policy. Next, the input data used to develop the model is presented. After the results calculated from the developed model are shown, the model is applied to the Eurasian continental area to

Global Logistics Network Modelling and Policy. https://doi.org/10.1016/B978-0-12-814060-4.00016-2

simulate the potential impact of policies related to BRI, such as freight discount from container block trains and stabilisation of a ferry service across the Caspian Sea.

International container railway services to/from China

The regular container train service (China–Europe Railway Express, or CERE) connecting Chinese cities with European cities (including European Russia) started in 2011 between Chongqing, Moscow, and Duisburg. Fig. 16.1 shows the number of regular container trains and amount of containers transported as the CERE each year. In particular, since the proposal of the SREB by President Xi Jinping in September 2013 in Astana, Kazakhstan, the services have rapidly increased in terms of not only the number of trains and amount of containers but also the number of cities covered in both China and Europe. Similarly, the train services and amount of containers for return trips from Europe to China have also increased since 2014. It is generally pointed out that local government in China subsidises the freight charges of the CERE, behind the background of rapid growth.

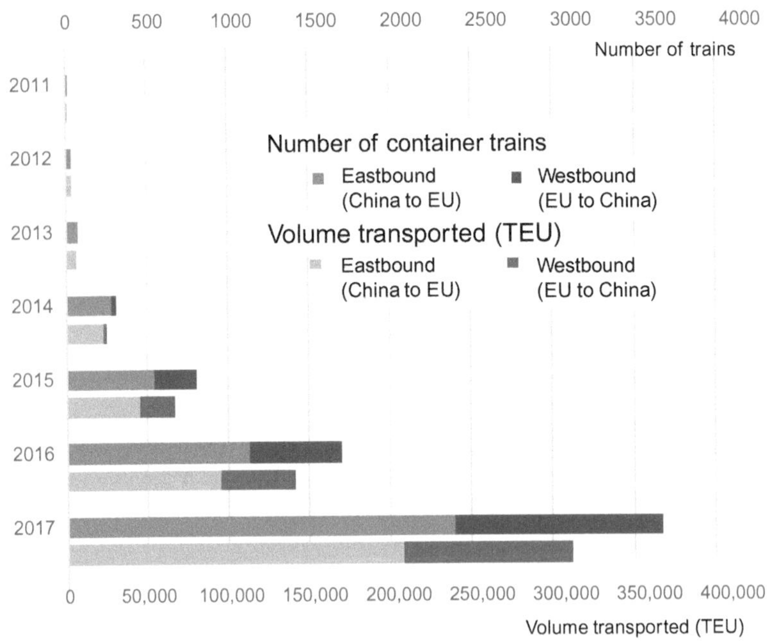

Fig. 16.1 Historical record of the CERE.
Source: Own compilation based on Eurasian Development Bank (Centre for Integration Studies), 2018. Silk Road Transport Corridors: Assessment of Trans-EAEU Freight Traffic Growth Potential, https://eabr.org/upload/iblock/0a8/EDB-Centre_2018_Report-49_Transport-Corridors_ENG.pdf (Accessed 29 April 2018) and Tsuji, H., 2018. Rapid Growth of Container Trains Between China and Europe. Russia and NIS Monthly Report, January 2018 (in Japanese).

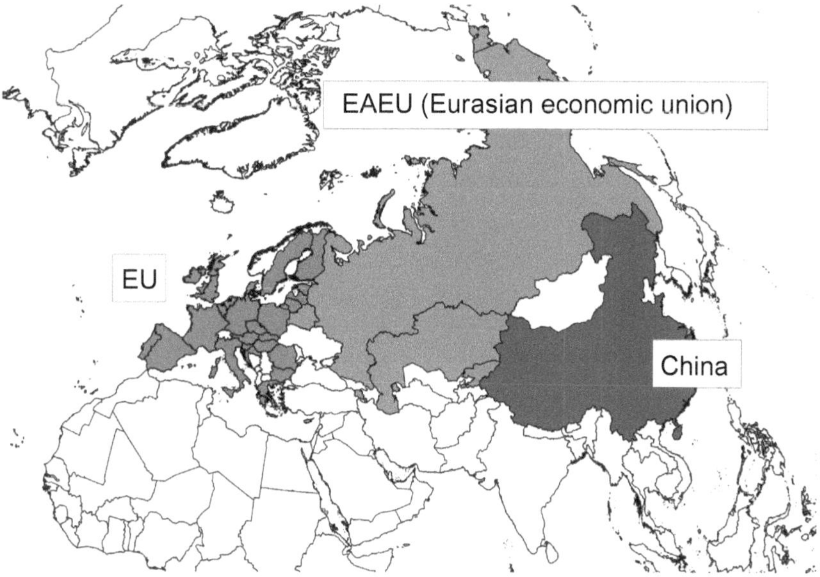

Fig. 16.2 Territory of Eurasian Economic Union.
Source: Own compilation.

Reduction in the barrier at national land borders also contributes to the growth of the CERE. For example, as also described in Chapter 13, the Eurasian Economic Union established in 2015 by reorganising the Eurasian Customs Cooperation would significantly reduce the barriers of national border such as customs procedure and document checking between the member countries (i.e. Russia, Belarus, Kazakhstan, Kyrgyz, and Armenia, see Fig. 16.2). Additionally, the facilities in some national borders have been improved and invested for realising smoother operation. Fig. 16.3 shows an example of the dry port developed in the border between China and Kazakhstan (Khorgos) in 2015, where containers are transshipped because of the different rail gauges.

As mentioned previously, the coverage of origin and destination of the CERE, not only the transport volume, has been also expanded. Fig. 16.4 shows the origin and destination of the CERE train that transited Kazakhstan in 2016. Major origins and

Fig. 16.3 Khorgos dry port (2017).
Source: Authors.

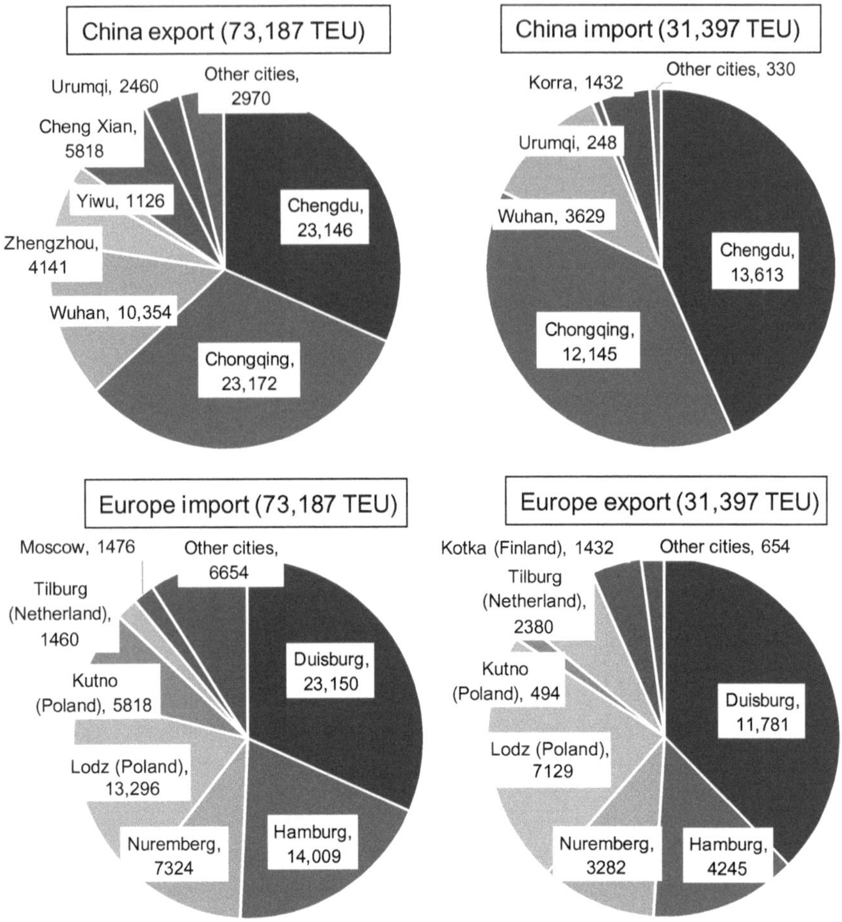

Fig. 16.4 Breakdown by origin and destination of CERE trains transiting Kazakhstan in 2016. *Source*: Own compilation based on the data provided by Kazakhstan Forwarder Association.

destinations of the CERE train in China are western inland cities such as Chengdu, Chongqing, and Wuhan, whereas major origins and destinations in Europe are German and Polish cities such as Duisburg, Hamburg, and Lodz. Note that the figure includes Russian and Belarussian cities other than EU cities but their cargo volume is small, because most of containers between China and Russia are transported via a direct border (i.e. Manzhouli), not transiting Kazakhstan, as indicated in Eurasian Development Bank (2018).

Moreover, the figure shows that some containers departed from Yiwu and Wuxi (included in 'other cities'), which are located in the coastal provinces (Zhejiang and Jiangsu Province) and famous as a wholesale and industrial city, respectively. Similarly, many European cities, in which Madrid (Spain) is included as the furthest city, are origin and destination of the CERE train, although the absolute numbers of containers are small.

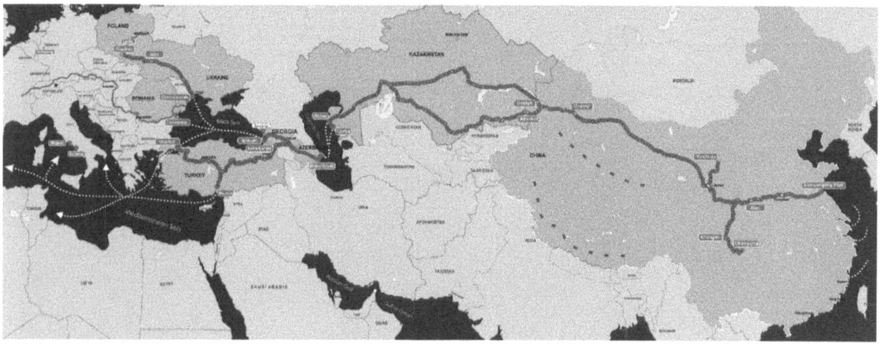

Fig. 16.5 Trans-Caspian International Transport Route and countries involved. *Source*: Trans-Caspian International Transport Route website.

Furthermore, several coastal and northeast cities in China are also considered to be the origin of the CERE train via Manzhouli (not via Kazakhstan, see Chapter 13). The container trains to Teheran (Iran) and Istanbul (Turkey) were also operated.

The directions of the access route to Iran and Turkey are different from that of the CERE. For example, an alternate route to access Europe from China by land is the 'Trans-Caspian International Transport Route' (TITR), which connects Kazakhstan with the South Caucasus countries and other Eastern European countries via the Caspian Sea. There are two key features of this route: access to Europe without passing through Russia (see Fig. 16.5); a shortcut to two Central Asian countries (i.e. Uzbekistan and Turkmenistan) compared to the conventional route. In other words, the TITR provides a reasonable alternative for countries in the Eurasian continent, apart from Russia, by reducing the number of countries necessary to transit. For example, the Kazakhstan government is committed to increasing the amount of containers transported along this route via Kazakhstan by 300,000 TEU in 2020, although only 200 TEU were transported in 2016.

For realising these goals, the Kazakhstan government has implemented many related policies including new rail construction across the land as well as port infrastructure investment along the Caspian Sea such as the Kuryk ferry complex (see Fig. 16.6). Additionally, some road, rail, and port infrastructure in other countries including Azerbaijan, Georgia, and Turkey have been improved or newly constructed.

Data

MCS and LT networks

The model in this chapter is developed based on 2013 data. Ports included in the model and liner service network is the same as those introduced in Chapter 13. The LT network covers 24 countries in the Eurasian continent (namely, five European countries, including Belarus, Germany, Moldova, the Netherlands, and Poland are added to the countries considered in Chapter 13), as shown in Fig. 16.7.

Fig. 16.6 Development plan of Kuryk port (above) and current situation (bottom, 2018). *Source*: Above: Kazakhstan Railways; bottom: authors.

Cargo shipping demand between regions (OD matrix)

The shipping demand for international cargo (OD matrix), as well as initial MCS demand, has been estimated using various existing data sources, as with the simulations in other chapters of this book. Because this study focuses mainly on the route competitions of international maritime containers and their relevant cargo, unlike other chapters, the OD matrix is made to include not only maritime containers, but also 'container-equivalent' land cargo, which is treated on a TEU basis.

The OD matrices obtained from the WTS data have been provided by transport mode (i.e. maritime and land), as well as by cargo type (container, dry bulk, liquid bulk, and neo bulk/general cargo) for maritime cargo. However, the matrix for land cargo, which even includes the cargo transported by pipeline, is not categorised by cargo type; therefore, the amount of container-equivalent land cargo is approximated by multiplying a maritime-based containerised rate by the commodity, which is estimated for each commodity as the share of maritime containers between all types of maritime cargo. Subsequently, the matrix of land container-equivalent cargo is converted on a TEU basis from a tonnage basis by dividing them into 10, by assuming an average weight of land container-equivalent cargo of 10 tons per TEU.

Fig. 16.7 LT network included in the model.
Source: Own compilation based on the ADC WorldMap.

The country-based OD matrix in the Eurasian continent is divided into subcountry levels (TAZ—traffic analysis zones), such as provinces, federal subjects, and oblasts, using the available statistics for the economy of each zone, similarly as shown in Chapter 13. China and Russia are differently divided into a second level of local municipalities (i.e. prefecture-city and federal subject level, respectively) using customs statistics data, which provides the trade amount by partner country and by commodity (which is only available for China), because the trade amount in each region of these countries is significantly larger than that in other countries.

Calculation results

Modal split in MCS and LT

The share of cargo transported by land between the countries in the Eurasian continent, which has been acquired from the WTS data, is 40.9% (4.2 million TEU), whereas that estimated in the model is slightly smaller, 38.1% (3.9 million TEU). Note that this observed and estimated share are calculated only between the countries where the LT network is considered in the model. Fig. 16.8 shows the comparison between the observed and the estimated amount of cargo transported by land by pairs of origin and destination country considered above. As shown in the figure, the amount and share of cargo transported by land are estimated well by the model, with a few exceptions.

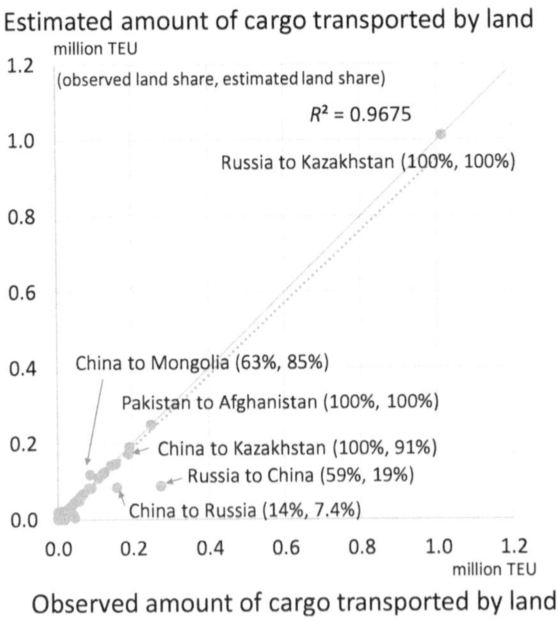

Fig. 16.8 Comparison between observed versus model-estimated amount of cargo transported by land.
Source: Own compilation.

The most significant differences observed are underestimations of those between China and Russia, mainly because the model cannot consider the difference in trade-partner countries for each region in Russia (e.g. the actual share of China in the regions of east Russia are considered larger than those in west Russia) due to data availability. Another underestimation is observed in the amount and share between China and Central Asian countries, including Kazakhstan, partly because the cargo transported by *domestic* MCS between Chinese seaports has to be categorised as 'maritime cargo' in the model estimation due to the structure of the network, whereas most are categorised as 'land cargo' in the WTS data. On the other hand, an overestimation of the amount and share to/from Mongolia is considered due to the 'observed' demand of LT in Mongolia being estimated by proportionally dividing that in 'other Northeast Asia' in the WTS data, including North Korea and Macau.

Figs 16.9 and 16.10 show the estimated share of cargo transported by land for each destination zone from Shanghai City and the Xinjiang Uygur Autonomous Region (XUAR), respectively. Fig. 16.9 reveals that for the cargo from Shanghai, only those to east Kazakhstan, Kyrgyzstan, and part of central Russia are transported by land (in other words, those going to any other regions from Shanghai are first transported using MCS), whereas Fig. 16.10 reveals that for the cargo from the XUAR, most of those going to all regions in Russia, Central Asia, the South Caucasus, Pakistan, and Iran are transported by land. These figures imply that the share of the cargo transported by land to each destination region of the Eurasian countries is significantly different from the origin city in China.

Container throughput at each port

Fig. 16.11 compares the observed and model-estimated annual export and import laden container throughputs for the gateway seaports across the Eurasian continent. The results shown in this figure seem similar to those shown in Figs 13.5 and 13.6, but this is

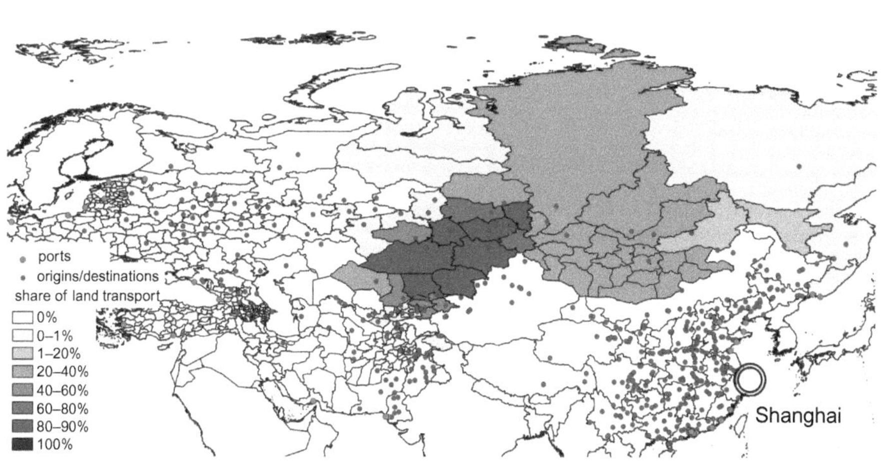

Fig. 16.9 Estimated share of cargo transported by land by destination from Shanghai. *Source*: Own compilation.

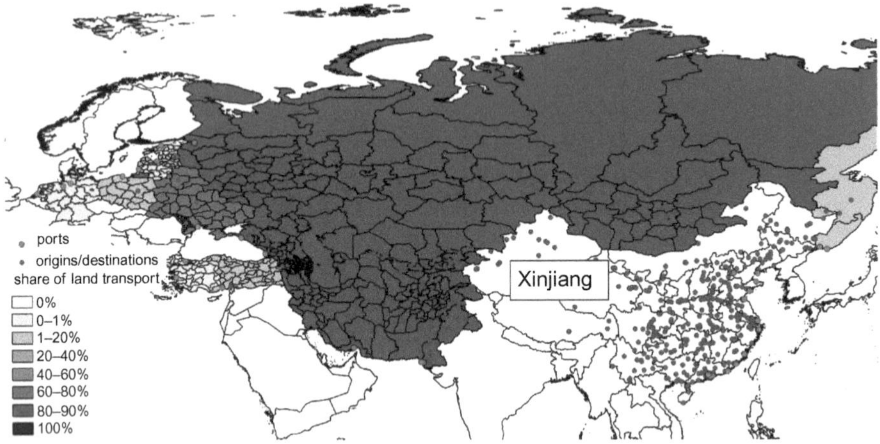

Fig. 16.10 Estimated share of cargo transported by land by destination from the XUAR.
Source: Own compilation.

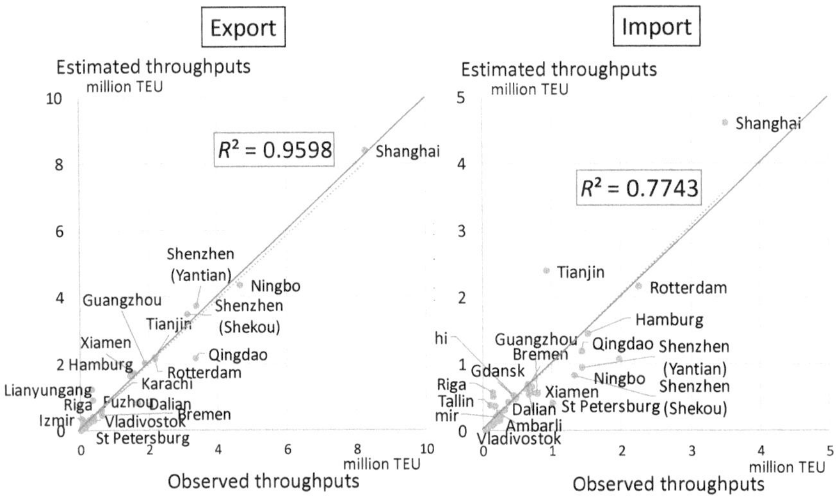

Fig. 16.11 Comparison between observed versus model-estimated annual export and import laden container throughputs of gateway seaports.
Source: Own compilation.

slightly improved as a whole by detailed zoning in China and Russia. From the figure, on the whole, the model estimates the observed throughputs of export and import laden containers well for each gateway seaport. The observed and model-estimated throughputs differ for several Chinese ports, such as Lianyungang–Qingdao for export and Shanghai–Ningbo for import. These should be further improved by focusing more on Chinese cargo.

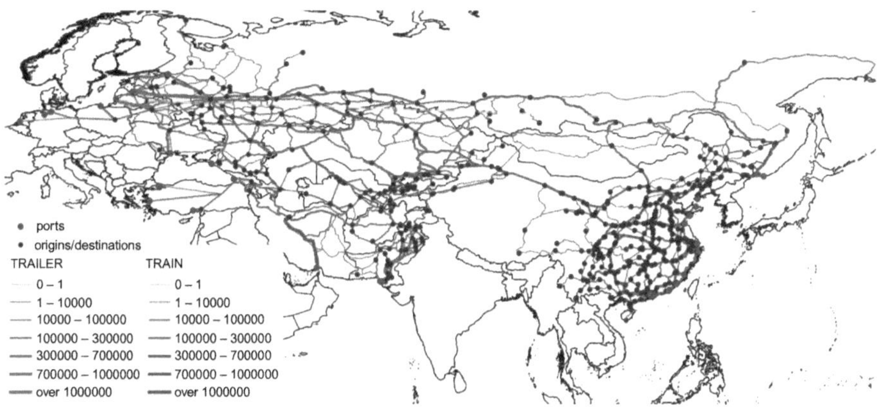

Fig. 16.12 Estimated annual cargo flow on the LT network.
Source: Own compilation.

Land cargo flow

Fig. 16.12 shows the estimated annual cargo flow in the LT network. Although it cannot be compared with the observed flow because of imperfect data availability, some calculated results quantitatively agree with the findings obtained from the authors' interview survey; for example, railways are mainly used for most long-distance transport between East Asia (i.e. China and Far East Russia) and Europe (including central/west Russia and Central Asia), whereas trucks are often used for regional transport within Europe, including central/west Russia and Central Asia.

Policy simulations

Promoting the CERE

The first simulation (scenario 1) examines the impact of improvement in level of service (LOS) of the CERE, including frequency, train speed, border barriers, and freight charges, on the amount of containers transported.

In the following scenarios, the authors assume the discount rate of freight charges from each Chinese city to each European city (i.e. cities in Poland, Germany, and the Netherlands) as (A) 25%, (B) 50%, and (C) 75%, from FL_{od} ($o \in$ Chinese city, $d \in$ European city), which is calculated in the LT submodel. It should be noted that the freight charge is discounted on a path basis for specific OD pairs, not on a link basis. The other improvements in LOS are assumed on a link basis, independent of freight discount; that is, the authors assume (i) no other improvements; (ii) increasing the frequency to four trains per day; (iii) increasing train speed to 40 km/hour, in addition to scenario (ii); and (iv) decreasing a barrier across the border between China and Kazakhstan by half, in addition to scenario (iii), respectively.

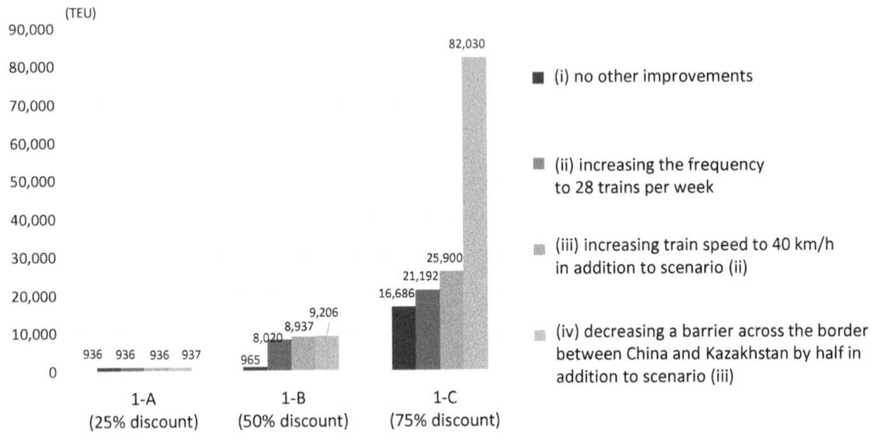

Fig. 16.13 Estimated annual amount of laden containers transported by land from China to Europe in the CERE scenarios.
Source: Own compilation.

Fig. 16.13 shows the estimated annual amount of laden containers transported by land from China to Europe by combining the above scenarios. It is found that if the freight discount rate is 25%, the amount of laden containers transported by land is very small (less than 1000 TEU/year), and this does not change if other improvements in LOS are assumed. Furthermore, if the freight discount rate is 50%, the amount of laden containers transported by land is expected to increase if the frequency is increased (in scenario 1-B-ii) to more than 8000 TEU/year. Further, if the freight discount rate is 75%, the amount of laden containers transported by land becomes significantly large; in particular, if the border barrier between China and Kazakhstan is decreased (in scenario 1-C-iv), the amount of laden containers is expected to be more than 80,000 TEU/year. Compared with the observed amount described in Fig. 16.1, the policies to promote the use of the CERE in recent years (i.e. from 2014 to 2016) likely correspond to somewhere in between the scenarios 1-C-ii and 1-C-iv.

Promoting the TITR

As shown earlier, the TITR could be one attractive option for accessing Europe by land from China, however, it has yet to be used. The most important key to encouraging the use of the TITR is to increase the LOS in shipping across the Caspian Sea. Currently, a ferry connecting Aktau (Kazakhstan) and Baku (Azerbaijan) is irregularly operated. The Kazakhstan government invests to improve the port facilities in Aktau, as well as constructing a new ferry port in Kuryk, which is located near Aktau, as mentioned earlier (see Fig. 16.6).

In simulations in scenario 2, as the first step of the TITR, the authors fix the discount rate of freight charges on the container trains from each Chinese city to the South Caucasian countries and Iran at 75%. In addition, the frequency is increased to three per day from two per week, and the freight charge of ferry is decreased by

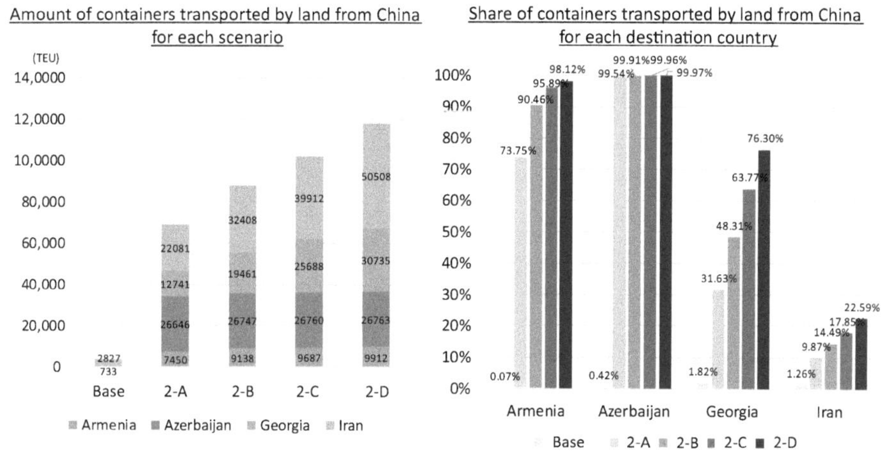

Fig. 16.14 Estimated annual amount of Chinese laden containers transported by land for each TITR scenario (left) and shares of LT for each destination country (right).
Source: Own compilation.

one-third by reflecting the future stabilisation of the ferry service. These changes are input into the model as scenario 2-A. Furthermore, the authors prepare several scenarios based on a decreasing rate of the barrier across the border between China and Kazakhstan, assumed at (B) 50%, (C) 75%, and (D) 100% (no barriers), respectively, in addition to the conditions considered in scenario 2-A.

The left figure of Fig. 16.14 describes the estimated annual amount of laden containers transported by land from China to destination countries. The amount of laden containers transported by land is expected to significantly increase in scenario 2-A from a base case (with the result shown in 'Calculation Results' section) for each country. Additionally, the amount of laden containers to Georgia and Iran is estimated to increase as the border barrier between China and Kazakhstan decreases, as described in the results of scenarios 2-B to 2-D, whereas those to Armenia and Azerbaijan are almost the same in each scenario.

The right figure of Fig. 16.14 describes the share of laden containers transported by land to each destination country from China. As shown in the figure, in scenario 2-A, the shares of laden containers transported by land to Iran and Georgia are less than half; in other words, the majority of laden containers are transported to these countries using MCS via their own seaports (i.e. Poti in Georgia and Bandar Abbas in Iran). Furthermore, the share of LT to Georgia is expected to increase as the border barrier between China and Kazakhstan decreases, whereas the share to Iran is still small (around 20%) even in scenario 2-D. The difference in both countries owes to the difference in the distance of MCS; if a vessel goes to a Georgian port from China, it has to navigate a long distance, including the Red Sea, the Suez Canal, the East Mediterranean Sea, and the Black Sea, which is different from that if it accesses an Iranian port at the mouth of the Persian Gulf.

Conclusion

This chapter applied the intermodal network assignment model as of 2013 for simulating the impact of China's BRI to promote LT across the continent, such as the CERE and the TITR, on the competitiveness of LT in container transport between China and the western countries of the Eurasian continent, including Europe, the South Caucasus, and Iran. For consideration of the competition between MCS and LT across the continent, the model was modified to include the cargo for which MCS is not used.

The first simulations on the CERE revealed that if the freight charge of a container train is reduced by 75% from the current level, around 20,000 TEU/year laden containers from China to Europe are expected to shift to LT from MCS, which is similar to the amount observed in 2014. Furthermore, if the border barrier between China and Kazakhstan is reduced by half of the current level, around 80,000 TEU/year laden containers are expected to shift to LT, which is slightly smaller than those observed in 2016. Considering that the border facilities between China and Kazakhstan have improved significantly in recent years, such as a dry port opened in 2015 in Khorgos, the finding from the simulation results that the significant discount of rail freight charge (by around 75% or more) should be provided by each Chinese local government is realistic, although the actual rate or amount of freight discount is often not identified.

Another simulation on the TITR revealed that if the ferry service across the Caspian Sea is stabilised, in addition to a 75% discount of the freight charge from China to the South Caucasian countries and Iran, the amount of laden containers is expected to be more than 20 times as large as the base case. In addition, if the border barrier between China and Kazakhstan is significantly reduced, even in the countries that have a seaport (namely, Georgia and Iran), the share of LT is expected to increase. These results suggest a potential for an alternate route connecting China with the western countries in the Eurasian continent without passing through Russia, by providing a shortcut route that can reduce the number of countries required to transit. Providing a stable trans-Caspian ferry service is key to the success of the southern route of the SREB.

References

Eurasian Development Bank (Centre for Integration Studies), 2018. Silk Road Transport Corridors: Assessment of Trans-EAEU Freight Traffic Growth Potential. https://eabr.org/upload/iblock/0a8/EDB-Centre_2018_Report-49_Transport-Corridors_ENG.pdf. (Accessed April 29, 2018).

Qiu, X., Wong, E.Y.C., Lam, J.S.L., 2018. Evaluating economic and environmental value of liner vessel sharing along the maritime silk road. Marit. Policy Manage. 45 (3), 336–350. https://doi.org/10.1080/03088839.2018.1437285.

Ren, J., Dong, L., Sun, L., 2018. Competitiveness prioritisation of container ports in Asia under the background of China's belt and road initiative. Transp. Rev. 38 (4), 436–456. https://doi.org/10.1080/01441647.2018.1451407.

Shao, Z.-Z., Ma, Z.-J., Sheu, J.-B., Gao, H.O., 2018. Evaluation of large-scale transnational high-speed railway construction priority in the belt and road region. Transport. Res. E 117, 40–57. https://doi.org/10.1016/j.tre.2017.07.007.

Sheu, J.B., Kundu, T., 2018. Forecasting time-varying logistics distribution flows in the One Belt-One Road strategic context. Transport. Res. E 117, 5–22. https://doi.org/10.1016/j.tre.2017.03.003.

Wang, S., Yang, D., Lu, J., 2018. A connectivity reliability-cost approach for path selection in the maritime transportation of China's crude oil imports. Marit. Policy Manage. 45 (5), 567–584. https://doi.org/10.1080/03088839.2018.1445308.

Wei, H., Sheng, Z., 2018. Logistics connectivity considering import and export for Chinese inland regions in the 21st-century maritime silk road by dry ports. Marit. Policy Manage. 45 (1), 53–70. https://doi.org/10.1080/03088839.2017.1403052.

Wei, H., Sheng, Z., Lee, P.T.-W., 2018. The role of dry port in hub-and-spoke network under Belt and Road initiative. Marit. Policy Manage. 45 (3), 370–387. https://doi.org/10.1080/03088839.2017.1396505.

Yang, D., Pan, K., Wang, S., 2018. On service network improvement for shipping lines under the one belt one road initiative of China. Transport. Res. E 117, 82–95. https://doi.org/10.1016/j.tre.2017.07.003.

Zeng, Q., Wang, G.W.Y., Qu, C., Li, K.X., 2018. Impact of the Carat Canal on the evolution of hub ports under China's Belt and Road initiative. Transport. Res. E 117, 96–107. https://doi.org/10.1016/j.tre.2017.05.009.

Zhao, L., Zhao, Y., Hu, Q., Li, H., Stoeter, J., 2018. Evaluation of consolidation center cargo capacity and locations for China railway express. Transport. Res. E 117, 58–81. https://doi.org/10.1016/j.tre.2017.09.007.

Conclusion

This book mainly focuses on the global container movement in both the maritime and land transport networks. The simulation model to describe the global container movement on the intermodal transport network was developed and applied to various regions of the world through Chapters 10–16, mainly developing countries, such as Central America, Southeast Asia (Lower Greater Mekong Subregion), South Asia, Central Asia, the Pacific Islands, Southern Africa, and the entire Eurasian continent.

The current trend to focus on the geopolitics stimulated by China's Belt and Road Initiative (BRI) reminds us of the importance of geographical consideration relating to the global logistics network. In particular, securing multiple trade route alternatives (e.g. multiple accesses to different gateway seaports or multiple transport modes to the same seaport) is critical for cargo owners. This is especially critical for inland cities and landlocked countries that have no coastlines, but also for 'sea-locked' countries that are far from the main sea lane of global maritime shipping (see Chapter 14). It is said that Central Asia, which is a part of the Eurasian continent, and the African continent are the two largest landlocked regions of the world (see Chapters 13, 15, and 16). The capital cities of India (Delhi) and Cambodia (Phnom Penh) are also located in inland areas that are far from the gateway seaports (see Chapters 11 and 12). Historically, these inland cities and countries, as well as the 'sea-locked' countries grew because they had some advantages, for example, from demographic or meteorological viewpoints. However, in the modern era of globalisation, they are largely handicapped from a logistics viewpoint because of the relatively expensive cost of land (or feeder) transport compared with maritime shipping costs. The importance of the policies that can reduce the land (or feeder) transport costs to/from these handicapped regions/countries, such as infrastructure investment along the trade corridor as well as border-crossing facilitation, was verified with the simulation model developed in each chapter of Part 3. In particular, many simulation results validated that reducing the border-crossing cost and increasing the level of service (e.g. capacity and frequency) in rail and inland waterway transport are critical factors for landlocked countries and inland cities because they have a relative advantage in terms of the cost of long-distance transport compared with that of road transport.

Moreover, the impacts of securing multiple trade routes were also examined with the model simulations in each chapter of this book. The new trade routes, which are available through the policies mentioned above, including the construction of new seaports and railways, may contribute to reducing the vulnerability of the trade route. Moreover, multiple trade routes are likely to encourage competition between routes, resulting in improving the level of service of each route, such as shipping cost, time, and frequency. For example, gateway seaport competition on the west coast of India that has increased because of new emerging ports (Mundra and Pipavav) may accelerate capacity expansion in JNP, which is currently heavily congested (see Chapter 12). In other words, the monopoly of the trade route may raise the shipping cost higher in ordinary times and cause a fatal crisis at a time of emergency, such as the COVID-19 pandemic. Furthermore, if the new route provides a shortcut to the global market, the impact would be much larger. For example, if Pakistani ports are connected with Central Asia through sufficient infrastructure, they can provide the shortest route for the Central Asian cargo in terms of hinterland transport distance (see Chapter 13). Another example in Chapter 11 indicated that the new navigational route in river shipping, which can accommodate larger vessels to decrease shipping costs, is likely to increase the share of inland waterway transport although the shipping distance of the new route is increased.

Another feature emphasised in this book was that containers are moved in the intermodal transport network on a global scale. The maritime container shipping network and land transport network often function complementary to the intermodal service network; however, they sometimes compete with each other. Central America, as discussed in Chapter 10, is an isthmus surrounded by two coastlines with different characteristics in terms of global maritime container shipping along the trunk route (Pacific coast) and in the feeder area (Caribbean coast). Therefore, international hinterland transport across national borders is occasionally useful to access the seaports on the opposite coast. Such complicated problems should be solved in the intermodal transport network. Chapter 16 extended the model to consider the competition between land transport and maritime shipping across the Eurasian continent. Moreover, the simulation result in Chapter 14 revealed that there are both complementary and competitive relationships, even between seaports in the maritime shipping network.

In contrast, several issues should be discussed if promoting policies to use multiple trade routes on the intermodal freight transport network. For example, providing too many alternatives will deconcentrate the resource for investment in each route, the impact of which cannot be neglected because mass freight transport generally has a significant scale of economy. In Chapter 12, we simulated that constructing many container ports in a state of India will cause a decrease in the total container throughput in the state owing to the decentralisation of port competitiveness. The simulation results in Chapter 14 also raised another important issue: it is difficult to pursue simultaneously both effective and regionally equalised investment. Furthermore, many other results in many chapters have shown a trade-off in the transport volume between the trade routes under the fixed demand of cargo shipping; for example, between road, rail, and inland waterways (see Chapters 11 and 12) and between access routes to different gateway seaports. The results applied to Southern Africa (Chapter 15) and the Eurasian continent (Chapter 13) show how the range of the gateway seaport competition is spread continentally. However, if some stakeholders gave up following the competition and stopped the investment, the competitiveness of the route in question would significantly decrease. In particular, the seaports in the minor trade regions are heavily affected, as shown in Chapter 10 (in Central America) and Chapter 14 (in the Pacific Islands) because containerships become larger as time goes by, even in the minor routes of the world. Therefore, national and local governments try to take an opportunity to effectively increase the availability of alternative trade routes and raise their competitiveness. As described in Chapter 16, the Kazakhstani government intensively invests in the infrastructure of rail and inland waterway transport along the Trans-Caspian International Transport Route, which provides alternative access to the Middle and Near East, South Caucasus, and East Europe by making use of the momentum of China's BRI policy. Chinese local governments also compete in subsidising the freight charge of the China-Europe Railway Express, under the policy promoted by the central government, to encourage the use of land transport across the Eurasian continent. The impact of China's BRI policy and its expected response should be understood and discussed in these contexts.

Finally, we propose the following eight recommendations based on the above discussions:

(1) *Think globally.* It is clear that the world is increasingly globalised and that each country's economy has become more strongly connected to each other. For the maritime container shipping network, in particular, being affected by the global trend of shipping cannot be avoided; this includes enlargement of vessels, even in the minor regions of the world. Thus, geographical thinking is very important in global logistics research. The authors expect that this trend will not be largely changed even in the post-COVID-19 era.

(2) *Analyse rationally and quantitatively by model simulations.* The authors believe that all decisions by any stakeholders—even governments—involved in the logistics industry should be made rationally (i.e. those that can be rationally explained) in principle because logistics is a part of economic activity under the condition that its sunk costs are not negligible (i.e. it depends heavily on the historical background). The authors also consider as experts in the logistics field that geopolitical issues, including China's BRI policy, can also be understood in this context. Therefore, a quantitative modelling approach to cover globally the logistics network is very effective in supporting the decision-making of logistics policy.

(3) *Consider as the intermodal network.* The strategies and policies on global logistics should be considered comprehensively in the intermodal network, including both maritime shipping and land transport. Enhancing connectivity by reducing the transfer cost and shortening the lead time between the different modes in the seaports and other nodes is a key issue.

(4) *Improve connectivity in inland cities, landlocked countries, and sea-locked countries by combining infrastructure investment and cross-border facilitation.* Handicaps in terms of logistics in these regions should be intensively mitigated for their economic development. This may often be realised through the international cooperation framework.

(5) *Secure multiple trade routes.* Strategies to secure multiple access routes to different gateway seaports (and multiple transport modes to the same seaport, if possible) from the network perspective are critical for ensuring the redundancy of trade and economics of the region/country, especially for inland cities, landlocked countries, and sea-locked countries.

(6) *Enhance the competition between the routes, whereas necessarily being coordinated.* The competition between multiple trade routes enables the enhancement of the level of service and reduces transport cost and time. However, too many alternatives are likely to weaken its competitiveness owing to diseconomy of scale. Therefore, proper coordination of the competition ('coopetition' policy) of the trade routes by the central government or other bodies is necessary after comprehending the complementary and competitive relationships of each trade route through model simulations.

(7) *Take care of the trade-off between the efficiency of investment and regional equality.* It is often difficult to pursue simultaneously both effective and regionally equalised investment. A multiobjective optimisation approach is one possible solution.

(8) *Mind a gap in decision-making between infrastructure investment and others.* Because infrastructure needs a significant amount of time until completion, future shipping demand should be well predicted. In other words, infrastructure policy should be planned to target future demand. Another difficulty in infrastructure planning is inflexibility against the change in demand. Although cargo owners and shipping companies can easily change shipping routes, the infrastructure management bodies cannot change the intended location. Sharing the risk in management amongst these stakeholders, such as concluding concessions, is important for providing a sufficient level of logistics infrastructure although it is not sufficiently considered in the model simulations in this book.

As summarised above, this book contributed to simulating the global container cargo flow on the intermodal network and to evaluating the impact of many policy options to enhance the level of service and competitiveness of each trade route. Some simulation results were used in the project studies conducted by the Japan International Cooperation Agency and local governments to examine their feasibility. The model that the authors developed would be particularly useful in the case of the impact of the project being considered simultaneously in both maritime and land transport networks, and its benefit is expected to spread widely across national borders.

However, the model should be further improved to increase its applicability to various kinds of logistics policy simulations, as discussed in the latter part of Chapter 5. This includes geographical expansion to every region of the world and detailed simulations at the city level, dynamic simulations including seasonal fluctuations and incorporation of other types of cargo (i.e. air, dry bulk, liquid bulk, and pipeline) as well as empty container repositioning. Model expansion to cover the entire supply chain—including many stakeholders as well as commercial distribution, relationships with financial markets, and investors' behaviour in the international transport market—is also worth considering. Moreover, many kinds of applications of the developed model are conceivable, including the impacts of introducing new technology, vulnerability analysis of the transport network, historical analysis of network development, and impacts of differences in policy systems and regulations. Furthermore, the economic impact of these logistics policies and the feedback to the cargo shipping demand (i.e. the induced demand) should be incorporated in the entire simulation system, as partly discussed in Chapters 4 and 9. The authors will continue to apply the model (being improved daily) to each region of the world, in addition to Japan, China, the entire land area of the ASEAN countries and Eastern Africa, which it has already been applied to, other than the examples introduced in this book.

Author Index

Note: Page numbers followed by *f* indicate figures, *np* indicate footnotes, and *t* indicate tables.

Subject Index

Note: Page numbers followed by *f* indicate figures and *t* indicate tables.

G

H

I